W9-BNO-455

Selected Papers of Nanjing Institute of Meteorology Alumni in Commemoration of Professor Jijia Zhang

Observation, Theory and Modeling of Atmospheric Variability

WORLD SCIENTIFIC SERIES ON METEOROLOGY OF EAST ASIA

Vol. 1: East Asia and Western Pacific Meteorology and Climate: Selected Papers of the Fourth Conference
edited by C.-P. Chang, Guoxiong Wu, Ben Jou & C. Y. Lam

Vol. 2: East Asian Monsoon
edited by C.-P. Chang

Vol. 3: Observation, Theory, and Modeling of Atmospheric Variability Selected Papers of Nanjing Institute of Meteorology Alumni in Commemoration of Professor Jijia Zhang
edited by Xun Zhu, Xiaofan Li, Ming Cai, Shuntai Zhou, Yuejian Zhu, Fei-Fei Jin, Xiaolei Zou & Minghua Zhang

WORLD SCIENTIFIC SERIES ON METEOROLOGY OF EAST ASIA

Selected Papers of Nanjing Institute of Meteorology
Alumni in Commemoration of Professor Jijia Zhang

Observation, Theory and Modeling of Atmospheric Variability

edited by

Xun Zhu (Chief Editor)
Applied Physics Laboratory, Johns Hopkins University

Xiaofan Li
NOAA/NESDIS/Office of Research and Application

Ming Cai
Florida State University

Shuntai Zhou
NOAA/NCEP/Climate Prediction Center

Yuejian Zhu
NOAA/NCEP/Environmental Modeling Center

Fei-Fei Jin
University of Hawaii

Xiaolei Zou
Florida State University

Minghua Zhang
State University of New York, Stony Brook

NEW JERSEY • LONDON • SINGAPORE • SHANGHAI • HONG KONG • TAIPEI • CHENNAI

Published by

World Scientific Publishing Co. Pte. Ltd.
5 Toh Tuck Link, Singapore 596224
USA office: Suite 202, 1060 Main Street, River Edge, NJ 07661
UK office: 57 Shelton Street, Covent Garden, London WC2H 9HE

British Library Cataloguing-in-Publication Data
A catalogue record for this book is available from the British Library.

OBSERVATION, THEORY, AND MODELING OF ATMOSPHERIC VARIABILITY
Selected Papers of Nanjing Institute of Meteorology Alumni in Commemoration of Professor Jijia Zhang

ISBN 981-238-704-8

Printed in Singapore by Mainland Press

Professor Jijia Zhang (章基嘉 教授, 1930 - 1995)

CONTENTS

PREFACE

This collection of selected papers addresses a variety of topics concerning the atmospheric variability. The authors are from various institutions mostly in the United States, along with several from Canada and China. They all have one thing in common: they graduated from Nanjing Institute of Meteorology of China and were taught by a great teacher, the late Professor Jijia Zhang, to whose memory this book is dedicated.

Professor Jijia Zhang is probably best known for his leadership in the modernization of China's meteorological enterprise. When he was the Deputy Administrator of China's State Meteorological Administration in 1980s, he led the development of modern technology such as meteorological satellites and numerical forecast models in national and regional observation and forecast systems. Envisioning the potential impacts of long-term global climate change, he directed the drafting of the first National Climate Plan and the first National Climate Bluebook of China. In China's "Seventh Five-year Plan", Professor Zhang led a challenging research project on medium to long-term weather forecasting and mesoscale disastrous weather prediction. In recognition of his tremendous effort and achievement, he was elected to the Chinese Engineering Academy in 1994, the first meteorologist to receive such honor.

Professor Jijia Zhang devoted his life and inspired many others to study atmospheric sciences. He was an industrious scholar, a respected scientist, a great teacher and mentor. Professor Zhang received his doctorate in 1958 from the Leningrad Hydro-meteorological Institute in the former Soviet Union (now the Russian State Hydro-meteorological University). He worked as a consultant to the Vietnam's Hydro-Meteorological Service and taught at Hanoi University in early 1960s. He was honored with the "Ho Chi Minh Medal" by the Vietnamese government, and was also awarded the First Prize by the Chinese government for his assistance to the Vietnamese people. In his legendary life, Professor Zhang spent twenty years at the Nanjing Institute of Meteorology, as a Professor, Department Chairman and Vice President. He was among the pioneers in building up the Institute from scratch. His leading role in teaching and research has shaped successfully the largest meteorological college in the world. He taught and influenced generations of students, who are now active in operational and research frontiers of the world meteorological stage. Part of their research activity is reflected in this memorial book.

This book contains relatively long tutorial articles as well as relatively short introductory notes, general review reports and specific research letters. For clarity we have organized these papers into four chapters. Chapter 1 covers the dynamics of atmospheric variability from the point of view of basic theory and data analysis. The topics include global scale variations of ENSO and extratropical general circulations, which are related to large-scale lower boundary forcing, air-sea interaction, troposphere-stratosphere exchange, sea ice distribution, and thermal forcing of Tibetan Plateau. They also include synoptic scale variations like storm tracks and extratropical cyclogenesis, which are associated with local instability. Chapter 2 consists of a series of papers addressing physical and mathematical problems in climate modeling and numerical weather prediction. Those problems include parameterization of sub-grid scale clouds and convection, direct simulation by cloud-resolving models, incorporation of turbulence-microphysics interaction, cloud-climate feedback, and coupling of land processes and mesoscale air-sea interaction into weather and climate models. A short-term regional climate model and a diagnostic technique calibrating climate model uncertainties are briefly introduced. Conventional and advanced concepts and methods in data assimilation and ensemble forecasting are also introduced in this chapter. In Chapter 3, various theories of atmospheric radiative transfer and their applications in satellite remote sensing are introduced and reviewed. The topics range from fundamental radiative transfer theories to current debate of cloud anomalous absorption. The utilization of satellite data (including solar radiation, temperature profile, cloud and precipitation properties, ozone, aerosol, etc.) in atmospheric research, education and operational forecast is also presented. The papers in Chapter 4 deal with mathematical and statistical theories and methods which can be used in atmospheric data analysis and modeling.

We hope that this book can serve two purposes: to be used by undergraduate and graduate students majoring in atmospheric sciences as an introduction to various research areas; and by researchers and educators as a general review or quick reference in their fields of interest. In editing this book, we were fortunate to receive help from many people in both the Western and Eastern Hemispheres. They include former colleagues and students of Professor Jijia Zhang, teachers of the Nanjing Institute of Meteorology, guest reviewers, and board members of the World Scientific Series on Meteorology of East Asia, which is chaired by Professor C.-P. Chang of the U. S. Navel Postgraduate School. We are grateful to Professor K. K. Phua, Chairman of World Scientific Publication Company for his generous support. We also thank Drs. Kan Liou, Steven Lloyd, and Elsayed Talaat from The Johns Hopkins University Applied Physics Laboratory for their editorial assistance.

--- Editorial Committee of the Professor Jijia Zhang's Memorial Paper Collection

List of Reviewers

(1)	Akio Arakawa	University of California at Los Angeles, USA
(2)	Jian-Wen Bao	NOAA Environmental Technology Laboratory, USA
(3)	Howard W. Barker	Meteorological Service of Canada, Canada
(4)	W. John Braun	University of Western Ontario, Canada
(5)	Fei Chen	National Center for Atmospheric Research, USA
(6)	Hong-Bin Chen	CAS Institute of Atmospheric Physics, China
(7)	Daniel C. Chin	Johns Hopkins University, USA
(8)	Ming-Dah Chou	NASA Goddard Space Flight Center, USA
(9)	William R. Cotton	Colorado State University, USA
(10)	Merritt N. Deeter	National Center for Atmospheric Research, USA
(11)	Mark Demaria	NOAA/NESDIS Office of Research and Application, USA
(12)	Yihui Ding	China Meteorological Administration, China
(13)	Leo J. Donner	NOAA Geophysical Fluid Dynamics Laboratory, USA
(14)	J. Gavin Esler	University of Cambridge, United Kingdom
(15)	Kristin Findell	NOAA Geophysical Fluid Dynamics Laboratory, USA
(16)	Norman Grody	NOAA/NESDIS Office of Research and Application, USA
(17)	John R. Gyakum	McGill University, Canada
(18)	Richard Haberman	Southern Methodist University, USA
(19)	Dennis L. Hartmann	University of Washington, USA
(20)	Fred J. Hickernell	Hong Kong Baptist University, China
(21)	Stacy D. Hill	Johns Hopkins University, USA
(22)	James R. Holton	University of Washington, USA
(23)	Lon L. Hood	University of Arizona, USA
(24)	Yimin Ji	George Mason University, USA
(25)	Eugenia Kalnay	University of Maryland, USA
(26)	In-Sik Kang	Seoul National University, Korea
(27)	Yoram J. Kaufman	NASA Goddard Space Flight Center, USA
(28)	Stanley Q. Kidder	CIRA Colorado State University, USA
(29)	Praveen Kumar	University of Illinois at Urbana-Champaign, USA
(30)	Christian D. Kummerow	Colorado State University, USA
(31)	William Lakin	University of Vermont, USA
(32)	Stephen E. Lang	NASA Goddard Space Flight Center, USA
(33)	Lance M. Leslie	University of Oklahoma, USA
(34)	Chongyin Li	CAS Institute of Atmospheric Physics, China
(35)	Wei Li	University of Toledo, USA
(36)	Bing Lin	NASA Langley Research Center, USA
(37)	Ping Liu	University of Hawaii, USA
(38)	Yimin Liu	CAS Institute of Atmospheric Physics, China
(39)	Yong-Qiang Liu	USDA Forest Service, USA
(40)	Steven A. Lloyd	Johns Hopkins University, USA
(41)	Dag Lohmann	NOAA Environmental Prediction Center, USA
(42)	Mankin Mak	University of Illinois at Urbana-Champaign, USA
(43)	Timothy P. Marchok	NOAA Geophysical Fluid Dynamics Laboratory, USA
(44)	Allan D. McQuarrie	Johns Hopkins University, USA

(45) Jungang Miao — University of Bremen, Germany
(46) Chin-Hoh Moeng — National Center for Atmospheric Research, USA
(47) Michael C. Morgan — University of Wisconsin, USA
(48) Mu Mu — CAS Institute of Atmospheric Physics, China
(49) Hisashi Nakamura — University of Tokyo, Japan
(50) Art B. Owen — Stanford University, USA
(51) Steven P. Pawson — NASA Goddard Space Flight Center, USA
(52) Roger A. Pielke, Sr. — Colorado State University, USA
(53) Jordan G. Powers — National Center for Atmospheric Research, USA
(54) Jinhuan Qiu — CAS Institute of Atmospheric Physics, China
(55) David W. Rusch — University of Colorado, USA
(56) Paul S. Schopf — George Mason University, USA
(57) Shaowen Shou — Nanjing Institute of Meteorology, China
(58) Joanne Simpson — NASA Goddard Space Flight Center, USA
(59) Kai-Sheng Song — Florida State University, USA
(60) James C. Spall — Johns Hopkins University, USA
(61) Darrell F. Strobel — Johns Hopkins University, USA
(62) Gunilla Svensson — Stockholm University, Sweden
(63) William H. Swartz — Johns Hopkins University, USA
(64) Elsayed R. Talaat — Johns Hopkins University, USA
(65) Wei-Kuo Tao — NASA Goddard Space Flight Center, USA
(66) Huug van den Dool — NOAA Climate Prediction Center, USA
(67) John E. Walsh — University of Illinois at Urbana-Champaign, USA
(68) Bin Wang — University of Hawaii, USA
(69) Xiaoqun Wang — University of New South Wales, Australia
(70) Ming Wei — Nanjing University, China
(71) Duming Weng — Nanjing Institute of Meteorology, China
(72) Fuzhong Weng — NOAA/NESDIS Office of Research and Applications, USA
(73) Roger Terry Williams — Naval Postgraduate School, USA
(74) Rongsheng Wu — Nanjing University, China
(75) Qin Xu — National Severe Storm Laboratory, USA
(76) Xiaofeng Xu — China Meteorological Administration, China
(77) Yongkang Xue — University of California at Los Angeles, USA
(78) Michio Yanai — University of California at Los Angeles, USA
(79) Song Yang — NOAA Climate Prediction Center, USA
(80) Zong-Liang Yang — University of Texas at Austin, USA
(81) Yuk L. Yung — California Institute of Technology, USA
(82) Da-Lin Zhang — University of Maryland, USA
(83) Ying Zhang — Canada Centre for Remote Sensing, Canada
(84) Yongliang Zhang — Johns Hopkins University, USA
(85) Keyun Zhu — Chengdu University of Information Technology, China

List of NIM Alumnus Authors and Academic Years Entering NIM

An,	Shu	(安　庶，1977)
Cai,	Ming	(蔡　鸣，1977)
Fang,	Zhi-Fang	(方之芳，1960)
Ji,	Yimin	(嵇驿民，1978)
Jin,	Fei-Fei	(金飞飞，1977)
Kong,	Fanyou	(孔繁铀，1978)
Li,	Xiaofan	(李小凡，1977)
Li,	Zhanqing	(李占清，1979)
Liu,	Guosheng	(刘国胜，1978)
Liu,	Quanhua	(刘全华，1978)
Liu,	Yimin	(刘屹岷，1981)
Liu,	Yong-Qiang	(刘永强，1977)
Rui,	Hualan	(芮华兰，1977)
Xia,	Youlong	(夏友龙，1980)
Xie,	Lian	(谢立安，1978)
Wang,	Shouping	(王首平，1978)
Weng,	Fuzhong	(翁富忠，1978)
Wu,	Guo-Xiong	(吴国雄，1961)
Yang,	Runhua	(杨润华，1974)
Yang,	Song	(杨　松，1979)
Yang,	Zong-Liang	(杨宗良，1980)
Yu,	Jun	(余　军，1977)
Yue,	Rong-Xian	(岳荣先，1977)
Zhang,	Guang Jun	(张广俊，1977)
Zhang,	Minghua	(张明华，1978)
Zhao,	Yiqiang Q.	(赵以强，1977)
Zhou,	Shuntai	(周顺泰，1977)
Zhou,	Xuelong	(周学龙，1985)
Zhu,	Tong	(朱　彤，1984)
Zhu,	Xun	(朱　迅，1977)
Zhu,	Yuejian	(朱跃建，1977)
Zou,	Xiaolei	(邹晓蕾，1977)

ACRONYMS

2D	Two-dimensional
3D	Three-dimensional
4D	Four-dimensional
ADM	Angular Dependence Model
AERONET	AEROsol NETwork
AFD	Adjoint of Finite-difference
AMIP	Atmospheric Model Inter-comparison Program
AMSR	Advanced Microwave Scanning Radiometer
AO	Arctic Oscillation
APS	Aerosol Polarimetry Sensor
ARESE	ARM Enhanced Shortwave Experiment
ARM	Atmospheric Radiation Measurement
ARMAR	Airborne Mapping Radar
ARME	Amazon Region Micrometeorology Experiment
AVHRR	Advanced Very High Resolution Radiometer
AWS	Automatic Weather Stations
BATS	Biosphere-Atmosphere Transfer Scheme
BFGS	Broyden, Fletcher, Goldfard, and Shannon
BOREAS	Boreal Ecosystem-Atmosphere Study
BRDF	Bidirectional Reflectance Distribution Function
BSI	Bayesian Stochastic Inversion
BSRN	Baseline Surface Radiation Network
CAA	Cloud Absorption Anomaly
CAD	Cold-Air Damming
CAMEX	Convection And Moisture EXperiment
CAPE	Convective Available Potential Energy
CAS	Chinese Academy of Sciences
CCM	Community Climate Model
CCN	Cloud Condensation Nuclei
CERES	Clouds and the Earth's Radiant Energy System
CHAMP	Challenging Minisatellite Payload
CHASM	CHAmeleon Surface Model
CIRA	Cooperative Institute for Research in the Atmosphere
CIRES	Cooperative Institute for Research in Environmental Sciences
CKD	Correlated-*k* Distribution
CLASS	Canadian Land Surface Scheme
CLASS	Cross-chain Loran Atmospheric Sounding System
CLM	Community Land Model
CMIP	Coupled Model Inter-comparison Program
CMIS	Conical Microwave Imager Sounder
CMPE	Cloud-microphysics Precipitation Efficiency
COARE	Coupled Ocean-Atmosphere Response Experiment

COLA	Center for Ocean-Land-Atmosphere Studies
CPT	Cold Point Tropopause
CPT-T	Cold Point Tropopause Temperatures
CPU	Central Processing Unit
CRF	Cloud Radiative Forcing
CSH	Convective/Stratiform Heating algorithm
CWT	Continuous Wavelet Transform
DAAC	Distributed Active Archive Center
DAS	Data Assimilation System
DFP	Davidon-Fletcher-Powell
DISORT	Discrete-Ordinate Radiative Transfer
DMSP	Defense Meteorological Satellite Program
ECMWF	European Center for Medium Range Weather Forecasts
EMD	Empirical Mode Decomposition
ENSO	El Niño-Southern Oscillation
ENVISAT	European Environment Satellite
EOF	Empirical Orthogonal Functions
EOS	Earth Observation System
EPS	Ensemble Prediction System
ERB	Earth Radiation Budget
ERBE	Earth Radiation Budget Experiment
ERS	European Earth Resource Satellite
ESMR	Electrically Scanning Microwave Radiometer
ESSIC	Earth System Science Interdisciplinary Center
EV	Economic Value
FAO	Food and Agriculture Organization (of the United Nations)
FAST	Fourier amplitude sensitivity test
FD	Finite Difference
FDA	Finite-difference of Adjoint
FFT	Fast Fourier Transform
FGGE	First GARP Global Experiment
FIFE	First ISLSCP Field Experiment
FSU	Florida State University
GACP	Global Aerosol Climatology Project
GALE	Genesis of Atlantic Lows Experiment
GAME	GEWEX Asian Monsoon Experiment
GAPP	GEWEX Americas Prediction Project
GARP	Global Atmospheric Research Program
GATE	Global Atmospheric Research Program Atlantic Tropical Experiment
GCE	Goddard Cumulus Ensemble Model
GCIP	GEWEX Continental-Scale International Project
GCM	General Circulation Model
GEOS	Goddard Earth Observation System
GEWEX	Global Energy and Water Cycle Experiment
GFDL	Geophysical Fluid Dynamics Laboratory

GFS	Global Forecast System
GISS	Goddard Institute for Space Studies
GKS	Generalized Kuramoto-Sivashinsky
GLASS	GEWEX/Global Land-Atmosphere System Study
GOALS	Global-Ocean-Atmosphere-Land-System
GOES	Geostationary Operational Environmental Satellites
GPM	Global Precipitation Mission
GPROF	Goddard Profiling algorithm
GPS	Global Positioning System
HAPEX-MOBILHY	Hydrological Atmospheric Pilot Experiment 1986 MOdelisation du BILan HYdrique
GPS/MET	meteorological applications of the Global Positioning System
GSF	Gulf Stream Front
GSFC	Goddard Space Flight Center
GWEC	Global Water and Energy Cycle Program
HH	Hydrometeor Heating algorithm
IFA	Intensive Flux Array
IMF	Intrinsic Mode Function
IOP	Intensive Observation Period
IP	Inertial Period
IPCC	Inter-Government Program for Climate Changes
IR	Infrared
ISBA	Interaction between Soil, Biosphere and Atmosphere
ISCCP	International Satellite Cloud Climatology Project
ISLSCP	International Satellite Land Surface Climatology Project
ITCZ	Intertropical Convergence Zone
IWC	Ice Water Content
JMA	Japanese Meteorological Agency
KdV	Korteweg-de Vries
LAI	Leaf Area Index
LASG	Laboratory of Atmospheric Sciences and Geophysical Fluid Dynamics
L-BFGS	Limited-memory Quasi-Newton that follow BFGS
LEO	Low Earth Orbiting
LIS	Lightning Imaging Sensor
LLJ	low level jet
LSM	Land Surface Model
LSPE	Large-scale Precipitation Efficiency
LTE	Local Thermodynamic Equilibrium
LWC	Liquid Water Content
MAB	Middle Atlantic Bight
MABL	Marine Atmospheric Boundary Layer
MC	Multi-Criteria
MJO	Madden Julian Oscillation
MM4/MM5	Mesoscale model, version 4/5

MODIS	Moderate Resolution Imaging Spectroradiometer
MPI	Maximum Potential Intensity
MRF	Medium-Range Forecast
MRI	Meteorological Research Institute
MSU	Microwave Sounding Unit
NAO	North Atlantic Oscillation
NASA	National Aeronautics and Space Administration
NASDA	National Space Development Agency
NCAR	National Center for Atmospheric Research
NCEP	National Centers for Environment Prediction
NDVI	Normalized Difference Vegetation Index
NESDIS	National Environmental Satellite, Data, and Information Service
NMVOCs	Non-methane Volatile Organic Compounds
NOAA	National Oceanic and Atmospheric Administration
NPO	North Pacific Oscillation
NPOESS	U.S. National Polar-Orbiting Environmental Satellite System
NSIDC	National Snow and Ice Data Center
NWP	Numerical Weather Prediction
NWS	National Weather Service
OGCM	Ocean General Circulation Model
OI	Optimal Interpolation
OLR	Outgoing Longwave Radiation
PAR	Photosynthetic Active Radiation
PBL	Planetary Boundary Layer
PE	Precipitation Efficiency
PMW	Passive Microwave
POES	Polar Orbiting Environmental Satellites
POLDER	POLarization and Directionality of the Earth Reflectance
PPD	Posterior Probability Density
PQPF	Probabilistic Quantitative Precipitation Forecast
PR	Precipitation Radar
QBO	Quasi-Biennial Oscillation
QPF	Quantitative Precipitation Forecast
RAMS	Regional Atmospheric Modeling System
RegCM	Regional Climate Model
RMOP	Relative Measure of Predictability
RMS	Root Means Square
RTE	Radiative Transfer Equation
SAB	South Atlantic Bight
SABER	Sounding of the Atmosphere using Broadband Emission Radiometry
SAGE	Stratospheric Aerosol and Gas Experiment
SAR	Synthetic Aperture Radar
SC	Successive Correction
SCAT	Sea winds Scatterometer

SCF	Squared Covariance Fraction
SCSMEX	South China Sea Monsoon Experiment
SED	Solar Energy Disposition
SHAP	Sensible Hear driven Air Pump
SiB	Simple Biosphere Model
SLH	Spectral Latent Heating ALgorithm
SLP	Sea Level Pressure
SMMR	Scanning Multichannel Microwave Radiometer
SPCZ	Southern Pacific Convergence Zone
SRB	Surface Radiation Budget
SSI	Spectral Statistical Interpolation
SSiB	Simplified Simple Biosphere
SSM/I	Special Sensor Microwave Imager
SST	Sea Surface Temperature
SSTA	Sea Surface Temperature Anomalies
STE	Stratosphere-Troposphere Exchange
SVD	Singular Value Decomposition
TE	Thermodynamic Equilibrium
TIROS	Television Infrared Observation Satellite
TMI	TRMM Microwave Imager
TOA	Top of the Atmosphere
TOGA	Tropical Ocean Global Atmosphere
TOMS	Total Ozone Mapping Spectrometer (TOMS)
TOVS	TIROS Operational Vertical Sounder
TPW	Total Precipitable Water
TRMM	Tropical Rainfall Measuring Mission
TSDIS	TRMM Science Data Information System
UKMO	United Kingdom Meteorological Office, Bracknell, England
UNESCO	United Nations Educational, Scientific and Cultural Organization
USDA	United States Department of Agriculture
UV	Ultraviolet
UW-NMS	University of Wisconsin-Numerical Modeling System
VCW	Virginia coastal water
VDISORT	Vector Discrete-Ordinate Radiative Transfer
VIRS	Visible/Infrared Scanner
VIS	Visible
VLF	Very Low-Frequency
WKB	Wentzel-Kramers-Brillouin
WMO	World Meteorological Organization
WVWV	Water Vapor Wind Vector

Part I

Dynamics of Atmospheric Variability

LOCAL INSTABILITY DYNAMICS OF STORM TRACKS

MING CAI
Department of Meteorology
University of Maryland, College Park, MD 20742, USA
E-mail: cai@atmos.umd.edu

(Manuscript received 30 December 2002)

This paper reviews local instability theory and its application to storm track dynamics. It begins with a brief introduction to the concepts pertinent to storm track dynamics: convective versus absolute instability, propagating versus stationary wave-packet resonance, local growth due to energy conversions versus spatial redistributions, stochastic non-modal instability, and localization due to the basic state's deformation. A brief review and relative merits of local energetics analysis, feedback diagnostics, and diagnostics using 3-D E-P flux vectors derived from pseudo-energy conservation relations are discussed.

The physical processes that are important to local instability can be delineated from a local energetics analysis. Because the availability of background energy sources is zonally inhomogeneous, the peak amplitude location becomes an intrinsic instability property of a local mode. The storm track location is determined by an optimal balance between the local growth due to energy extracted from the basic flow and the local growth due to energy redistributions. In particular, the advection always acts to spread perturbations downstream, implying the preferred location of a storm track would be downstream of the basic jet core. The stronger the advection is, the farther downstream the storm track and the slower the growth rate of perturbations would be. This explains why the domain-averaged basic wind, rather than the domain-averaged barotropic/baroclinic wind shear, is one of the most determining factors for instability of a local mode. The energy propagation associated with the ageostrophic flow also tends to cause downstream development although it may not be as dominant as the advective processes. Another important factor for determining the storm track length is the basic deformation field in the jet exit region where meridionally elongated baroclinic eddies tend to decay barotropically.

1. Introduction

The longitudinal inhomogeneity of atmospheric statistics has been well documented in a large number of observational studies (e.g., Blackmon 1976; Blackmon et al. 1977; Lau 1978; Lau and Wallace 1979). In the Northern Hemisphere, the most salient features on a temporal variance map of high frequency eddies that have time scales shorter than 10 days or so are the two zonally elongated local maxima extending from the east coasts of Asia and North America cross the Pacific and Atlantic oceans (Fig. 1). The local maxima on high frequency transient variance maps largely coincide with the track density maxima of individual cyclones (Hoskins and Hodges 2002). In the literature, they are referred to as the Pacific and Atlantic storm tracks, respectively. With reference to the mean circulation, these two storm tracks are located downstream of the two major localized jet streams associated with the eastern Asian and North American climatological mean upper-level troughs. The seasonality and interannual to interdecadal variations of storm tracks have been documented in the literature (e.g., Nakamura 1992, Christoph et al. 1997, and Chang 2001 for the annual cycle of storm tracks, particularly the "midwinter suppression" of the Pacific storm track and its interannual variability; Lau 1988 and Nakamura et al. 2002 for wintertime interannual variability in general; Hoerling and Ting 1994, Chen and van den Dool 1995, 1997, Trenberth and Hurrell 1994, Rogers 1997, Straus and Shukla 1997, Honda and Nakamura 2001, and Honda et al. 2001 for interannual variability related to the El-Niño-Southern oscillation and the north Atlantic oscillation phenomena; Ebisuzaki and Chelliah 1998, Chang and Fu 2002 and Nakamura et al. 2002 for decadal variability).

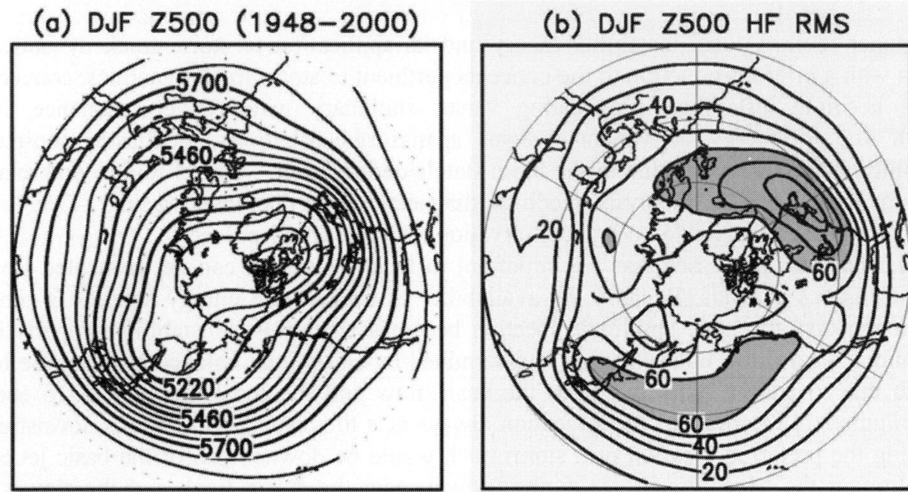

Figure 1 (a) December-January-February mean 500 hPa geopotential height (CI = 60m). (b) Standard deviation of high-frequency (< 10 days) transient eddies in the 500 hPa geopotential height field (CI = 10m). The data are derived from the 1948-2000 NCEP/NCAR Reanalysis. (courtesy of Dr. Song Yang).

The climatological zonally localized jet streams in the Northern Hemisphere are ultimately caused by the presence of zonal asymmetries in the lower boundary conditions for the atmosphere, such as large-scale orography and land-ocean thermal contrasts. The existence of zonally localized storm tracks, however, does not necessarily hinge upon a zonal asymmetry on the earth's surface per se. The linear/nonlinear local instability theory[1] attempts to explain the existence of zonally localized storm tracks primarily as a result of the zonal inhomogeneity in the background flow from which synoptical scale transient eddies

[1] Generally, one can find both "local" and "global" modes in an instability analysis of a zonally varying basic flow. The latter exists exclusively only in a cyclic domain that facilitates "recycling perturbation" repetitively through a jet core. Very often, local instability refers to instability property of local modes, which do not require a streamwise cyclic boundary condition for their existence. Here, the author takes the liberty to use the notion of "local instability" to refer to an instability analysis of a zonally varying basic flow without distinguishing explicitly "local" versus "global" modes. In the WKB theory, the term "unstable local modes" has been used exclusively to describe those arising from an absolute instability. In this paper, "unstable local modes" are referred to those that do not require a streamwise cyclic boundary condition for their existence without concerning "absolute" versus "convective" instability.

draw energy. A local instability analysis typically excludes the zonal inhomogeneity in surface boundary condition although the surface diabatic latent and sensible heat fluxes (Hoskins and Valdes 1990;Mak 1998; Black 1998; Zhu et al. 2001; and Chang et al 2002) and orographic forcing (e.g., Frederiksen and Bell 1987; Lee 1995a; Lee and Mak 1996) may act as additional energy sources to the development of weather systems. Cai and Mak (1990b) demonstrated that storm tracks can form preferentially downstream of the troughs of an eddy-driven traveling planetary scale wave without any zonal asymmetry in the basic flow and in the lower boundary conditions. They found that the traveling storm tracks that are localized with respect to a companion traveling planetary wave are symbiotically coupled to the traveling planetary wave itself. Cai and van den Dool (1991, 1992) and Cuff and Cai (1995) further confirmed from data the existence of traveling storm tracks in both Northern and Southern Hemispheres, explaining the belt of large temporal variability of high-frequency eddies looping around the middle latitudes.

The majority of papers on local instability focus solely on the zonal inhomogeneity of horizontal gradients of the basic state (vertical wind shear is related to horizontal temperature gradient via the thermal wind relation). Mak (1993) discussed the local baroclinic instability resulting from the zonal inhomogeneity in the basic stratification, suggesting that storm tracks tend to form in the regions where the background static stability is weaker. Lee and Mak (1993) found that the static stability over the continents is larger than over the oceans and that between the two oceans the static stability over the Atlantic is weaker. Therefore, it appears that the spatial variations of atmospheric static stability at least partially explain the locations of the observed storm tracks as well as their spatial variations, particularly the Atlantic storm track being stronger than the Pacific storm track (Fig. 1).

This paper reviews storm track dynamics from the viewpoint of local instability with a focus on the quasi-geostrophic local energetics dynamics, which has not received adequate attention in a recent review paper on storm track dynamics by Chang et al. (2002). Readers are advised to consult with Chang et al. (2002) for a broader overview on storm track dynamics. This paper is organized as follows. The next section briefly reviews the basic theoretic considerations on storm track dynamics, loosely in chronological order as they first appeared in the literature. Section 3 is devoted to discussions of quasi-geostrophic local energetics. Also discussed in this section are the generalized necessary conditions derived from local energetics prospective for barotropic and baroclinic instabilities of a zonally varying basic flow. Examples of local instability and local energetics analysis are presented in section 4. The concluding remarks, including brief review and discussions on the relative merits of local energetics analysis, feedback diagnostics, and diagnostics using 3-D E-P flux vectors derived from pseudo-energy conservation relations, are given in section 5.

2. Theoretical development of local instability theory

Lorenz (1972) pioneered the instability analysis of a zonally varying basic state by considering a monochromatic wavy basic state. However the more intriguing consequence of the presence of zonal inhomogeneity in a basic flow did not become evident until the instability of a localized jet stream was examined in an attempt to explain the variance/covariance statistics of atmospheric transient eddies (e.g., Tupaz et al. 1978; Frederiksen 1978, 1979, 1983; and Simmons et al. 1983). As summarized in Frederiksen and Bell (1987), a whole class of unstable modes of the Northern Hemisphere wintertime basic flow exhibits a spatial variation of synoptic scale eddy statistics similar to the observed Pacific and Atlantic storm tracks.

Local instability theory seeks answers to the following single most pertinent question: what are the most essential factors of a zonally varying background flow that would organize disturbances to grow preferentially downstream of a jet stream? A zonal inhomogeneity by itself implies a zonal variation of the instability criterion evaluated locally. However, this factor alone obviously would not explain why the instability does not preferentially take place at the jet core where the background baroclinicity is largest. The answers rest on other factors that are not necessarily exclusively due to a zonal inhomogeneity in a basic flow. These factors are either inconsequential or exist only in an extreme limit for a zonally homogeneous background flow and become relevant only in the presence of a zonal inhomogeneity in a basic flow.

2.1. *Convective versus absolute instability*

Merkine (1977) first introduced the concept of convective versus absolute instability in geophysical fluid dynamics by considering the asymptotic instability property of a zonally uniform basic flow in a quasi-geostrophic two-layer f-plane *infinitely long* channel model. Absolute instability refers to continuous amplification of perturbations at a longitude sector after an unstable wave packet has passed through. The absolute growth rate is much smaller than the most unstable normal mode growth rate. Convective instability refers to a situation in that temporal amplification of perturbations at a longitude only takes place when an unstable wave packet is passing through although the unstable wave packet itself continuously amplifies as it propagates in the zonal direction with an asymptotic growth rate equal to the growth rate of the most unstable normal mode.

The theoretical implication of absolute instability was not fully recognized until Pierrehumbert (1984, referred to as P84 hereafter) put the concept of absolute instability in the context of a zonally varying baroclinic flow. For a zonally uniform basic state in a finite length domain where a cyclic boundary condition is applied, there is no chance for absolute instability to be dominant because the rapidly growing wave packet with an asymptotic growth rate equal to the most unstable normal mode would soon overpower the absolute growth by passing through the streamwise cyclic boundaries repetitively. However, absolute instability can exist naturally in a zonally inhomogeneous flow in a finite length domain. After moving out of the region where the background baroclinicity is largest, an unstable wave packet arising from a convective instability would immediately suffer a significant reduction of growth because its environmental baroclinicity becomes weaker as it moves downstream of the jet core. The perturbation may be completely damped out by dissipations before it can reach the jet core again through the cyclic boundary condition. The absolute growth, however, remains in the region where the background baroclinicity is still sufficient large. Convective instability can still be dominant if the growing perturbation can reenter the jet core through the streamwise cyclic boundaries.

From a dispersion relation, D(ω, k) = 0, one can infer the asymptotic growth of a wave packet along a particular ray, $V = x/t$, by finding the root k_V satisfying

$$\left.\frac{\partial \omega}{\partial k}\right|_{k=k_v} = V, \tag{1}$$

where, k and ω are complex wavenumber and frequency, respectively. Convective instability corresponds to the solution along the ray of $V = C_g$ with C_g being the group velocity. Along this ray, the wave packet grows with the most unstable growth rate and travels with the speed of the group velocity evaluated at the most unstable wavenumber. Absolute growth is the solution along the ray of $V = 0$, a saddle point of the dispersion relation in the complex wavenumber plane that yields both a zero group velocity and a maximum growth rate. Readers can consult the papers by Merkine (1978), Pierrehumbert (1984, 1986), and, Lin and Pierrehumbert (1993) for elaborated discussions on absolute instability.

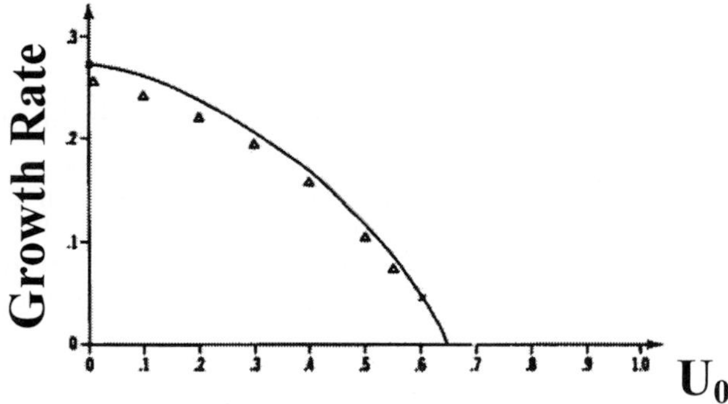

Figure 2. Growth rate of the most unstable local mode. The solid curve is the absolute instability growth rate evaluated at the jet core where the background baroclinicity is the maximum and the triangles are obtained by numerical integrations with the zonally inhomogeneous flow (after Pierrehumbert, 1984).

Figure 2 summarizes the main results of P84. It shows that the local mode growth rate is very close to the absolute instability evaluated at the jet core. In other words, the local mode growth rate is determined primarily by the maximum baroclinicity in the domain but not by the average baroclinicity. This particularly implies that the growth rate of the local mode is not sensitive to the local baroclinicity far away from the jet core. It is also seen that the local mode growth rate is largest for $U_0 = 0$ and gradually becomes zero when U_0 is sufficiently large (U_0 is the constant part of the basic zonal wind). This is consistent with the absolute instability theory that requires the ratio of DU (baroclinicity) to U_0 to be larger than a critical value (Merkine 1977). The maximum amplitude location of the local mode is downstream of the jet core and moves farther downstream as U_0 increases (see Figs. 6 and 8 in P84). Again this feature can be explained by the absolute instability analysis of a zonally inhomogeneous flow under the WKB approximation. Fig. 3 shows that the location of the maximum perturbation amplitude (represented by the "0" contour) is just slightly downstream of the jet core for $U_0 = 0$ and it moves farther downstream (a smaller *DU*) as U_0 increases. The degree of localization (measured by the gradient of solid contours with respect to *DU* along a constant U_0 in Fig. 3) is strongest when $U_0 = 0$ and becomes weaker as U_0 increases. A global mode, which requires cyclic zonal boundary conditions for its existence, exhibits a pronounced sensitivity to the local baroclinicity far away from the jet core as one might expect. However, the growth rate of a global mode is relatively insensitive to U_0 and so is its maximum amplitude position (see Fig. 13 in P84), as the case of a zonally parallel flow.

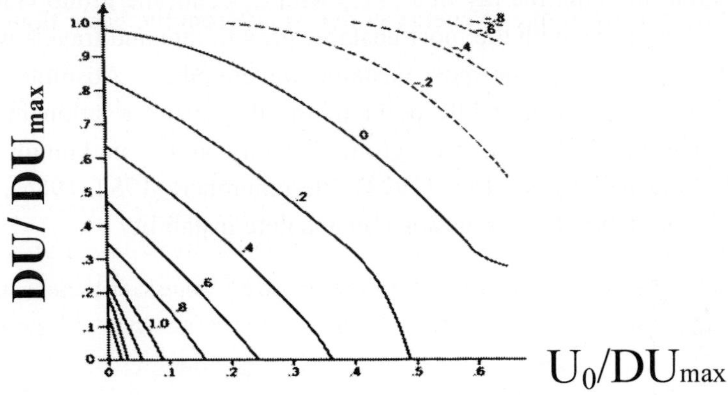

Figure 3. Contours of exponential downstream decay (solid contours) or amplification (dashed contours) rate of the absolute instability solution evaluated at the jet core. The ratio DU/DU_{max} is indicative of the downstream location in reference to the jet core where $DU/DU_{max} = 1$. The smaller the ratio is, farther downstream the location is from the jet core. The contour marked with "0" corresponds to the location of the peak amplitude of the local mode (after Pierrehumbert, 1984).

Although the absolute instability analysis in the WKB limit explains the growth rate and peak location relative to the jet core of the most unstable local mode quite successfully, it is yet to be demonstrated theoretically that the absolute growth evaluated at the jet core could explain the most unstable local mode when the WKB approximation is no longer valid. Furthermore, there are typically a large number of unstable modes for the observed wintertime mean state and a significant portion of them appears very localized in the zonal direction as the case of local modes (e.g., Frederiksen and Bell 1987; Lee and Mak 1995). It is natural to ask whether different local unstable modes could correspond to the absolute growths evaluated at different locations. Lee and Mak (1995) made an instability analysis of the observed 1982-83 wintertime mean flow and found two unstable eigenmodes that are reminiscent of the observed Pacific and Atlantic storm tracks. And yet these two modes are associated with convective instability rather than absolute instability.

2.2. *Local growth due to energy conversions versus redistributions*

Mak and Cai (1989) and Cai and Mak (1990a) (referred to as MC89 and CM90, hereafter) applied local energetics analysis to elucidate the factors that are inconsequential for a zonally parallel flow but they become important for a zonally varying basic flow in a zonally bounded domain. Particularly, local energetics analysis would delineate the roles played by the perturbation energy redistribution processes due to (i) the advection and (ii) the ageostrophic fluxes in the context of a zonally varying basic flow. Although the dynamic processes associated with spatial energy redistributions do not provide net energy to perturbations regardless of whether the basic flow is zonally uniform or inhomogeneous, perturbation energy redistribution processes in a zonally varying flow can indirectly determine the perturbation growth by controlling the preferred location(s) for perturbations to grow. In other words, the preferred location of disturbances becomes an important factor in a zonally varying basic flow because of the zonal variations of background energy sources. The balance between the local growth due to energy conversions from the basic flow and the local growth due to energy redistributions determines the location of the maximum perturbation amplitude and thereby the overall perturbation growth rate. It follows that one can easily attribute the growth rate of different local modes in the same zonally varying basic state to their relative positions with respect to the jet core, as demonstrated in MC89 and CM90. It should be emphasized that the alternative physical interpretation from the local energetics prospective does not invalidate nor validate the absolute instability theory. It is supplementary to the instability theory by invoking less abstract and more physically intuitive dynamical concepts that are readily applicable regardless of whether the WKB approximation is valid or not.

The downstream development of disturbances due to spatial energy redistributions is akin to that resulting from energy dispersion of Rossby waves, which was first, studied by Rossby (1945) and Yeh (1949) in a barotropic β-plane model and was later extended to the spherical domain by Hoskins et al. (1977). Simmons and Hoskins (1979) studied subsequent development of a localized initial perturbation in a baroclinically unstable domain. They found that both upstream and downstream developments relative to the mean zonal wind are plausible. Their vorticity budget analysis indicates that the advection term attributes to the development immediately downstream from the initial vorticity perturbation. The developments of new disturbances upstream and farther downstream are triggered by the stretching term (note that the stretching term is associated with ageostrophic wind in a QG model). Orlanski and Katzfey (1991) applied the local kinetic energy budget analysis derived in a primitive equation model to study the life cycle of an explosive cyclone observed over the South Pacific. They concluded that the downstream energy propagation by the ageostrophic geopotential fluxes was the primary mechanism causing the cyclone to decay and triggering a new disturbance to grow in the downstream direction. Orlanski and Chang (1993) found that the energy propagation associated with the ageostrophic flow indeed could be both upstream and downstream as in Simmons and Hoskins (1979). They showed evidence suggesting that downstream development is predominantly in the upper troposphere while upstream development dominates in the lower troposphere. The analysis of Chang (1993) further showed that the successive developments of perturbations downstream of the existing perturbations could be explained by the downstream energy propagation associated with the ageostrophic flow. Such a successor in the downstream direction then grows vigorously by extracting energy from the basic flow although its predecessor has already ceased to grow. The newly developed perturbation triggers another successor farther downstream. This series of subsequent downstream developments essentially forms a coherent wave-packet wave train propagating in the zonal direction, as first documented by Lee and Held (1993) in the Southern Hemisphere and in a GCM experiment driven by a zonally symmetric radiative equilibrium temperature. The downstream development due to the ageostrophic flow has also been applied to explain the nonlinear dynamics of baroclinic eddies (Cai and Mak 1990b; Chang and Orlanski 1993).

Energy redistributions due to the ageostrophic flow in general can be both upstream and downstream as found in Simmons and Hoskins (1979), CM90, and Orlanski and Chang (1993). It is impossible to definitely know which of the two is more dominant a priori because the ageostrophic component itself is part of the perturbation solution. In contrast, the advection always directs growing perturbations downstream of the location where the local growth due to energy conversions from the basic flow to perturbations is largest. To illustrate this point, let us consider a generic form of time mean local energy equation as in MC89:

$$\frac{\overline{\partial E}}{\partial t} = \sigma \overline{E} = -U_0 \frac{\partial \overline{E}}{\partial x} + \overline{G}, \tag{2}$$

where E is the meridional (and vertical if applicable) mean of perturbation energy as a function of longitude x and time t; the overbar stands for a time mean operator and σ can be interpreted as the local growth rate. The first term on the RHS of (2) represents the energy advection by the uniform part of zonal wind, U_0, and G stands for the meridional mean of the sum of all perturbation energy generation terms and other energy redistribution terms. As far as the following reasoning is concerned, the detailed forms of E and G are inconsequential. Taking the derivative of (2) with respect to x yields

$$\sigma \frac{\partial \overline{E}}{\partial x} = -U_0 \frac{\partial^2 \overline{E}}{\partial x^2} + \frac{\partial \overline{G}}{\partial x} \quad . \tag{3}$$

Let x_G and x_E be the longitudes where $\overline{G}$ and $\overline{E}$ have local maxima, respectively. It follows that, by definition, we have

$$\sigma\left[\partial \overline{E}/\partial x\right]_{x=x_G} = -U_0\left[\partial^2 \overline{E}/\partial x^2\right]_{x=x_G} \text{ and } U_0\left[\partial^2 \overline{E}/\partial x^2\right]_{x=x_E} = \left[\partial \overline{G}/\partial x\right]_{x=x_E} . \tag{4}$$

One can easily prove that for an unstable local mode ($\sigma > 0$), we have $\left[\partial \overline{E}/\partial x\right]_{x=x_G} > 0$ and therefore $x_E > x_G$ when $U_0 > 0$ (or when $U_0 < 0$, $\left[\partial \overline{E}/\partial x\right]_{x=x_G} < 0$; equivalently $x_E < x_G$). In other words, the peak longitude of an unstable local mode has to be downstream of the maximum $\overline{G}$. One can also easily reach the above conclusion from the energetics equation evaluated at $x = x_E$. According to the second equation in (4), we have $\left[\partial \overline{G}/\partial x\right]_{x=x_E} < 0$ for U_0

> 0 or $\left[\partial\bar{G}/\partial x\right]_{x=x_E} > 0$ for $U_0 < 0$. This immediately tells us that the peak longitude of unstable perturbations has to be located downstream of the maximum value of G, the sum of all of the perturbation energy generation and other energy redistribution terms. The longitudinal position of the maximum G may be conservatively expected to coincide with the jet core. Therefore, the maximum perturbation activities would have to be downstream of the jet core. Intuitively, one may envision that the farther the peak perturbation is away from the jet core, the less the amount of background energy would be available for perturbations to extract, thereby the smaller growth rate would be. It follows that the dependency of the growth rate and the peak location of the most unstable local mode to the constant part of the zonal flow found in P84 can be qualitatively explained by this simple intuitive argument. The results of MC89 and CM90 support this simple intuitive argument.

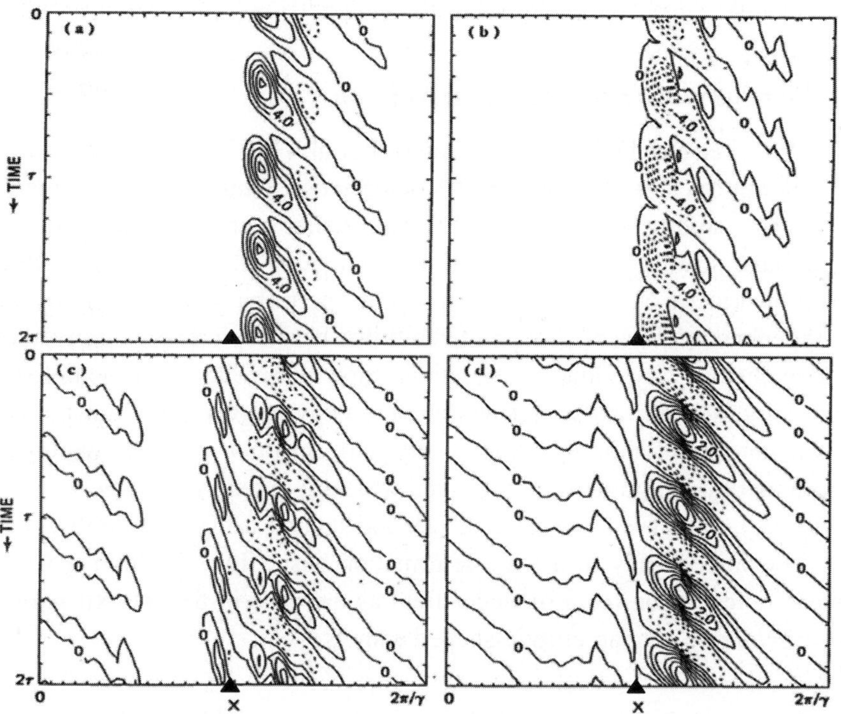

Figure 4. Hovmöller diagram of various (nondimensional) energetics budget terms for the most unstable local mode described in Mak and Cai (1989). (a) energy conversion rate minus the dissipation rate; (b) ageostrophic pressure work; (c) energy advection by the basic wind; and (d) the net local energy change rate or the sum of (a)-(c). The contour interval for (a) and (b) is 2 and 1 for (c) and 0.5 for (d). The solid triangle indicates the longitude (x) position of the basic jet core (after Mak and Cai 1989).

Figure 4 illustrates how the "competition" between the local growth due to energy conversions and that due to energy redistributions regulates the temporal variation of overall growth of a traveling unstable local mode in a barotropic model. The energy conversion from the basic flow to perturbations is positive at any given time except in the region farther downstream of the jet core where the energy conversion rate is much smaller with alternating signs within each period of the unstable mode (Fig. 4a). The energy redistribution terms associated with the mean flow advection and the ageostrophic perturbation flow immediately transfer perturbation energy downstream (Fig. 4b-c). As a result, the net local growth over the region where perturbations extract energy from the basic flow becomes negative after a short time delay. The net local growth (Fig. 4d) in fact follows more closely to the sum of the local growth due to these two energy redistribution terms than the energy conversion term itself although the latter is the only source that feeds the perturbation in a barotropic model. As the perturbation moves out of the unstable region, it gets less energy from the basic flow and it loses the "battle" to its successor generated near the jet core in terms of accessibility to the energy reservoir stored in the basic flow. The energy redistribution terms again act to move the perturbation out of the unstable region. It is the net balance between the growth near the jet core due to the energy conversion and the growth due to the downstream energy propagation that determines the location where the time mean perturbation peaks.

2.3. *Localization due to the basic state's deformation*

Lee (1995b), Whitaker and Dole (1995), Swanson et al. (1997) investigated the localization mechanism of storm track by the basic state's deformation. Shutts (1983) pointed out that the transfer of energy among eddies with different scales can be related to the straining mechanism of large-scale deformation fields. Simmons et al. (1983) and Wallace and Lau (1985) derived the kinetic energy budget equation for perturbations in a zonally varying basic flow in a spherical coordinate. MC89 pointed out that barotropic energy conversion to eddies is determined exclusively by eddies' orientation/shape relative to the basic deformation field. The basic vorticity is irrelevant as far as kinetic energy conversion from the basic state to perturbations is concerned although in a zonally uniform flow the deformation and vorticity associated with meridional gradient of the zonal wind are indistinguishable.

Farrell (1989b), Cai (1992, hereafter C92), and Iacono (2002) illustrated the non-modal instability of a localized eddy in a pure deformation basic flow. Following C92, for a pure stretching basic deformation (or a diffluent field as the case in the jet exit region), the non-dimensional growth rate σ of an elliptic-shaped eddy is,[2]

$$\sigma = -\underbrace{\frac{1-r^2}{1+r^2}}_{S} + 2\underbrace{\frac{3(1-r^4)}{3(1+r^4)+2r^2}}_{A} = \frac{1-r^2}{1+r^2}\left(1+\frac{8r^2}{3(1+r^4)+2r^2}\right), \tag{5}$$

[2] C92 made several mistakes in deriving his Eqs. (14), (16), and (18). One should multiply the right hand sides of his Eqs. (14) and (16) by a factor of "2" and "4", respectively, in order to correct the mistakes in the two equations. Using these two corrected equations, one can then correct his Eq. (18), which is Eq. (5) in this paper. Due to this correction, Fig. 3 in C92 should be replaced by Fig. 5 in this paper. It should be pointed out that Iacono (2002) noticed the possible errors and made a correction to C92's Eq. (18) from C92's Eq. (17) without using the incorrect Eqs. (14) and (16) in C92, leading to Eq. (5) in this paper.

where r is the eddy eccentricity parameter equal to the ratio of the meridional length of an elliptic-shaped eddy to the zonal length. The growth rate of an elliptic-shaped eddy in a confluent field as in the jet entrance region is the same as (5) except with a reversed sign. C92 obtained (5) under the assumption that the initial development of the amplitude and eccentricity of an elliptic-shaped eddy is a function of time only. The term with an underlining label "S" is the growth rate due to the change in the perturbation kinetic energy associated with the change in eccentricity of the eddy alone. The term labeled "A" corresponds to the growth rate due to the amplitude change of the eddy which exactly balances the change of the perturbation enstrophy associated with the change in the eddy eccentricity resulting in a net zero change in the total enstrophy. These two terms have an opposite sign. Because the kinetic energy change due to the amplitude change is at least twice as large as that due to the change in eccentricity, the net growth rate is about half of that due to the amplitude change. Iacono (2002) derived an exact analytical expression for the growth rate of an elliptic-shaped eddy caused by the straining of a pure deformation field without invoking any assumptions about how the eddy evolves in time. This refined expression is

$$\sigma = (1 - r^2)/(1 + r^2)\,. \tag{6}$$

It is seen that (5) differs from (6) by a factor that is slightly larger than 1.

Whitaker and Dole (1995) pointed out that localization due to the basic deformation has two effects. First, the zonal variation of basic deformation field itself can lead to a localized development of a storm track in the entrance region of a pure barotropic jet superimposed onto a zonally uniform baroclinic shear flow. Secondly, the basic deformation plays an important role in determining the length of a storm track. These two effects can be illustrated by examining the initial temporal growth rate of an elliptic-shaped eddy caused by the straining of the basic deformation as expressed in (5) or (6). Figure 5 plots the growth rate, as well as the terms "S" (the short-long dashed curve) and "A" (the long dashed curve), as a function of the eccentricity parameter, r for a diffluent basic flow as in the jet exit region. It is seen that in the jet exit region, a meridionally elongated eddy ($r > 1$) would have a net loss of kinetic energy to the basic flow as it is strained out by the basic diffluent flow. As the eccentricity continues to increase, the meridionally elongated eddy would experience an accelerated kinetic energy loss. It follows that being meridionally elongated, the longer (shorter) the baroclinically unstable eddies stay in the jet exit region, the shorter (longer) the storm track would be (Whitaker and Dole 1995). The results of local energetics analysis by CM90 showed that in the jet exit region, the meridionally elongated baroclinic eddies lose kinetic energy to the basic flow due to the straining mechanism (see Fig. 10b in this paper). Lee (1995b) studied the modulation mechanism of storm tracks by a zonal variation of the barotropic component of the basic flow in much less idealized situations, confirming analytically that the zonal wavelength of an eddy has to become smaller as it moves from a region of stronger zonal flow to a region of weaker zonal flow. The barotropic decay of synoptic scale eddies in the jet exit region has been confirmed with the observed data as well (e.g., Black and Dole 2002; Chang et al. 2002).

One could also infer the growth rate of an elliptic-shaped eddy in a confluent field (or in the jet entrance region) from Fig. 5 by simply replacing "*r*" with "*R*", where "*R*" denotes the ratio of the zonal length of an elliptic-shaped eddy to the meridional length (i.e., $R = 1/r$). This is because the growth rate of "$R < 1$" (meridionally elongated eddies) and "$R > 1$" (zonally elongated eddies) in a confluent flow is identical to the growth rate of "$r > 1$" and "$r < 1$" in a diffluent flow. It follows that meridionally elongated eddies ($R < 1$) in the jet entrance region have a positive growth rate under the basic straining field. This explains why a storm track tends to form in the entrance region of a purely barotropic jet stream superimposed onto a zonally uniform baroclinic shear flow (Whitaker and Dole 1995).

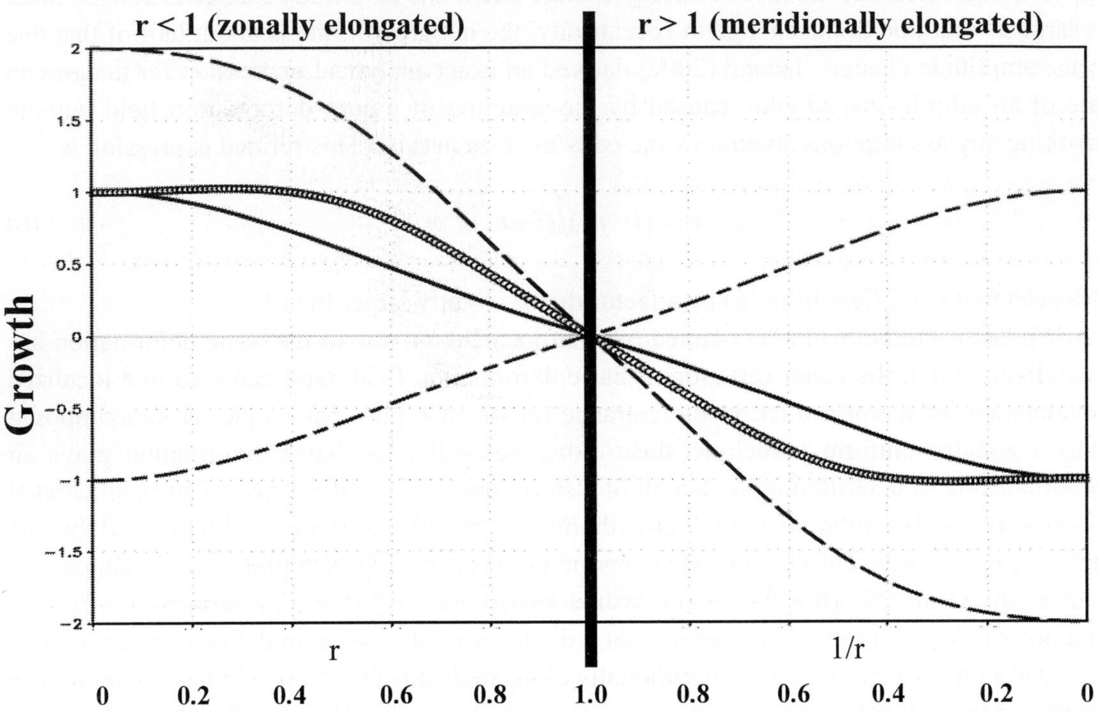

Figure 5. Non-dimensional growth rate of an elliptic-shaped eddy in a diffluent field as a function of the eccentricity parameter r. The thick solid vertical line separates the area of $r < 1$ (on the left, a zonally elongated eddy) from that where $r > 1$ (on the right, a meridionally elongated eddy). The abscissa for the region of $r < 1$ and $r > 1$ is "r" and "$1/r$", respectively. The curve with open circles and the solid curve are calculated based on Eqs. (5) and (6), respectively. The long-dashed and short-long-dashed curves correspond to the terms labeled with "A" and "S" in Eq. (5), respectively. The sum of the two curves is equal to the curve with open circles.

2.4. *Stochastic non-modal instability*

Linear eigenmode (or modal) instability analysis of the observed climatological mean flow appears to be able to explain qualitatively some key features of the observed storm tracks, including the peak amplitude location, the preferred (horizontal and vertical) orientations of eddies, the energetic processes, and the spatial variations of heat/momentum fluxes (Lee and Mak, 1995). However, it would be rather daunting task for one to infer the observed time-mean transient eddy statistics from a modal instability analysis. The non-orthogonal property of eigenmodes is a mathematical barrier for uniquely partitioning the observed variance using these eigenmodes. Moreover, there are typically a large number of unstable modes for the observed mean state and they appear similar to one another in certain aspects and yet differ from one another subtly. Particularly, the most unstable modes are often not substantially more unstable than other modes. It would take a much longer time for the most unstable modes to be dominant than the *e*-folding time scales of these modes.

Non-modal instability analysis offers an alternative approach to identify unstable perturbations. Instead of seeking for an exponential growth, a non-model instability analysis enables one to determine the fastest possible transient growths (thereby referred as "optimal modes") measured by a norm $||N||$ within a specific time interval τ (e.g., Farrell 1988, 1989ab, 1989b, Borges and Hartmanm 1992, Lee and Mak 1995; Chang and Mak 1995). The transient growths associated with optimal modes often are several factors faster than the exponential growths of unstable eigenmodes. Optimal modes have been argued to be more relevant to the development of observed transient eddies. In addition, one could explain the observed variance using a set of optimal modes because they are orthogonal to one another (Chang and Mak 1995). By definition, optimal modes are sensitive to the choices of the norm $||N||$ and the time interval τ. Particularly, for a larger value of τ, optimal modes become unstable eigenmodes.

Branstator (1995) attempted to explain the variance of synoptic scale eddies as the short-time ensemble statistics of initial random perturbations that evolve linearly about a climatological mean flow. As long as the integration time is not sufficiently long (say 5 days), the ensemble statistics of such a linear model bears a resemblance to the statistics of synoptic scale eddies in the fully non-linear integration. Effectively, Branstator's approach is a natural way of generating non-modal instability. The short-time ensemble statistics corresponds closely to the statistics of non-modal instability, rather than few fastest growing individual optimal modes. Presumably, the ensemble statistics of linear model would become the most unstable eigenmode for a sufficiently long integration.

The notion of stochastic non-modal instability has been introduced to the literature in a series of publications of Farrell and his collaborators (Farrell and Ioannou1993, 1994, 1995, 1996a,b; DelSole and Farrell 1995, DelSole 1996). In this framework, stochastically generated perturbations only grow temporally by extracting energy from the basic flow that is stable in the sense of modal instability. Because of the lack of modal instability, an individual perturbation would be asymptotically damped out due to dissipations. Such a linear system essentially mimics in a crude way two of the most important aspects of nonlinear dynamics: saturation and excitation. Whitaker and Sardeshmukh (1998) and Zhang and Held (1999) applied this idea to explain the storm tacks in the context of a zonally varying basic flow. They found that many salient features of the observed storm tracks, such as the midwinter suppression of the Pacific storm track and a relatively stronger storm track over the Atlantic, could be simulated by the statistics of the stochastically forced perturbations that are superimposed on the observed zonally varying basic flow. With reference to Branstator (1995)'s work, the stochastic forcing acts mechanically to prevent initially random perturbation from collapsing into the dominant eigenmodes (i.e., the least stable modes for a stable basic flow) by continuously "refreshing" the existing perturbations. As a result, non-modal transient growths are excited continuously and the statistics of the linear model can be regarded as the statistics of non-modal instability. The basic flow acts as an energy source to the stochastically generated perturbations and organizes the perturbations downstream of the jet cores via energy redistributions by advection and ageostrophic fluxes.

2.5. *Propagating versus stationary wave-packet resonance*

Mak (2002) proposed a complementary view on the underlying dynamics of local instability, namely, "propagating versus stationary wave-packet resonance" as opposed to "convective versus absolute instability". For a zonally parallel flow, the classic normal mode instability can be succinctly interpreted in terms of wave resonance (e.g., Bretherton 1966; Baines and Mitsudera 1994), which views the normal mode instability as a result of mutual reinforcement of two waves of the same wavelength with the same phase speed. Wave-packet resonance is the generalized interpretation for the instability of a zonally varying basic flow.

Whether the instability is due to propagating or stationary wave-packet resonance is determined by many factors. There are three primary factors dictating wave packet propagation along closed vorticity contours in a barotropic model: (a) planetary beta effect, (b) advection by basic flow, (c) dynamical beta effect due to the basic vorticity gradient. As in Mak (2002), let us consider a pair of wave packets that are in resonance (traveling with the same speed) looping around along the closed basic vorticity contours (Fig. 6). The wave-packet along the north/south rim of the maximum positive/negative vorticity patch would propagate eastward because of negative meridional gradient of the basic vorticity. These two wave packets would move westward when they are through the jet core where the meridional gradient of the basic vorticity is positive. In other words, the wave packet in the north (south) of the jet would move clockwise (counterclockwise) along the closed vorticity contours due to the dynamical beta effect of the basic flow. The planetary beta effect counters the eastward propagation. The westward propagation is opposed by the advection of a westerly basic mean flow. When these factors counterbalance one another, the wave packet resonance becomes stationary. The two wave-packets mutually reinforce each other by advecting the basic absolute vorticity. The resonance growth can be achieved by a self-reseeding mechanism without requiring recycling of perturbations through the global domain. Therefore, it corresponds to an unstable local mode. When these factors are off balance, the wave packets are no longer looping around the closed basic vorticity contours exactly. As a result, the resonance becomes propagating and the self-reseeding mechanism becomes weaker. For a sufficiently large propagation speed, the recycling of perturbation through the streamwise cyclic boundaries would overpower the self-reseeding mechanism. As a result, global modes become dominant. The stationary versus propagating wave-packet resonance offers an intuitive interpretation why instability properties of a zonally varying basic flow are generally sensitive to seemingly minor features, such as the domain-averaged wind and detailed spatial distribution of the basic vorticity gradient field.

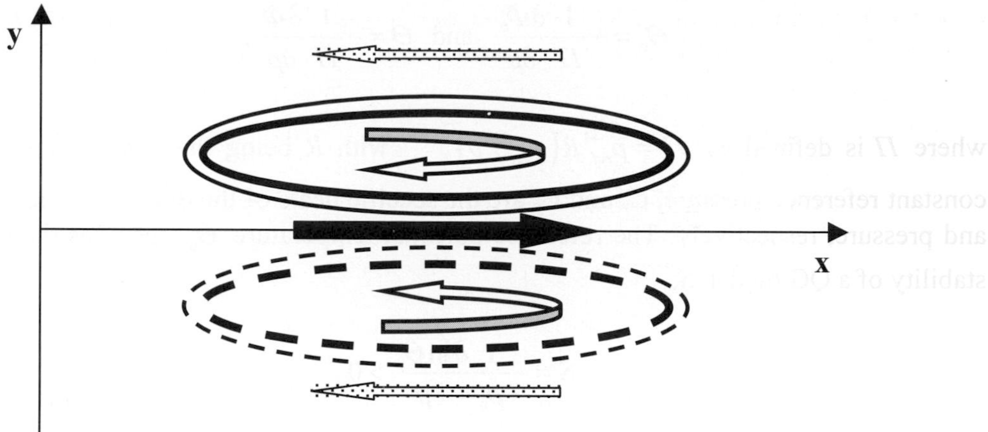

Figure 6. Schematic diagram illustrating the three primary factors dictating wave-packet resonance propagation in a barotropic model as discussed in Mak (2002). The ellipses represent contours of the relative vorticity of the basic flow. Solid (dashed) contours are positive (negative) vorticity. The thicker contours have a large absolute value. The curved arrows indicate the propagation of wave packets looping around the vorticity contours. The thick arrow indicates the propagation due to advective effect of the basic flow and shaded arrows the propagation due to the planetary beta effect.

3. Quasi-geostrophic framework of local energetics

This section presents a complete set of local energetics equations for disturbances superimposed onto a prescribed basic state in a continuous quasi-geostrophic (QG) β-plane model with pressure as the vertical coordinate. The derivation of local energetics equations in this paper is an extended version shown in CM90 with inclusion of nonlinear terms.

3.1. *Basic State Specification*

The QG dynamics can be modeled by a single prognostic equation, namely, the potential vorticity equation, which governs the evolution of one unknown, geopotential. Therefore, it suffices to define the basic state of a QG model with a single spatially varying field, namely the basic geopotential $\tilde{\Phi}(x,y,p)$,

$$\tilde{\Phi}(x,y,p)=\Phi_0(p)+\Phi(x,y,p), \tag{7}$$

where x and y stand for longitude and latitude and p is the pressure. In (7), $\Phi_0(p)$ is the reference geopotential, defining the basic stratification of a QG model and $\Phi(x,y,p)$ is the part predicted by the QG model. The corresponding basic potential temperature field $\tilde{\Theta}(x,y,p)=\Theta_0(p)+\Theta(x,y,p)$ is related to the basic geopotential through the hydrostatic equation, viz.

$$\Theta_0 = -\frac{1}{\Pi}\frac{\partial \Phi_0}{\partial p} \text{ and } \Theta = -\frac{1}{\Pi}\frac{\partial \Phi}{\partial p}, \tag{8}$$

where Π is defined as $\Pi = p_{ref}^{-1} R\left(p_{ref}/p\right)^{C_v/C_p}$ with R being the gas constant and p_{ref} a constant reference pressure, C_v and C_p are the specific heats of the dry air at constant volume and pressure, respectively. The reference potential temperature Θ_0 specifies the basic static stability of a QG model, S,

$$S = -\frac{1}{\rho_0}\frac{\partial \ln \Theta_0}{\partial p} > 0,$$

which is treated as one of a QG model's parameters with ρ_0 as the air density of a QG model. It should be added that $|\partial\Theta_0/\partial p| >> |\partial\Theta/\partial p|$ under the QG approximation.

For an easy reference, Φ and Θ are referred to as the basic geopotential and potential temperature, respectively, although they are really the departures of the basic flow from the reference geopotential and potential temperature (Φ_0 and Θ_0) that specify the basic stratification of a QG model. The basic geopotential $\Phi(x,y,p)$ is maintained by a forcing $H(x,y,p)$ so that

$$J(\Phi,\overline{q}) = H \text{ and } \overline{q} = \frac{1}{f_0}\nabla^2\Phi + \beta y + f_0\frac{\partial}{\partial p}\left(\frac{1}{S}\frac{\partial \Phi}{\partial p}\right). \tag{9}$$

where J is the Jacobian operator and $\overline{q}$ the basic potential vorticity. In (9), the vertical component of Coriolis parameter, $2\Omega\sin\varphi$, is approximated to be in the form of $(f_0 + \beta y)$ under the QG beta plane approximation, where f_0 and β are evaluated at a latitude $\varphi = \varphi_0 = 45°$ and y is the latitude coordinate centered at φ_0. The basic wind vector (U, V) can be determined from the basic geopotential Φ geostrophically. The vertical motion of the basic flow can be diagnosed from the basic geopotential Φ and the diabatic heating part, Q, of the basic external forcing H using the thermodynamic equation. In a QG model, the basic vertical motion is inconsequential to the evolution of disturbances superimposed onto the basic flow.

3.2. *The governing equations*

Unlike the global energetics budget analysis, it is impossible to derive a budget equation for a complete local energetics analysis from the potential vorticity equation without introducing errors. Particularly, the ageostrophic effect might not be fully recognized when working with the potential vorticity equation alone (MC89 and CM90). Although part of the ageostrophic flow is inconsequential to the development of the quasi-geostrophic potential vorticity, that variable is needed to obtain an exact local energetics budget for the perturbations in a QG model. In order to obtain an exact local energetics equation in the QG dynamics framework, one has to start from the quasi-geostrophic momentum, thermodynamics, hydrostatic, continuity equations with an explicit inclusion of the ageostrophic flow. The ageostrophic flow consists of (i) ageostrophic wind $\vec{v}^{(1)} = (u^{(1)}, v^{(1)})$ and (ii) ageostrophic geopotential $\phi^{(1)}$ (or ageostrophic pressure in the z coordinate). Following CM90, it is useful to decompose the ageostrophic wind into a rotational part $\vec{v}_r^{(1)}$ and an irrotational part $\vec{v}_d^{(1)}$, which can in turn be expressed as[3]

$$\vec{v}^{(1)} = \vec{v}_r^{(1)} + \vec{v}_d^{(1)} = \vec{k} \wedge \nabla\xi + \nabla\chi, \tag{10}$$

where $\vec{k}$ is the vertical unit vector; ξ and χ, respectively, are the streamfunction and velocity potential associated with the rotational and irrotational parts of the ageostrophic wind. Similarly, the ageostrophic geopotential $\phi^{(1)}$ can be partitioned as

$$\phi^{(1)} = f_0\xi + \phi_a^{(1)}. \tag{11}$$

Using (10) and (11), the nonlinear quasi-geostrophic governing equations for the disturbance departure from the basic state can be written as follows:

$$\frac{\partial u}{\partial t} + (U+u)\frac{\partial u}{\partial x} + (V+v)\frac{\partial u}{\partial y} = -u\frac{\partial U}{\partial x} - v\frac{\partial U}{\partial y} - \frac{\partial \phi_a^{(1)}}{\partial x} + f_0\frac{\partial \chi}{\partial y} + \beta y v, \tag{12a}$$

$$\frac{\partial v}{\partial t} + (U+u)\frac{\partial v}{\partial x} + (V+v)\frac{\partial v}{\partial y} = -u\frac{\partial V}{\partial x} - v\frac{\partial V}{\partial y} - \frac{\partial \phi_a^{(1)}}{\partial y} - f_0\frac{\partial \chi}{\partial x} - \beta y u, \tag{12b}$$

$$\frac{\partial \theta}{\partial t} + (U+u)\frac{\partial \theta}{\partial x} + (V+v)\frac{\partial \theta}{\partial y} = -u\frac{\partial \Theta}{\partial x} - v\frac{\partial \Theta}{\partial y} - \omega\frac{\partial \Theta_0}{\partial p}, \tag{12c}$$

$$\nabla^2\chi + \frac{\partial \omega}{\partial p} = 0, -\frac{\partial \phi}{\partial x} + f_0 v = 0, -\frac{\partial \phi}{\partial y} - f_0 u = 0, -\frac{\partial \phi}{\partial p} - \Pi\theta = 0. \tag{12d–g}$$

In the equations above, (u, v) are the zonal and meridional components of perturbation velocity that are related to the perturbation geopotential ϕ via the geostrophic balance as stated in (12e) and (12f); θ is the perturbation potential temperature, which is related to ϕ via the hydrostatic balance (12g); ω is the perturbation vertical velocity in p-coordinate; and ∇^2 is the horizontal Laplacian operator. There are 7 equations for 7 unknowns (u, v, ϕ, θ, ω, χ, $\phi_a^{(1)}$). Therefore, we have a close set of governing equations for disturbances

[3] In the literature, two extreme definitions of geostrophy are often used: one defines the wind that is in balance with the observed pressure gradient and the other defines the pressure gradient that balances with the observed wind. Should one of the two definitions be used in a QG model, one would not be able to obtain an energetically consistent QG momentum equation as in Andrews et al (1987,pg120-123) and Holton (1992,pg149-158). The inconsistency results in an unbalanced irrotational wind tendency in the momentum equation of geostrophic wind. The correct way of partitioning is to formally apply a perturbation expansion method with the Rossby number as the small parameter as in Pedlosky (1987, pg345-351) and MC90, which leads to equations (10) and (11) in this paper. The resulting QG momentum equations (12.a-b) are complete and energetically consistent.

superimposed onto a prescribed basic flow Φ. Note that the terms associated with dissipation and diabatic heating are not included in (12) for simplicity. This set of equations can be reduced to a single prognostic equation, the potential vorticity equation, in terms of one unknown, the perturbation geopotential, ϕ. The perturbation potential vorticity can be determined from ϕ according to

$$q = \frac{1}{f_0}\nabla^2\phi + f_0\frac{\partial}{\partial p}(\frac{1}{S}\frac{\partial\phi}{\partial p}).$$

The perturbation potential vorticity equation is

$$f_0\frac{\partial q}{\partial t} + J(\Phi, q) + J(\phi, \overline{q}) + J(\phi, q) = 0. \tag{13}$$

Note that the rotational part of the ageostrophic wind (ξ) does not appear in (12). Therefore, it is inconsequential to the evolution of both the momentum and vorticity of the quasi-geostrophic flow. This particularly implies that the rotational part of ageostrophic wind does not affect the energetics of a quasi-geostrophic disturbance. The irrotational part of the ageostrophic wind (χ) appears explicitly in the momentum equation. It also implicitly affects the vorticity tendency of a quasi-geostrophic disturbance through the stretching, implying that its effect is recorded in the evolution of the quasi-geostrophic perturbation potential vorticity. The ageostrophic geopotential $\phi_a^{(1)}$, on the other hand, is the part of the ageostrophic flow that is inconsequential to the development of the perturbation potential vorticity but pivotal to the local kinetic energetics budget of a quasi-geostrophic disturbance because it appears in the momentum equation explicitly. This implies that as long as one's analysis is solely based on the potential vorticity in a QG model, one would not even be conscious of the presence or absence of the effects of the ageostrophic geopotential. From the energetics viewpoint, the ageostrophic flow produces "pressure work" causing the change of the geostrophic wind kinetic energy. The pressure work associated with the ageostrophic wind can be rewritten into two parts: one has a net non-zero global mean value representing the transfer from potential to kinetic energy and the other has a net zero global mean value. However, the global mean of pressure work associated with the ageostrophic potential is always zero.

It might sound puzzling that the ageostrophic geopotential (or the ageostrophic pressure in z-coordinate) has no impact on the evolution of the vorticity of a quasi-geostrophic disturbance and yet it affects the perturbation kinetic energetics (not globally but locally). Physically, the terms, $(\partial\phi_a^{(1)}/\partial x, \partial\phi_a^{(1)}/\partial y)$, act to balance the tendency of irrotational wind in the momentum equations that arise from advection terms. When the momentum equations are rewritten in terms of the vorticity and divergence equations, the tendency of irrotational wind, by definition, only appears in the divergence equation but not in the vorticity equation. Under the QG approximation, the tendency of irrotational wind due to the advection of rotational wind by rotational wind is neglected automatically although it implicitly exists at any given time. When considering the local energetics budget, these terms have to be considered explicitly in order to get an exact energy balance at each grid point as shown in MC89 and CM90. In a primitive equation model, it is no longer necessary to define the ageostrophic potential $\phi_a^{(1)}$ as long as one defines the "geostrophic wind" as the part of the wind balancing with the pressure gradient force, as in Orlanski and Chang (1993), because the tendency of irrotational wind is always explicitly present in a primitive equation model. In this case, the only the ageostrophic terms that would redistribute energy spatially are the terms associated with the ageostrophic wind, which includes both rotational and irrotational winds as in (11).

Among the seven unknowns, (u, v, θ) are determined by prognostic equations (12a-c). The geopotential ϕ can be obtained from one of the equations (12e-g). The vertical motion ω can be determined from the omega equation with appropriate boundary conditions

$$(\nabla^2 + \frac{f_0^2}{S}\frac{\partial^2}{\partial p^2})\omega = \frac{f_0}{S}G, \tag{14}$$

where

$$\begin{aligned}
G = {} & 2[\frac{\partial^2 U}{\partial x \partial p}(\frac{\partial v}{\partial x}+\frac{\partial u}{\partial y}) - \frac{\partial U}{\partial x}(\frac{\partial^2 v}{\partial x \partial p}+\frac{\partial^2 u}{\partial y \partial p}) + \frac{\partial U}{\partial p}\nabla^2 v - \frac{\partial V}{\partial p}\nabla^2 u] \\
& + 2[\frac{\partial^2 u}{\partial x \partial p}(\frac{\partial V}{\partial x}+\frac{\partial U}{\partial y}) - \frac{\partial u}{\partial x}(\frac{\partial^2 V}{\partial x \partial p}+\frac{\partial^2 U}{\partial y \partial p}) + \frac{\partial u}{\partial p}\nabla^2 V - \frac{\partial v}{\partial p}\nabla^2 U] \\
& + 2[\frac{\partial^2 u}{\partial x \partial p}(\frac{\partial v}{\partial x}+\frac{\partial u}{\partial y}) - \frac{\partial u}{\partial x}(\frac{\partial^2 v}{\partial x \partial p}+\frac{\partial^2 u}{\partial y \partial p}) + \frac{\partial u}{\partial p}\nabla^2 v - \frac{\partial v}{\partial p}\nabla^2 u] + \beta\frac{\partial v}{\partial p}.
\end{aligned}$$

We have made use of $S = -\Pi(\partial \Theta_0 / \partial p)$ and assumed it to be a constant in deriving (14). The ageostrophic velocity potential χ and the corresponding irrotational ageostrophic wind $\vec{v}_d^{(1)} = \nabla\chi$ can then be determined from ω by solving (12d). After taking $\partial / \partial x$ of (12a) and $\partial / \partial y$ of (12b) and summing the resulting expressions, we obtain a diagnostic equation for the ageostrophic geopotential as

$$-\nabla^2 \phi_a^{(1)} = 4\frac{\partial U}{\partial x}\frac{\partial u}{\partial x} + 2\frac{\partial U}{\partial y}\frac{\partial v}{\partial x} + 2\frac{\partial u}{\partial y}\frac{\partial V}{\partial x} + 2\left(\frac{\partial u}{\partial x}\right)^2 + 2\frac{\partial u}{\partial y}\frac{\partial v}{\partial x} - \frac{\partial(\beta y v)}{\partial x} + \frac{\partial(\beta y u)}{\partial y}. \tag{15}$$

The fact that both basic and perturbation wind are non-divergent is useful in simplifying the right hand side terms in (15).

3.3. *Local energetics equations*

The kinetic energy K and potential energy P at each grid point are defined as

$$K=\frac{1}{2}\left(u^2+v^2\right),\ P=\frac{1}{2S}\left(\frac{\partial\phi}{\partial p}\right)^2. \tag{16}$$

Now the local energetics equations can be derived from (12.a,b,c) using (12.g) as

$$\frac{\partial K}{\partial t}=-(\vec{V}+\vec{v})\cdot\nabla K-\nabla\cdot\left(\phi_a^{(1)}\vec{v}+\phi\vec{v}_d^{(1)}\right)-\frac{\partial(\omega\phi)}{\partial p}+\vec{E}\cdot\vec{D}-(\vec{k}\cdot\vec{F}_3)(\vec{k}\cdot\vec{T}_3), \tag{17}$$

$$\frac{\partial P}{\partial t}=-(\vec{V}+\vec{v})\cdot\nabla P+\vec{F}_3\cdot\vec{T}_3, \tag{18}$$

where

$$\vec{E}=\left(\frac{1}{2}(v^2-u^2),\ -uv\right),\qquad \vec{D}=\left(\frac{\partial U}{\partial x}-\frac{\partial V}{\partial y},\ \frac{\partial V}{\partial x}+\frac{\partial U}{\partial y}\right), \tag{19a, b}$$

$$\vec{F}_3=\Pi\left(u\theta,\ v\theta,\ -\omega\theta\right),\qquad \vec{T}_3=\left(-\frac{\partial\Theta_0}{\partial p}\right)^{-1}\left(-\frac{\partial\Theta}{\partial x},\ -\frac{\partial\Theta}{\partial y},\ \frac{\partial\Theta_0}{\partial p}\right). \tag{19c, d}$$

Note that the vectors $\vec{F}_3$ and $\vec{T}_3$ are three-dimensional and all other vectors are two-dimensional. Finally, the sum of (17) and (18) yields the equation for the total disturbance energy $(K+P)$,

$$\frac{\partial(K+P)}{\partial t}=-(\vec{V}+\vec{v})\cdot\nabla(K+P)-\nabla\cdot(\phi_a^{(1)}\vec{v}+\phi\overline{v_d^{(1)}})-\frac{\partial\omega\phi}{\partial p}+\vec{E}\cdot\vec{D}+\vec{F}_2\cdot\vec{T}_2. \tag{20}$$

In (20), $\vec{F}_2$ and $\vec{T}_2$ are the horizontal components of the vectors $\vec{F}_3$ and $\vec{T}_3$, respectively.

3.4. *Physical interpretations*

(i) Local growth due to spatial energy redistributions

There are three spatial energy redistribution terms in (20): the horizontal energy advection (the first term), the horizontal energy flux convergence due to the ageostrophic flow (the second term), and the vertical energy flux convergence due to the vertical motion (the third term). The latter two terms are the part of "pressure work" that becomes zero when averaged over the whole domain. Therefore, these two terms are labeled as "redistribution" terms instead of source terms. The part of the "pressure work" that has a non-zero global mean value, thereby being called as a source term, is the last term in (17). This term will be discussed shortly in the context of baroclinic energy conversion.

The importance of these energy redistribution terms in the context of a zonally varying basic flow has been discussed in section 2. It suffices to reiterate here that although the domain mean of each of these three terms is equal to zero, implying that these terms have no contribution to the global energetics of disturbances, they indirectly dictate the instability properties of a local mode by determining the location where perturbations extract energy from a zonally varying background flow most vigorously. Mathematically, in the context of a zonally homogeneous flow, the time-mean and domain-mean operators are equivalent to one another as far as a linear eigenmode is concerned. For the case of a zonally varying basic flow, the time mean is no longer the same as the domain mean even for a linear eigenmode because of the spatial variation of perturbation amplitude. Particularly, the time mean of an energy redistribution term can be non-zero although its global mean has to be zero.

(ii) Local growth due to barotropic energy conversion

The inner product of $\vec{E}$ and $\vec{D}$ is the kinetic energy conversion rate between the mean state and disturbance. The E-vector, consisting of elements of the anisotropic part (trace free) of the instantaneous horizontal velocity correlation tensor, measures the disturbance's local orientation/shape, as pointed out in Hoskins et al. (1983).[4] The x-component and y-component of $\vec{D}$ are the stretching and shearing deformations of the background flow, respectively. One can easily verify that the magnitude of both $\vec{E}$ and $\vec{D}$ is invariant under a coordinate transformation by rotation about the vertical axis. However, both $\vec{E}$ and $\vec{D}$ still cannot be regarded as the true vectors because the direction of each of the two vectors changes by an angle of -2λ if the coordinate system is rotated by λ. Nevertheless, the angle between the two vectors is invariant with respect to a rotation coordinate transform. It follows that the inner product of these two vectors is invariant under a rotation coordinate transform. The inner product of $\vec{E}$ and $\vec{D}$ can be rewritten as

$$\vec{E}\cdot\vec{D}=\left|\vec{E}\right|\left|\vec{D}\right|\cos\left[2(\varepsilon-\gamma)\right], \tag{21}$$

$$2\varepsilon=\arctan\left[-uv\Big/\left(\frac{1}{2}(v^2-u^2)\right)\right],\ -\frac{\pi}{2}\le\varepsilon\le\frac{\pi}{2} \tag{22}$$

$$2\gamma=\arctan\left[\left(\frac{\partial V}{\partial x}+\frac{\partial U}{\partial y}\right)\Big/\left(\frac{\partial U}{\partial x}-\frac{\partial V}{\partial y}\right)\right],\ -\frac{\pi}{2}\le\gamma\le\frac{\pi}{2} \tag{23}$$

where (2ε) and (2γ) are the angles of the vectors $\vec{E}$ and $\vec{D}$ with respect to the x-axis, respectively. The local orientation of eddies with respect to the x-axis, α, can be deduced from the direction of the E-vector according to

---

[4] The E-vector defined in Hoskins et al. (1983) neglects the factor of ½ in the x-component. As pointed out in MC89, this seemingly innocent constant factor is important not only because it appears naturally in the kinetic energy equation but also the E-vector would preserves its amplitude under a coordinate rotation transform only when this constant factor is included. The inclusion of the factor (1/2) is also necessary for a straight-forward connection of the x-component of the E-vector to the zonal component of the group velocity relative to the mean flow (Plumb 1986). The advantage of defining the E-vector without the factor is to express the barotropic energy conversion in terms of the gradient of zonal wind, as shown in Hoskins et al. (1983) and Simmons et al. (1983).

$$\alpha = \varepsilon + \frac{\pi}{2} \, . \tag{24}$$

Since ε is in the range from $-\pi/2$ to $\pi/2$, α is in the range from 0 to π, covering all possible directions in terms of orientation. The angle γ corresponds to the orientation of the axis of dilatation of the basic flow. Let δ be the angle of the contraction axis of the basic deformation with respect to the x-axis, which is equal to $(\gamma - \pi/2)$. It follows that $(\varepsilon - \gamma) = (\alpha - \delta)$. Therefore, we have

$$\begin{cases} \vec{E}\cdot\vec{D} > 0 \ \textit{for}\ (\alpha - \delta) \in (-\pi/4,\ \pi/4), \\ \vec{E}\cdot\vec{D} = 0 \ \textit{for}\ (\alpha - \delta) = \pm\pi/4\ , \\ \vec{E}\cdot\vec{D} < 0 \ \textit{for}\ (\alpha - \delta) \in [-\pi/2, -\pi/4),\ \textit{or}\ (\alpha - \delta) \in (\pi/4,\ \pi/2], \end{cases} \tag{25}$$

As a special case of (25), one can conclude that *to optimally extract kinetic energy from the basic flow, a disturbance must be orientated along the axis of contraction of the basic flow* ($\alpha = \delta$) *and on the other hand, if a disturbance is orientated along the axis of dilatation of the basic flow* $\alpha - \delta = \pm\pi/2$), *it would lose kinetic energy to the basic flow at the fastest possible rate*. It follows that the classic statement about the barotropic energy conversion, namely that "*perturbation must lean against the shear*" (e.g., Pedlosky 1987, pg499-508), has to be rephrased to that "*a barotropically unstable perturbation has to be elongated along the contraction axis of the basic deformation*". For a zonally uniform barotropic jet, the "against shear" orientation is the same as the contraction axis, but for a zonally varying barotropic jet this is ***not*** the case.

Iacono (2002) showed that in the natural coordinate system that follows the basic wind, the kinetic energy conversion rate from the basic flow to perturbations can be expressed in terms of an "effective shear" vector, which exactly coincides with the shear vector in the natural coordinate of an irrotational flow. The magnitude of the effective shear vector gives rise to the upper bound of the growth rate of a localized perturbation whereas the direction of the effective shear vector is an indicative of the optimal orientation of the unstable perturbations. Specifically, the optimal orientation, measured by the local averaged dominant perturbation wind orientation, is perpendicular to the bisector of the angle between the effective shear vector and the basic wind. In the case of a zonally uniform flow, the effective shear vector is always perpendicular to the basic wind and it points to the left (right) of the basic wind for a westerly (easterly) shear. This implies that the bisector of the angle between the effective shear vector and the basic wind is a SW-NE (NW-SE) orientation on the south (north) side of a westerly jet. Therefore, the preferred dominant perturbation wind orientation aligns along the NW-SE (SW-NE) direction on the south (north) side of the jet. Obviously, this is the familiar "leaning against shear" statement. For a non-parallel basic flow, the preferred orientation still appears always "leaning against the shear" to an observer who follows the basic wind.

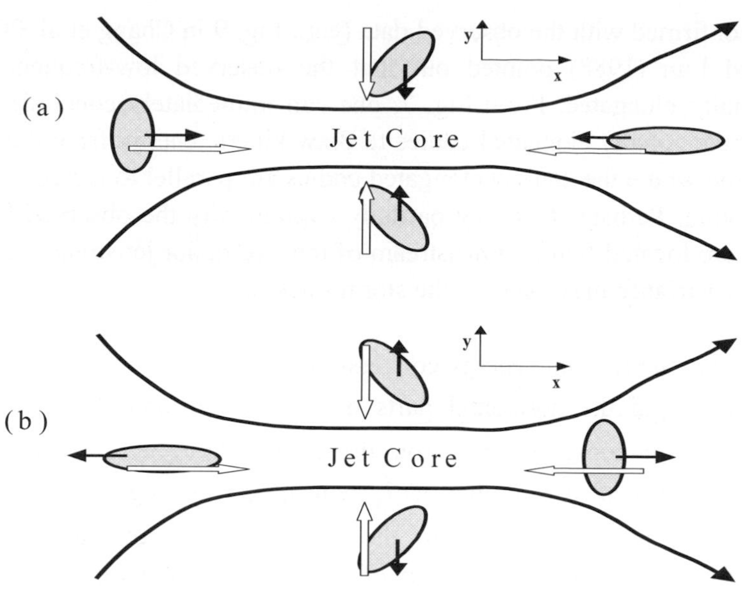

Figure 7. Schematic diagram illustrating (a) the optimal and (b) the worst-possible eddy orientations for extracting kinetic energy from the basic flow in streamwise and traverse directions. The curves represent the basic streamfunction. The block arrows are the basic *D*-vectors at different locations of the localized jet stream. The shaded ellipses represents eddies' orientations with the solid arrows indicating their E-vectors. The contraction axis of the basic flow has a NW-SE (SW-NE) orientation on the south (north) side of the jet core and is parallel to the y-axis (x-axis) in the jet entrance (exit) region.

Figure 7 shows schematically the optimal and worst possible orientations of eddies in terms of their ability to extract kinetic energy from a localized jet stream in both streamwise and cross-streamwise directions. It is known the observed high-frequency eddies have a preferred SW-NE (NW-SE) orientation south (north) of the jet core. As indicated in Fig. 7b, they are the worst possible orientations for eddies to draw kinetic energy from a jet stream. Therefore, the observed high-frequency eddies act to maintain the mean jet core by losing kinetic energy to the mean flow. In other words, high-frequency eddies have a positive feedback on the time mean jet by depositing westerly momentum to the mean flow (e.g., Hoskins et al. 1983; Lau 1988; Cai and van den Dool 1992). It is also seen from Fig. 7a that the entrance region of a barotropic jet with a zonally uniform baroclinicity would be the favorite location for meridionally elongated baroclinic eddies to grow because they would be able to extract energy baroclinically and barotropically. As the meridionally elongated eddies move to the jet exit region, they would begin to lose energy to the basic flow barotropically because their orientation is along the dilatation axis in the jet exit region (Fig. 7b). Subsequently, they would be deformed to be more meridionally elongated and lose energy to the basic flow barotropically at the fastest possible rate (Fig. 5), explaining the barotropic decay of baroclinically unstable eddies in the jet exit region (e.g., CM90; Whitaker and Dole 1995; and Lee 1995b). The mechanism of barotropic growth (decay) in the jet entrance (exit) region has been confirmed with the observed data (e.g., Fig. 9 in Chang et al. 2002). Hoskins et al. (1983) and Lau (1988) pointed out that the observed low-frequency eddies are preferentially zonally elongated. From Fig. 7, one can immediately conclude that the only favorite place for the zonally elongated eddies to draw kinetic energy from the mean flow is in the jet exit region where the zonally elongated eddies are parallel to the contraction axis of the basic deformation. Perhaps, this may partially explain why the observed low-frequency variance maxima are located farther downstream of the two major jet streams compared with the high-frequency variance maxima (i.e., the storm tracks).

(iii) Local growth due to baroclinic energy conversion

The inner product of the horizontal parts of vectors $\vec{F}_3$ and $\vec{T}_3$ (i.e., $\vec{F}_2 \cdot \vec{T}_2$) is the available potential energy conversion rate from the mean state to perturbations. The product of the vertical component of $\vec{F}_3$ and that of $\vec{T}_3$ with a negative sign is the conversion rate from the perturbation potential energy to the perturbation kinetic energy. This term corresponds to the part of the pressure work due to the ageostrophic wind that has a non-zero global mean value. The full inner product of these two three-dimensional vectors $\vec{F}_3$ and $\vec{T}_3$ itself gives rises to a net perturbation potential energy growth due to the baroclinic energy conversion. The vector $\vec{F}_3$ is the three-dimensional heat flux vector of perturbation flow. The vector $\vec{T}_3$ approximately represents the vector $\vec{N}_{\tilde{\Theta}}$, the normal vector of a basic isentropic surface pointing to the down-gradient direction. This becomes clear after rewriting $\vec{T}_3$ as

$$\vec{T}_3 \approx \vec{N}_{\tilde{\Theta}} = \left[\frac{\partial(-p)}{\partial x}, \frac{\partial(-p)}{\partial y}, -1\right]_{\tilde{\Theta}=\mathrm{Const}} . \tag{26}$$

Particularly, the two-dimensional vector $\vec{T}_2$ (the horizontal part of $\vec{T}_3$) is parallel to the horizontal gradient of the basic $\tilde{\Theta}$ fields pointing to the down-gradient direction.

The generalized necessary condition for baroclinic instability in the context of a zonally varying basic flow can be written, in a local area average sense, as

$$\vec{F}_2 \cdot \vec{T}_2 > -(\vec{k} \cdot \vec{F}_3)(\vec{k} \cdot \vec{T}_3) > 0 \quad \text{or} \quad \left(-\frac{\partial \Theta_0}{\partial p}\right)^{-1} \left(-u\theta \frac{\partial \Theta}{\partial x} - v\theta \frac{\partial \Theta}{\partial y}\right) > -\omega\theta > 0 . \tag{27}$$

It is said according to (27) that the potential energy conversion from the basic flow to perturbations resulting from the *down-gradient horizontal heat fluxes* of a baroclinically unstable perturbation has to be larger than the energy conversion from the perturbation potential energy to the perturbation kinetic energy associated with the *up-gradient vertical heat fluxes* of a baroclinically unstable perturbation. As a result, the unstable perturbation has a net gain of both the available potential energy and kinetic energy. In the subsequent nonlinear development, the horizontal gradient of the basic $\tilde{\Theta}$ field would be reduced and the vertical gradient of the basic $\tilde{\Theta}$ field would be increased (the latter would increase the static stability, which is neglected under QG approximation). Figure 8 sketches the relation between $\vec{T}_3$ and $\vec{F}_3$ for an unstable perturbation on an edge-wise cross-section normal to the $\tilde{\Theta}$ surfaces passing through the point under consideration.

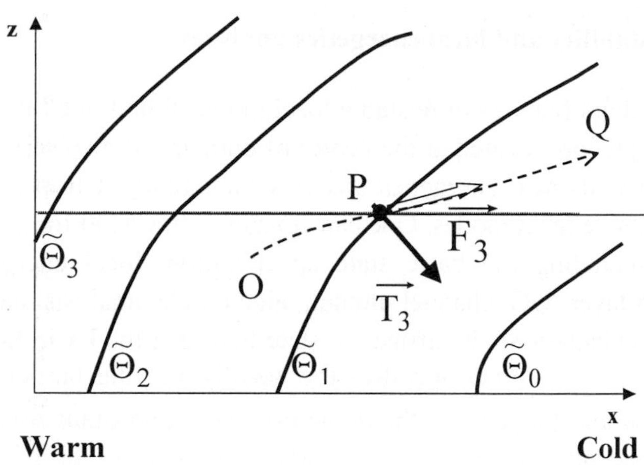

Figure 8. Schematic diagram illustrating the relation between a baroclinically unstable parcel trajectory and the basic isentropic surfaces. The thick curves represent isentropic surfaces $\tilde{\Theta}_i$ with that $\tilde{\Theta}_i > \tilde{\Theta}_j$ for $i > j$. The solid arrow is the 3-D T-vector at point P, pointing to the down-gradient direction of the isentropic surfaces. The point O is on the warmer and Q colder side of the $\tilde{\Theta}_1$ surface. The trajectory OQ penetrates the isentropic surface $\tilde{\Theta}_1$ at point P. The block arrow represents the 3-D F-vector associated with the unstable trajectory at the point P and it is parallel to the trajectory at P at the time when the parcel passes through the point P.

Dividing both sides of (27) by $(-\omega\theta)$, which is positive for an unstable perturbation, leads to a relation between the trajectory of an unstable parcel and the slope of basic isentropic surfaces, namely,

$$\frac{\delta x}{\delta z}\frac{\Delta z}{\Delta x}+\left[\frac{\delta y}{\delta z}\frac{\Delta z}{\Delta y}\right]_{\tilde{\Theta}=\text{Const}}>1, \tag{28}$$

where δx, δy, and δz $(= -\delta p/\rho_0 g)$ are displacements of the parcel relative to the basic flow whereas Δx, Δy, and Δz $(= -\Delta p/\rho_0 g)$ are increments between two points on a basic isentropic surface. It is said according to (28) that the trajectory of a baroclinically unstable parcel must be within the triangle wedge between a horizontal plane (the thin line in Fig. 8) and the local isentropic surface (the thick contour line labeled with $\tilde{\Theta}_1$ in Fig. 8) through which the parcel penetrates with a warm/cold air parcel (the thick dashed curve OPQ in Fig. 8, representing a warm air parcel) moving down-gradient/up-gradient direction of the basic isentropic surface. The gentler trajectory slope compared to the basic isentropic surface would ensure that the local temperature anomaly due to the horizontal advection exceeds that is required to overcome the static stability and therefore leads to a net acceleration forward, implying an instability. This statement generalizes the physical interpretation about baroclinic instability put forward by Eady (1949) to the case of a zonally varying basic flow. As in the conventional baroclinic instability problem, the criterion (28) is a necessary but not sufficient condition for local baroclinic instability. It is the generalized form for a zonally uniform baroclinic basic flow [e.g., Eq. (7.6.9) in Pedlosky 1987].

4. Examples of local instability and local energetics analyses

This section reviews the salient features of unstable local modes found in CM90 for a zonally varying basic state with a jet core located at the center of domain. It also serves the purpose of demonstrating the utility of local energetics analysis for gaining a better understanding about the behaviors of unstable local modes. One can consult with CM90 for the details about the instability analysis, including the basic state specification, local energetics analysis implementation in a two-layer QG channel model, eigenmode analysis and diagnostics procedures, and feedback calculations. It suffices to state here that the basic flow consists of two parts: (U_S, V_S) and U_0. The former specifies the zonally varying baroclinic/barotropic shear part of the basic flow and the latter is the domain-invariant constant zonal wind. As in the case of P84, it is intrinsically important to understand why the instability properties of a local mode would change by just varying the constant part of the zonal wind without alternating the baroclinic/barotropic shear part.

Figures 9 shows the growth rate of unstable modes as a function of U_0 with the same amount of baroclinic/barotropic shear specified by (U_S, V_S). It is seen that the growth rate of an unstable local mode is large when U_0 is small and gradually diminishes as U_0 increases. As shown in Fig. 6 in CM90, a local mode consists of a group of localized waves with their total amplitude peaking downstream of the jet core. As U_0 increases, the location where the amplitude of an unstable local mode peaks moves further downstream of the jet core. The unstable global modes, on the other hand, remain unstable for a large value of U_0. An unstable global mode also consists of a group of waves but has a single dominant monochromatic wave component. As a result, the time-mean amplitude of an unstable global mode has a weaker spatial variation with the peak position located far away downstream of the jet core. Moreover, the peak location of an unstable global mode shows little sensitivity with respect to U_0. Another important difference between local and global modes is that the instability of a local mode does not requires a "recycling" of perturbations through the cyclic channel domain which is necessary for instability of a global mode (see Fig. 5 of CM90).

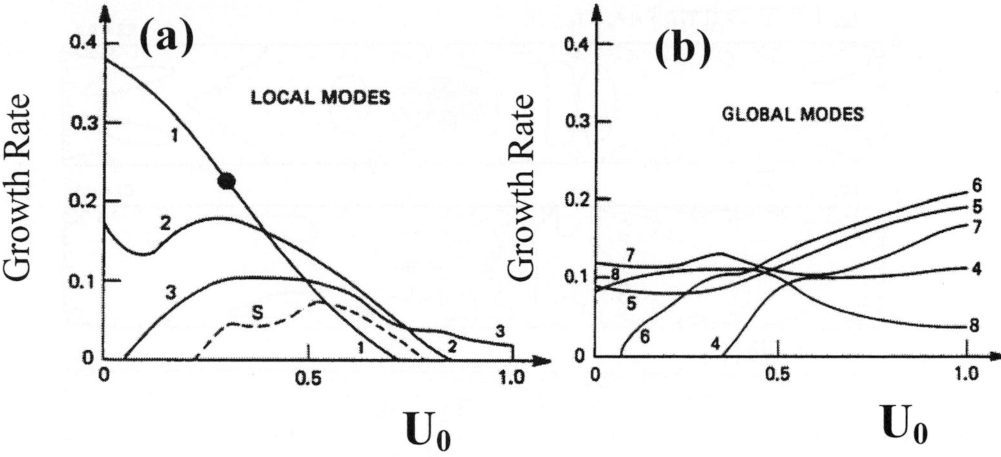

Figure 9. Growth rate of the unstable modes as a function of U_0. (a) local modes and (b) global modes. The solid curves are traveling modes and dashed curve stationary (zero phase velocity relative to the ground) mode (adapted from Cai and Mak 1990).

Figure 10 summarizes the local energetics calculation of the most unstable local mode (indicated by a "dot" in Fig. 9a). The maximum amplitude of this unstable local mode is located at the longitude "1.0" (1 tick mark from the jet core). Some of the salient features shown in Fig. 10 are:

- The potential energy conversion from the basic flow to perturbations is rather localized. The maximum conversion is located at the longitude "1.0", immediate-downstream of the jet core (Fig. 10a). It is seen that the spatial distribution of the potential energy conversion rate is fairly symmetric about its maximum center. Most of the potential energy extracted from the basic flow is immediately converted locally to the perturbation kinetic energy (Fig. 8a of CM90).
- Because of being meridionally elongated, baroclinically unstable eddies lose kinetic energy in the exit region of the basic jet stream (Fig. 10b). The negative center is located at the longitude "2.0". This demonstrates the "barotropic decay" of the meridionally elongated synoptical scale eddies in the jet exit region.
- The energy redistribution terms (Figs. 10c-e) overall act to remove perturbation energy from the immediate-downstream region to a region further downstream of the jet core. This explains the asymmetry of the perturbation energy of this local mode with respect to its maximum center showing a much-elongated termination of storm track in the downstream direction (Fig. 10f). In particular, the advection terms play a significant role in causing the downstream development of perturbation activities. The energy flux associated with the ageostrophic potential also contributes to the downstream development. But the energy flux by the ageostrophic wind exhibits upstream energy propagation. The net of these two ageostrophic energy redistribution terms contributes a stronger positive center upstream and a weaker positive center downstream.

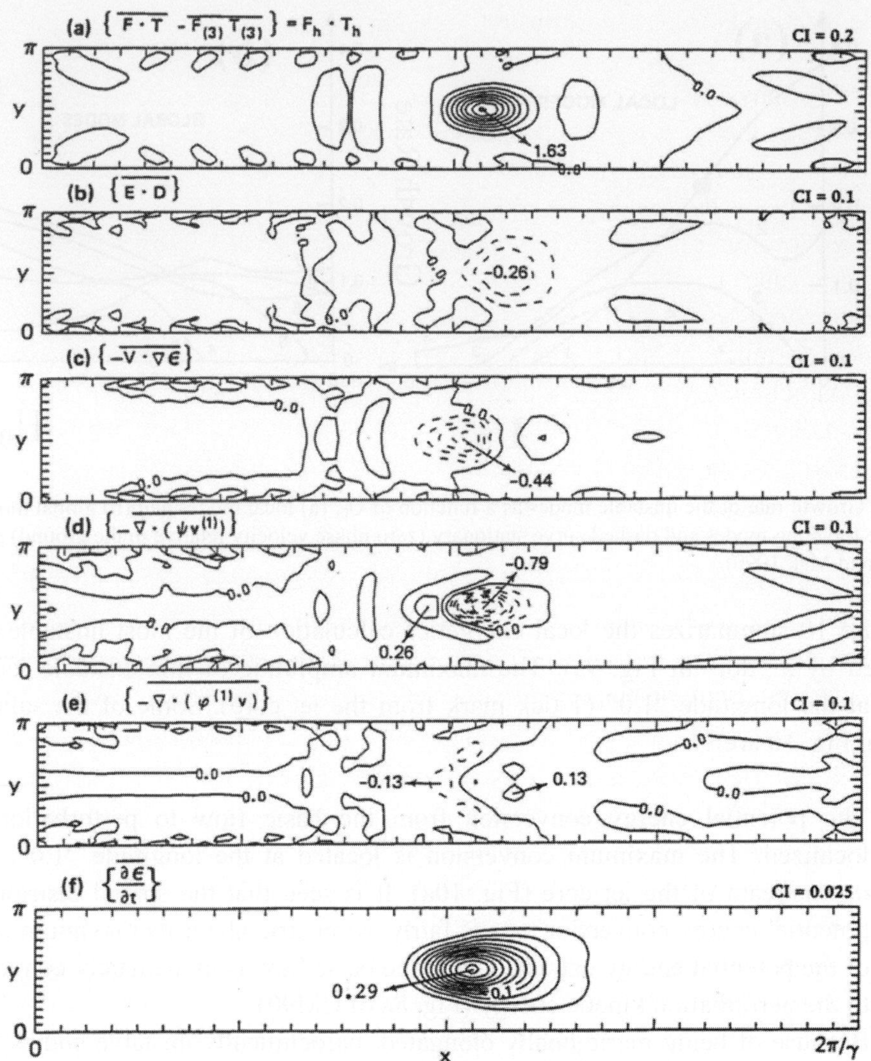

Figure 10. Local energetics of the unstable mode #1 shown in Fig. 9a. (a) Potential energy conversion rate. (b) Kinetic energy conversion rate. (c) Energy advection by the basic flow. (d) Convergence of energy flux associated with the divergent part of the ageostrophic flow. (e) Convergence of energy flux associated with the ageostrophic geopotential. (f) Net local energy change rate. Spatial distribution of the total energy of the unstable mode is equal to panel (f) divided by the growth rate which is equal to 0.22 as marked with the solid circle in Fig. 9a (after Cai and Mak 1990).

5. Concluding remarks

This paper reviews the local instability theory of a zonally varying basic flow focusing on explaining the storm track location with respect to the background jet core. Because the availability of background energy sources is zonally inhomogeneous, the peak perturbation amplitude location becomes an intrinsic factor that reflects the instability properties of a local mode.

In terms of physical processes, the peak perturbation location is determined by an optimal balance between the local growth due to energy extracted from the basic flow and the local growth due to energy redistributions. In particular, the energy redistribution term due to advection always leads to energy propagation in the downstream direction. This implies that the preferred location of a storm track has to be downstream of the basic jet core. The energy redistribution due to the ageostrophic fluxes generally leads to both upstream and downstream energy propagations. Numerical calculations reveal that the downstream energy propagation is much stronger than the upstream one. This is particularly the case for nonlinear baroclinic wave packets, causing successive development of perturbations in the downstream direction.

In terms of dynamical mechanisms, the weaker background baroclinicity away from the basic jet core inhibits the global mode growth by limiting "reseeding" through cyclic boundaries. As a result, local modes that do not require "reseeding" through cyclic boundaries may prevail as long as the downstream spread of perturbations by advection is not sufficiently strong. The stationary wave-packet resonance provides a "self-reseeding" mechanism for local modes to grow. When the wave-packet resonance becomes propagating, self-reseeding mechanism becomes secondary and the repetitive reseeding through cyclic boundary becomes dominant. This particularly implies that local modes would diminish when the zonal mean wind speed exceeds a critical value.

Local energetics analysis has been proved to be a valuable tool to gain additional insight into the underlying dynamics and physical processes for storm tracks. It can be and has been generically applied to linear/nonlinear and modal/nonmodal instability analyses. As discussed in section 3.4, with the aid from local energetics analysis, the classic interpretations for the barotropic/baroclinic instability can be straightforwardly and unambiguously generalized for a zonally varying basic flow. In terms of baroclinic instability, the classic interpretation for the baroclinic instability of a zonally uniform flow (Eady, 1949) remains qualitatively valid if applied locally for a zonally varying basic flow. The relation between the perturbation orientation and the basic deformation field rather than the basic vorticity determines the barotropic energy conversion from the basic flow to perturbations. For a zonally uniform barotropic jet, the "against shear" orientation is the same as the contraction axis, but this is ***not*** the case for a zonally varying barotropic jet. Therefore, the statement that "*a barotropically unstable eddy has to lean against shear*" should be rephrased as "*a barotropically unstable perturbation has to be elongated along the contraction axis of the basic deformation*". The classic "leaning against shear" instability picture is meaningful only to an observer who follows a zonally non-parallel basic wind. Because the basic deformation field reverses its polarity near a jet core in both streamwise and traverse directions, a perturbation would experience a change of its stability/instability property as it moves from the jet entrance to the exit region if its orientation remains unchanged. Particularly, the meridionally elongated baroclinic eddies would grow (decay) barotropically in the jet entrance (exit) region. The barotropic decay rate in the jet exit region partially determines the storm track length. Moreover, a down-gradient momentum transport is no longer the necessary condition for barotropic instability of a zonally varying flow. A meridionally (zonally) elongated non-tilting perturbation can be unstable in the jet entrance (exit) region without having a net momentum transport in either zonal or meridional direction.

In terms of wave-mean flow interaction, local energetics analysis explicitly separates energy "source" terms (i.e., those conversion terms discussed in subsections (ii) and (iii) in section 3.4) from energy propagation terms (i.e., spatial energy redistribution terms discussed in subsection (i) in section 3.4). The source terms depict wave-mean flow energy exchange locally and energy propagation terms describe how perturbations disperse spatially without interacting with the mean flow. In other words, energy propagation terms only indicate where perturbations would develop but not where perturbations would interact with the mean flow. The latter is described by energy source terms from the perturbation perspectives. A local energetics analysis, however, does not directly yield information about how unstable disturbances would modify the mean state. A complementary diagnostics analysis, namely, "feedback analysis" pioneered by Lau and Holopainen (1984), can be carried out to infer the potential modification to the basic state (i.e., the tendencies induced by heat/vorticity fluxes or potential vorticity flux of unstable disturbances) as demonstrated in Pierrehumbert (1985) and CM90 for local baroclinic instability. The feedback calculation depicts wave-mean flow interaction from the mean flow perspectives. The same dynamical quantities (i.e., eddies' momentum/heat or potential vorticity fluxes) are involved in both diagnostics but with opposite polarity. Therefore, energy source terms are complementary to feedback diagnostics (e.g., a positive energy conversion from the mean flow to perturbations would imply a negative feedback to the mean flow or vice versa). For a zonally averaged flow, the wave-mean flow interaction can be diagnosed using the Eliassen-Palm (E-P) flux defined in the meridional-vertical plane based on a conservation relation (e.g., Edmon et al. 1980). The E-P flux itself, being proportional to the group velocity, is indicative of the direction of wave activity propagation and its convergence measures the forcing to the zonal-mean flow. For a zonally varying basic flow, the 2-D E-P flux vector can be extended to a 3-D flux vector that satisfies a conservation relation of a pseudo energy (e.g., Andrews, 1983; Plumb, 1986; Swanson et al. 1997). The diagnostics using a 3-D flux vector in a conservation relation may look more appealing in comparison with the two-step diagnostics (i.e., local energetics and feedback analyses) because the flux vector itself is indicative of wave activity propagation and its divergence is a measure of the down-gradient flux of potential vorticity by eddies which implies a net forcing to the mean flow. Therefore, in principle, such a wave-activity flux vector describes not only propagation of wave activities but also their sources/sinks (or locations where eddies interact with the mean flow). However, it remains to be demonstrated that such a unity of energy source and propagation terms would be really appropriate in the context of a zonally varying basic flow because of somewhat arbitrariness in deriving them. One of the main disadvantages of using such flux vectors is the potential ambiguity to interpret the results because such a conservation relation is built on the divergence of a flux vector rather than the flux vector itself, resulting in the non-uniqueness in defining a flux vector (Plumb 1985; Chang and Orlanski 1994). Local energetics analysis, on the other hand, is derived directly from the governing equations without adding/deleting any non-divergent terms into the equations. In that regards, there is no ambiguity to interpret the direction of eddy activity propagation based on local energetics analysis once the time-mean flow and transient eddies are defined. Chang and Orlanski (1994) showed that the energy flux vector (i.e., the resultant vector of kinetic energy advection by the mean flow and the vectors inside the divergence operator in Eq. (17) in this paper) is a better approximation to group velocity in comparison with the flux vector defined by Plumb (1986).

It is the author's view that the existence of unstable local modes may not necessarily be a manifestation of absolute instability at the maximum baroclinicity or any other locations. The strong dependence of instability properties on the domain-averaged wind can be simply explained by the advective effect of the basic flow, as shown in MC89 and CM90. The self-reseeding mechanism may not be necessarily required for the existence of local modes in nature. Lee and Mak (1995) argued that the Atlantic storm track could be maintained by global recycling of perturbations whereas the upstream self-reseeding mechanism associated with convergence of the ageostrophic energy fluxes could explain the Pacific storm track. This upstream self-reseeding mechanism essentially is a manifestation of the stationary/weak-propagating wave-packet resonance as discussed recently in Mak (2002). Furthermore, Lin and Pierrehumbert (1993) proposed two alternative "reseeding" mechanisms for unstable local modes, namely, "stochastic atmospheric noises" and "reseeding from upstream" through global recycling of debris of old baroclinic eddies. In the former case, the exponential growth may not even be present, as demonstrated in the publications by Farrell and his collaborators and by Whitaker and Sardeshmukh (1998) and Zhang and Held (1999). The latter requires nonlinear constructive/destructive interferences among several global modes to generate spatially localized time-mean perturbation statistics. Perhaps, non-modal instability also serves as a seeding mechanism for perturbations to grow preferentially downstream of a jet stream.

Although the exponential growth associated with modal instability may not necessarily be required to produce storm tracks, the underlying physical/dynamic mechanisms that are pertinent to modal instability should also be important ingredients of non-modal instability. Particularly, the basic flow would act as an energy source to perturbations and organize perturbations downstream of the jet cores via energy redistributions by advection and ageostrophic fluxes regardless of the source of perturbations. The stochastic dynamics seems to explain the variance of the observed synoptic scale eddies very reasonably, including "midwinter suppression" of the Pacific storm track and a stronger storm track over the Atlantic. However, linear stochastic models have a less skill in simulating the covariance statistics of synoptic scale eddies, such as heat and momentum fluxes, and the feedback effects. In terms of dynamics, the covariance statistics of transients is far more important than the variance itself because the covariance records the interactions among transients and the basic flow. Particularly, the covariance statistics is indicative of the preferred horizontal/vertical orientation of eddies in additional to their amplitude. It is relatively easier to explain the variance than the covariance statistics. One may further argue that there might be a fundamental difference between (modal or non-modal) local instability theory and stochastic dynamics. For the former case, the energy used to sustain perturbation growths would ultimately come from the basic baroclinic/barotropic shear itself. For the latter case, it does not have to be the case since the stochastic forcing itself is an energy source to perturbations. It might be possible that eddies are generated stochastically and they are organized into storm tracks simply by advection and ageostrophic fluxes and by a barotropic growth in the jet entrance region and decay in the jet exit region. If that were the case, one would conjecture that the preferred orientations of eddies might not be relevant at all or purely coincidental and the formation of storm tracks in a linear stochastic model would primarily be a mechanic process and instability (or energy conversion from the basic flow to perturbations) is of secondary importance.

Acknowledgements. The author is greatly appreciative of the insightful comments and suggestions from Drs. Mankin Mak and Huug van den Dool during the course of this study. The detailed suggestions from Dr. Hisashi Nakamura and two anonymous reviewers are extremely helpful for clarifying several important issues and have led a significant improvement to the early version of the paper. This work was supported by grants from the NASA Seasonal-to-Interseasonal Prediction Project (NASA-NAG-55825) and NOAA OGP CLIVAR Pacific Program (NA15GP2026).

References

Andrew, D. G., 1983: A conservation law for small-amplitude quasi-geostrophic disturbances on a zonally asymmetric basic flow. *J. Atmos. Sci.*, **40**, 85-90.

Baines, P. G., and H. Mitsudera, 1994: On the mechanism of shear flow instabilities. *J. Fluid Mech.*, **276**, 327-342.

Black, R. X., 1998: The maintenance of extratropical intraseasonal transient eddy activity in the GEOS-1 assimilated dataset. *J. Atmos. Sci.*, **55**, 3159-3175.

___, and R. M. Dole, 2000: Storm tracks and barotropic deformation in climate models. *J. Climate*, **13**, 2712-2728.

Blackmon, M. L., 1976: A climatological spectral study of the 500 mb geopotential height of the Northern Hemisphere wintertime circulation. *J. Atmos. Sci.*, **33**, 1607-1623.

___, J. M. Wallace, N.-C. Lau, and S. L. Mullen, 1977: An observational study of the Northern Hemisphere wintertime circulation. *J. Atmos. Sci.*, **34**, 1040-1053.

Borges, M. D., and D. L. Hartmann, 1992: Barotropic instability and optimal perturbations of observed nonzonal flows. *J. Atmos. Sci.*, **49**, 335-354.

Branstator, G. W., 1995: Organization of stormtrack anomalies by recurring low-frequency circulation anomalies. *J. Atmos. Sci.*, **52**, 207-226.

___, and I. M. Held, 1995: Westward propagation normal modes in the presence of stationary background waves. *J. Atmos. Sci.*, **52**, 247-262.

Bretherton, E. P., 1966: Critical layer instability in baroclinic flows. *Quart. J. Roy. Meteor. Soc.*, **92**, 325-334.

Cai, M., 1992: A physical interpretation for the stability property of a localized disturbance in a deformation flow. *J. Atmos. Sci.*, **49**, 2177-2182.

___, and M. Mak, 1990a: On the basic dynamics of regional cyclogenesis. *J. Atmos. Sci.*, **47**, 1417-1442.

___, and M. Mak, 1990b: Symbiotic relation between planetary and synoptic scale waves. *J. Atmos. Sci.*, **47**, 2953-2968.

___, and ___ , 1991: Low-frequency waves and traveling storm tracks. Part I: Barotropic component. *J. Atmos. Sci.*, **48**, 1420-1436.

___, and H. M. van den Dool, 1992: Low-frequency waves and traveling storm tracks. Part II: Three-dimensional structure. *J. Atmos. Sci.*, **49**, 2506-2524.

Chang, E. K. M., 1993: Downstream development of baroclinic waves as inferred from regression analysis. *J. Atmos. Sci.*, **50**, 2038-2053.

___, 2001: GCM and observational diagnoses of the seasonal and interannual variations of the Pacific storm track during the cool seasons. *J. Atmos. Sci.*, **58**, 1784-1800.

___, and, I. Orlanski, 1993: On the dynamics of a storm track. *J. Atmos. Sci.*, **50**, 999-1015.

___, and ___, 1994: On energy flux and group velocity of waves in baroclinic flows. *J. Atmos. Sci.*, **51**, 3823-3828.

___, and Y. Fu, 2002: Interdecadal variations in Northern Hemisphere winter storm track intensity. *J. Climate*, **15**, 642-658.

___, S. Lee, and, K. L. Swanson, 2002: Storm track dynamics. *J. Climate*, **15**, 2163-2183.

Chang, J.-C. and M. Mak, 1995: Nonmodal barotropical dynamics of the intraseasonal disturbances. *J. Atmos. Sci.*, **52**, 896-914.

Chen, W-B and H. M. van den Dool, 1995: Low-frequency variabilities for widely different basic flows. *Tellus*, **47A**, 526-540.

___, and ___, 1996: Asymmetric impact of tropical SST anomalies on atmospheric internal variability over the North Pacific. *J. Atmos. Sci.*, **54**, 725-740.

Christoph, M., U. Ulbrich, and P. Speth, 1997: Midwinter suppression of Northern Hemisphere storm track activity in the real atmosphere and in GCM experiments. *J. Atmos. Sci.*, **54**, 1589-1599.

Cuff, T. J., and M. Cai, 1995: Interaction Between the Low- and High-Frequency Transients in the Southern Hemisphere Winter Circulation. *Tellus*, **47A**, 331-350.

DelSole, T., 1996: Can quasi-geostrophic turbulence be modeled stochastically? *J. Atmos. Sci.*, **53**, 207-226.

____, and B. E. Farrell, 1995: A stochastically excited linear system as a model for quasi-geostrophic turbulence: Analytic results for one- and two-layer fluids. *J. Atmos. Sci.*, **52**, 2531-2547.

Eady, E. T., 1949: Long waves and cyclone waves. *Tellus*, 1, 33-52.

Ebisuzaki, W. and M. Chelliah, 1998: ENSO and inter-decadal variability in storm tracks over North America and vicinity. *Proc. 23rd Annual Climate Diagnostics and Prediction Workshop,* Miami, FL, NOAA, 243-246.

Edmon, H. J., B. J. Hoskins, and M. E. McIntyre, 1980: Eliassen-Palm cross section for the troposphere. *J. Atmos. Sci.*, **37**, 2600-2616.

Farrell, B. E., 1984: Modal and non-modal baroclinic waves. *J. Atmos. Sci.*, **41**, 668-673.

___, 1988: Optimal excitation of neutral Rossby waves. *J. Atmos. Sci.*, **45**, 163-172.

___, 1989a: Optimal excitation of baroclinic waves. *J. Atmos. Sci.*, **46**, 1193-1206.

___, 1989b: Transient development in confluent and diffluent flow. *J. Atmos. Sci.*, **46**, 3279-3288.

___, and P. J. Iannou, 1993: Stochastic dynamics of baroclinic waves. *J. Atmos. Sci.*, **50**, 4044-4057.

___, and ___, 1994: A theory for the statistical equilibrium energy spectrum and heat flux produced by transient baroclinic waves. *J. Atmos. Sci.*, **51**, 2685-2698.

___, and ___, 1995: Stochastic dynamics of the midlatitude atmospheric jet. *J. Atmos. Sci.*, **52**, 1642-1656.

___, and ___, 1996a: Generalized stability theory: Part I: Autonomous operators. *J. Atmos. Sci.*, **53**, 2025-2040.

___, and ___, 1996b: Generalized stability theory: Part II: Non-autonomous operators. *J. Atmos. Sci.*, **53**, 2041-2053.

Frederiksen, J. S., 1978: Instability of planetary waves and zonal flows in two-layer models on a sphere. *Quart. J. Roy. Meteor. Soc.*, **104**, 841-872.

___, 1979: The effects of long planetary waves on the regional cyclogenesis. *J. Atmos. Sci.*, **36**, 195-204.

___,1983: Disturbances and eddy fluxes in Northern Hemisphere flows: Instability of three-dimensional January and July flows. *J. Atmos. Sci.*, **40**, 836-855.

___, and R. C. Bell, 1987: Teleconnection patterns and the roles of baroclinic and topographic instability. *J. Atmos. Sci.*, **40**, 2200-2218.

Hoerling, M. P., and M. Ting, 1994: Organization of extratropical transients during El-Niño. *J. Climate,* **7**,745-766.

Honda, M., and H. Nakamura, 2001: Interannual seesaw between the Aleutian and Icelandic lows. Part II: Its significance in the interannual variability over the wintertime Northern Hemisphere. *J. Climate*, **14**, 4512–4529.

___, H. Nakamura, J. Ukita, I. Kousaka, K. Takeuchi, 2001: Interannual seesaw between the Aleutian and Icelandic lows. Part I: Seasonal dependence and life cycle. *J. Climate*, **14**, 1029–1042.

Holton, J. R., 1992: *An Introduction to Dynamic Meteorology*, Third Edition, Academic Press. 511pp.

Hoskins, B. J., A. J. Simmons, and D. G. Andrews, 1977: Energy dispersion in a barotropic atmosphere. *Quart. J. Roy. Meteor. Soc.*, **103**, 553-567.

___, I. N. James, and G. H. White, 1983: The shape, propagation and mean-flow interaction of large-scale weather systems. *J. Atmos. Sci.*, **40**, 1595-1612.

___, P. J. Valdes, 1990: On the existence of storm tracks. *J. Atmos. Sci.*, **47**, 1854-1864.

___, and K. I. Hodges, 2002: New Perspectives on the Northern Hemisphere Winter Storm Tracks. *J. Atmos. Sci.*, **47**, 1041–1061.

Huang, H. P., and W. A. Robinson, 1995: Barotropic model simulations of the North Pacific retrograde disturbances. *J. Atmos. Sci.*, **52**, 1630-1641.

Iacono, R., 2002: Local energy generation in barotropic flows. *J. Atmos. Sci.*, **59**, 2153-2163.

Lau, N.-C., 1978: On the three-dimensional structure of the observed transient eddy statistics of the Northern Hemisphere wintertime circulation. *J. Atmos. Sci.*, **35**, 1900-1923.

___, 1988: Variability of the observed midlatitude storm tracks in relation to low-frequency changes in circulation pattern. *J. Atmos. Sci.*, **45**, 2718-2743.

___, and J. M. Wallace, 1979: On the disturbance of horizontal transports by transient eddies in the Northern Hemisphere wintertime circulation. *J. Atmos. Sci.*, **36**, 1844-1861.

Lin, S.-J., and R. T. Pierrehumbert, 1993: Is the midlatitude zonal flow absolutely unstable? *J. Atmos. Sci.*, **50**, 505-517.

Lee, S., 1995a: Linear modes and storm tracks in a two-level primitive equation model. *J. Atmos. Sci.*, **52**, 1841-1862.

___, 1995b: Localized storm tracks in the absence of local instability. *J. Atmos. Sci.*, **52**, 977-989.

___, and I. M. Held, 1993: Baroclinic wave packets in models and observations. . *J. Atmos. Sci.*, **50**, 1413-1428.

Lee, W. –J., and Mak, 1994: Observed variability in the large scale static stability. *J. Atmos. Sci.*, **51**, 2137-2144.

___, and ___, 1995: Dynamics of storm tracks: A linear instability perspective. *J. Atmos. Sci.*, **52**, 697-723.

___, and, ___, 1996: The role of orography in the dynamics of storm tracks. *J. Atmos. Sci.*, **53**, 1737-1750.

Lorenz, E. N., 1972: Barotropic instability of Rossby wave motion. *J. Atmos. Sci.*, **29**, 258-264.

Mak, M., 1993: Local baroclinic instability induced by inhomogeneous stratification. *J. Atmos. Sci.*, **50**, 1629-1642.

___, 1998: Influence of surface sensible heat flux on incipient marine cyclogenesis. *J. Atmos. Sci.*, **55**, 820-834.

___, 2002: Wave-packet resonance: Instability of a localized barotropic jet. *J. Atmos. Sci.*, **59**, 823-836.

___, and M. Cai, 1989: Local barotropic instability. *J. Atmos. Sci.*, **46**, 3289-3311.

Merkine, L., 1977: Convective and absolute instability of baroclinic eddies. *Geophys. Astrophys. Fluid Dyn.*, **9**, 129-157.

Nakamura, H., 1992: Midwinter suppression of baroclinic wave activity in the Pacific. *J. Atmos. Sci.*, **49**, 1629-1642.

____, T. Isumi, T. Sampe, 2002: Interannual and decadal modulations recently observed in the Pacific storm track activity and East Asian winter monsoon. *J. Climate*, **15**, 1855-1874.

Orlanski, I., and J. Katzfey, 1991: The life cycle of a cyclone wave in the Southern Hemisphere. Part I: Eddy energy budget. *J. Atmos. Sci.*, **48**, 1972-1998.

___, and, E. K. M. Chang, 1993: Ageostrophic geopotential fluxes in downstream and upstream development of baroclinic waves. *J. Atmos. Sci.*, **50**, 212-225.

Pedlosky, J., 1987: *Geophysical Fluid Dynamics.* Second Edition. Springer-Verlag. 710 pp.

Pierrehumbert, R. T., 1984: Local and global instability of zonally varying flow. *J. Atmos. Sci.*, **40**, 2141-2162.

___, 1985: The effects of local baroclinic instability on zonal inhomogeneities of vorticity and temperature. *Advances in Geophysics,* **29**, Academic Press, 165-182.

___, 1986: Spatially amplifying modes of the Charney baroclinic instability problem. *J. Fluid Mech.*, **170**, 293-317.

Plumb, R. A., 1985: An alternative form of Andrew's conservation law for quasi-geostrophic waves on a steady, nonuniform flow. *J. Atmos. Sci.*, **42**, 298-300.

___, 1986: Three-dimensional propagation of transient quasi-geostrophic eddies and its relationship with the eddy forcing of the time-mean flow. *J. Atmos. Sci.*, **43**, 1657-1678.

Rogers, J. C., 1997: North Atlantic storm track variability and its association to the North Atlantic oscillation and climate variability of Northern Europe. *J. Climate*, **10**, 1635–1647.

Rossby, C.-G.. 1945: On the propagation of frequencies and energy in certain types of oceanic and atmospheric waves. *J. Meteor.*, **2**, 187–204.

Simmons, A. J., and B. J. Hoskins, 1979: The downstream and upstream development of unstable baroclinic waves. *J. Atmos. Sci.*, **36**, 1239-1254.

___, J. M. Wallace, and G. W. Branstator, 1983: Barotropic wave propagation and instability, and the atmospheric teleconnection patterns. *J. Atmos. Sci.*, **40**, 1362-1392.

Straus, D. M., and J. Shukla, 1997: Variations of midlatitude transient dynamics associated with ENSO. *J. Atmos. Sci.*, **54**, 777-790.

Shutts, G. J., 1983: The propagation of eddies in diffluent jetstreams: eddy vorticity forcing of "blocking" flow fields. *Quart. J. Roy. Meteor. Soc.*, **109**, 737-761.

Swanson, K. L., P. J. Kushner, and, I. M. Held, 1997: Dynamics of barotropic storm tracks. *J. Atmos. Sci.*, **54**, 790-810.

Trenberth, K. E., and J. W. Hurrell, 1994: Decadal atmosphere-ocean variations in the Pacific. *Climate Dyn.*, **9**, 303-319.

Tupaz, J. B., R. T. Williams, and C.-P. Chang, 1978: A numerical study of barotropic instability in a zonally varying easterly jet. *J. Atmos. Sci.*, **35**, 1265-1280.

Wallace, J. M. and N-C Lau, 1985: On the role of barotropic energy conversion in general circulation. *Issues in Atmospheric and Oceanic Modeling. Part A: Climate Dynamics*, S. Manabe, Ed., Academic Press, 33-74.

Whitaker, J. S., and R. M. Dole, 1995: Organization of storm tracks in zonally varying flows. *J. Atmos. Sci.*, **35**, 1265-1280.

___, and P. D. Sardeshmukh, 1998: A linear theory of extratropical synoptic eddy statistics. *J. Atmos. Sci.*, **55**, 237-258.

Yeh, T-C., 1949: On energy dispersion in the atmosphere. *J. Meteor.*, **6**, 1–16.

Zhang, Y., and I. M. Held, 1999: A linear stochastic model of a GCM's midlatitude storm tracks. *J. Atmos. Sci.*, **56**, 3416-3435.

Zhu, W-J, Z-B Sun, and B. Zhou, 2001: The impact of Pacific SSTA on the interannual variability of northern Pacific storm track during winter. *Adv. in Atmos. Sci.*, **40**, 259-268.

UNDERSTANDING THE COUPLED OCEAN-ATMOSPHERE DYNAMICS OF ENSO

FEI-FEI JIN

Department of Meteorology, University of Hawaii at Manoa, Honolulu HI 96822, USA
E-mail: jff@hawaii.edu

(Manuscript received 17 January 2003)

Basic theories of the El Niño-Southern Oscillation (ENSO) phenomenon are briefly reviewed in this paper. The main focus is on the dynamics of ENSO within coupled dynamic frameworks of limited complexity. The simple coupled models capture the growth mechanism related to the positive feedback of tropical ocean-atmosphere interaction proposed by Bjerknes. Yet, different considerations of coupled processes have led to different hypotheses for the phase-transition mechanisms of ENSO-like coupled oscillations. The relevance of these theories to the understanding of the nature of the complex ENSO phenomenon will be discussed.

1. Introduction

The El Niño-Southern Oscillation (ENSO) phenomenon is known as the most dramatic and best defined interannual climate variability. ENSO affects many global climatic and ecological systems. El Niño and its counterpart La Niña are events of remarkable increases and deceases in the equatorial Pacific sea surface temperature (SST). The Southern Oscillation (SO) describes a large-scale seesaw of sea level pressure between the eastern and western sides of the tropical Pacific. The sea level pressure (SLP) is normally relatively high in the southeastern Pacific (e.g. Tahiti) and relatively low over the western tropical Pacific (e.g., Darwin), thus driving the easterly surface winds. Bjerknes (1969) first recognized the connection between the Southern Oscillation and El Niño. He noticed that the SST values in the eastern Pacific cold tongue region are much lower than those in the western Pacific warm pool region. This SST gradient generates a thermal circulation called the Walker circulation in the equatorial atmosphere along the equatorial Pacific. The equatorial SLP gradient associated with SO and the equatorial easterly is partly due to this Walker circulation. Bjerknes thus first hypothesized that a positive feedback of tropical ocean-atmosphere interaction can amplify SST perturbations of the cold tongue to sustain either a warm or a cold phase of ENSO. On one hand, the easterly trade winds force the thermocline depth, the layer of sharp vertical temperature gradient that separates the upper ocean from the abyssal deep ocean, to be shallower in the equatorial eastern Pacific than in the western Pacific. The trade winds also induce the equatorial Ekman upwelling due to Coriolis effects, which effectively brings cold water from the subsurface to the surface layer to generate a cold tongue in the eastern Pacific. On the other hand, the atmospheric zonal pressure gradient caused by the east-west contrast of the SST drives an equatorial Walker circulation, which enhances the surface easterlies over the Pacific basin and thus strengthens the cold tongue. This provides an explanation for the occurrence of the cold tongue in the eastern equatorial Pacific and for the association of the weak Southern Oscillations with El Niño events.

The recognition that tropical ocean-atmosphere interactions produce the ENSO phenomenon has led to great progress in understanding the coupled dynamics of ENSO. (e.g., Bjerknes 1969; Wyrtki 1975; Rasmusson and Carpenter 1982; Anderson and McCreary 1985; Cane and Zebiak 1985; Cane et al. 1986; Graham and White 1988; Philander 1990; Barnett et al. 1991; Philander et al. 1992; Neelin et al 1994; Jin 1996, 1997; Neelin et al 1998). In addition to the Bjerknes coupled positive feedback hypothesis, a number of phase transition mechanisms were proposed. (Schopf and Suarez 1988; Battisti and Hirst 1989; Neelin 1991; Jin and Neelin 1993; Wang and Weisberg 1996; Jin 1996, 1997; Jin and An 1999; Picaut et al 1997). In fact, it is the hypotheses proposed by Wyrtki (1985) and Cane-Zebiak (1985) that first emphasized the importance of the discharge and recharge of the equatorial heat content in the ENSO phase transition. It was found that during warm (cold) ENSO phases, the equatorial heat content is often draining out (building up) as the result of the mass exchange between the equatorial belt and off-equatorial regions through ocean dynamical adjustment. The discharge/recharge process of the equatorial heat content, which is out of phase with SST anomalies of ENSO, was shown (Jin 1996, 1997; and Jin and An 1999*)* to be responsible for ENSO phase transitions with simple coupled models. The combined Bjerknes-Wyrtki-Cane-Zebiak (BWCZ) hypothesis indicated that ENSO is a natural basin-wide oscillation of the tropical Pacific ocean-atmosphere system. Both the positive feedback of the coupled interaction of the ocean-atmosphere system of the tropical Pacific and the memory of the system in the subsurface ocean dynamical adjustment are essential for ENSO. This recharge-oscillation mechanism is at the heart of ENSO theory pointing to the instability of the tropical Pacific climate state.

Today, theories for the ENSO phenomenon have approached a mature stage (Neelin et al. 1994, 1998). ENSO modeling has advanced to a point where predictions are made on a regular basis. Yet, there are still many unresolved issues. In this paper, I shall discuss the fundamental aspects of ENSO dynamics and some interesting directions for further research.

2. Some observed features of ENSO

Although solar heating is zonally uniform, SST in the tropics is far from zonally symmetric as shown in Fig. 1a. In particular, the tropical Pacific features the largest zonal contrast in SST along the equator, with a warm pool in the west and a cold tongue in the east. The equatorial SST gradient is collocated with the equatorial trade winds. Associated with the easterly wind on the equator, there exists a strong zonal gradient in ocean thermocline depth (Fig1.c). The latter marks the layer of sharp vertical temperature gradient that separates the upper ocean from the abyssal deep ocean.

During El Niño, the normal tropical climate state is altered significantly. As an example, the peak of the 1997/1998 ENSO is shown in Fig. 1b. At the mature phase of the event, the warm pool expands so far to the east that the climatologic cold tongue (Fig. 1a) almost vanishes (Fig. 1b). The strong tilt of the thermocline depth in the climate state (Fig. 1c) reverses (Fig. 1d). Therefore, the anomalies in SST and thermocline depth are very strong. During La Nina, the opposite situation happens (not shown).

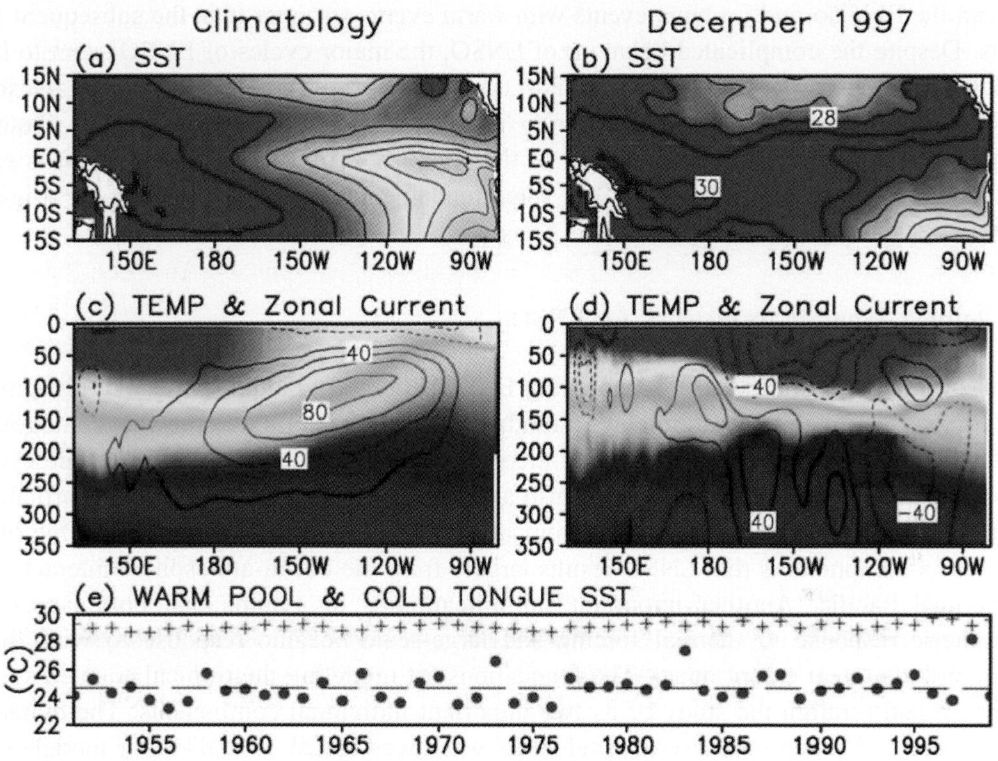

Figure 1. (a) Sea surface temperatures (SST) (°C); (c) upper ocean temperature (°C) (in color) and zonal currents (cm s^{-1}) (in contours). (a) and (c) are December means averaged from 1978-1998; (b) and (d) represent the 1997 December fields of (a) and (b); (e) Winter seasonal mean (November to January) SST in the warm pool (+) (averaged over the area 5°S to 5°N, 130° to 170°E) and in the cold tongue (averaged over the area 5°S to 5°N, 120° to 80°W). The large El Nino events of 1982/83 and 1997/98 are characterized by very small zonal temperature gradients (after Jin et al. 2003).

Because strong SST anomalies occur in the eastern Pacific cold tongue region, an area-averaged SST anomaly is often used as an index to represent the time scales and intensity of ENSO as shown in Fig. 1e. The temporal behavior of ENSO is highly irregular and complicated. Its time scales and strengths seem to evolve from decade to decade and from event to event. The two biggest events (1982-1983, 1997-1998) occurred in the past two decades. It is not yet clear whether these changes in ENSO are simply part of the chaotic nature of ENSO or if they are modulated by the slow background changes in the climate state of the tropical Pacific. The spatial structures of each ENSO are related to its temporal behavior and they evolve significantly as well. For instance, most ENSO events before 1980 showed some westward propagation tendency in the SST anomalies (Rasmusson and Carpenter 1982). Since then ENSO events are dominated by stationary fluctuations in SST anomalies. In fact, some (1982-83 events) even showed some eastward propagation. It seems that each El Niño and La Niña is different. Even in the past two decades, each ENSO event is quite different. In the earlier 1990s, some ENSO events occurred near the dateline rather than in the cold tongue region. These events seem to have short cycles while the major ENSO events occurred across the central to eastern Pacific. There is also substantial asymmetry between the El Niño and La Niña events with warm events stronger than the subsequent cold events. Despite the complicated behavior of ENSO, the major cycles of ENSO seem to have periodicity on an order of 3-5 years. For each warm (cold) event, equatorial westerly (easterly) winds anomalies always occur to its west, and the thermocline tilt is flattened (steepened). Therefore, even in the midst of the complex evolution of ENSO, there is some substantial commonality that points to the some fundamental common causes of the origin of ENSO.

3. Coupled dynamic framework for ENSO

The basic dynamic framework for modeling ENSO emerged gradually after the recognition that large-scale air-sea interaction plays a fundamental role in ENSO. Somewhat analogous to the development of the quasi-geostrophic theory for the dynamics of middle latitude cycles, the basic dynamic framework for ENSO also adopts a number of simplifications, particularly the filtering of the fast processes of the tropical atmosphere and ocean dynamics. One of the important assumptions is that ENSO results largely from the ocean-atmosphere interaction in the tropical Pacific. Another important simplification is to assume that both large-scale atmospheric response to thermal forcing and large-scale oceanic response to wind stress forcing are to a great extent linear. The foundations for modeling the tropical coupled system were then laid through the study of its two important individual components. The dynamics of the equatorial ocean response to wind stress was investigated in shall-water models (e.g., Moore 1968; McCreary 1976; Cane and Sarachick 1981), and oceanic GCMs (Philander and Pacanowski 1980; Philander 1981). The tropical atmospheric dynamics and the atmospheric dynamical response to tropical heating associated with equatorial SST anomalies were also simulated with linear shallow-water modes (Mastuno 1966; Gill 1980; Zebiak 1986). A number of coupled models combining the two components through further considerations of the process in generating equatorial SST anomalies (e.g., Zebiak and Cane 1987). The Zebiak and Cane model is now one of the best known among the coupled models with limited complexity.

3.1. *A simple atmospheric model*

The so-called Gill-type atmospheric models were established based on the dynamic framework for the equatorial wave dynamics developed by Matsuno (1966). Two very different approaches were taken by Gill (1980) and Lindzen and Nigam (1987) involving rather different hypotheses on how the atmosphere responds to SST anomalies. These two approaches were proven to be equivalent in terms of the final model formulation (Neelin 1989) and were unified later by a reduced gravity interpretation (Battisti et al. 1999). We shall follow this reduced gravity interpretation (Battisti et al. 1999) and consider a reduced gravity model for the lower atmosphere:

$$\varepsilon u_a - \beta y v_a + \partial_x \phi_a = 0,$$
$$\beta y u_a + \partial_y \phi_a = 0, \quad (1)$$
$$\varepsilon \phi_a + C_a^{\,2}(1-b)\nabla \vec{u}_a = -\varepsilon \Gamma T'.$$

Here (u_a, ϕ_a) denote the anomalies of low level winds and geopotential, ε is the linear friction constant, C_a is the gravity wave speed with the reduced gravity parameter, b is the factor that takes the convective heating into the consideration, Γ is a constant, and finally T' is SST anomaly. This model describes dynamical balances in the low troposphere as among the pressure gradient, Coriolis force, and the boundary-layer friction. The thermal dynamical balance includes diabetic heating and divergence flow acting in the moisture convergence and adiabatic cooling. On the atmospheric response time scale, the associated thermal forcing is considered as steady forcing because SST varies slowly due to the large heat inertia of the ocean surface layer and slow ocean dynamics. The details of the derivation of this equation can be found in Battisti et al. (1999). Furthermore, for simplicity, we neglected the friction term in the second equation of system (1) by adopting the so-called long wave approximation.

This simple model framework can successfully describe the equatorial atmospheric Kelvin wave and Rossby wave response to a quasi-steady thermal forcing associated with SST anomalies. For simplicity, we consider an SST anomaly T' with a simple meridional structure:

$$T' = T(x)\exp(-y^2/2L_a^2), \quad (2)$$

where $L_a^2 = C_a(1-b)^{1/2}/\beta$ is the atmospheric Rossby radius of deformation and $T(x)$ describes the distributions of SST anomalies along the equator. In this case, the forcing only excites the Kelvin wave and the first symmetric Rossby wave. The zonal wind stress anomalies associated with the forced solution thus can be expressed as

$$\tau(x,y) = A_0\Big\{3(1-y^2/L_a^2)\exp(-y^2/2L_a^2)\exp(3\varepsilon x/L)\int_x^L T(x)\exp(-3\varepsilon x/L)dx/L$$
$$-\exp(-y^2/2L_a^2)\exp(-\varepsilon x/L)\int_0^x T(x)\exp(\varepsilon x/L)dx/L\Big\}, \quad (3)$$

where L is the width of the Pacific basin, A_0 is constant (details for its evaluation can be found in Jin and Neelin (1993)). The first term is for the Rossby wave response. It corresponds to a westerly wind stress anomaly from heating (warm SST anomaly) region and toward the west. The response gradually fades away from the forcing region due to the atmospheric damping. The second term is the Kelvin wave response. It associates with an easterly wind stress anomaly from the forced region and decays eastward.

Using this Gill-type atmospheric model with some further elaborations regarding the convective heating, Zebiak (1986) showed that in the deep tropics this kind of model gives rise to realistic simulations of equatorial wind anomalies associated with ENSO. This kind of model fails to simulate the response in higher latitudes because it does not capture the dynamics for equivalent barotropic Rossby and its propagation. Nevertheless, to the zeroth order approximation, the simple model (1) and its simple solution (3) give a qualitatively correct representation of the zonal wind stress field forced by equatorial SST anomalies associated with El Niño. Therefore, the simple Gill-type tropical atmospheric model is adapted to serve as a basic component for the coupled dynamics of the tropical Pacific in simulating the basic features of ENSO.

3.2. *A simple model for tropical ocean dynamics*

After the classic paper of Matsuno (1966) for tropical atmospheric wave dynamics, it was quickly realized that the same dynamics operate in the tropical ocean. A parallel work was done by Moore (1968) for the unforced solutions of a shallow-water-like model for the tropical ocean in an ocean basin. The usefulness of this kind model for simulating wind driven tropical ocean variability was subsequently demonstrated (e.g., Cane and Sarachik 1981; Busalacchi and O'Brine 1981). Here, we will again use the reduced gravity interpretation for the shallow-water-like model for the tropical upper ocean dynamics. We assume that an active and warm upper layer of ocean is on top of a motionless deep layer. Then the linear equations of the motion for tropical upper ocean can be written as

$$\begin{aligned}(\partial_t + \varepsilon_m)u - \beta yv + g'\partial_x h &= 0,\\ \beta yu + g'\partial_y h &= 0,\\ (\partial_t + \varepsilon_m)h + H(\partial_x u + \partial_y v) &= 0.\end{aligned} \tag{4}$$

Here, u and v are the vertically averaged zonal and meridional ocean currents, respectively, h is the departure of the upper layer thickness from the reference depth H (150 m) and represents a thermocline depth anomaly field, and g' is the reduced gravity parameter which gives rise to an internal gravity speed as $C_o = \sqrt{g'H}$ = 2.5 m s^{-1}. The suitable boundary conditions under the long-wave are as follows

$$u(L, y, t) = 0\,, \quad \int_{-\infty}^{\infty} u(0, y, t)dy = 0\,. \tag{5}$$

System (4) can be nondimensionalized by the oceanic Rossby deformation radius $L_o = \sqrt{C_o / \beta}$ for y, the ocean basin width L for x, the Kelvin wave crossing time L/C_o (set as 2 months) for t, the reference depth H for h, and the Kelvin wave speed C_o for u and v respectively. Expanding the variable in terms of hyperbolic cylinder functions $\{\psi_{2n}(y),\ n = 0,1,2,...\}$ (c.f., Battisti 1988), and considering only hemispheric symmetric solutions, one can obtain the following eigen-solutions (details see Jin et al. 2001)

$$h_{j,k}(x,y,t)=\sum_{n=0}^{N}(q_{2n}^{j,k}(x,t)+p_{2n}^{j,k}(x,t))\psi_{2n}(y)/2,$$

$$u_{j,k}(x,y,t)=\sum_{n=0}^{N}(q_{2n}^{j,k}(x,t)-p_{2n}^{j,k}(x,t))\psi_{2n}(y)/2,$$

where

$$q_{2n}^{j,k}=B\sqrt{(2n-1)!!/(2n)!!}\exp(\sigma_{j,k}(4n(x-1)-x+t)-\varepsilon_m t),$$
$$p_{2n}^{j,k}=\sqrt{(2n+2)/(2n+1)}\,q_{2n+2}^{j,k},$$
$$\sigma_{j,k}=(-(\ln\rho_j-ij2\pi/N)\pm ik2\pi)/4;\;\; j=0,1,...,N-1; k=0,1,2,... \quad (6)$$

Here k is the node number of the eigenfunction in the zonal direction near the equator, B is an arbitrary constant, j is an index and N is a truncation number. All $\ln\rho_j$ are positive but small, indicating that the eigen modes are all damped. For N = 101, the highest damping rate is about 0.06, which corresponds to an e-folding time of about 2.5 years. This damping rate can be further cut to about half if the truncation is doubled. When $N\to\infty$, $\sigma_{0,k}=\pm ik\pi/2$ (k = 1,2,3,...) and corresponds to the so-called ocean basin modes (Cane and Moore 1981). The gravest ocean basin mode (k = 1, j =0) has the lowest frequency of $\pi/2$, which corresponds to a period about 8 months.

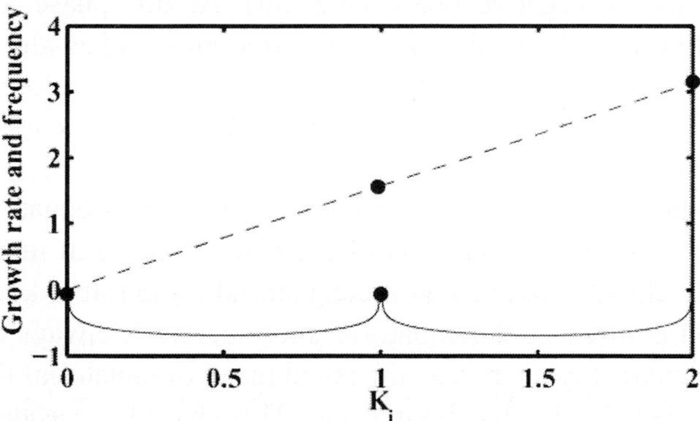

Figure 2. The eigenvalue distribution with respect to the index $k_j=k+j/N$. The straight line is the dependence of the imaginary part (frequency) of the eigen value on k_j and the curve is the dependence of the real part (growth rate) of the eigen values on k_j. The real part is multiplied by 10 to display its variations more clearly. Heavy dots indicate the eigen values for the ocean basin modes (k_j = 0,1,2, ...) (after Jin 2001).

The real and imaginary parts of the eigenvalues can be plotted as the function of a combined index defined as $s=k+j/N$ (k = 0,1,2,... and j = 0,1,2,..., N). The results are shown in Fig. 2. The modes found by Cane and Moore (1981) now correspond to the least damped modes with j=0 in the truncated case (s = 0,1,2, as indicated by the heavy dots in Fig. 1). However, the modes found by Cane and Moore (1981) are only a small part of the complete and almost continuous spectrum of the system (4) under the boundary condition (5). The modes near the zero frequency are called very low-frequency (VLF) modes ($k = 0, j << N$). This class of modes was implicitly included in the solutions of Moore (1968).

These VLF modes may be of vital importance for interannual and perhaps decadal climate variability in the tropics and subtropics. They show zonal-uniform deepening and shoaling of the equatorial thermocline. These vacillations in the equatorial thermocline are associated with the slow westward progression of the off equatorial Rossby waves. The periods for the VLF modes shown ($k = 0, j = 5$ and 14, s about 0.04 and 0.14 respectively) are about 14 and 4 years. The oscillations can be understood by relating the slope of the thermocline field and the meridional geostrophic velocity in the off-equatorial region and the associated discharge/recharge of the equatorial thermocline through the oscillation. For instance, at the maximum phase and outside of the relative uniform equatorial thermocline anomaly (Fig. 3a), there is a clear zonal slope in the latitude band about 5-10N, whereas at higher latitudes, the thermocline slope alternates crossing the basin in the zonal direction. The slope of the thermocline in the latitude band adjacent to the equatorial region indicates a strong net meridional transport of mass out of the equatorial region due to the Rossby wave activity, whereas alternating slopes at the higher latitude result in little net meridional transport across the whole basin. It is the discharge process in this effective latitude band next to the equatorial wave-guide that causes the shoaling of the equatorial thermocline and makes the positive equatorial thermocline anomaly diminish at the transition phase (Fig. 3c). At this time, the sign of the major part of the zonal slope still remains and thus the equatorial thermocline goes on to the negative phase (Fig. 3d). At this phase, the slope is already reversed in the effective latitude band and the recharge process has already begun; this will bring the equatorial thermocline back to the maximum phase. Therefore, the deepening and shoaling of the equatorial thermocline associated with the VLF mode can be understood as a free recharge oscillation.

The nearly zonal-uniform shoaling and deepening of the near equatorial thermocline is a unique feature of the VLF modes, which is of great significance in understanding the low-frequency thermocline climate variation in the equatorial region. It is known that, associated with ENSO, the zonal contrast of the equatorial thermocline is always in quasi-equilibrium with zonal integrated wind stress, whereas the zonal mean of equatorial thermocline anomaly lags behind the wind stress forcing by about a year. This phase lag was demonstrated as being critical for the phase transition of the ENSO and is at the heart of the coupled recharge oscillator paradigm of ENSO (Jin 1996, 1997a,b; Jin and An 1999) and will be discussed later in section 5. The shoaling and deepening of the nearly zonal-uniform thermocline associated with the VLF modes are related to the equatorial ocean heat content recharge and discharge. Thus, the observed coupled recharge and discharge of equatorial heat content associated with ENSO result from the slow dynamics of the tropical ocean dynamics.

The VLF mode on the decadal time scale (Fig. 4) can be understood as an even slower recharge oscillator in the same manner as the mode on the interannual time scale. The only difference between the interannual and decadal modes is the location of the effective latitude band of the discharge/recharge of the equatorial heat content by the Rossby wave activity. For the decadal modes, the effective latitude band is about 10-25 degrees from the equator. The existence of the VLF modes on the decadal time scale provides to a new possible mechanism for the decadal climate variability in the tropical-subtropical regions (Wang et al. 2003a, b).

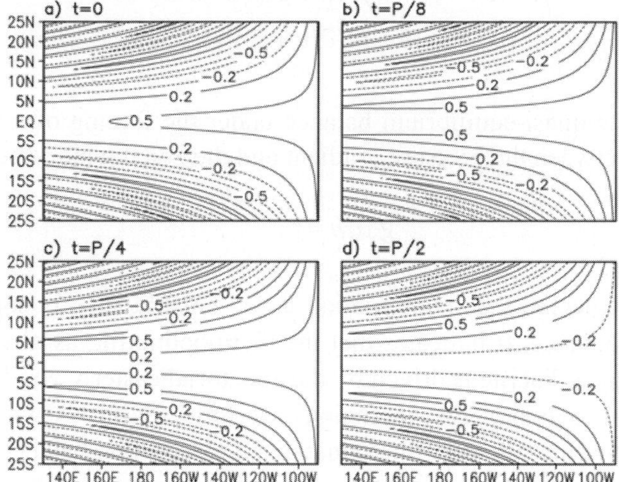

Figure 3. The thermocline of the solution in Fig. 2 b (k_j = 0.14) at the different phases of the oscillation throughout a half period. The truncation number N is 101 (after Jin 2001).

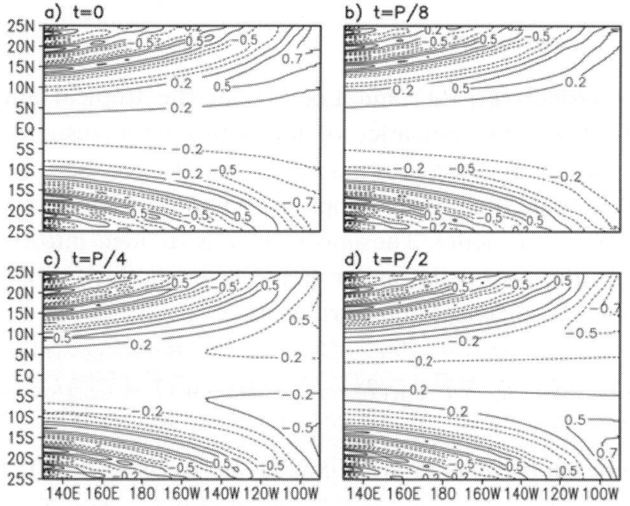

Figure 4. The Same as in Fig. 3 except for (k_j = 0.04) (after Jin 2001).

The existence of the VLF spectrum of tropical ocean dynamics not only provides a necessary element for ENSO to oscillate on an interannual time scale, but also provides a case to understand how the tropical ocean responds to the anomalous wind stress forcing. Under the long wave approximation, the contribution from meridional wind stress forcing can be ignored and the forced model thus can be written as

$$(\partial_t + \varepsilon_m)u - \beta yv + g'\partial_x h = \tau$$
$$\beta yu + g'\partial_y h = 0 \qquad (7)$$
$$(\partial_t + \varepsilon_m)h + H(\partial_x u + \partial_y v) = 0$$

Along the equator, the quasi-equilibrium balance under the forcing of a very slowly varying wind stress is basic between the pressure gradient and the zonal wind stress forcing,

$$g'\partial_x h = \tau\,. \qquad (8)$$

Here we have neglected the small friction term. This balance is sometimes also called equatorial Sverdrup balance. It indicates that the thermocline tilt, as observed in the climate mean state of the tropical Pacific is driven by the equatorial easterly wind stress. For instance, during the El Niño (La Niña), the anomalous westerly (easterly) wind stress diminishes (enhances) the thermocline tilt along the equator. However, this balance only determine the thermocline tilt for a given wind stress. It does not determine the zonal mean of the equatorial thermocline anomalies. In general, the zonal mean thermocline response depends on the excitations of the VLF modes, which are response for the slow adjustment of the zonal mean thermocline depth anomalies along the equator.

3.3. *ZC-type coupled model for ENSO*

The reduced gravity models Eq. (1) and Eq. (4) for the tropical atmosphere and ocean respectively, describe the basic dynamics of the atmospheric response to equatorial SST anomalies and upper ocean response to wind anomalies. To consider the thermodynamics of SST changes, Zebiak proposed a creative approach by embedding a mixed layer model into the upper layer of the ocean model. The upper layer is divided into two layers: a constant depth mixed-layer and the rest of the upper active layer. Following ZC approach, the SST equation of the surface mixed layer can be written as

$$\partial_t T' = -\bar{\vec{u}}_1 \cdot \nabla T' - \vec{u}_1' \cdot \nabla \bar{T} - \bar{w}(T' - T'_{sub})/H_1 - w'(\bar{T} - \bar{T}_{sub})/H_1 + Q'\,. \qquad (9)$$

Here (u_1, v_1) are the ocean currents in the mixed layer, H_1 is its depth, and T_{sub} is the ocean temperature in the subsurface. It was shown in ZC that the surface layer anomalous current can be expressed as a combination of the upper layer mean currents (u, v, w) and the Ekman currents (u_s, v_s, w_s):

$$u_1 = u + u_s, \qquad v_1 = v + v_s, \qquad w_1 = w + w_s \qquad (10)$$

where the Ekman flow can be determined by

$$\varepsilon_s u_s - \beta yv_s = \tau(H - H_1)/(\rho HH_1), \quad \varepsilon_s v_s - \beta yu_s = 0.. \qquad (11)$$

The meridional component of the wind stress should be added into the above equation when it is not negligible. The vertical motion associated with the upper layer mean currents and Ekman currents can be calculated using the continuity equation.

By assuming that the subsurface temperature is related to the thermocline depth,

$$T'_{sub} = T'_{sub}(h), \quad Q' = -\varepsilon_T T'. \tag{12}$$

Equations (3), (7), (9), (11) and (12) form a so-called intermediate coupled model for tropical Pacific. This kind of model, first constructed by ZC, produced the first realistic ENSO simulations. It has also served as the framework for much of the theoretical analysis of the basic dynamics of ENSO.

4. ENSO-like coupled modes

Although the ZC model was the first coupled model to successfully simulate and predict ENSO, a number of models of similar complexity had been used in development of the ENSO theory. A thorough review of ENSO theories developed up to the earlier and middle 1990's was given by Neelin et al. (1994, 1998). To a large extent, ENSO theories were developed based on the hypothesis that ENSO may be viewed as a leading mode of the coupled tropical ocean-atmosphere system. Various ENSO-like coupled mode were found in simple coupled models (e.g. Lau 1981; Philander et al 1984; Anderson and McCreay 1985; Cane and Zebiak 1985; Zebiak and Cane 87; Gill 1985; Hirst 1986, 1988; Graham and White 1988; Battisti 1988; Battisti and Hirst 1989; Schopf and Suarez 1988; Yamagata and Masumoto 1989; Xie et. al. 1989; Wakata and Sarachik 1991; Neelin 1991; Jin and Neelin 1993a, b; Chang et al 1994; Wang and Weisberg 1996; Jin 1996, 1997a,b; Jin and An 1999; An and Jin 2001; Cai 2003). Similar modes were simulated by most coupled GCMs (e.g. Philander et al. 1992, Collins 2000, Yu and Mechoso 2001).

A unified linear stability theory for ENSO that delineates the relations of different coupled modes and their dependence on parameter regime was first proposed by Jin and Neelin (1993a,b) using a stripped–down version of the ZC type model. The stripped-down model, first proposed by Neelin 1991, was motivated by the fact that the observed strong surface temperature response to changes in upwelling and advection is confined to a narrow band along the equator. With this simplification, Jin and Neelin (1993) solved the eigen solutions of the coupled system (3, 7, 9, 11, 12) in various parameter regimes analytically and mapped the regimes of different coupled mode in the full parameter space numerically.

In Jin and Neelin (1993a,b), three free parameters (μ, δ, δ_s) are introduced to map out the regimes of different coupled modes relevant to ENSO. The relative coupling coefficient μ is introduced as a multiplier to A in the equation (3). The relative time scale coefficient δ is introduced as a multiplier to the time derivatives of the equations for ocean wave dynamics (7). The surface-layer coefficient δ_s is introduced as a multiplier to the Ekman flows in equation (10). Here we provide a brief summary of the main finding of this research. A schematic regime diagram in the parameter space is given in Fig. 5.

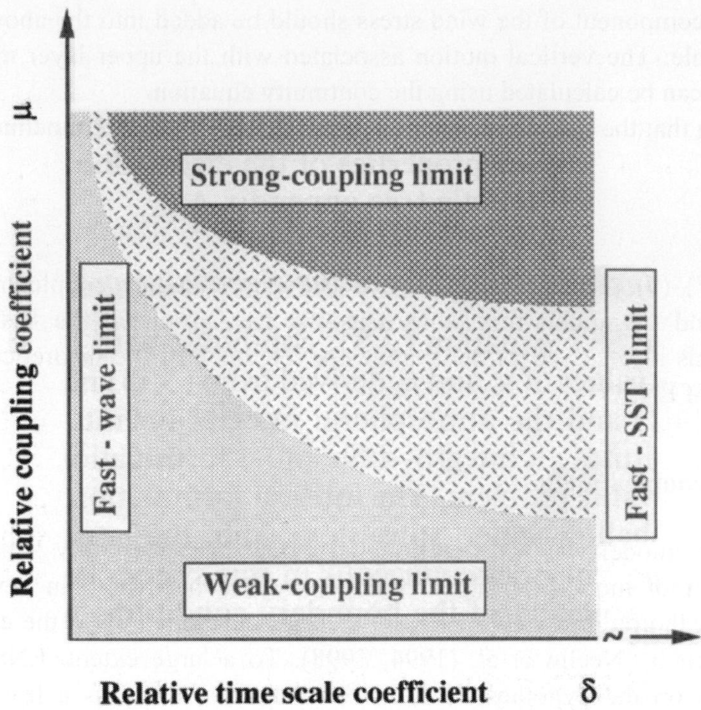

Figure 5. Schematic regime diagram of the (μ, δ) parameter space showing regions of validity of various limits (after Neelin and Jin 1993).

(1) When μ is small, the dynamical coupling of ocean and atmosphere in the tropics is relatively weak. A useful idealization of the so-called weak coupling limit is proposed. In this limit, it was shown that uncoupled ocean dynamical modes can be destabilized. In particular, the gravest ocean basin mode was found being destabilized effectively through the two terms in the equations for SST anomalies: the anomalous zonal advection $u\partial_x \overline{T}$, and anomalous vertical advection of subsurface temperature anomalies $\overline{w}(\partial T'_{sub} / \partial h)h$ (Neelin and Jin 1993). The SST anomalies generated by these two terms produce the equatorial zonal wind anomalies that drive the zonal current and equatorial thermocline anomalies. Thus, the coupled feedbacks through these two terms are some time referred as zonal advective feedback and thermocline feedback. In the weak coupling limit, the thermocline feedback are also found to destabilize ocean dynamics modes with much lower frequencies (Jin and Neelin 1993a,b; Wang et al. 2003a, b).

(2) When δ is small, the ocean adjustment is fast relative to the change rate in SST anomalies. This is the so-called fast wave limit. In this case, ocean dynamics is more or less in quasi-equilibrium with the wind stress forcing and does not serve as a source of the memory of the coupled system. The oscillatory modes exist only due to the spatial phase differences between SST anomalies and SST tendency owing to the anomalous zonal and vertical advection or heat fluxes. The oscillatory modes in these regimes are called SST modes as was first proposed by Neelin (1991). Analytical solutions of eastward and

westward propagating in the Pacific ocean basin are given in Jin and Neelin (1993b). The physics of the eastward propagating mode can be explained as follows. For a warm (cold) SST anomaly, zonal wind response is westerly (easterly) over and to the west of the warm (cold) SST anomaly and easterly (westerly) to the far east of the warm (cold) SST anomaly. The phase of zero zonal wind is thus to the east of warmest (coldest) SST anomaly. Due the quasi-balance of Eq. (7) at the fast wave limit, the phase of maximum positive (negative) thermocline anomalies is also located to east of warmest (coldest) SST anomalies. The SST tendency term $\overline{w}(\partial T'_{sub}/\partial h)h$ tends to give the maximum warming (cooling) to the east of warmest (coldest) SST anomalies. Therefore thermocline feedback will favor an eastward propagating SST mode through thermocline feedback at the fast wave limit. The physics of the westward propagating SST can be understood in the same manner if we only consider the anomalous Ekman pumping (δ_s is very large). In that case, the SST tendency term $w'_s\partial_z\overline{T}$ tends to give a westward propagating SST mode because on the equator, the anomalous Ekman upwelling is negatively proportional to the equatorial zonal wind stress anomalies. The westward SST mode has since been recognized as being responsible for the generation of the equatorial annual cycle (Xie 1994). Due to the existence of the ocean basin boundary, the eastward propagating mode exists in a rather limited parameter space, and away from the fast wave limit, it is merged with coupled modes originating from the destabilized ocean dynamics mode.

(3) When coupling is strong, the positive feedback dominates the ocean adjustment process. In this case, the oscillatory mechanism from the ocean dynamic adjustment is again lost and non-oscillatory growing modes tend to dominate this regime. This regime implies that the coupled system will exhibit a climate state drift to a permanent warm or cold state when nonlinearity is included. To a large extent this climate drift is built in through the anomalous coupling used in Eqs. (3) and (11) so that only the anomalous wind stress and heat flux are determined by the coupled model. The anomalous coupling implies a flux correction to maintain the prescribed climate state. By eliminating the flux correction of the coupled models, it was shown by Neelin and Dijkstra (1995) that the climate drift due to the non-oscillatory coupled instability is removed.

(4) When δ is very large, the ocean adjustment is slow. This is the so-called slow wave or fast SST limit. One of the special cases of this limit was explored by a number of simple models by directly relating the SST anomalies to thermocline depth anomalies. Cane et al (1990) showed analytically that the ocean basin mode is destabilized into a coupled wave oscillator.

(5) The most realistic regime is in the middle of the parameter regime and is bounded by above extreme limits where analytical solutions are accessible. In the more realistic regime, the leading coupled mode is a mixture of ocean wave-SST mode. Numerical results of Jin and Neelin (1993) shown that in this regime there is a standing oscillatory coupled mode. Two prototype models have been developed as an approximation to describe this mixed wave-SST mode. One is the delayed-oscillator model proposed by

Suarez and Schopf (1988) and Battisti and Hirst (1989) and the other is the recharge oscillator model proposed by Jin (1996,97a,b; Jin and An 1999). We will discuss in detail the development of these two models in next section.

The parameters discussed above provide convenient means for defining various extreme cases to obtain analytical solutions of coupled modes. It indicates the sensitivity of the leading coupled mode of the tropical ocean-atmosphere system to the various parameters. However, it is hard to translate this dependence of the coupled mode on the parameters into the sensitivity of the coupled to the changes in the climate mean state. An and Jin (2000) and Fedrove and Philander (2000) recognized that there is a need to investigate the dependence of the coupled mode on the changes in the basic state with a fixed set of parameters. These new efforts of analyzing the coupled mode under different climate mean states have cast new insight on the modulation of the ENSO by climate state (An and Jin 2000) and behaviors of ENSO under paleoclimate settings (Fedrove and Philander 2000).

5. Conceptual models of ENSO

To refine the linear stability theory of ENSO into a simple prototype model has been challenging issue. The linear stability analysis indicates that the coupled mode resembles ENSO in the most realistic way lies in the parameter regime where both time scales in the wave dynamics and SST changes are important and the coupling is moderate. The approaches for analytical solutions proposed by Jin and Neelin (1993b) fail in this region. Nevertheless, based on physical intuitions, simple conceptual models, such as the delayed oscillator model and recharge oscillator model, were proposed through heuristic arguments. In fact these conceptual models were later deduced from the coupled model framework through systematic simplifications (e.g. Jin 1997b). In this section, we will briefly describe the conceptual delayed-oscillator and recharge oscillator models.

5.1. *SSBH delayed-oscillator model*

With the earlier simple coupled model successfully simulating ENSO-like behaviors, a simple and clear conceptual model was put forth by Suarez and Schopf (1988) and Battisti and Hirst (1989). It combines the SST mode dynamics that describe Bjerknes feedback, with ocean adjustment of slow Rossby propagation. Therefore it is one of possible ways to represent the mixed wave-SST mode as discussed in the previous section. However, the development of the delayed oscillator comes before the recognition of the relations between different coupled modes in different parameter regimes. Rather, the delayed oscillator is proposed to explain the simulated ENSO-like oscillations in the coupled models (Schopf and Suarez 1988; Battisti and Hirst 1989). Following the arguments from Battisti and Hirst (1989), the SST anomalies in the eastern equatorial Pacific depend largely on anomalous vertical advection of the subsurface temperature into the surface layer. Thus, the SST anomalies $\mathbf{T}_{\mathbf{E}}'$ in the equatorial eastern Pacific can be approximately expressed as following:

$$\partial_t T_E' = -\varepsilon_T T_E' - \overline{w}(T_E' - \gamma h_E) / H_m, \tag{13}$$

where h_E is the thermocline anomaly in the equatorial eastern Pacific. We used the linearized relation between subsurface temperature anomalies and the thermocline depth anomalies in the eastern equatorial Pacific with $\gamma = \partial T'_{sub} / \partial h$. The thermocline anomalies in the eastern Pacific consist of two parts: the thermocline anomalies in the western equatorial Pacific plus the difference between the thermocline anomalies of the eastern and western Pacific. This zonal thermocline difference along the equator is proportional to the zonal integrated wind stress forcing as indicated by the quasi-balance equation (8). Since the zonal wind stress anomalies in the central Pacific are proportional to SST anomalies in the eastern Pacific as indicated by the equation (3), the thermocline anomalies in the eastern Pacific thus equals the thermocline anomalies in the western equatorial Pacific plus a term proportional to the equatorial SST anomalies:

$$h_E = h_w + L\hat{\tau} = h_w + aT'_E . \tag{14}$$

Battisti and Hirst (1989) argued that the wind stress anomalies in the equatorial Pacific generated the forced equatorial Rossby waves that take some time to reach the western boundary, where they are then reflected as the equatorial Kelvin waves. Therefore the thermocline at the equatorial western Pacific is controlled by the Rossby waves forced by the wind stress anomalies of the central Pacific. A westerly (easterly) wind stress anomaly generates Rossby of negative (positive) thermocline anomalies. Consider the forced Rossby propagating freely from the central Pacific to the western boundary in a time interval of η, then Battisti and Hirst (1989) propose that the thermocline at the western Pacific can be approximated as

$$h_w = -bT'_E(t-\eta) . \tag{15}$$

In other words, the thermocline at the western Pacific is a delayed negative response to the SST anomalies in the central to eastern equatorial Pacific. This heuristic argument leads to a very simple delayed oscillator model for ENSO:

$$\partial_t T'_E = -(\varepsilon_T + \varepsilon_w)T'_E + \mu\varepsilon_w\gamma(aT'_E - bT'_E(t-\delta\eta)) . \tag{16}$$

Here we set $\varepsilon_w = \overline{w}/H$ and we added nondimensional relative coupling and relative time scale parameters (μ, δ) introduced by Jin and Neelin (1993a,b) into this equation so to relate the delayed oscillator model to the coupled model regimes analyzed by Jin and Neelin (1993a,b). When the system is uncoupled, there is only a damped SST mode representing the natural recovery to climatological SST after an anomalous SST occurs. In the weakly coupled limit, it can be easily shown that there is another real damped mode and a number of damped oscillatory modes with periods on the order of the time delay and its harmonics. These modes are distorted ocean adjustment modes brought in through the Rossby wave delay. When coupling becomes stronger, the SST mode and the real mode of the ocean adjustment from the Rossby wave propagation merge into a leading coupled oscillatory mode while all other oscillatory modes are damped. In the strong coupling regime with large μ, the leading oscillatory mode breaks into two real modes again with one of them being a very strong pure growing mode. This pure growing mode is related to the pure growing SST mode at the fast wave limit ($\delta = 0$). In the fast SST limit, it can be shown the leading nonoscillatory mode becomes unstable. In the middle of the parameter space, there is a mixed SST-ocean adjustment mode that is the ENSO-like delayed oscillation mode.

The delayed oscillator model is a useful prototype model that gives a simple explanation of ENSO. Using the delayed-oscillator model, we show that the result from this simple model is qualitatively consistent with the analysis of Jin and Neelin (1993) in terms of regime behavior of the coupled in the parameter space. However, it is not immediately clear how to relate the phase-transition mechanism in the delayed-oscillator model to the hypothesis by Wyrtki and ZC about ENSO phase transition. Moreover, the delayed-oscillator model also ignores the wave reflection at the eastern boundary, which has been recognized as playing an important role in ENSO (Zebiak and Cane 1987; Picaut et al 1997).

5.2. *The recharge oscillator model for ENSO*

The physics of the recharge oscillator are schematically illustrated in Fig. 6. An initial positive SST anomaly induces a westerly wind forcing over the central to western Pacific. The anomalous slope of the equatorial thermocline is promptly set up to be proportional to the wind stress and thus to the SST anomaly. The deepening of the thermocline in the eastern Pacific results in a positive feedback process that amplifies this anomaly and brings the warm anomaly to a mature phase as shown in Fig. 6. While the warming due to the positive thermocline depth anomaly in the eastern Pacific is balanced by the damping process of the SST, the wind stress leads to a negative zonal mean thermocline depth across the Pacific as the result of the divergence of zonal integrated Sverdrup transport. This process can be viewed as the discharge of the zonal mean equatorial heat content because thermocline depth is highly related to the dynamical part of the upper ocean heat content (e.g., Zebiak 1989). This process gradually reduces thermocline depth in the western and eastern Pacific and leads to a cooling trend for the SST anomaly. Thus, the warm phase evolves to the transition phase as shown in Fig. 6. At this time, the SST anomaly cools to zero. The east-west thermocline tilt diminishes because it is always in a quasi-equilibrium balance with the zonal wind stress that has disappeared following the SST anomaly. However, the entire equatorial Pacific thermocline depth and thus the eastern Pacific thermocline depth is anomalously shallow because of the discharge of the equatorial heat content during the warm phase. Subsurface ocean temperature is anomalously cold across the equatorial Pacific in the aftermath of the warm phase. The corresponding negative zonal mean thermocline anomaly is not in quasi-equilibrium with the wind stress and slowly diminishes as a result of the ocean adjustment if the cold phase does not set in. However, it is this anomalous shallow thermocline depth at the transition that allows anomalous cold water to be pumped into the surface layer by climatological upwelling; the SST anomaly thus slides into a negative phase. Once the SST anomaly becomes negative, the cooling trend proceeds because the negative SST anomaly will be further amplified through the positive thermocline feedback. That is, the enhanced trades in response to the cold SST anomaly deepen the thermocline depth in the western Pacific and lift the thermocline depth up in the east. Thus the oscillation develops into its mature cold phase as shown in Fig. 6. At the same time, the zonal mean thermocline depth over the equatorial Pacific is deepening as the result of the recharging of the equatorial heat content due to the strengthened trades. This reverses the cooling trend after the mature cold phase and brings it to another transition phase as shown in Fig. 6. When the cold SST anomaly reduces to zero, the positive zonal mean thermocline depth generated by the recharging process will lead the SST anomaly going back to another warm phase.

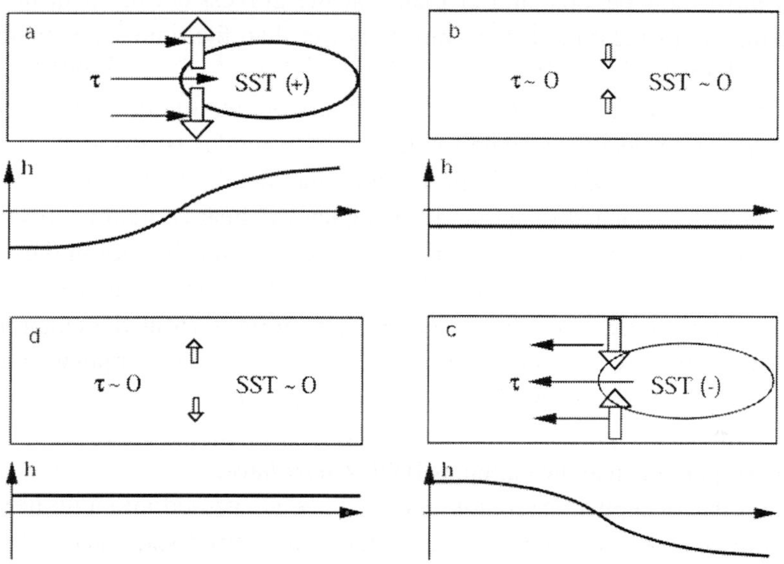

Figure 6. Schematic panels of the four phases of the recharge oscillation: (a) the warm phase, (b) the warm to cold transition phase, (c) the cold phase, and (d) the cold to warm transition phase. The rectangle box represents the equatorial Pacific basin, the elliptical circle represents the SST anomaly, thin and filled arrows represent wind stress anomaly associated with the SST anomaly, and the thick unfilled arrows represent the recharge/discharge of equatorial heat content. Each panel also shows the distribution of the thermocline depth anomaly (h) along the equator (after Jin 1997a).

During the cycle, a positive western Pacific thermocline depth anomaly leads to a warm SST anomaly in the eastern Pacific. This can also be viewed as the thermocline depth anomaly of the western Pacific being negatively proportional to the SST anomaly of the eastern Pacific with a time-lag. Thus the non-equilibrium between zonal mean thermocline depth and the wind stress forcing, because of the slow basin-wide ocean adjustment, serves as the memory of the coupled system. The discharge of the equatorial heat content in the warm phase and recharge at the cold phase serve as the phase transition mechanism from a warm event to a cold event and vice versa.

First, the recharge oscillator model for ENSO captures the Bjerknes positive feedback. A warm (cold) SST anomaly in the central to eastern Pacific induces equatorial westerly (easterly) winds and thus an anomalous thermocline zonal contrast. The latter gives rise to a positive (negative) thermocline anomaly in the eastern equatorial Pacific and thus enhances the warm (cold) SST by anomalous warm (cold) vertical advection by climatological upwelling. The coupling of relatively fast atmospheric and oceanic adjustment processes can quickly establish this positive dynamical ocean-atmosphere feedback. The atmospheric wind responses to the forcing generated by SST anomalies are nearly instant. The anomalies in the equatorial zonal contrast of the thermocline and the zonal wind stress anomalies are also always in quasi-equilibrium which can be easily understood from the dispersion diagram (2). Because the thermocline zonal contrast corresponds to a zonal spatial scale with the pseudo-zonal wave number greater than 1, the ocean adjustment modes that contribute to establishing the quasi-equilibrium between equatorial thermocline zonal contrast and the zonal wind stress anomalies will have their periods shorter than that the gravest ocean basin mode. The time scales of these fast adjustment modes are much shorter than the time scale characterizing the slow ENSO periodicity.

Secondly, the oscillation mechanism is from the slow non-equilibrium dynamics of the zonal mean thermocline depth associated with the VLF modes of tropical ocean dynamics. The oscillations of the zonally uniform thermocline depth of the VLF modes can be understood by the discharge and recharge of the equatorial heat content as discussed in Jin et al. (2001a, b). The conceptual recharge oscillator model thus has its root in the slow spectrum of the tropical ocean dynamics. It naturally captures the slow adjustment process of the equatorial zonal mean thermocline depth anomalies to forcing from the equatorial wind and wind stress curl anomalies. By separating the fast and slow adjustment processes of the tropical ocean dynamics, this conceptual basin-wide coupled recharge oscillator model combines the hypotheses put forward by Bjerknes-Wyrtki-Cane-Zebiak (BWCZ). Thus recharge oscillator perhaps may be called as BWCZ oscillator.

In principle, this conceptual oscillator model can be simplified into a harmonic oscillator. This was accomplished by Jin (1996,1997a,b). Here only a brief summary of the formulation of the model is given. Under a slow time scale, the quasi-equilibrium relationship (9) still holds, for instance, within one oceanic Rossby radius of deformation from the equator. This relation approximately holds for the anomalies associated with the ENSO, because ENSO-related wind stress anomalies change slowly in time relative to the time scale of the equatorial oceanic Kelvin waves. However, this relation only constrains the east-west contrast of the thermocline depth. The absolute depth at the western Pacific or the mean thermocline depth over the equatorial band is not constrained by this balance. The mean thermocline depth depends on the mass adjustment of the entire tropical Pacific ocean and is not in equilibrium with the slowly varying wind forcing. The non-equilibrium between the zonal mean thermocline depth and the wind stress forcing resides in the memory of the subsurface ocean dynamics. This non-equilibrium makes the fast-wave limit approximation inappropriate for an ENSO-like mode that relies on the subsurface ocean memory (e.g., Cane 1992b).

There are two equivalent qualitative approaches to this adjustment process. One is from the wave propagation point of view. The ocean mass adjustment is completed through the oceanic Kelvin and Rossby waves that propagate in opposite directions and are forced by the basin boundaries to change into opposite wave characteristics through reflections. The damping effect due to the leakage of energy via the boundaries and other damping processes allows this adjustment to settle into a quasi-equilibrium state in the equatorial region. In this view, one of the two unknowns of the thermocline anomalies in the eastern and western equatorial Pacific can be related to a boundary condition that determines wave reflections as shown in a number of theoretical studies (e.g., Cane and Sarachik 1981). One possible approach is the determination of the thermocline anomalies in the western equatorial Pacific in the delay oscillator. Another is to consider that the equatorial wave propagation process is relatively fast for establishing the thermocline slope that extends to the off-equatorial region as a result of the broadness of the atmospheric wind system. The Coriolis force becomes important off the equatorial band and therefore there will be Sverdrup transport either pumping the mass in or out of the equatorial region depending on wind forcing, as hypothesized by Cane and Zebiak (1985) and Wyrtki (1986). Under linear shallow water dynamics, this Sverdrup transport is accomplished by the Rossby waves. The zonally integrated effect of this Sverdrup transport of mass or, equivalently, heat content (in the context of shallow water dynamics) results in the deepening or shoaling of the Pacific thermocline depth. Thus, although the thermocline tilt along the equator is set up quickly to balance the equatorial wind stress as expressed by (8), the thermocline depth of the warm pool takes time to adjust to the zonally integrated meridional transport, which is related to both the wind stress and its curl off the equatorial band.

Using the second approach and assuming that the adjustment time scale is much longer than that for a Kelvin wave crossing the basin, this adjustment process can be symbolically described by the following equation

$$dh_W / dt = -rh_W - F_\tau . \tag{17}$$

This equation focuses on thermocline depth changes averaged over the western equatorial Pacific during basin-wide ocean dynamic adjustments because the tropical wind anomalies associated with ENSO are largely over the western to central Pacific (Deser and Wallace 1990). The first term on the right-hand side represents the ocean adjustment. It is simply characterized by a damping process with a rate r which collectively represents the damping of the upper ocean system through mixing and the equatorial energy loss to the boundary layer currents at the east and west sides of the ocean basin. This was shown clearly in Jin (1997b) where this relation is derived from basic dynamical principles through a number of simplifications. The wind forcing F_τ is related to the zonally integrated wind stress and its curl. It represents the Sverdrup transport across the basin. To a large extent F_τ is also proportional to the value of $\hat{\tau}$ alone. For example, $F_\tau = \alpha\hat{\tau}$, which approximately holds for the ENSO-related wind-stress anomalies with a broad meridional scale. One also expects that F_τ is a weak forcing because only a part of the wind stress forcing is involved in the slow adjustment process whereas the other part is in the quasi-Sverdrup balance. A small r and a weak forcing F_τ are consistent in describing the slow basin-wide ocean adjustment process. The minus sign for the wind forcing term comes from the fact that a westerly wind-stress anomaly will lead to a shallower thermocline over the western Pacific, whereas a strengthened trade will result in a build-up of the warm pool as suggested by Wyrtki (1975, 1986). The equation (17) can be rewritten as

$$dh_W / dt = -rh_W - \alpha\hat{\tau} . \tag{18}$$

Equations (14) and (18) give a gross description of the basin-wide oceanic adjustment under anomalous wind-stress forcing. The equatorial Sverdrup balance is assumed and the explicit wave time scale is omitted but the role of these waves in achieving the quasi-equilibrium adjustment process is accounted for. The whole basin-wide distribution of heat content, residing in the memory of subsurface ocean dynamics, is crudely but explicitly taken into consideration. The slow buildup of the western Pacific warm pool in terms of thermocline depth, or the recharging of the entire equatorial heat content, takes place during the phase of strengthened trade winds ($\hat{\tau} < 0$), whereas a weakened trade results in the gradual discharging of the western Pacific warm pool and the reducing of the entire equatorial thermocline depth. Thus Equation (18) gives a simple and clear description of the equatorial heat content recharge / discharge process which is the essential phase-transition mechanism of ENSO as suggested by Cane and Zebiak (1985) and Wyrtki (1986). Mathematically, the zero frequency mode in the equations (14 and 18) serves as an approximation of the VLF mode. Clearly without wind forcing the ocean dynamics mode of this simple model has only one free solution with the zonally uniform equatorial thermocline as its eigen vector and $-r$ as its eigen value. In other words, although drastically simplified, the simple model indeed captures the slow ocean adjustment of the zonally uniform thermocline and the fast adjustment of zonal contrast of thermocline to equatorial wind forcing.

The variation of SST during ENSO is largely confined within the central to eastern equatorial Pacific. The SST anomaly in this region strongly depends on the local thermocline depth, which determines the temperature of the subsurface water as it is pumped up into the surface layer by the climatological upwelling associated with the climatological trade wind along the equator. Changes in the trade-wind intensity in response to the SST anomaly may also further reinforce the SST anomaly by altering upwelling and horizontal advection. The mean climatological upwelling and heat exchange between the atmosphere and ocean tend to damp out the SST anomaly. Although the details of all these processes can be complicated (e.g., Jin and Neelin 1993a,b), they can be roughly depicted in a simple equation similar to equation (13) for the SST anomaly T'_E, averaged over the central to eastern equatorial Pacific:

$$\partial_t T'_E = -cT'_E + \varepsilon_w \gamma h_E + \delta_s \tau_E \,. \tag{19}$$

The first term on the right-hand side is the relaxation of SST anomaly toward climatology (or zero anomaly) caused by the damping processes with a collective damping rate $c = \varepsilon_T + \varepsilon_w$. The second and third terms are the thermocline and upwelling feedback processes respectively; τ_E is wind stress averaged over the domain where the SST anomaly occurs.

As discussed above, atmospheric response to a warm SST anomaly of the central to eastern Pacific is a westerly wind over the central to western equatorial Pacific and an easterly anomaly to the east of the SST anomaly. There is an overall westerly (easterly) anomaly for a positive (negative) SST anomaly averaged over the entire basin of the equatorial band but a much weaker easterly (wasterly) anomaly averaged over the eastern half of the basin (e.g., Deser and Wallace 1990). This allows the simple approximate relations of the wind stress and SST anomalies:

$$\hat{\tau} = aT_E', \quad \tau_E = -\tilde{a}T_E'. \tag{20}$$

Combining the equations (14), and (18)-(20), one obtains a simple linear coupled system is obtained with both the subsurface ocean-adjustment dynamics and the surface-layer SST dynamics:

$$\frac{dT_E'}{dt} = RT_E' + \varepsilon_w \gamma h_W, \quad \frac{dh_W}{dt} = -rh_W - \alpha a T_E'. \tag{21}$$

Here $R = a\varepsilon_w\gamma - \tilde{a}\delta_s - c$ collectively describes the Bjerknes positive feedback hypothesis of the tropical ocean-atmospheric interaction. This system therefore combines the mechanisms in the BWCZ hypotheses to describe a decaying or growing coupled recharge oscillator depending on the sign of $(R-r)/2$ (growth rate) when $\omega = \sqrt{\alpha a\varepsilon_w\gamma - (r+R)^2/4}$ (frequency) is real. Both growth rate and the frequency depend heavily on the coupling factor $a\varepsilon_w\gamma$ because the Ekman upwelling feedback parameter $-\tilde{a}\delta_s$ is a relatively small negative value due to the weak local wind stress averaged over the region of SST anomaly. For simplicity, this term will be ignored throughout the rest of the paper by setting $\tilde{a}\delta_s = 0$ or so that $R = a\varepsilon_w\gamma - c$.

To facilitate some quantitative analysis, an estimate of the values of the parameters in the coupled system (21) is needed. The collective damping rate c is dominated by the mean climatological upwelling, which yields a typical damping time scale of about two months; $\varepsilon_w\gamma$ is related to both the mean climatological upwelling and the sensitivity of subsurface ocean temperature to the thermocline depth. It is chosen to give an SST change rate of 1.5 degrees over two months (which is upwelling time scale) per 10 m of thermocline depth anomaly over the eastern Pacific. This value of $\varepsilon_w\gamma$ is close to that in the ZC model. The collective damping rate r in the ocean adjustment includes the weak linear damping (about 1/(2.5 years) in the ZC model), and the damping effect due to the loss of energy to the boundary currents of the west and east boundary layers. The latter could lead to a much shorter damping time scale. Parameter r is thus set as 1/(8 months), which still gives a damping time scale much longer than the typical 2-month crossing time scale of the first baroclinic Kelvin wave. Assuming that for a given steady wind-stress forcing the zonal mean thermocline anomaly of this linear system is about zero at the equilibrium state, i.e., $h_E + h_W = 0$, then from (16) and (18), one finds that c, r, and α are found to be about half of r. Finally, parameter a is a measure of the thermocline slope that is in balance with the zonal wind stress produced by the SST anomaly. It is chosen to give 50 m of east-west thermocline depth difference per one degree of the SST anomaly. This is a high-end estimation of the parameter, whereas for a typical ENSO event the value may be somewhat smaller. For convenience, a so-called relative coupling coefficient μ is introduced as

$$a = a_0 \mu , \tag{22}$$

where a_0 is the high-end estimation of parameter a, and μ is referred to as a relative coupling coefficient which will be varied in the range (0, 1.5) embracing the uncoupled and strongly coupled cases. System (21) is hereafter nondimensionalized by scales of $[h]$ = 150 meters, $[T]$ = 7.5°C, and $[t]$ = 2 months for anomalous thermocline depth, SST, and the time variable respectively. Accordingly, parameters c, r, and α are scaled by $[t]^{-1}$, and $\varepsilon_w \gamma$, a_0 by $[h][t]/[T]$, $[T]/[h]$. Their dimensionless values are c = 1, $\varepsilon_w \gamma$ = 0.75, r = 0.25, α = 0.125, and a_0 = 2.5, respectively.

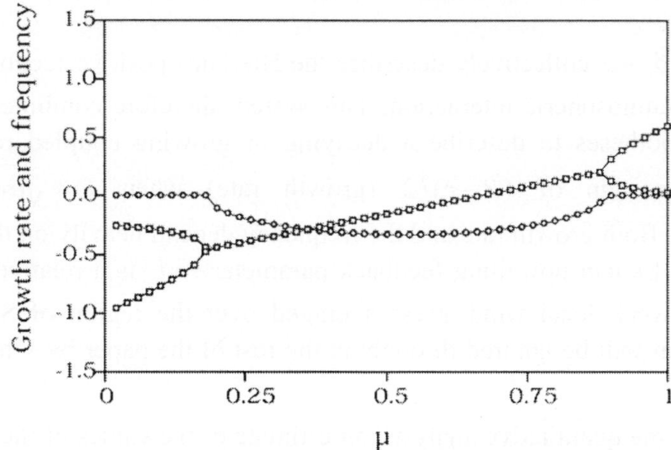

Figure 7. Dependence of eigenvalues on the relative coupling coefficient. The curves with dots represents the growth rates and the curve with circles is frequency when the real modes merge as a complex mode (corresponding periods in years equal $\pi/3$ divided by the frequencies) (after Jin 1997a).

The detailed analyses of the model sensitivity to this set of parameters can be found in Jin (1997b). With the values of the parameters given above, the dependence of the eigen modes of the coupled system (21) on the relative coupling coefficient can be analytically solved as

$$\sigma_{1,2} = -\frac{15}{16}\left(\frac{2}{3} - \mu \pm i\sqrt{(\mu - \mu_1)(\mu_2 - \mu)}\right), \quad \mu_1 = (8 - \sqrt{28})/15, \mu_2 = (8 + \sqrt{28})/15. \tag{23}$$

The result is also shown in Fig. 7. When the relative coupling coefficient μ is weak ($\mu < \mu_1$), the system has two decaying modes that can be identified at μ = 0 as the uncoupled SST model and the ocean-adjustment mode, respectively. These two modes eventually merge into an oscillatory mode as the coupling coefficient increases to $\mu > \mu_1$. When the coupling is further increased to $\mu > \mu_2$, the oscillator breaks down to give two modes: One is a purely growing mode because the strong coupling through Bjerknes feedback results in a growth rate being too fast to be linearly checked by the slow ocean adjustment process. The other is a real mode whose growth rate decreases rapidly and becomes negative when μ is larger than

μ_2. For a wide range of moderate coupling between (μ_1, μ_2), the system does support an oscillatory mode with a period mostly in the 3-5 year range. This oscillatory mode is supercritical if $\mu > \mu_c = 2/3$ and subcritical if the coupling coefficient is smaller than the critical value μ_c. In the supercritical oscillatory regime, adding nonlinearity to the coupled system will limit the linear growth to a self excited coupled oscillation (Jin 1997a). In the subcritical oscillatory regime, a coupled oscillatory solution can be also sustained by stochastic forcing. An analytical neutral solution for $\mu = \mu_c$ is shown in Fig. 8.

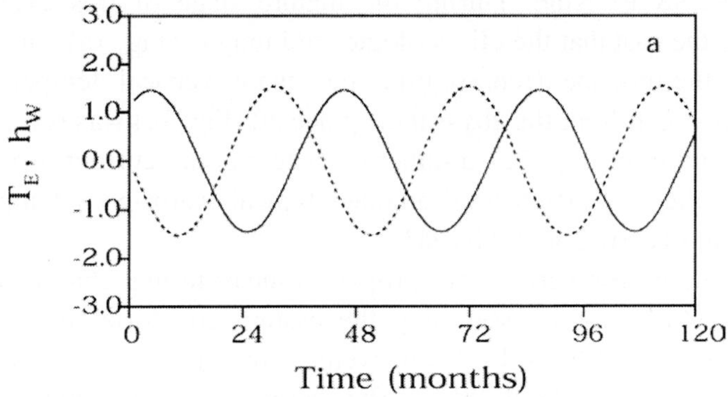

Figure 8. The time series for the SST (in °C, solid line) and thermocline depth in the western Pacific (in 10 meters, dashed line) (a) with initial conditions of SST at 1.125 °C and thermocline depth in the western Pacific is zero (after Jin 1997a).

The relevance of the recharge oscillator model of ENSO has been examined by a number of studies with observational data (Meinen and McPhaden 2000) and GCM simulations (e.g. Collins 2000, Yu and Mechoso 2001). The advantage of this simple conceptual model for ENSO is that it can be easily tested by examining the equatorial mass or heat content budgets without following the details of forced and free equatorial waves and their reflections. One of the critical elements of the recharge oscillator model for ENSO is the slow adjustment of the zonally averaged thermocline depth anomalies along the equator. Thus there is a phase difference of about 90 degrees between the variations of the zonally uniform thermocline depth along the equator and the SST anomalies in the eastern equatorial Pacific. This critical element is responsible for making the ENSO oscillate on a 3-4 year time scale.

6. Nonlinear dynamics of ENSO

Theories based on the linear dynamics of the coupled ocean-atmospheric models provide an understanding of the mechanisms for its oscillatory nature and periodicity of about 3-5 years. In general, nonlinear processes may become important when the ENSO perturbations become intense. In addition to the simple role of nonlinearity to control the exponential growth of unstable ENSO mode in the coupled system, it is known that the nonlinear dynamics of ENSO can produce ENSO irregularity, frequency and phase locking, amplitude modulations (e.g. Jin et al. 1994; Tzipermann et al. 1994; Timmermann et al. 2003). However, the relevance of the nonlinear dynamics in generating the observed feature of ENSO is still a unsettled subject. It has been suggested that linear and marginally stable coupled dynamics with stochastic forcing are sufficient to explain the overall feature of ENSO (Tomperson and Battisti 2001). Here, I shall focus on a basic question: how to measure the ENSO nonlinearity and to reveal evidence of the importance of that nonlinear dynamics in shaping ENSO.

6.1. *Maximum Potential Intensity (MPI) for ENSO*

First, we will examine the strongest El Niño event ever recorded instrumentally (McPhaden 1999) — the 1997/98 El Niño. During the mature stage of this event, the warm pool expanded so far to the east that the climatologic cold tongue (Fig. 1a) vanished (Fig. 1b). The mean tilt of the thermocline (representing the sharp vertical temperature gradient that separates the upper ocean from the abyssal deep ocean) (Fig. 1c) was reversed (Fig. 1d). Even the equatorial undercurrent (Fig. 1c), a rather persistent ocean current, was strongly disrupted (Fig. 1d). In fact, the 1997/1998 SST attained typical warm-pool temperatures (Fig. 1e). Similar strong event occurred in 1982/1983.

Motivated by these observations, we propose a measure that characterizes the maximum potential intensity (MPI) for ENSO using the eastern equatorial Pacific SST. The upper bound of this SST, the MPI of El Nino events, corresponds to the radiative-convective equilibrium temperature of about 30°C. The warm pool SST attains values close this temperature. The low bound of the equatorial SST in the eastern Pacific, the MPI for La Niña events, corresponds to a complete surface outcropping of the thermocline and is about 20°C. Because the average SST in the cold tongue region is about 25°C, the MPI of ENSO measured by SST anomalies averaged in the cold tongue region is about 5°C.

The 1982/1983 and 1997/1998 El Niño events were so strong that they nearly reached the MPI, which is also clear from Fig. 1e. The La Niña events, however have never reached MPI, at least for the data shown in Fig. 1e. For those warm events nearly reaching the MPI, the anomalies are too large to be viewed as small perturbations to the climate mean state. In other words, the nonlinear processes are important for these events.

The above definition of MPI for ENSO is based on the current climate state of the tropical Pacific. The paleoclimate states of the tropical Pacific could be quite different. For a relatively cold tropical Pacific, such as during glacial times, the range of MPI could be reduced, which might also limit the ENSO intensity. On the contrary, in the warmer climate as simulated in the global warming simulations, the MPI may further increase to allow strong ENSO activity (Timmermann et al. 1999).

6.2. *Nonlinear dynamical Heating and ENSO asymmetry*

The fact that the strong 1982/1983 and 1997/1998 events reached the MPI for ENSO gives one measure of the ENSO nonlinearity. Another possible measure for the ENSO nonlinearity is the dominance of the NDH terms in the heat budget of the upper ocean. To quantify this second measure, we calculated the heat budget in the uppermost 50 m of the tropical Pacific, using the NCEP ocean assimilation data.

The heat budget of the ocean surface layer is calculated using the following SST equation

$$\frac{\partial T'}{\partial t} = -(u'\partial_x \overline{T} + v'\partial_y \overline{T} + w'\partial_z \overline{T} + \overline{u}\partial_x T' + \overline{v}\partial_y T' + \overline{w}\partial_z T') \\ -(u'\partial_x T' + v'\partial_y T' + w'\partial_z T') + R' . \tag{24}$$

In (24), T, u, v, w are SST, zonal, meridional, and vertical velocities; overbar and prime denote the climatologic mean and anomalies, respectively. The contributions from heat fluxes and subgrid scale processes are denoted by the residual term R'. The term for NDH is in the bracket of the second line of the equation. As shown in Fig. 9a the 1997/1998 warm event is characterized by a large NDH anomaly of about 2°C/month that is located in the center of the El Niño SST anomaly. It is comparable in magnitude to the linear heat advection terms throughout much of the 1997/1998 warm event. The subsequent La Niña event from 1998/99 was also characterized by a positive NDH anomaly of about 2°C/month. The overall effect of the strong NDH anomaly is to amplify El Niño events and to damp La Niña events. This leads to an asymmetry in the magnitude of El Niño and La Niña in consistency with the observations (Burgers and Stephenson 1999). The nonlinear advection of heat is negligible for modest ENSO events such as the 1986/87 El Niño and the subsequent La Niña state (Fig. 9b). In this case, El Niño and La Niña attained similar absolute magnitudes, also lending support to the assertion that NDH is responsible for the asymmetry.

6.3. *Eastward propagating El Nino and its strong intensity*

The nonlinear advection of heating depends on particular temporal and spatial phase differences between the temperature and current fields. This is shown for the 1997/98 El Niño in Fig. 3. During the mature phase, westerly zonal wind stress anomalies occur in the central to western equatorial Pacific, whereas easterly wind anomalies can be seen in the eastern Pacific (Fig. 10a). This wind pattern is consistent with the linear atmospheric dynamic response to the SST anomalies (Gill 1980) (Fig. 10c). The westerly (easterly) wind stress anomalies near the equator induce anomalous downwelling (upwelling) (Fig. 10b) due to the Coriolis effect. The reduction in the integrated zonal wind stress across the equatorial Pacific flattens the tilt of the equatorial thermocline. The deepening of the thermocline in the eastern equatorial Pacific leads to an adiabatic warming in the subsurface ocean (Fig. 10d) that exceeds the surface warming (Fig. 10c) throughout the development phase of the El Niño event (Fig. 10e). At the same time, an enhanced upwelling (Fig. 10b) due to easterly wind anomalies leads to an enhanced vertical advection of anomalously warm waters, thereby accelerating the surface warming.

Similarly, the transition to the La Niña phase involves anomalous cooling in the subsurface ocean resulting in a positive vertical temperature gradient. At the same time, there is reduced upwelling in the eastern equatorial Pacific due to the prevailing westerly wind anomalies. The result is that upwelling of anomalously cold subsurface waters into the surface layer is prevented, thereby slowing down surface cooling. Therefore, the out-of-phase relationship between the vertical temperature gradient and the upwelling in the eastern equatorial Pacific gives rise to positive NDH throughout the 1997 to 1999 ENSO cycle.

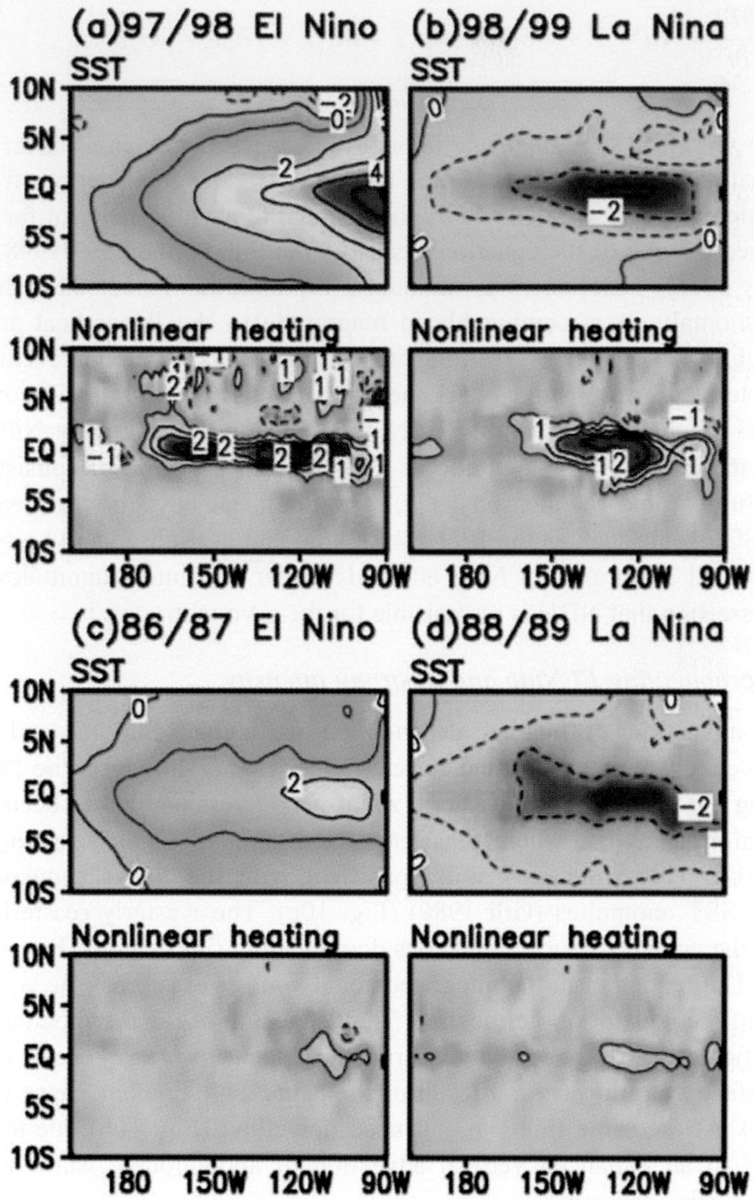

Figure 9. December SST anomaly (°C) and rate of change in SST (°C month^{-1}) due to the nonlinear dynamic heating terms *(23)* computed for (a) the El Niño event in 1997 and (b) the La Niña event in 1998. (c) SST and nonlinear heating for the El Niño event in March 1987 and (d) the mature La Niña situation in December 1988. The data were prefiltered with 11-month running mean. The anomalies were obtained based on the climatology of 1978-1998 (after Jin et al. 2003).

The nonlinear warming serves as a strong positive feedback for the El Niño event and as a strong negative feedback for the following La Niña event. Similar results (not shown) were obtained for the 1982/83 El Niño and its subsequent La Niña phase. As illustrated above, a prerequisite for this type of nonlinear heating is an eastward movement of the anomalous wind-stress. Before the 1976 climate shift, ENSO events were characterized by westward propagating anomalies. A heat budget analysis of another ocean assimilation data set (Carton et al. 2000) covering the period from 1950-1999 (Figure 11) confirms that the pre-1976 era exhibited much less nonlinear heating during ENSO cycles and hence smaller amplitudes of El Niño events than that of the post-1976 era. Therefore, the direction into which ENSO events propagate may serve as a useful indicator to estimate the potential for nonlinear amplification and hence for the probability to generate strong El Niño events. Further studies are needed to show the linkage between the eastward propagation of an El Nino event and the preferred nonlinear dynamical amplification. Firmly establishing this linkage will help to predict the amplitudes of strong El Niño events, which so far have been very difficult to predict (Landsea and Knaff 2000).

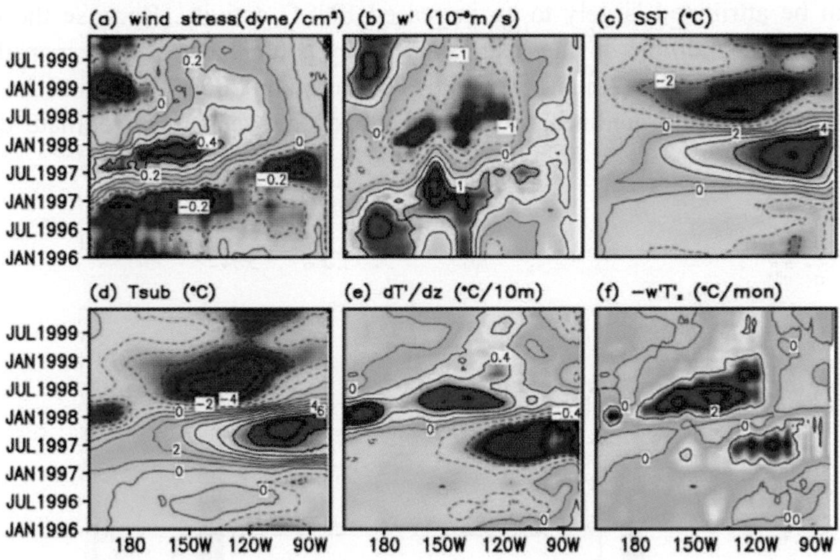

Figure 10. Time-longitude plots of (a) wind stress (dyne cm^{-2}), (b) upwelling velocity (10^{-5} m s^{-1}), (c) ocean temperature anomaly in the surface layer (°C), (d) subsurface ocean temperature (obtained at 65 m depth) (°C), (e) vertical temperature difference (between surface layer and the subsurface) (0.1 °C m^{-1}), and (f) nonlinear vertical heat advection (°C month^{-1}) along the equator. Anomalies are computed with respect to the 1978-1998 climatology (after Jin et al. 2003).

6.4. *Nonlinear dynamic heating and tropical warming*

In Fig. 11 we show the NDH anomaly for the last 2 and 5 decades from NECP (Ji et al. 1995) and SODA (Carton et al. 2000) data sets. There has been an increase in the net warming starting from about 1976 with an average value of about 0.2°C/month, which was significantly higher than the average value of the pre-1976 period (Fig. 4). Half of the increased warming after the 1976 climate shift may be attributable to an increased El Niño activity associated with 1982/83 and 1997/98 events. To estimate the response to this dynamical heating, we adopt the simple ENSO recharge-oscillator equation (Eq. 21) and add a NDH forcing term denoted as N' to the SST anomaly equation. For a rough estimate of the mean temperature changes generated by the term N', we consider the stationary solution in a realistic parameter regime allowing a slightly damped oscillation. Following the choices of other parameters in Jin (1997) and setting the relative coupling coefficient at $\mu = 0.6$, the system is slightly subcritical. In this case, the change in mean temperature is $\tau_c \overline{N'}$. Here $\overline{N'}$ is the time mean of the heating term. The response time scale of the coupled model to steady forcing, τ_c, is about 4.5 months. With the excessive mean warming in the past two decades due to strong ENSO of about 0.1°C/month, the simple coupled model estimation for a steady response is thus about 0.45 °C. This estimation is perhaps is a bit on the high side because the nonlinear heating term is not steady and the non-steady response will be less sensitive, particularly for a 20-year mean. Yet, this estimation is not far from the observed tropical Pacific warming of about 0.3-0.4 °C throughout the last decades. This implies that the mean warming can be attributed largely to an increased ENSO activity. Because the changes in ocean background conditions in the last few decades could have been responsible for the change in the ENSO activity (An and Jin 2000; Timmermann 2001; Wang and An 2001), our result suggests a possible nonlinear positive feedback between mean climate change and ENSO variability.

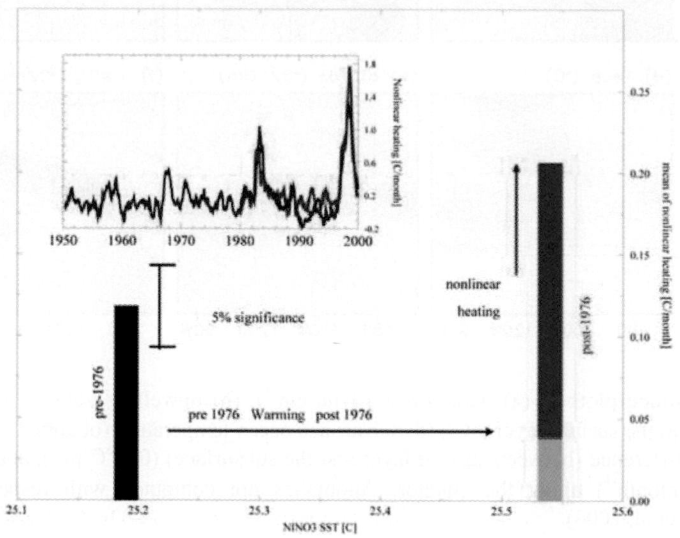

Figure 11. The inserted plot shows the time series of the nonlinear dynamic heating rates (°C month^{-1}) in the central-eastern equatorial Pacific (averaged over the area 2.5°S to 2.5°N, 150° to 100°W) based on NCEP (red) and SODA (black) data sets. The bar plot shows the mean nonlinear dynamic heating rates (°C month^{-1}) averaged from 1950 to 1976 (black bar) and from 1976 to 2000 (red and green bar, where red indicates the contribution from two strong ENSO events and the green is from the rest) together with the mean SST for these two periods in the same region. The 5% significance level for mean changes of the nonlinear heating is computed from a t-test using the variances of the pre-1976 and post-1976 period (after Jin et al. 2003).

7. Beyond a simple oscillator paradigm

So far, ENSO has been portrayed as a coupled oscillator of the tropical Pacific. However, observed spectra of the tropical Pacific climate variability, characterized by index time series such as the "Niño 3" sea surface temperature anomalies (SSTA) and the Southern-Oscillation index, exhibit much more structure than just a single spectral peak at interannual timescales associated with ENSO. In addition to the 3-5 year main periodicity of ENSO, there is significant variance on various other time scales, and in particular on decadal and on near annual periods. Recently it was shown that there is a fast coupled mode in the tropical Pacific atmosphere-ocean system (Jin et al. 2003). Evidence for the co-existence of different coupled modes in the tropical climate system was in fact found in a number of coupled models. For instance, in the Zebiak and Cane model, a fast mode, called the "mobile mode" (Mantua and Battisti 1995; Perigaud and Dewitte 1996), can be identified that is characterized by a 9-12 month period and that can co-exist with the simulated interannual ENSO mode. Similar co-existences between fast and slow modes have been found also in more complex models (Neelin 1990; Philander et al. 1992). So far, however, the relevance of these simulated fast modes has not been recognized. The potential co-existence of this fast with the main interannual mode will enrich the coupled variability in the tropics and poses a challenge for ENSO predictions. When the background in the equatorial central to eastern Pacific is warm, such as in the early 1990s, this fast mode variability may surface as near-annual mini-El Niño events, whereas during cold background the fast mode is expected to lead to near-annual La Niña events such as in the past few years. Clearly, further studies are needed for a better understanding of the fast mode on the interannual time scale and its interaction with the slow ENSO mode. Moreover, decadal variability related to coupled dynamics of the tropical Pacific was also simulated in coupled GCMs (Manabe, Yokimoto etc). In addition to an interannual coupled mode of a 3-4 year period, a co-exiting coupled decadal mode is also revealed in a simple coupled model of the tropical Pacific (Wang et al. 2003a, b). Whether this kind of tropical decadal coupled mode could really exist in the tropical Pacific and how it interacts with the main ENSO mode are questions that needs further investigation. Further studies along these lines may lead to a more complete picture about ENSO: its irregular and eventful features may be combinations of a team of the modes with periods of near annual, interannual and decadal scales.

Acknowledgement. This work was partly supported by NSF grants ATM-9615952 and ATM-0226141 and NOAA grants GC99234 and GC01246. The author thanks Drs. Jingxia Zhao and S.-I. An for valuable comments and Diane Henderson for editing the manuscripts. SOEST # xxxx.

References:

An, S.-I, and F.-F. Jin, 2000: An eigen analysis of the interdecadal changes in the structure and frequency of ENSO mode. *Geophys. Res. Lett.*, **27**, 2573-2576.

An, S.-I. and F-F. Jin, 2001: Collective role of zonal advective and thermocline feedbacks in ENSO mode. *J. Climate*, **14**, 3421-3432.

Anderson, D. L. T., and J. P. McCreary, 1985: Slowly propagating disturbances in a coupled ocean-atmosphere model. *J. Atmos. Sci.* **42**, 615-629.

Barnett, T. P., N. Graham, M. Cane, S. Zebiak, S. Dolan, J. O'Brien, and D. Legler, 1988: On the prediction of El Niño of 1986--1987, *Science*, **241**, 192-196.

Barnett, T. P., M. Latif, E. Kirk, and E. Roeckner, 1991: On ENSO physics. *J. Climate,* **4,** 487-515.

Battisti, D. S., 1988: The dynamics and thermodynamics of a warming event in a coupled tropical atmosphere/ocean model. *J. Atmos. Sci.*, **45**, 2889-2919.

Battisti, D. S., and A. C. Hirst, 1989: Interannual variability in the tropical atmosphere/ocean system: influence of the basic state and ocean geometry. *J. Atmos. Sci.*, **46**, 1687-1712.

Battisti, D. S., E. S Sarachik, and A. C Hirst 1999: A consistent model for the large-Scale steady surface atmospheric circulation in the tropics. *J. Climate,* **12**, 2956-2964.

Bjerknes, J., 1969: Atmospheric teleconnections from the equatorial Pacific. *Mon. Wea. Rev.*, **97**, 163-172.

Busalacchi, A. J., J. J. O'Brine, 1981: Interannual variability of the equatorial Pacific in the 1960s. *J. Geophys. Res.*, **86**, 10,901-10,907.

Burgers, G., and D. B. Stephenson, The normality of El Niño, 1999: *Geophys. Res. Lett.*, **26**, 1027-1030.

Cai, M., 2003: On the west-east asymmetry and ENSO variability in the equatorial Pacific basin. *J. Climate,* in press.

Cane M. A., 1992a: Tropical Pacific ENSO models: ENSO as a model of the coupled system in *Climate System Modeling*, 583-616. Trenberth, Ed., Cambridge University Press, 788 pp.

Cane M. A., 1992b: Comments on "The fast wave limit and interannual oscillations". *J. Atmos. Sci.*, **49**, 1947-1949.

Cane, M. A., and D. W. Moore, 1981: A note on low-frequency equatorial basin modes. *J. Phys. Oceanogr.*, **11**, 1578-1584.

Cane, M. A., M. Münnich, and S. E. Zebiak, 1990: A study of self-excited oscillations of the tropical ocean-atmosphere system. Part I: Linear analysis. *J. Atmos. Sci.*, **47**, 1562-1577.

Cane M., and E. S. Sarachik, 1981: The periodic response of a linear baroclinic equatorial ocean. *J. Mar. Res.*, **39**, 651-693.

Cane, M. A., and S. E. Zebiak, 1985: A theory for El Niño and the Southern Oscillation. *Science*, **228**, 1084-1087.

Cane, M. A., S. E. Zebiak, and S. C. Dolan, 1986: Experimental forecasts of El Niño. *Nature*, **321**, 827-832.

Carton, J., G. Chepurin, X. Cao, and B. Giese, 2000, A simple ocean data assimilation analysis of the global upper ocean 1950-1995, Part 1: methodology, *J. Phys. Oceanogr.*, **30**, 294-309.

Chang, P, B. Wang, T. Li, and J. Lin, 1994: Interaction between the seasonal cycle and ENSO-frequency entrainment and chaos in a coupled ocean-atmosphere model. *J. Atmos. Sci.*, **52**, 2353-2372.

Collins, M., 2000: Understanding uncertainties in the response of ENSO to greenhouse warming, *Geophys. Res. Lett.*, **27**, 3509-3513.

Deser, C., and J. M. Wallace, 1990: Large-scale atmospheric circulation features of warm and cold episodes in the tropical Pacific, *J. Climate*, **3**,1254-1281.

Fedorov, A. V., and S. G. H. Philander, 2000: Is El Niño changing?. *Science.*, **288**, 1997–2002.

Gill, A. E., 1980: Some simple solutions for heat induced tropical circulation. *Quart. J. Roy. Met. Soc.*, **106**, 447-462.

Gill, A. E., 1985: Elements of coupled ocean-atmosphere models fo the tropics. In "coupled ocean-atmosphere models" (J. Nihoul, ed.), Elsevier, Oceanogr. Sci., **40**, Amsterdam, 303-327.

Graham, N. E., and W. B. White, 1988: The El Niño cycle: Pacific ocean-atmosphere system, *Science*, **240**, 1293-1302.

Hirst, A. C., 1986: Unstable and damped equatorial modes in simple coupled ocean-atmosphere models. *J. Atmos. Sci.*, **43**, 606-630.

Hirst, A. C., 1988: Slow instabilities in tropical ocean basin-global atmosphere models. *J. Atmos. Sci.*, **45**, 830-852.

Ji, M., A. Leetmaa, and J. Derber, 1995: An ocean analysis system for seasonal to interannual climate studies. *Mon. Wea. Rev.*, **123**, 460-488.

Jin, F.-F., 1996: Tropical ocean-atmosphere interaction, the Pacific cold tongue, and the El Niño Southern Oscillation. *Science,* **274**, 76-78.

Jin, F.-F., 1997a: An equatorial ocean recharge paradigm for ENSO. Part I: Conceptual model. *J. Atmos. Sci.*, **54** 811-829.

Jin, F.-F., 1997b: An equatorial ocean recharge paradigm for ENSO. Part II: A stripped-down coupled model. *J. Atmos. Sci.*, **54**, 830-847.

Jin, F.-F., 1997c: A theory of interdecadal climate variability of the North Pacific ocean-atmosphere system. *J. Climate*, **10**, 1821-1834.

Jin. F.-F., 2001: Low-frequency modes of tropical ocean dynamics. *J. Climate*, **14**, 3874-3881.

Jin F.-F., and S.-I. An, 1999: Thermocline and zonal advective feedbacks within the equatorial ocean recharge oscillator model for ENSO. *Geophys. Res. Lett*, **26**, 2989-2992.

Jin F.-F., S. An, A. Timmermann, and J. Zhao, 2003: Strong El Nino events and nonlinear dynamical heating, *Geophys. Res. Lett,* in press.

Jin, F.-F., and Co-authors, 2001a: Dynamical and cloud-radiation feedbacks in *El Niño* and Greenhouse warming. *Geophys. Res. Lett.,* **28,** 1539-1542.

Jin F-F, M. Kimoto, and X.-C. Wang, 2001b: Decadal extratropical-tropical ocean-atmosphere interaction and its modulation on ENSO. *Geophys. Res. Lett.* **28**, 1531-1534.

Jin F.-F., J. Kug, S. An, 2003: A near-annual coupled mode in the equatorial Pacific ocean. *Geophys. Res. Lett*, in press.

Jin, F.-F., and J. D. Neelin, 1993a: Modes of interannual tropical ocean-atmosphere inter-action—a unified view. Part I: Numerical results. *J. Atmos. Sci.,* **50**, 3477-3502

Jin, F.-F., and J. D. Neelin, 1993b: Modes of interannual tropical ocean-atmosphere inter-action—a unified view. Part III: Analytical results in fully coupled cases. *J. Atmos. Sci.* **50**, 3523-40.

Jin, F.-F., J. D. Neelin and M. Ghil, 1994: El Niño on the Devil's Staircase: annual subharmonic steps to chaos. *Science*, **206**, 70-72.

Jin, F.-F., J. D. Neelin and M. Ghil, 1996: The interaction of ENSO and the annual cycle: Subharmonic frequency-locking and ENSO aperiodicity. *Physica D*, **98**, 442-465.

Knutson, T. R. and S. Manabe, 1998: Model assessment of decadal variability and trends in the tropical Pacific Ocean, J. Climate, **11**, 2273-2296.

Landsea, C. W., and J. A. Knaff, 2000: How much skill was there in forecasting the very strong 1997-98 El Niño? *Bull. Am. Meteorol. Soc.*, **81**, 2107-2120.

Latif, M., T.P. Barnett, M.A. Cane, M. Flügel, N.E. Grahem, H. von Storch, J.-S. Xu, and S.E. Zebiak, 1994: A review of ENSO prediction studies. *Climate Dynamics,* **9,** 167-179.

Latif, M., and Co-authors, 1998: A review of the predictability and prediction of ENSO, *J. Geophys. Res.*, **103**(C7), 14,375-14,393.

Lau, K. M., 1981: Oscillations in a simple equatorial climate system. *J. Atmos. Sci.*, **38**, 248-261.

Lau, K. M., 1985: Elements of a stochastic-dynamical theory of the long-term variability of the El Niño-Southern Oscillation. *J. Atmos. Sci.*, **42**, 1552-1558.

Lau, N. C., S. G. H. Philander, and M. J. Nath, 1992: Simulation of El Niño/Southern Oscillation phenomena with a low-resolution coupled general circulation model of the global ocean and atmosphere. *J. Climate*, **5**, 284-307.

Lindzen, R. S., and S. Nigam, 1987: On the role of sea surface temperature gradients in forcing low level winds and convergence in the tropics. *J. Atmos. Sci.*, **44**, 2440-2458.

Mantua N. J. and D. S. Battisti, 1995; Aperiodic variability in the Zebiak-Cane coupled ocean-atmosphere model: Air-sea interaction in the western equatorial Pacific. *J. Climate*, **8**, 2897-2927.

Matsuno, T., 1966: Quasi-geostrophic motions in the equatorial area, *J. Meteorol. Soc. Jp,* **44**, 25-43.

McCreary, J. P., 1983: A model of tropical ocean-atmosphere interaction, *Mon. Weather Rev.*, **111**, 370-387.

McPhaden M. J., and Coauthors, 1998: The tropical ocean global atmosphere (TOGA) observing system: a decade of progress. *J. Geophys. Res.* **104**, 14,169-14,240.

McPhaden, M.J., 1999: The child prodigy of 1997-1998, *Nature*, **398**, 559-562.

McWilliams, J. and P. Gent, 1978: A coupled air-sea model for the tropical Pacific. *J. Atmos. Sci.*, **35**, 962-989.

Meinen, C. S. and M. J. McPhaden, 2000: Observations of warm water volume changes in the equatorial Pacific and their relationship to El Niño and La Niña. *J. Climate,* **13**, 3551-3559.

Moore, D. W., 1968: *Planetary-gravity waves in an equatorial ocean*. Ph.D. thesis, Harvard University, Cambridge, Mass., 207 pp.

Neelin, J. D. 1989, On the interpretation of the Gill model, *J. Atmos. Sci.,* **46**, 2466-2468.

Neelin, J. D., 1990: A hybrid coupled general circulation model for El Niño studies. *J. Atmos. Sci.*, **47**, 674-693.

Neelin, J. D., 1991: The slow sea surface temperature mode and the fast-wave limit: Analytic theory for tropical interannual oscillations and experiments in a hybrid coupled model. *J. Atmos. Sci.*, **48**, 584-606.

Neelin, J. D., D. S. Battisti, A. C. Hirst, F.-F. Jin, Y. Wakata, T. Yamagata, S. Zebiak, 1998: ENSO theory. *J. Geophys. Res.,* **104**, 14,262-14,290.

Neelin, J. D., and H. A. Dijkstra, 1995: Ocean-atmospheric interaction and the tropical climatology, Part I The danger of Flux-Correction, *J. Climate*, **8,** 1325-1342.

Neelin, J. D., and F.-F. Jin, 1993: Modes of interannual tropical ocean-atmosphere interaction—a unified view. Part II: Analytical results in the weak-coupling limit. *J. Atmos. Sci.*, **50**, 3504-3522.

Neelin, J. D., M. Latif, and F.-F. Jin 1994: Dynamics of coupled ocean-atmosphere models: the tropical problem. *Ann. Rev. Fluid Mech.* **26,** 617-59.

Penland, C., and L. Matrosova, 1994: A balance condition for stochastic numerical model with application to the El Niño-Southern Oscillation. *J. Climate*, **7**, 1352-1372.

Penland, C., and P. Sardeshmuhk, 1995: The optimal growth of the tropical sea surface temperature anomalies. *J. Climate*, **8**, 1999-2024.

Perigaud C., and B.Dewitte,1996:E1 Niño-La Niña events simulated with Cane and Zebiak's model and observed with satellite or in situ data,Part I:model data comparison. *J.Climate*, **9**, 66-84.

Philander, S. G. H., 1981: The response of the equatorial oceans to a relaxation of the trade winds, *J. Phys. Oceanogr.*, **11**, 176-189.

Philander, S. G. H., 1990: El Niño, La Niña, and the Southern Oscillation. Academic Press, San Diego, 293 pp.

Philander, S. G. H., and R. C. Pacanowski, 1980: The generation of equatorial currents, *J. Geophys. Res.,* **85**, 1123-1136.

Philander, S. G. H., R. C. Pacanowski, N. C. Lau, and M. J. Nath, 1992: A simulation of the Southern Oscillation with a global atmospheric GCM coupled to a high-resolution, tropical Pacific ocean GCM. *J. Climate*, **5**, 308-329.

Philander, S. G. H., T. Yamagata, and R. C. Pacanowski, 1984: Unstable air-sea interactions in the tropics. *J. Atmos. Sci.*, **41,** 604-613.

Picaut J. M. Ioualalen, c. Mekes, T. Delcroix, and M. J. McPhaden, 1996: Mechanism of the zonal displacements of the Pacific Warm Pool: Implications for ENSO, *Science*, **274**, 1486-1489.

Rasmusson, E. M., and T. H. Carpenter, 1982: Variations in tropical sea surface temperature and surface wind fields associated with the Southern Oscillation/El Niño. *Mon. Wea. Rev.*, **110**, 354-384.

Schneider E. K., B. Huang and J. Shukla, 1995: Ocean wave dynamics and El Niño, *J. Climate*, **8**, 2415-2439.

Schopf, P. S., and M. J. Suarez, 1988: Vacillations in a coupled ocean-atmosphere model. *J. Atmos. Sci.*, **45**, 549-566.

Suarez, M. J., and P. S. Schopf, 1988: A delayed action oscillator for ENSO. *J. Atmos. Sci.*, **45**, 3283-3287.

Sun D. and Z. Liu, 1996: Dynamic ocean-atmosphere coupling: A thermostat for the tropics. *Science*, **272,** 1148-1150.

Timmermann, A., 2001: Changes in ENSO stability due to greenhouse warming, *Geophys. Res. Lett.*, **28**, 2061-2064.

Timmermann A., and Co-authors, 1999: Increased El Niño frequency in a climate model forced by future greenhouse warming. *Nature,* **357,** 230.

Timmermann A., F.-F. Jin, and Jan Abshagen, 2003: A nonlinear theory for El Niño bursting. *J. Atmos. Sci*, **60**, 152-165.

Thompson, C. J., and D. S. Battisti, 2000: A linear stochastic dynamical model of ENSO. Part I: model development. *J. Climate,* **13**, 2818-2832.

Trenberth, K., 1990: Recent observed interdecadal climate changes in the Northern Hemisphere, *Bull. Am. Meteorol. Soc*, **71**, 988-993.

Tziperman, E., M.A. Cane, and S. Zebiak, 1995: Irregularity and locking to the seasonal cycle in an ENSO prediction model as explained by the quasi-periodicity route to chaos. *J. Atmos. Sci.*, **50**, 293-306.

Tziperman, E. , L. Stone. M.A. Cane, and H. Jarosh, 1994: El Niño chaos: overlapping of resonances between the seasonal cycle and the Pacific ocean-atmosphere oscillator, *Science*, **263**, 72-74.

van der Vaart, P. C., H. Dijkstra and F.-F. Jin, 2000: The Pacific cold tongue and the ENSO mode, A unified theory within the Cane-Zebiak model. *J. Atmos. Sci.* **55,** 967-988.

Wakata, Y., and E. S. Sarachik, 1991: Unstable coupled atmosphere-ocean basin modes in the presence of a spatially varying basic state. *J. Atmos. Sci.*, **48**, 2060-2077.

Wallace, J. M., E. M. Rasmusson., T. P. Mitchell., V. E. Kousky., E. S. Sarachik and H. von Storch, 1998: The structure and evolution of ENSO-related climate variability in the tropical Pacific: Lessons from TOGA. *J. Geophys. Res.,* **103,** 14,241-14,259.

Wang, B. and Z. Fang, 1996: Chaotic oscillations of tropical climate: a dynamic system theory for ENSO. *J. Atmos. Sci.*, **53**, 2786-2802.

Wang, B., and S.-I. An, 2001: Why the properties of *El Niño* changed during the late 1970s. *Geophys. Res. Lett.*, **28**, 3709-3712.

Wang, C., and R. H. Weisberg, 1996: Stability of equatorial modes in a simplified coupled ocean-atmosphere model. *J. Climate*, **12**, 3132-3148.

Wang, X,-C., F,-F Jin and Y.-Q. Wang, 2003a: Tropical ocean recharge mechanism for climate variability. Part I: Equatorial heat content changes induced by the off-equatorial wind. *J. Climate*, in revision.

Wang, X,-C., F,-F Jin and Y.-Q. Wang 2003b: Tropical ocean recharge mechanism for climate variability. Part II: A unified theory for decadal and ENSO modes. *J. Climate*, in revision.

Wyrtki, K., 1975: El Niño—the dynamic response of the equatorial Pacific Ocean to atmospheric forcing. *J. Phys. Oceanogr.*, **5**, 572-584.

Wyrtki K., 1986: Water displacements in the Pacific and the genesis of El Niño cycles, *J. Geophys. Res.*, **91**, 7129-7132.

Xie, S.-P., A. Kubokawa and K. Hanawa, 1989: Oscillations with two feedback processes in a coupled ocean-atmosphere model, *J. Climate,* **2**, 946-964.

Xie, S.-P., 1994: On the genesis of the equatorial annual cycle. *J. Climate,* **7**, 2008-2013.

Yamagata, T., and Y. Masumoto, 1989: A simple ocean-atmosphere coupled model for the origin of warm El Niño Southern Oscillation event. *Phil. Trans. Roy. Soc. London*, **A329**, 225-236.

Yu, J.-Y, and C. R. Mechoso, 2001: A coupled atmosphere–ocean GCM study of the ENSO cycle. *J. Climate*, **14**, 2329-2350.

Yukimoto, S., M. Endoh, and A. Noda, 2000, ENSO-like interdecadal variability in the Pacific ocean as simulated in a coupled general circulation model. *Geophys. Res. Lett.*, **105**, 13,945 –13,963.

Zebiak, S. E., 1989: Ocean heat content variability and El Niño cycles. *J. Phys. Oceanogr.*, **19**, 475-486.

Zebiak, S. E., 1986: Atmospheric convergence feedback in a simple model for El Niño, *Mom. Wea. Rev.*, **114**, 1263-1271.

Zebiak, S. E. and M. A. Cane, 1987: A model El Niño Southern Oscillation. *Mon. Wea. Rev.*, **115**, 2262-2278.

MONTHLY AND SEASONAL VARIABILITY OF THE LAND-ATMOSPHERE SYSTEM

YONG-QIANG LIU*

Georgia Institute of Technology
Atlanta, GA 30332, USA
E-mail: liuy@eas.gatech.edu

(Manuscript received 28 January 2003)

The land surface and the atmosphere can interact with each other through exchanges of energy, water, and momentum. With the capacity of long memory, land surface processes can contribute to long-term variability of atmospheric processes. Great efforts have been made in the past three decades to study land-atmosphere interactions and their importance to long-term variability. This paper reviews studies on monthly and seasonal variability of the land-atmosphere system. Issues to be addressed include the importance of land surface processes, time scales, persistence, coupled patterns of soil moisture and precipitation, and prediction. A perspective on future studies is given.

1. Introduction

Land surface condition is a major factor for the partitioning of radiative energy absorbed on the ground into sensible and latent heat fluxes, which in turn affects the development of the atmospheric planetary boundary layer (PBL) and clouds. On the other hand, atmospheric processes, especially radiation and precipitation, are driving forces of land surface thermal and hydrological variability. Therefore, the land surface and atmosphere are closely related to each other. The importance of interaction between the two climate system components has been emphasized (e.g., Shukla and Mintz 1982; Yeh et al. 1984; Mintz 1984; Xue and Shukla 1993; Avissar 1995; Betts et al. 1996).

Land-atmosphere interaction has emerged as one of the most active research areas in atmospheric and hydrological sciences in the past three decades, partly due to the increasing attention to human activity related to regional environmental changes. Landscape in Amazonia, the Sahel, and Northwest China has changed dramatically since the 1970s as a result of deforestation and over-cultivation. It might be responsible for some regional climate disasters such as the prolonged drought in northern Africa during the 1970s (Charney 1975; Xue and Shukla 1993; Zeng et al. 1999).

Great efforts have been made in developing parameterizations and datasets, and in understanding physical mechanisms. Several land surface schemes (e.g., Dickinson et al.

* Present address: Forestry Sciences Laboratory, USDA Forest Service, 320 Green St., Athens, GA 30602, USA. email: yliu@fs.fed.us

1993; Sellers et al. 1986; Xue et al. 1991; Ji and Hu 1989) were developed. Parameterization schemes were also developed for heat and water transfer and convections induced by landscape heterogeneity (e.g. Chen and Avissar 1994; Zeng and Pielke 1995; Avissar and Liu 1996; Liu et al. 1999). Several datasets of decade-long soil moisture measurements have been archived (Robock et al. 2000), which are valuable for the studies of monthly and seasonal variability of the land-atmosphere system because of their long length of record. A number of feedback mechanisms were proposed (e.g., Rodriguez-Iturbe et al. 1991; Entekhabi et al. 1992; Eltahir 1998).

Similar to oceans, land has the capacity to retain anomalous signals over a much longer period than the atmosphere. This suggests that land surface processes could contribute to long-term atmospheric variability by passing their relatively slow anomalous signals to the atmosphere. A large number of studies have provided evidence for this contribution (e.g., Yeh et al. 1984; Dickinson and Handerson-Sellers 1988; Xue and Shukla 1993; Delworth and Manabe 1989; Vinnikov et al. 1996; Liu and Avissar 1999; Koster and Suarez 2001; Robock et al. 2003).

Understanding the importance of soil moisture memory to monthly and seasonal atmospheric variability has been a goal of the Global Energy and Water Cycle Experiment (GEWEX). Measurements of land-atmosphere variability were made and models were developed to study the relationships between soil and atmospheric processes in the GEWEX Continental-Scale International Project (GCIP) and GEWEX Asian Monsoon Experiment (GAME), which were conducted in the Mississippi River area of the USA and the Huaihe River area of China, respectively. The GEWEX America Prediction Project (GAPP) has established a goal to develop and demonstrate a capacity to make reliable monthly and seasonal predictions of precipitation and land-surface hydrologic processes using soil moisture memory.

The purpose of this paper is to review studies on monthly and seasonal variability of the land-atmosphere system, including those that the author participated in. The importance of various land properties is first discussed in Section 2. Four specific issues of time scales, persistence, spatial relations, and prediction are then discussed in Sections 3-6, with the focus on soil moisture. A perspective on future research is given in Section 7. (For a review of land-surface processes and their representation in weather and climate models, readers are referred to the paper by Yang 2003.)

2. The Importance of Land Properties

2.1. *Soil Moisture*

Soil moisture of the root zone layer (about 1-2 m deep) and the surface layer (about 0.1 m deep) can affect long-term and short-term atmospheric processes, respectively. Observational evidence for the importance of the root-zone layer soil moisture has been obtained. Vinnikov et al. (1996) found significant lag autocorrelation over a few months using the soil moisture measurements from Russia. Similar results were obtained for summer months from the soil moisture measurements in Illinois of the U.S. (Findell and Eltahir 1997). Eltahir (1998) provided an observational basis for a soil moisture-rainfall feedback mechanism, which linked changes in soil moisture, PBL moist static energy, and rainfall. Koster et al. (2003) provided evidence for the feedback by which precipitation-induced soil moisture anomalies affect subsequent precipitation. Also, evidence of the dominant role of soil moisture in persistence of the land-atmosphere system was presented by analyzing soil moisture measured in China (Liu and Avissar 1999a). However, only a very limited number of observational studies on the monthly and seasonal variability of the land-atmosphere system have been conducted due to the lack in systematic measurements of long-term soil moisture and the difficulty in isolating the signals of land's impacts.

Climate models have been a major tool for studies of monthly and seasonal variability of the land-atmosphere system. Early simulations with atmospheric general circulation models (GCM) coupled with simple land-surface schemes have showed significant impacts of soil moisture on the surface air temperature and precipitation (Mintz 1984). Some of them further indicated that the impacts of initial soil moisture anomalies could last for several months (e.g., Walker and Rowntree 1977; Rind 1982; Rowntree and Bolton 1983; Yeh et al. 1984).

Soil moisture conditions may also contribute to floods and droughts. Both GCM and regional climate models (RegCM) have been used to investigate the role of soil moisture and associated mechanisms. Giorgi et al. (1996) proposed a soil moisture-rainfall feedback which might contributes to the 1988 drought and the 1993 flood in the Midwest U.S. Hong and Pan (2000) found a positive feedback which affects low level structure, and Pal and Eltahir (2001) proposed a positive feedback mechanism linking soil moisture, moist static energy in the PBL, and the frequency and magnitude of convective rainfall processes. Bosilovich and Sun (1999) and Schar et al. (1999) suggested a mechanism linking soil moisture anomaly, change in the low-level jet, and atmospheric moisture change within the flood regions.

2.2. *Soil Temperature*

Soil thermal condition is a less important factor for monthly and seasonal variability in comparison with soil moisture and vegetation because of the relatively small heat capacity of soil. Anomalies in soil temperature last less than one month in a system without hydrological interactions (Liu and Avissar 1999b). However, it is possible that the soil thermal condition affects seasonal atmospheric variability through a mechanism other than modifying land-surface energy and water balances. Tang and Zhong (1984) proposed a mechanism of self-sustained oscillation in the land-atmosphere coupled thermal system, in which the oscillation period was found to be about half a year in a soil layer of more than 10 meters deep.

2.3. *Vegetation*

Vegetation can influence long-term atmospheric variability by modifying land-atmosphere energy and water balances. In comparison with bare soil, vegetated soil has a lower albedo and therefore receives more solar radiation for sensible and latent heat transfer. Vegetation affects the water balance by intercepting precipitation, extracting soil water from deep layers through transpiration, and resisting runoff. With larger roughness, momentum and turbulent transfers on the surface are changed.

Charney (1975) proposed a bio-geophysical feedback mechanism to explain the prolonged drought in northern Africa. A reduction in vegetation cover in the Sahel due to agricultural activities would lead to increased albedo, which would lead to a reduction in heat transfer from the surface and cooling in the atmosphere. The air would be forced to subside and therefore rainfall is reduced. Simulations with GCMs have indicated that degradation of vegetation conditions such as deforestation and desertification would lead to an increase in the surface temperature and a decrease in soil moisture, evaporation, runoff, and possibly precipitation in most cases (Dickinson and Henderson-Sellers 1988; Xue 1997). Vegetation interaction may also contribute to climate variability at interannual and decadal scales (Zeng et al. 1999; Guillevic et al. 2000).

A limitation in simulating the roles of vegetation in seasonal variability is that vegetation is prescribed or constant in most land-surface models. Solutions to this problem have been explored by imposing prescribed seasonal patterns of the evolution of vegetation based on the previously observed cycle of vegetation climatology, or incorporating dynamic vegetation into a land-surface model (GAPP 2000). Ji (1995) developed a climate-vegetation interaction model to simulate the seasonal variations of biomass, energy, and water fluxes for temperate forest ecosystems in northeastern China. Dickinson et al. (1998) added an interactive canopy model to the Biosphere-Atmosphere Transfer Scheme (BATS, Dickinson et al. 1993) to describe the seasonal evolution in leaf area needed in atmospheric models and to estimate carbon fluxes and net primary productivity. Lu et al. (1999) developed and implemented a coupled RAMS/CENTURY modeling system.

2.4. *Snow Cover*

When winter snow melts in the following spring or summer, land-surface heat and water balances can be changed due to heat consumption and liquid water release. This way snow is connected to seasonal atmospheric variability. In addition, snow can directly affect surface radiative balance by increasing the surface albedo.

Robock et al. (2003) investigated the relationship between interannual variations of the monsoon strength and snow cover over Eurasia. They found that the Indian summer monsoon precipitation is negatively correlated with snow cover intensity over Europe in the previous winter and over western Asia in the previous spring, and positively correlated with snow cover over Tibet. The relationship with snow cover over Europe and western Asia is consistent with Bamzai and Shukla (1999).

2.5. *Joint Roles*

Soil moisture, vegetation, and snow cover are closely related. Two or all of them could be involved in a land surface process at the same time. Vegetation often acts in a way similar to soil moisture. For example, soil moisture is reduced due to rainfall interception by vegetation, which leads to less significant persistence (Scott et al. 1995, 1997). Also, both wet and vegetated surfaces have a large water exchange rate, which leads to a short length of the seasonal scale in the land-atmosphere system (Liu and Avissar 1999b). The possible role of soil moisture in the effect of snow cover on the subsequent Indian summer monsoon precipitation was investigated by Robock et al. (2003). For the case examined in that study, no evidence for a significant role was found.

Time and space scales of anomalous processes are among the important properties determining relative contributions of soil moisture and other land surface factors. Soil moisture may play more important roles in monthly and seasonal variability than interannual one. It is more important to seasonal precipitation variability at regional scales than continental ones (Liu 2002). Vegetation, on the other hand, may be important in both seasonal and interannual variability. The GAPP science and implementation plan (GAPP 2000) describes in great detail the individual and joint roles of various land properties, development of modeling tools, and major issues to be investigated.

2.6. *Relative Contributions of Land and Ocean Processes*

There are some substantial differences between land and ocean processes. Physically, soil moisture (a hydrological factor) is of major importance for land process, while SST (a thermal factor) is of major importance for ocean processes, which, of course, is due to the fact that the amount of water available for evaporation is unlimited in the ocean, and its thermal capacity is very large. Their relative contributions to seasonal-interannual atmospheric variability depend on geographic and climate regimes (Koster and Suarez 1995; Koster et al. 2000). GCM simulations showed that land surface processes contribute mostly to the variance of precipitation out of the Tropics and in the regions where the strength of land-atmosphere feedback is controlled largely by the relative availability of energy and water there.

Findell and Eltahir (2003a, b) used the convective triggering potential and a low-level humidity index to distinguish between three types of early-morning atmospheric conditions (those favoring moist convection over dry soils, those favoring moist convection over wet soils, and those that will allow or prevent deep convective activity, independent of the surface flux partitioning). Analyses of the two measures from radiosonde stations across the contiguous 48 United States reveal that during the summer months positive feedbacks between soil moisture and moist convection are likely in much of the eastern half of the country. Over the western half of the country, atmospheric conditions and the likelihood of moist convection are largely determined by oceanic influences, and land surface conditions in the summer are unlikely to impact convective triggering.

3. Time Scales of Land-Atmosphere System Variability

A time scale concerned here represents a period over which an anomalous event spans. It is measured by the length of the period during which the anomalies remain a same sign for non-oscillatory variability, or by the half-length of its time period for oscillatory variability. Atmospheric variability occurs at many time scales, in addition to the sun-forced diurnal and seasonal cycles. Some scales identified include monthly-to-seasonal fluctuations, quasi-biannual fluctuations, and the Southern Oscillation (SO) (2-5 years). Abnormal weather events like floods and droughts mostly happen at monthly and seasonal scales.

The atmospheric fluctuations and weather anomalies at these time scales result from interactions between the atmosphere and other components of the climate system. It has been revealed, for example, that ocean-atmospheric interaction is the cause for the SO (Cane 1992). Studies on the time-scales in the land-atmosphere system are aimed at identifying the major scales at which land surface processes can contribute to long-term atmospheric variability and anomalous weather events.

3.1. *Theoretical Framework of Time Scales*

The definition and physical interpretation of time scales can be illustrated using the soil water balance equation, a first-order model (Delworth and Manabe 1988),

$$dw(t)/dt = P - E - R, \tag{1}$$

where *w* is soil moisture; *P, E,* and *R* are precipitation, evapotranspiration, and runoff. Snowmelt is neglected in the model. *P* is an external forcing. Although varied levels of parameterization schemes are available for *E* and *R* (see Yang 2003 for detail), some simple schemes are usually used for scale analyses. *R* can be related to *P,* and *E* can be calculated by

$$E = [w(t)/w_{fc}]E_p, \tag{2}$$

where E_p and w_{fc} are potential evaporation, which is determined by meteorological conditions, and soil water field capacity. Precipitation usually has a much faster pace than soil moisture, suggesting that variations of precipitation and soil moisture could be expressed as red noise, *y*(*t*), and white noise, *z*(*t*), respectively

$$dy(t)/dt = -\lambda y(t) + z(t). \tag{3}$$

This represents a system possessing an inherent exponential damping with a rate of $\lambda = E_p / w_{fc}$. The reciprocal of the rate is the *e*-folding time = $1/\lambda = w_{fc} / E_p$, that is, the time of a period during which a disturbance is reduced to e^{-1} times of its initial amplitude. This time is used to measure time scale of soil moisture variability. λ can be related to time-lag autocorrelation function

$$r(\tau) = -e^{\lambda\tau}, \tag{4}$$

where r and τ are autocorrelation and time lag, respectively. It can be seen from Eqs. 2 and 3 that major contributing factors for restoring soil moisture to its normal status are potential evaporation and soil water field capacity. Eq. 4 indicates that the longer the time scale is, the more closely consequent soil moisture is related its to initial anomaly.

A fourth-order model can be developed based on the energy and water conservation equations of the soil and atmosphere, which determines damping rates of soil moisture, soil temperature, air humidity, and air temperature with the effects of various interactions (e.g., Liu and Avissar 1999b). By assuming no disturbance with soil moisture in the fourth-order model, a third-order model can be obtained. There are no interactions between soil moisture and other model variables. Similarly, three other third-order models can be obtained by assuming no disturbance with soil temperature, air humidity, or air temperature in the fourth-order model. Four second-order models can be obtained by assuming no disturbance with any two of the four system variables. Finally, by assuming no interactions among the four system variables, the fourth-order model becomes four independent first-order models, each of which contains only one of the four variables. In these cases, the variation of the disturbance in the models is caused by self-feedback.

3.2. *Scales in Land-Atmosphere System*

Using the first-order model described by Eqs. (1-4), Delworth and Manabe (1988) obtained a globally averaged scale of soil moisture variability of about one to two months based on the simulations with the Geophysical Fluid Dynamics Laboratory (GFDL) GCM. Using the same technique, a value of three months was obtained based on the soil moisture measurements in Russia (Vinnikov et al. 1996) and about two and half months based on the soil moisture measurements in North China (Entin et al. 2000).

Interactions between soil moisture and other variables included in some higher-order models (Liu et al. 1993; Liu and Avissar 1999b; Liu and Avissar 2003) can significantly increase the length of a scale. In a fourth-order model (Liu and Avissar 1999b), the atmosphere is assumed to consist of an air column and the soil is assumed to consist of a thermally active layer and of a hydrologically active layer. Radiation, cloud and precipitation, sensible and latent heat (evaporation) fluxes are calculated based on various parameterization schemes. The time scales represented by the four solutions of the fourth-order model are of the orders of one day, one week, two months, and eight months (Table 1). The two longer scales represent monthly- and seasonal-scale processes, respectively. The seasonal scale appears only in those third- or second-order models that include disturbance of soil moisture. The longest scale varies between about 190 and 240 days. This emphasizes that the soil moisture feedback, and its interactions with the other variables, are the primary cause for the scale. In the model without soil moisture disturbance, the maximum time scale is only about two months. In addition, among the various interactions, the one between soil moisture and air humidity is the predominant: excluding this interaction results in a reduction in the length of the seasonal scale by more than one month. Without any interactions (results of first–order models), the longest scale caused by soil moisture self-feedback is only about three and a half months.

These studies suggested that the major scales at which land surface processes could affect the long-term atmospheric variability are monthly and seasonal ones. Because of the lack of long-term soil moisture measurements, it is difficult to find observational evidence for the scale features obtained from the simple models of water and energy conservation equations.

However, global climate models have provided some numerical evidence (e.g., Yeh et al. 1984; Liu et al. 1992 a, b; and Liu and Avissar 1999a). Yeh et al (1984) conducted a numerical experiment of initial soil moisture forcing with a simplified GFDL GCM, and showed that the induced anomalies in evaporation, precipitation, and soil moisture last for three to five months. The time scales calculated using a 10-year simulation with the NCAR Community Climate Model version 2 (CCM2) coupled with BATS (Dickinson et al. 1993) range between about two months in the tropics and over eight months at high latitudes (Liu and Avissar 1999a).

Table 1 Damping times of the fourth- and third-order models (in days) (From Liu and Avissar 1999b)

Model Order	Variables Inactivated	Scales 1	2	3	4
4th	None	232	58	6	1
3rd	Air Temperature	229	7	1	
	Air Humidity	196	19	1	
	Soil Temperature	236	29	6	
	Soil Moisture	58	6	1	

3.3. *Impacts of Physical Parameters*

A technique called Fourier Amplitude Sensitivity Test (FAST) was used to identify which parameters mostly affect the scales obtained from the fourth-order land-atmosphere model (Liu and Avissar 1999b). This technique was introduced by Cukier et al. (1973) and was used, for example, by Liu and Avissar (1996) to examine sensitivity of shallow convective precipitation to atmospheric dynamic and cloud microphysical parameters.

A number of parameters were found to have a very strong impact on the seasonal scale. Basically, the smaller the fluxes of water in the land-atmosphere system (i.e., evaporation, runoff, and underground diffusion), the longer the scale is. As a result, a moister soil, a higher soil temperature, more solar radiation, larger potential evaporation, or stronger turbulent activity (thus a larger actual evaporation rate) leads to a smaller length of the scale.

4. Persistence

It has long been recognized that atmospheric anomalies can persist over relatively long time periods (i.e., months to seasons), a feature known as persistence of atmospheric disturbances. The significance of this feature is that, if a variable has strong persistence, the variable itself can be a good predictor for predicting its variability. Using autocorrelations between adjacent monthly or seasonal atmospheric variables, Namias (1952) demonstrated the existence of such persistence.

Land can contribute to the atmospheric persistence through its long memory and interaction with the atmosphere. For example, following a dry spring, soil would likely be desiccated during the summer. This would result in a relatively large sensible heat flux injected in the atmosphere at the ground surface, which could perhaps maintain anticyclonic circulations. Under such circumstances, one could expect reduced summer rainfall. A theoretical framework of the role of soil moisture memory in precipitation statistics was presented by Koster et al. (2000), which indicates that a larger soil moisture memory would lead to a larger correlation between initial precipitation and its subsequent variability. Studies of the persistence issue are aimed at understanding the features of persistence in the land-atmosphere system and the contribution of land surface processes.

4.1. *Features of Soil Moisture Persistence*

Persistence of the land-atmosphere system has been investigated by analyzing the natural variability of long-term simulations produced with GCMs (e.g., Delworth and Manabe 1988, 1989; Manabe and Delworth 1990). Using multi-year simulations produced with the GFDL GCM, Delworth and Manabe (1988, 1989) analyzed the variability in soil moisture and atmospheric variables, and tried to identify relationships between them. They found that persistence of soil moisture is statistically significant, and is more intensive at high latitudes and during winter seasons.

Persistence also depends on climate regime: it is more significant in drier geographic regions. Figure 1 presents one-month lag autocorrelation coefficients of four variables obtained from measurements in China (Liu and Avissar 1999a). Because of the unavailability of soil moisture data in North China during winter time due to frozen soil, one-month lag autocorrelations were calculated only for the months of April to October. For soil moisture, the one-month lag autocorrelation coefficients are less than 55% in the moist Southeast China, and over 80% in the dry Northwest China, indicating increased significance of persistence from the moist region to the dry one. Persistence of other three variables (soil temperature, air humidity and temperature) has the same dependence on climate regime as soil moisture. In addition, a global analysis (Liu and Avissar 1999a) based on the CCM2 simulation (Bonan 1994) obtained relatively large autocorrelation coefficients in dry northern Africa, and the strong contrast between this region and the moist tropics, where autocorrelation coefficients are relatively low.

4.2. *Physical Mechanisms*

Actual evaporation was found to be a major factor determining the seasonal persistence in the land-atmosphere system (Liu and Avissar 1999b). Persistence of soil moisture is inversely proportional to evaporation. At high latitudes and in dry regions, actual evaporation is low, resulting in strong persistence. Furthermore, because water exchanges between the land surface and its overlying atmosphere are dominant in dry regions, as indicated by a large ratio of evaporation to precipitation (or small runoff) obtained in these regions, soil moisture has a significant impact on atmospheric persistence. By contrast, soil moisture has a smaller impact in the tropics and at high latitudes, where water exchanges between the atmosphere in these regions and its surroundings are more important.

It was indicated that soil moisture plays a critically important role in the persistence of the land-atmosphere system (Liu and Avissar 1999a). Soil moisture has the strongest persistence among various land and atmospheric variables. The one-month autocorrelation coefficients are over 30% anywhere on the globe, with a global average of about 60%.

Persistence of soil temperature was also found in most parts of the globe. However, its intensity is much weaker than that of soil moisture. Persistence of atmospheric variables is found only in some regions.

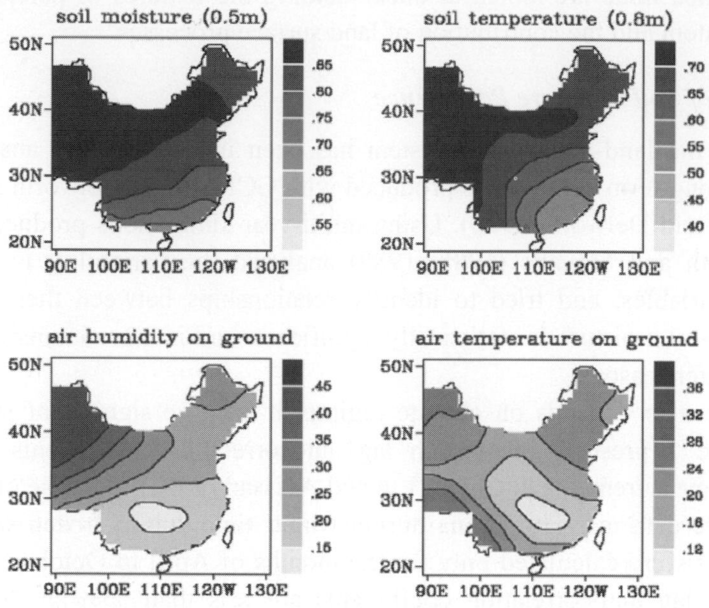

Figure 1. One-month lag correlation coefficients (After Liu and Avissar 1999a).

5. Spatial Relations

5.1. *Singular Value Decomposition (SVD) Patterns*

Unlike soil moisture, whose status is determined by local hydrological processes (rainfall reaching the surface, evapotranspiration, and runoff), precipitation is controlled by both large-scale circulation patterns and local land-atmosphere exchanges. How strong land surface can affect precipitation depends on the relative importance of the two processes. It is assumed that the local processes may play more important roles for certain spatial patterns of soil moisture and precipitation.

Relations in spatial patterns between soil moisture and precipitation have been investigated by applying techniques such as principal component analysis (PCA). Wang and Kumer (1998), for example, used empirical orthogonal function (EOF) to analyze soil moisture patterns and the effects on variability of the surface climate in the USA. Another PCA technique called singular value decomposition (SVD) (*Bretherton* et al. 1992) was used to identify soil moisture patterns closely connected to precipitation variability in China (Liu 2002). This technique realizes separation of each of the two fields into space patterns and time coefficients (expansion coefficients). Sorted in declined order of singular values, response between the first pair of spatial patterns is the largest, that between the second pair is the second largest, and so on. The first several pairs of modes are regarded as SVD leading modes.

The soil moisture and precipitation data used in the SVD analysis were simulated with the NCAR regional climate model (RegCM) (Dickinson et al. 1989; Giorgi and Bates 1989; Giorgi et al. 1999). This model is a tool for studying regional features of climate and land-surface processes at geographic regions of interest. It was developed based on NCAR / Pennsylvania State University Mesoscale Model Version 4 (MM4) (Anthes et al. 1987). Some detailed parameterization schemes were incorporated, including BATS land-surface physics (*Dickinson* et al. 1993) and the NCAR radiative transfer model. RegCM was able to reproduce some important high-resolution spatial characteristics of climate over East Asia (e.g., Liu et al. 1994, 1996; Liu and Ding 1995).

Four leading modes of soil moisture (Figure 2) were identified by the SVD analysis. The spatial pattern of the 1st soil moisture mode consists of a pair of positive and negative anomalies separated between 32-35°N. The anomalous regions are basically zonally oriented, each of which is about 30 degrees long and 15 degrees wide. The 2nd mode represents a pattern of more or less meridionally oriented anomalies. There is a large area of negative anomaly spreading from Mongolia and Northeast China down southwestward to Central China. The 3rd mode is similar to the 2nd mode in terms of its meridional orientation, but the negative region in the north does not spread over Mongolia, and its intensity is much weaker. The last mode is featured by varied anomalies which are negative in Northeast China and become positive southwestward down to Central China.

Based on the corresponding atmospheric patterns, the SVD leading modes of soil moisture were divided into two types. Type I consists of the 1st and 4th soil moisture SVD modes, whose corresponding atmospheric anomalous patterns are featured by strong relations in the middle latitudes of 500 hPa and by the same signs of anomalies between the middle and low troposphere; Type II consists of the 2nd and 3rd modes, whose corresponding atmospheric patterns are featured by comparable importance between the atmospheric systems in the middle and low latitudes and by opposite signs between the middle and low troposphere in some regions.

5.2. *Temporal Relations*

The two types of soil moisture SVD modes have different importance to subsequent precipitation. The time-lag correlation of the SVD expansion coefficient series between soil moisture and precipitation (Figure 3) is more significant for the Type II than Type I SVD patterns of soil moisture. For both patterns of the 2nd and 3rd modes, the correlation coefficients with soil moisture leading precipitation are significant for the lag-time up to six months at the 99.9% confidence level (the critical correlation value of 30%). In contrast, those for the patterns of the 1st and 4th modes are only barely significant for the lag-time of one month at the confidence level.

The above results indicate that, among the SVD leading patterns of soil moisture, only those of Type II have close relations to subsequent variability of precipitation. Some explanations were obtained based on the features of their corresponding atmospheric patterns, that is, opposite signs of anomalies between the middle and low troposphere, and significant anomalies at both middle and low latitudes. The first feature usually indicates a weak control of an atmospheric system. As a result, soil moisture plays a relatively important role in variability of precipitation. In contrast, the same signs of atmospheric anomalies between the two heights for the Type I SVD patterns usually indicate a strong control of an atmospheric system throughout the entire troposphere. In this case, soil moisture's role is less significant. The second feature may reflect a low westerly index, i.e., anomalously strong planetary-wave activities. This favors the development of synoptic systems which have small scales and move fast. Their effects on long-term atmospheric processes are very limited. As a result, the role of soil moisture becomes relatively important.

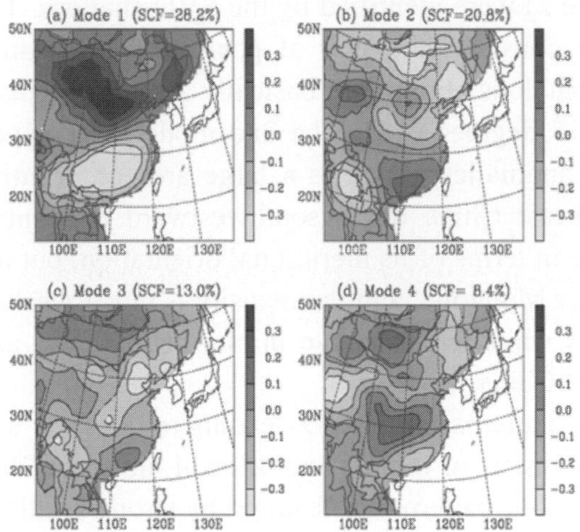

Figure 2. Soil moisture SVD patterns coupled with precipitation (After Liu 2002).

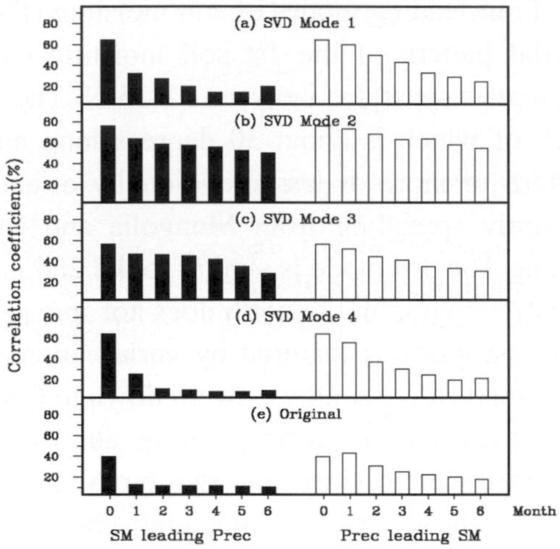

Figure 3. Time-lag correlation coefficients between soil moisture (SM) and precipitation (Prec) (After Liu 2002).

5.3. *Predictive Significance*

The significance of the SVD patterns to monthly and seasonal prediction was illustrated by comparing the time-lag correlation with soil moisture leading precipitation between the SVD modes (Fig. 3a-d) and the original data series (Fig. 3e). The coefficients in Fig. 3e were obtained by first calculating time-lag correlation coefficients at each location and then obtaining space average of their absolute values. Although the simultaneous correlation coefficient (zero-lag month) of 40% is comparable to those of the SVD leading modes, the correlation coefficients of the original data series with soil moisture leading precipitation by one month or longer are only about 15%.

This result suggests that, if using soil moisture as a predictor to forecast monthly and seasonal precipitation variability, better predictability could be achieved by using the SVD patterns than original data at individual locations of the predictor. There are also differences in the time-lag correlation between precipitation and subsequent soil moisture variability between the original data series and the SVD expansion series, though not as large as those in the time-lag correlation between soil moisture and subsequent precipitation variability.

6. Prediction

The importance of land-surface conditions to short- and medium-term weather forecasts has been recognized. A number of operational and research models, including the NCEP ETA model (Black 1994), NCAR/Penn State MM5 (Grell et al. 1994), Regional Atmospheric Modeling System (RAMS) (Pielke et al. 1992), have been coupled with land-surface parameterization schemes to include the impacts of land-surface conditions on sensible and latent heat fluxes, PBL, atmospheric stability, and clouds and precipitation. There is also evidence, including that described above, for the importance of soil moisture to long-term weather forecast. As described below, dynamic and statistical methods have been used to estimate predictability and to develop forecast techniques.

6.1. *Dynamical Method*

This method basically tests significance of simulated responses in the surface air temperature, precipitation, and soil moisture to initial soil moisture forcing with land-atmosphere coupled climate models. The response could be measured by the differences in the variables between simulations with and without the initial soil moisture forcing. The role of ocean processes is excluded by using prescribed SST. The ensemble technique is often used to compare two or more groups of multiple simulations. Studies by Schlosser and Milly (2002) and Dirmeyer (2003) showed generally higher predictability of the surface air temperature, consistent with the resulted obtained from reanalysis data (e.g., Huang et al. 1996). Predictability of precipitation was found to be much lower, but for some specific situations such as transition zones between dry and humid climates, or the regions with a tendency toward large initial soil moisture anomalies and a strong precipitation-evaporation-soil moisture connection, it could be significant (Koster et al 2000; Koster and Suarez 2003).

6.2. *Statistical Method*

This method uses statistical techniques like regression to build relations between initial soil moisture and subsequent variability of precipitation or other variables. Soil moisture could be a sole predictor or combined with others like SST. Traditionally a regression relation has been built for a specific location. Because precipitation at a given location is determined by the combined effects of systematic relationships, which mostly are of large spatial scale, identifying its spatial patterns and utilizing them as prediction factors can enhance predictive skill at individual locations (Barnston 1994). Some tools like PCA have been applied to obtaining spatial patterns of soil moisture and precipitation aimed at improving skills of prediction models (e.g., Mo 2002).

Predictability of monthly and seasonal precipitation could be improved by using SVD patterns between soil moisture and precipitation instead of their values at individual locations.

A recent study provided predictive evidence for this suggestion by comparing skills of two statistical prediction models based on the coupled SVD patterns and local relationships (Liu 2003). The data used for model development and validation were obtained from the simulation over East Asia with RegCM (Liu 2002). The results showed much improved skill with the prediction model using the coupled SVD patterns. The seasonal prediction skill is higher than the monthly one. The most remarkable contribution of soil moisture to the prediction skill is found in warm seasons, opposite to that of SST.

7. Discussion

The major findings from the studies reviewed in this paper are as follows:

a. Land processes have significant contribution to long-term variability of the land-atmosphere system at time scales up to seasons. Self-feedback of soil moisture is a major contributor to the seasonal scale and the interactions between soil moisture and other components of the system can remarkably increase the length of the seasonal scale.
b. Land processes are more important than ocean processes to seasonal variability of precipitation over the mid-latitude continents. Vegetation could contribute to interannual and decadal variability of precipitation.
c. Persistence of soil and atmospheric variables, especially soil moisture, is a basic feature of long-term variability of the land-atmosphere system. It is more significant in high latitudes, during winter seasons, and in arid regions. The rate of actual evaporation, which is dependent on soil water content, is a major factor determining the persistence.
d. Among various spatial SVD patterns of soil moisture in East Asia, only those corresponding to the atmospheric anomalous patterns with opposite signs between the middle and low troposphere and significant anomalies at both middle and low latitudes have close relations to subsequent monthly and seasonal precipitation variability.
e. There is high predictability of monthly and seasonal variability in the surface air temperature when soil moisture is used as a predictor. Predictability of monthly and seasonal variability in precipitation is generally low, but it could be significant in certain climatic regions. Application of spatial relations like soil moisture SVD patterns can improve statistical predictability of precipitation.

The lack of soil moisture data has been a major difficulty in studying the issues discussed above. This problem will remain for quite a while. A number of research projects like GCIP have made efforts to obtain more frequent and higher-resolution measurements of soil moisture data. This data is valuable for land-surface model development and water and energy balance analysis, but their relatively short length presents a problem to the studies of long-term variability. The application of remote sensing is believed to be one possible solution for global coverage of high-resolution soil moisture data, which is useful for obtaining initial fields for simulation and prediction. One limitation with the technique is that only soil moisture of a thin layer can be detected, which is much less valuable than soil moisture of a deep layer to studies on long-term variability.

Techniques have been under development to create alternative data of soil moisture measurements by using land-surface hydrological models with observed/assimilated precipitation. The application of such precipitation is expected to produce more reliable soil moisture. Among various types of data are the NCEP global reanalysis (Kalnay et al. 1996), which has been used in the issues discussed here (e.g., Huang et al. 1996), the NCEP regional reanalysis (Mesinger et al. 2003), which will be available soon, and the GEWEX/Global Land-Atmosphere System Study (GLASS) outputs (Dirmeyer 2002).

Most studies on the role of land processes in monthly and seasonal variability of the land-atmosphere system have been conducted using land-atmosphere coupled models. Our understanding of the variability, therefore, is affected by uncertainties with the models. One example of model uncertainty is coupling strength (Koster et al. 2002). There is a need for continuously improving calculation schemes of various land-surface processes, including snow pack, frozen soil, vegetation, and landscape heterogeneity.

An objective of developing and demonstrating the capacity of making reliable monthly and seasonal predictions of precipitation and land-surface hydrologic processes using soil moisture memory was recently developed with GAPP. This raises some new issues about the monthly and seasonal variability of the land-atmosphere system. It presents some new challenges and, at the same time, new opportunities for scientists for many years to come.

Acknowledgments I would like to thank three anonymous reviewers whose comments significantly improved the manuscript. One of the reviewers kindly checked English grammar and usage. I also thank Dr. Xun Zhu for his encouragement, and Tim Giddens for editorial check. This research was supported by the National Aeronautics and Space Administration under grant NAG8-1513.

References:

Anthes, R. A., E.-Y. Hsie, and Y.-H. Kuo, 1987: *Description of the Penn State/NCAR Mesoscale Model Version 4 (MM4)*, Technical Note, NCAR/TN-282+STR, National Center for Atmospheric Research, Boulder, Colorado, 66 pp.

Avissar, R., 1995: Recent Advances in the representation of land-atmosphere interactions in global climate models, *Rev. Geophys.*, **33**, 1005-1010.

Avissar, R., and Y. Liu, 1996: A three-dimensional numerical study of shallow convective clouds and precipitation induced by land-surface forcings, *J. Geophys. Res.*, **101**, 7499-7518.

Bamzai, A. S., and J. Shukla, 1999: Relation between Eurasia snow cover, snow depth, and the Indian summer monsoon: An observational study, *J. Clim.*, **12**, 3117-3132.

Betts, A. K., J. H. Ball, and Coauthors, 1996: The land-surface atmosphere interaction: A review based on observational and global modeling perspectives, *J. Geophys. Res.*, **101**, 7209-7226.

Black, T., 1994: The new NMC mesoscale Eta model: description and forecast examples, *Wea. Forecasting*, **9**, 265-278.

Bonan, G. B., 1994: Comparison of the land surface climatology of the NCAR CCM2 at R15 and T42 resolutions with implications for sub-grid land surface heterogeneity. *J. Geophys. Res.*, **99**, 10357-10364.

Bosilovich, M. G., and W.-Y. Sun, 1999: Numerical simulation of the 1993 Midwestern flood: land-atmosphere interactions, *J. Clim.*, **12**, 1490-1505.

Bretherton, C. S., C. Smith, and J. M. Wallace, 1992: An intercomparison of methods for finding coupled patterns in climate data, *J. Clim.*, **5**, 541-560.

Cane, M. A., 1992: Tropical Pacific ENSO models: ENSO as a mode of the coupled system, in "*Climate System Modeling*" (Ed. K.E. Trenberth), The Press of the Uni. Of Cambridge, 788 pp.

Charney, J. B., 1975: Dynamics of deserts and drought in the Sahel, *Q.J.Roy. Meteor. Soc.*, **101**, 193-202.

Chen, F. and R. Avissar, 1994: Impact of land--surface moisture variability on local shallow convective cumulus and precipitation in large-scale models, *J. Appl. Meteor.*, **33**, 1382-1401.

Cukier, R. I., C. M. Fortuin, and Coauthors, 1973: Study of the sensitivity of coupled reaction systems to uncertainties in rate coefficients. I. Theory. *J. Chem. Phys.*, **59**, 3873-3878.

Delworth, T. and S. Manabe, 1988: The influence of potential evaporation on the variabilities of simulated soil wetness and climate, *J. Clim.*, **1**, 523-547.

Delworth, T. and S. Manabe, 1989: The influence of soil wetness on near-surface atmospheric variability. *J. Clim.*, **2**, 1447-1462.

Delworth, T. and S. Manabe, 1993: Climate variability and land-surface processes, *Adv. Water Resour.*, **16**, 3-20.

Dickinson, R. E., and A. Henderson-Sellers, 1988: Modeling tropical deforestation: A study of GCM land-surface parameterizations, *Quart. J. Roy. Meteor. Soc.*, **114**, 439-462.

Dickinson, R. E., R. M. Errico, F. Giorgi, and G. T., Bates, 1989: A regional climate model for the western U.S., *Clim. Change*, **15**, 383-422.

Dickinson, R. E., A. Henderson-Sellers, and P. J.Kennedy, 1993: *Biosphere-Atmosphere Transfer Scheme (BATS) Version 1E as Coupled to the NCAR Community Climate Model*, NCAR Tech. Note/TN-387, National Center for Atmospheric Research, Boulder, CO., 72pp.

Dickinson, R. E., M. Shaikh, R. Bryant, and L. Graumlich, 1998: Interactive canopies for a climate model. *J. Clim.*, **11**, 2823-2836.

Dirmeyer, P. A., 2002: Second GEWEX/GLASS Global Soil Wetness Project (GSWP2), in "Mississippi River Climate & Hydrology Conference", May 13-17, New Orleans, LA, P.54.

Dirmeyer, P. A., 2003: The role of the land surface background state in climate predictability, *J. Hydrometeor.*, **4**, 599-610.

Eltahir, E. A. B., 1998: A soil moisture-rainfall feedback mechanism. 1. Theory and observations, *Water Resour. Res.*, **34**, 765-785.

Entekhabi, D., I. Rodriguez-Iturbe, and R. I. Bras, 1992: Variability in large-scale water balance with land surface-atmosphere interaction, *J. Clim.*, **57**, 798-813.

Entin, J. K., A. Robock, and Coauthors, 2000: Temporal and spatial scales of observed soil moisture variations in the extratropics. *J. Geophys. Res.*, **105**, 11,865-11,877.

Findell, K. L. and E. A. B. Eltahir, 1997: An analysis of the soil moisture-rainfall feedback, based on direct observations from Illinois, *Water Resour. Res.*, **33**, 725-735.

Findell, K. L., and E. A. B. Eltahir, 2003a: Atmospheric controls on soil moisture–boundary layer interactions. Part I: Framework development, *J. Hydrometeorol.*, **4**, 552-569.

Findell, K. L., and E. A. B. Eltahir, 2003b: Atmospheric controls on soil moisture–boundary layer interactions. Part I: Feedbacks within the continental United States, *J. Hydrometeorol.*, **4**, 570-583.

GAPP, 2000: GEWEX Americas Prediction Project (GAPP) Science Plan and Implementation Strategy, 160pp.

Giorgi, F., and G. T. Bates, 1989: The climatological skill of a regional model over complex terrain. *Mon. Wea. Rev.*, **117**, 2325-2347.

Giorgi, F., L. O. Means, C. Shields, and L. Mayer, 1996: A regional model study of the importance of local versus remote controls of the 1988 drought and the 1993 flood over the Central United States, *J. Clim.*, **9**, 1150-1162.

Giorgi, F., Y. Huang, K. Nishizawa, and C. Fu, 1999: seasonal cycle simulation over eastern Asia and its sensitivity to radiative transfer and surface processes, *J. Geophys. Res.*, **104**, 6403-6424.

Grell, A. G., J. Dudhia, and D. R. Stauffer, 1994: *A Description of the Fifth-Generation Penn State/NCAR mesoscale Model (MM5)*, NCAR Tech. Note, 398, 122pp.

Guillevic, P., R. D. Koster, and Coauthors, 2002: Influence of the interannual variability of vegetation on the surface energy balance-A global sensitivity study, *J. Hydrometeor.*, **3**, 617-629.

Hong, S.-Y., and H.-L. Pan, 2000: Impact of soil moisture anomalies on seasonal summertime circulation over North America in a regional climate model, *J. Geophys. Res.*, **105**, 29625-29634.

Huang, J., H. M. van den Dool, and K. P. Georgakakos, 1996: Analysis of model calculated soil moisture over the United States (1931-1993) and applications to long-range temperature forecasts, *J. Clim.*, **9**, 1350-1362.

Kalnay, E., M. Kanamitsu, and Coauthors, 1996: The NCEP/NCAR 40-year reanalysis project, *Bull. Amer. Met. Soc.*, **77**, 437-471.

Koster, R. D. and M. J. Suarez, 1995: Relative contributions of land and ocean processes to precipitation variability, *J. Geophys. Res-Atoms.*, **100** (D7), 13775-13790.

Koster, R. D. and M. J. Suarez, 1996: The influence of land surface moisture retention on precipitation statistics, *J. Clim.*, **9**, 2551-2567.

Koster, R. D., M. J. Suarez, and M. Heiser, 2000: Variance and predictability of precipitation at seasonal-to-interannual timescales, on precipitation, *J. Hydrometeor.*, **1**, 26-46.

Koster, R. D. and M. J. Suarez, 2001: Soil moisture memory in climate models, *J. Hydrometeor.*, **2**, 558-570.

Koster, R. D., P. A. Dirmeyer, and Coauthors, 2002: Comparing the degree of land-atmosphere interaction in four atmospheric general circulation models, *J. Hydrometeor.*, **3**, 363-375.

Koster, R. D. and M. J. Suarez, 2003: Impact of land surface initialization on seasonal precipitation and temperature prediction, *J. Hydrometeor.*, **4**, 408-423.

Koster, R. D., M. J. Suarez, and Coauthors, 2003: Observational evidence that soil moisture variations affect precipitation, *Geophys Res. Lett.,* **30**(5), art. No. 1241.

Ji, J. J. and Y. C. Hu, 1989: A simple land surface process model for use in climate study, *Acta, Meteorologica Sinica*, **3**, 342-351.

Ji, J. J., 1995: A climate-vegetation interaction model: simulating physical and biological processes at the surface, *Journal of Biogeochemistry*, **22**, 2063-2068.

Liu, Y.-Q., D. Z. Ye, and J. J. Ji, 1992a: Influence of soil moisture and vegetation on climate. I: A theoretical analysis on persistence of short-term climatic anomalies. *Science in China*, **35**, 441-448.

Liu, Y.-Q., D. Z. Ye, and J. J. Ji, 1992b: Influence of soil moisture and vegetation on climate changes induced by thermal forcing, *Acta Meteor. Sinica*, **6**, 58-69.

Liu, Y.-Q., D. Z. Ye, and J. J. Ji, 1993: Influence of soil moisture and vegetation on climate. II: numerical experiments on persistence of short-term climatic anomalies. *Science in China*, **36B**, 102-109.

Liu, Y.-Q., F. Giorgi, and W. M. Washington, 1994: Simulation of summer monsoon climate over East Asia with an NCAR regional climate model, *Mon. Wea. Rev.*, **122**, 2331-2348.

Liu, Y.-Q., and Y. H. Ding, 1995: A review of the study on simulation of regional climate, *Q.J. Appl. Meteor.*, **6**, 228-239 (in Chinese).

Liu, Y.-Q. and R. Avissar, 1996: Sensitivity of shallow convective precipitation induced by land surface heterogeneities to dynamical and cloud microphysical parameters, *J. Geophys. Res.*, **101**, 7477-7497.

Liu, Y.-Q., R. Avissar, and F. Giorgi, 1996: A simulation with the regional climate model (RegCM2) of extremely anomalous precipitation during the 1991 East-Asia flood: An evaluation study. *J. Geophy. Res.*, **101**, 26199-26215.

Liu, Y.-Q., and R. Avissar, 1999a: A study of persistence in the land-atmosphere system using a general circulation model and conservations, *J. Clim.*, **12**, 2139-2153.

Liu, Y.-Q., and R. Avissar, 1999b: A study of persistence in the land-atmosphere system with a fourth-order analytical model, *J. Clim.*, **12**, 2154-2168.

Liu, Y.-Q., C. P. Weaver, and R. Avissar, 1999: Toward a parameterization of mesoscale fluxes and moist convection induced by landscape heterogeneity, *J. Geophy. Res.*, **104**, 19515-19534.

Liu, Y.-Q., 2002: Spatial patterns of soil moisture connected to monthly-seasonal precipitation variability in a monsoon region, *J. Geophy. Res.* (in press).

Liu, Y.-Q., 2003: Prediction of monthly-seasonal precipitation using coupled SVD patterns between soil moisture and subsequent precipitation, *J. Geophys. Lett.*, 30 (15), 1827, doi:10.1029/2003GL017709.

Liu, Y.-Q., and R. Avissar, 2003: Modeling of the global water cycle - analytical models (1st, 2nd, 3rd, 4th order, in "The Encyclopedia of Hydrological Sciences", John Wiley & Sons (in preparation).

Lu, L., R.A. Pielke, and Coauthors, 2001: Implementation of a two-way interactive atmospheric and ecological model and its application to the central United States. *J. Clim.*, **13**, 900-919.

Manabe, S., and T. Delworth, 1990: The temporal variability of soil wetness and its impact on climate. *Clim. Change*, **16**, 185-192.

Mesinger, F., G. DiMego, and Coauthors, 2003: NCEP regional reanalysis, Symp. on Observing and Understanding the Variability of Water in Weather and Climate, AMS Annual Meeting, Long Beach, CA, Feb. 2003.

Mintz, Y., 1984: *The sensitivity of numerically simulated climate to land-surface boundary conditions*, The Global Climate, Houghton, J. T., Ed., Cambridge University Press, 79-105.

Namias, J., 1952: The annual course of month-to-month persistence in climatic anomalies. *Bull. Amer. Meteor. Soc.*, **33**, 279-285.

Pal, J. S., and E. A. B. Eltahir, 2001: Pathway relating soil moisture conditions to future summer rainfall within a model of the land-atmosphere system, *J. Clim.*, **14**, 1227-1242.

Pielke, R. A., W. R. Cotton, and Coauthors, 1992: A comprehensive meteorological modeling system - RAMS, *Meteorol. Atmos. Phys.,* **49**, 69-91.

Rind, D., 1982: The influence of ground moisture conditions in North America on summer climate as modeled in the GISS GCM, *Mon. Wea. Rev.*, **110,** 1487-1494.

Robock, A., K. Y. Vinnikov, and Coauthors, 2000: The global soil moisture data bank, *Bull. Amer. Meteor. Soc.*, **81**, 1281-1299.

Robock, A., M. Q. Mu, K. Y. Vinnikov, and D. Robinson, 2003: Land surface conditions over Euraia and Indian summer monsoon rainfall, *J. Geophys. Res.*, **108**(D4), art no. 4131.

Rodriguez-Iturbe, I., D. Entekhabi, and R.I. Bras, 1991: Nonlinear dynamics of soil moisture at climate scales: I. Stochastic analysis, *Water Resour. Res.*, **27**,1899-1906.

Rowntree, P.R. and J.A. Bolton, 1983: Simulation of the atmospheric response to soil moisture anomalies over Europe, *Quart. J.R.Met.Soc.*, **109**, 501-526.

Schar, C., D. Luthi, U. Beyerle, and E. Heise, 1999: The soil-precipitation feedback: a process study with a regional climate model, *J. Clim.*, **12**, 722-741.

Schlosser, C. A., and P. C. D. Milly, 2002: A model-based investigation of soil moisture predictability and associated climate predictability, *J. Hydrometeor.*, **3**, 483-501.

Scott, R., R. D. Koster, D. Entekhabi, and M. J. Suarez, 1995: Effect of a canopy interception reservoir on hydrological persistence in a general circulation model, *J. Clim.,* **8**, 1917-1928.

Scott, R., D. Entekhabi, R. D. Koster, and M. J. Suarez, 1997: Timescales of land surface evapotranspiration response, *J. Clim.,* **10**, 559-566.

Sellers, P. J., Y. Mintz, Y. C. Sud, and A. Dalcher, 1986: A simple biosphere model (SiB) for use within general circulation models, *J. Atmos. Sci.*, **43**, 505-531.

Shukla, J. and Y. Mintz, 1982: Influence of land-surface evapotranspiration on the earth's climate. *Science*, **215**, 1498-1501.

Tang, M. C. and Q. Zhong, 1984: Self-sustained vertical oscillation in a simple climate system, *Arch. Met. Geoph. Biocl., Ser.B*, **34**, 21-37.

Vinnikov, K., A. Robock, N. A. Speranskaya, and C. A. Schlosser, 1996: Scales of temporal and spatial variability of midlatitude soil moisture, *J. Geophys. Res.*, **101**, 7163-7174.

Walker, J. M. and P. R. Rowntree, 1977: The effect of soil moisture on circulation and rainfall in a tropical model, *Quart. J. Roy. Meteor. Soc.*, **103**, 29-46.

Wang, W. Q. and A. Kumar, 1998: A GCM assessment of atmospheric seasonal predictability associated with soil moisture anomalies over North America, *J. Geophy. Res.*, **103**, 28637-28646.

Xue, Y.-K., P. J. Sellers, J. L. Kinter and J. Shukla, 1991: A simplified biosphere model for global climate studies. *J. Clim.*, **4**, 345-364.

Xue, Y. and J. Shukla, 1993: The influence of land surface properties on Sahel climate. Part I: Desertification. *J. Clim.,* **6**, 2232-2245.

Xue, Y., 1997: Biosphere feedback on regional climate in tropical North Africa, *Q. J. Roy. Met. Soc.,* **123** B, 1483-1515.

Yang, Z.-L., R. E. Dickinson, A. Robock, and K. Y. Vinnikov, 1997: On validation of the snow submodel of the biosphere-Atmosphere Transfer Scheme with Russian snow cover and meteorological observational data, *J. Clim.*, **10**, 353-373.

Yang, Z.-L., 2003: Modeling land surface processes in short-term weather and climate studies. *This volume.*

Yeh, T. C., R. T. Wetherald, and S. Manabe, 1984: The effect of soil moisture on the short-term climate and hydrology change-A numerical experiment. *Mon. Wea. Rev.*, **112**, 474-490.

Zeng, N., J. D. Neelin, K. M. Lau, C. J. Tucker, 1999: Enhancement of interdecadal climate variability in the Sahel by vegetation interaction, *Sciences*, **286** (5444), 1537-1540.

Zeng, X and R. A. Pielke, 1995: Landscape-induced atmospheric flow and its parameterization in large-scale numerical models, *J.Clim.*, **8,** 1156-1177.

ADAPTATION OF THE ATMOSPHERIC CIRCULATION TO THERMAL FORCING OVER THE TIBETAN PLATEAU

GUOXIONG WU, YIMIN LIU, JIANGYU MAO, XIN LIU, AND WEIPING LI

State Key Laboratory of Atmospheric Sciences and Geophysical Fluid Dynamics (LASG)
Institute of Atmospheric Physics, Chinese Academy of Sciences, Beijing 100029, China
E-mail: gxwu@lasg.iap.ac.cn

(Manuscript received 15 November 2002)

The theory of Ertel's potential vorticity is applied to investigating the adaptation of the atmospheric circulation to diabatic heating. Results show that such atmospheric thermal adaptation depends critically on the vertical distribution of diabatic heating. Data diagnosis shows that in summer over the Tibetan Plateau, although the column-integrated convective condensation heating overwhelms diffusive sensible heating, it is the sensible heating that determines the vertical profile of the total diabatic heating over the Tibetan Plateau and forces the prominent and unique atmospheric circulation patterns over the plateau and the surrounding area.

The strong near-surface diabatic heating over the Tibetan Plateau increases the lower layer convergence and the concavity of isentropic surfaces that intersect the plateau. Such heating, in combination with surface friction, makes the Tibetan Plateau region a strong source of negative vorticity in the atmosphere. The sensible heating-induced ascent along the lateral boundaries of the Tibetan Plateau draws in the surrounding air in the lower troposphere and pumps the air out of the region into the upper troposphere. Results obtained from numerical experiments reveal the effectiveness of such a sensible heat-driven air pump (SHAP). The SHAP not only regulates the climate in the surrounding area but also affects the circulation anomalies over the Northern Hemisphere in the form of the Rossby wave rays.

1. Introduction

The Tibetan Plateau covers nearly two and half million square kilometers, about one quarter of the Chinese territory. Its height reaches the middle of the troposphere. Its center is located near 90°E and 30°N in the subtropics. Such a huge land mass will inevitably affect the atmospheric circulation. Since the end of the 1940s, the dynamic effects of the plateau during the winter have been examined (e.g., Charney and Eliassen 1949; Yin 1949; Bolin 1950; Yeh 1950; Wu 1984; Rodwell and Hoskins 2001). In the period prior to the mid-1950s, studies of the plateau meteorology were mainly concerned with its dynamical influence. In the mid-1950s, Yeh et al. (1957) and Flohn (1957) found independently that a huge heat source exists over the Tibetan Plateau in summer. Since then the thermal influence of the plateau has also become the subject of intense study. One of the basic questions concerns how the plateau heating contributes to the maintenance of the circulation patterns in the surrounding area. Figure 1 shows the July mean cross sections of geopotential height deviation from the corresponding zonal mean (a) along 30°N and (b) along 90°E. The data are from the National

Centers for Environment Prediction/National Center for Atmospheric Research (NCEP/NCAR) Reanalysis (Kalnay et al. 1996), ranging from 1980 to 1997. A unique feature over the Tibetan Plateau is clearly shown: over the Tibetan Plateau in summer, there is a strong but very shallow lower tropospheric low underlying a strong and deep high in the upper troposphere. Continuous efforts have been made since 1957 to understand the mechanism linking the diabatic heating over the Tibetan Plateau with such peculiar large-scale circulation (Ye and Gao 1979; Luo and Yanai 1984; Zhang et al., 1988; Yanai et al. 1992; Ye and Wu 1998). By employing the barotropic vorticity equation, Holton and Colton (1972) and Ye and Zhang (1974) found that a reasonable circulation can be simulated (or the balance of the vorticity equation obtained) only if an extremely large diffusion coefficient is used. Using observations and the European Center for Medium-Range Weather Forecasts (ECMWF) FGGE-IIIb data, Yang et al. (1992) interpreted such a large diffusion coefficient as the effects of sub-grid-scale convection that transports positive vorticity from the boundary layer to the upper troposphere, thus maintaining the cyclonic system near the surface and the anticyclonic system in the upper troposphere. However, the circulation pattern persists from day to day in summer regardless of the variation of weather systems. Its maintenance mechanism deserves further investigation.

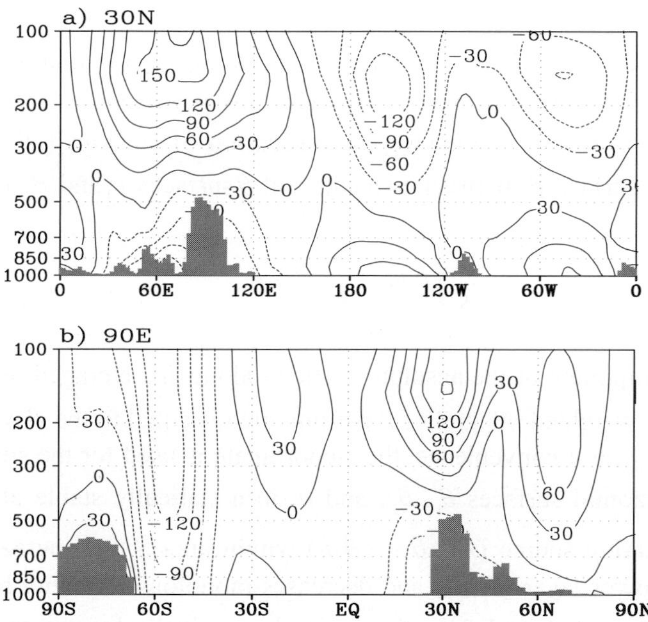

Figure 1. July mean cross sections of geopotential height deviation from the corresponding zonal mean (a) along 30°N and (b) along 90°E, in units of geopotential meter (gpm).

In this study, the theory of Ertel's potential vorticity (Ertel 1942; Hoskins et al. 1985) is used to study the adaptation of the atmospheric circulation to diabatic heating. The thermal adaptation theory developed in Section 2 is then examined with numerical experiments in Section 3. The heating feature over the Tibetan Plateau and its influence on the surrounding circulation are explored in Sections 4 and 5, respectively. The concept of a sensible heat-driven air pump (SHAP) is proposed in Section 6 to express how the heating over the plateau can affect the climate in the surrounding area and over the Northern Hemisphere. Some concluding remarks are presented in Section 7.

2. Adaptation of the Atmospheric Circulation to Diabatic Heating

2.1 *The PV-θ View*

Ertel's potential vorticity P is defined as

$$P = \frac{1}{\rho}\vec{\zeta}_a \cdot \nabla\theta , \tag{1}$$

where $\vec{\zeta}_a$ is the three-dimensional absolute vorticity, ρ is air density, and θ is potential temperature. In a θ-coordinate system, where $\nabla_\theta \theta \equiv 0$, (1) becomes

$$P = \frac{1}{\rho}\vec{\zeta}_a \cdot \vec{\mathbf{n}} \frac{\partial\theta}{\partial z} , \tag{2}$$

where $\vec{\mathbf{n}}$ is the unit vector along the gradient of θ. We define the unit area on a constant-θ surface as $ds = dxdy$. The mass contained in a unit box between two constant-θ surfaces of thickness dz is $dm = \rho dxdydz$. Following the discussion of Hoskins (1991) and integrating (2) over an area S at surface θ_1 between two adjacent θ surfaces θ_1 and θ_2 lead to

$$\frac{1}{S}\int P dm = \int_{\theta_1}^{\theta_2} \zeta_{an} d\theta , \tag{3}$$

where ζ_{an} is the component of the absolute vorticity along $\vec{\mathbf{n}}$, averaged over S. Equation (3) shows that the mass-weighted integral of P per unit area is equal to the θ-weighted integral of absolute vorticity ζ_{an}. For convenience, the z-coordinate is used for the study in this section. Consider three horizontal surfaces θ_1, θ_2, and θ_3 in a statically stable atmosphere with an external diabatic heating source Q ($= \dot{\theta} = d\theta/dt$) within area S that increases monotonically with height in the lower layer (θ_1, θ_2) and decreases monotonically with height in the upper layer (θ_2, θ_3) (Fig. 2a). T, A, and B are three parcels, originally located at the θ_3, θ_2, and θ_1 surfaces, respectively. Thermodynamically, the heating at θ_2 causes the θ-surface to sink, increasing stability and P below and decreasing stability and P above the surface (Fig. 1b). Dynamically, the heating-induced ascent of parcel A results in convergence in the lower layer and divergence in the upper layer. Since Pdm increases in the lower layer but decreases in the upper layer, according to (3), cyclonic vorticity in the lower layer and anticyclonic vorticity in the upper layer are thus generated (Wu and Liu 2000). Notice that for the convenience of following study, the z-coordinate and a fixed area S are chosen for the present discussion, using the Eulerian viewpoint. In the study of Hoskins (1991), the θ-coordinate and the

Lagrangian viewpoint are used, and the ascent at location A is considered a result of the rapid geostrophic adjustment towards balance with mass flowing inwards in the lower layer and outwards in the upper layer, leading to cyclonic flow below and anticyclonic flow above.

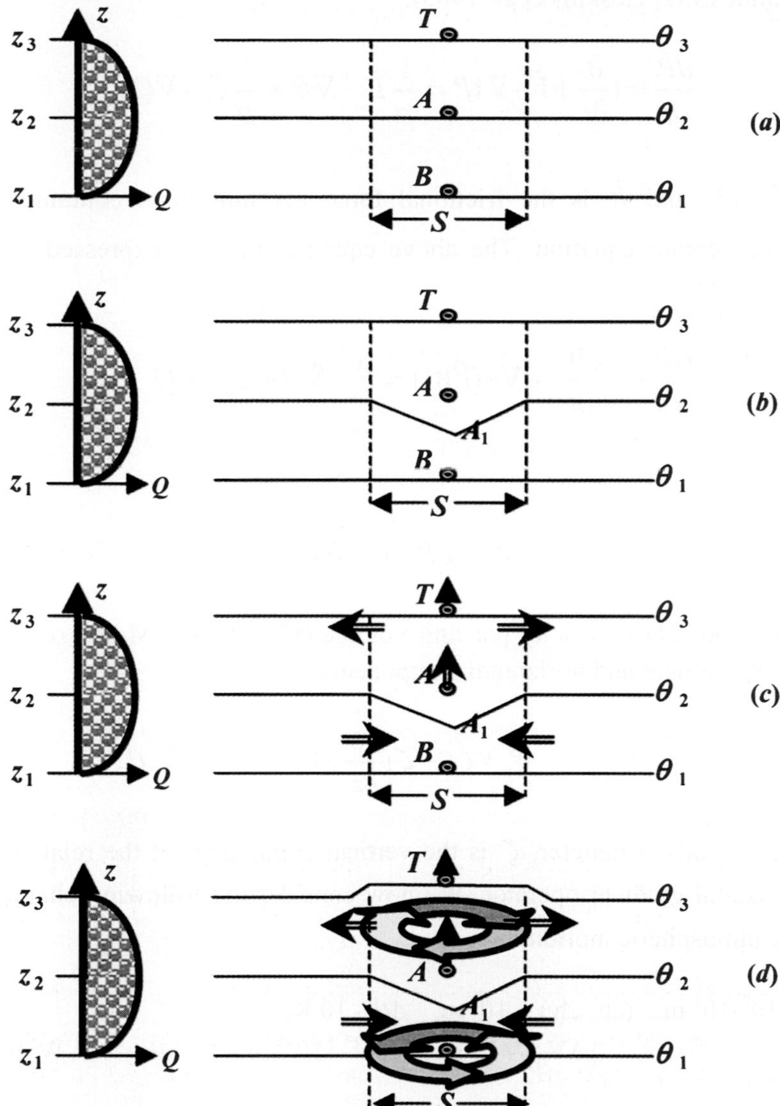

Figure 2. Diagram showing the thermal adaptation of the atmospheric circulation (arrows) to external heating Q. The left panel indicates the vertical distribution of heating with height within the area S. T, A, and B denote parcels initially at the θ_3, θ_2, and θ_1 surfaces, respectively. (a) Initial state. (b) Thermal impact: heating increases θ within S, the θ_2 surface descends, and potential vorticity increases in the lower layer and decreases in the upper layer. (c) Dynamic impact: heating causes ascent of parcel A, air converges in the lower layer and diverges in the upper layer, leading to an increase of mass m in the lower layer and a decrease in the upper layer. (d) Due to the combined effects of (b) and (c) and according to (3), cyclonic circulation develops in the lower layer with anticyclonic circulation in the upper layer (see also Fig. 2 of Hoskins 1991).

2.2 *Potential Vorticity Budget and Boundary Symmetric Instability*

The potential vorticity budget associated with such heating-induced circulation can be explored by employing the following Ertel potential vorticity equation (Ertel 1942; Eliassen and Kleinschmidt 1957; Hoskins et al. 1985):

$$\frac{dP}{dt} = (\frac{\partial}{\partial t} + \vec{V} \cdot \nabla)P = \frac{1}{\rho}\vec{F}_\zeta \cdot \nabla\theta + \frac{1}{\rho}\vec{\zeta}_a \cdot \nabla Q , \tag{4a}$$

where $\vec{F}_\zeta = \nabla \times \vec{F}$ and $\vec{F}$ is the frictional force per unit mass contained in the three-dimensional momentum equation. The above equation may be expressed in the following equivalent flux form:

$$\frac{DW}{Dt} = \frac{\partial W}{\partial t} + \nabla \cdot (\vec{V}W) = \vec{F}_\zeta \cdot \nabla\theta + \vec{\zeta}_a \cdot \nabla Q \quad , \tag{4b}$$

where

$$W = \rho P = \vec{\zeta}_a \cdot \nabla\theta \tag{5}$$

is the amount of potential vorticity per unit volume (Haynes and McIntyre 1987). W can be separated into its vertical and horizontal components, i.e.:

$$W = W_v + W_h , \; W_v = (f + \zeta)\frac{\partial\theta}{\partial z} , \; W_h = \nabla \times \vec{\mathbf{V}} \cdot \nabla_h\theta , \tag{6}$$

where f is the Coriolis parameter, ζ is the vertical component of the relative vorticity, and ∇_h is the horizontal gradient operator. We now consider the following characteristic values for large-scale atmospheric motions:

$$\Delta z \sim 10^3\text{-}10^4 \text{ m}, \; (\Delta x, \Delta y) \sim 10^6 \text{ m}, \; \Delta\theta \sim 10 \text{ K},$$
$$|\Delta\vec{\mathbf{V}}| \sim 10 \text{ m s}^{-1}, \; w \sim 10^{-3} \text{ m s}^{-1}, \; \Delta F \sim 10^2 \text{ W m}^{-2}, \tag{7}$$

where F is the heat flux of the atmosphere, which usually has a magnitude of 10^2 W m^{-2}. Thus,

$$Q = -\frac{\theta}{c_p T \rho}\frac{\partial F}{\partial z} \sim 10^{-5}\text{-}10^{-4} \text{ K s}^{-1}.$$

The magnitude of W is then estimated as

$$W = W_v + W_h = (f+\zeta)\frac{\partial\theta}{\partial z} + \nabla\times\vec{\mathbf{V}}\cdot\nabla_h\theta \sim 10^{-6}\ \mathrm{K\ s^{-1}\ m^{-1}}\{(10^{-1}\text{-}10^{0}) + (10^{-2}\text{-}10^{-12})\}. \quad (8)$$

Defining the following units for the total potential vorticity,

$$1\ \mathrm{TU} = 10^{-6}\ \mathrm{K\ s^{-1}\ m^{-1}},$$

we can rewrite (8) as

$$W = W_v + W_h \sim (10^{-1}\text{-}10^{0})\mathrm{TU} + (10^{-2}\text{-}10^{-1})\mathrm{TU}. \quad (9)$$

Equation (9) indicates that the vertical component W_v is one order of magnitude greater than the horizontal component W_h.

If the left-hand side of (4) is expanded with Δz being taken as 10^3 m, then the magnitudes of the various terms in (4) can be estimated as

$$\begin{array}{ccccc} \dfrac{\partial W}{\partial t} & +\nabla_h\cdot(\vec{\mathbf{V}}_h W) & +\dfrac{\partial}{\partial z}(wW) & =\vec{\mathbf{F}}_\zeta\cdot\nabla\theta & +\vec{\zeta}_a\cdot\nabla Q, \\ \text{(a)} & \text{(b)} & \text{(c)} & \text{(d)} & \text{(e)} \\ & 10^{-11} & 10^{-12} & 10^{-11} & 10^{-11} \end{array} \quad (10)$$

where $\vec{\mathbf{V}}_h$ is the horizontal wind. Equation (10) indicates that the horizontal divergence of the horizontal flux of W is one order of magnitude greater than the vertical divergence of its vertical flux. In other words, the vertical divergence of the vertical flux of W is not large enough to balance the generation of potential vorticity induced by diabatic heating.

Now let us consider the potential vorticity budget in the lower layer (z_1, z_2) of the heated column (Fig. 3a). In this layer, $\partial Q/\partial z > 0$, so term (e) in (10) is positive. According to Fig. 2, cyclonic circulation is generated in the lower layer. At steady state, $\partial W/\partial t$ vanishes. Since there is convergence in the interior of S, term (b) in (10) becomes negative and cannot balance term (e). Although term (c) is positive, its magnitude is too small and cannot balance term (e) either. Therefore the generation of potential vorticity near the surface due to diabatic heating must be balanced, to a major degree, by the effects of friction, i.e.,

$$\vec{\mathbf{F}}_\zeta\cdot\nabla\theta \cong -\vec{\zeta}_a\cdot\nabla Q. \quad (11)$$

Then, the rate of vorticity dissipation in the vertical direction

$$F_z \cong -(f+\zeta)\frac{\partial Q}{\partial z}\bigg/\frac{\partial\theta}{\partial z} < 0 \quad (12)$$

can be estimated, based on (7), to be as high as 10^{-9} s^{-1}. Therefore, a heating rate of 100 W m^{-2} in the boundary layer will force a surface cyclone with an intensity of 10^{-5} s^{-1} within 2 to 3 hours. However, the growth rate of the surface cyclone is limited because the heating causes the intersection of the θ-surface with the Earth's surface, resulting in the $\nabla_h\theta$ along the boundary Γ pointed inward. At the same time, the cyclonic circulation in the lower layer and the anticyclonic circulation in the upper layer create a vertical wind shear, resulting in the horizontal vorticity along Γ pointed outward. Therefore, along the boundary Γ of area S,

$$W_h = \nabla \times \vec{\mathbf{V}} \cdot \nabla_h \theta < 0 \text{, at } \Gamma. \tag{13}$$

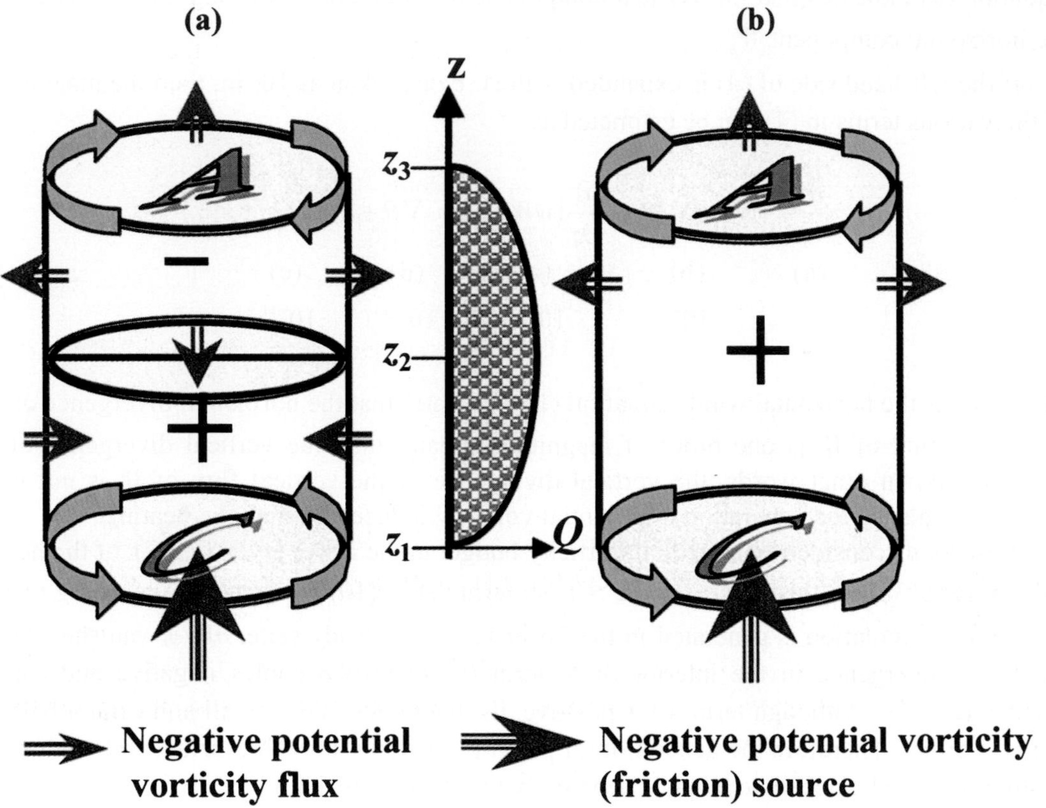

Figure 3. Budget of total potential vorticity W during atmospheric thermal adaptation. (+) and (–) indicate the source and sink of W, respectively. Arrows indicate the direction of the flux of negative W. (a) In the lower layer of a two-layer model, generation of positive W due to heating is balanced mainly by frictional dissipation and cross-lateral boundary flux of negative W; the generation of negative W in the upper layer resulting from the decreasing heating with height is balanced mainly by the cross-lateral boundary flux of negative W. (b) In a column model, the negative W generation due to friction at the lower boundary is partly balanced by the heating-generated positive W ($\zeta\partial Q/\partial z$), and mainly by the cross-boundary negative W flux.

Assuming a horizontal scale of the boundary (Δx, Δy) of $10^4 \sim 10^5$ m and a vertical scale Δz of 10^3 m, by referring to (7), we can estimate the magnitude of W along the boundary as:

$$W_v \sim 10^{-1}\text{-}10^{0}\ \text{TU},\ \ W_h \sim 10^{0}\text{-}10^{1}\ \text{TU, at the boundary } \Gamma. \tag{14}$$

Equations (13) and (14) suggest that, along the boundary,

$$|W_h| > |W_v| \tag{15}$$

and

$$W = W_v + W_h < 0 . \tag{16}$$

The condition (16) implies the existence of symmetric instability at the boundary of the heating region S. Integrating (10) from z_1 to z_2 over area S and employing the following Gaussian law

$$\int_S \nabla_h \cdot (\vec{\mathbf{V}}_h A) ds = \oint_\Gamma (A\vec{\mathbf{V}}_h) \cdot \vec{\mathbf{m}} d\ell \tag{17}$$

lead to

$$\underset{(\mathrm{b})>0}{\int_{z_1}^{z_2} dz \oint_\Gamma W\vec{\mathbf{V}}_h \cdot \vec{\mathbf{m}} d\ell} + \underset{(\mathrm{c})>0}{\int_S W_2 w_2 ds} = \int_{z_1}^{z_2} \int_S \left[\underset{(\mathrm{d})<0}{F_z \frac{\partial \theta}{\partial z}} + \underset{(\mathrm{e})>0}{(f+\zeta)\frac{\partial Q}{\partial z}} \right] ds dz\ , \tag{18}$$

where $W_2 w_2$ is the vertical potential vorticity flux at level z_2, Γ is in the counter-clockwise direction, and $\vec{\mathbf{m}}$ is the unit vector perpendicular to Γ and pointed outward. Since there is convergence across the lower layer boundary ($\vec{\mathbf{V}}_\Gamma \cdot \vec{\mathbf{m}} < 0$) and since W is negative along the boundary (refer to Eq. (16)), the sign of term (b) in (10) or (18) becomes positive. This means that, due to the intersection of the θ-surface with the Earth's surface and the appearance of symmetric instability along the boundary, negative potential vorticity is pumped across the boundary into the heating area, thus diluting the generation of positive potential vorticity due to diabatic heating. As a result, the signs of terms (b) through (e) in (18) become (+), (+), (–), and (+), respectively. Therefore, the heating-induced generation of positive potential vorticity in the lower layer is balanced by surface frictional dissipation, transport across the upper boundary, and dilution of negative potential vorticity flux across the transverse boundary. A steady state is then maintained (Fig. 3a).

Since $\partial Q/\partial z < 0$ in the upper layer (z_2, z_3), negative potential vorticity generation and anticyclonic circulation are anticipated (Fig. 3). In this layer, frictional dissipation can be ignored, and the generation of negative potential vorticity is balanced mainly by the horizontal divergence of potential vorticity flux. Because $\nabla_h \theta$ and the horizontal vorticity vector possess opposite signs as they do in the lower layer, the discussions presented from

(13) to (16) for the lower layer are applicable to the upper layer. Since the airflow is divergent in the layer, $\vec{\mathbf{V}}_{\Gamma} \cdot \vec{\mathbf{m}} > 0$. Integration of (10) from z_2 to z_3 over S then leads to

$$\underset{(b)<0}{\int_{z_2}^{z_3} dz \oint_{\Gamma} W\vec{\mathbf{V}}_h \cdot \vec{\mathbf{m}} d\ell} - \underset{(c)<0}{\int_S W_2 w_2 ds} = \underset{(e)<0}{\int_{z_2}^{z_3} \int_S (f+\zeta) \frac{\partial Q}{\partial z} dsdz} \quad . \tag{19}$$

Equation (19) shows that, in the upper layer, the generation of negative potential vorticity due to the decreasing with height of diabatic heating is partly diluted by the upward transfer of positive potential vorticity across the lower boundary from below, but mainly by the outward transport of negative potential vorticity across the lateral boundary Γ of the heating area (Fig. 3a). Here we also observe the importance of the symmetric instability along the boundary in the maintenance of the potential vorticity balance within the heating area.

If the integration of (10) is taken from z_1 to z_3 over S, the potential vorticity balance over the whole column can then be estimated, and the result can be obtained by adding (18) and (19):

$$\underset{(b)<0}{\int_{z_1}^{z_3} dz \oint_{\Gamma} W\vec{\mathbf{V}}_h \cdot \vec{\mathbf{m}} d\ell} = \int_{z_1}^{z_3} \int_S \left[\underset{(d)<0}{F_z \frac{\partial \theta}{\partial z}} + \underset{(e)>0}{\zeta \frac{\partial Q}{\partial z}} \right] dsdz \quad . \tag{20}$$

The term for vertical transfer (c) vanishes. The larger part, i.e., $f\partial Q/\partial z$, of the generation term for the lower and upper portions of the region cancel each other, leaving only a small term, $\zeta\partial Q/\partial z$, in (e). The frictional dissipation term (d) remains unchanged. The sign of term (b) should be the same as that in the upper layer. This is because, based on (18) and (19), the generation rate in the upper layer that is approximated as $(f\partial Q/\partial z)$ is not dissipated by friction as it is in the lower layer. Also, the scale analyses presented in (6), (7), and (10) indicate that the magnitude of the integrand (d) in (20) is larger than that of the integrand (e); thus the sign of term (b) on the left-hand side of (20) follows the sign of term (d). The (b), (d), and (e) terms in (20) possess signs of (–), (–), and (+), respectively. This implies that for the total-column integration, the negative vorticity pumped into the column due to surface friction is partly canceled by the heating-generated positive vorticity. The rest is pumped out of the column in the upper layer, affecting the circulation outside the heating region (Fig. 3b).

3. Numerical Experiments

We now conduct a numerical experiment that is designed to examine the thermal adaptation theory established in the preceding section. The model used is the climate model Global-Ocean-Atmosphere-Land-System (GOALS) that was developed at LASG, IAP (Wu et al. 1997; Zhang et al. 2000). Its atmosphere component (Wu et al. 1996) has 9 levels in the vertical and is rhomboidally truncated at wave number 15 in the horizontal. The ocean component (Zhang et al. 1996) has 20 layers in the vertical with a horizontal resolution of 4° latitude by 5° longitude. The land component uses the simplified Simple Biosphere (SSiB) model (Xue et al. 1991; Liu and Wu 1997). The K-distribution scheme developed by Shi (1981) is used for parameterizing radiation processes. This model has been used in the Atmospheric Model Inter-comparison Program (AMIP), Coupled Model Inter-comparison Program (CMIP), and Task I of the Inter-Governmental Program for Climate Changes (IPCC, 2001). For the present purpose, the ocean and land components are switched off, and an aqua-planet is assumed. The zonal mean sea surface temperature is imposed as the lower boundary condition. The solar angle is fixed to its value on July 15. Other variables, including CO_2, aerosol, cloud amount, and atmospheric variables, all assume their corresponding July zonal means. The initial wind and horizontal gradient of temperature $\nabla_h T$ are set to zero. For simplicity, an axially symmetric surface heating source is imposed at the equator, centered at 176°W bounded in the region 11°S-11°N and 160°E-150°W. The intensity of the heating source is 100 W m^{-2} at the center and decreases gradually in cosine form to zero at the boundary. The heating region bounded by the 1-W m^{-2} heating isoline is shown in Fig. 4 with heavy curves. The heating is set to 100 Wm^{-2} for the experiment (the surface sensible heat flux over equatorial Africa and Latin America is commonly above this value in July). To concentrate on the atmospheric response to such imposed diabatic heating, latent heating and the sensible heating over other parts of the world are switched off in the thermodynamic equation.

Figures 4a and 4d show the vertical distributions of diffusive sensible heating over the heating region on Day 1 and Day 3 of the model integration, respectively. After Day 3 the model asymptotic responses are similar to those demonstrated in Figs. 4 and 5 for Day 3 and will not shown here. The maximum heating is located at the lowest level ($\sigma = 0.991$), with an intensity of 8.5 K day^{-1} on Day 3. The heating decreases with height and approaches zero near 800 hPa. Figures 4c and 4f present the wind and vorticity fields at the $\sigma = 0.991$ level on Day 1 and Day 3, respectively. The diabatic heating results in cyclonic vorticity and horizontal convergence in the lower layer, the so-called "Gill pattern" (Gill 1980). On the contrary, such heating generates anticyclonic circulation and divergence at the upper level $\sigma = 0.664$ (Figs. 4b and 4e). The intensity of the vorticity is on the order of 10^{-6} s^{-1} outside the equator. The vertical cross sections of potential temperature and the wind vector composed of the zonal and vertical components on (a) Day 1 and (b) Day 3 are demonstrated in Fig. 5. After one day of heating, the isentropic surfaces become concave in the heating region. The airflow converges in the lower layer, penetrates the isentropic surfaces over the heating region, and diverges at the upper level $\sigma = 0.664$. These processes intensify on Day 3. All of these results are similar to those shown in Fig. 2.

The spectacular phenomenon appears at the western as well as eastern boundaries of the heating region, i.e., around 165°E and 157°W, where the isentropic surfaces intersect with the Earth's surface. According to Figs. 4 and 5, the magnitudes of the terms relevant to the calculation of W along the boundary can be estimated as follows:

$$(f+\zeta) \sim 10^{-6}\text{-}10^{-5}\ \text{s}^{-1}, \quad \frac{\Delta\theta}{\Delta z} \sim 10^{-4}\text{-}10^{-3}\ \text{K m}^{-1},$$

$$\left(\frac{\Delta u}{\Delta z}, \frac{\Delta v}{\Delta z}\right) \sim 10^{-2}\ \text{s}^{-1}, \quad \left(\frac{\Delta\theta}{\Delta x}, \frac{\Delta\theta}{\Delta y}\right) \sim 5\times 10^{-6}\ \text{K m}^{-1}. \tag{21}$$

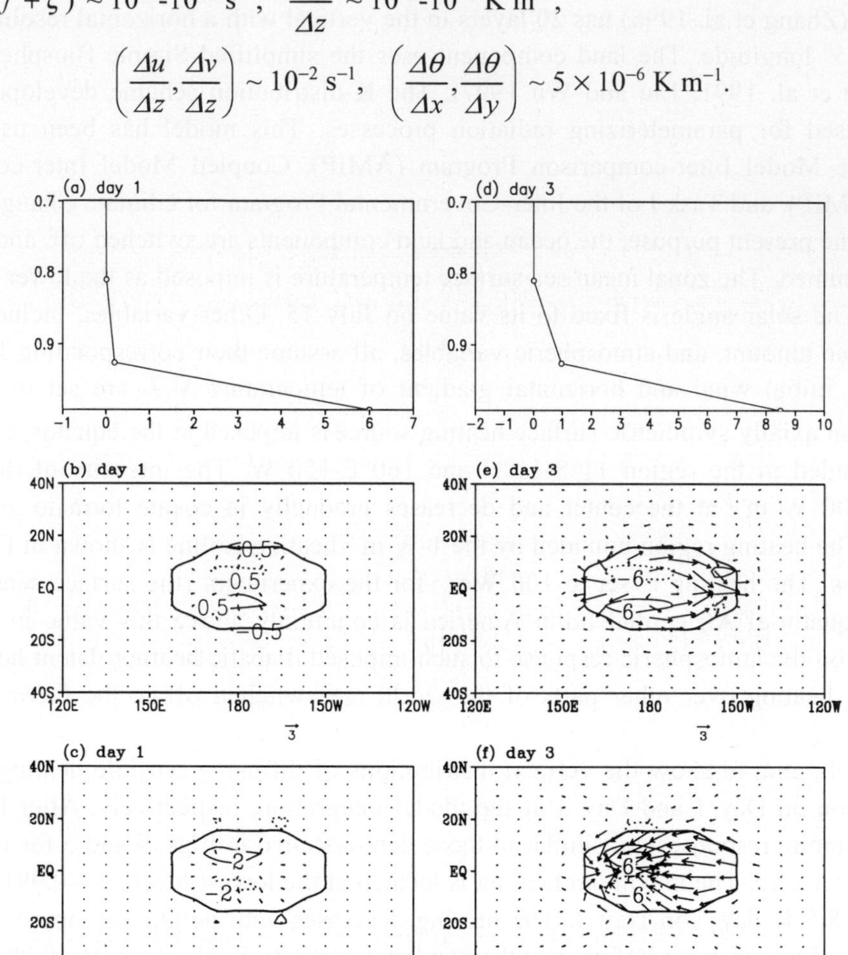

Figure 4. Numerical experiment on the thermal adaptation of the atmospheric flow field and vorticity field (10^{-6} s^{-1}) to a prescribed surface sensible heating at Day 1 (a, b, and c) and Day 3 (d, e, and f) at the upper level $\sigma = 0.664$ (b and e) and lower level $\sigma = 0.991$ (c and f). (a) and (d) present the heating rate (K day^{-1}) at Day 1 and Day 3, respectively. Solid and dashed curves denote positive and negative values, respectively, and the heavy solid curve bounds the region with heating greater than 1 W m^{-2}.

It is evident that even for a region with a horizontal scale of 4×10^{6} m and an average heating rate 64 W m^{-2} ($200/\pi$), the heating-induced horizontal vorticity (the vertical wind shear) along the boundary is about three orders of magnitude larger than the vertical vorticity. According to (21), the magnitudes of W_{v} and W_{h} can be estimated as

$$W_v = (f+\zeta)\frac{\partial\theta}{\partial z} \sim 10^{-4}\text{-}10^{-2}\ \text{TU}, \quad W_h = \nabla\times\vec{\mathbf{V}}\cdot\nabla_h\theta \sim 10^{-2}\text{-}10^{-1}\text{TU}. \tag{22}$$

It becomes apparent that along the boundary of the heating region, $|W_h| > |W_v|$. Since $W_h < 0$, the validity of (15) and (16) and the occurrence of symmetric instability ($W < 0$) along the boundary are proved. These then support our analysis of the vorticity balance as presented in Fig. 3. The only difference is that the magnitude of W shown in (22) is smaller than the previous scale analysis. This is because in the numerical experiment, the heating source is located at the equator, where f vanishes.

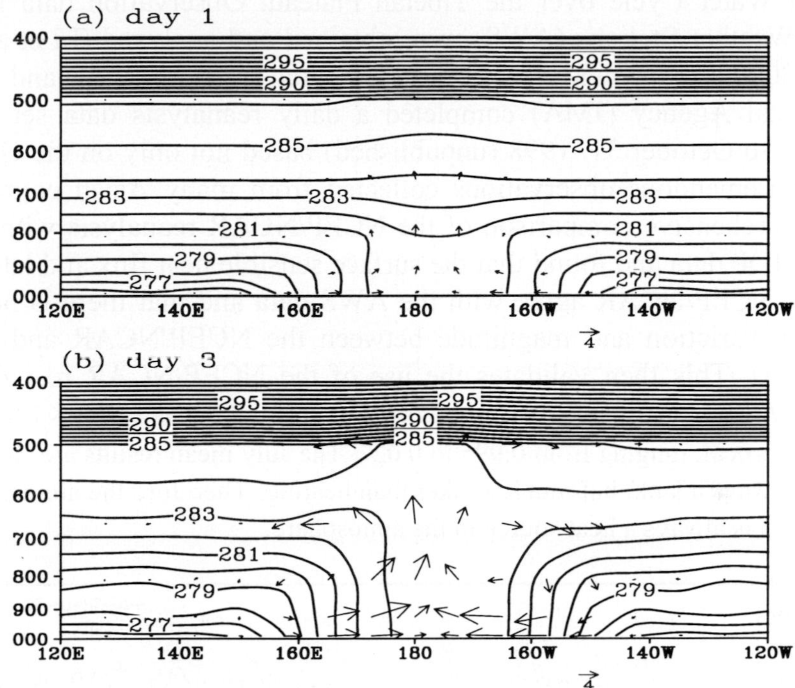

Figure 5. Vertical cross sections at the equator of potential temperature (K) and velocity (units: u in m s^{-1}, ω in −40 Pa s^{-1}) based on the numerical experiment of thermal adaptation of the atmosphere to a prescribed surface sensible heating. (a) Day 1; (b) Day 3.

4. Heating Feature over the Tibetan Plateau

Let us consider the characteristics of the heating over the Tibetan Plateau in summer. At any particular level, the total heating Q is defined as the sum of diffusive sensible heating (SH), latent heating (LH), and radiative cooling RC (RC is negative), i.e.,

$$Q(\sigma) = \mathrm{SH}(\sigma) + \mathrm{LH}(\sigma) + \mathrm{RC}(\sigma). \tag{23}$$

Integration over the atmospheric column then produces the total atmospheric heating:

$$H = \frac{c_p p_s}{g} \int_0^1 Q(\sigma) d\sigma, \tag{24}$$

where c_p is the specific heat at constant pressure, p_s is surface pressure, and g is the gravitational acceleration of the Earth. The plateau region is defined here as the region where the surface elevation is above 3000 m. The heating rate data considered here were obtained from the NCAR/NCEP reanalysis from 1958 to 1997. Such data are model dependent but not observation dependent. Therefore caution is required before these data are used. From July 1993 to March 1999, China and Japan organized a Field Experiment of Surface Energy and Water Cycle over the Tibetan Plateau. Observation data from the six Automatic Weather Stations (AWS) were obtained and analyzed (Li et al., 2001). In September 2000, the Japan Meteorological Research Institute (MRI) and the Japanese Meteorological Agency (JMA) completed a daily reanalysis data set GAME-IOP from April 1 to October 31, 1998 (unpublished) based not only on the GTS data, but also on the tremendous observations collected from many Asian countries. Duan[1] made a comprehensive comparison of the NCEP/NCAR reanalysis with these AWS and GAME-IOP data and found that the surface sensible heat flux and latent heat flux provided by NCEP/NCAR agree with the AWS data and that there is no significant difference in variation and magnitude between the NCEP/NCAR and GAME-IOP daily data sets. This then validates the use of the NCEP/NCAR reanalysis for the present study. This analysis possesses a horizontal resolution of $1.875^{\circ} \times 1.875^{\circ}$ and 28 σ-layers in the vertical, ranging from 0.995 to 0.027. The July mean results are shown in Fig. 6. Radiation is always a sink, but much weaker than heating. Therefore the atmosphere over the Tibetan Plateau is always a heat source to the atmosphere.

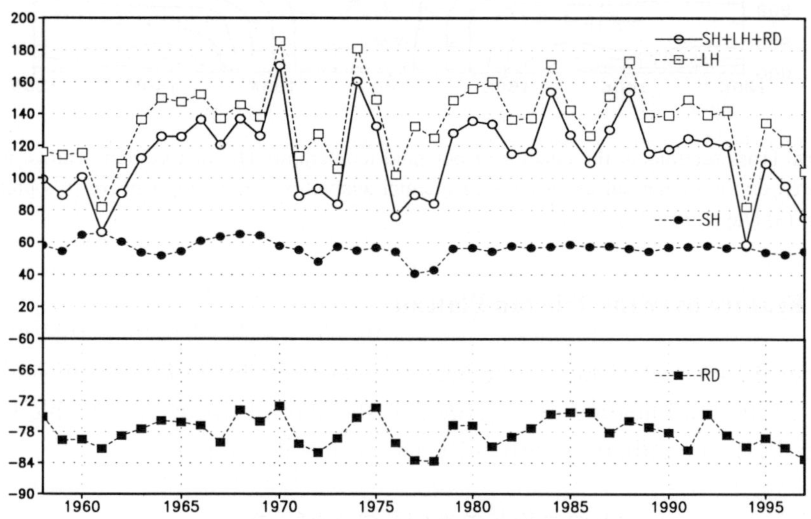

Figure 6. Evolution of the July mean diabatic heating from 1958 to 1997 averaged over the Tibetan Plateau where the altitude is higher than 3 km above sea level. Open circles and squares indicate total heating and latent heating, respectively; and solid circles and squares represent sensible heating and radiative cooling, respectively. The units are W m^{-2}.

[1] Duan, A., Chapter 2, Ph.D. Thesis, Chinese Academy of Sciences (unpublished). Private communication.

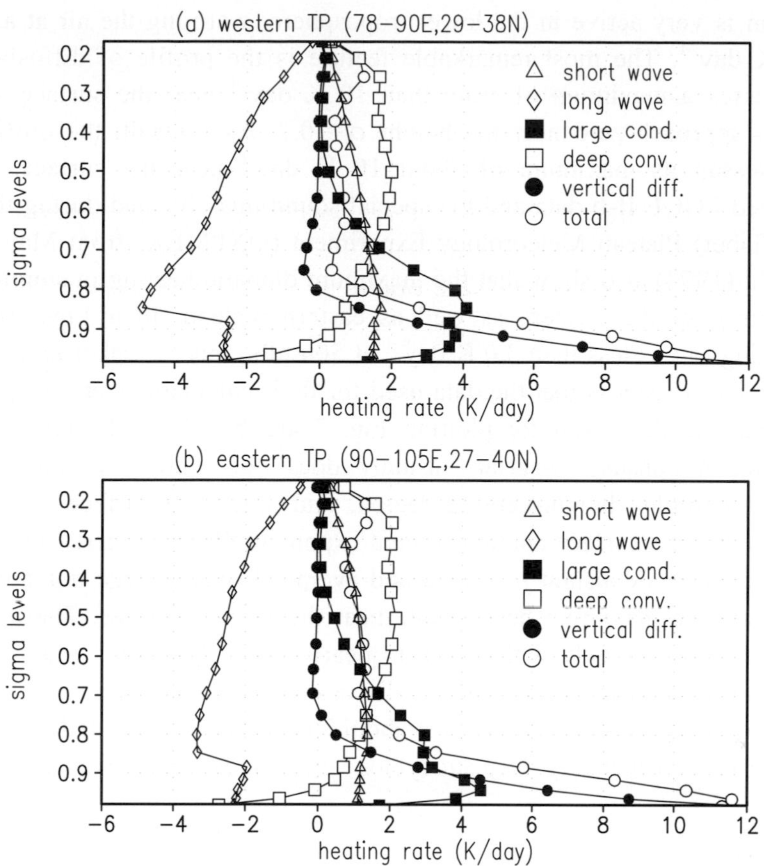

Figure 7. The vertical profiles of the July (1986-1995) mean diabatic heating over (a) the western Tibetan Plateau (78°E-90°E, 29°N-38°N) and the eastern Tibetan Plateau (90°E-105°E, 27°N-40°N), in units of K day^{-1}. The (◊) sign demonstrates longwave radiation cooling rate; the Δ and solid circles represent the shortwave radiation and vertical diffusive diabatic heating rate, respectively; the solid and open squares represent the large-scale and deep convective latent heating rate, respectively; and the open circles indicate the total heating rate.

Diffusive sensible heating in July is positive, but with a weak interannual variation. Latent heat release, which determines the interannual variation of the total heating, is the dominant heating mechanism over the Tibetan Plateau. If the vertical profiles of heating are considered, however, the result is totally different. In Fig. 7, we show the vertical profiles of the July mean diabatic heating for the period 1986-1995 over the western and eastern Tibetan Plateau (Liu et al. 2001). The domain of the western Tibetan Plateau for the calculation is 78°E-90°E and 29°N-38°N, and that of its eastern part is 90°E-105°E and 27°N-40°N. The boundary of these domains follows the 3000-m contour of the plateau very well. The figure shows that both the western and eastern parts of the plateau share similar characteristics in the distributions of diabatic heating except that the longwave radiation cooling over the western part is somehow stronger. Shortwave radiative heating is relatively uniform in the vertical. Deep convection warms the upper troposphere by about 2 K day^{-1}, whereas large-scale shallow convection is very active in the lower troposphere, warming the air at a maximum rate of about 4 K day^{-1}. The most remarkable feature is the profile of diffusive sensible heating that possesses a maximum of more than 10 K day^{-1} near the surface, decreasing upward rapidly to approach zero near the height σ= 0.7. As a result, the profile of total heating also possesses its maximum of about 10 K day^{-1} near the surface. Using the objectively analyzed FGGE II-b data and the special soundings obtained during the Chinese Qinghai-Xizang (Tibet) Plateau Meteorology Experiment (QXPMEX) from May to August 1979, Yanai and Li (1994) also show that the maximum diabatic heating in summer appears in the lower layers over the western as well as eastern plateau (their Figs. 16 and 17). However, the heating rate is less than 3.0 K day^{-1} at 500 hPa, much weaker than that shown in Fig. 7. One possible reason is that the data used for their calculations are based on the p-coordinate, whereas the data used for plotting Fig. 7 are based on the terrain-following σ-coordinate. Since a constant p-surface usually does not follow the Earth's surface, particularly over a complex height-varying region, and since the surface heating rate is maximized near the ground surface and evanescent exponentially with increasing height, the heating rate calculated at a constant p-surface and averaged over a large domain then may largely underestimate the maximum heating rate near the ground surface. The total heating shown in Fig. 7 decreases quickly with height to nearly about 1 K day^{-1} near σ = 0.7 and changing little at higher altitudes. Following the discussion in Section 2 and presented in Fig. 2, such a profile of total heating in summer should generate a very shallow and strong cyclone near the surface, with a very deep anticyclone aloft. This is what we have observed in Fig. 1.

5. Vorticity Source

Figure 8 shows cross sections along (a) 30°N and (b) 90°E of the July mean potential temperature and wind vector projected on the corresponding cross sections in which the vertical velocity w has been multiplied by 10^3. The two cross sections pass through the Tibetan Plateau, based on data used from the NCEP/NCAR reanalysis between 1986 and 1995. During summer, the warmest center of potential temperature is just over the Tibetan Plateau. This is in agreement with the existence of the warm temperature center in July over the plateau (Yanai et al. 1992). Strong ascent prevails in the area from the plateau to eastern China on the east and to the Bay of Bengal to the south, penetrating the isentropic surfaces almost perpendicularly and indicating the existence of a strong heating source over the area in summer. In addition to this well-known fact, another spectacular phenomenon is also observed in the figure: the isentropic surfaces between 315 K and 335 K intersect with the Tibetan Plateau while remaining continuous surfaces over the other parts of the world.

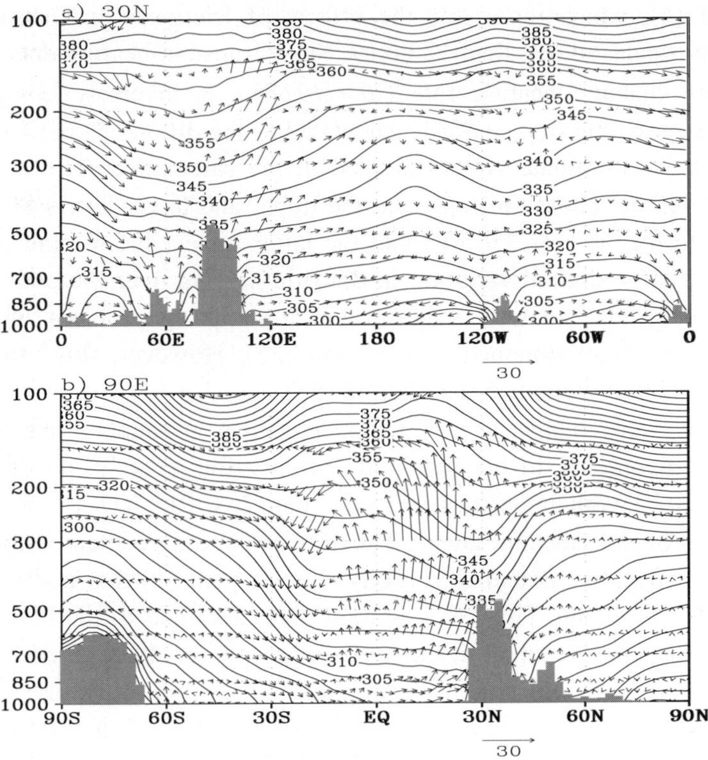

Figure 8. July mean (1986-1995) cross sections of potential temperature (contours, interval is 10 K) and vertical circulation (vectors), (a) along 30°N and (b) along 90°E.

A study by Haynes and McIntyre (1987) has shown that there is no net creation and destruction of potential vorticity between two complete isentropic surfaces. The intersection of those θ-surfaces with the plateau then implies that *the Tibetan Plateau is an important source of potential vorticity for atmospheric motions in summer.* As a matter of fact, if the σ-coordinate is used, the Earth's surface would become flat and the intersecting θ-surfaces in Fig. 8 would be similar to what we have observed in Fig. 5. Therefore, it would be instructive to employ the thermal adaptation theory just demonstrated in the preceding section to gain some new insights on the maintenance of the circulation pattern shown in Fig. 1. For this purpose, we take the monthly average of (4) and perform diagnosis by using the 12-hour (00Z and 12Z) NCEP/NCAR reanalysis data for July from 1986 to 1995. In such circumstances, the local variation is small and can be omitted. Let the over bars indicate the monthly mean and the primes indicate the deviations from it. Equation (4) can then be written as

$$\nabla \cdot \overline{\vec{\mathbf{V}} W} = \overline{\vec{\zeta}_a \cdot \nabla \dot{\theta}} + R \tag{25}$$

(Liu et al. 2001), where

$$R = \overline{\vec{\mathbf{F}}_\zeta \cdot \nabla \theta} + \overline{\vec{\zeta}'_a \cdot \nabla \dot{\theta}'} - \nabla \cdot \overline{\vec{\mathbf{V}}' W'} \tag{26}$$

is the residual of (4), which represents the effects of frictional dissipation and transient processes with time scales of less than 12 hours. Since transient heating data are absent from the reanalysis, the diabatic heating rate ($\dot{\theta} = d\theta / dt$) is inversely computed from the distribution of θ at each time step, and R is calculated as the difference between the two other terms in (25). The σ coordinate system is used in the calculation. The two levels of σ = 0.4357 and σ= 0.995 are chosen to represent the upper and lower layers of the atmosphere, respectively. The former is near 250 hPa and the latter is near the ground over the Tibetan Plateau. Due to the large diurnal variation in diabatic heating over the plateau (Yanai et al. 1992), use of the 12-hour analysis may present a sampling problem in calculating the magnitudes of those terms contained in (25) and (26). However, this will not affect the evaluation of the potential vorticity balance from (25).

The potential vorticity budget at σ= 0.4357 is shown in panels *a* and *b* of Fig. 9. The distribution of the divergence of the potential vorticity flux (Fig. 9, panel *a*) exhibits two strong negative centers with intensities of more than 2×10^{-6} TU s^{-1} on the eastern and northwestern flanks of the Tibetan Plateau. They are in good agreement with the distribution of the generation of negative potential vorticity (Fig. 9, panel *b*) that is due to the decrease with height of diabatic heating. Thus, we can infer that the effects of the frictional dissipation and the transient processes are small in the upper troposphere, although cumulus friction can exist when cumulus convection is active (Tung and Yanai 2002). The result obtained here looks different from that of Yang et al. (1992). They used the classical vorticity equation to diagnose the balance of vorticity, with the conclusion that in July over the Tibetan Plateau the divergence of the vorticity flux is balanced by the contribution of sub-grid-scale processes and the residual term (see Fig. 1 in Yang et al. (1992)). Since there is no diabatic heating in the classical vorticity equation, it is hard to compare our result with theirs. However, Yang et al. found that the contribution of the sub-grid-scale processes mainly comprised convective activity, a substantial part of the diabatic heating over the Tibetan Plateau in July. From this viewpoint, the implication of their research is consistent with the results of this study, although in an indirect sense.

The potential vorticity budget at the σ= 0.995 level is shown in panels *c* and *d* of Fig. 9. The strong center of positive potential vorticity generation is due to the increase of diabatic heating with height on the southern side of the Tibetan Plateau, from 25°N to 31°N (Fig. 9, panel *d*), revealing the importance of near-surface sensible heating in the maintenance of the lower-layer cyclone over the Tibetan Plateau. On the other hand, the divergence of vorticity flux (Fig. 9, panel *c*) is too small to balance the vorticity generation. The heating term must be balanced mainly due to the frictional and transient processes (Fig. 9, panel *e*). The results obtained from the above analysis not only confirm the thermal adaptation theory but also provide an interpretation for the formation of the atmospheric circulation over the Tibetan Plateau. It is mainly due to the strong surface sensible heating over the plateau that the shallow and strong surface cyclone and deep anticyclone aloft are maintained.

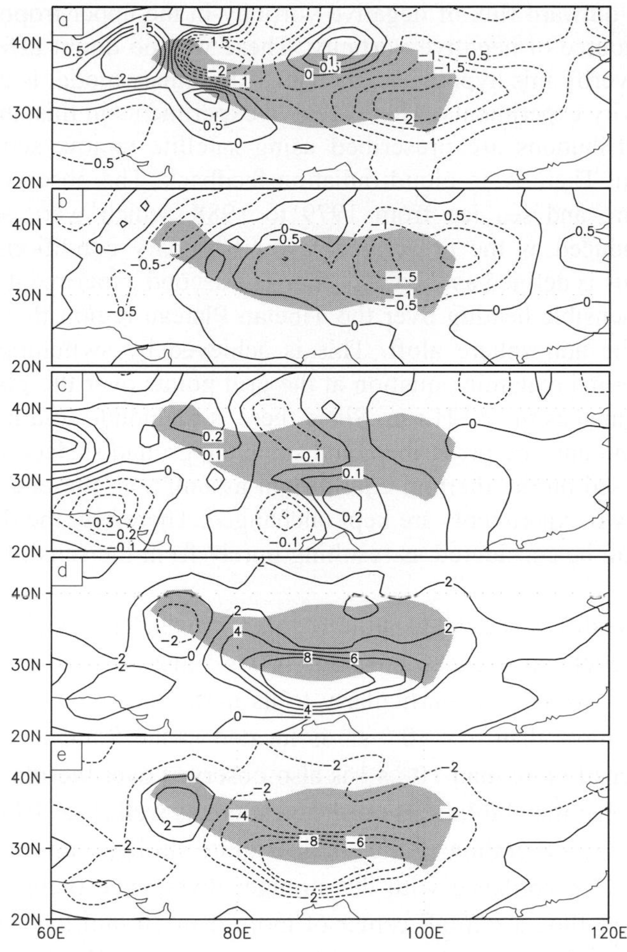

Figure 9. The July (1986-1995) mean potential vorticity budget calculated from the potential vorticity equation (25) at the $\sigma = 0.4357$ level ((a) and (b)) and at the $\sigma = 0.995$ level ((c), (d), and (e)). Regions higher than 3000 m above sea level are shaded. Panels (a) and (c) are the divergence of the W flux; (b) and (d) are the generation of W due to diabatic heating; and (e) is the residual term R. The units are 10^{-12} K m^{-1} s^{-2}.

6. Impacts of the Sensible Heat Driven Air Pump over the Tibetan Plateau

Due to the thermal adaptation of the atmospheric circulation to the diabatic heating over the plateau, strong ascent with lower layer convergence and upper layer divergence occurs over the plateau. Were there no diabatic heating along the lateral boundaries of the plateau, the impinging air parcels should stay on the same potential temperature surface and go around the plateau. The strong ascending flow at the lateral boundaries and from lower to higher potential temperature as demonstrated in Fig. 8 therefore suggests that it is the surface sensible heating along the lateral boundaries of the plateau that forces the near-surface ascending flow. This works as an air pump that sucks the air from below and pumps it outward to the surrounding area in the upper layer. Because such an air pump is mainly sensible heat-driven, it can be termed a sensible heat-driven air pump (Wu et al. 1998; 2002).

In addition, the strong outward flux of negative vorticity in the upper troposphere makes the plateau an important source of negative vorticity. These will no doubt affect the circulation over a larger area. To verify this hypothesis, the GOALS climate model is again employed to initiate a set of sensitivity experiments. There are two experiments in this analysis. In the first experiment, cloud distributions are prescribed using satellite remote sensing data for the calculation of radiation. There is no cloud-radiation feedback. The observed distributions of sea surface temperature and sea ice from 1979 to 1988, which were set for the AMIP experiments, are introduced as the prescribed lower boundary conditions to integrate the model for 10 years. This is defined as the CON run. The second experiment is the same as the CON except that the sensible heating over the Tibetan Plateau region that is above 3 km is not allowed to heat the atmosphere aloft. This is achieved by switching off the sensible heating term in the thermal dynamic equation at the grid points over the plateau region. This experiment is then defined as the NSH run. Since there is no cloud-radiation feedback in the model and the cloud amounts are prescribed, and since the ground surface temperature in the model is calculated based on the thermal equilibrium assumption, the energy budgets at the ground surface in the two experiments are kept unchanged. Therefore, the difference between the two experiments can be considered as resulting purely from the sensible heating over the plateau.

Figure 10 shows the difference fields at the $\sigma = 0.991$ level. The sensible heating over the plateau causes lower level convergence of about 10^{-5} s^{-1} just above the plateau (Fig. 10a), which is surrounded by divergence, particularly to its north and south. Also, another strong convergence center of more than 6×10^{-6} s^{-1} is located in the South China Sea region. A positive vorticity center of more than 10^{-5} s^{-1} is also observed over the plateau, accompanied by negative vorticity generation both to its north and south as well (Fig. 10b).

The negative vorticity generation in the south is quite strong, with an intensity of more than -6×10^{-6} s^{-1} and in accordance with the stronger divergence there. Due to the strong convergence of moisture flux, a rainfall center of more than 10 mm day^{-1} is forced over the plateau that extends northeastward to central and northern China (Fig. 10c). Reduced rainfall is observed in the area ranging from northwest India, the Bay of Bengal, the Indochina Peninsula, to South China. The effects on local climate of the SHAP over the plateau become prominent. In Figs. 11a and 11b, the streamlines are shown at 200 hPa (a) from the NCEP/NCAR reanalysis and (b) from the CON experiment averaged over the 10-year AMIP period. Despite the difference in details, the GOALS/LASG simulation of the climate basic state is reasonable. Two anticyclones are over the Tibetan Plateau and the south of North America. Two upper tropospheric troughs are over the two oceans. The difference between CON and NSH shown in Fig. 11c demonstrates that the sensible heating over the Tibetan Plateau does generate negative vorticity in the upper troposphere. More importantly, such a sensible heat-induced vorticity source over the plateau forces a series of geopotential height anomalies in the form of a ray of Rossby waves. This implies that the sensible heating over the Tibetan Plateau affects not only the climate anomaly in the surrounding area but also the circulation over the Northern Hemisphere. Although this result is obtained through numerical experiments, it is also supported by observational evidence. This will be shown in the study of the plateau's impact on the interannual climate variations in an accompanying paper.

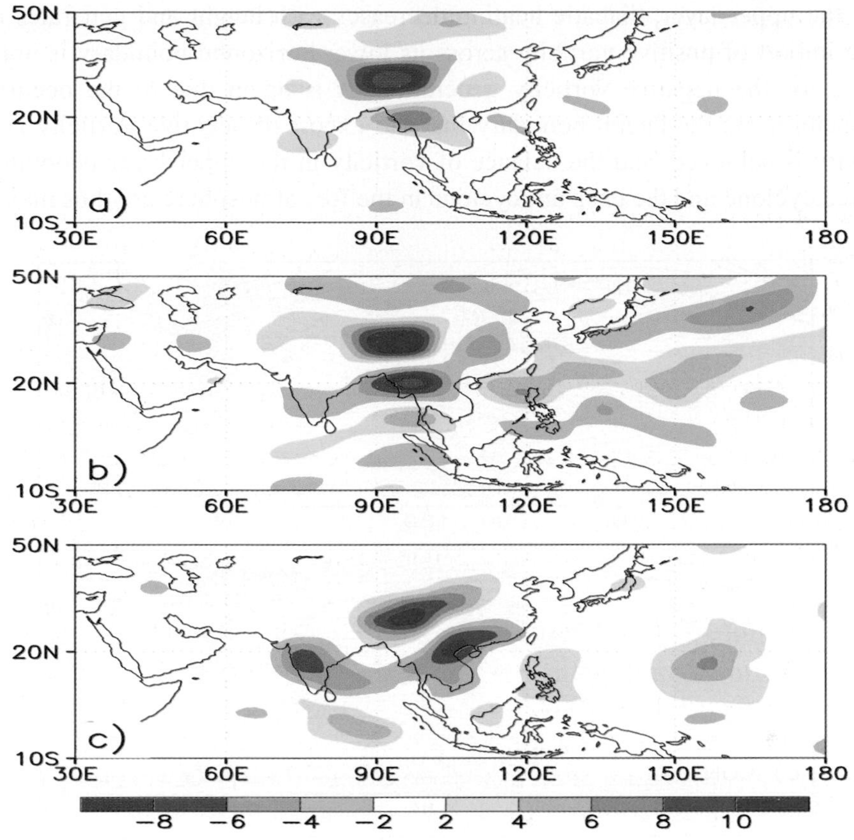

Figure 10. The July mean differences between CON and NSH experiments of (a) the wind divergence, (b) vorticity at $\sigma = 0.991$, and (c) precipitation. The contour interval is $2 \times 10^{-6}\mathrm{s}^{-1}$ in (a) and (b) and 2 mm day^{-1} in (c).

7. Concluding Remarks

This study has shown that even in midsummer, the shallow surface cyclone and the deep anticyclone in the middle and upper troposphere over the Tibetan Plateau are mainly forced by the surface sensible heating over the plateau. Based on the theory of potential vorticity, a theory of the adaptation of the atmospheric circulation to a given diabatic heating source is developed. It shows that the vertical profile of diabatic heating is important in determining the forced circulation patterns. In the case of surface sensible heating, the maximum heating level is so close to the surface that the isentropic surfaces at the boundary of the heating region intersect the surface of the Earth, resulting in the occurrence of symmetric instability along the boundary. Negative vorticity flux is then transported into the heating region across the boundary. The positive vorticity generation in the lower layer due to the increase with height of diabatic heating is then balanced by frictional dissipation at the surface, the negative vorticity import across the lateral boundary, and the export of positive vorticity at the upper boundary. In the upper layer, diabatic heating decreases with height and generates negative vorticity. The import of positive vorticity across its lower horizontal boundary is not enough to compensate for the negative vorticity generation. It is again due to the occurrence of symmetric instability at the lateral boundary that the export of negative vorticity across the lateral boundary is achieved, and the balance of vorticity in the upper layer is obtained. The shallow surface cyclone and the deep anticyclone in the free atmosphere are thus maintained.

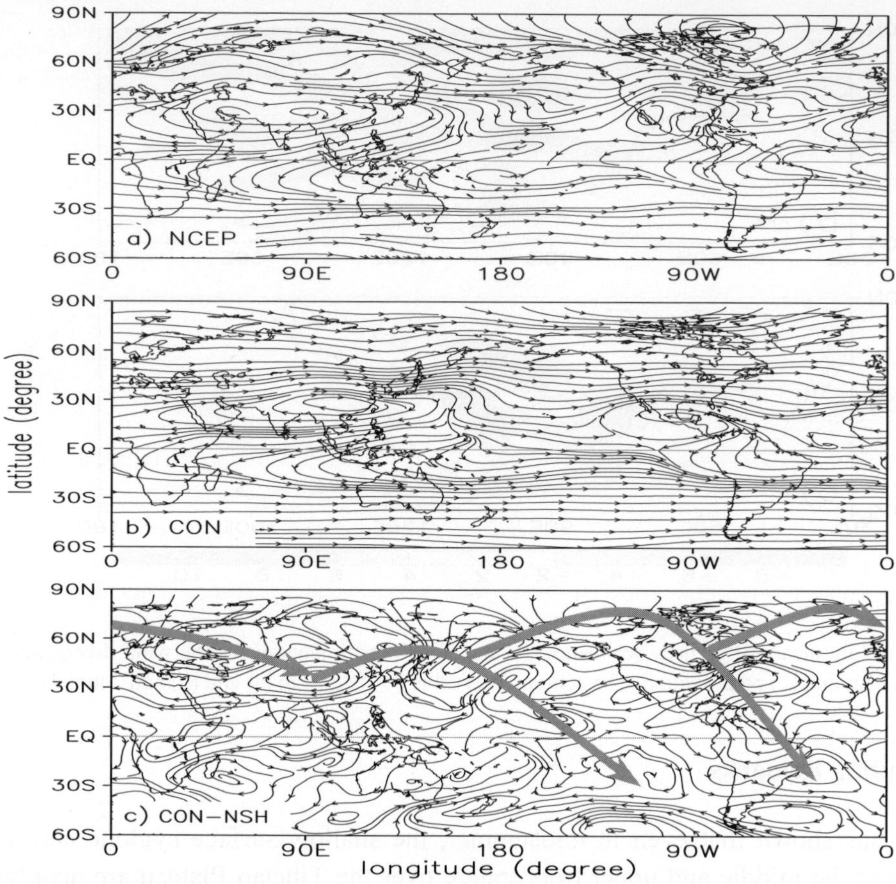

Figure 11. July mean streamlines at 200 hPa. (a) Calculation from the NCEP/NCAR reanalysis average from 1979 to 1988; (b) numerical simulation from the CON experiment; and (c) difference between the CON and NSH experiments. Refer to text for details.

The strong ascent induced by the heating over the plateau works as an air pump. It is mainly sensible heat-driven and is defined as a sensible heat-driven air pump over the Tibetan Plateau. The SHAP affects not only the climate in the neighboring regions but also the circulation anomaly over the Northern Hemisphere.

The Tibetan Plateau is a heat source only in the summer months and is a heat sink in winter. The transition of the thermal feature of the plateau from heat sink to source also has profound impacts on the seasonal transition of the general circulation of the atmosphere. This will be discussed in a separate study.

Acknowledgments. This paper is devoted to the late Professor Zhang Jijia who contributed to the study of the thermal status of the Tibetan Plateau and its impact on climate. We are in debt to Prof. M Yanai for his valuable suggestions. This study was jointly supported by the Chinese Academy of Sciences under the projects ZKCX2-SW-210, the Excellent Ph. D. Thesis Award, and by the Natural Science Foundation of China under the projects 40135020, 40023001 and 40221503.

References

Bolin, B., 1950: On the influence of the earth's orography on the westerlies. *Tellus*, **2**, 184-195.

Charney, J. G., and A. Eliassen, 1949: A numerical method for predicting the perturbations of the middle latitude westerly. *Tellus*, **1**, 38-54.

Eliassen, A. and E. Kleinschmidt, 1957: Dynamic meteorology. *Handb. Phys., Berlin*, **48**, 71-72.

Ertel, H., 1942: Ein neuer hydrodynamische wirbdsatz. *Meteorology. Z. Braunschweig.* **59**, 277-281.

Flohn, H., 1957: Large-scale aspects of the summer monsoon in South and East Asia. *J. Meteor. Soc. Japan*, **75**, 180.

Gill, A. E., 1980: Some simple solutions for heat-induced tropical circulation. *Quart. J. Roy. Meteor. Soc.,* **106**: 447~662.

Haynes, P. H., and M. E. Mclntyre, 1987: On the evolution of vorticity and potential vorticity in the presence of diabatic heating and frictional or other forces. *J. Atmos. Sci.*, **44**, 828-841.

Holton, J. R., and D. Colton, 1972: A diagnostic study of the vorticity balance at 200 mb in the tropics during the northern summer. *J. Atmos. Sci.,* **29**, 1124-1128.

Hoskins, B. J., M. E. Mclntyre, and A. W. Robertson, 1985: On the use and significance of isentropic potential vorticity maps. *Quart. J. Roy. Meteor. Soc.*, **111**, 877-946.

Hoskins, B. J., 1991: Towards a PV-θ view of the general circulation, *Tellus*, **43AB**, 27-35.

IPCC, 2001: Chap. 8, *Climate Change 2001: The Scientific Basis. Contribution of Working Group I to the Third Assessment Report of the Intergovernmental Panel on Climate Change*. Houghton, J. T. et al. ed., Cambridge University Press, Cambridge and New York, 881 pp.

Kalnay, E., M. Kanamitsu, and Coauthors, 1996: The NCEP/NCAR 40-year reanalysis project. *Bull. Amer. Meteor. Soc.,* **77**, 437-471.

Li, G. P., T. Y. Duan, S. Haginoya, et al., 2001: Estimates of the bulk transfer coefficients and surface fluxes over the Tibetan Plateau using AWS data. *J. Meteor. Soc. Japan,* **79**(2), 625~635.

Liu, H., and G.-X. Wu, 1997: Impacts of land surface on climate of July and onset of summer monsoon: A study with an AGCM plus SSiB. *Adv. Atmos. Sci.,* **14(3)**, 289-308.

Liu, X., G.-X. Wu, W. Li, and Y. Liu, 2001: Thermal adaptation of the large-scale circulation to the summer heating over the Tibetan Plateau, *Progress in Natural Science*, **11(3)**, 207-214.

Luo, H. B., and M. Yanai, 1984: The large-scale circulation and heat sources over the Tibetan Plateau and surrounding area during the early summer of 1979. Part II: heat and moisture budgets. *Mon. Wea. Rev.*, **112**, 966-989.

Rodwell, M. R., and B. J. Hoskins, 2001: Subtropical anticyclones and monsoons. *J. Climate*, **14**, 3192-3211.

Shi, G. Y., 1981: An accurate calculation and the infrared transmission function of the atmospheric constituents. Ph.D. Thesis, Dept. of Sci., Tohoku University of Japan. 191 pp.

Tung, W.-W., and M. Yanai, 2002: Convective momentum transport observed during the TOGA COARE IOP. Part I: General features. *J. Atmos. Sci.*, **59**, 1857-1871.

Wu, G. X, 1984: The nonlinear response of the atmosphere to large-scale mechanical and thermal forcing. *J. Atmos. Sci.*, **41**, 2456-2476.

Wu, G. X, H. Liu, Y. Zhao, and W. Li, 1996: A nine-layer atmospheric general circulation model and its performance, *Adv. Atmos. Sci.*, **13**, 1-18.

Wu, G. X, X. Zhang, and Coauthors, 1997: The LASG global ocean-atmosphere-land system model GOALS/LASG and its simulation study. *App. Meteor.,* **8**(spec.), 15-28.

Wu, G. X, W. Li, H. Guo, H. Liu, J. Xue, and Z. Wang, 1998: Sensible heat driven air-pump over the Tibetan Plateau and its impacts on the Asian Summer Monsoon. *Collections on the Memory of Zhao Jiuzhang*, Ye Duzheng ed., Chinese Science Press, Beijing, pp. 116-126.

Wu G. X. and Y. M. Liu, 2000: Thermal adaptation, overshooting, dispersion and subtropical anticyclone, I. Thermal adaptation and overshooting, *Chinese J. Atmos.*, **24**(4), 433-446.

Wu G. X., L. Sun, and coauthors, 2002: Impacts of land surface processes on summer climate. *Selected Papaers of the Fourth Conference on East Asia and Western Pacific Meteorology and Climate*. C. P. Chang et al. ed., World Scientific, Singapore, pp. 64-76.

Xue, Y. K., P. J. Sellers, J. L. Kinter, and J. Shukla, 1991: A simplified biosphere model for global climate studies. *J. Climate*, **4**, 345-364.

Yanai M., C. F. Li, and Z. S. Song, 1992: Seasonal heating of the Tibetan Plateau and its effects on the evolution of the Asian Summer Monsoon, *J. Meteor. Soc. Japan*, **70**, 319-351.

Yanai, M., and C. Li, 1994: Mechanism of heating and the boundary layer over the Tibetan Plateau. *Mon. Wea. Rev.*, **122**, 305-323.

Yang, W, D. Z. Ye, and G. X. Wu, 1992: Diagnosis study on the heating and circulation over the Tibetan Plateau in summer season. III Mechanism for the maintenance of circulation. *Chinese J. Atmos.*, **16(4)**, 409-426.

Ye, D. Z., and C. C. Zhang, 1974: The preliminary annulus simulation of the influence of the heating effect of the Tibetan Plateau on the atmospheric general circulation over East Asia in summer. *Scientia Sinica*, 301-320.

Ye, D. Z., and Y. Gao, 1979: *Meteorology of the Qinghai-Xizang Plateau.* Beijing, Chinese Science Press, 278 pp.

Ye, D. Z., and G.-X. Wu, 1998: The role of the heat source of the Tibetan Plateau in the general circulation. *Meteor. Atmos. Phys.*, **67**, 181.

Yeh, T. C., 1950: The circulation of the high troposphere over China in the winter of 1945-1946. *Tellus*, **2**, 173-183.

Yeh, T. C., S.-W. Lo, and P.-C. Chu, 1957: The wind structure and heat balance in the lower troposphere over Tibetan Plateau and its surroundings. *Acta Meteor. Sinica*, **28,** 108-121.

Yin, M. T., 1949: A synoptic-aerologic study of the onset of the summer monsoon over India and Burma. *J. Meteor.*, **6**, 393-400.

Zhang, J., et al., 1988: *Advances in the Qinghai-Xizang Plateau Meteorology-- The Qinghai-Xizang Meteorology Experiment (QXPMEX, 1979) and Research.* Chinese Science Press, Beijing, 268 pp.

Zhang X. H., K. M. Chen, and Coauthors, 1996: Simulation of thermohaline circulation with a twenty layer oceanic general circulation model. *Theo. Appl. Climatol.*, **55**, 65-88.

Zhang X. H., G. Y. Shi, H. Liu, and Y. Yu, 2000: *IAP Global Ocean-Atmosphere-Land System Model.* Science Press, Beijing, New York, 252 pp.

TROPOPAUSE VARIATIONS IN THE EQUATORIAL REGION

XUELONG ZHOU
Institute for Terrestrial and Planetary Atmospheres
State University of New York at Stony Brook
Stony Brook, NY 11794 USA
E-mail: xzhou@notes.cc.sunysb.edu

(Manuscript received 30 September 2002)

Stratosphere-troposphere exchange is dominated by the meridional circulation, with upward motion in the tropics and downward motion in the polar regions. When tropospheric air enters the stratosphere across the tropical tropopause, it is dehydrated by the cold tropopause temperatures. Thus, the tropical tropopause temperatures are essential for explaining observed distribution pattern and variations of stratospheric water vapor. Investigations about the tropical tropopause also provide information about the coupling between the stratosphere and troposphere in the tropics. In this review article, the author summarizes recent research on the tropical tropopause

The tropical tropopause was found to be influenced by the stratospheric quasi-biennial oscillation (QBO), the tropospheric El Niño southern oscillation (ENSO), and the Madden Julian oscillation (MJO). The MJO in the cold point tropopause temperatures (CPT-T) shows Kelvin wave features in the deep tropics and Rossby wave features in the subtropics. The QBO in the CPT-T is mainly zonally symmetric and is associated with downward propagating temperature anomalies accompanying the QBO meridional circulation. The influence of the ENSO on the tropical tropopause shows east-west dipole and north-south dumbbell features. A cooling trend of the tropical CPT-T was found. This cooling trend could not explain the observed positive trend of stratospheric water vapor if the tropical tropopause temperature were the only factor that determines stratospheric water vapor concentration. The three-dimensional structure of the tropical "cold trap" is also discussed.

1. Introduction

Variations of the tropical tropopause are of interest for quantifying climate variability and for understanding mechanisms of the coupling between the stratosphere and the troposphere. Water vapor experiences phase change at the tropical tropopause. This is an advantage of water vapor compared with other trace gases in studies of the coupling between the stratosphere and the troposphere. Brewer (1949) noted that the distribution of stratospheric water vapor could be explained only if there exists a meridional overturning circulation that transports water vapor upward into the stratosphere through the cold tropical tropopause with poleward and downward transport at high latitudes. Dobson (1956) pointed out that a similar circulation is required to explain the observed distribution of stratospheric ozone. To honor their work, this circulation is called the "Brewer-Dobson" circulation. It should be regarded as a Lagrangian mean circulation and can be approximated by the residual circulation of the transformed Eulerian mean equations (Dunkerton 1978). The source of most stratospheric water vapor is methane oxidation (e.g., LeTexier et al. 1988). Methane oxidation and the transport effect of the Brewer-Dobson circulation are important in determining the climatological pattern of stratospheric water vapor, with more water vapor in the upper stratosphere than in the lower stratosphere, and more in the polar stratosphere than in the tropical stratosphere. Contrary to intuition, the troposphere is a sink of stratospheric water vapor rather than a source. The crucial factor in determining stratospheric water vapor is the cold tropical tropopause temperatures. In summary, three factors control stratospheric water vapor concentration: dehydration at the tropical tropopause, methane oxidation, and the transport effect of the residual circulation.

Because tropospheric air enters the stratosphere mainly through the tropical tropopause, investigations of variations and structure of the tropical tropopause are of fundamental importance for stratospheric water vapor. This paper reviews recent studies on the tropical tropopause. There are several definitions of the tropopause, e.g., lapse rate tropopause, cold point tropopause (CPT), transition layer tropopause, etc. (Highwood and Hoskins 1998). Different tropopause definitions were applied in the literature for different purposes. The CPT definition of the tropical tropopause, the position of the coldest temperature in the tropical vertical temperature profile, was used by the author in studies on tropopause variations, with emphasis on the dehydration effect of the cold tropopause temperatures on stratosphere water vapor.

This paper is organized as follows. In section 2, the seasonal cycle of the tropical tropopause is presented. Intraseasonal variations of the tropical tropopause temperatures are described in section 3. In section 4, signatures of the quasi-biennial oscillation (QBO) and the El Niño and southern oscillation (ENSO) are reviewed. In section 5, the trend of the tropical tropopause temperature is discussed. Preliminary discussion about the structure of the "cold trap" is presented in section 6. The last section is the summary.

2. Seasonal Cycle of Tropical CPT Temperatures

Failure of "freeze-drying" at the mean tropopause temperature to dehydrate the stratosphere to the observed water vapor mixing ratio led Newell and Gould-Stewart (1981) to propose the "stratospheric fountain" hypothesis. They showed that sufficiently cold temperatures were frequently found over the maritime continent warm pool region in January and were also found, though less frequently, over the Indian summer monsoon region during July. They suggested that tropospheric air enters the stratosphere mostly at these preferred times and locations. This is the stratospheric fountain hypothesis. Atticks and Robinson (1983) were the first to look at the Brewer hypothesis by examining tropical radiosondes. Frederick and Douglass (1983) extended the analysis of Newell and Gould-Stewart using 8 years of rawinsonde data from eight tropical stations. Their results are generally consistent with those of Newell and Gould-Stewart (1981).

Figure 1 shows the tropical CPT temperatures (CPT-T) in January and July averaged over the 15-year European Center for Medium Range Weather Forecasts (ECMWF) reanalysis period (1973-1993). The CPT was obtained by applying cubic spline fitting to ECMWF reanalysis daily data. Comparison between the CPT calculation based on reanalysis and high-resolution sounding data indicated that the ECMWF-based CPT has a warm bias about 2 K and the CPT based on National Centers for Environment Prediction (NCEP) reanalysis data has a warm bias about 4 K (Zhou 2000; Randel et al. 2000). However, the bias is almost longitudinally uniform and exists mainly in the seasonal cycle. The western Pacific, the stratospheric fountain area, is the coldest region in the tropics. In summer, the cold CPT area extends to the region over Tibet plateau though the coldest region is still over the western Pacific. This is an indication of the influence of the summer monsoon on the tropical tropopause. It can be seen in Figure 1 that the tropical CPT is about 4 K colder in winter than in summer (Zhou 2000). Consistently, the CPT is higher in altitude and lower in pressure in winter than in summer (figures not shown). This annual cycle is mainly caused by the annual variation of the extratropical wave forcing (Yulaeva et al. 1994) and is consistent with the "tape recorder" signal in stratospheric water vapor (Mote et al. 1996). However, this consistency does not mean that the air parcel across the tropical CPT is dehydrated at the monthly and/or zonal mean temperatures. Generally, the monthly and/or zonal tropopause temperature is too warm to dehydrate water vapor content to the entry value of the mixing ratio (Zhou et al. 2001a). Figure 1 also indicates a zonally asymmetric feature, i.e., the western Pacific is colder than other regions. This is a result of zonal asymmetry in the distribution of diabatic heating (Highwood and Hoskins 1998).

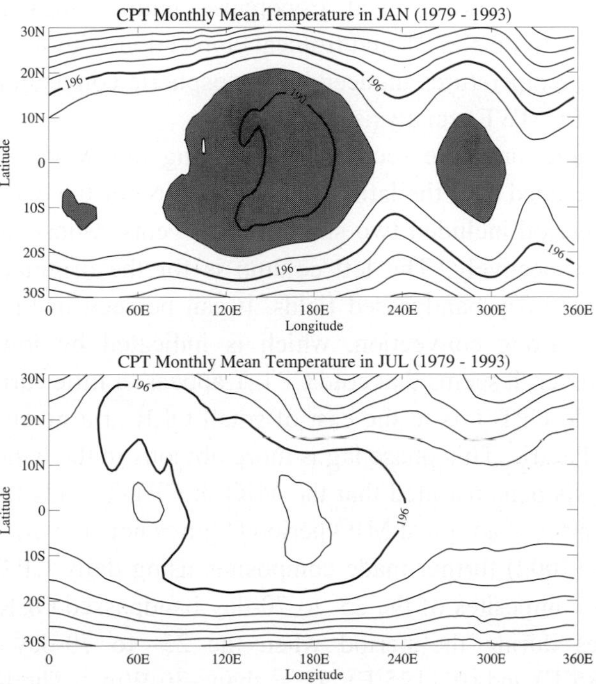

Figure 1. Monthly mean CPT temperature (top, January; bottom, July) averaged over the ECMWF reanalyses period (1979-1993). The contour interval is 2 K. Areas with values below 192 K are shaded. Note that an overestimate of about 2 K in ECMWF-based CPT-T calculations was not deducted (after Zhou et al. 2001a).

Freeze-drying to the saturation mixing ratio characteristics of the exceptionally cold tropopause temperature over the western Pacific is obviously required to explain the observed low stratospheric water vapor concentration. Such dehydration would appear to require slow upwelling in this region to produce condensation and allow ice crystals to have enough time to sediment. However, wind analysis suggested that the mean vertical wind might be downward (Sherwood 2000; Gettelman et al. 2000). To resolve this paradox, Holton and Gettelman (2001) suggested a new term, "cold trap," to replace the term "fountain" and proposed a mechanism model. In their model, air parcels, which may first enter the tropopause transition layer at longitudes other than the western Pacific, encounter the cold trap and get dehydrated due to fast zonal motion.

3. Intraseasonal Variations of the Tropical CPT Temperature

The Madden Julian oscillation (MJO) is a major component of tropical low-frequency variations. The oscillation is a result of large-scale circulation cells oriented in the equatorial plane that move eastward from the Indian Ocean to the central Pacific Ocean (Madden and Julian 1994). The oscillation has a range of periods of about 30-60 days and is manifest in eastward-propagating complex convective regions. Zhou and Holton (2002) analyzed the influence of the MJO on the tropical tropopause temperatures and documented the relationship between the tropical CPT and tropical convection with focus on wave patterns in the tropical tropopause temperatures induced by tropospheric convection, using daily tropical CPT derived from the ECMWF reanalysis (1979-1993).

Figure 2 shows time-longitude sections of outgoing longwave radiation (shaded) and CPT-T (contoured) averaged over the latitude band 5°S-5°N for the period November 1992 to February 1993. This period included two strong MJO events, whose start points are labeled by letters A and B, respectively. The left column is for the original fields, and the right column for the 25- to 70-day bandpassed fields. It can be seen in Figure 2a that CPT-T is colder when there is strong convection, which is indicated by low outgoing longwave radiation (OLR). However, it seems that cold CPT-T appears a little earlier than low OLR at a given longitude, or cold CPT-T is to the east of small OLR at a given time from the Indian Ocean to the western Pacific. This phase lag is more obvious in the bandpassed fields (Figure 2b). Correlation analysis demonstrated that the MJO in CPT-T leads the MJO in OLR by 8-12 days, about one quarter of a typical MJO period (figures not shown).

Zhou and Holton (2002) further made composites using daily CPT-T data during 1979-1993. Figure 3 shows composites of the 25- to 70-day bandpassed OLR (shaded) and CPT-T (contoured) anomalies during the period when the 25- to 70-day bandpassed OLR at reference points (0°, 85°E) and (0°, 125°E) is less than -30 Wm^{-2}. The Kelvin wave pattern is obvious in the deep Tropics, and there is a Rossby wave pattern in the subtropics. The maximum CPT-T anomalies lead the OLR anomalies by one-quarter wavelength over both the Indian Ocean and the western Pacific. Hendon and Salby (1996) simulated the composite life cycle of the MJO using the diabatic heating that was prescribed from the anomalous OLR in the life cycle of the MJO constructed by Hendon and Salby (1994). The composites of

CPT-T shown in Figure 3 show patterns, if the sign is reversed, generally similar to the temperature patterns on day 5 and day 15 shown in Figure 2 of Hendon and Salby (1996), which is for simulated tropospheric mean temperature anomalies.

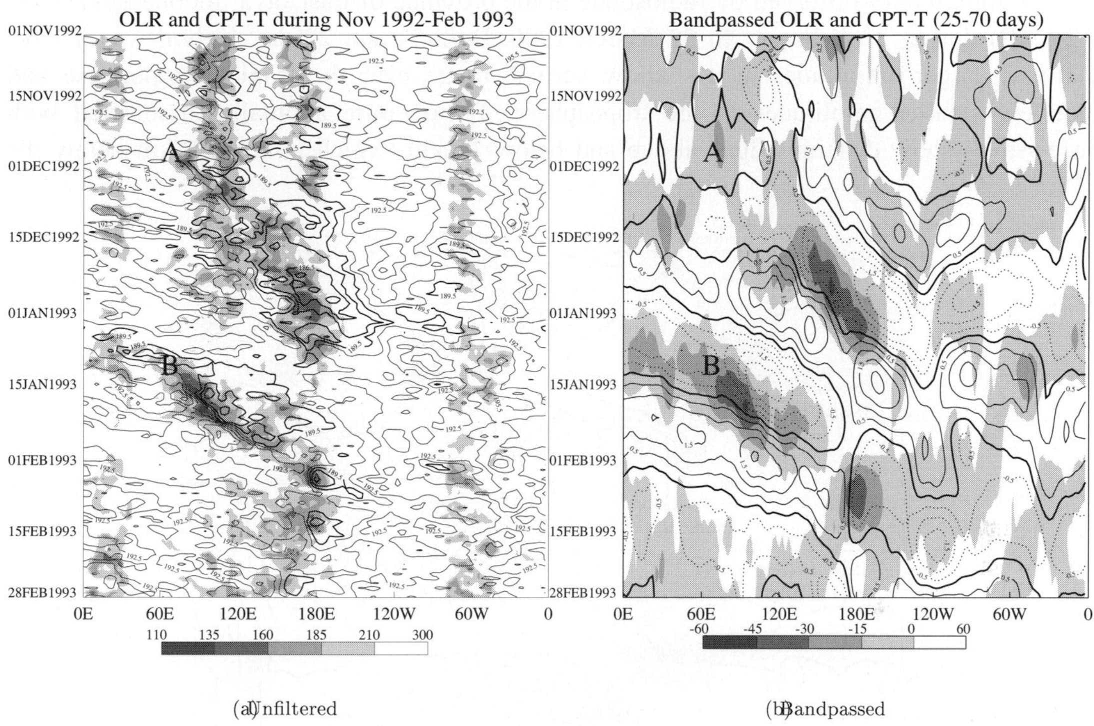

Figure 2. Time-longitude sections of (a) unfiltered and (b) 25- to 70-day bandpassed OLR (shaded) and CPT-T (contoured) averaged over the latitude band 5°S-5°N. The shading scale for OLR (Wm^{-2}) is indicated by the bar at the bottom. The contour interval for CPT-T is 1.5 K in a and 0.5 K in b. Contours less than 190 K are indicated by thick curves in a (after Zhou and Holton 2002).

Zhou and Holton (2002) also analyzed the 6- to 25-day variation, which is an important component of tropical intraseasonal variations (Vincent et al. 1998). The 6- to 25-day variation in CPT-T leads the 6- to 25- day variation in OLR by one-quarter period, and there are indications of a Kelvin wave in the deep tropics and a Rossby wave in composites like Figure 3.

Wheeler et al. (2000) found large tropopause temperature perturbations associated with equatorial waves coupled to convection. These signals lead those in OLR for the convectively coupled Kelvin and eastward and westward inertio-gravity waves and could be interpreted as a vertically propagating response excited by a moving equatorial heat source. Using radiosonde and the NCEP reanalysis data and the lapse-rate tropopause definition, Kiladis et al. (2001) showed upward- and eastward-tilting temperature anomalies, indicating that the tropopause temperature disturbance leads the convection center.

The Kelvin wave pattern in the tropical CPT is also consistent with some sounding observations. Boehm and Verlinde (2000) used sounding data observed in the summer of 1999 at the Republic of Nauru to find that the tropical tropopause is affected by stratospheric waves with a timescale of several days and amplitude of up to 8 K in temperature near the tropopause. These waves have been identified as Kelvin waves, which have their origin in the troposphere (Holton et al. 2001). Tsuda et al. (1994) reported Kelvin waves with periods of about 7 and 20 days observed by radiosonde in the province of East Java, Indonesia (7.57°S, 112.68°E) during 27 February to 22 March 1990. Both 7- and 20-day Kelvin waves have relatively larger amplitudes in a shallow vertical range near the tropical tropopause and relatively smaller amplitudes in the troposphere. Temperature anomalies associated with these waves show different signs above and below 15 km, which is about 2 km below the CPT.

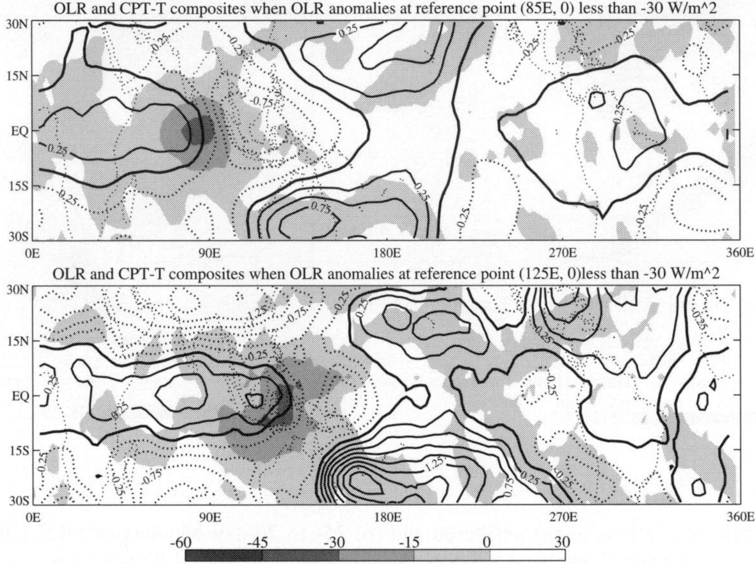

Figure 3. Composites of 25- to 70-day bandpassed OLR (shaded) and CPT-T (contoured) anomalies during the period when the 25- to 70-day bandpassed OLR at reference points (top) (0°, 85°E) and (bottom) (0°, 85°E) are less than –30 Wm^{-2}. The shading scale for OLR anomalies is indicated by the bar at the bottom. Contour interval for the CPT-T composites is 0.25 K (after Zhou and Holton 2002).

4. Interannual Variations of the Tropical CPT Temperatures

The stratospheric QBO and the tropospheric ENSO are two of the dominant interannual variations in the tropics. Some features and the influence of the QBO and ENSO on the tropical tropopause were reported before the year 2000, limited to sparse tropical stations (e.g., Reid and Gage 1981, 1985, 1996; Frederick and Douglass 1983; Gage and Reid 1987). Only recently were features of tropopause over the entire tropics revealed (Randel et al. 2000; Zhou 2000; Zhou et al. 2001a, 2001b; Seidel et al. 2001).

4.1. *QBO and ENSO Signatures in the Tropical CPT*

To extract the QBO and ENSO signatures in the tropical CPT, Zhou et al. (2001b) used the zonal wind shear at 50 mb (difference between the winds at 40 mb and 70 mb) over

Singapore and the sea surface temperature anomaly (SSTA) in the Niño3.4 region as reference indices in the bivariate regression. Mathematically, the SSTA and the QBO are not independent due to the overlap in their frequency domain. A QBO in SST has been noted although it is not clear whether it is physically related to the stratospheric QBO (Xu 1992; Meehl 1993; Geller et al. 1997). In order to separate the influences of the stratospheric QBO and the tropospheric ENSO on the tropical CPT, Zhou et al. (2001b) made two independent QBO and SSTA time series using a Butterworth bandpassed filter (Hamming 1989). The response function of the bandpass filter has the value 1 at a period of 28 months, and ½ at a period of about 22 and 34 months. This filter was used to extract the quasi-biennial variability from the wind shears at 50 mb over Singapore. For the SSTA, the bandpassed time series was subtracted from the SSTA. The above process can be expressed by the following

$$QBO(t)' = QBO(t)_{22-34m}, \tag{1}$$

$$SSTA(t)' = SSTA(t) - SSTA(t)_{22-34m}. \tag{2}$$

Zhou et al. (2001b) found that the CPT is simultaneously correlated to the ENSO and that the CPT has the strongest correlation to the QBO at a time lag. The westerly wind shears at 50 mb lead warm CPT-T anomalies by about 6 months. Taking this time lag into account, the bivariate regression can be written as follows:

$$CPTT(t) = a \times QBO(t-6)' + b \times SSTA(t)' + \gamma(t), \tag{3}$$

in which *CPTT* is the CPT-T anomalies (subtract the seasonal cycle from the monthly CPT-T) and γ is the residual. Figure 4 shows the composites of CPT-T anomalies associated with the QBO, based on regressed time series, i.e., $a \times QBO(t-6)'$, which is adjusted in time. The zonally symmetric features of the QBO in CPT properties are obvious. During the westerly shear period, the composite CPT is warmer by 0.2 to 0.3 K, and during the easterly shear conditions it is colder by 0.2 to 0.4 K. There seems to be a stationary wave with wavenumber 2 superposing on a zonally homogeneous field. For example, there are minima over the western Pacific and southern America. However, this asymmetry might be due to the fact that the influence of the QBO and ENSO on the tropical CPT cannot be perfectly separated. Figure 5 shows the composites of the CPT-T anomalies for two El Niños and La Niñas that occurred during 1979–1993 (the two El Niños occurred during April 1982–July 1983 and August 1986–February 1988, and the two La Niñas during September 1984–June 1985 and May 1988–January 1989). Generally, the temperature anomalies associated with the ENSO are larger than those associated with the QBO. Figure 5 shows three distinct features. First, there is an east–west (E–W) dipole over the tropical Pacific. The second feature is that there are three north–south (N–S) dumbbells in the tropics. The strongest dumbbell is over the central to eastern Pacific. One dumbbell pattern is clear over the Atlantic, and similar features are also seen over the eastern Indian Ocean and the western Pacific, though less clearly. The third feature is that there is a maximum (in absolute value) over the equatorial western Pacific. These three features can be explained by changes of tropical convection activity associated with ENSO, as will be discussed. Similar features of the CPT variability associated with QBO and ENSO also can be seen in the pressure and height composites (figures not shown), although those features are not as clear as in the temperature composites. Similar results were found by Randel et al. (2000) using the NCEP reanalyses and lapse rate tropopause definition.

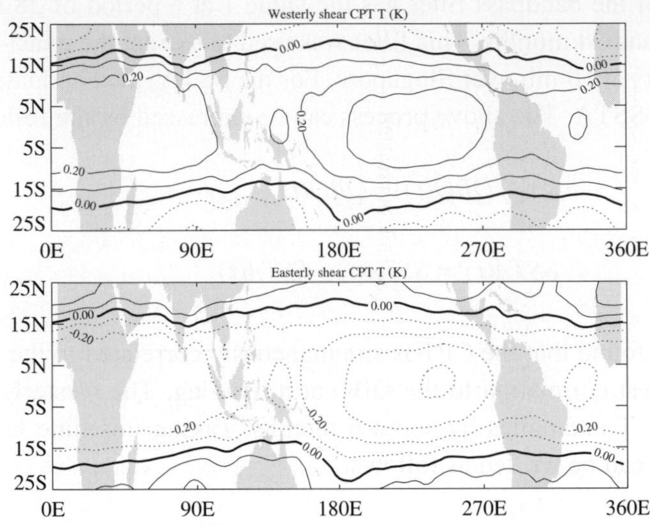

Figure 4. Composite analysis of CPT-T under westerly (upper) and easterly (lower) stratospheric zonal wind shear conditions. Note that the shears lead the CPT-T by 6 months. Contour interval is K (after Zhou et al. 2001b).

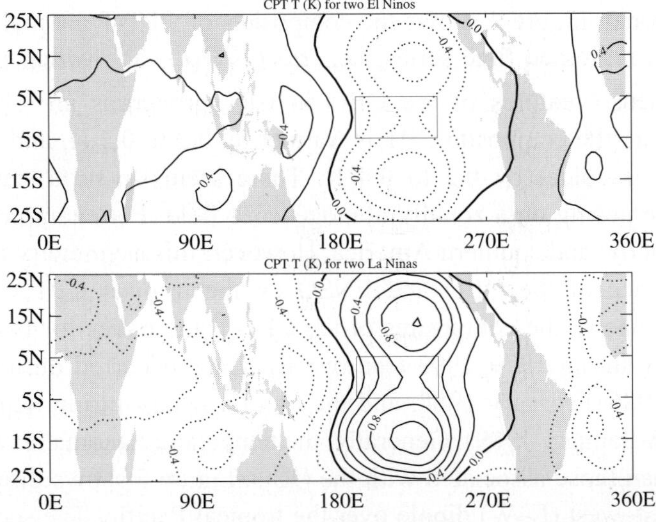

Figure 5. Composite analysis of CPT-T during two ENSO events that occurred during 1979–1993. See the text for the selection of ENSO events. Square boxes show the Niño3.4 region. Contour interval is K (after Zhou et al. 2001b).

Geller et al. (2002) used a two-dimensional model to simulate the interannual variability of stratospheric water vapor. They indicated that the QBO modulation of stratospheric water vapor results from two causes. Dynamical redistribution of water vapor from the QBO-induced mean meridional circulation dominates the observed variability in the middle and upper stratosphere. In the lower tropical stratosphere, the QBO water vapor variability is dominated by a "tape recorder" that results from the dehydration signal accompanying the QBO variation of the tropical CPT. It is suggested that another low-frequency tape recorder exists as a result of ENSO modulation of the tropical CPT.

4.2. *Mechanism for the QBO Signature in the Tropical CPT*

Tropical stratospheric QBO is a significant low-frequency process and has been intensively investigated for decades (e.g., Reed et al,. 1961; Veryard and Ebdon 1961; Plumb and Bell 1982; Naujoket 1986). Zhou et al. (2001b) indicated that the westerly shears at 50 mb, which are accompanied by warm temperature anomalies (Plumb and Bell 1982), lead the tropical CPT-T by about 6 months and are positively correlated with the tropical CPT-T. It takes about 3–4 months for the westerly shear at 50 mb to reach 100 mb, and takes about 7 months for the easterly shear to propagate from 50 mb to 100 mb (Naujoket 1986), which gives an average of about 5–6 months. This time lag is consistent with the analyses by Zhou et al. (2001b). The amplitude of QBO temperature perturbation can be estimated using

$$f_0\bar{u}_z + H^{-1}R\bar{T}_y = 0\,, \tag{4}$$

which is equation (7.1.1d) of Andrews et al. (1987). Given a wind change over a scale height $\Delta\bar{u} = 10$ m/s and a latitudinal scale L = 1000 km, Eq. (4) gives an estimate of about 0.79 K for temperature change over a distance L. The amplitude of the CPT-T anomalies associated with the QBO is about 0.3–0.5 K (Figure 4). Consistency in time lag and amplitude suggests that the QBO signature in the tropical CPT temperatures is probably due to the stratospheric QBO temperature anomalies that accompany the downward-propagating QBO meridional circulation (Plumb and Bell 1982).

4.3. *Mechanism for the ENSO Signature in the Tropical CPT*

Yulaeva and Wallace (1994) investigated the signature in global temperature and precipitation fields derived from the Microwave Sounding Unit (MSU). The ENSO-associated temperature anomalies in the troposphere (MSU-2) and the lower stratosphere (MSU-4) are consistent with the ENSO signature in the tropical CPT, in terms of the shape and positions of the E–W dipole and N–S dumbbells and the maximum anomaly over the western Pacific. The ENSO has a coherent vertical structure in the tropical atmospheric temperature that reverses sign at certain altitude below the CPT. For instance, during El Niño events, the tropical troposphere (1000- to 200-mb layer) over the eastern Pacific is warmer and the tropopause and the lower stratosphere is colder. Zhou and Sun (1994) extended the Gill model (Gill 1980), including a cooling over the western domain and a heating over the eastern domain to study the tropical tropospheric nonlinear steady response to two heating sources of contrasting nature. OLR observations indicate positive anomalies (negative diabatic heating anomalies) over the eastern Indian Ocean and the western Pacific and negative anomalies (positive diabatic heating anomalies) over the central to eastern Pacific (Yulaeva and Wallace 1994). Thus, the distribution of the two idealized heat sources considered by Zhou and Sun (1994) represents the longitudinal distribution of major diabatic heating anomalies in the tropics during El Niño events. The steady geopotential height response at the top of the model domain to these two heat sources of contrasting nature shows that there is a positive dumbbell over the eastern-half domain and a negative dumbbell over the western-half domain. These positive and negative dumbbells form an E–W (positive–negative) dipole. In addition, there is a negative maximum geopotential height anomaly to the east of the western cooling source, and this maximum anomaly is a result of wave–wave interaction (Zhou and Sun 1994). Thus, the geopotential height response to W–E negative-positive paired heating sources is hydrostatically consistent with Figure 5a, which shows the El Niño associated tropical CPT-T anomalies. However, the response of the temperature anomaly was assumed to have the same vertical structure as the heating (Gill 1980; Zhou and Sun 1994). The Gill model needs to be extended further to explain the ENSO signature in the tropical CPT-T more reasonably though results from this model have shown that the E–W dipole feature in the ENSO signature in the tropical CPT is caused by the changes in tropical convection during ENSO events, the N–S dumbbell feature is due to rotation of the Earth, and the maximum over the equatorial western Pacific is caused by nonlinear wave-wave interaction (Zhou and Sun 1994).

5. Long-term Trend of the Tropical CPT Temperatures

Zhou et al. (2001a) analyzed the trend of the tropical CPT using tropical daily sounding data. Figure 6 shows tropical monthly mean CPT-T during 1973–1998. A cooling trend (–0.57 ± 0.06 K per decade during 1973–1998) in the tropical CPT-T was found. This number is close to the cooling rate (–0.6 K/decade) found by Simmons et al. (1999) in the global mean 100 mb temperatures. Independent calculation by Seidel et al. (2001) showed a cooling of about –0.5 K/decade at the tropical tropopause. Zhou et al (2001) also calculated the trends for each tropical region and found that the trend of the CPT over the western Pacific is about –0.33 ± 0.08 K/decade, and the trend is a little stronger in other regions.

The trend may be influenced by inhomogeneities in the sounding data. To examine the impact of the inhomogeneity problem, Zhou et al. (2001a) included a few binary terms in the regression equation corresponding to changes in the station log and calculated the trend station by station. They found that nearly all tropical stations have a cooling trend. The trend is sensitive to the inhomogeneity of the sounding data (Gaffen et al. 2000). However, no statistically significant warming trend has been found.

Zhou et al. (2001a) proposed that the cooling of the tropical tropopause might be caused by changes in tropical convection. If convection occurs more frequently and/or is more intense than in earlier years, then the tropical CPT tropopause will be getting higher and colder. Calculations indicated a warming trend in SST and a negative trend in OLR almost everywhere in the tropics (Waliser and Zhou 1997). As mentioned above, the tropical tropopause is cooling almost everywhere in the tropics. These three trends are consistent with one another. The warming SST tends to destabilize the static stability of the troposphere, and convection will occur more frequently and/or be more intense. Convective clouds will reach higher altitude or cover a larger area. As an indication of this, the OLR shows a negative trend. Waliser and Zhou (1997) indicated that the negative OLR trend corresponds closely to the positive trend observed directly from the *in situ* rainfall observation. Stronger convection and more precipitation produce larger diabatic heating in the tropics, which forces a higher tropopause, so the pressures and temperatures of the tropical tropopause become lower and colder.

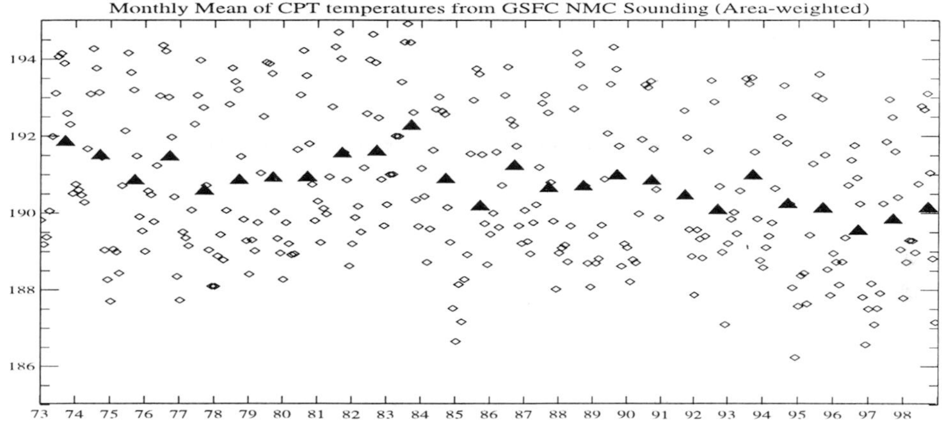

Figure 6. Monthly mean temperatures over the entire tropical cold point tropopause (CPT) in 1973–1998 calculated from operational sounding profiles (diamonds) and their annual means (triangles). Temperature is in Kelvin (after Zhou et al. 2001a).

Observations showed a positive trend in stratospheric water vapor in the past decades (Rosenlof et al. 2001). Three factors influence stratospheric water vapor: methane oxidation, dehydration at the tropical tropopause, and transport by the meridional circulation. The increase in methane emission is too small to explain the observed trend of stratospheric water vapor (Nedoluha et al. 1998). Thus, the cooling trend in the CPT temperatures implies that the positive trend in stratospheric water vapor is probably a result of changes in the dynamic circulation. For instance, if more tropospheric air enters the stratosphere through the tropopause in the subtropics where the temperatures are warmer than in the deep tropics, the water vapor flux into the stratosphere would be larger.

6. Spatial Structure of the "Cold Trap"

In section 2, it was pointed that a paradox exists about the western Pacific. Air moving from the troposphere into the stratosphere needs to be dehydrated in this region, but the wind analysis indicated a downward motion in the region. To solve this paradox, Holton and Gettelman (2001) proposed a mechanism model in which the zonally moving air encounters the "cold trap" in response to radiative cooling of subvisible cirrus lying above deep convective anvil clouds. According to this mechanism, the three-dimensional structure of the cold trap and its variations are important.

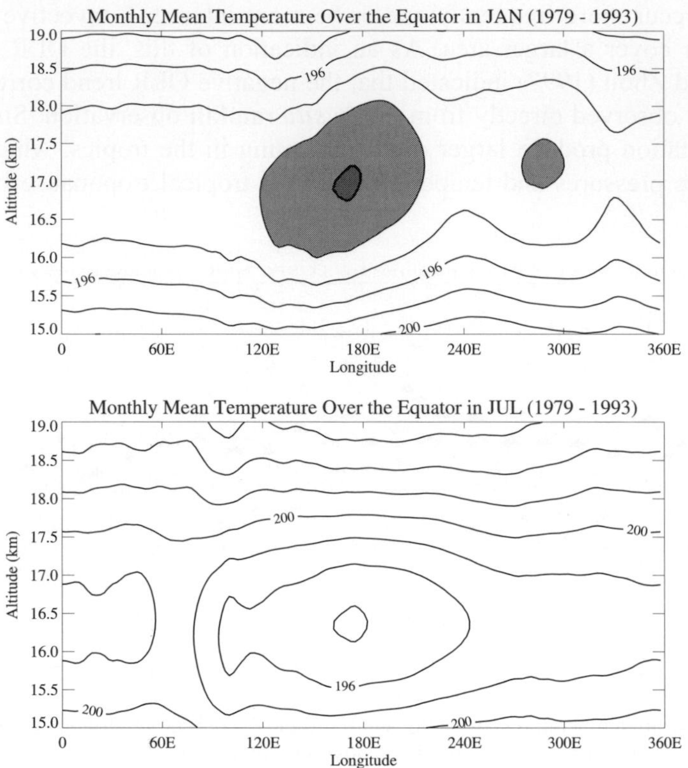

Figure 7. Vertical sections of monthly mean temperature near the tropical tropopause (upper, January; lower, July) averaged over the ECMWF reanalyses period (1979–1993). The contour interval is 2 K. Areas with values below 192 K are shaded.

Figure 7 shows a vertical section of the 15-year mean temperature in January and July over the equator, which was based on the ECMWF reanalysis monthly temperature field. Cubic spline fitting was applied to obtain temperatures near the tropopause. It can be seen that the cold trap is about 4 K colder in winter than in summer and 0.5 km higher in winter. The horizontal section of the temperature field at a given altitude near the tropopause, e.g., 17 km in winter, looks like Figure 1. The saturation water vapor mixing ratio near the tropopause strongly depends on the temperature and slightly depends on the pressure. If we define the cold trap to be the volume enclosed by the surface of 190 K, one can conceive the general structure of the wintertime cold trap by combining Figures 1a and 7a. In Figures 1a and 7a, values below 192 K instead of below 190 K were shaded since the ECMWF reanalysis overestimates the tropopause temperatures by about 2 K. The height of the cold trap is about 1.5 km. The upwelling velocity in the tropical lower stratosphere is about 0.2 to about 0.4 mm/s (Plumb and Eluszkiewicz 1999). It takes about 2 months for an air parcel to climb 1.5 km in the tropical lower stratosphere. During this period, the air parcel travel about

1.3 rounds along the equator, assuming a zonal wind speed of 10 m/s. Thus, most air parcels, which enter the stratosphere at regions other than the western Pacific, have a chance to encounter the cold trap. Air parcels that enter the stratosphere about 1 month earlier also can encounter the cold trap and get dehydrated.

Zhou et al. (2003) found that dehydration regions with extremely cold temperatures and large sizes occur when cold temperature anomalies associated with the QBO arrive at the tropical tropopause layer in wintertime while the tropical tropopause layer is at the coldest phase of the annual cycle and under La Niña conditions. La Niña events have a more dramatic influence on the dehydration than El Niño events.

7. Summary

Much progress about the tropical tropopause has been made in recent years. It includes progress in studies of the seasonal cycle, intraseasonal variations, interannual variations, and the long-term trend of the tropical tropopause. Tropical convection plays an important role in tropopause variations. For the timescale of a few hours, Teitelbaum et al. (2000) found an almost simultaneous response of the tropical CPT to strong convection in limited observations. For the intraseasonal timescale, the CPT temperature anomalies in the deep tropics lead the OLR anomalies by about one quarter major period, and are likely the Kelvin wave excited by convection that occurs to their west (Zhou and Holton 2002). The annual cycle of the tropopause is mainly driven by the extratropical stratospheric wave forcing (Yulaeva et al. 1994), but the zonal asymmetry of the tropical tropopause is attributed to the asymmetry of the distribution of tropical convection (Highwood and Hoskins 1998). At interannual timescales, the tropical tropopause is affected by the QBO and the ENSO. The signature of the QBO in the tropopause is mainly zonally symmetric, but the ENSO signature shows dipole and dumbbell patterns (Zhou 2000; Randel et al. 2000; Zhou et al. 2001a). These patterns are believed to be stationary Rossby and Kelvin waves associated with tropical convection (Zhou et al. 2001b). Zhou et al. (2001a) found a cooling trend in the tropical CPT, and this trend is consistent with trends in tropical SST, OLR, and *in situ* precipitation. It was suggested that the trend in the tropical tropopause is caused by changes in tropical convection.

The general picture of stratosphere-troposphere exchange (STE) is clear. Tropospheric air enters the stratosphere across the tropical tropopause where it is dehydrated by the cold temperature. The air is transported to high altitudes and high latitudes by the residual circulation, while methane oxidation increases water vapor concentration. Finally, it returns to the troposphere at the polar region, where polar stratospheric air experiences a second dehydration as a result of the cold temperature inside the polar vortex. However, our knowledge about STE is far from complete. Many uncertainties lie mainly in the tropical tropopause region. For simplicity, the tropical tropopause is often treated as if it were a surface without thickness. In fact, the tropopause is better thought of as a transition layer (Highwood and Hoskins 1998; Sherwood and Dessler 2001). Studies about the transition layer tropopause are very limited due to its complexity. However, understanding the structure and variations of the cold trap and many processes (sedimentation and/or re-evaporation of ice crystals, cooling by overshooting convective turrets, etc.) in the transition layer is crucial for studies on STE.

Acknowledgments. This research is supported by NASA under SAGE grant NAS 196006 to SUNY at Stony Brook.

References

Andrews, D. G., J. R. Holton, and C. B. Leovy, 1987: Middle atmospheric dynamics, Academic Press.

Atticks, M., and G. Robinson, 1983: Some features of the structure of the tropical tropopause, *Quart. J. Roy. Meteor. Soc.,* **109**, 295-308.

Boehm, M. T., and J. Verlinde, 2000: Stratospheric influence on upper tropospheric tropical cirrus, *Geophys. Res. Lett*, **27**, 3209-3212.

Brewer, A. M., 1949: Evidence for a world circulation provided by the measurements of helium and water vapor distribution in the stratosphere, *Quart. J. Roy. Meteor. Soc.,* **75**, 351-363.

Dobson, G. M. B., 1956: Origin and distribution of polyatomic molecules in the atmosphere, *Pro. R. Soc. London, Ser.* **A236**, 187-193.

Dunkerton, T. J., 1978: On the mean meridional mass motions of the stratosphere and mesosphere, *J. Atmos. Sci.*, **35**, 2325-2333.

Frederick, J. E., and A. R. Douglass, 1983: Atmospheric temperatures near the tropical tropopause: Temporal variations, zonal asymmetry and implications for stratospheric water vapor, *Mon. Wea. Rev.*, **111**, 1397-1401.

Gaffen, D. J., M. A. Sargent, R. E. Habermann, and J. R. Lanzante, 2000: Sensitivity of the tropospheric and stratospheric temperature trends to radiosonde data quality, *J. Climate*, **13**, 1776-1796.

Gage, K. S., and G. C. Reid, 1987: Longitudinal variations in tropical tropopause properties in relation to tropical convection and ENSO events, *J. Geophys. Res.*, **92**, 14197-14203.

Geller, M. A., W. Shen, M. Zhang, and W. Tan, 1997: Calculation of the stratospheric quasi-biennial oscillation for time-varying forcing, *J. Atmos. Sci.*, **54**, 883-894.

Geller, M. A., X.-L. Zhou, and M. Zhang, 2002: Simulations of the interannual variability of stratospheric water vapor, *J. Atmos. Sci.*, **59**, 1076-1085.

Gettelman, A., A. R. Douglass, and J. R. Holton, 2000: Simulations of water vapor in the upper troposphere and lower stratosphere, *J. Geophys. Res.*, **105**, 8317-9003-9023.

Gill, A. E., 1980: Some simple solutions for heat induced tropical circulation, *Quart. J. Roy. Meteor. Soc.*, **106**, 447-462.

Hamming, R. W., 1989: Digital filter, Prentice-Hall, Inc., 3rd ed., p. 284.

Hendon, H. H., and M. L. Salby, 1994: The life cycle of the Madden-Julian oscillation, *J. Atmos. Sci.*, **51**, 2225-2237.

Hendon, H. H., and M. L. Salby, 1996: Planetary-scale circulations forced by intraseasonal variations of observed convection, *J. Atmos. Sci.,* **53**, 1751-1758.

Highwood, E. J., and B. J. Hoskins, 1998: The tropical tropopause, *Quart. J. Roy. Meteor. Soc.*, **124**, 1579-1604.

Holton, J. R., and A. Gettelman, 2001: Horizontal transport and the dehydration of the stratosphere, *Geophys. Res. Lett.,* **28,** 2799-2802.

Holton, J. R., M. J. Alexander, and M. T. Boehm, 2001: Evidence for short vertical wavelength Kelvin waves in the DOE-ARM Nauru99 radiosonde data, *J. Geophys. Res*., **106**, 20 125-20 129.

Kiladis, G. N., K. H. Straub, G. C. Reid, and K. S. Gage, 2001: Aspects of interannual intraseasonal variability of the tropopause and lower stratosphere, *Quart. J. Roy. Meteor. Soc.*, **127**, 1961-1984.

LeTexier, H. S., H. S. Solomon, and R. R. Garcia, 1988: The role of molecular hydrogen and methane oxidation in water vapor budget of the stratosphere, *Quart. J. Roy. Meteor. Soc.*, **114**, 281-295.

Madden, R. A., and P. R. Julian, 1994: Observations of the 40-50 day tropical oscillation — A review, *Mon. Wea. Rev*., **122**, 814-837.

Meehl, G. A., 1993: A coupled air-sea biennial mechanism in the tropical Indian and Pacific regions: role of the ocean, *J. Climate*, **6**, 31-40.

Mote, P. W., K. H. Rosenlof, M. E. McIntyre, E. S. Carr, J. C. Gille, J. R. Holton, J. S. Kinnersley, H. C. Pumphrey, J. M. Russell III, and J. W. Waters, 1996: An atmospheric tape recorder: The imprint of tropical tropopause temperatures on stratospheric water vapor, *J. Geophys. Res*., **101**, 3989-4006.

Naujokat, B., 1986: An update of the observed quasi-biennial oscillation of the stratospheric winds over the Tropics. *J. Atmos. Sci*., **43**, 1873-1877.

Nedoluha, G. E., R. M. Bevilacqua, R. M. Gomez, D. E. Siskind, B. C. Hicks, J. M. Russell III, and B. J. Conner, 1998: Increases in middle atmospheric water vapor observed by the Halogen Occultation Experiment and the ground-based water vapor millimeter-wave spectrometer from 1991-1997, *J. Geophys. Res*., **103**, 3531-3543.

Newell, R. E., and S. Gould-Stewart, 1981: A stratospheric fountain, *J. Atmos. Sci*., **38**, 2789-2795.

Plumb, R. A., and R. C. Bell, 1982: Equatorial waves in steady zonal shear flow, *Quart. J. Roy. Meteor. Soc.*, **108**, 313-334.

Plumb, R. A., and J. Eluszkiewicz, 1999: The Brewer-Dobson circulation: dynamics of the tropical upwelling, *J. Atmos. Sci*., **56**, 868-890.

Randel, W. J., F. Wu, and D. J. Gaffen, 2000: Interannual variability of the tropical tropopause derived from radiosonde data and NCEP reanalyses, *J. Geophys. Res*., **105**, 15 509-15 523.

Reed, R. J., W. J. Campbell, L. A. Rasmussen, and D. G. Rogers, 1961: Evidence of downward-propagating annual wind reversal in the equatorial stratosphere, *J. Geophys. Res*., **66**, 813-818.

Reid, G. C., and K. S. Gage, 1981: On the annual variation in eight of the tropical tropopause, *J. Atmos. Sci.*, **38**, 1928-1938.

Reid, G. C., and K. S. Gage, 1985: Interannual variation in the height of the tropical tropopause, *J. Geophys. Res*., **90**, 5629-5635.

Reid, G. C., and K. S. Gage, 1996: The tropical tropopause over the western Pacific: Wave driving, convection and the annual cycle, *J. Geophys. Res*., **101**, 21 233-21 242.

Rosenlof K. H., S. J. Oltmans, D. Kley, J. M. Russell III, E.-W. Chiou, W. P. Chu, D. G. Johnson, K. K. Kelley, H. A. Michelsen, G. E. Nedoluha, E. E. Remsberg, G. C. Toon, and M. P. McCormick, 2001: Stratospheric water vapor increases over the past half-century, *Geophys. Res. Lett*., **28**, 1195-1198.

Seidel, D. J., R. J. Ross, J. K. Angell, and G. C. Reid, 2001: Climatological characteristics of the tropical tropopause as revealed by radiosondes, *J. Geophys. Res*., **106**, 7857-7878.

Sherwood, S. C., 2000: A stratospheric "drain" over the maritime continent, *Geophys. Res. Lett*., **27**, 677-680.

Sherwood, S. C., and A. E. Dessler, 2001: A model for transport across the tropical tropopause, *J. Atmos. Sci.*, **58**, 765-779.

Simmons, A. J., A. Untch, C. Jakob, P. Kallberg, and P. Unden, 1999: Stratospheric water vapor and tropical tropopause temperatures in ECMWF reanalyses and multi-year simulations, *Quart. J. Roy. Meteor. Soc.*, **125**, 353-386.

SPARC, 2000: SPARC assessment of upper tropospheric and stratospheric water vapor, Eds.: D. Kley, J. M. Russell III and C. Phillips, WMO/TD No. 1043.

Teitelbaum, H., M. Moustaoui, C. Basdevant, and J. R. Holton, 2000: An alternative mechanism explaining the hygropause formation in tropical region, *Geophys. Res. Lett*, **27**, 221-224.

Tsuda, T., Y. Murayama, H. Wiryosumarto, S. W. B. Harijino, and S. Kato, 1994: Radiososnde observations of equatorial atmospheric dynamics over Indonesi. 1. Equatorial waves and diurnal tides, *J. Geophys. Res.*, **99**. 10,491-10,505.

Veryard, R. G., and R. A. Ebdon, 1961: Fluctuations in tropical stratospheric winds. *Meteor. Mag.*, **90**, 125-143.

Vincent, D. G., A. Fink, J. M. Schrage, and P. Speth, 1998: High and low-frequency intraseasonal variations of OLR on annual and ENSO timescales, *J. Climate*, **11**, 968-986.

Waliser, D. E., and W. F. Zhou, 1997: Removing satellite equatorial crossing time bias from the OLR and HRC datasets, *J. Climate*, **10**, 2125-2146.

Wheeler, M., G. N. Kiladis, and P. J. Webster, 2000: Large-scale dynamical fields associated with convectively coupled equatorial waves, *J. Atmos. Sci.*, **57**, 613-640.

Xu, J. S., 1992: On the relationship between the stratospheric quasi-biennial oscillation and the tropospheric southern oscillation, *J. Atmos. Sci.*, **49**, 725-734.

Yulaeva, E., J. R. Holton, and J. M. Wallace, 1994: On the cause of annual cycle in the tropical lower stratospheric temperature, *J. Atmos. Sci.*, **51**, 169-174.

Yulaeva, E., and J. M. Wallace, 1994: The signature of ENSO in global temperature and precipitation fields derived from the microwave sounding unit, *J. Climate*, **7**, 1719-1736.

Zhou, X.-L., 2000: The tropical cold point tropopause and stratospheric water vapor, Ph.D. dissertation, State University of New York, 121 pp.

Zhou, X.-L., and Z. Sun, 1994: Tropical atmospheric nonlinear steady response solution under effects of paired heat sources of contrasting nature, *ACTA Meteor. Sin.*, **8**, 356-364.

Zhou, X.-L., M. A. Geller, and M. Zhang, 2001a: The cooling trend in the tropical cold point tropopause temperatures and its implications, *J. Geophys. Res.*, **106**, 1511-1522.

Zhou, X.-L., M. A. Geller, and M. Zhang, 2001b: Tropical cold point tropopause characteristics derived from ECMWF reanalyses and sounding, *J. Climate*, **14**, 1823-1838.

Zhou, X.-L., and J. R. Holton, 2002: Intraseasonal variations of tropical cold point tropopause temperatures, *J. Climate*, **15**, 1460-1473.

Zhou, X.-L., M. A. Geller, and M. Zhang, 2003: Temperature fields in the tropical tropopause transition layer, *J. Climate,* submitted.

STATISTICAL RELATIONSHIP BETWEEN THE NORTHERN HEMISPHERE SEA ICE AND ATMOSPHERIC CIRCULATION DURING WINTERTIME

ZHI-FANG FANG

Chengdu University of Information Technology
Chengdu, Sichuan 610041,China
E-mail: Fangzf@public.cd.sc.cn

(Manuscript received 6 November 2002)

The wintertime relationship between the Northern Hemisphere sea-ice concentration, 500-hPa height, sea level pressure and 1000-500-hPa thickness is examined. The Northern Hemispheric sea ice extent exhibits a strong sensitivity to the climatic variation of atmospheric circulation anomalies. The sea-ice extent has reduced in the Barents Sea, Greenland Sea and Labrador Sea since 1990. Particularly, the reduction of sea ice extent in the Barents Sea and Greenland Sea became evident as early as 1968. The Northern Hemispheric sea ice extent also exhibits a strong signal of decadal variability except the Greenland Sea where the downward trend is more pronounced. The sea ice variability is characterized by a dipole pattern in both the Atlantic and Pacific sectors. Its temporal variability is strongly coupled to the North Atlantic Oscillation (North Pacific Oscillation) in the Atlantic (Pacific) sector. The relationship is strongest when the atmosphere leads the sea ice by 1-2 weeks.

1. Introduction

Sea ice has been known to play an important role in the earth climate system. The climate variability of sea ice extent reflects the atmospheric circulation variability. The North Atlantic Oscillation (NAO) or the Arctic Oscillation (AO) and North Pacific Oscillation (NPO) are the prominent atmospheric circulation anomaly patterns in the Northern Hemisphere winter season (e.g., Rogers 1981; Wallace and Gutzler 1981; Thompson and Wallace 1998). These two patterns have been related to the variations of sea ice in both Pacific and Atlantic basins (e.g., Walsh and Johnson 1979; Rogers and van Loon 1979; Kelly et al. 1987; Fang and Wallace 1994; Fang, et al. 2002). Furthermore, the movement of individual sea ice floes tends to follow the surface wind at an angle between 10° and 45° or more for a wind speed from 10m/s to a low wind speed (Thorndike and Colony 1982). On the other hand, the sea-ice extent may act as additional thermal forcing via thermodynamic fluxes at air-sea interface that can lead to changes in circulation. Particularly, the extent of sea-ice affects the radiation budget of the earth-atmosphere system because of the high reflecting power to the solar radiation of sea ice. This partially explains the fact that the secular temperature variation in high latitudes is more significant than in low-middle latitudes (Deser and Blackmon 1993). Also the change in albedo alone due to a small perturbation in sea ice extent may lead to an accelerated and amplified change in sea ice via albedo feedback (North 1975; Huang and Bowman 1990). Moreover, the energy exchange between sea ice and the ocean underneath and the atmosphere above due to sea-ice melting and condensation provides a memory to the land-ocean-atmosphere climate system that regulates not only the seasonal but also the inter-annual/decadal variability the atmosphere.

This paper is a summary report of a series of publications by the author and collaborators on the Northern Hemisphere sea ice climate variability and its relation to atmospheric circulation anomalies (Fang et al. 1987; Fang 1990; Fang et al. 1991; Fang and Wallace 1993, 1994, 1996, 1998; Fang et al. 1998; and Fang et al. 2002). The data and methods are discussed in the next section. Section 3 devotes a brief description of the prominent climate variation patterns of the sea ice extent over the high latitudes of the Pacific and Atlantic oceans. Section 4 presents the temporal relationship between the prominent climate variation patterns of the sea ice extent and the dominant circulation anomaly patterns. The key findings are summarized in section 5.

2. Data and methods

The sea ice concentration data are from three different datasets. One is the weekly sea ice concentration data in the North Hemisphere derived from satellite images (Knight 1984). The weekly dataset is archived in the US National Snow and Ice Data Center (NSIDC) at the University of Colorado at Boulder. The weekly sea ice dataset has a latitude-dependent resolution ranging from 0.25°× 0.25° equatorward of 60°N to 2° × 2° at 85°N. The record encompasses the period from 3 January 1972 through 26 December 1989. The other two are monthly ice concentration data. One of the monthly sea ice concentration datasets is obtained from the Hadley Centre which were derived from satellite images for the period of 1968-98 covering the area poleward of 40°N. In this study, we have converted the Hadley Centre's monthly sea-ice data from 1°(latitude)×1°(longitude) resolution to 1°(latitude)×3°(longitude) with a total number of 6000 grid points. The other monthly dataset is from the collection of Professor Walsh at the University of Illinois at Champaign-Urbana for the period of 1953 through 1991 (referred to as the LAUI dataset). This dataset, first used in Walsh and Johnson (1979), has a resolution of 60 n.mi × 60 n.mi rectangle grids covering the Arctic, the North Atlantic and the most part of the northern Pacific except over the Sea of Okhotsk with a total number of 4640 grid points.

Three atmospheric circulation variables, namely, the 500-hPa geopotential height, sea level pressure, and 1000-500 hPa thickness, are analyzed. The monthly data are from the NCEP/NCAR reanalysis dataset for the period from 1958 to 1997. The weekly data are derived from the U.S. National Meteorological Center operational daily analyses which have been archived by the NCAR Data Library. Following a procedure described in Kushnir and Wallace (1989), we first obtained 10-day low-pass-filtered anomalies on a reduced 445-point Gaussian grid covering the Northern Hemisphere poleward of 20°N. Values were then extracted from the low-pass-filtered anomalies at 7-day interval corresponding to the timing of weekly sea ice data.

The primary analysis tools are the Empirical Orthogonal Function (EOF) and Singular Value Decomposition (SVD) techniques. Bretherton et al. (1992) introduced the SVD method and Wallace et al. (1992) have discussed the two leading SVD modes of the North Hemispheric 500 hPa height field and SST field over the North Pacific and North Atlantic. The temporal covariance matrix between normalized sea-ice concentration at each grid point and an atmosphere variable at each grid point are constructed. The output of SVD analysis consists of two matrices, the "left" matrix for sea ice and the "right" matrix for the atmosphere variable, together with the same number of singular values. Singular values are proportional to the squared covariance fraction (SCF) accounted for by the corresponding singular value vectors. The temporal correlation between the expansion coefficients of the left and right fields measures the coupling strength between a pair of singular value vectors. The heterogeneous correlation pattern for the left field is a correlation map obtained by correlating the time series of the expansion coefficient of the right field with grid point values of the left field, and vice versa.

3. Inter-annual/decadal variability of sea ice

Fang and Wallace (1994) examined the prominent sea ice anomaly patterns in the North Atlantic and North Pacific sectors using the weekly NSIDC data in winter season. Figure 1a shows the leading EOF of sea ice concentration over the Atlantic sector (the explained variance is 22.3%). It is seen that the sea ice extent in the Greenland/Barents seas is negatively correlated with that over the Davis Strait/Labrador Sea. In a similar manner, the leading EOF of sea ice concentration in the Pacific sector (the explained variance is 35.0%) exhibits an out-of-phase relation between Bering Sea and the Sea of Okhotsk (Figure 1b).

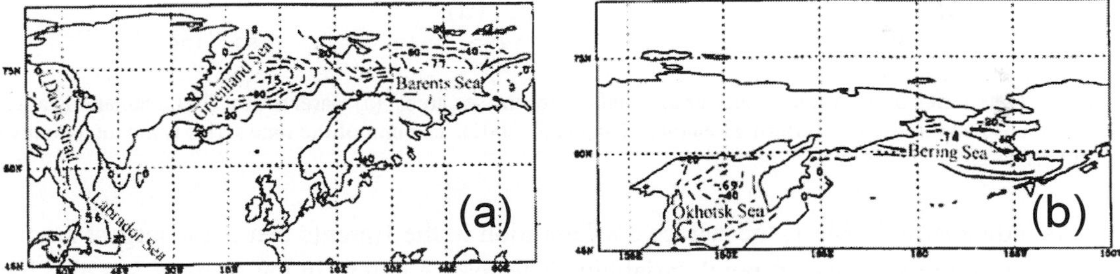

Figure 1. Horizontal structure of the leading EOF of temporal covariance matrix of the weekly sea ice concentration data during the winter season for (a) the Atlantic and (b) the Pacific sector. The contours are the temporal correlation between the time series of the leading EOF and sea ice concentration at each grid point. Contour interval is 20%; negative contours are dashed (after Fang and Wallace 1994).

From the monthly sea ice datasets, we derived sea-ice indices for the four sea areas: the Greenland Sea, Barents Sea, Davis Strait/Labrador Sea and the Bering Sea. A sea-ice index for a particular area is equal to the total number of grid points covered by ice in that area. A standard grid size of 60 n.mi × 60 n.mi is used to calculate the sea-ice indices from the two monthly sea ice datasets. It follows that the total area of sea ice concentration in a particular region can be obtained by multiplying its sea-ice index with the standard grid size. Using linear regression method, we have merged sea-ice indices derived from the two monthly sea ice datasets so that the time series of the sea-ice indices are prolonged to a 46-year span from 1953 to 1998. The period of 1968-1991 was covered by the both datasets. The monthly correlation between the two datasets during the common period exceeds 0.89 in each of the four sea areas. Particularly, the correlation between the two datasets is above 0.97 in the Barents Sea. Therefore, the likelihood of introducing heterogeneities to the merged dataset would be minimal. It is seen in Figure 2 that the Barents Sea sea-ice index is negatively correlated with that for the Davis Strait/Labrador Seas and each of them shows four maximum peaks and minimum valleys. The Barents Sea sea-ice index appears to have a long-term downward trend since 1968. The sea ice concentration in the Davis Strait/Labrador Sea apparently does not suffer a long-term downward trend. The Bering Sea sea-ice index exhibits three pronounced peaks without an obvious long-term trend. The Greenland Sea shows little interannual variability but a pronounced downward trend after 1968. The reduction of sea ice concentrations in the four sea areas are observed in 1990s.

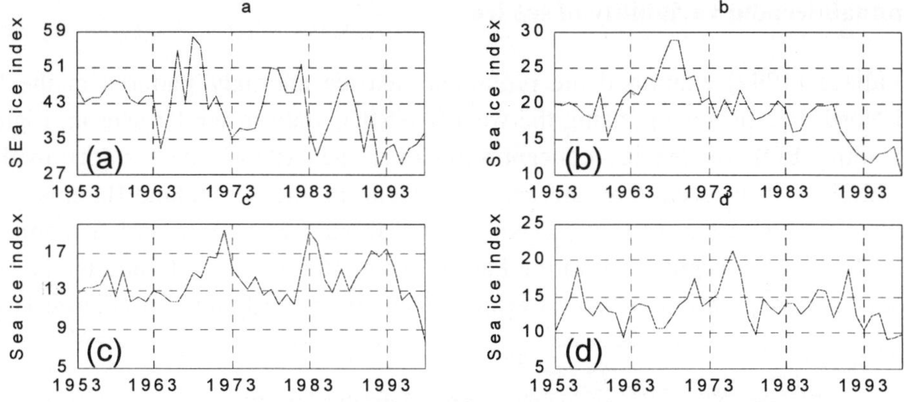

Figure 2. Time series of the sea ice concentration indices for the sea areas: (a) Barents Sea; (b) Greenland Sea; (c) Davis Strait / Labrador seas; (d) Bering Sea (after Fang et al. 2002). The unit of the time series is the total number of grids covered by sea ice in each region.

As indicated in Table 1, the sea ice concentration in the Barents Sea is the largest among the four seas and so is its temporal variability. The sea ice extent in the Barents Sea also has the largest downward trend which is equivalent to a reduction of about 972 square n.mi per year. This downward trend accounts for about 27% of the total variance of the Barents Sea sea-ice index and exceeds 99% significant level based on the F-test. On top of this long-term downward trend, the Barents Sea sea-ice index exhibits a pronounced decadal variability with a time scale of about 11.5 years. The 11.5-year cycle explains about 20% of the total variance of this index and exceeds the 95% significant level based on the F-test. The Greenland Sea has the second largest sea ice coverage among the four seas. It shows a significant downward trend (exceeding 99% level based on the F-test). However there is no pronounced decadal variability in the Greenland Sea sea-ice extent. The Davis Strait/Labrador Sea and the Bering Sea, however, show little long-term trend. Nevertheless, there are still significant amounts of decadal variability of sea ice in the two seas, accounting for about 20% of the total variance in each of the two regions.

Table 1. Statistics of the sea-ice indices. The mean value has a unit of "number of grid points". The variance parameters have a unit of "square of number of grid points". The F-test scores marked with "*" and "**" are 95% and 99% significant, respectively. [1] (adapted from Fang et al. 2002).

	Barents	Greenland	Labrador	Bering
Mean	41.91	19.08	14.04	13.53
Variance	49.78	16.41	4.94	7.59
Trend Coefficient (number of grids per year)	-0.27	-0.18	0.02	-0.02
Variance associated with the trend	13.43	5.81	0.06	0.08
F-test score of the trend estimate	16.26**	24.11**	0.57	0.49
The dominant period (years)	11.5	46.0	11.5	15.3
Variance associated with the dominant period	10.90	8.91	1.01	1.99
F-test score of the dominant period estimate	6.03*	25.54**	5.54*	7.64**

4. Effect of atmospheric circulation on sea-ice distribution

Rogers and Van Loon (1979), Walsh et al. (1979) and Kelly et al. (1987) showed that there exists a statistically significant relationship between the NAO and the sea ice extent in the Baltic Sea and the Labrador shelf as well as the coastal areas of Iceland. Walsh et al.(1986) further showed that the NAO index can be used to predict the number of icebergs that penetrate southward of 48°N along the Labrador shelf in subsequent months.

To elucidate the impact of atmospheric circulation anomalies to sea ice variations over the Atlantic, we define the Atlantic sea ice index as the difference between total sea ice concentration in the region of 110°W-20°W and the region of 20°W-70°E, based on the LAUI dataset. The bottom panel of Figure 3 shows the time series of the Atlantic sea ice index. The simultaneous correlation between the sea-ice index of the Atlantic section and the 500-hPa geopotential height exhibits a large-scale circulation anomaly pattern that bears a resemblance to the NAO pattern (Figure 4), showing an out-of-phase relation between Greenland and the North Atlantic. The maximum variability center over Greenland corresponds to one of the most robust patterns derived from cluster analyses of the 700-hPa and 500-hpa height fields (Kimoto and Ghil 1993; Cheng and Wallace 1992), which is referred to as the Greenland Blocking. Following Kushnir et. al (1989), the Greenland

[1] The F-test scores of the trend estimate are calculated based on the statistic $F = U / [(n-2)Q]$, where U and Q are regression variance and residual variance, n is sample size. The F-test score of the dominant period estimate are calculated based on the statistic $F = \{(a_k^2 + b_k^2)/2\} / \{[(s_y^2 - (a_k^2 + b_k^2)/2]/(n-2-1)\}$, where s_y^2 is total variance for the sea ice index, $(a_k^2 + b_k^2)$ is the amplitude of K wave.

Blocking index is constructed as the average of 500-hPa height over the Greenland area (60-80°N and 20-80°W), which is equivalent to the NAO/AO index but emphasizing more on the northern center of the NAO dipole pattern. The correlation between the Greenland Blocking and the NAO/AO indices is –0.85/–0.89, respectively. The top panel of Figure 3 shows the time series of the Greenland Blocking index. The correlation between the two time series shown in Figure 3 is –0.68, which exceeds 99% significance level. It follows that the Greenland blocking anomalies tend to be accompanied by a reduction of the Atlantic sea-ice index, representing the case of increasing sea-ice concentration in the Baffin/Davis Strait, and decreasing sea-ice in the Barents Sea.

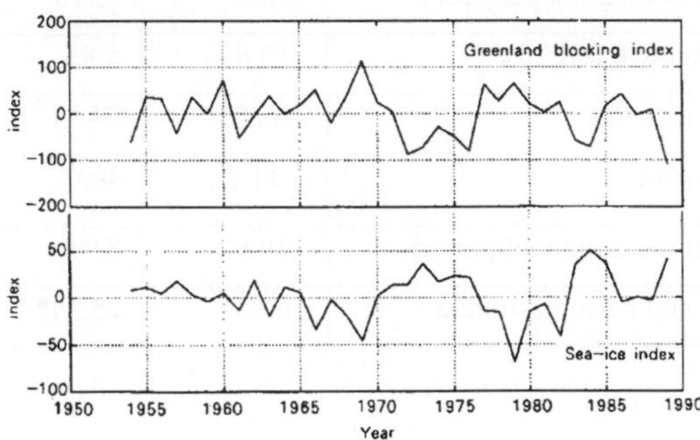

Figure 3. Time series of the Greenland Blocking and sea ice index in the winter season (DJF) (after Fang and Wallace 1993).

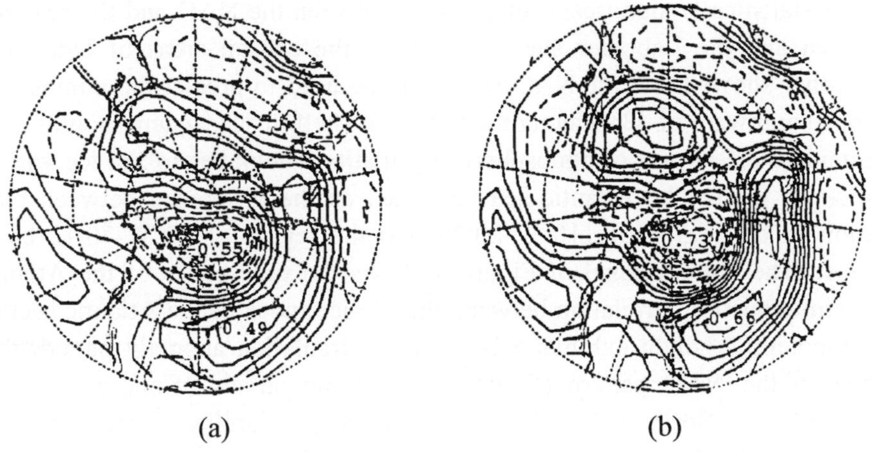

Figure 4. The simultaneous correlation pattern between wintertime sea-ice index over the north Atlantic sector and 500-hPa height field for (a) monthly data and (b) seasonal data. Contour interval is 0.1, Negative contours are dashed (after Fang and Wallace 1993).

Fang and Wallace (1994) applied SVD technique to study the relations between fluctuations in wintertime sea ice concentration field over the North Atlantic and the 500hPa height, sea level pressure (SLP), and 1000-500-hPa thickness fields using the weekly datasets. In order to allow for the possibility that the strongest relationship might be observed, not only in simultaneous correlations, but in the correlations with one field lagged by some time interval relative to the other, the analysis was performed for a sequence of different lag intervals, ranging from –7 weeks to +7 weeks, where negative values denote sea ice leading the atmospheric variables and positive values denote sea ice lagging. The statistics of the leading modes are summarized in Table 2. Both the SCF and the correlation coefficient (R) measure the coupling strength between the two fields of the SVD analyses. It is seen from Table 2 that both of these indicators reach their peak values with sea ice lagging the atmospheric variables by 2-3 weeks. The relationships are of comparable strength for all three atmospheric fields. With the atmosphere lagging sea ice, the correlations are uniformly weak, whereas with the atmosphere leading, the correlations remain strong out to a month or longer.

Table 2. Statistics for the leading mode in the SVD expansion of wintertime sea ice concentrations over the North Atlantic sector at 7-day intervals and the hemispheric 500-hPa height, sea level pressure (SLP) and 1000-500-hPa thickness fields. Lag time is the number of 7-day time steps that the sea ice data were lagged relative to the atmospheric data. SCF is the squared covariance fraction between the two fields explained by the leading mode; R is the correlation coefficient between the time series of the expansion coefficients for the leading mode. The shaded row represents the simultaneous relation. The bold numbers are the maximum values (adopted from Fang and Wallace 1994).

	SCF(%)			R(%)		
Lag	500-hPa	SLP	Thickness	500-hPa	SLP	Thickness
-2	25	26	36	35	35	41
-1	29	31	45	38	36	48
0	42	38	53	48	42	58
1	50	47	56	58	52	**64**
2	53	53	57	**59**	**56**	**64**
3	**55**	56	**58**	57	**56**	61
4	**55**	58	55	56	**56**	58
5	**55**	**59**	53	55	54	57
6	52	57	51	54	53	56
7	49	57	49	52	52	55

To assess the statistical significance of these results, the SVD analysis was repeated with a series of scrambled 500-hPa height data by a Monte Carlo random number generator. The scrambling is applied only at the year level. As a result, the autocorrelation of the weekly data at each grid point within each individual year is preserved. Also the reordering has no effect on the variance or the EOFs of the 500-hPa height field because the temporal scrambling does not change the spatial structure of the data at any given time. With these scrambled data, the values of R and SCF are substantially lower than in the calculation with the unscrambled data, implying that the results in Table 2 are statistically significant.

The heterogeneous correlation pattern for sea ice (Figure 5a) is virtually identical to the correlation pattern for the leading EOF of wintertime sea ice over the North Atlantic shown in Figure 1a, exhibiting out-of-phase fluctuations between Greenland/Barents seas and the Davis Strait/Labrador sea. The temporal correlation between the expansion coefficient of the leading EOF and the time series of the SVD sea ice expansion coefficient is 0.99. The corresponding 500-hPa height pattern (Figure 5b) resembles the correlation pattern shown in Figure 3 over the North Atlantic and Greenland regions, showing a negative phase pattern of the NAO. In other words, the negative NAO pattern tends to be accompanied by decreasing sea ice edge in the Davis Strait and along the edge of the Labrador Shelf, and increasing sea ice concentration in the Barents Sea and Greenland Sea. The changes in sea ice over the Davis Strait/Labrador Shelf region can be understood as a response to surface wind and temperature anomalies. For example, positive 500-hPa anomalies over southern Greenland are accompanied by anomalous easterly or southeasterly (geostrophic) surface winds (Figure 5c), which would cause the sea ice boundary along the Labrador Shelf to retreat northwestward, and this tendency would be reinforced by the positive 1000-500-hPa thickness anomalies over this region (Figure 5d). The northeastward geostrophic surface winds are observed over the Greenland Sea at these times, which would cause the sea ice boundary to advance and extend the sea-ice range.

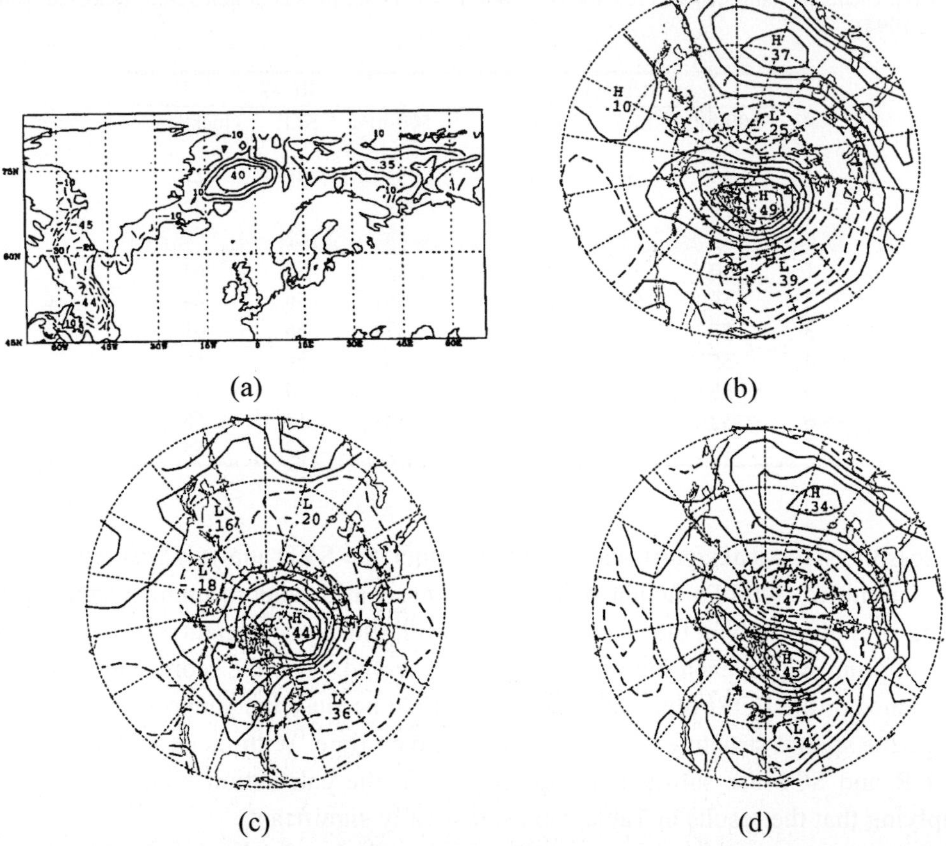

Figure 5. Heterogeneous correlation pattern of the leading mode derived from SVD of the temporal covariance matrix between wintertime sea ice concentration and atmospheric variables at two weeks earlier. (a)The Atlantic sea ice concentration; (b) the 500-hPa height; (c) SLP; and (d) 1000-500-hPa thickness. Contour interval is 0.1. Negative contours are dashed (adopted from Fang and Wallace 1994).

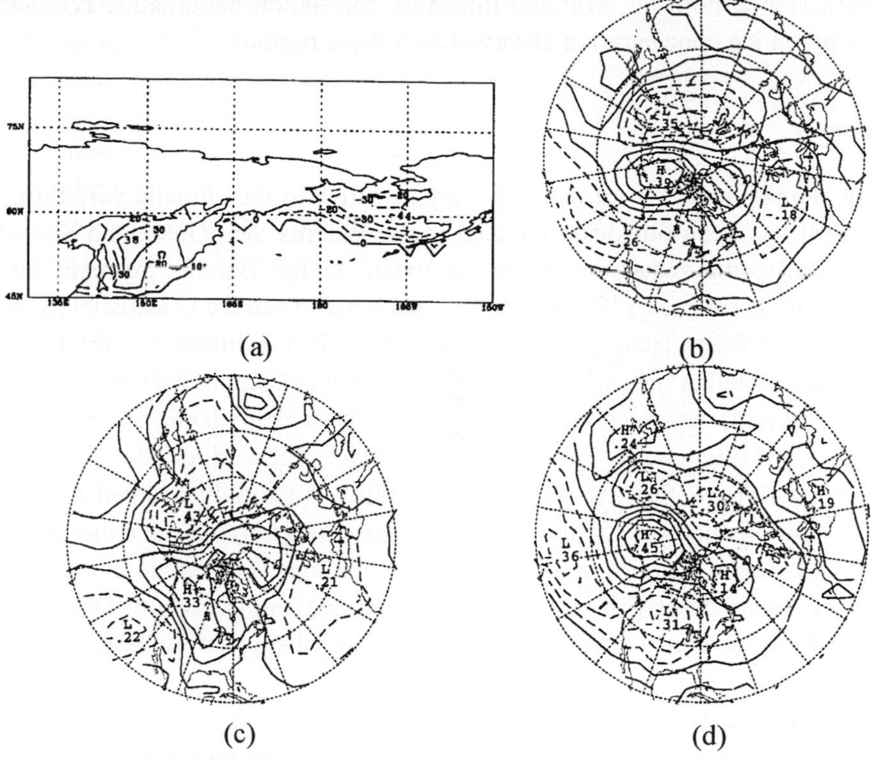

Figure. 6 As in Figure 5 except for the Pacific sea ice concentration and the atmospheric variables are one week earlier than the sea ice field. (adopted from Fang and Wallace 1994).

Figure 6 is the counterpart of Figure 5 for the Pacific sector. The correlations between the North Pacific sea ice concentration and the atmospheric circulation anomalies are not as strong as over the North Atlantic. Nevertheless, they still stand out well above the corresponding populations derived from Monte Carlo simulations analogous. The strongest correlations are observed when the sea ice concentration lags 500-hPa height by one week. The pattern of sea ice concentration (Figure 6a) is very similar to the leading EOF shown in Figure 1b, displaying an out-of-phase relation between anomalies over the Bering Sea and those over the Sea of Okhotsk. The corresponding 500-hPa height pattern (Figure 6b) is characterized by a dominant positive centers of action over the Alaska, flanked by weaker negative centers over the west pacific and Siberia. It bears a resemblance to the NPO and Alaska Blocking discussed by Wallace and Gutzler (1981) and Cheng et. al (1993). However, the primary positive center of action is located father northeast than the NPO pattern and the negative center of action over the Sea of Okhotsk is much more prominent. The corresponding SLP correlation pattern (Figure 6c) is indicative of that the southerly surface wind anomalies over the Bering Sea and Bering strait would cause the sea ice boundary to retreat northward. The westerly surface wind anomalies over the Sea of Okhotsk would cause the sea ice boundary to advance eastward. These conditions are accompanied by positive thickness anomalies over the Bering Sea and negative anomalies over the Sea of Okhotsk

(Figure 6d). Hence both the SLP and thickness correlation patterns are consistent with the anomalies in sea ice concentration observed over these regions.

5. Summary

The North Hemispheric sea ice extent is very sensitive to the climatic variation. Since 1990, sea ice extent has been obviously reduced in the Barents Sea, Greenland Sea and Labrador Sea. Particularly, the reduction of sea ice extent in the Barents Sea and Greenland Sea became evident as early as 1968. The downward trend of sea ice is statistically significant. In addition, the Northern Hemisphere sea ice extent also exhibits a statistically significant decadal variation with a time scale of about 11-15 years except in the Greenland Sea..

The spatial patterns of sea ice variability over the north Atlantic are characterized with a dipole with opposing centers of action in the Davis Straits/Labrador Sea region and the Greenland and Barents Seas. Its temporal variability is strongly coupled to the atmospheric NAO. The relationship between the two patterns is strongest with the atmosphere leading the sea ice by two week. A dipole pattern of sea ice variations is observed in the North Pacific as well, with opposing centers of action in the Bering Sea and the Sea of Okhotsk. The Pacific sea ice variability is related to the NPO, with the atmosphere leading the sea ice by one week.

Acknowledgments. This summary work was supported by the National Science Foundation of China and LASG, the Chinese Academy of Science. The author greatly thanks Dr. Ming Cai for his editorial assistance that significantly improved the quality of this paper. The comments and suggestions from two anonymous reviewers are helpful in clarifying several important issues.

References:

Bretherton, C. S., C. Smith, and J. M. Wallace, 1992: An intercomparison of methods for finding coupled patterns in climate data. *J. Climate*, **5**, 541-560.

Cheng, X-H., and J. M. Wallace, 1992: Cluster analysis of the Northern Hemisphere wintertime 500-hPa height field: Spatial patterns. *J. Atmos. Sci.*, **49**, 2674-2696.

Deser, C., and M. L. Blackmon, 1993: Surface climate variations over the north Atlantic ocean during winter. *J. Climate,* **6**, 1743-1753.

Fang, Z-F., 1987: Interaction between subtropical high and polar ice in North Hemisphere. *Science Bulletin*, **32**, 330-335.

Fang, Z-F., Y-Z. Gao and M. Dai, 1987: A probable avenue to effect of polar ice on North Pacific subtropical high. *ACTA OCEANOLOGICA SINICA*, **6**, 190-195.

Fang, Z-F., 1990: Effect of the Arctic sea ice in January on the Asia-Western Pacific circulation in June. *Chinese Journal of Atmospheric Sciences,* **14**, 113-119.

Fang, Z-F., Y-B. Tan, and X-H Sui, 1991: Teleconnection pattern between the Arctic ice area and the 500hPa geopotential height field during the Northern hemispheric summer. *Chinese Journal of Atmospheric Sciences,* **15**, 53-60.

Fang, Z-F and J. M. Wallace, 1993: The Relationship between the wintertime Blocking over the Greenland and the sea-ice distribution over the North Atlantic. *Advances in Atmospheric Sciences*, **10**, 453-464.

Fang, Z-F. and J. M. Wallace, 1994: Arctic Sea ice variability on a time-scale of weeks and its relation to atmospheric forcing. *J. Climate,* **7**, 1897-1914.

Fang, Z-F., and J. M. Wallace, 1996: The atmospheric circulation to forcing of the Sea-ice in the North Pacific and Kuroshio SST. *Chinese Journal of Atmospheric Sciences*, **20**, 541-546.

Fang, Z-F., R-C Yu, X-Z Jin and X-H Zhang, 1998: Comparison of Arctic Sea-ice variation during 1966-1991 between an Ocean-Sea Ice model calculations and observations. *Chinese Journal of Atmospheric Sciences,* **22**, 149-162.

Fang, Z-F., and J. M. Wallace, 1998: North Pacific sea ice and Kuroshio SST variability and its relation to the winter monsoon, *Polar Meteorology and Glaciology Jpn.,* **12**, 58-67。

Fang, Z-F.,Y-F. Guo, Q. Qiao, and J-H. Fang, 2002: Decrease of Arctic sea-ice and its relationship with the sea surface pressure and temperature in high latitude. *Plateau Meteorology,* **21,** 565-575.

Huang, J. and K. P. Bowman, 1992: The small ice cap instability in seasonal energy balance models. *Climate Dynamics*, *7*, 205-215.

Kelly, P. M., C. M. Goodess, and B. S. G. Cherry, 1987: The interpretation of the Icelandic sea-ice record. *J. Geophys. Res*., **92C**, 10835-10843.

Kimoto, M., and M. Ghil, 1993: Multiple flow regimes in the northern hemisphere winter. Part 1: Methodology and hemispheric regimes. *J. Atmos. Sci*., **50**, 2625-2643.

Knight, R. W., 1984: Introduction to a new sea-ice database. *Ann.Glaciol.*, **5,** 81-84.

Kushnir,Y., and J. M. Wallace, 1989: Low-frequency variability in the northern hemisphere winter: Geographical distribution, structure, and time scale dependence. *J. Atmos. Sci*., **46**, 3122-3142.

North, G. R., 1975: Theory of Energy-Balance Climate Models. *J. Atmos. Sci.*, *32*, 2033–2043.

Rogers, J. C., and H. Van loon, 1979: The sea-saw in winter temperatures between Greenland and northern Europe. Part 2: Some oceanic and atmospheric effects in middle and high latitudes. *Mon. Wea. Rev.*, **107**, 509-519.

Rogers, J. C., 1981: The North Pacific Oscillation. *J. Climate*, **1**, 39-57.

Thorndike, A. S., and R. Colony, 1982: Sea ice motion in response to geostrophic winds. *J. Geophys. Res*., **87,** 5845-5852.

Thompson, D. W. .J., and J. M. Wallace, 1998: The Arctic Oscillation signature in the wintertime geopotential height and temperature fields. *Geophys. Res. Lett.*, **25**, 1297-1300.

Walsh, J. E. and C. M. Johnson, 1979: Interannual atmospheric variability and associated fluctuations in Arctic sea ice extent. *J. Geophys. Res*., **84,** 6915-6928.

Walsh, J. E., W. I. Wittmann, L. H. Hester, and W. S. Dehn, 1986: Seasonal prediction of iceberg severity in the Labrador Sea. *J. Geophys. Res.*, **91**, 9683-9692.

Wallace, J. M., and D. S. Gutzler, 1981: Teleconnection in the geopotential height field during the northern hemisphere winter. *Mon. Weather Rev.*, **109**, 784-812.

Wallace, J. M., C. Smith, and C. S. Bretherton, 1992: Singular value decomposition of wintertime sea surface temperature and 500-mb height anomalies. *J. Climate*, **5**, 561-576.

A PV STUDY OF AN EXPLOSIVE EXTRATROPICAL CYCLOGENESIS EVENT

FANYOU KONG
Center for Analysis and Prediction of Storms, University of Oklahoma
100 E. Boyd St., Norman, OK 73019, USA
E-mail: fkong@ou.edu

(Manuscript received February 17, 2003)

The piecewise potential vorticity (PV) inversion technique is applied to the ERICA IOP2 storm case. The inversion results suggest that the mid and lower level diabatic condensation-derived PV anomaly contributes a major role to this cyclone's deepening, and the contribution increases during the most rapid deepening phase. It also shows that the diabatic processes appear to augment the positive contribution of the lower boundary thermal perturbation but to reduce the relative importance of the upper level PV anomaly. Sensitivity experiments reveal that, by removing part or all of the upper level positive PV anomaly from the preconditioned atmosphere but retaining the lower level perturbation and diabatic processes, the cyclogenesis might weaken substantially. For this storm, however, the impact of the upper level PV anomaly removal seems less significant relative to diabatic processes since a dry simulation results even weaker cyclogenesis. The absence of surface thermal anomaly in the initial atmosphere could severely hamper the subsequent cyclogenesis but only if there is lack of upper level PV anomaly, implying that the thermal anomaly is an internal factor and is largely controlled by the upper level PV perturbation.

1. Introduction

Rapidly intensifying (or explosive) extratropical cyclogenesis, defined when the surface cyclone central pressure fall averages at least 1 hPa h^{-1} for 24 h, is the most common maritime weather phenomenon in winter season and has been broadly and intensively investigated over the decades (Sanders and Gyakum 1980; Bosart 1981; Gyakum 1983a,b; Anthes et al. 1983; Uccellini 1986; Kuo and Reed 1988; Shapiro and Keyser 1990). It is widely recognized that upper level forcing, latent heat release due to condensation, surface fluxes of heat and moisture, and effective static stability are important components of the cyclogenesis, but how the sequence of events interacts in producing the intensity of a particular storm is case dependent and unclear.

Evidence suggests that initial conditions can be significant in explosive cyclogenesis. Gyakum et al. (1992) suggested that for a given low level convergence the surface vorticity spinup in the period prior to a cyclone's maximum intensification (so called antecedent vorticity development) acted as an important dynamical conditioning process for explosive cyclogenesis. They stressed the importance of accurate forecasts during the storm's antecedent phase of development. Their research also suggested that the explosive cyclone development is typically characterized by a nonlinear interaction between two cyclonic disturbances in the lower and upper troposphere. Kuo et al. (1991a) studied the role of surface heat and moisture fluxes at the antecedent stage of surface cyclone development through a set of sensitivity experiments on several western Atlantic explosive cyclogenesis events using the PSU/NCAR mesoscale model. They concluded that the surface energy fluxes preceding the rapid deepening phase had substantial impacts on the subsequent deepening rate, while the fluxes occurring during the rapid cyclogenesis stage had only small effect.

The effects of latent heat release associated with condensation within frontal rainbands are relatively poorly understood, especially at upper levels. Because of insufficient data resolution both in time and space, it is not feasible to adequately investigate the influence of latent heat release on the upper level structure. The issue is better addressed through numerical simulations. Kuo et al. (1991b) examined the relative contributions of baroclinic and diabatic processes to explosive cyclogenesis. Their research suggested a strong nonlinear interaction (positive feedback) between the two processes in rapid cyclone development. Numerous other studies also addressed the roles of convection and precipitation processes on the dynamics of explosive cyclogenesis (Gyakum 1983b; Leslie et al. 1987; Liou and Elsberry 1987; Kuo and Reed 1988; Davis et al. 1993; Balasubramanian and Yau 1994), and their results showed enormous case-to-case variability.

The potential vorticity (PV) inversion technique, especially the piecewise PV inversion, proposed by Davis and Emanuel (1991), is considered to be the most useful diagnostic tool for its ability to isolate circulations associated with individual perturbations and to further examine their interactions. Davis and Emanuel (1991) and Davis (1992a,b) demonstrated its usefulness by applying the PV inversion technique to an observational dataset to investigate extratropical cyclogenesis. Davis et al. (1993) extended the diagnostic system to numerical simulation output from the Penn State/NCAR mesoscale model (MM4) to illustrate the integrated influence of latent heating on the structure and evolution of several simulated extratropical cyclones. Balasubramanian and Yau (1994, 1996) applied the technique to idealized extratropical oceanic cyclone simulations from a hydrostatic primitive equation model to study the physics of convective-cyclogenetic interaction. Huo et al. (1998, 1999) made novel use of piecewise PV inversion by altering the upper level PV component in their study of a superstorm development and by rebuilding the initial condition fields in an attempt to improve the numerical prediction of the storm.

In this study, one typical explosive cyclogenesis case (ERICA IOP2) has been simulated using MC2 (Canadian Mesoscale Compressible Community Model) (Benoit et al. 1997), and the Davis piecewise PV inversion technique is applied to the model data to examine the relative importance of different physics processes and their interactions, especially the condensation-related diabatic process. The preconditioned model atmosphere is then modified by removing some or all of the different PV perturbation components to conduct sensitivity experiments in an effort to understand how the factors affect the subsequent cyclogenesis.

2. Case Description and Numerical Simulations

The case selected in this study is the IOP2 storm during the Experiment on Rapidly Intensifying Cyclone over the Atlantic (ERICA). It is also known as the *Rowan Gorilla Rapid Deepener* because of its role in capsizing the deep-sea drilling rig Rowan Gorilla in the Nova Scotia offshore oil fields. Details on the storm can be found in Harnett et al. (1989) and Sanders (1990). Only a brief description is given here.

Occurring over the eastern Atlantic between 35°N and 40°N on 13-15 December 1988, the ERICA IOP2 storm recorded a 24 h central mean sea level (MSL) pressure fall of 43 hPa from 1800 UTC 13 to 1800 UTC 14, reaching a minimum pressure of 957 hPa, the second most rapid deepening cyclone observed during the ERICA season. The storm environment in its early developing stage was characterized by a complex multi-center surface low pattern located within a broad zone of surface cyclonic vorticity and evolved into a rapidly deepening phase when the second and stronger of the two upper level troughs moved offshore in Virginia and North Carolina after 1200 UTC 13 and then passed over the surface cyclone at about 0000 UTC 14. The upper-levels were characterized by a jet streak and an amplifying short-wave trough. A single center formed after 0600 UTC 14 (Harnett et al. 1989; Roebber 1993; Reuter and Yau 1993). By 1200 UTC 14 the cyclone was cut off, with a near vertical close circulation center extending from the sea surface up to 250 hPa.

The Canadian mesoscale model MC2 is used to simulate this explosive cyclone case. MC2 is a three-dimensional nonhydrostatic model making use of advanced semi-implicit time differencing, semi-Lagrangian advection algorithm, and a full physics package. Detail model features and numerical and physics descriptions are presented in Benoit et al. (1997). Two sets of model simulations are conducted. The regular simulations are performed with a horizontal resolution of 25 km and at 23 irregularly spaced vertical levels within a 25 km model lid. With the combination of Kuo cumulus parameterization (Kuo 1974) and Kong-Yau explicit microphysics (Kong and Yau 1997), the model simulated surface low center is in very good agreement with the observation, with a 24 h deepening rate of 41 hPa, *vs.* 43 hPa observed and a location error less than 35 km (Kong and Yau 1997: Fig.2 - Fig.4). No repetition will be given here other than referring readers to Kong and Yau (1997) in the interest of brevity.

The second set of simulations, configured to accommodate the PV inversion and diagnostics, uses 50 km horizontal resolution and 19 equally spaced pressure levels. According to Davis et al. (1993), the PV inversion for a dataset of high resolution simulation is not very efficient since "*the PV inversion process naturally smooths out small-scale features*". In their work, they even coarsen the 45 km grid data by a factor of 3 and performed the PV inversions on a 135 km grid. By comparing the inversion results of the two grid sets they suggested "*the coarsening of PV didn't affect the solution on scales greater than 200 km*". In this study, rather than coarsening data from the 25 km runs, the PV analysis data are obtained by directly rerunning the case on a 50 km grid. Another important consideration in rationalizing the use of a 50 km grid in the PV diagnostics is the need for a set of sensitivity experiments in Section 4, launched with the reinitialized atmosphere based directly on the PV inversion.

One moist run with Kuo cumulus parameterization scheme and a simple resolvable scale condensation (KC in short), and one run without diabatic processes (no condensation at all) have been conducted. The latter may be thought of as a kind of dry simulation (DRY in short). Both runs proceed 36 h starting from 1200 UTC 13. The initialization and lateral boundary conditions are provided from the CMC (Canadian Meteorological Centre) operational objective analysis.

The moist simulation (KC) with 50 km grid deepens less vigorously than its 25 km counterpart. It still however has a 24 h deepening rate of 33.9 hPa from 6 h to 30 h into the simulation, falling well into the explosive cyclone category. The surface pressure and temperature pattern and the storm track do not exhibit any significant differences from the fine grid run. The simulation without diabatic processes (DRY), on the contrary, exhibits a far weaker deepening rate, only 9.5 hPa in the same 24 h period or one third of that from the diabatic run during the same period. Surface isobars show a generally similar pattern with a much shallower surface low and weaker baroclinic frontal zones for the dry cyclone (Figures not shown).

3. Piecewise PV Inversion

The PV diagnostic system equations and the computation procedure are described in detail in the literature (Davis and Emanuel 1991; Davis 1992a; Davis et al. 1993) and are not outlined here. To conduct the PV inversion, a mean flow is obtained for each time from a five day average of the twice-daily CMC (Canadian Meteorological Center) global analysis centered around that time. From the perspective of PV thinking, three main components of positive PV anomaly contribute to the extratropical cyclogenesis: 1) upper level dry PV anomaly (UPV in short), which originated from the upper-tropospheric and lower-stratospheric perturbation, 2) low to mid level PV anomaly (LPV), which is diabatically produced by the condensation in and near the frontal cloud bands through vertical and/or slantwise convection, and 3) the surface potential temperature anomaly, or surrogate PV, which is generated primarily from the horizontal northward advection of surface warm air and also from the sensible heat flux over the ocean surface (Reed et al. 1992). Following the partitioning method of Davis and Emanuel (1991), in this study the UPV comprises the PV anomalies at the grid levels of 500 hPa and above, the LPV consists of the anomalies between 950 hPa and 550 hPa, and the surface potential temperature anomaly, θ'_B, is the average of the 1000 hPa and 950 hPa anomalies.

The relative contributions associated with PV anomalies in low level cyclone deepening are examined by performing the PV inversion for each individual anomaly. Table 1 lists the 900 hPa geopotential height perturbations at the surface low centers at successive simulation times for both moist and dry simulations. It is interesting to note that while in the dry cyclone the upper level perturbation and the surface potential temperature advection contribute to the 900 hPa height decrease, the mostly convection-related lower level positive PV anomaly in the moist simulation plays a major role to the IOP2 cyclone's deepening, along with the surface thermal perturbation. Moreover, the contribution from this condensation-driven PV anomaly increases significantly through the

Table 1: 900 hPa Geopotential Height Perturbation from PV inversion (unit: dm)

	Moist - Mean				
	06 h	12 h	18 h	24 h	30 h
θ'_B	-5.30	-10.60	-12.67	-19.08	-19.44
LPV	+3.31	-3.30	-10.16	-11.30	-11.43
UPV	-3.33	-3.97	-8.64	-2.49	+0.91
Total*	-5.32 (-5.71)	-17.9 (-15.6)	-31.5 (-31.1)	-32.9 (-31.1)	-30.0 (-30.0)
	Dry - Mean				
	06 h	12 h	18 h	24 h	30 h
θ'_B	-2.64	-5.85	-10.32	-12.50	-10.21
LPV	+3.26	+7.72	+10.47	+11.30	+9.01
UPV	-4.19	-7.76	-7.52	-5.42	-3.01
Total*	-3.57 (-3.60)	-5.9 (-6.6)	-7.3 (-7.1)	-6.6 (-6.5)	-4.2 (-4.6)

* The figures in brackets refer to the simulation values.

cyclone's most rapid deepening phase until 24 h of the simulation (1200 UTC 14). The LPV contributes to increase the low level geopotential height at the early stage (before 6 h) instead. The perturbation associated with the UPV, on the other hand, follows the increasing trend initially but contributes less and less after 18 h of the simulation (0600 UTC 14), and finally reverses course around 30 h. The most noteworthy feature might be the counter-effect in the LPV contribution for the dry cyclone vs. the moist one. It is this difference that leads the dry cyclone to exhibit a much weaker cyclogenesis for this case. Inspecting the 18 h typical upper level and low level PV anomalies for the moist and dry simulations, respectively (Figures not shown), reveals that the diabatic condensation processes in the moist simulation produce a large positive PV anomaly along the warm and bent-back fronts with the maximum magnitude of more than 3.5 PVU (1 PVU $= 10^{-6}$ m^2 K kg^{-1} s^{-1}). This is in good agreement with many observations (Manabe 1956; Eliassen and Kleinschmidt 1957; Boyle and Bosart 1986; Whitaker et al. 1988; Davis and Emanuel 1991; Nieman et al. 1993) and simulations (Kuo and Reed 1988; Hoskins and Berrisford 1988; Kuo et al. 1991b) of various extratropical cyclone cases. By examining the sources of each perturbation, Davis and Emanuel (1991) pointed out that this intense low level anomaly was not of stratospheric origin but instead generated by condensation. The dry storm in contrast shows a broad region of marginal negative potential vorticity anomaly over the western Atlantic Ocean, except for a small area of weak positive anomaly near the surface cyclone center. Considering there is no latent heat release for the dry case, it is a natural hypothesis that the negative anomaly is attributed to the lower level neutral or weakly unstable thermal structure associated with the upward heat fluxes that stemmed from the ocean surface. On the other hand, both moist and dry storms show similar magnitudes of positive UPV perturbation over the surface lows.

Table 1 also reveals that in both simulations the lower boundary thermal advection alone contributes the largest amount in the cyclone's low level deepening. Nevertheless, the LPV anomaly in the dry simulation offsets nearly all the positive contribution from the surface thermal perturbation, leaving the UPV the main net contributor for the dry cyclone's surface deepening.

Table 2: 900 hPa Geopotential Height Perturbation from PV inversion (unit: dm)

	Moist - Dry				
	06 h	12 h	18 h	24 h	30 h
θ'_B	-3.48	-5.98	-12.61	-11.96	-10.74
LPV	-0.75	-14.70	-18.17	-21.71	-24.41
UPV	+1.40	+7.05	+10.22	+10.95	+9.98

Inversions of the PV perturbations between the moist and dry simulations (Table 2) provide further understanding of the net effects of the diabatic moist microphysics processes on the overall cyclogenesis and the upper level signature. In this situation, the flow in the dry simulation is taken to be the '*mean*' at each specific inversion time. Since the dry simulation exhibits all dynamic processes in the cyclone development, the inversion has the ability to isolate to a large degree the behavior of the lower to mid level latent heat related diabatic processes in the rapid intensification of surface cyclone, and its possible modification on the upper level perturbation and the surface thermal advection. Three conclusions can be drawn from Table 2: (1) The mainly condensation-derived LPVs do play a major role in this explosive cyclogenesis case. (2) The diabatic processes in the lower to mid troposphere possibly augment the positive contribution of the surface thermal perturbation on the rapid deepening of the surface cyclone, and (3) they, however, seem to be in a position to significantly weaken the UPV's relative contribution, for the anomalies between the moist and dry dynamics now contribute negatively (filling) to the surface deepening. This final point does not mean the UPV contribution is actually weakened but just reflects the fact that the local diabatic heating weakens the UPV (Fig. 1a) and increases the LPV.

The difference of the surface potential temperature between the two simulations (Fig. 1b) provides further explanation to the second remark above. The diabatic processes in the moist simulation produces a broad region of additional and intense warming with magnitudes up to 2.5 K over the cyclone center and along the frontal areas. This surface warming has a role on the surface cyclogenesis equivalent to a positive PV anomaly (Davis 1992b; Davis et al. 1993). Two possible factors are believed to contribute to the additional low level warming. The first factor is the increase of the horizontal warm advection associated with the intensifying surface circulation in the moist simulation, and the second is the increase of the surface energy fluxes, owing to the strong surface wind field to which the fluxes are directly proportional.

4. Sensitivity to the Preconditioned Atmosphere

The instantaneous PV inversion in the previous section reveals at each stage of the cyclone's life cycle specific details about the relative importance of different aspects of a PV disturbance to a balanced synoptic circulation. However, one remaining question is how these perturbations interact each other in the nonlinear real atmosphere. More specifically, the issue addressed in this section is how significantly different PV anomalies in the preconditioned atmosphere (or antecedent stage) affect the subsequent rapid cyclogenesis in an integrated way.

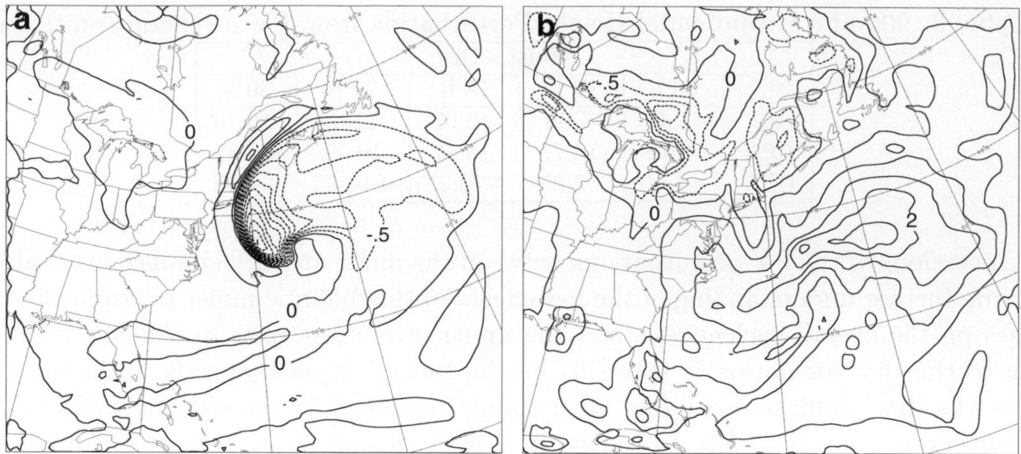

Figure 1: Deviations of the moist simulation (KC) vs. dry (DRY) at 18 h. (a) PV perturbation at 350 hPa (0.5 PVU interval), and (b) surface potential temperature perturbation (0.5 K interval), with negative values dashed.

4.1. *Methodology*

The following procedure, similar to Huo et al. (1999), is employed to re-construct the IOP2's preconditioned atmosphere: First determine the portion of the PV anomaly which will be removed from the initial atmosphere. Then employ the piecewise inversion technique for the anomaly to obtain a balanced fragment of the (partial) flow (containing perturbations of the geopotential height, air velocity, and temperature), in association with that anomaly. Finally, reinitialize the atmosphere by taking out the fragment of the balanced flow from the original initial field. Unlike Huo et al. (1999), who only studied the impacts of different UPV anomalies, this study focus on the interaction of the UPV anomalies and the surface thermal perturbation by modifying both in the storm preconditioned atmosphere.

The initial time of the simulations is 1200 UTC 13, when the upper-troposphere featured the passage of a weak short-wave trough followed by a digging polar trough, and the surface featured a complex pressure field with three separate cyclonic vorticity maxima located within a board zone of surface cyclonic vorticity. Explosive cyclogenesis was poised to commence 6 h later. Fig. 2 depicts the initial geopotential height, potential vorticity and its anomaly from the mean state at the level of 350 hPa. The intense PV maximum associated with the digging polar trough is located over the eastern U.S. region south of the Great Lakes, with a tongue stretching southeastward to join the weak southwest-northeast-orientated maximum, reflecting the short-wave signature above the surface baroclinic zone. The vertical cross section (Fig. 3) along the heavy dash line of Fig. 2a further reveals that this tongue of weak PV anomaly extends down to the surface layer. This is consistent with the result of Roebber (1993, Fig.13a and 14a). Fig. 4 depicts the initial surface wind vector, potential temperature and its anomaly from the mean. It can be seen that a synoptic-scale thermal wave (baroclinic zone) with cyclonic flow is coincident with a region of weak PV anomaly centered around 32N and 74W (Fig. 4a). This region corresponds to a surface warm perturbation with a

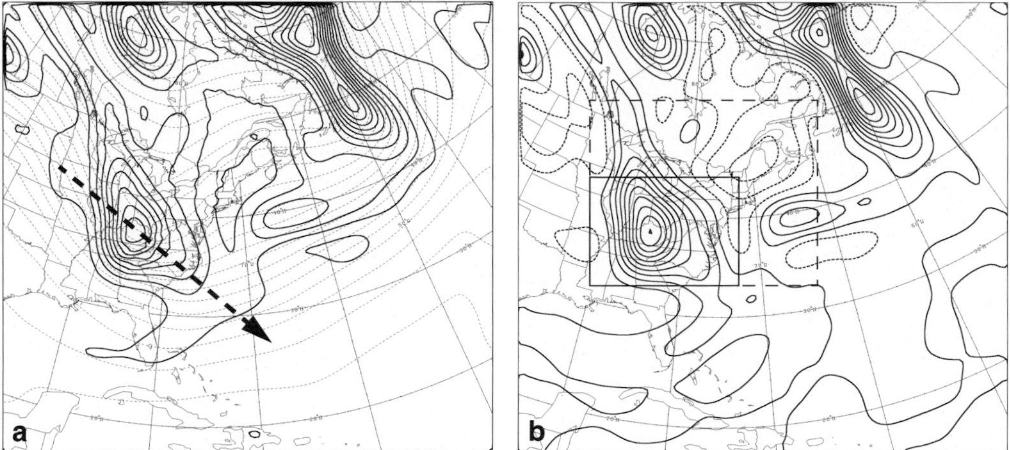

Figure 2: 350 hPa potential vorticity (solid, 0.5 PVU) and geopotential height (dashed, 6 dm) (a), and PV anomaly from mean (0.5 PVU) (b) at initial time (1200 UTC 13 December). The dashed arrow in (a) labels the cross section in Fig. 3. Dashed lines in (b) refer to negative anomaly, and the boxes mark the UPV removal regions.

magnitude of up to 2.5 K (Fig. 4b), which is believed to contribute to the early surface pressure fall of the IOP2 cyclone.

Six experiments are conducted with the configurations and results briefly summarized in Table 3. Four of the experiments relate to the removal of UPV anomalies above 500 hPa from the initial field. 50% (PV05) and 100% (PV10) of the positive UPV are removed from the small rectangular box in Fig. 2b to represent a weak or nonexistent initial tropopause PV intrusion. 50% (PV05BIG) and 100% (PV10BIG) of the entire UPV are removed from the large dash rectangular area in Fig. 2b, which effect a removal of the initial downstream ridge. The other two experiments involve the modification associated with the lower boundary thermal perturbation. In THETA, the surface potential temperature anomaly is removed from the initial atmosphere. TH10 is a combination of PV10 and THETA, in which both the positive UPV and the surface thermal perturbation are removed from the initial fields. For purpose of comparison, Table 3 includes results of the moist control run (KC) and the dry simulation (DRY).

Fig. 5 shows an example of the fragment of balanced nondivergent flow and the geopotential height deviation obtained from the inversion of positive UPV (in PV10). After subtracting this fraction of flow from the initial fields, the rebuilt initial atmosphere is shown in Fig. 6c and d, which is used to initiate PV10. Also shown in Fig. 6a and b is the unmodified original analysis at the initial time, which is used in the KC and DRY simulations. It is evident that the sharp upper level digging polar trough over the Great Lakes and eastern U.S. region is efficiently removed, retaining only a much weaker wave in the rebuilt initial atmosphere. This upper level smoothness is also clearly indicated in Fig. 3b. The removal of the positive UPV results in a somewhat shallower low level trough, leading to certain decrease of the low level cyclonic vorticity over the region. The general flow pattern, however, is not altered.

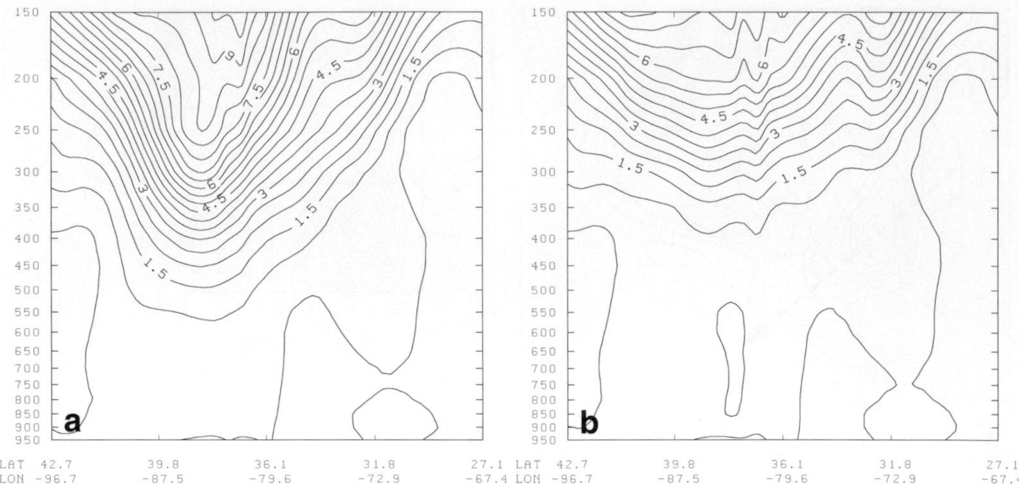

Figure 3: Vertical cross sections along the heavy dash arrow line in Fig. 2a. (a) PV without modification, and (b) PV with removal of 100% positive UPV within the small box in Fig. 2b

4.2. *General Results*

Table 3 summarizes the 24 h deepening rates from 1200 UTC 13 December through 1200 UTC 14 December for all sensitivity simulations. The following conclusions can be drawn from the table: 1) The removal of a portion of the UPV anomalies from the initial atmosphere expectantly results in weaker deepening rates, with nearly a half less deepening for the entire removal runs (PV10 and PV10BIG) and about one-quarter less deepening for the half removal runs (PV05 and PV05BIG); 2) The positive UPV within the upstream digging polar trough region (small box in Fig. 2b) accounts for almost the entire contribution of the upper-troposphere perturbation to the surface cyclogenesis, since the two experiments (PV05BIG and PV10BIG), with the UPV removal region

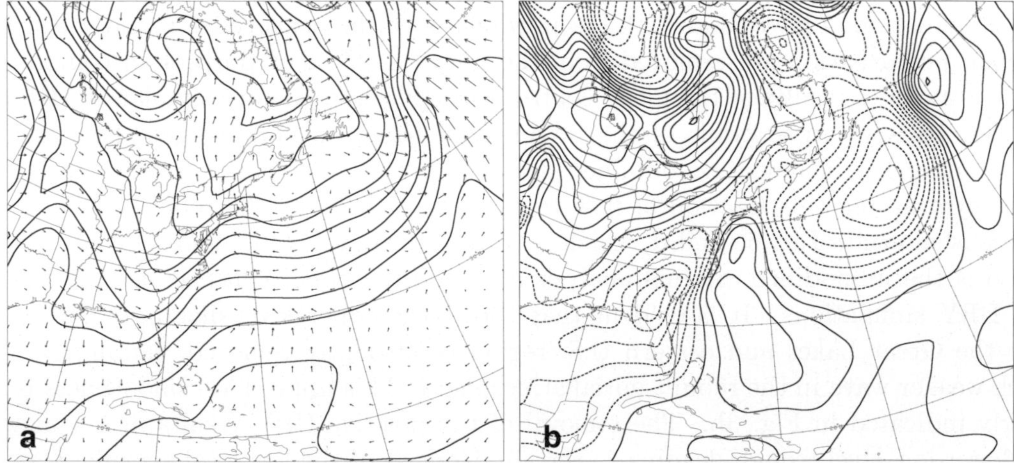

Figure 4: Surface wind vectors and potential temperature (5K interval) (a), and potential temperature anomaly from mean at initial time (1K interval, with negative values dashed) (b)

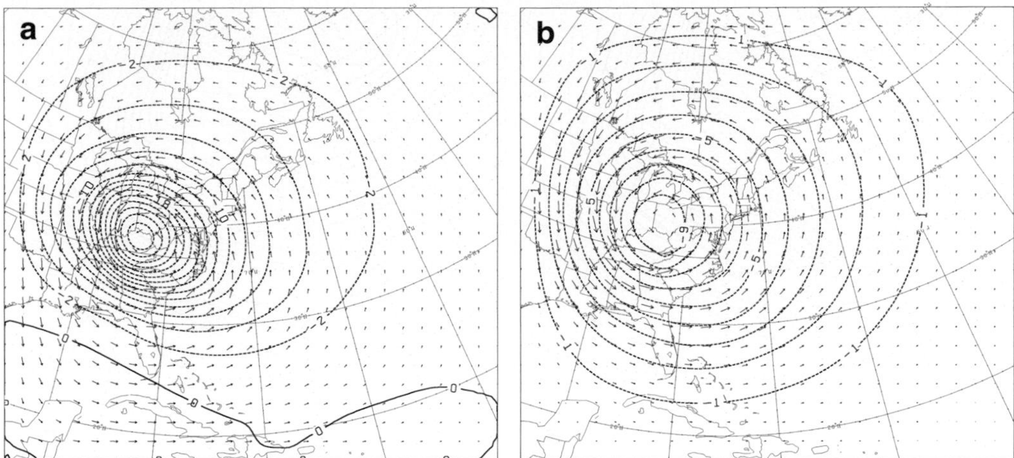

Figure 5: Geopotential height perturbation and nondivergent wind associated with the positive UPV (PV10) at 350 hPa (a) and 850 hPa (b), respectively. The contour interval is 2 dm in (a) and 1 dm in (b). Negative perturbations are dashed.

extending to cover the downstream ridge, have only small differences in the 24 h deepening rates comparing with PV05 and PV10, respectively; 3) It seems the elimination of the initial surface thermal perturbation alone (THETA) does not make any significant difference to the subsequent cyclogenesis, for its 24 h deepening still reaches 36.6 hPa, only 2 hPa less than the control run and well within the scope of explosive deepening extratropical cyclones. However, the simulation with both positive UPV and surface potential temperature perturbation removed (TH10), a combination of PV10 and THETA, strikingly eliminates the explosive cyclogenesis with a deepening rate equivalent to the dry cyclone. Obviously, the joint impact of the two perturbations exceeds the sum of impacts in association with each single perturbation. More analysis presented later in this section for the last two experiments suggest that among those major perturbations that contribute to rapid cyclogenesis the surface thermal wave would be an internal factor in that it could be efficiently controlled by other perturbations during the cyclone's life cycle.

Fig. 7 shows the 24 h sea surface pressure and 950 hPa temperature from the six sensitivity experiments. In general, an absence of the initial UPV produces less spinup

Table 3: Sensitivity tests with modifications of initial anomaly

case	24 hr deepening (hPa)*	description
KC	38.8	moist run (control) with Kuo scheme
DRY	11.0	dry run, without condensation
PV05	30.2	remove half positive UPV (small box)
PV10	20.6	remove all positive UPV (small box)
PV05BIG	29.1	remove half UPV (big box)
PV10BIG	19.5	remove all UPV (big box)
THETA	36.6	remove surface potential temperature anomaly
TH10	11.5	PV10 + THETA

* From 1200 UTC 13 to 1200 UTC 14 December 1988

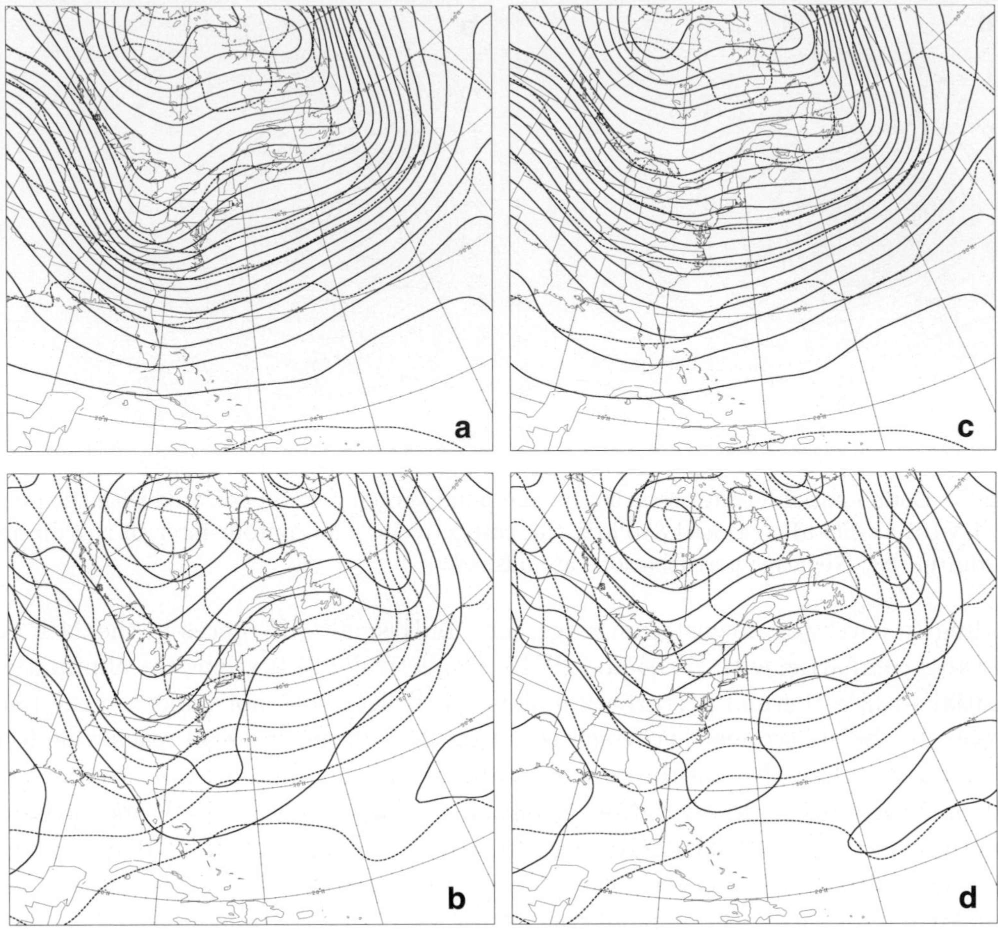

Figure 6: Initial geopotential height (solid lines) and temperature (dashed) of the unmodified initial atmosphere (a and b) and the rebuilt initial atmosphere with removal of the positive UPV (in PV10) (c and d). The upper panels (a and c) are at 350 hPa level, and the bottom panel (b and d) are at 850 hPa level.

of the surface cyclones with the low centers located somewhat southeastward and much weaker baroclinic zones along the warm front and bent-back front. This response feature is more obvious in the two fully removal runs (PV10 and PV10BIG). For PV10, the 24 h low center is located 200 km southeast of that in the control run. The low center for PV10BIG is another 150 km eastward. In both cases, the northern portion of the surface cyclone is shaped like a sharp inverted trough stretching north-northwest toward Maine and Nova Scotia, which correspondingly causes the isotherms in this region to bulge out and reduces the baroclinicity in the warm front region. Among all experiments, THETA has the most similar surface pattern to the control run, except for a southeastward offset of 90 km for the surface low center. On the other hand, the biggest difference exists when both upper level and surface perturbations are eliminated in the initial fields (TH10). This is like removing the background of vertical and/or slantwise convection development, so its result is similar to the dry run. Though its 24 h deepening rate is comparable to the dry cyclone, TH10 has a quite different surface pattern, with a largely modified deformation field and a weak thermal wave. The surface low is elongated along the northeast-southwest orientation, with a low center shifted southeastward about 350 km away.

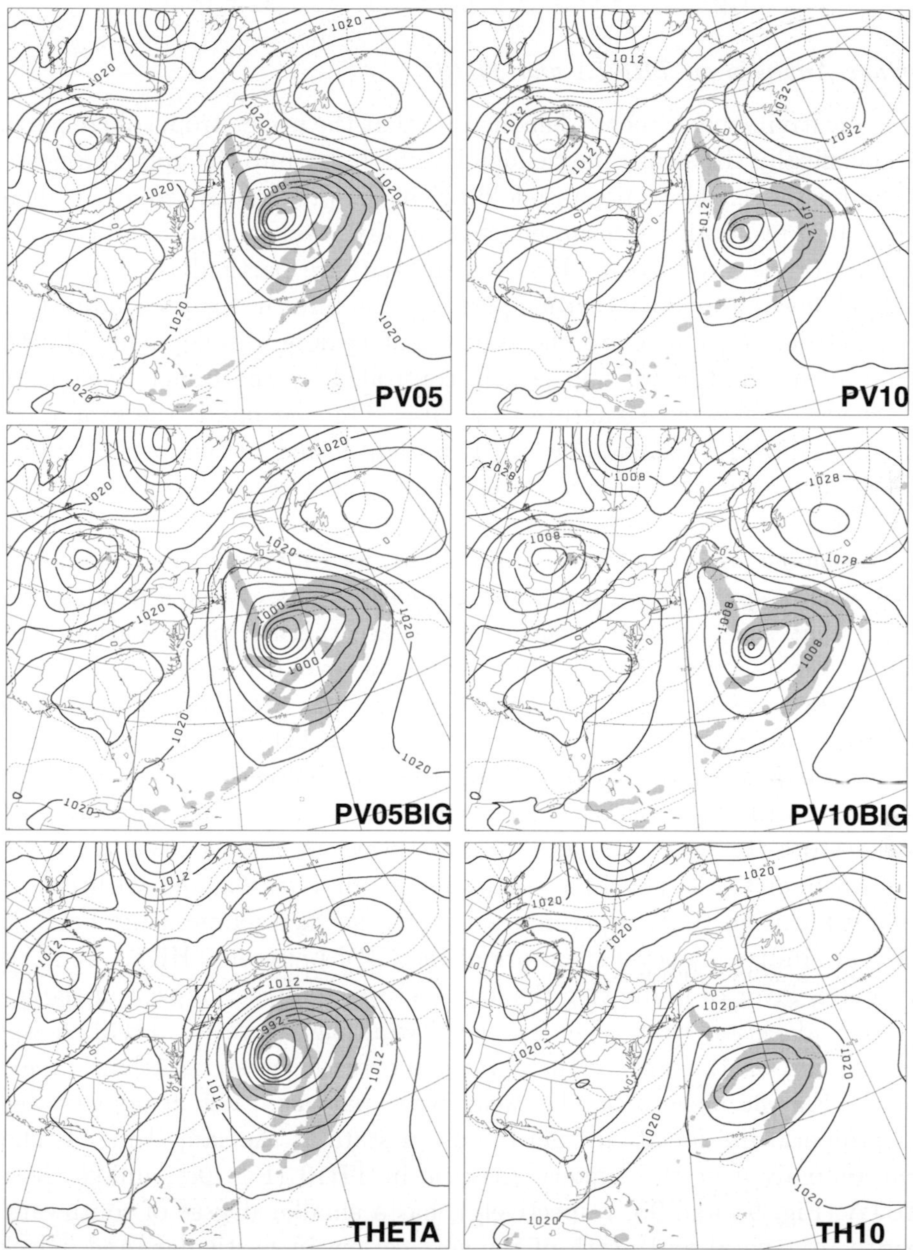

Figure 7: 24 h MSL isobars (solid lines, 4 hPa interval) and 950 hPa isotherms (dashed, 5 K interval) for sensitivity experiments with the modifications of initial atmosphere. Stippled areas refer surface precipitation rates over 1 mm h^{-1}.

Table 4: 900 hPa Geopotential Height Perturbation from PV inversion (unit: dm)

	THETA - Mean				
	06 h	12 h	18 h	24 h	30 h
θ'_B	-4.91	-8.45	-6.98	-10.16	-16.12
LPV	+1.82	-1.99	-9.54	-10.36	-10.81
UPV	-4.05	-5.44	-11.63	-7.22	-2.86

4.3. *Further Analysis on THETA and TH10*

The above sensitivity experiments demonstrate that the upper-troposphere perturbation (or more specifically the stratospheric high PV intrusion into the upper-troposphere, or, in some circumstance, the tropopause folding) in the preconditioned atmosphere does a significant job in promoting the subsequent explosive cyclogenesis in an integrated way. Implications from the results of THETA and TH10 are particularly impressive. It appears that the lower boundary thermal perturbation built in the preconditioning period could exert no lasting impact to the subsequent cyclone development. Furthermore, it could be an internal factor and largely controlled by the upper level perturbation through its low level wind signature (Fig. 5b). On one hand, the removed thermal perturbation in THETA seems to be swiftly re-established and supports the rapid deepening of the surface cyclone. On the other hand, the absence of UPV in TH10 makes this re-establishment process extremely inefficient. In order to quantitatively verify the foregoing statement, instantaneous PV inversions for different portions of perturbation in THETA with respect to the mean are also carried out. Table 4 lists the results. It can be seen that the initial removal of the surface thermal perturbation indeed does not affect its role as a major contributor to the low level deepening, the only exception being for the 18 h result which shows relatively weaker deepening associated with θ'_B. Comparison of Table 4 and Table 1 indicates that for THETA the relative importance of UPV increases while that of θ'_B decreases to some degree.

Fig. 8 and 9 provide a further elucidation to the different behavior of how THETA and TH10 restore the lower boundary thermal wave in the cyclone's early developing stage. Fig. 8 depicts the 900 hPa nondivergent winds associated with the UPV 6 h after the initialization time, superimposed on the surface temperature field. It clearly reveals a very similar low level signature of the UPV between THETA and the control run, which is in favor of promoting the low level warm advection east of the low center and the cold advection west. The low level wind signature for TH10, however, has much less impact on the surface thermal wave over the western Atlantic region. The isotherms in Fig. 8c show a similar weak wave pattern. Fig. 9 depicts the surface potential temperature deviation from the mean state at the same time. It shows that the warm anomaly over the western Atlantic in TH10 (Fig. 9c), unlike those in KC and THETA (Fig. 9a and 9b, respectively), has a smaller degree of southward extent, less broadth in general and north of the surface low in particular, and a lack of any northern maxima. In addition, the southwestern negative anomaly, which is obvious both in Fig. 9a and 18b, is nonexistent in Fig. 9c.

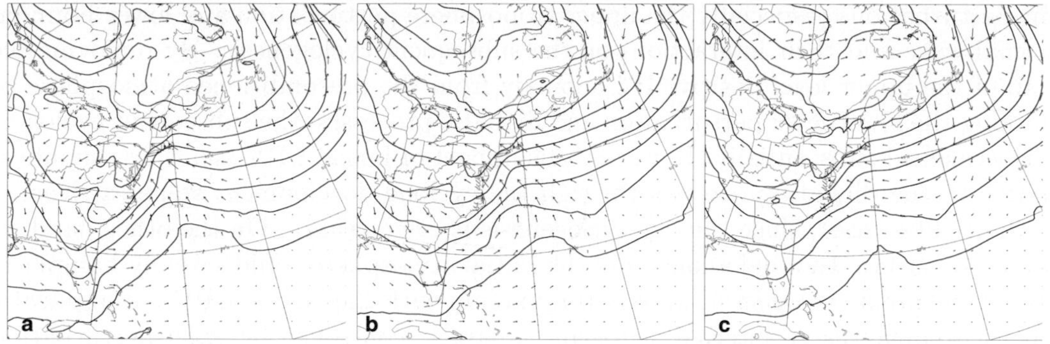

Figure 8: Surface isotherms (solid, 5 K) and 900 hPa nondivergent winds associated with UPV at 1800 UTC 13 December 1988 (6 h of simulation time) for KC (a), THETA (b), and TH10 (c), respectively.

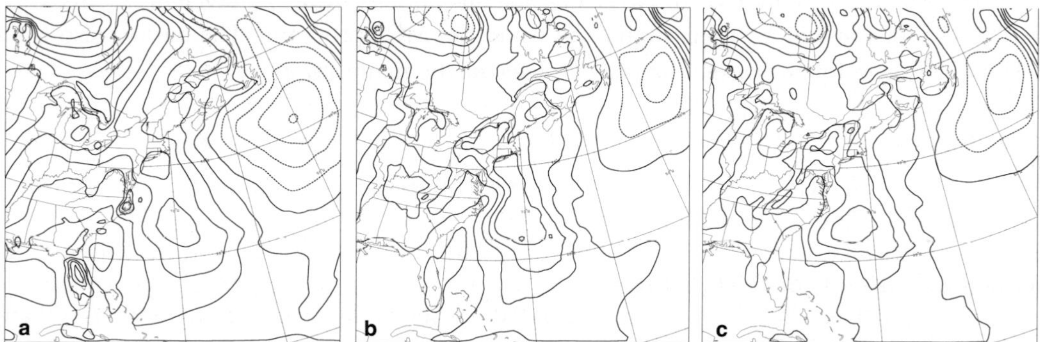

Figure 9: Surface potential temperature deviation from mean state at 1800 UTC 13 December (6 h of simulation time) for KC (a), THETA (b), and TH10 (c). The contour interval is 1 K and dashed contours refer to negative deviation.

5. Conclusions

The Canadian MC2 model is used to simulate an extratropical explosive marine cyclogenesis case ERICA IOP2, and PV diagnostics are employed to the simulation data. The goal is to understand the relative importance of different physics processes to the explosive cyclogenesis and how they interact each other. The inversions of different portions of PV anomaly from the simulation dataset suggest that the mid and lower level diabatic condensation-derived PV anomaly (LPV), attributed largely to the vertical and/or slantwise convections within the frontal rainbands, and the thermal advection within the lower boundary contribute major roles to the cyclone's deepening. The contributions grow during the most rapid deepening phase. It also shows that the diabatic processes appear to augment the positive contribution of the lower boundary thermal perturbation while to relatively weaken the UPV contribution by reducing the upper layer absolute circulation (UPV).

Sensitivity experiments are conducted using a methodology of modifying (or rebuilding) the initial conditions by means of piecewise PV inversion technique in order to further reveal the impacts of a preconditioned atmosphere through several important perturbations (especially the upper-troposphere perturbation and lower boundary thermal wave and their interaction) on the subsequent cyclogenesis. By removing part or all of the upper level positive PV anomaly from the preconditioned atmosphere but retaining the lower level perturbation and diabatic processes, the subsequent cyclogenesis is weakened substantially. However, this impact of initial UPV seems less significant relative to the diabatic processes through the cyclone development stage (at least for this particular case), since the dry cyclogenesis is even weaker. On the other hand, the absence of surface thermal anomaly in the initial atmosphere could severely hamper the subsequent rapid cyclogenesis only in conjunction with the lack of UPV, an implication that the anomaly is an internal factor and is largely controlled by other perturbations, especially by UPV.

Acknowledgments

The author wishes to thank Dr. M. K. Yau for his full support in this study, and the MC2 team at RPN/AES of Canada for their support in using the MC2 program. The author also thanks Dr. Davis Mechem for his careful language review of the manuscript.

References

Anthes, R. A., Y.-H. Kuo and J. R. Gyakum, 1983: Numerical simulations of a case of explosive cyclogenesis. *Mon. Wea. Rev.*, **111**, 1174-1188.

Balasubramanian, G., and M. K. Yau, 1994: The effects of convection on a simulated marine cyclone. *J. Atmos. Sci.*, **51**, 2397-2417.

——, and ——, 1996: The life cycle of a simulated marine cyclone: Energetics and PV diagnostics. *J. Atmos. Sci.*, **53**, 639-653.

Benoit, R., M. Desgagne, P. Pellerin, S. Pellerin, and Y. Chartier, 1997: The Canadian MC2: A semi-Lagrangian, semi-implicit wideband atmospheric model suited for finescale process studies and simulations. *Mon. Wea. Rev.*, **125**, 2382-2415.

Bosart, L. F., 1981: The Presidents' Day snowstorm of 18-19 February 1979: A subsynoptic-scale event. *Mon. Wea. Rev.*, **109**, 1542-1566.

Boyle, J. S., and L. F. Bosart, 1986: Cyclone-anticyclone couples over North America. Part II: Analysis of a major cyclone event over the eastern United States. *Mon. Wea. Rev.*, **114**, 2432-2465.

Davis, C. A., 1992a: Piecewise potential vorticity inversion. *J. Atmos. Sci.*, **49**, 1397-1411.

——, 1992b: A potential-vorticity diagnosis of the importance of initial structure and condensational heating in observed extratropical cyclogenesis. *Mon. Wea. Rev.*, **120**, 2409-2428.

——, and K. E. Emanuel, 1991: Potential vorticity diagnostics of cyclogenesis. *Mon. Wea. Rev.*, **119**, 1929-1953.

——, M. T. Stoelinga, and Y.-H. Kuo, 1993: The integrated effect of condensation in numerical simulations of extratropical cyclogenesis. *Mon. Wea. Rev.*, **121**, 2309-2330.

Eliassen, A., and E. Kleinschmidt, 1957: *Dynamic meteorology.* Handbuch der Physik, Vol 48, Springer Verlag, 1-154.

Gyakum, J. R., 1983a: On the evolution of the QE II storm. I: Synoptic aspects. *Mon. Wea. Rev.*, **111**, 1137-1155.

——, 1983b: On the evolution of the QE II storm. II: Dynamic and thermodynamic structure. *Mon. Wea. Rev.*, **111**, 1156-1173.

——, P. J. Roebber, and T. A. Bullock, 1992: The role of antecedent surface vorticity development as a conditioning process in explosive cyclone intensification. *Mon. Wea. Rev.*, **120**,

1465-1489.

Harnett, E., G. Forbes, and R. Hadlock, 1989: *ERICA Field Phase Summary.* ERICA Data Center, Drexed University, Philadelphia PA 19104, 217pp.

Hoskins, B. J., and P. Berrisford, 1988: A potential vorticity perspective of the storm of 15-16 October 1987. *Weather*, **43**, 122-129.

Huo, Z., D.-L., Zhang, and J. Gyakum, 1998: An application of potential vorticity inversion to improve the numerical prediction of the March 1993 superstorm. *Mon. Wea. Rev.*, **126**, 426-436.

——, ——, and ——, 1999: Interaction of potential vorticity anomalies in extratropical cyclogenesis. Part II: Sensitivity to initial perturbation. *Mon. Wea. Rev.*, **127**, 2563-2575

Kong, F.-Y., and M. K. Yau, 1997: An explicit approach of microphysics in MC2. *Atmosphere-Ocean*, **35**, 257-291.

Kuo, H. L., 1974: Further studies on the parameterization of the influence of cumulus convection on large-scale flow. *J. Atmos. Sci.*, **31**, 1232-1240.

Kuo, Y.-H., and R. J. Reed, 1988: Numerical simulation of an explosively deepening cyclone in the eastern Pacific. *Mon. Wea. Rev.*, **116**, 2081-2105.

——, ——, and S. Low-Nam, 1991a: Effects of surface energy fluxes during the early development and rapid intensification stage of seven explosive cyclones in the western Atlantic. *Mon. Wea. Rev.*, **119**, 457-476.

——, M. A. Shapiro, and E. G. Donall, 1991b: The interaction between baroclinic and diabatic processes in a numerical simulation of a rapid intensifying extratropical marine cyclone. *Mon. Wea. Rev.*, **119**, 368-384.

Leslie, L. M., G. J. Holland, and A. H. Lynch, 1987: Australian east-coast cyclones. Part II: Numerical modeling study. *Mon. Wea. Rev.*, **115**, 3037-3054.

Liou, C. S., and R. L. Elsberry, 1987: Heat budgets analyses and forecasts of an explosively deepening maritime cyclone. *Mon. Wea. Rev.*, **115**, 1809-1824.

Manabe, S., 1956: On the contribution of heat released by condensation to the change in pressure pattern. *J. Meteor. Soc. Jap.*, **34**, 12-24.

Neiman, P. J., M. A. Shapiro, and L. S. Fedor, 1993: The life cycle of an extratropical marine cyclone. Part II: Mesoscale structure and diagnostics. *Mon. Wea. Rev.*, **121**, 2177-2199.

Reed, R. J., and M. T. Stoelinga, and Y.-H. Kuo, 1992: A model-aided study of the origin and evolution of the anomalously high potential vorticity in the inner region of a rapidly deepening marine cyclone. *Mon. Wea. Rev.*, **120**, 893-913.

Reuter, G. W., and M. K. Yau, 1993: Assessment of slantwise convection in ERICA cyclones. *Mon. Wea. Rev.*, **121**, 375-386.

Roebber, P. J., 1993: A diagnostic case study of self-development as an antecedent conditioning process in explosive cyclogenesis. *Mon. Wea. Rev.*, **121**, 976-1006.

Sanders, F., 1990: Surface analysis over the oceans - Searching for sea truth. *Weather and Forecasting*, **5**, 596-612.

Sanders, F., and J. R. Gyakum, 1980: Synoptic-dynamic climatology of the "bomb". *Mon. Wea. Rev.*, **108**, 1589-1606.

Shapiro, M. A., and D. Keyser, 1990: *Fronts, jet streams and the tropopause. Extratropical Cyclones: The Erik Palmen Memorial Volume*, C. W. Newton and E. Holopainen, Eds., Amer. Meteo. Soc., 167-191.

Uccellini, L. W., 1986: The possible influence of upstream upper-level baroclinic processes on the development of the QE II storm. *Mon. Wea. Rev.*, **114**, 1019-1027.

Whitaker, J. S., L. W. Uccellini, and K. F. Brill, 1988: A model-based diagnostic study of the rapid development phase of the Presidents' Day cyclone. *Mon. Wea. Rev.*, **116**, 2337-2365.

Part II

Climate Modeling and Numerical Weather Prediction

CLOUD-CLIMATE FEEDBACK: HOW MUCH DO WE KNOW?

MINGHUA ZHANG
Institute for Terrestrial and Planetary Atmospheres
Stony Brook University, Stony Brook, NY 11794-5000, USA
E-mail: mzhang@notes.cc.sunysb.edu

(Manuscript received 6 March 2003)

This paper introduces the concept of cloud-climate feedback along with its role in the sensitivity of the climate system. It reviews available cloud feedback diagnostic methods and representative results in three-dimensional atmospheric general circulation models. A case study is presented to analyze physical processes responsible for cloud feedbacks that control the sensitivity of the NCAR Community Climate Model. The paper further discusses weaknesses of current numerical models and areas of required research on this subject.

1. Introduction

Potential global warming caused by increasing greenhouse gases in the atmosphere is currently a major environmental concern worldwide. Given the complexity of the climate system, three-dimensional numerical models are the only tools to reliably project future climate changes in response to human activities.

One of the unsettling issues from using these models is the large discrepancy in the magnitude of the simulated global warming in different models. In response to a doubling of CO_2 concentration in the atmosphere, coupled general circulation models (GCMs) projected warming that ranged from 1.9 to 5.2 °C in 1990 (Mitchell et al. 1990). In 1995, this range was 2.1 to 4.6 °C (Kattenberg et al. 1996). In the latest IPCC report, this range is 2.0 to 5.1 °C (Cubasch et al. 2001). This spread of model projections casts large uncertainties to regional climate changes and prompts indeterminable societal responses. There is therefore the need to understand and narrow this model difference.

Cloud-climate feedback emerged as the leading cause of model uncertainties. I use Figure 1 to schematically show the physical concept. Clouds, consisting of water or ice particles, affect radiative transfer. Clouds reflect solar (shortwave) radiation to space, thus serving as a cooling agent to the Earth atmosphere system. This is analogous to an umbrella in the summer to shield the Earth from sunlight. Clouds also act as a greenhouse agent to the infrared radiation (longwave). This role of clouds is analogous to a blanket that keeps the Earth warm. The question is: how do these umbrella and blanket effects of clouds vary as a result of climate change? In a climate warming scenario, if the cooling umbrella becomes larger while the greenhouse blanket does not change, the variation of clouds would offset the warming, and thus a negative feedback. If the greenhouse blanket is larger while the cooling umbrella is fixed, variation of clouds would amplify the initial warming. In reality, clouds are an internal variable of the climate system. The size and thickness of both the umbrella and the blanket vary with the climate. The exact cooling and warming effects of clouds depend on the height, location, amount, and the microphysical and radiative properties of clouds, as well as their appearance of time with respect to the seasonal and diurnal cycles of the incoming solar radiation. Simple theoretical calculations, with hypothetical yet reasonable assumption on cloud variation, show that clouds can indeed either significantly reduce or amplify a global warming projection.

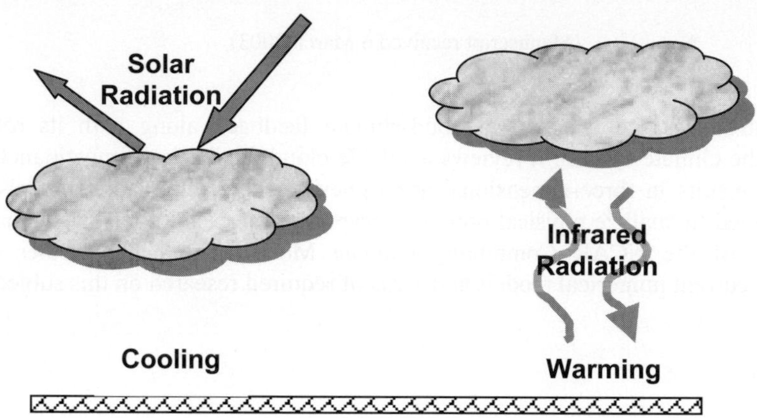

Figure 1. Schematic illustration of the solar and infrared effects of clouds on surface temperature.

Cloud-climate feedback has therefore risen to the list of highest priority in the U.S. Global Change Program. Several satellite and surface measurement programs were specifically established to narrow model uncertainties of cloud feedbacks. After about 15 years of intensive research, however, this issue is still evasive.

The objective of this paper is to appraise where we stand in the research of cloud-climate feedback, and to point out remaining challenges. The paper is organized as follow. Section 2 introduces the definition of cloud feedback and diagnostic methods in climate models. Section 3 reviews representative cloud feedback results that highlight differences among models. Section 4 reports a case analysis by using the latest version of the NCAR Community Atmospheric Model Version 2 (CAM2). Section 5 further evaluates simulated clouds in the CAM2 by using observations. The last section contains a summary and a discussion on challenges and related science issues.

2. Definition and Diagnostic Methods

2.1. *Terminology*

The quantitative definition of feedback was first used in electrical signal control systems (Bode 1945). Given forcing ΔQ to a system, if the system responds by one measure ΔT_0, and if this response does not impact the initial forcing, the system is said to be without feedback.

This is schematically shown in Figure 2a. When the relationship between the response and the forcing is written as

$$\Delta T_0 = G_0 \Delta Q, \tag{1}$$

G_0 is defined as the zero-feedback gain. It describes the system response per unit forcing. The system does not have to be linear. Nevertheless, it is more convenient to consider both the forcing and response as small perturbations, thus the linear approximation holds, so that G_0 is independent of the magnitude of the forcing.

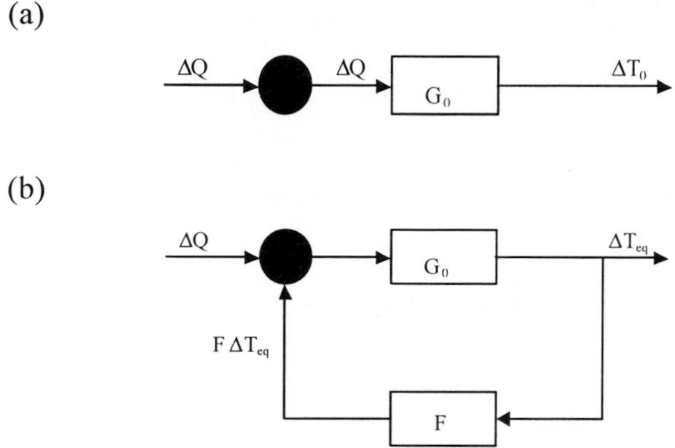

Figure 2. System response to a forcing: (a) without feedback, (b) with feedback.

If the system response also induces changes in the forcing, as schematically shown in Figure 2b, and if the induced forcing is written as proportional to the final response ΔT_{eq} by a factor of F, namely $F\Delta T_{eq}$, then F is called the feedback. Feedback therefore is *the induced forcing from a unit response*. The total forcing to the system at the final state is $\Delta Q + F\Delta T_{eq}$. As a result,

$$\Delta T_{eq} = G_0(\Delta Q + F\Delta T_{eq}). \tag{2}$$

It follows from the above that

$$\Delta T_{eq} = \frac{G_0 \Delta Q}{1 - G_0 F}. \tag{3}$$

This equation can be also written in terms of the zero feedback response ΔT_0 as

$$\Delta T_{eq} = \frac{\Delta T_0}{1-f}, \tag{4}$$

where

$$f = G_0 F. \tag{5}$$

The dimensionless parameter f is called the feedback factor. It scales the feedback against the zero-feedback gain.

Since there could be more than one process that can induce changes in the forcing, Equation (2) can be expanded to:

$$\Delta T_{eq} = G_0 (\Delta Q + \sum_i F_i \Delta T_{eq}). \tag{6}$$

Equation (4) then becomes

$$\Delta T_{eq} = \frac{\Delta T_0}{1-\sum_i f_i}, \tag{7}$$

where

$$f_i = G_0 F_i. \tag{8}$$

While feedbacks and feedback factors are additive, the system response to additional feedback is not. Theoretically, the denominator in (4) and (7) could be zero or negative. In the case of negative value, the final response of the system is opposite to what is initially forced. In the zero case, the system is unstable.

2.2. *Application to climate models*

If we write the net *upward* radiative flux at the top-of-the-atmosphere (TOA) as N, then

$$N = N(Solar\ and\ GHG\ Forcing,\quad T_s, H_2O, Clouds, Snow, T_{air}\ \cdots,\quad) \tag{9}$$

where *GHG* represents greenhouse gases" such as CO_2 and methane etc., T_s is the surface temperature, other variables are as commonly used. With a small climate change from one equilibrium to another caused by a forcing, since $\Delta N = 0$, one has

$$-\Delta N_{forcing} = \frac{\partial N}{\partial T_s}\Delta T_s + \frac{\partial N}{\partial (H_2O)}\frac{d(H_2O)}{dT_s}\Delta T_s + \frac{\partial N}{\partial (Clouds)}\frac{d(Clouds)}{dT_s}\Delta T_s + \ldots \tag{10}$$

In the above, $-\Delta N_{forcing}$ is the change of net *downward* radiation at TOA as a result of a direct forcing (since N is defined as the upward flux), such as from an increase in solar radiation or increase of CO_2. The right hand side of the above equation describes the increase of upward radiation as a result of the induced climate change. Equation (10) is therefore a simple balance of forcing and response.

Without internal feedbacks --- no contributions from water vapor, clouds and other terms on the right hand side of (10) except for the surface temperature, Equation (10) becomes

$$-\Delta N_{forcing} = \frac{\partial N}{\partial T_s}\Delta T_s \, . \tag{11}$$

When the simple forcing and response format of Equation (1) is used, the above is written as

$$\Delta Q = G_0^{-1}\Delta T_s \, , \tag{12}$$

where the zero-feedback gain is

$$G_0 = \left(\frac{\partial N}{\partial T_s}\right)^{-1} . \tag{13}$$

G_0 can then be estimated with the Stefan-Boltzman law using an effective atmospheric emissivity ε through

$$\frac{\partial N}{\partial T_s} = \varepsilon 4\sigma T_s^3 = \frac{4\times \mathrm{OLR}}{T_s} .$$

where σ is the Stefan-Boltzman constant, and OLR is the outgoing longwave radiation at TOA. With 240 W m^{-2} for OLR, and 280 K for T_s, the zero feedback gain is estimated to be about 0.3 K/(W m^{-2}). For a doubling of CO_2, the greenhouse forcing is about 4 W m^{-2}, which corresponds to a direct warming of 1.2 K.

With feedbacks from water vapor, clouds and other processes, (10) can be written as

$$\Delta Q = \frac{1}{G_0}\Delta T_s + \frac{\partial N}{\partial (H_2O)}\frac{d(H_2O)}{dT_s}\Delta T_s + \frac{\partial N}{\partial (Clouds)}\frac{d(Clouds)}{dT_s}\Delta T_s + \ldots . \tag{14}$$

or

$$\Delta T_s = \frac{G_0 \Delta Q}{1 - G_0 F} , \tag{15}$$

where the total feedback is

$$F = -\frac{\partial N}{\partial(H_2O)}\frac{d(H_2O)}{dT_s} - \frac{\partial N}{\partial(Clouds)}\frac{d(Clouds)}{dT_s} - \ldots = -\frac{\delta_{H_2O}N}{dT_s} - \frac{\delta_{Clouds}N}{dT_s} - \ldots \quad (16)$$

The individual feedbacks are thus the partial differentiation of the net downward radiative flux at TOA with respect to the individual physical process (water vapor variation, cloud variation etc.) accompanying a unit change of surface temperature. For example, the first term on the right hand side of (16) is the water vapor feedback, and the second term is the cloud feedback.

This definition of feedbacks for climate models is also used in Wetherald and Manabe (1988) and Schlesinger (1988). The feedback has a unit of W m^{-2} K^{-1}.

Some caveats are pointed here. First, the original definition of feedback was introduced for a zero-dimensional system. For the three-dimensional climate system, it is customary to use the globally averaged net radiation at TOA as the forcing and the globally averaged surface temperature as the response. The partial differentiations, however, should be evaluated from spatially varying physical quantities. Second, since air temperature is strongly coupled to surface temperature, in the zero-feedback calculation, the atmospheric temperature change is often assumed to be the same as that of the surface. The remaining air temperature variation can be considered as a temperature lapse rate feedback. The decomposition of feedbacks is therefore subjective that should be guided by the gain of physical insights. Third, feedback analysis does not directly lead to improved models. It, however, helps to pinpoint why a model is sensitive or insensitive to a forcing. This is similar to a physician diagnosing a disease without directly curing it.

2.3. *Diagnostic methods*

Based on the above discussion, there are two ways to calculate feedbacks in climate models. One is to impose forcing and calculate the response in a model. Feedback processes are introduced into the model one at a time by arbitrarily holding other physical quantities fixed. The coefficient terms of ΔT_s in Equation (14) are then obtained by differencing results from two experiments. This approach is feasible for feedback studies using simple climate models since it requires multiple climate change integrations of the model.

The second method is to directly calculate the partial differentiations of radiative flux in Equation (16) by using a single climate change simulation. The partial differentiations with respect to various physical quantities are calculated offline. This method was used in Wetherald and Manabe (1988) and in Zhang et al. (1994).

An elegant implementation of the second method, specific for diagnosing model cloud feedbacks, was introduced by Cess and Potter (1988). Instead of obtaining a climate change by imposing a forcing and calculating the system response, they prescribed a simple climate change at the sea surface and calculated the induced radiation perturbation at the top of the atmosphere. The method thus imposes ΔT_s and calculates the required forcing ΔQ in

Equation (14). The partial differentiation of radiative fluxes with respect to clouds is calculated from the change of cloud-radiative forcing (CRF), which is a standard model diagnostics after ERBE (Ramanathan 1987). The cloud feedback in Equation (16) therefore becomes a simple diagnostic of

$$F_{clouds} = \frac{\Delta CRF}{\Delta T_s}. \tag{17}$$

In Cess and Potter (1988), a uniform plus or minus 2 degree SST perturbation was used. It is noted that cloud feedback can be different under different forcing conditions. Yet, the feedback analysis from this type of surrogate climate change has provided considerable insights about climate model sensitivities. Furthermore, as has been recently shown with several GCMs, cloud feedback diagnosed from this surrogate climate change is consistent with those diagnosed directly from global warming simulations.

3. Model Results

3.1. *Results from early models*

Early studies of cloud feedbacks were carried out through zero-dimensional energy balance models and one-dimensional radiative-convective models (RCM). Since clouds are not explicitly calculated in these models, empirical relationships between clouds and temperature had to be used. Representative works include Budyko (1969), Schneider (1972), and Cess (1975) among others. A review of these studies can be found in Schlesinger (1988). In the present paper, we restrict our discussion to cloud feedback analysis in GCMs.

Hansen et al. (1984) were the first to diagnose cloud feedback from a GCM. They combined the GISS GCM output from a CO_2 climate change simulation with a one dimensional RCM. The RCM was used to calculate the change of surface temperature in response to a doubling of CO_2, with variations of water vapor, clouds, and snow replaced by those from the GCM simulations. In the case of cloud feedback, Hansen et al. (1984) separated it into cloud amount feedback and cloud height feedback. The first was obtained by inserting a globally averaged total cloud change in the GCM into the whole column of the RCM. The cloud height feedback was calculated as a residual of the total cloud feedback and the cloud amount feedback. Hansen et al. (1984) reported that both feedbacks were positive as a result of the following process. In a warmer climate, the model had a reduction of clouds and a shift of clouds to higher altitude. Reduction in low and middle clouds has larger impact on solar radiation than on the infrared radiation, and thus the net cloud cooling becomes smaller, a positive cloud amount feedback. On the other hand, the raised cloud altitude enhances the cloud greenhouse effect of clouds, also producing a positive feedback. The diagnosed cloud feedbacks from the two processes are listed in Table 1. With a zero-feedback gain of 0.3 K /(W m^{-2}) as discussed earlier, the cloud feedback of 0.73 (W m^{-2} K^{-1}) translates to a feedback factor of $0.73 \times 0.3 = 0.22$, which alone amplifies the global warming by a factor of $1/(1-f) = 1/0.78 = 1.28$.

Table 1. Cloud feedback results

	Cloud amount feedback (W/m^2/K)	Cloud height feedback (W/m^2/K)	Total cloud feedback (W/m^2/K)
Hansen et al. (1984)	0.33	0.40	0.73
Wetherald and Manabe (1988)	0.12	0.25	0.37

Wetherald and Manabe (1988) calculated the partial differentials of the TOA net downward radiation with respect to various physical quantities, by using the GFDL GCM, to derive the feedback components in Equation (16). The cloud amount feedback was obtained by scaling the model cloud radiative forcing with the variation of cloud amount. The cloud height feedback was derived as a residual similarly to Hansen et al. (1984). Modest positive cloud amount and cloud height feedbacks were reported. They are also listed in Table 1. These feedbacks were also attributed to reduction of middle and low clouds and increase of high clouds.

3.2. *Model intercomparisons*

The similarity of diagnosed positive cloud feedbacks between the early GISS GCM and the GFDL GCM was later found to be a coincidence rather than a true physical consistency. Cess et al. (1990) used the surrogate climate change of plus and minus 2 K over the oceans and diagnosed the cloud feedback through Equation (17) in 19 GCMs. Figure 3 shows the total cloud feedback from these models. It is seen that they range from negative to strongly positive. These feedback values, along with a zero-feedback gain of 0.3 K/(W m^{-2}), correspond to a feedback factor of –0.2 to 0.4. This would imply a difference of a factor of two difference in the temperature response to a prescribed forcing.

The Cess et al. (1990) study was updated in Cess et al. (1996) and it was found that the models continue to show serious physical disagreements as measured by cloud feedbacks from infrared and solar components separately, even though the range of the net cloud feedbacks appeared to be smaller.

The next interesting set of GCM cloud feedbacks was reported by Senior and Mitchell (1993, 1996). They used the same version of the unified GCM at the UK Met Office (UKMO) with only small modifications to the cloud scheme. CO_2 climate change simulations were carried out and cloud feedbacks were diagnosed. Table 2 lists cloud feedbacks from three experiments. Experiment B is from the standard UKMO GCM using the cloud scheme as described in Smith (1990). Experiment A differs from the standard model in that cloud particle size is calculated based on the cloud liquid amount through an empirically observed relationship. Experiment C differs from B in that the assumed subgrid scale distribution of total water within a GCM grid is changed from a triangle distribution to a top-hat distribution.

It is seen that the model exhibited quite large changes in cloud feedbacks. The magnitudes of the simulated global warming are also listed in the table and they are very different. This sensitivity has been also demonstrated in other GCMs (Le Treut et al. 1994).

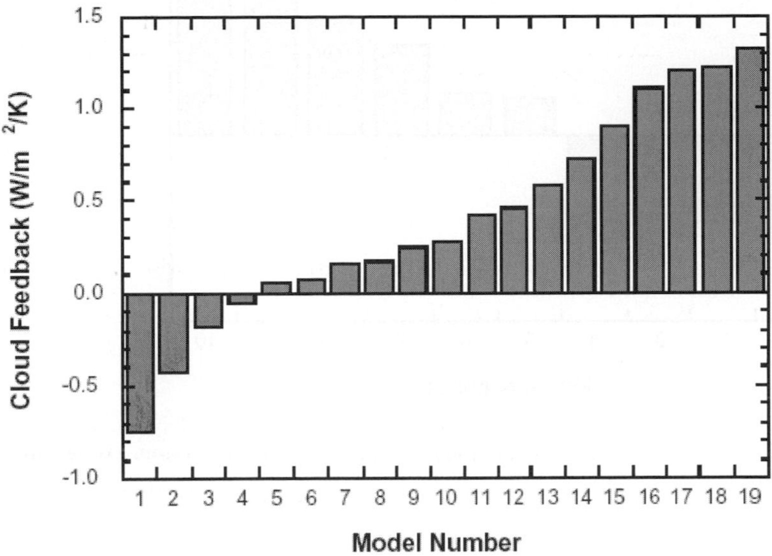

Figure 3. Cloud feedback (W m^{-2} K^{-1}) in 19 General Circulation Models. Adapted from Cess et al. (1990).

Table 2, Climate change simulation and cloud feedbacks in the UKMO GCM

Experiment ID	A	B	C
ΔT_S (°C)	1.9	3.4	5.5
ΔCRF (W/m^2)	−1.04	0.93	3.64
Cloud Feedback (W/m^2/K)	−0.55	0.27	0.66

Figure 4 shows an updated model intercomparison of the change of CRF from doubling CO_2 simulations in 10 GCMs as reported in Stocker et al. (2001). Since the direct radiative perturbation from a doubling of CO_2 is about 4 W m^{-2}, cloud feedbacks in the figure add up to a range of 2.8 W m^{-2} to over 7.0 W m^{-2} of radiative imbalance in the models, which is sufficiently large to explain the spread in simulated global warming as reported in Cubasch et al. (2001). Figure 4 suggests that cloud feedback uncertainties in current climate models are still about as large as they were fifteen years ago.

At the writing of this paper, an international model intercomparison project is being initiated to carry out a systematic assessment of cloud feedbacks in climate models and why they differ from each other (McAveney and Le Treut, 2003, personnel communication).

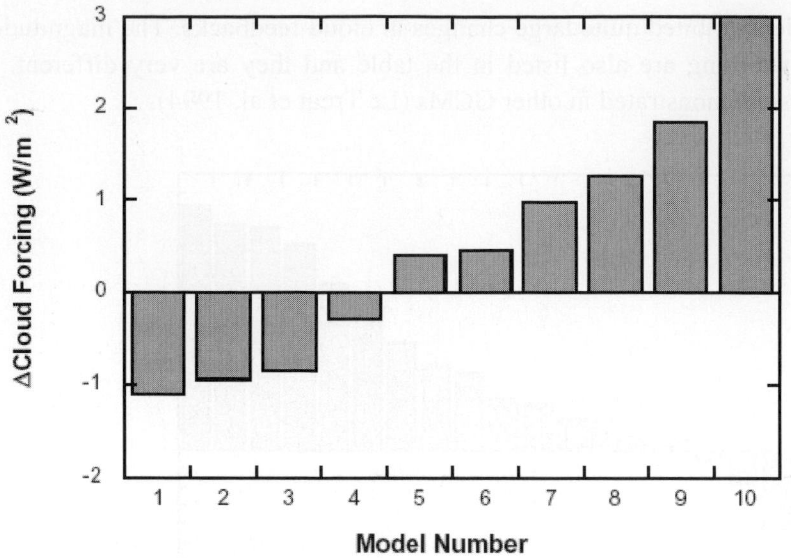

Figure 4. Variation of cloud radiative forcing (W m^{-2}) in doubling CO_2 simulations from 10 General Circulation Models. Adapted from Stocker et al. (2001).

4. Case Study Using the NCAR Community Atmospheric Model

The NCAR Community Climate Model (CCM) is one of the most widely used atmospheric GCMs for climate simulation studies. The model has evolved over the years with continuing modifications and enhancements in its physical components (Williamson et al. 1987; Hack et al. 1993; Kiehl et al. 1996; Collins et al., 2003). Figure 5 shows the evolution of diagnosed cloud feedbacks in CCM0, CCM1, CCM2, CCM3, and the latest Community Atmospheric Model Version 2 (CAM2). It is seen that cloud feedback in this model started from a strong positive feedback to a modest negative feedback since the introduction of the CCM2.

To understand the physical mechanism behind these diagnostics, a brief description of the essential changes to the model cloud parameterization is given here. CCM0 and CCM1 had several important commonalities in their cloud parameterizations. In these models, stratiform clouds were assigned a fixed cloud amount of 95% when there was large-scale condensation in a grid box. Convective clouds were assigned a 30% total amount and the whole convective column was assumed to have randomly overlapping clouds. CCM0 required eighty percent relative humidity for large-scale condensation to occur, while CCM1 required one hundred percent humidity. Cloud radiative properties were specified by using condensed water from large-scale saturation and a moist adiabatic adjustment convection scheme (Ramanathan et al. 1983). Increased cloud amount in the upper troposphere in a warmer climate is likely responsible for the strong positive feedback in these two versions of the model.

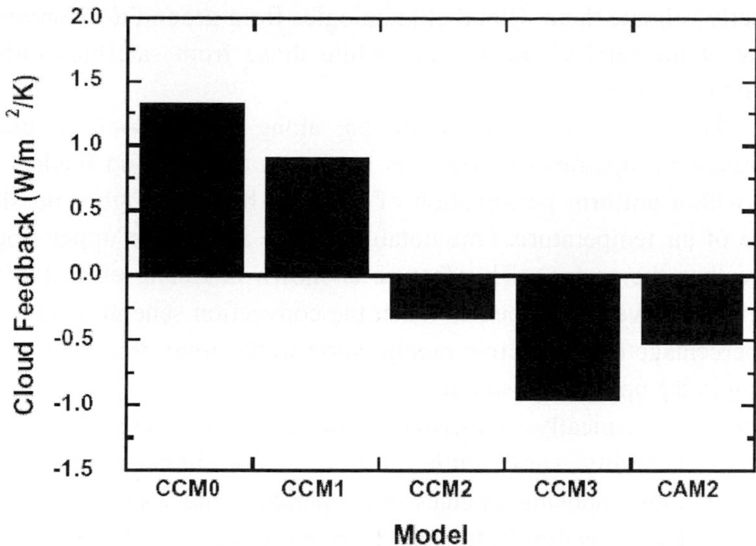

Figure 5. Evolution of cloud feedbacks (W m^{-2} K^{-1}) in different versions of the NCAR Community climate model. CCM0 and CCM1 results are derived from Cess et al. (1990). CCM2 result is from Zhang et al. (1994). Feedbacks from CCM0 to CCM3 are derived from perpetual July surrogate climate change. Feedback from CAM2 is derived from prescribing sea surface temperature variation from a doubling CO_2 experiment in the UKMO GCM.

CCM2 allowed the calculation of fractional cloudiness following Slingo (1987). Parameterized relationships between grid-scale relative humidity and cloud amount were used. Vertical velocity and stability were also used as input. Convective clouds were parameterized based on convective precipitation from the Hack (1993) mass flux scheme.

CCM3 used essentially the same cloud parameterization as in CCM2, except that convective cloud amount was parameterized based on the deep convective mass flux in the Zhang and McFarlane (1995) scheme. The Hack (1993) scheme is still used to simulate shallow convection, but it does not directly generate clouds.

The latest CAM2 contains three main changes related with clouds. First, boundary layer stratus clouds are calculated to be proportional to the atmospheric vertical stability between 700 mb and surface following the observations of Klein and Hartmann (1993). Second, CAM2 uses a microphysical prognostic cloud condensate scheme of Rasch and Kristjansson (1998) and a macrophysical formulation of Zhang et al. (2003). Third, falling rain is allowed to evaporate in the Zhang and McFarlane (1993) convection scheme.

We will contrast cloud feedbacks in the CCM2 and in the CAM2 to gain some insight in these models. Figure 6a shows the latitude-pressure distribution of the zonally averaged cloud amount in the CCM2 in a perpetual July simulation. It is characterized by two minimum regions of clouds in the subtropics that are associated with the descending branches of the Hadley circulations, and three maximum cloud regions associated with the Inter Tropical Convergence Zone (ITCZ) and middle latitude frontal clouds in the two hemispheres. Note that while these features of vertical cloud distributions qualitatively agree with our understanding of the general circulation of the atmosphere, there are no observational measurements to directly validate them. Cloud climatologies from ground measurements only give an upward view of the total cloud amount, while those from satellites only give a downward view of the total cloud.

The variation of the cloud pattern in Figure 5a, along with associated changes in microphysical and radiative properties of clouds, is important to the cloud feedback. When the CCM2 is forced with a uniform perturbation of the SST by 4 K, Figure 6b shows the corresponding change of air temperature. One notable feature is that the upper troposphere warms up much more than the surface. This feature is shown in Zhang et al. (1994) to be dependent on the cumulus convection scheme. When the convection scheme is less rigorous, as measured by the percentage of convective precipitation in the total precipitation, there is less amplified warming in the upper troposphere.

Warmer temperature is typically associated with more moisture. It is generally understood that relative humidity varies little in a changed climate, since there is a cancellation between two large opposing changes in temperature and water vapor. However, the residual after the cancellation matters in the cloud variation and cloud feedback. Figure 6c shows the simulated change of clouds in the CCM2. There is a broad reduction of clouds in the middle and upper troposphere, which is closely related with the broad warming in Figure 6b. There is also an increase of cloudiness near the tropopause, associated with increased convection.

The reduction of clouds above 500 mb in the model turns out to dominate the model cloud feedback. Because of the relatively high altitude of cloud variations, the longwave effect overrides the shortwave effect, which permits more longwave radiation to escape to space, and thus a negative feedback in the model. Figure 7 shows the decomposed cloud feedbacks from infrared and solar radiation. The net negative cloud feedback is driven by the negative infrared cloud feedback.

The corresponding feedback components in the CAM2 are also plotted in Figure 7. While both CCM2 and CAM2 have negative net cloud feedbacks, it is seen that in CAM2 it is the negative solar component that dominates the cloud feedback, i.e., the enhanced reflection of solar radiation by clouds. This increase of solar reflection in CAM2 is a result of increased low-level stratus that can be traced to the modifications of cloud parameterizations. Figure 8 shows the latitudinal distribution of CRF variation for the southern summer season. The reduction of cloud forcing is primarily in the shortwave component at around 65°S and in the subtropics of the two hemispheres. Figure 9 shows the geographical distribution of the shortwave CRF in the control simulation and in the warmer simulation in the northern winter. The change of shortwave shortwave CRF is distinct around the sea ice line in the southern hemisphere. As the surface warms up, the atmosphere warms up globally. As a result, near the sea ice, the vertical stability of the lower atmosphere is increased, which in turn produces more low clouds in the CAM2 to reflect solar radiation.

The CAM2 modification of rain evaporation in the cumulus convection scheme also weakens the model convection, which is speculated to reduce the magnitude of temperature and cloud variations in the upper troposphere. This should contribute to the disappearance of negative longwave cloud feedback in CAM2.

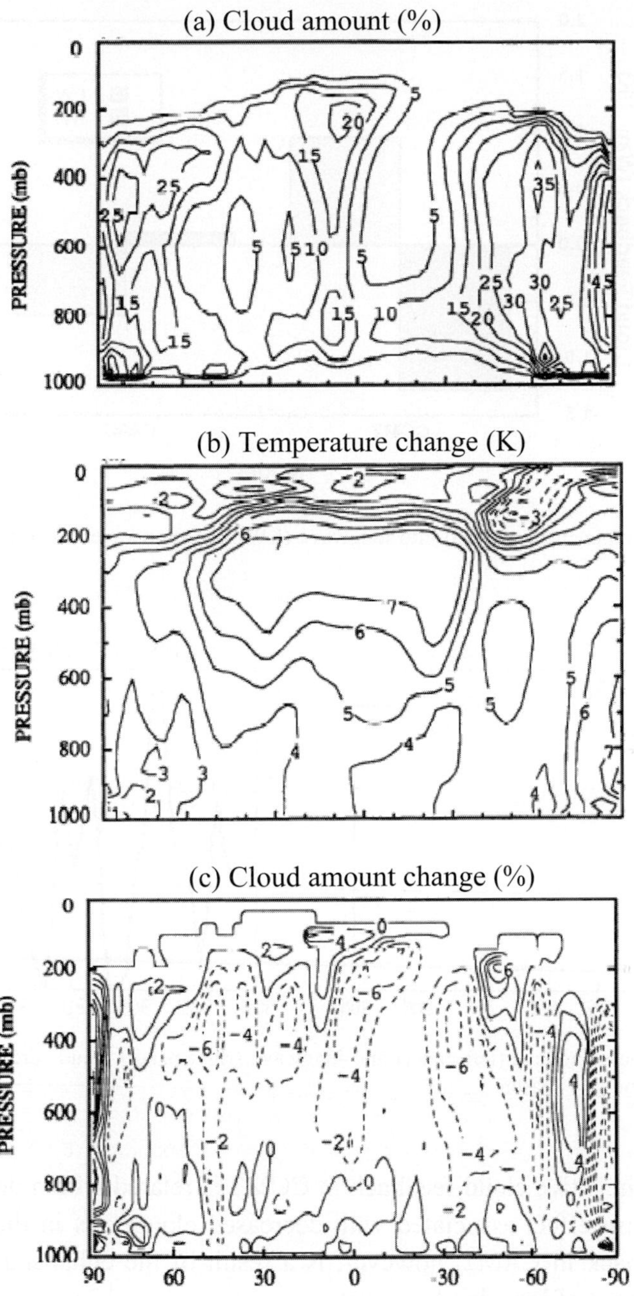

Figure 6. Zonal-pressure distributions of: (a) clouds, (b) temperature variation in perpetual July surrogate climate change, and (c) cloud variation.

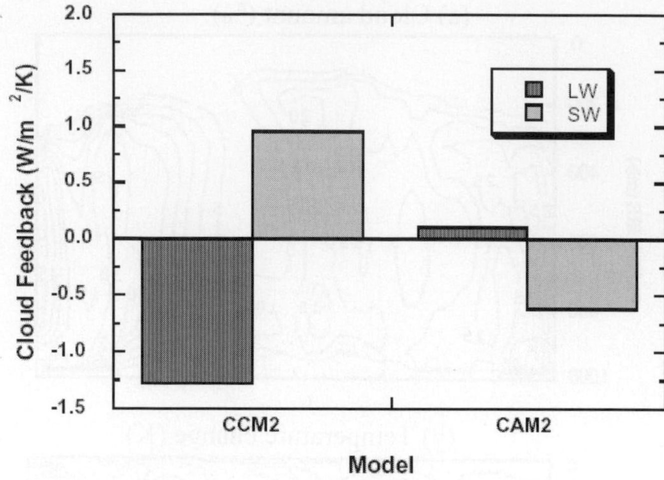

Figure 7. Feedbacks components separated into longwave (LW) and shortwave (SW) radiation in the CCM2 and in CAM2.

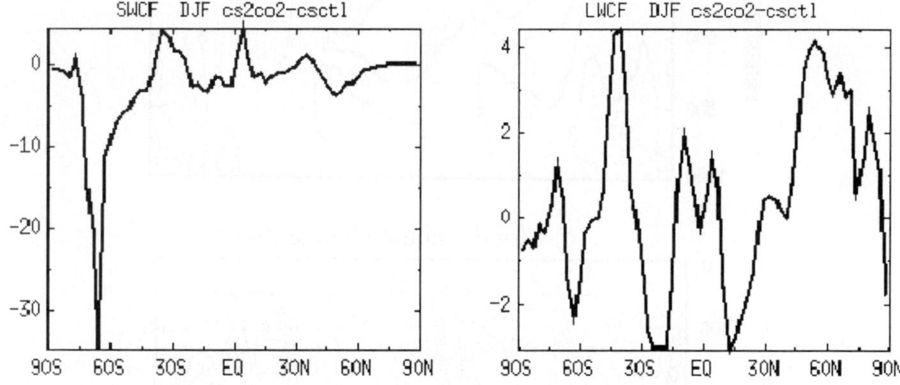

Figure 8. Change of the shortwave (left panel) and longwave (right panel) cloud forcing (W m^{-2}) in a climate change simulation for the DJF season.

Therefore, the negative cloud feedback in CCM2 is related with a dominant reduction of the cloud greenhouse effect associated with decreased cloudiness in the upper troposphere. The negative feedback in CAM2, however, is a result of the enhanced solar cooling effect from increased amount of low clouds.

5. Further Evaluation of Model Clouds

Given the above discussion, a reasonable question to ask is whether cloud feedbacks in current models have any fidelity to the real climate system. Since observations of cloud feedbacks are not directly available, a natural step is to evaluate model simulated clouds against available observations.

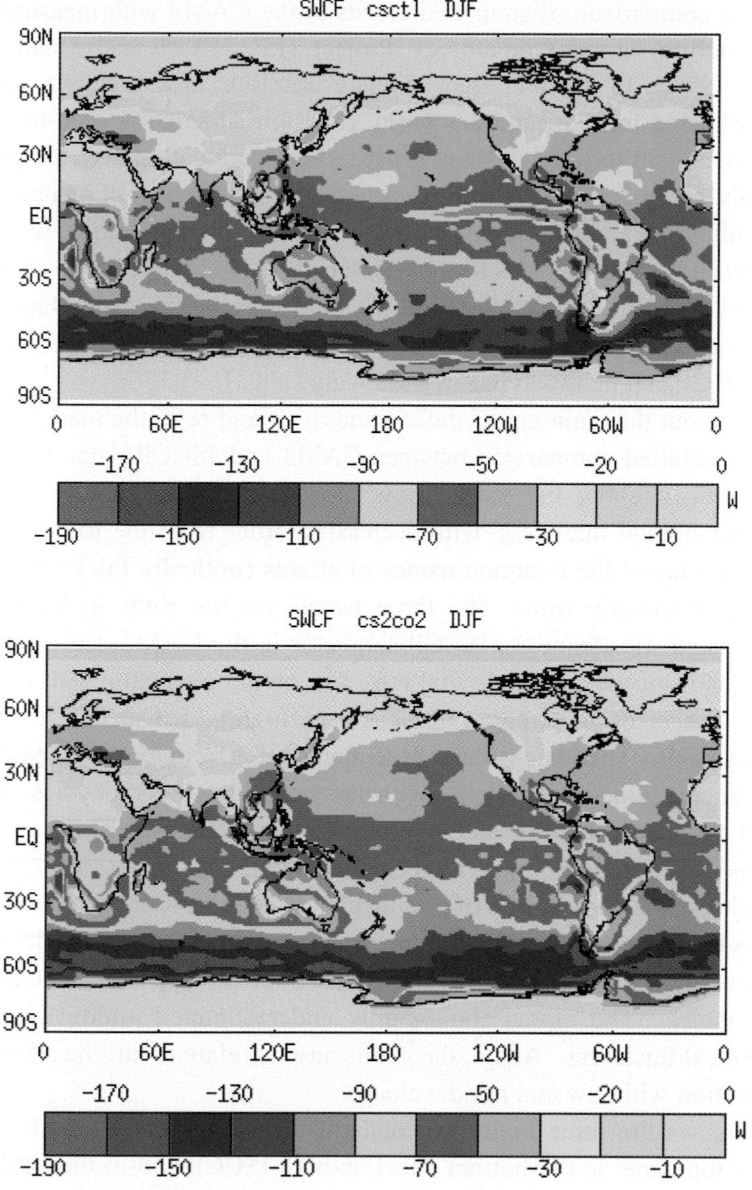

Figure 9. Geographical distribution of the January shortwave cloud forcing (W m^{-2}) in the control climate (upper panel) and warmer climate (lower panel).

A common practice in the climate modeling community is to use CRF measurements from ERBE as model validation datasets (e.g., Zhang et al. 2003). The same CRF at TOA, however, does not necessarily mean the same vertical structures of clouds, which is more directly related with cloud changes in response to a climate forcing. It is therefore desirable to use as many independent measurements as possible.

We present a comparison of simulated clouds in the CAM2 with measurements from the International Satellite Cloud Climatology Project (ISCCP) (Rossow and Schiffer 1991; Rossow et al. 1996). In ISCCP, geostationary satellite measurements of narrow band radiances are collected from around the world. A visible channel is used to derive the cloud optical thickness, and an infrared channel is used to derive the cloud top temperature and thus pressure. Clouds are then categorized according to their optical thickness and cloud top heights. To facilitate the comparison of model fields with ISCCP clouds, we implemented an ISCCP simulator in the CAM2, which was developed by Drs. Steve Klein at GFDL and Mark Webb at UKMO. The ISCCP simulator allows us to retrieve the same diagnostics from the CAM2 as what are available in ISCCP. The implementation method and discussions on the limitations of ISCCP can be found in Lin and Zhang (2003).

We only point out the main model deficiencies here and refer the reader to Lin and Zhang (2003) for more detailed comparison between CAM2 and ISCCP data. The three panels on the left in Figure 10 show the ISCCP low cloud amount in January 1983 categorized according to their optical thickness, with decreasing order from the top panel to the bottom panel. They are assigned the common names of stratus (optically thick), stratocumulus, and shallow cumulus (optically thin). The three panels on the right in Figure 10 show the corresponding diagnostics from the ISCCP simulator in the CAM2. Low clouds are defined as those with tops from the surface to 700 mb. The model overestimated optically thick low clouds, especially near 60°S, a region that has been highlighted in Figure 9 to describe the model cloud feedback. This bias is due to the model stratiform cloud scheme that tries to do the job of producing clouds associated with the boundary layer physics. The model also underestimated the optically thin and intermediate low clouds. This is because shallow convection does not directly produce clouds in the model.

Figure 11 shows the comparison of middle clouds in observations and in the model. They have cloud tops between 700 mb and 400 mb. The three types of middle clouds are given the names of nimbostratus (optically thick, top panels), altostratus, and altocumulus (optically thin, bottom panels). The model significantly underestimated middle clouds of thin and intermediate optical thickness. Again, the main cause is related with the arbitrary decoupling of model convection with low and middle clouds.

Figure 12 shows the three high-top clouds of deep convective clouds, cirrus, and thin cirrus, from the top panel to the bottom panel, in both ISCCP and in the CAM2. The model overestimated both optically thick and optically thin high clouds. The overestimation of high thin cirrus is caused by an inaccurate algorithmic relationship between the detrainment convective mass flux and cirrus anvils (Rasch and Kristjansson 1998). The overestimation of optically thick high clouds is likely due to biases in the model microphysical cloud scheme associated with the conversion rate of cloud droplets to precipitation.

These multiple opposing errors in high and middle clouds jointly produce a reasonable longwave cloud forcing at TOA. At the meanwhile, the opposing errors in optically thick and thin/intermediate clouds offset to produce a reasonable shortwave cloud radiative forcing at the TOA.

It is unlikely that a model can produce a reliable cloud variation and thus cloud feedback if its basic state is problematic. The overestimation of low stratus in the CAM2 has direct bearings on the negative feedback in the model. At the writing of this paper, the CAM2 is undergone significant revisions in its cloud scheme to improve the above-mentioned biases, so that it can be more faithfully used to project the sensitivity of the climate system for use in the next IPCC report scheduled for the year 2005.

We are therefore still far from getting a confident cloud feedback in climate models. Constructive process-oriented analysis is needed to understand the physical causes of model deficiencies. Figure 13 is used to illustrate this point. It shows two snapshots of one day apart of the observed surface pressure field, 500 mb height, and the infrared cloud image over North America on September 26 and September 27, 2002 respectively.

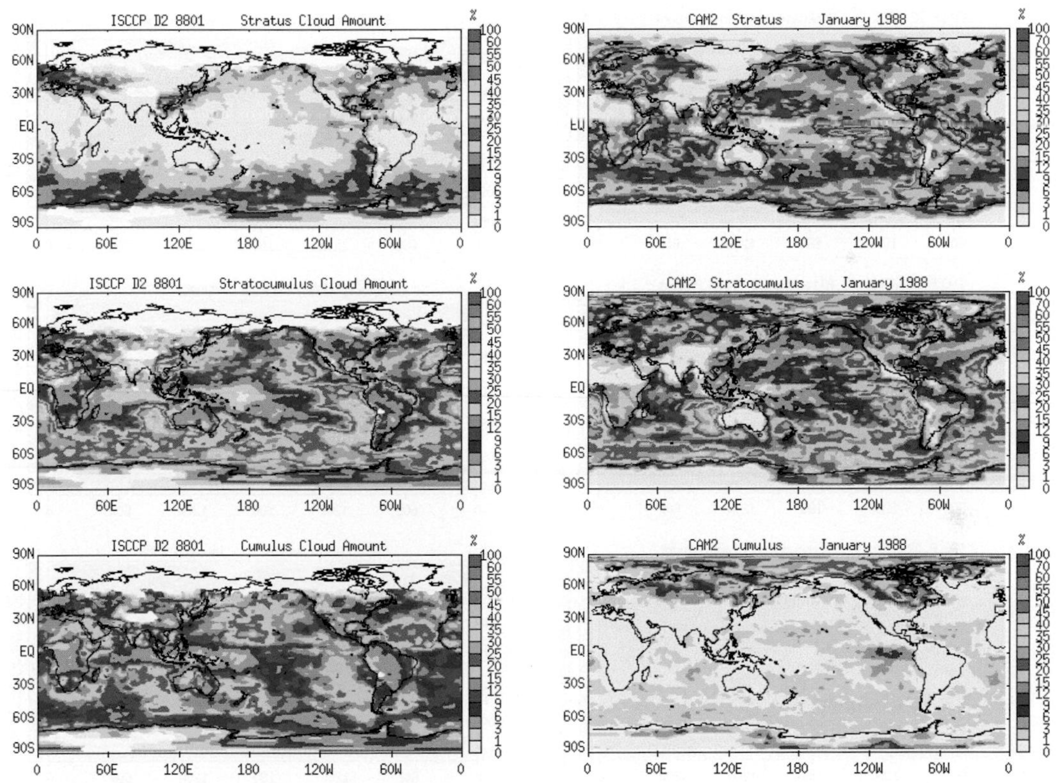

Figure 10. Low clouds in January 1988. Left panels: ISCCP observations with optical thickness decreasing from the top panel to the bottom panel. Right panels: CAM2 simulations.

Figure 14a shows the 12-hour CAM2 forecast corresponding to Figure 13a when it is initialized from the operational analysis. Plotted in the image is the model high cloud amount. The model is able to capture the cloud band associated with a hurricane. But the model overestimates high clouds in low latitudes, consistent with the ISCCP comparison in Figure 12. This confirms an algorithmic cause of error in the parameterization of high cloud amount since the model thermodynamic fields are close to observations and yet the high clouds are not.

A completely different inference can be made, however, for the model cloud behavior in the 36 hour forecast shown in Figure 14b. When compared with Figures 13b, aside from the high cloud deficiencies at low latitudes, model clouds are very different from observations over the United States. This is because the hurricane in the model does not move northward as in observation. This failure is mainly the result of model resolution rather than initialization errors. Thus, the deficiency in simulated clouds is a reflection of errors in the model dynamics rather than in its cloud parameterizations. Any tuning of the cloud scheme to match simulated clouds with observed clouds could deteriorate the model rather than improve it. Correct simulation of clouds in this case thus requires a deeper effort in improving other aspects of the model.

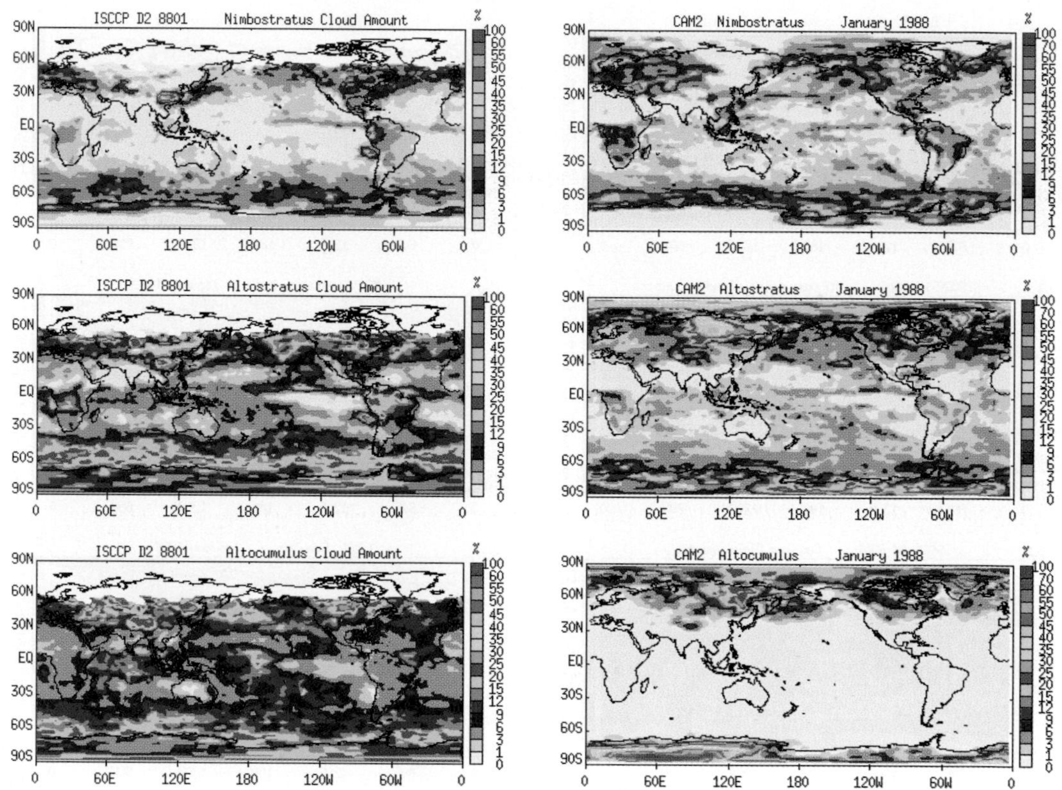

Figure 11. Same as Figure 10 except for middle clouds.

6. Summary and Discussions

Clouds are an integral part of the moist geophysical dynamics. They are strongly coupled with both grid scale and sub-grid scale processes. Cloud feedback is a result of aggregated change of cloud radiative forcing associated with chaotic transient atmospheric circulation.

This paper reviewed the concepts of cloud feedback and its role in defining the sensitivity of a climate model. It also discussed methods that have been used to diagnose the cloud feedbacks in climate models. I have attempted to present what cloud feedbacks are in current models, what processes caused the cloud feedbacks, and how model clouds compare with observations. From the discussions presented, it can be concluded that we are still far away from confidently simulating model clouds and their climate feedbacks.

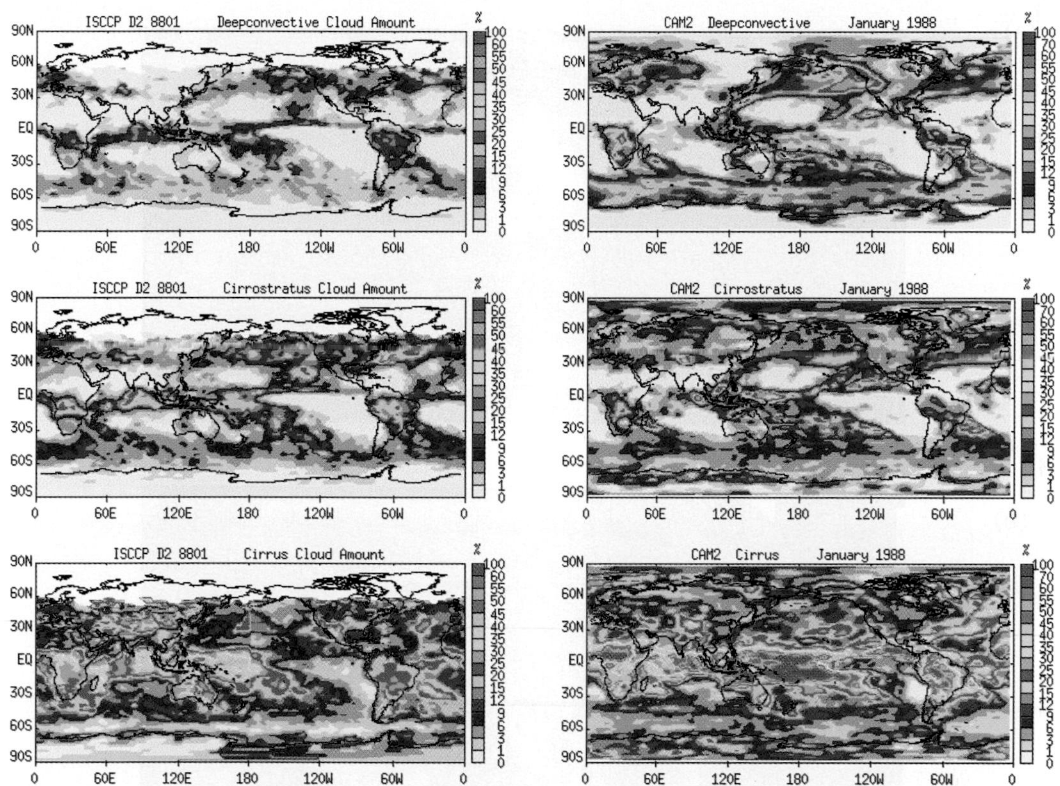

Figure 12. Same as Figure 10 except for high clouds.

The cloud feedback problem could become at least conceptually more tractable if we divide it into three different tasks. One is the microphysical cloud calculation with given dynamical circulation features. With detailed knowledge of aerosol distribution, the dynamical circulation provides the information on the generation of super saturation. This can then be used to estimate the number of cloud droplets nucleated, as in the simplified method of Ghan and Easter (1992) or more elaborate calculation using CCN spectral information as in Kogan (1991). The spectral size distribution of cloud droplets can then be calculated. This procedure itself incurs considerable demand on computational resource. This is one direction several modeling groups are currently pursuing.

The second aspect is the specification of the atmospheric dynamics on the subgrid scale. Clouds are mostly generated by subgrid scale processes, which have to be parameterized in climate models. To describe the variability of the atmospheric thermodynamic and dynamical structures within a grid, statistical description is needed. Yet, these statistical relationships should be based on realistic physical principles. At the present time, these subgrid models are either from empirical relationships or from highly intuitive conceptualizations. This deficiency in the subgrid scale dynamics has prompted Randall et al. (2003b) to use cloud resolving models inside a climate model to replace the parameterization package.

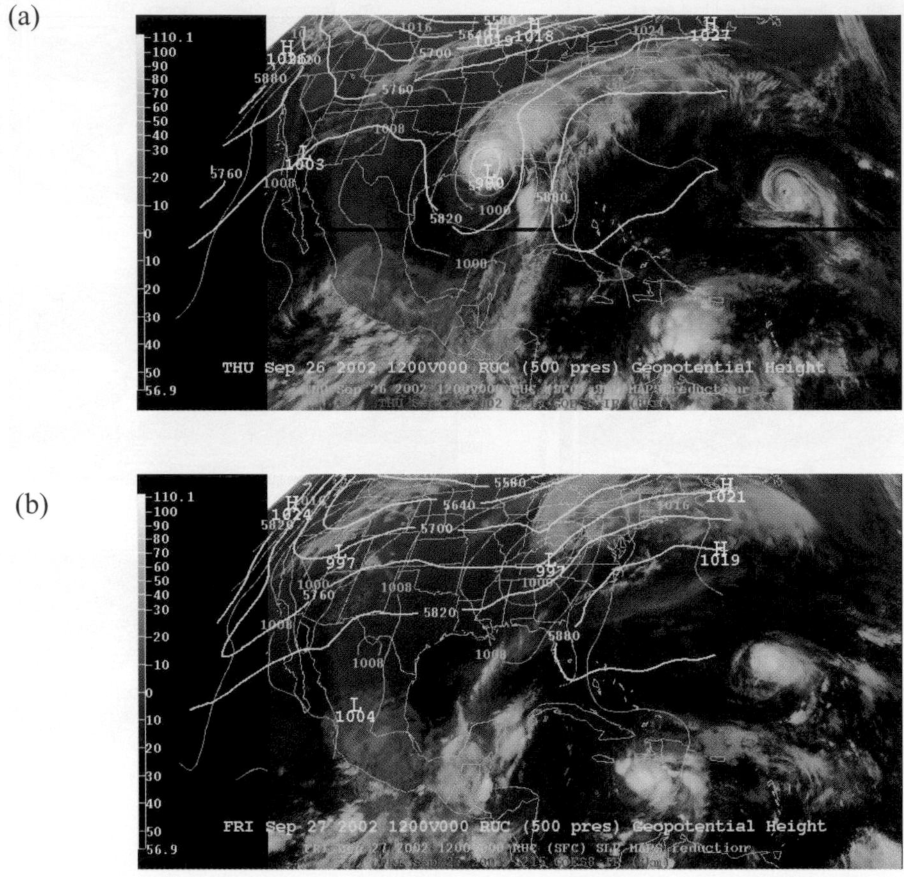

Figure 13. Observations of infrared clouds, 500mb height (cyan), and surface pressure (green). (a) 9/26/2002 12:00 GMT. (b) 9/27/2002 12:00 GMT.

The third issue is the abstraction of the coupling of subgrid scale dynamics with the subgrid scale cloud processes into a practical parameterization formulation. It is not clear whether this abstraction is possible. This issue is still valid even if spatial resolution of current models is reduced by an order of magnitude. On the practical side, very few existing convective schemes even include a component of the cloud microphysics.

The study of cloud-climate feedback is therefore rooted in the subgrid-scale dynamics and physics. This problem poses challenges and opportunities for constructive use of observations and calls for a breakthrough of parameterization methodology. Recent coordinated research activities in this regard include GCSS (GEWEX Cloud System Studies) (Randall et al., 2003a) and the DOE ARM (Atmospheric Radiation Measurement) program (Xie et al., 2002; Xu et al., 2002). Several subsequent papers in this volume describe specific examples of issues involved in the parameterization of subgrid scale physics and dynamics. The benefit of these activities will not just be to resolve the cloud-climate feedback uncertainty, but also to improve numerical modeling and numerical prediction of weather and climate in general.

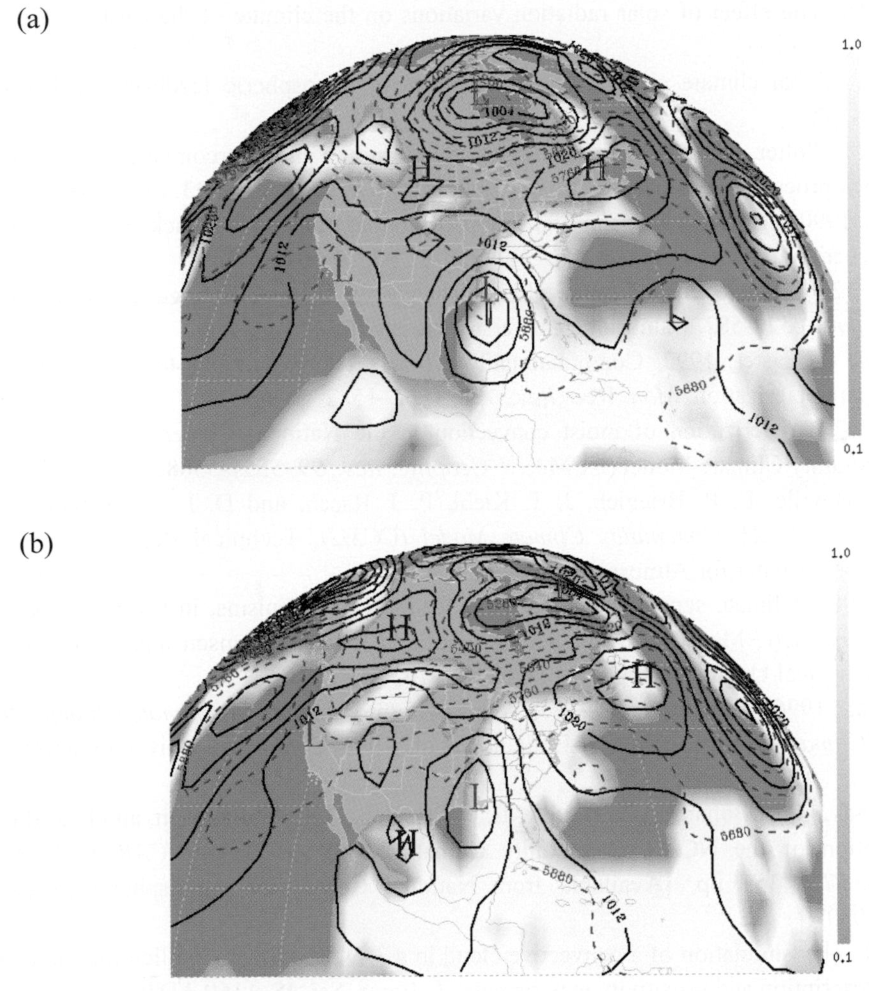

Figure 14. CAM2 simulated high clouds, 500mb height (red dashed), and surface pressure (black). (a) 9/26/2002 12:00 GMT. (b) 9/27/2002 12:00 GMT.

Acknowledgements. The author wishes to acknowledge the contribution of Dr. Wuyin Lin for his help in carrying out some of the model simulations reported in the paper, and Drs. Steve Klein at GFDL and Mark Webb at UKMO for making their ISCCP Simulator available to this study. This research was supported by the U. S. National Science Foundation under grant ATM901950, by the Department of Energy under grant EFG0298ER62570, and by NASA under its TRMM and GPM program, to the Stony Brook University.

References:

Bode, H. W., 1945, *Network analysis and feedback amplifier design*, 551 pp. Van Nostrand, New York.

Budyko, M. I., 1969: The effect of solar radiation variations on the climate of the earth. *Tellus*, **21**, 611-619.

Cess, R. D., 1975: Global climate change: an investigation of atmospheric feedback mechanisms. *Tellus*, **27**, 193-198.

Cess, R. D., and G. L. Potter, 1988: A methodology of understanding and intercomparing atmospheric climate feedback processes in general circulation models, *J. Geophys. Res.*, **93**, 8305-8314.

Cess, R. D. et al., 1990: J Intercomparison and interpretation of climate feedback processes in 19 atmospheric general circulation models. *J. Geophys. Res.*, **95**, 16,601-16,615.

Cubasch, U., et al., 2001: Projection of future climate change. In *Climate Change, the Third IPCC Assessment Report*, p528-582, Cambridge University Press, 2001.

Ghan, S. J., and R. C. Easter, 1992: Computationally efficient approximations to stratiform cloud microphysics parameterization. *Mon. Wea. Rev.*, **120**, 1572–1582.

Hack, J. J., 1994: Parameterization of moist convection in the National Center for Atmospheric Research Community Climate Model (CCM2), *J. Geophys. Res.*, **99**, 5551-5568.

Hack, J. J., B. A. Boville, B. P. Briegleb, J. T. Kiehl, P. J. Rasch, and D. L. Williamson,1993: *Description of the NCAR Community Climate Model (CCM2)*, Technical Report NCAR/TN-382+STR, National Center for Atmospheric Research, 120 pp.

Hansen, J., et al., 1984: Climate sensitivity: analysis of feedback mechanisms, in *Climate Processes and Climate Sensitivity*, Maurice Ewing Series, 5, Edited by J. E. Hansen and T. Takahashi, American Geophysical Union, Washington, D. C., 130-163.

Kattenberg, A., et al., 1996: Climate models–projections of future climate, in *Climate Change 1995, the Science of Climate Change*, edited by Houghton et al., 1996, Cambridge University Press, 572 pp.

Kiehl, J. T., J. J. Hack, G. B. Bonan, B. A. Boville, B. P. Briegleb, D. L. Williamson, and P. J. Rasch, 1996: Description of the NCAR Community Climate Model (CCM3). *NCAR Tech. Note, NCAR/TN-420+STR*, 151 pp. [Available from National Center for Atmospheric Research, Boulder, CO 80307.]

Kogan, Y. L., 1991: The Simulation of a convective cloud in a 3-D model with explicit microphysics. Part I: model description and sensitivity experiments. *J. Atmos. Sci*, **48**, 1160-1189.

Lin, W. Y., and M. H. Zhang, 2003: Evaluation of Clouds and Their Radiative Effects Simulated by the NCAR Community Atmospheric Model CAM2 Against Satellite Observations. *J. Climate,* submitted.

Ramanathan, V. et al., 1983: The response of a spectral general circulation model to refinements in radiative processes. *J. Atmos. Sci.*, **40**, 605-630.

Ramanathan, V., 1987: The role of earth radiation budget studies in climate and general circulation research, *J. Geophys. Res.*, **92**, 4075-4095.

Randall, D. A. et al., 2003a: Confronting models with data, the GEWEX cloud systems study. *Bull. Amer. Meteor. Soc.*, in press.

Randall, D. A., M. Khairoutdinov, A. Arakawa, and W. W. Grabowski, 2003b: Breaking the cloudparameterization deadlock. Submitted to *Bull. Amer. Meteor. Soc.*.

Rasch, P. J., and J. E. Kristjánsson, 1998: A comparison of the CCM3 model climate using diagnosed and predicted condensate parameterizations. *J. Climate*, **11**, 1587-1614.

Rossow, W.B., and R.A. Schiffer, 1991: ISCCP cloud data products. *Bull. Amer. Meteor. Soc.*, **72**, 2-20.

Rossow, W. B., A. W. Walker, D. E. Beuschel, and M. D. Roiter, 1996: *International Satellite Cloud Climatology Project (ISCCP) Documentation of New Cloud Datasets*. WMO/TD-No. **737**, World MeteorologicalOrganization, 115 pp.

Schlesinger, M. E., 1988: Quantitative analysis of feedbacks in climate model simulations of CO2 induced warming, in *Physically-based Modeling and Simulation of Climate and Climate Change*, NATO ASI series, edited by Schlesinger, M. E., pp. 653-735, Kluwer Academic Press, 1988.

Schneider, S. H., 1972: Cloudiness as a global climatic feedback mechanism: the effects on radiation balance and surface temperature of variations in cloudiness. *J. Atmos. Sci.*, **29**, 1413-1422.

Senior, C. A., and J. F. B. Mitchell, 1993: Carbon dioxide and climate: the impact of cloud parameterization. *J. Climate*, **6**, 5-21.

Senior, C. A., and J. F. B. Mitchell, 1996: cloud feedbacks in the unified UKMO GCM. In *Climate Sensitivity to Radiative Perturbations, Physical Mechanism and Their Validation*, edited by H. Le Treut, 331pp, Springer, 1996.

Slingo, J. M., 1987: The development and verification of a cloud prediction scheme for the ECMWF model. *Quart. J. Roy. Meteor. Soc.*, **113**, 899–927.

Smith, R. N. B., 1990: A scheme for predicting layer clouds and their water content in a general circulation model. *Quart. J. Roy. Meteor. Soc.,* **116**, 435-460.

Stocker, T. F., et al., 2001: Physical climate processes and feedbacks. In *Climate Change, the Third IPCC Assessment Report*, Cambridge University Press, 2001.

Wetherald, R. T., and S. Manabe, 1988: Cloud feedback processes in general circulation models, *J. Atmos. Sci.*, **45**, 1397-1415.

Xie, S. C., et al. 2002: Intercomparison and evaluation of cumulus parameterization under summertime midlatitude continental conditions. *Quart. J. Roy. Meteorol. Soc.*, **128**, 1095-1135.

Xu, K. M., et al. 2002: An intercomparison of cloud resolving models with the ARM summer 1997 Intensive Observation Period data. *Quart. J. Roy. Meteorol. Soc.*, **128**, 593-624.

Zhang, G. J., and N. A. McFarlane, 1995: Sensitivity of climate simulations to the parameterization of cumulus convection in the Canadian Climate Centre general circulation model. *Atmos.-Ocean*, 33, 407-446.

Zhang, M. H., J. J. Hack, J. T. Kiehl and R. D. Cess, 1994: Diagnostic study of climate feedback processes in atmospheric general circulation models. *J. Geophys. Res.,* **99**, 5525-5537.

Zhang, M., W. Lin, C. Bretherton, J. J. Hack, and P. J. Rasch, 2003: A modified formulation of fractional stratiform condensation rate in the NCAR Community Atmospheric Model (CAM2), *J. Geophys. Res.*, **108**(D1), 4035, doi:10.1029/2002JD002523.

PARAMETERIZATION OF CONVECTION IN GLOBAL CLIMATE MODELS

GUANG JUN ZHANG

Scripps Institution of Oceanography
La Jolla, CA 92093-0221, USA
E-mail: gzhang@ucsd.edu

(Manuscript received 12 January 2003)

Atmospheric convection is an important energy source for the global circulation. It has a typical spatial scale of a few kilometers to a few tens of kilometers. Thus, in global climate models (GCMs), which have a horizontal resolution of 200 to 300 kilometers, convection must be parameterized. This study reviews some widely used convective parameterization schemes in GCMs. Emphasis will be placed on the mass flux type of schemes and associated closure conditions. Recent development on a fundamental issue related to closure will be discussed. An Example will be presented to demonstrate the role of closure using the single column version of the National Center for Atmospheric Research Community Climate Model CCM3.

1. Introduction

The vertical transport of mass, heat, moisture and momentum in the atmosphere plays a fundamental role in all areas related to climate. For example, vertical transport of moisture determines how latent heating is deposited in the atmosphere, which in turn determines the strength of the Hadley circulation, the mid-latitude jet stream, etc. It also determines the vertical distribution of clouds and moisture, which in turn determines the radiation budget of the earth's climate system. Three dimensional global climate models (GCM) have to treat this phenomenon, but the major stumbling block is that the vertical transport happens on scales of a few kilometers (individual convective cells) to hundreds (meso-scale convective clusters) and thousands of kilometers (tropical disturbances such as cyclones, etc). The challenge then is to understand from observations the nature of this multi-scale transport and incorporate the subgrid-scale transport in the climate model, the so-called parameterization problem.

The parameterization of the vertical transport of mass, heat and moisture by atmospheric moist convection has been a long-standing issue in global climate modeling and numerical weather prediction. It is an extremely difficult problem since convection strongly interacts with clouds, atmospheric water vapor and radiation, involving processes operative at space and time scales spanning several orders of magnitudes. Over the years, a range of parameterization schemes has been developed, varying from moist convective adjustment schemes (Manabe et al. 1965, Betts 1986) to the more sophisticated moisture-convergence-based (Kuo 1965, 1974, Tiedke 1989) and convective-instability-based mass flux representation (Arakawa and Schubert 1974, Emanuel 1991, Donner 1993, Zhang and McFarlane 1995, and many more). Accompanying these are extensive studies on the role of convection in various aspects of climate simulations, ranging from the mean tropical climate to Asian summer monsoon and intraseasonal variability of tropical precipitation (e.g. Slingo et al. 1988, 1994, Zhang 1994, Eitzen and Randall 1999, Maloney and Hartmann 2001). In spite of the tremendous efforts made over the past forty years, accurately representing convection and understanding its interaction with other processes in climate remain to be the most challenging problems in today's GCM development. This, to a large extent, reflects the complexity of the problem and the deficiencies in our understanding of convective processes. This paper will give a comprehensive review of the development of convective parameterization. We will concentrate on deep, precipitating atmospheric convection. Shallow convection such as subtropical trade cumulus is excluded to focus on the subject of interest. Section 2 will introduce the basic problem of convective parameterization. Section 3 will review a number of popular convective parameterization schemes. In section 4, closure conditions needed for convective parameterization are discussed. Section 5 will present an example of recent development in convective parameterization from personal viewpoint. Section 6 will conclude the paper.

2. Formalism of the convective effects on the large-scale fields

Atmospheric convection occurs on spatial scales of tens of kilometers. For typical GCMs with horizontal resolution of ~300 km, convection is a subgrid scale phenomenon. The governing equations for the large-scale temperature and moisture fields are given by (Yanai et al. 1973)

$$\frac{\partial \bar{s}}{\partial t} + \nabla \cdot (\bar{\mathbf{v}}\bar{s}) + \frac{\partial \bar{\omega}\bar{s}}{\partial p} = \bar{Q}_R + L(c-e) + L_d \Upsilon_{ds} + L_f \Upsilon_{fm} - \nabla \cdot (\overline{\mathbf{v}'s'}) - \frac{\partial \overline{\omega' s'}}{\partial p}, \tag{1}$$

$$\frac{\partial \bar{q}}{\partial t} + \nabla \cdot (\bar{\mathbf{v}}\bar{q}) + \frac{\partial \bar{\omega}\bar{q}}{\partial p} = -(c-e) - \Upsilon_{ds} - \nabla \cdot (\overline{\mathbf{v}'q'}) - \frac{\partial \overline{\omega' q'}}{\partial p}, \tag{2}$$

where s (= $C_pT + gz$) is the dry static energy, q is the specific humidity, **v** is the horizontal wind vector, ω is the vertical p-velocity, L is the latent heat of vaporization, L_d is the latent heat of deposition, and L_f is the latent heat of freezing. The overbar represents averaging over the large-scale domain or a GCM grid and the prime represents the deviation from the mean. The perturbation product terms on the right hand side represent the effect of subgrid scale transport (in this case, convection) on the large-scale or grid mean fields. Q_R is the radiative heating rate. $c - e$ represents the net condensation (condensation minus evaporation) within the GCM grid, Υ_{ds} is the net deposition (deposition minus sublimation) and Υ_{fm} is the net freezing (freezing minus melting). In the original derivation of (1) and (2) by Yanai et al. (1973) the Υ terms associated with ice phase change were ignored. Thus for simplicity, these terms will be dropped in the presentation from now on. However, they can be easily incorporated if desired, as has been done in Donner (1993). Typically, the horizontal divergence of the perturbation flux is much smaller than the vertical divergence term and can be neglected. For the convenience of presentation, we use the short hand $(\partial T/\partial t)_c$ and $(\partial q/\partial t)_c$, respectively, to denote the convective effects on the large-scale temperature and moisture fields, i.e.:

$$C_p\left(\frac{\partial \overline{T}}{\partial t}\right)_c = L(c-e) - \frac{\partial \overline{\omega' s'}}{\partial p}, \tag{3}$$

$$\left(\frac{\partial \overline{q}}{\partial t}\right)_c = (e-c) - \frac{\partial \overline{\omega' q'}}{\partial p}. \tag{4}$$

The task of convective parameterization is to represent the collective effects of convection on the right hand side of Eqs. (3) and (4) in terms of the resolved fields.

3. Representation of convective effects

3.1. *Adjustment schemes*

One of the earliest convective parameterization schemes for use in GCMs was the moist convective adjustment scheme developed by Manabe et al. (1965). The essence of the scheme is to adjust the temperature and moisture profiles following a moist adiabat when the atmosphere is saturated and moist statically unstable over two adjacent model layers. While this approach can effectively remove the atmospheric instability in the lower troposphere, it does a poor job in the upper troposphere where the atmosphere is locally stable. As a result, the simulated troposphere using moist convective adjustment is too cold in the upper troposphere (Zhang and McFarlane 1995). Using the adjustment concept, Betts (1986) and Betts and Miller (1986) developed a penetrative convective adjustment scheme generally known as the Betts-Miller scheme. According to their work, the convective effects on the temperature and moisture fields can be represented through convective adjustment.

$$(\partial T/\partial t)_c = (T_r - \overline{T})/\tau, \tag{5}$$

$$(\partial q/\partial t)_c = (q_r - \overline{q})/\tau, \tag{6}$$

where T_r and q_r are the reference temperature and moisture toward which the actual temperature and moisture adjust, τ is the adjustment timescale. The reference thermodynamic profiles are empirically determined from observations. The reference temperature profile nearly follows a virtual moist adiabat up to the freezing level, and a slightly more stable profile above. This specification is based on two observational studies (Betts 1982, 1986).

Making use of these observations, the Betts-Miller scheme assumes that the reference saturation equivalent potential temperature (which is a function of temperature and pressure only) decreases linearly with pressure from the cloud base level to the freezing level, then increases to the environmental saturation equivalent potential temperature at the cloud top level. The rate of decrease in the lower troposphere is such that the potential temperature change with pressure of the reference profile is 85% of that along the moist adiabat. Thus, the reference temperature profile is slightly unstable with respect to the virtual moist adiabat of the near surface air. The moisture profile is specified by utilizing the saturation pressure deficit, which is the difference of pressure between a parcel's lifting condensation level and its original level. The pressure deficit at three levels, the cloud base, the freezing level and the cloud top, are specified, with values in the intervening levels obtained by linear interpolation. The specified temperature and moisture profiles serve as a first guess. The final reference profiles must satisfy certain energy conservation constraints. Since convective processes conserve moist static energy, the vertically integrated moist static energy over the convection layer at the observed state must be the same as that of the reference state after convection, that is,

$$\int_{p_t}^{p_b} (h_r - h) dp = 0 . \tag{7}$$

Thus correction to the first guess moist static energy can be made by calculating

$$\Delta h = \frac{1}{p_b - p_t} \int_{p_t}^{p_b} (h - h_r^{(1)}) dp , \tag{8}$$

where $h_r^{(1)}$ is the first guess reference moist static energy. This correction is then added to the reference moist static energy profile. The partitioning between temperature and moisture correction is done such that the specified pressure deficit at each level remains unchanged after the correction. The precipitation rate on the surface is just the column integrated moisture change due to this adjustment, that is,

$$P = -\frac{1}{g} \int_{p_t}^{p_b} \left(\frac{q_r - q}{\tau} \right) dp . \tag{9}$$

In summary, in the Betts-Miller scheme, the large-scale processes act to pull the temperature and moisture profiles away from the reference profiles and convection acts to push them back to the reference profiles within an adjustment timescale. Precipitation is a by-product of this scheme. At the same time it also serves as a closure. If precipitation calculated using Eq. (9) is positive, the scheme is activated and the ensuing temperature and moisture changes are added to the thermodynamic equations. If the calculated precipitation is negative, convection is not allowed.

The Betts-Miller scheme has three tunable parameters: the adjustment timescale, the potential temperature lapse rate below freezing level with respect to moist adiabat of near surface air, and the saturation pressure deficit at the cloud base, the freezing level and the cloud top. It is simple to use and it gives a reasonable simulation of the climate (Slingo et al. 1994). However, it is difficult to justify that a particular choice of these parameters is universally valid. Vaidya and Singh (1997) found that the simulated Indian monsoon is very sensitive to the choice of these parameters. Furthermore, the reference profiles are based on the observations of the current climate state. In a changed climate, it is not clear if the reference atmospheric states remain the same, especially the moisture profile. Thus, the usefulness of the Betts-Miller scheme in climate change simulations is questionable.

3.2. *Kuo Scheme*

Kuo (1965) developed a convective parameterization scheme in an effort to incorporate the effect of latent heat release by convection on the intensification of tropical cyclones. The scheme was based on an observational fact that convection is highly correlated with the low-level moisture convergence and the CISK (convective instability of the second kind) concept proposed by Charney and Eliassen (1964). It assumes that (i) convection occurs in a region where the atmosphere is conditionally unstable and there is low-level moisture convergence; (ii) convective clouds originate from the boundary layer and the cloud temperature and moisture profiles can be characterized by a pseudo-moist adiabat typical of the boundary layer air; and (iii) clouds extend from the lifting condensation level of the boundary layer air to the neutral buoyancy level of this air. With these assumptions, the Kuo scheme starts with Eq. (2) for the large-scale moisture field. Integrating over the atmospheric column gives:

$$\int_0^{p_s} \frac{\partial \overline{q}}{\partial t} dp = -Pg - \int_0^{p_s} \nabla \cdot (\overline{\mathbf{v}}\overline{q}) dp + gF_{LH} = g(M_t - P), \tag{10}$$

where

$$M_t = -\frac{1}{g} \int_0^{p_s} \nabla \cdot (\overline{\mathbf{v}}\overline{q}) dp + F_{LH} \tag{11}$$

is the total moisture supply to the atmospheric column, F_{LH} is surface evaporation. P is precipitation on the surface, which equals the column integral of the net condensation:

$$P = \frac{1}{g} \int_0^{p_s} (c - e) dp. \tag{12}$$

Kuo assumed that a small fraction of the moisture supply (bM_t) is used to moisten the atmosphere and the rest of it is precipitated out as rain, i.e.

$$P = (1 - b) M_t \tag{13}$$

and

$$\frac{1}{g}\int_0^{p_s}\frac{\partial \overline{q}}{\partial t}dp = bM_t\,. \tag{14}$$

Here b is a tunable parameter. Thus, knowing the total moisture supply M_t, which can be computed from the large-scale fields or GCM output, one can compute the surface precipitation P, thus the vertical integral of latent heating through Eqs. (12) and (13). One needs to know the vertical distribution of the heating in order to determine the effect of condensational heating on temperature field at each GCM level. Kuo (1974) assumed that it is distributed according to the temperature difference between the cloud air following a pseudo-moist adiabat and the environmental air. With this, one can readily show that:

$$L(c-e) = \frac{gLP}{p_b - p_t}\frac{T_c - \overline{T}}{< T_c - \overline{T} >}\,, \tag{15}$$

where the angle brackets represent the vertical average over the cloud layer. p_b and p_t are the pressure levels of the cloud base and cloud top, respectively. Outside the cloud layer latent heating is zero.

Equation (15) represents the effect of latent heating from convection on the large-scale temperature field in terms of the large-scale moisture supply (through moisture convergence and surface evaporation) and the cloud-environment temperature difference, the latter of which is determined by following a pseudo-moist adiabat of the boundary layer air. In order for the scheme to be activated, the moisture supply must be positive.

Kuo's scheme has been used successfully in modeling tropical rainfall (e.g. Krishnamurti et al. 1980). However, the CISK concept, on which Kuo's scheme was based, together with the lack of theoretical rigor of the scheme has been criticized by Emanuel (1994, pp 527-529). In addition, one important defect of the scheme is the fact that by prescribing the amount of moisture used for moistening the atmosphere, the predictability of the moisture field is lost. This is particularly serious for GCMs since moisture is a fundamental quantity, which affects the atmospheric greenhouse effect, clouds, radiation and the hydrological cycle.

3.3. *Mass flux schemes*

The mass flux form of convective parameterization has been widely used since Arakawa and Schubert (1974) developed their scheme. One important advantage of this type of scheme over others is the ease of incorporating convective transport of tracers, which requires the knowledge of vertical mass flux within convective drafts. In addition, detrained mass of hydrometeor from convective cores can be incorporated into large-scale cloud parameterization. Most of the parameterization schemes in use nowadays in GCMs are in mass flux form. Thus, we will devote more space to this class of schemes.

The large-scale temperature and moisture budget equations with the incorporation of the effect of convection can be written as (Tiedtke 1989):

$$\frac{\partial \bar{s}}{\partial t}+\nabla\cdot(\bar{\mathbf{v}}\bar{s})+\frac{\partial(\bar{\rho}\overline{ws})}{\bar{\rho}\partial z}=\bar{Q}_R+L(c-e)+\frac{\partial}{\bar{\rho}\partial z}[M_u(s_u-\bar{s})+M_d(s_d-\bar{s})], \quad (16)$$

$$\frac{\partial \bar{q}}{\partial t}+\nabla\cdot(\bar{\mathbf{v}}\bar{q})+\frac{\partial(\bar{\rho}\overline{wq})}{\bar{\rho}\partial z}=-(c-e)+\frac{\partial}{\bar{\rho}\partial z}[M_u(q_u-\bar{q})+M_d(q_d-\bar{q})], \quad (17)$$

where subscripts u and d denote variables in updrafts and downdrafts, $M_u=\rho\sigma_u w_u$ and $M_d=\rho\sigma_d w_d$ are mass fluxes in convective updrafts and downdrafts, respectively. The above equations could have been written in the p-coordinate system. However, for the convenience of presentation we will use z-coordinate hereafter.

From Eqs. (16) and (17), it is clear that to parameterize the convective effects on the large-scale temperature/dry static energy and moisture fields, one needs to know the vertical profiles of the cloud mass flux, in-cloud temperature and moisture, as well as condensation and evaporation inside convective updrafts and downdrafts. This is accomplished through the introduction of simple updraft and downdraft models.

Arakawa and Schubert (1974) introduced a spectral cloud model (updrafts only). The cloud population was considered as consisting of a spectrum of entraining plumes of different sizes, characterized by the fractional entrainment rate. Here we will follow the bulk cloud approach instead for its simplicity (Tiedtke 1989, Zhang and McFarlane 1995). For steady-state clouds, the bulk equations for mass flux, heat, moisture and cloud condensate within updrafts are:

$$\frac{\partial M_u}{\rho\partial z}=\varepsilon_u-\delta_u, \quad (18)$$

$$\frac{\partial M_u s_u}{\rho\partial z}=\varepsilon_u\bar{s}-\delta_u\hat{s}_u-\delta_u l+L(c-e), \quad (19)$$

$$\frac{\partial M_u q_u}{\rho\partial z}=\varepsilon_u\bar{q}-\delta_u\hat{q}_u-(c-e), \quad (20)$$

$$\frac{\partial M_u l}{\rho\partial z}=-\delta_u l+(c-e)-R_r/\rho, \quad (21)$$

where ε_u and δ_u are the mass entrainment and detrainment. s_u and q_u are dry static energy and specific humidity in the updrafts. Air within updrafts is assumed saturated, thus:

$$q_u=q_s(s_u). \quad (22)$$

$\hat{s}_u$ and $\hat{q}_u$ are the dry static energy and moisture detrained into the environment. At the detrainment level, it is often assumed that the air has the same temperature as that of its environment and is saturated, i.e.,

$$\hat{s}_u = \overline{s}\,, \tag{23}$$

$$\hat{q}_u = q_s(\overline{s})\,, \tag{24}$$

Furthermore, l is the cloud liquid water detrained into the environment and it is assumed to be the same as the mean updraft liquid water content at that level. R_r is the conversion rate of cloud liquid water to rain and can be set proportional to l as a first order approximation (Lord 1982):

$$R_r = c_0 M_u l \tag{25}$$

with $c_0 = 2\times10^{-3}\,\mathrm{m}^{-1}$.

The vertical profile of the updraft mass flux is specified through mass fractional entrainment and detrainment rates. Depending on the cloud model used, they are treated differently. Tiedtke (1989) partitioned the entrainment into organized inflow and turbulent entrainment. Similarly, mass detrainment was also partitioned into organized outflow and turbulent detrainment. Turbulent entrainment and detrainment were assumed to be proportional to the updraft mass flux similar to the plume model (see below). The organized inflow is assumed to occur in the lower part of the cloud layer where there is large-scale moisture convergence. On the other hand, the organized outflow is assumed to exist only in the top layer of the deepest clouds.

In ensemble plume models, only turbulent entrainment is considered. To illustrate how the bulk mass flux is specified in this case, we follow the approach of Zhang and McFarlane (1995), which makes simplification of the Arakawa-Schubert (1974) spectral plume model. For each cloud type with fractional entrainment rate λ, the variation of mass flux with height is given by:

$$\frac{\partial m_u(\lambda,z)}{\partial z} = \lambda m_u(\lambda,z)\,. \tag{26}$$

Integrating from the cloud base z_b to z gives:

$$m_u(\lambda,z) = m_b(\lambda)\exp[\lambda(z-z_b)]\,, \tag{27}$$

where $m_b(\lambda)$ is the updraft mass flux at the cloud base level for clouds spanning unit interval of fractional entrainment rates characterized by λ. Thus, the mass flux for each subensemble of cloud increases exponentially with height. Integrating over all possible λ's that contribute to the mass flux at level z gives:

$$M_u(z) = \int_0^{\lambda_D(z)} m_b(\lambda) \exp[\lambda(z - z_b)] d\lambda\,, \tag{28}$$

where $\lambda_D(z)$ is the fractional entrainment rate of the updrafts that detrain at height z. Here it is implicitly assumed that $\lambda_D(z)$ decreases monotonically with height. Thus, clouds with $\lambda > \lambda_D(z)$ have no contribution to mass flux at height z. In the Arakawa-Schubert (1974) scheme, $m_b(\lambda)$ was solved for by applying a closure condition to each cloud type. Zhang and McFarlane (1995), on the other hand, introduced additional simplifying assumptions. Noting that the bulk cloud base mass flux, denoted by M_b, is given by:

$$M_b = \int_0^{\lambda_0} m_b(\lambda) d\lambda\,, \tag{29}$$

where λ_0 is the maximum fractional entrainment rate corresponding to the shallowest of the updraft plume ensemble, they assume that the cloud base mass flux for each cloud type is independent of the cloud type for the large range of deep clouds considered. Thus,

$$m_b(\lambda) = M_b / \lambda_0\,. \tag{29a}$$

Substituting Eq. (29a) into Eq. (28) and carrying out the integration yield:

$$M_u(z) = \frac{M_b}{\lambda_0 (z - z_b)} \{\exp[\lambda_D(z)(z - z_b)] - 1\}\,. \tag{30}$$

The quantity λ_D is determined by the requirement that the temperature of the clouds detraining at height z is the same as that in its environment, which is ensured by requiring that

$$h_b - h^*(z) = \lambda_D(z) \int_{z_b}^{z} [h_u(\lambda, z') - h_b] dz'\,, \tag{31}$$

where h_u is the moist static energy in the updraft with fractional entrainment rate λ and h^* is the saturation moist static energy. The procedure of calculating λ_D can be found in Zhang and McFarlane (1995). To determine the mass detrainment from the subensemble of updrafts with tops at z, the same procedure that was used in Arakawa and Schubert (1974) can be applied here, i.e.

$$\delta_u(z) = -m_u(\lambda_D(z), z) \frac{d\lambda_D(z)}{\rho dz} = -\frac{M_b}{\lambda_0} \exp[\lambda_D(z)(z - z_b)] \frac{d\lambda_D(z)}{\rho dz}\,. \tag{32}$$

With this, the total updraft mass entrainment ε_u can be obtained from the mass continuity equation (18).

The equations of mass flux, temperature and moisture for downdrafts can be written as:

$$\frac{\partial M_d}{\rho \partial z} = -\delta_d + \varepsilon_d \tag{33}$$

$$\frac{\partial M_d s_d}{\rho \partial z} = -\delta_d \hat{s}_d + \varepsilon_d \overline{s} \tag{34}$$

$$\frac{\partial M_d q_d}{\rho \partial z} = -\delta_d \hat{q}_d + \varepsilon_d \overline{q} \tag{35}$$

The downdraft mass flux is often related to the updraft mass flux at the downdraft initial level with a proportionality constant based on Johnson (1978). Furthermore, detrainment of downdraft mass is often ignored until the subcloud layer, and the entrainment rate is either specified or related to the updraft entrainment. With an assumption similar to that in Eq. (22), the temperature and moisture profiles in downdrafts can be determined.

Substitution of Eqs. (18) – (20) and (33) – (35) into (16) and (17), together with Eq. (23) yields:

$$\frac{\partial \overline{s}}{\partial t} + \nabla \cdot (\overline{\mathbf{v}} \overline{s}) + \frac{\partial (\overline{\rho} \overline{w} \overline{s})}{\overline{\rho} \partial z} = \overline{Q}_R - \delta_u L l + M_c \frac{\partial \overline{s}}{\overline{\rho} \partial z}, \tag{16a}$$

$$\frac{\partial \overline{q}}{\partial t} + \nabla \cdot (\overline{\mathbf{v}} \overline{q}) + \frac{\partial (\overline{\rho} \overline{w} \overline{q})}{\overline{\rho} \partial z} = \delta_u (q_s - \overline{q} + l) + M_c \frac{\partial \overline{q}}{\overline{\rho} \partial z}. \tag{17a}$$

These are the same equations as Eqs. (74) and (75) in Arakawa and Schubert, except that here $M_c = M_u + M_d$ is the net mass flux.

By assuming that the cloud base mass flux for each cloud type is independent of the cloud type, the bulk mass flux profile has an analytic form of Eq. (30), leaving only the bulk cloud base mass flux to be determined. However, this assumption also leads to some serious limitations. It limits the cloud population that constitutes the bulk cloud to deep convection only. Thus a separate shallow convection scheme is required when the Zhang and McFarlane (1995) scheme is used. In the National Center for Atmospheric Research (NCAR) Community Atmospheric Model, Hack (1994) scheme is used for shallow convection to supplement the Zhang-McFarlane scheme (Zhang et al. 1998).

4. The closure conditions

4.1. *Moisture flux closure*

To determine the convective effect in mass flux schemes, the cloud base mass flux needs to be specified. This is often done through the so called "closure conditions". In Betts-Miller scheme, the closure is through requirement on surface precipitation (see discussions following Eq. (9)). In the Kuo scheme, the closure is provided by relating the low-level moisture convergence to precipitation. Tiedtke (1989) modified this closure to relate the low-level moisture convergence to the cloud base mass flux in his mass flux scheme. He assumed that moisture convergence in the subcloud layer together with the surface turbulent flux is balanced by the convective transport through the cloud base:

$$[M_u(q_u-\overline{q})+M_d(q_d-\overline{q})]_{z=z_b}=-\int_0^{z_b}\left[\overline{\mathbf{v}}\cdot\nabla\overline{q}+\overline{w}\frac{\partial\overline{q}}{\partial z}\right]\overline{\rho}dz+(\overline{\rho}\overline{w'q'})_{tur}, \tag{36}$$

where the last term on the right hand side represents the surface turbulent moisture flux. Since the downdraft mass flux is related to the updraft mass flux by design, the above equation determines the cloud base mass flux, thus closes the parameterization.

4.2. *Arakawa-Schubert quasi-equilibrium closure*

The Arakawa-Schubert parameterization scheme is closed through the quasi-equilibrium assumption, which states that the stabilization of the atmosphere by convection is in quasi-equilibrium with the destabilization by the large-scale processes. Arakawa-Schubert introduced the "cloud work function", which is the vertical integral of the buoyancy of a cloud parcel lifted adiabatically, weighted by the normalized cloud mass flux:

$$A(\lambda)=\int_{p_t}^{p_b}R_d\eta(\lambda)(T_{vp}(\lambda)-T_{ve})d\ln p, \tag{37}$$

where p_b and p_t are the pressure at the parcel's initial level, i.e. the boundary layer, and the neutral buoyancy level, respectively, R_d is the gas constant of the dry air, T_{vp} and T_{ve} are the virtual temperature of the cloud parcel and its environment, respectively. $\eta(\lambda)$ is the normalized cloud mass flux for the subensemble of clouds characterized by the fractional entrainment parameter λ, with $\eta(\lambda)=1$ at the cloud base. Since Arakawa-Schubert used a spectral cloud model, the quasi-equilibrium closure was applied to each subensemble of clouds. Mathematically, the quasi-equilibrium can be written as:

$$\frac{dA(\lambda)}{dt}=\frac{dA_c(\lambda)}{dt}+\frac{dA_{ls}(\lambda)}{dt}\approx 0. \tag{38}$$

$dA_c(\lambda)/dt$ and $dA_{ls}(\lambda)/dt$ are the time rate of change of cloud work function due to convective and the large-scale processes, respectively. Thus, the convective stabilization is related to the grid-resolvable destabilization for each cloud type:

$$\frac{dA_c(\lambda)}{dt} = -\frac{dA_{ls}(\lambda)}{dt} \approx -\frac{A_{ls}^t(\lambda) - A^{t-\Delta t}(\lambda)}{\Delta t}, \tag{39}$$

where $A_{ls}^t(\lambda)$ is the cloud work function at time t after the large-scale forcing is applied, $A^{t-\Delta t}$ is the observed cloud work function at $t-\Delta t$ (after convection), Δt is the time interval of the observations. Lord (1982) estimated the observed cloud work function in the atmosphere using several different datasets in the tropics, and found that for a given cloud type they are in general invariant with time and can be replaced by a climatological value A_0. The values of A_0 for different λ are given in Lord (1982). Thus, Eq. (39) can be written as:

$$\frac{dA_c(\lambda)}{dt} \approx -\frac{A_{ls}^t(\lambda) - A_0(\lambda)}{\Delta t}. \tag{40}$$

Arakawa and Schubert (1974) showed that the change in cloud work function by convection is proportional to the cloud base mass flux:

$$\frac{dA_c(\lambda)}{dt} = -F(\lambda)m_b(\lambda), \tag{41}$$

where $F(\lambda)$ can be computed from the large-scale conditions and the spectral cloud model. Thus, the closure equation becomes:

$$m_b(\lambda) = \frac{1}{F(\lambda)} \frac{A_{ls}^t(\lambda) - A_0(\lambda)}{\Delta t}. \tag{42}$$

This means that the cloud base mass flux for each cloud type is proportional to the convective instability in the atmosphere after the large-scale forcing. This closure has been used in almost all of the implementations of the Arakawa-Schubert scheme.

4.3. *CAPE-based closure*

Following the concept of the Arakawa-Schubert quasi-equilibrium, Zhang and McFarlane (1995) explicitly used CAPE (convective available potential energy) removal as the closure, noting that CAPE is the same as the cloud work function for non-entraining air plumes. Since the large-scale temperature and moisture changes in both the cloud layer and the subcloud layer due to convective activity are linearly related to the cloud base mass flux, CAPE change due to convection can be written as

$$\frac{\partial}{\partial t}\text{CAPE} = -M_b K, \tag{43}$$

where K is the CAPE consumption rate by convection per unit cloud base updraft mass flux, and is determined by the large-scale thermodynamic profiles and the cloud model, similar to $F(\lambda)$ in Eq. (41). The closure condition is that the CAPE is removed at an exponential rate by convection with a characteristic adjustment time scale τ. Thus

$$M_b = \frac{\text{CAPE}}{\tau K}, \tag{44}$$

where τ is typically a few hours. This type of closure has also been used in the ECMWF Integrated Forecast System (Gregory et al. 2000) and the Hadley Centre climate model HadAM3 (Pope et al. 2000).

5. Recent work from personal research and remaining outstanding issues

5.1. *Refinement of closure conditions.*

The Arakawa-Schubert quasi-equilibrium assumption was developed using tropical soundings, and has been used for convective parameterization in global climate models. Its applicability in midlatitude continental convection environment was tested in a case study of mesoscale convective systems (Grell et al. 1991), and more recently using the concept of generalized convective available potential energy by Cripe and Randall (2001). These studies support the Arakawa-Schubert quasi-equilibrium assumption. On the other hand, recently Zhang (2002) examined the validity of the Arakawa-Schubert quasi-equilibrium in midlatitude land convection using 3-hr averaged data, and found that it does not hold well even in actively convective regime. He introduced a free tropospheric quasi-equilibrium as a refinement to the Arakawa-Schubert quasi-equilibrium. Zhang (2003) and Donner and Phillips (2003) further examined this issue using both tropical oceanic and midlatitude continental data. They find that at sub-diurnal timescales, the Arakawa-Schubert quasi-equilibrium is not an accurate assumption. As the timescale increases, the accuracy of the Arakawa-Schubert quasi-equilibrium approaches that of the free tropospheric quasi-equilibrium.

By definition, CAPE is the vertical integral of buoyancy of a parcel lifted from the boundary layer following the moist adiabat to its neutral buoyancy level, and is the same as the cloud work function of the non-entraining air plumes:

$$\text{CAPE} \equiv \text{A}(0) = \int_{p_t}^{p_b} R_d (T_{vp} - T_{ve}) d\ln p, \tag{37a}$$

For notational simplicity, A (with its argument dropped) will be used to denote CAPE in this section. Analogous to Eq. (38), CAPE change is due to two types of processes: convective processes and large-scale processes:

$$\frac{dA}{dt} = \frac{dA_c}{dt} + \frac{dA_{ls}}{dt}. \tag{38a}$$

We can rewrite Eq. (38a) as

$$\frac{dA_c}{dt} = -\frac{dA_{ls}}{dt} + \frac{dA}{dt} \tag{38b}$$

to diagnose CAPE change due to convection from the observed net CAPE change and CAPE change from the large-scale forcing. The Arakawa-Schubert quasi-equilibrium assumption requires that $|dA/dt| << |dA_{ls}/dt|$. When this assumption is valid, we expect the diagnosed CAPE change due to convection to be approximately balanced by CAPE change due to the large-scale forcing.

This quasi-equilibrium assumption can be tested using observational data. For this purpose, two datasets are used, one from the midlatitude continental environment and one from the tropical maritime environment. The midlatitude data are from the U. S. Department of Energy Atmospheric Radiation Measurement (ARM) Program in the Southern Great Plains (SGP). They cover a period of 29 days from June 19 to July 17, 1997. The details of the analysis procedures can be found in Zhang (2002). The tropical maritime data are from the TOGA COARE Intensive Observation Period (IOP), which covers 120 days from Nov. 1 1992 to Feb. 28 1993. The sounding data are averaged over the Intensive Flux Area at 6h intervals.

Figure 1 shows the scatter plots of convective removal of CAPE diagnosed from Eq. (38b) versus the large-scale generation of CAPE for the 29-day ARM SGP IOP (top) and the 120-day TOGA COARE IOP (bottom) for convective periods. If the Arakawa-Schubert quasi-equilibrium is valid, the points should fall close to the diagonal line. Clearly, although they tend to fall in the right direction in general, there is a significant degree of scatter in both the tropics and the midlatitude. Based on this figure, it would be difficult to state that the Arakawa-Schubert quasi-equilibrium is a good approximation in either case.

Another way to look at the convective quasi-equilibrium issue is to follow the approach of Emanuel et al. (1994). From Eq. (37a), the time rate of change of CAPE is given by:

$$\begin{aligned}\frac{dA}{dt} &= \frac{d}{dt}\left\{\int_{p_t}^{p_b} R_d(T_{vp} - T_{ve})\,d\ln p\right\} \\ &= \int_{p_t}^{p_b} R_d\left(\frac{dT_{vp}}{dt} - \frac{dT_{ve}}{dt}\right) d\ln p - R_d\left[T_{vp} - T_{ve}\right]_{p_t}\frac{dp_t}{dt}.\end{aligned} \tag{45}$$

The last term on the second line vanishes since the virtual temperature of the parcel at the neutral buoyancy level is the same as that of its environment. Thus, CAPE change consists of two parts, one due to the free tropospheric environmental virtual temperature change and one due to the parcel's virtual temperature change, that is,

$$\frac{dA}{dt}=\frac{dA^p}{dt}+\frac{dA^e}{dt}, \tag{46}$$

where

$$\frac{dA^p}{dt}=R_d\int_{p_t}^{p_b}\frac{dT_{vp}}{dt}d\ln p\,,\quad \frac{dA^e}{dt}=-R_d\int_{p_t}^{p_b}\frac{dT_{ve}}{dt}d\ln p \tag{47}$$

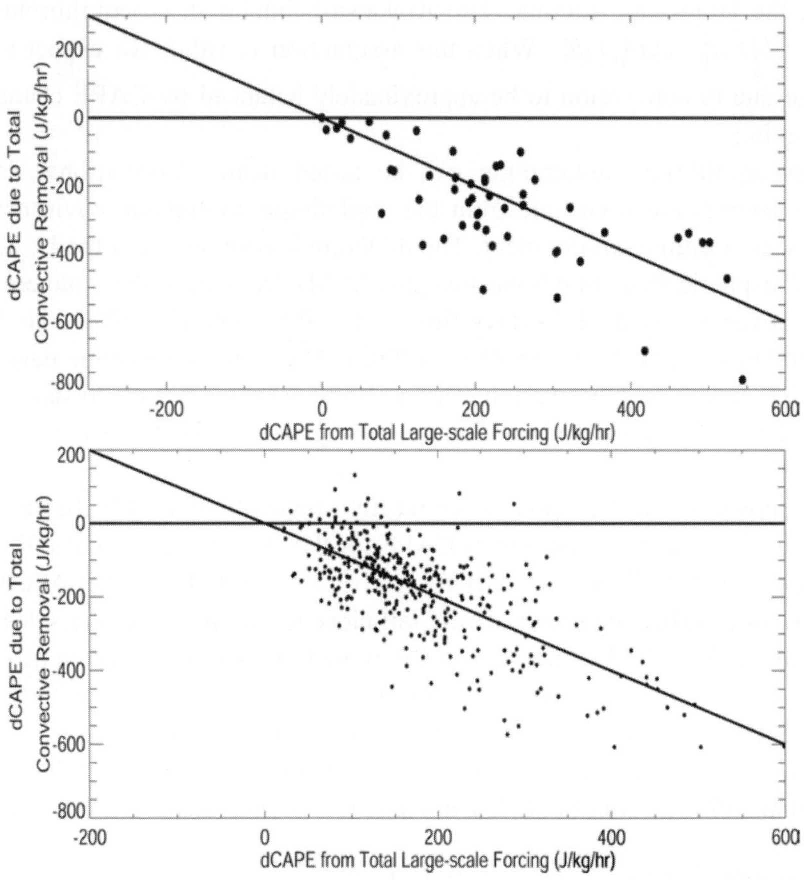

Fig. 1: Scatter plots demonstrating the validity of the Arakawa-Schubert quasi-equilibrium assumption in the midlatitude (top) and tropical (bottom) convection environment. Each point represents a 3 h average for the midlatitude data and 6 h average for the tropical data due to the data availability. The x-axis is the CAPE change due to the large-scale forcing, and the y-axis is the CAPE change due to convection diagnosed from Eq. (38a).

Figure 2 displays the scatter plots of the terms in Eq. (46) for the ARM SGP site data. The top frame shows the scatter plot of the net CAPE change versus CAPE change resulting from the parcel's temperature change. The bottom frame shows the scatter plot of CAPE change resulting from the ambient temperature versus the net CAPE change. Clearly, CAPE variations resulting from changes in the boundary layer temperature and moisture are largely reflected (about 90%) in the net atmospheric CAPE variations, and more importantly, the CAPE variations due to the ambient temperature changes above the parcel's source level are insignificant compared to the net CAPE change, that is, $dA^e / dt \approx 0$.

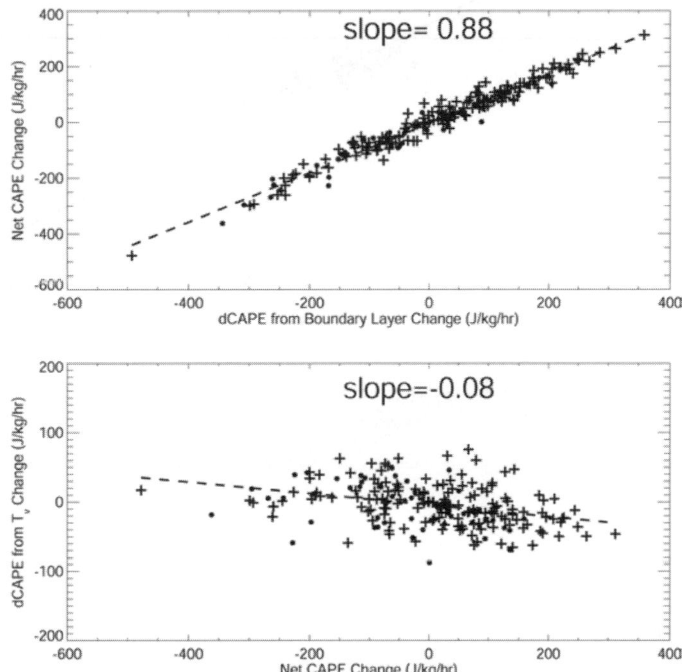

Fig. 2: Scatter plot of the net atmospheric CAPE change versus that due to the boundary layer temperature and moisture changes (top), and scatter plot of CAPE change due to parcel's ambient temperature change versus the net atmospheric CAPE change using data from the ARM SGP site. Dots are for convective periods and crosses are for non-convective periods. The slopes are from linear regression of all points.

As CAPE change due to contributions from the ambient air (or the free tropospheric air above the boundary layer) temperature change is a result of the large-scale and convective processes in analogy to Eq. (38a), we can write:

$$\frac{dA^e}{dt} = \frac{dA^e_c}{dt} + \frac{dA^e_{ls}}{dt} \tag{48}$$

or

$$\frac{dA_c^e}{dt} = -\frac{dA_{ls}^e}{dt} + \frac{dA^e}{dt}. \tag{48a}$$

Eq. (48a) is an alternative way to diagnose the convective effect from the large-scale forcing and the observed changes of temperature and moisture. Note that in this approach only fields and large-scale forcing above the parcel's source level (or the boundary layer) are involved.

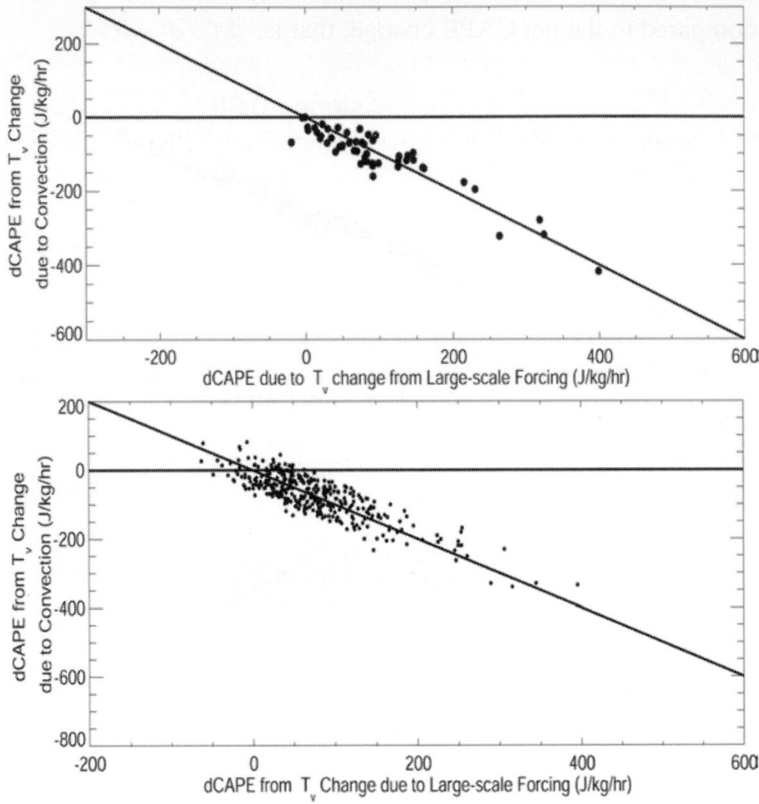

Fig. 3: Scatter plots of convective removal of partial CAPE contribution from the free tropospheric virtual temperature change versus its large-scale generation for the 29-day ARM Southern Great Plains IOP (top) and the 120-day TOGA COARE period (bottom).

Figure 3 shows the scatter plots of the diagnosed convective removal of CAPE due to changes of temperature and moisture in the free troposphere versus the large-scale forcing on the same fields. Similar to the philosophy in plotting Fig. 2, if convective stabilization can be diagnosed in this new way, the observations during convective periods should fall close to the diagonal line. Indeed, for both the midlatitude convection (top) and the tropical convection (bottom), the agreement between the diagnosed and "predicted" convective removal of CAPE is excellent. Comparing with Fig. 1, it is clear that the improvement in predicting convection using the new approach is significant. We call this "the free tropospheric quasi-equilibrium" since only the free tropospheric processes are involved. It works well in both tropical and midlatitude convection environment, and should be built into convective parameterization schemes.

Separating the boundary layer from the rest of the troposphere in the free tropospheric quasi-equilibrium assumption may make one wonder how the effect of the surface conditions such as sea surface temperature (SST) on convection can be represented. SST affects the thermodynamic properties of the boundary-layer air through surface sensible and latent heat fluxes. The variation of the boundary-layer air properties in turn determines the variation of CAPE. In both CAPE-based and Arakawa-Schubert quasi-equilibrium closures, convection responds to remove this instability. In the free tropospheric quasi-equilibrium assumption, SST affects convection in at least two ways. First, as above, it changes CAPE, providing a necessary (but not sufficient) condition for convection. Second, SST gradients interact with the large-scale dynamics through the pressure gradient force, as suggested by Lindzen and Nigam (1987). This dynamically induced circulation would give rise to upward motion in the warm SST region, leading to destabilization of the atmosphere and thus convection.

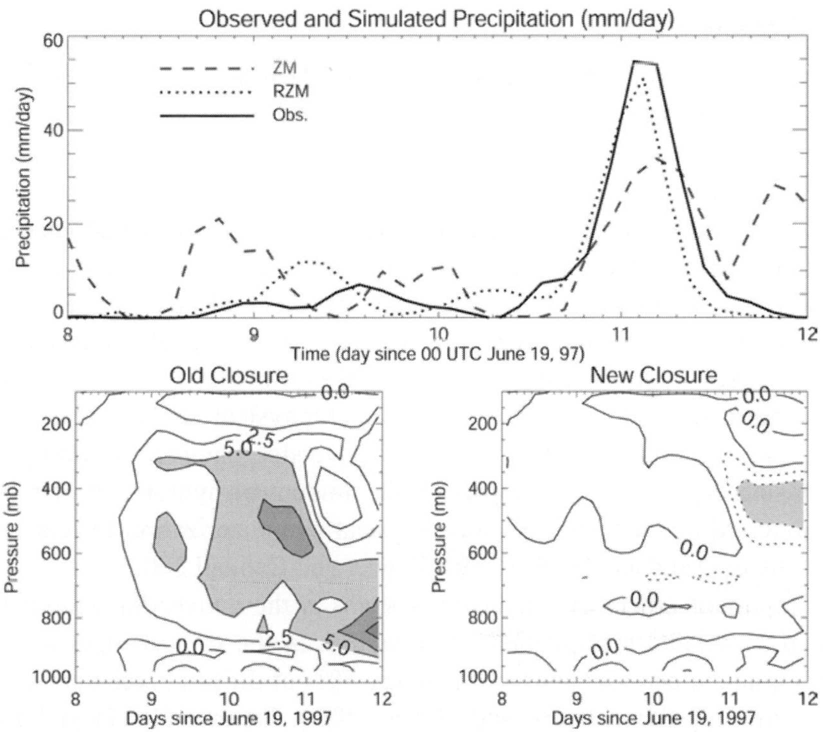

Fig. 4: Observed and simulated precipitation with the original and modified closures in the Zhang-McFarlane scheme (top), and the temperature biases in the simulations (bottom). Contour intervals are 2.5 K.

5.2. *Test of the modified quasi-equilibrium closure*

A modified quasi-equilibrium closure was developed based on Eq. (48a), aiming to improve the simulation of convection in GCMs (Zhang 2002). Here we use the single column version of the NCAR Community Climate Model CCM3 to demonstrate the impact of this new closure on the simulation of surface precipitation and tropospheric temperature at the ARM SGP site.

Figure 4 shows a 4-day time series of the observed and simulated precipitation, during which a strong convective system developed. Also shown are the temperature biases from the simulations. When the original Zhang-McFarlane scheme is used, the simulated precipitation occurs daily, compared to the observations. The intensity of the simulated precipitation is weak when heavy precipitation is observed. When the modified closure is used, both the magnitude and timing of the observed heavy precipitation event are well simulated. The simulated temperature field has a large warm bias during most of the simulation period when the original closure is used. This contrasts with a much smaller bias in magnitude when the new closure is used. Note that in the two model simulations, the only difference is the closure. This clearly demonstrates the improvement of the new closure on the simulation of convection. Obviously, the ultimate test of the new closure needs to be done with the 3-D GCM to fully determine its impact on climate simulation. This is currently underway.

5.3. *Outstanding issues*

In the early efforts of convective parameterization development, convective heating and drying were the main focus. The cloud water content inside convective cells was often crudely parameterized without affecting the convective heating and drying too much. However, convection, clouds and radiation in the climate system are highly interactive. Detrained cloud hydrometeor is an important source of cloud water and ice for large-scale cloud formation and evolution. These clouds in turn exert large radiative forcing on the atmosphere and the underlying ocean and land. As more elegant cloud parameterization schemes are developed, different treatment of the convective cloud condensate and its detrainment can have a large impact on the simulated climate (Fowler and Randall 2002). This requires more accurate treatment of the microphysical processes inside convective systems. For example, cloud water to rainwater conversion parameterization Eq. (25) is too simplistic, considering its potential impact on the amount of hydrometer detrainment to the convective anvils. More sophisticated microphysical parameterization for convective clouds has started to emerge (Sud and Walker 1999, Fowler and Randall 2002).

Another important aspect of convective systems is the convection-generated mesoscale circulation and associated precipitation. Observations in the tropics suggest that mesoscale stratiform precipitation can account for as much as 50% of the total precipitation in organized tropical convection systems (Leary and Houze 1980). Donner (1993) and Donner et al. (2001) have pioneered to incorporate the effects of mesoscale circulation into convective parameterization for GCMs with some success.

Finally, advancement in cloud-system-resolving models and computing power has made it possible to embed these models in GCMs under the so-called "superparameterization" framework (Grabowski 2001, Khairoutdinov and Randall 2001). While these models will not replace convective parameterization in any foreseeable future, they are excellent tools for evaluating and improving convective parameterization and are likely to be further pursued in this respect.

6. Summary

This paper reviews several widely used convection parameterization schemes and discusses their shortcomings. The emphasis is on the mass flux representation of convection in GCMs and associated closure conditions, since this type of schemes are advantageous in representing convective transport of chemical species.

While significant progress has been made in the past in developing improved convective parameterization schemes, some fundamental issues remain to be addressed. One such issue is the closure condition based on the quasi-equilibrium assumption of Arakawa and Schubert, the validity of which is often taken for granted. Results from recent field experiments are used to demonstrate that the Arakawa-Schubert quasi-equilibrium assumption can be refined to provide a more accurate relationship between convection and the large-scale fields. The modification based on the results yields a much better agreement between the predicted and diagnosed convective removal of the atmospheric convective instability. A single column version of the NCAR Community Climate Model is used to test the modified closure of the Zhang-McFarlane convection scheme. Significant improvement is achieved in simulating the timing and intensity of precipitation with the modified closure.

Some outstanding issues still facing today's convection parameterization developers are briefly outlined. Although methods to tackle these issues have started to emerge, they remain to be fully utilized by the GCM communities.

Acknowledgments. This work was supported by NSF grant ATM-0204798 and by the Office of Science (BER), U.S. Department of Energy, grant No. DE-FG03-03ER63532. The author thanks the reviewers for their constructive comments, which have led to improvement in the presentation.

References:

Arakawa, A., and W. H. Schubert, Interaction of a cumulus cloud ensemble with the large-scale environment. Part I. *J. Atmos. Sci.*, **31**, 674-701, 1974.

Betts, A. K., 1982: Saturation point analysis of moist convective overturning. *J. Atmos. Sci.*, **39**, 1484-1505.

Betts, A. K., 1986: A new convective adjustment scheme. I. Observational and theoretical basis. *Quart. J. Roy. Meteor. Soc.*, **112**, 677-691.

Betts, A. K., and M. J. Miller, 1986: A new convective adjustment scheme. Part II: Single column tests using GATE wave, BOMEX, ATEX and arctic air-mass data sets. *Quart. J. Roy. Meteor. Soc.*, **112**, 693–709.

Charney, J. G., and A. Eliassen, 1964: On the growth of the hurricane depression. *J. Atmos. Sci.*, **21**, 68-75.

Cripe, D. G., and D. A. Randall, 2001: Joint variations of temperature and water vapor over the midlatitude continents. *Geophys. Res. Lett.*, **28**, 2613-2616.

Donner, L. J., 1993: A cumulus parameterization including mass fluxes, vertical momentum dynamics, and mesoscale effects. *J. Atmos. Sci.*, **50**, 889-906.

Donner, L. J., C. J. Seman, R. S. Helmer, and S. Fan, 2001: A cumulus parameterization including mass fluxes, convective vertical velocities, and mesoscale effects: Thermodynamic and hydrological aspects in a general circulation model. *J. Climate*. **14**, 3444-3463.

Donner, L. J., and V. T. Phillips, 2003: Boundary-layer control on convective available potential energy: Implications for cumulus parameterization. *J. Geophys. Res.*, submitted.

Eitzen, M. A., and D. A. Randall, 1999: Sensitivity of the simulated Asian summer monsoon to parameterized physical processes. *J. Geophys. Res.*, **104**, 12,177-12,191.

Emanuel, K. A., 1991: A scheme for representing cumulus convection in large-scale models. *J. Atmos. Sci.*, **48**, 2313-2335.

Emanuel, K. A., 1994: Atmospheric convection. *Oxford University Press*, pp 580.

Emanuel, K. A., J. D. Neelin, and C. S. Bretherton, 1994: On large-scale circulation in convective atmospheres. *Quart. J. Roy. Meteor. Soc.*, **120**, 1111-1143.

Fowler, L. D., and D. A. Randall, 2002: Interactions between cloud microphysics and cumulus convection in a general circulation model. J. Atmos. Sci., 59, 3074-2098.

Grabowski, W. W., 2001: Coupling cloud processes with the large-scale dynamics using the cloud-resolving convection parameterization (CRCP). *J. Atmos. Sci.*, **58**, 978-997.

Gregory, D., J.-J. Morcrette, C. Jacob, A. C. M. Beljaars, and T. Stockdale, Revision of convection, radiation and cloud schemes in the ECMWF Integrated Forecasting System. *Quart. J. Roy. Meteor. Soc.*, **126**, 1685-1710, 2000.

Grell, G. A., Kuo, Y.-H., and Pasch, R. J., 1991: Semiprognostic tests of cumulus parameterization schemes in the middle latitudes. *Mon. Wea. Rev.*, **119**, 5-30.

Hack, J. J., 1994: Hack, J. J., 1994: Parameterization of moist convection in the National Center for Atmospheric Research community climate model (CCM2). *J. Geophy. Res.*, **99**, 5551-5568.

Johnson, R. H., 1978: Cumulus transports in a tropical wave composite for phase III of GATE. *J. Atmos. Sci.*, **35**, 484-494.

Khairoutdinov, M. F., and D. A. Randall, 2001: A cloud resolving model as a cloud parameterization in the NCAR Community System Model: Preliminary results. *Geophys. Res. Lett.*, **28**, 3617-3620.

Krishnamurti, T. N., Y. Ramanathan, H.-L. Pan, R. J. Pasch, and J. Molinari, 1980: Cumulus parameterization and rainfall rates I. *Mon. Wea. Rev.*, **108**, 465-472.

Kuo, H.-L., 1965: On the formation and intensification of tropical cyclones through latent heat release by cumulus convection. *J. Atmos. Sci.*, **22**, 40-63.

Kuo, H.-L., Further studies of the parameterization of the influence of cumulus convection on large-scale flow. *J. Atmos. Sci.,* **31**, 1232-1240, 1974.

Leary, C. A., and R. A. Houze, Jr., 1980: The contribution of mesoscale motions to the mass and heat fluxes of an intense tropical convective system. *J. Atmos. Sci.*, **37**, 784-796.

Lindzen, R. S., and S. Nigam, 1987: On the role of sea surface temperature gradients in forcing the low-level winds and convergence in the tropics. *J. Atmos. Sci.*, **44**, 2418-2436.

Lord, S., J., 1982: Interaction of a cumulus cloud ensemble with large-scale environment. Part III: Semi-prognostic test of the Arakawa-Schubert cumulus parameterization. *J. Atmos. Sci.*, **39**, 88-103.

Maloney, E. D., and D. L. Hartmann, 2001: The sensitivity of intraseasonal variability in the NCAR CCM3 to changes in convective parameterization. *J. Climate*, **14**, 2015-2034.

Manabe, S., J. Smagorinsky, and R. F. Strickler, 1965: Simulated climatology of a general circulation model with a hydrological cycle. *Mon. Wea. Rev.*, **93**, 769-798.

Pope, V. D., M. L. Gallani, P. R. Rowntree, and R. A. Stratton, 2000: The impact of new physical parameterizations in the Hadley Centre climate model: HadAM3. *Climate Dyn.*, **16**, 123-146.

Slingo, J. M., U. C. Mohanty, M. Tiedtke, and R. P. Pearce, 1988: Prediction of the 1979 summer monsoon onset with modified parameterization schemes. *Mon. Wea. Rev.*, **116**, 328-346.

Slingo, J. M., and Coauthors, 1994: Mean climate and transience in the tropics of the UGAMP GCM: Sensitivity to convective parameterization. *Quart. J. Roy. Meteor. Soc.,* **120,** 881–922.

Sud, Y. C., and G. K. Walker, 1999: Microphysics of clouds with the Relaxed Arakawa-Schubert Scheme (McRAS). Part I: Design and evaluation with GATE Phase III data. *J. Atmos. Sci.*, **56**, 3196-3220.

Tiedtke, M., 1989: A comprehensive mass flux scheme for cumulus parameterization in large-scale models. *Mon. Wea. Rev.*, **117**, 1779-1800.

Vaidya, S. S., and S. S. Singh, 1997: Thermodynamic adjustment parameters in the Betts-Miller scheme of convection. *Wea. Forecasting*, **12**, 819-825.

Yanai, M. S. Esbensen, and J. H. Chu, 1973: Determination of bulk properties of tropical cloud clusters from large-scale heat and moisture budgets. *J. Atmos. Sci.*, **30**, 611-627.

Zhang, G. J., 1994: Effects of cumulus convection on the simulated monsoon circulation in a general circulation model. *Mon. Wea. Rev.*, **122**, 2022-2038.

Zhang, G. J., and N. A. McFarlane, 1995: Sensitivity of climate simulations to the parameterization of cumulus convection in the Canadian Climate Centre general circulation model. *Atmosphere-Ocean*, **33**, 407-446.

Zhang, G. J., J. T. Kiehl, and P.J. Rasch, 1998: Response of climate simulation to a new convective parameterization in the National Center for Atmospheric Research Community Climate Model (CCM3), *J. Climate*, **11**, 2097-2115.

Zhang, G. J., 2002: Convective quasi-equilibrium in midlatitude continental environment and its effect on convective parameterization. *J. Geophys. Res.*, **107**, doi: 10.1029/2001JD001005.

Zhang, G. J., 2003: Convective quasi-equilibrium in the Tropical western Pacific: Comparison with midlatitude continental environment. *J. Geophys. Res.*, in press.

CLOUD MODELING IN THE TROPICAL DEEP CONVECTIVE REGIME

XIAOFAN LI
Joint Center for Satellite Data Assimilation
NOAA/NESDIS/Office of Research and Applications
5200 Auth Road, Camp Springs, MD 20746, USA
E-mail: Xiaofan@noaa.gov

(Manuscript received 5 September 2002)

Understanding cloud processes and the associated interactions with their environment is crucial for better predictions of tropical climate. The cloud-resolving model is demonstrated to be a powerful tool for process studies. In this paper, cloud modeling in the tropical deep convective regime is reviewed based on the author's research work. The review focuses on model setup, cloud-radiation interaction processes, convective-radiative processes associated with the diurnal variation of tropical oceanic convection, dominant cloud microphysical processes producing precipitation, precipitation efficiency, physical processes responsible for the phase relation between surface rain rate and convective available potential energy, and the effects of precipitation on the tropical upper ocean.

1. Introduction

Clouds play an important role in the hydrological and energy cycles in the tropical atmosphere. Cloud formation is determined by the large-scale thermal and moisture conditions of the environment. Once formed, clouds feedback to the environment by redistributing temperature, moisture, and momentum vertically through radiative, microphysical, and mixing processes. However, how to accurately simulate cloud processes and the associated interactions with their environment remains a big challenge in the climate modeling community. The intercomparison studies of atmospheric general circulation models indicated that the lack of understanding clouds represents one of the largest uncertainties in climate modeling and prediction (e.g., Cess et al. 1990; Gates 1992). Thus, understanding the interaction between clouds and their environment is fundamentally important for better prediction of tropical climate.

Due to the scarcity of observations and the nonlinear nature, the interaction between tropical convection and the large-scale circulation is often studied with cloud-resolving models that include explicit cloud-scale dynamics, detailed microphysics, and sophisticated radiative transfer calculations based on physically determined cloud optical properties. Sui et al. (1994) were the first to successfully integrate the cloud-resolving model to the equilibrium climate to study the tropical water and energy cycles and their role in the climate system. Recently, cloud-resolving models have been used to simulate the deep convective response to large-scale forcing observed in the Global Atmosphere Research Programme Atlantic

Tropical Experiment (GATE) (e.g., Xu and Randall 1996; Grabowski et al. 1996) and the Tropical Ocean Global Atmosphere (TOGA) Coupled Ocean-Atmosphere Response Experiment (COARE) (e.g., Wu et al. 1998; Li et al. 1999).

This paper will highlight the research contributions of the author which were made primarily through the use of a cloud-resolving model in recent years. The model setup will be discussed in section 2. The model includes radiation and explicit cloud microphysical parameterization schemes, which allow cloud-radiation interaction. The new cloud-infrared interaction mechanism from Li et al. (1999) is reviewed in section 3. Diurnal variation of tropical convection is an important variability in terms of the hydrological and energy cycles. The physical processes associated with the diurnal variation will be identified in section 4. The dominant cloud microphysical processes responsible for tropical precipitation are reviewed in section 5. Precipitation efficiency (PE) is an important physical parameter for measuring the interaction between clouds and their environment. The definition of PE and the possible physical parameters controlling it are discussed in section 6. In section 7, a new concept of moist available potential energy is introduced into the Eulerian framework, and a set of energetics equations are derived to explain the phase relationship between the surface rain rate and the convective available potential energy (CAPE). A coupled ocean-cloud-resolving atmosphere model is developed to examine the effects of small temporal and spatial scale precipitation on the tropical upper ocean in section 8. The conclusions are given in section 9.

2. Model and forcing

For cloud modeling studies, a scale separation between the large-scale and convective scale is assumed so that the cloud response to the large-scale forcing can be examined. Soong and Ogura (1980) were the first to develop ways to impose the observed large-scale variables into a cloud-resolving model to examine the "one-way" response of the model to the imposed "large-scale forcing". There are two ways to impose the large-scale forcing into a cloud model. The horizontally uniform and vertically varying vertical velocity can be imposed, as first introduced by Soong and Ogura (1980), or the horizontally uniform total advection of the heat and moisture can be imposed (e.g., Wu et al. 1998). The corresponding 2-D equations in an x-z framework are, respectively,

$$\frac{\partial A}{\partial t}=-\frac{\partial(u'A')}{\partial x}-\overline{u}^o\frac{\partial A'}{\partial x}-\frac{1}{\overline{\rho}}\frac{\partial(\overline{\rho}w'A')}{\partial z}-\overline{w}^o\frac{\partial A'}{\partial z}-w'\frac{\partial\overline{A}}{\partial z}+S_A+D_A-\overline{u}^o\frac{\partial\overline{A}^o}{\partial x}-\overline{w}^o\frac{\partial\overline{A}}{\partial z}, \quad (1)$$

$$\frac{\partial A}{\partial t}=-\frac{\partial(u'A)}{\partial x}-\frac{1}{\overline{\rho}}\frac{\partial(\overline{\rho}w'A)}{\partial z}+S_A+D_A-\overline{u}^o\frac{\partial\overline{A}^o}{\partial x}-\overline{w}^o\frac{\partial\overline{A}^o}{\partial z}. \quad (2)$$

Here, $A = (\theta, q_v)$, the potential temperature and specific humidity, respectively; u and w are the zonal and vertical wind components; ρ is air density; S_A and D_A are the thermodynamic forcing including radiation and cloud microphysics and the dissipation terms, respectively; overbars denote a zonal-mean; primes are derivations from the zonal-mean the superscript ^{o} denotes imposed observed variables in the model. The derivation and assumptions for the above equations can be found in Li et al. (1999).

Li et al. (1999) compared the simulations with the observations in terms of vertical temperature and moisture profiles, surface heat fluxes, and surface rain rates. The root-mean-square differences of temperature and specific humidity between the observation and simulation with the imposed vertical velocity over vertical and horizontal model domains are 1.44°C and 0.42 g kg^{-1}, respectively. The six-day-mean latent heat flux in the simulation (164 Wm^{-2}) is larger than the observation (145 Wm^{-2}). The simulated and observed surface rain rates are similar in both amplitude and phase when imposed vertical velocity is accurately derived from the observations. Li et al. (1999) also found that the root-mea-square temperature difference between the observation and simulation with the imposed vertical velocity (1.44°C) is smaller than that between the observation and simulation with imposed sources and sinks (2.25°C), indicating that the adjustment of the mean thermodynamic stability distribution by vertical advection in model with the imposed vertical velocity produces better simulations. Thus, the vertical-velocity forcing is used in the following discussions. The governing equations with an anelastic approximation can be expressed as follows:

$$\frac{\partial u'}{\partial x}+\frac{1}{\bar{\rho}}\frac{\partial(\bar{\rho}w')}{\partial z}=0, \tag{3}$$

$$\frac{\partial u'}{\partial t}=-\frac{\partial}{\partial x}(2u'\bar{u}^o+u'u')-\frac{1}{\bar{\rho}}\frac{\partial}{\partial z}\bar{\rho}(w'\bar{u}^o+\bar{w}^o u'+w'u'-\overline{w'u'})-c_p\frac{\partial(\bar{\theta}\pi')}{\partial x}+D_u-\bar{D}_u, \tag{4}$$

$$\begin{aligned}\frac{\partial w'}{\partial t}=&-\frac{\partial}{\partial x}(u'\bar{w}^o+\bar{u}^o w'+u'w')-\frac{1}{\bar{\rho}}\frac{\partial}{\partial z}\bar{\rho}(2w'\bar{w}^o+w'w'-\overline{w'w'})-c_p\frac{\partial(\bar{\theta}\pi')}{\partial z}\\&+g(\frac{\theta'}{\theta_b}+0.61q_v'-q_l')+D_w-\bar{D}_w,\end{aligned} \tag{5}$$

$$\begin{aligned}\frac{\partial \theta}{\partial t}=&-\frac{\partial(u'\theta')}{\partial x}-\bar{u}^o\frac{\partial\theta'}{\partial x}-\frac{1}{\bar{\rho}}\frac{\partial}{\partial z}\bar{\rho}w'\theta'-\bar{w}^o\frac{\partial\theta'}{\partial z}-w'\frac{\partial\bar{\theta}}{\partial z}+\frac{Q_{cn}}{\pi c_p}+\frac{Q_R}{\pi c_p}\\&-\bar{u}^o\frac{\partial\bar{\theta}^o}{\partial x}-\bar{w}^o\frac{\partial\bar{\theta}}{\partial z}+D_\theta,\end{aligned} \tag{6}$$

$$\frac{\partial q_v}{\partial t} = -\frac{\partial (u' q_v')}{\partial x} - \bar{u}^o \frac{\partial q_v'}{\partial x} - \frac{1}{\bar{\rho}} \frac{\partial}{\partial z} \bar{\rho} w' q_v' - \bar{w}^o \frac{\partial q_v'}{\partial z} - w' \frac{\partial \overline{q_v}}{\partial z} - (c - e + d - s)$$

$$- \bar{u}^o \frac{\partial \overline{q_v}^o}{\partial x} - \bar{w}^o \frac{\partial \overline{q_v}}{\partial z} + D_{qv}, \tag{7}$$

$$\frac{\partial C}{\partial t} = -\frac{\partial (uC)}{\partial x} - \frac{1}{\bar{\rho}} \frac{\partial}{\partial z} [\bar{\rho}(w - w_{TC})C] + S_C + D_C \tag{8}$$

Here, $C = (q_c, q_r, q_i, q_s, q_g)$, the mixing ratios of cloud water, raindrops, cloud ice, snow, and graupel, respectively; w_{TC} is the terminal velocity for cloud species C which is zero for cloud water and ice; $\pi = (p/p_o)^{\kappa}$; $\kappa = R/c_p$; R is the gas constant; c_p is the specific heat of dry air at constant pressure p, and p_o=1000 mb; c, e, d, and s denote vapor condensation, evaporation, deposition, and sublimation, respectively; $Q_{cn} = L_v(c - e) + L_s(d - s) + L_f(f - m)$ denotes the net heat released through phase changes among different cloud species, where f and m are fusion and melting, respectively; L_v, L_s, and L_f are heat coefficients due to phase changes; Q_R is the radiative heating rate due to the net flux convergence of solar and infrared radiation (Chou and Suarez 1994; Chou et al. 1991, 1997); S_C is the source and sink of various hydrometeors through the explicit cloud microphysical parameterization (Rutledge and Hobbs 1983, 1984; Krueger et al. 1995; Tao et al. 1989; Lin et al. 1983; also see comprehensive discussions in Li et al. 1999, 2002c). A complete description of the cloud-resolving model can be found in Tao and Simpson (1993) and Li et al. (1999).

Figure 1 shows an example the time evolution of the vertical distribution of large-scale atmospheric vertical velocity and zonal wind (Prof. M. Zhang, personal communication, 2000) and sea surface temperature (SST) (Weller and Anderson 1996) from 19 December 1992 to 9 January 1993 that have been imposed in the model. From 19-25 December 1992, upward motion dominates, indicating strong convection. From 26 December 1992 to 3 January 1993, downward motion becomes dominant, with only occasional upward motion, suggesting a relatively dry phase. In the last few days, there is moderate upward motion. Diurnal and two-day signals are evident in Fig. 1a as shown by Sui et al. (1997a) and Takayabu et al. (1996). Large-scale westerly winds increased significantly in the lower- and mid-troposphere and reached their maximum of 20 ms^{-1} at 600 mb around 3 January 1993 (Fig. 1b). Except for the first and the last four days, SST had only a weak diurnal variation with a slowly decreasing trend (Fig. 1c).

Figure 2 shows a typical case with the development of strong convection. The rainbands propagate westward before hour 20 then start to propagate eastward. The change in direction of the rainbands is a result of intensifying lower-tropospheric westerly winds and weakening mid-tropospheric easterly winds (Fig. 1b).

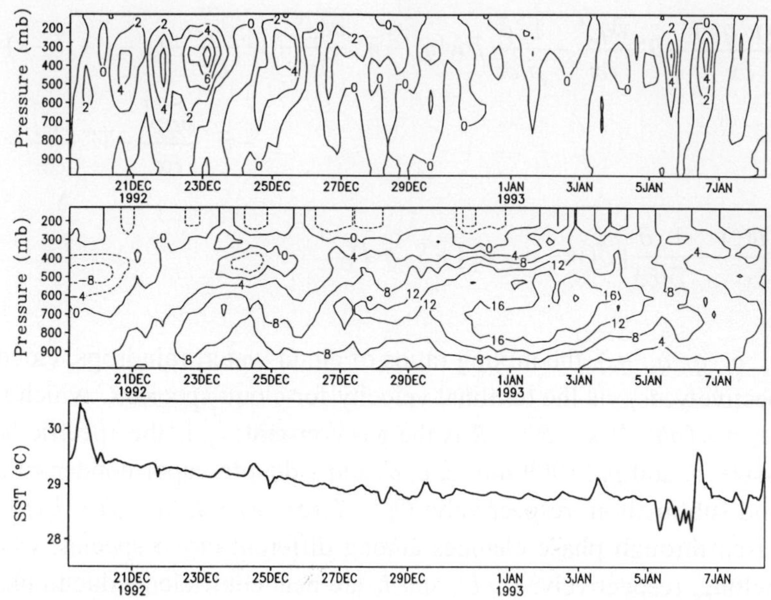

Fig. 1 Time evolution of (a) vertical velocity (cm s^{-1}), (b) zonal wind (m s^{-1}), and (c) sea surface temperature (°C) observed and derived during TOGA COARE for a 20-day period.

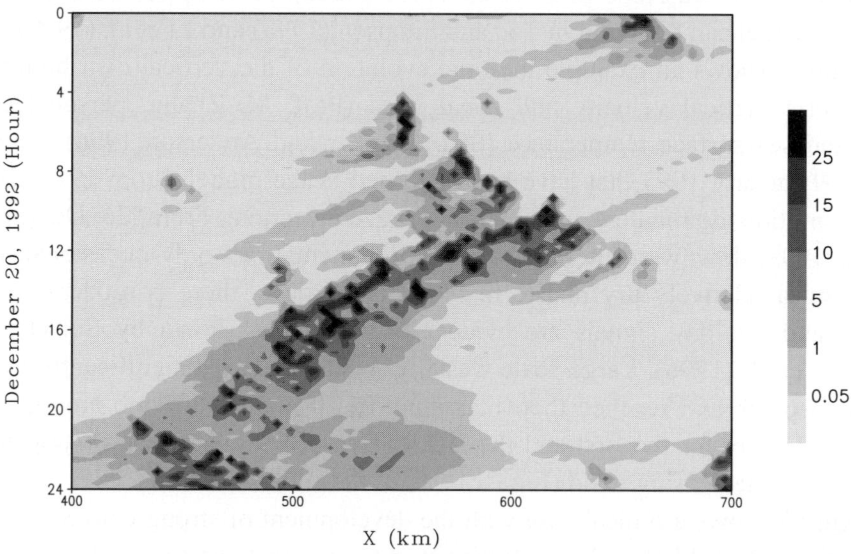

Fig. 2 Time evolution and zonal distribution of surface rain rate (mmh^{-1}) on December 20, 1992 as simulated by the cloud-resolving model.

3. Cloud-radiation interaction

Cloud-radiation interaction was examined by a comparison between two experiments. In one, cloud single scattering albedo and asymmetry factor varied with clouds and environmental thermodynamic conditions, in the other, they were fixed at 0.99 and 0.843, respectively (Li et al. 1999). A comparison of solar radiation calculations between the two experiments showed that the experiment with the varying single scattering albedo and asymmetry factor had stronger solar radiation absorption by ice clouds in the upper troposphere than does the experiment with the constant single scattering albedo and asymmetry factor. The difference in temperatures between the two experiments further showed that the temperature was 2°C warmer around 200 mb in the experiment with variable single scattering albedo and asymmetry factor than in the experiment with constant values.

A statistical analysis of the clouds and surface rain rates revealed that stratiform (convective) clouds contributed to 33 (67) % of the total rain in the experiment with the variable cloud optical properties and 40 (60) % in the experiment with the constant cloud optical properties. The fractional cover by stratiform clouds increased from 64% in the experiment with the variations to 70% in the experiment with the constants. These sensitivity tests show the cloud-radiation interaction process for stabilizing the atmosphere in which the change in the vertical heating gradient by solar radiation due to variations of cloud optical properties stabilizes the middle and upper troposphere and contributes to the reduction of stratiform clouds that further stabilizes the cloud system by reducing infrared cloud top cooling and cloud base warming.

4. Diurnal variation of tropical oceanic convection

The diurnal variation of tropical oceanic convection has been studied by observational analysis and numerical simulation. The dominant diurnal signal is the nocturnal peak in precipitation that occurs in the early morning. Kraus (1963) emphasized the role of radiative forcing in the diurnal variation and suggested that solar heating and IR cooling tends to suppress convection during daytime and enhance convection during nighttime respectively. Gray and Jacobson (1977) found that the diurnal variation of convection is a result of a synoptic-scale dynamic response to cloud radiative forcing (the radiational differences between cloudy regions and clear-sky regions). Such cloud radiative forcing drives low-level convergence into the cloudy regions causing upward motion and convection during nighttime. Tao et al. (1996) carried out a series of cloud-resolving simulations to study cloud-radiative mechanisms and emphasized that the increase of surface precipitation by IR cooling is due to the increased relative humidity. Recently, Sui et al. (1997a) analyzed the observational data from TOGA COARE and found that the primary peak in tropical convection on a diurnal time-scale appears during early morning with a secondary peak occurring in the early afternoon. Sui et al. (1998a) further conducted a series of numerical experiments using a cloud-resolving model and proposed a mechanism in which the nocturnal rainfall maximum is due to more available precipitable water during nighttime as a result of radiative cooling, which is similar to the mechanism proposed by Tao et al. (1996). The secondary rainfall maximum in the early afternoon results from the SST-induced instability associated with solar heating. Liu and Moncrieff (1998) found from their cloud resolving simulations that the simulated diurnal variation is primarily due to the direct interaction between radiation and convection with the mechanism proposed by Gray and Jacobson (1977) being a secondary factor. In the rest of the section, the results are from a recent analysis of the cloud-resolving simulation and provide some basic details to support the theory proposed by Sui et al. (1998a) in the cloud microphysics perspective.

The diurnal variation of the composite surface rain rate shows that rainfall peaks around hour 3 and hour 17, and rainfall is at a minimum at hour 13-14 (Fig. 3a). The hour-17 rainfall maximum results from the maximum in the sum of the surface evaporation and vertically-integrated horizontal and vertical specific humidity advection (large-scale moisture source) (Fig. 3b) that is due to the maximum imposed upward motion at hour 16 whose maximum center of 3 cms^{-1} is located at 350 mb (not shown, see Sui et al. 1997a for a discussion). SST does not significantly contribute to this rainfall peak because of the small amplitude in its diurnal variation (~0.2°C). Thus, external forcing plays a major role in the afternoon rainfall peak. The magnitude of the large-scale moisture source is much smaller around hour 3 than around hour 17. Thus, the moisture source cannot explain the maximum surface rain rate at hour 3.

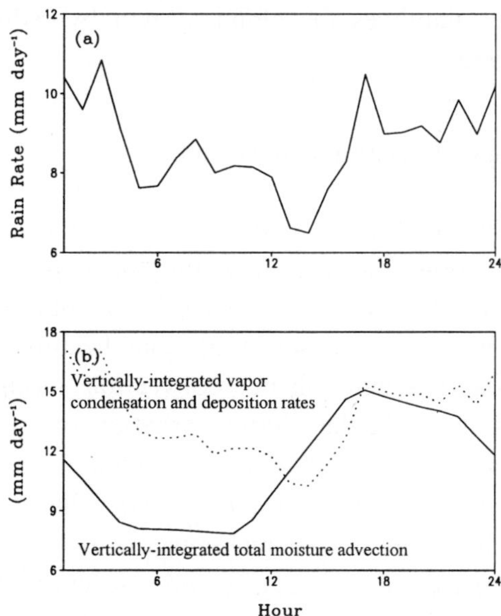

Fig. 3 Composite diurnal variations of (a) surface rain rate and (b) the sum of surface evaporation and the vertically-integrated horizontal and vertical moisture advection (solid line) and the sum of the vertically-integrated vapor condensation and deposition rates (dot line). Units are in mm d^{-1}.

Calculation of the cloud microphysics precipitation efficiency [defined as the ratio of the surface rain rate to the sum of the vapor condensation and deposition rates (Li et al. 2002a, also see the discussion in Section 6)] shows that the mean and standard deviation of the precipitation efficiency associated with the diurnal variation are, respectively, 63.8% and 3%. The small standard deviation in the precipitation efficiency indicates a linear relationship between the surface rain rate and the sum of the vapor condensation and deposition rates whereby larger (smaller) condensation and deposition rates cause larger (smaller) surface rain rates.

The diurnal variation of the sum of the vapor condensation and deposition rates is shown in Fig. 3b (dashed line) in comparison with the diurnal variation of the large-scale moisture source (solid line). Both variations show a similarity in the afternoon and evening, but they display large differences in the morning. The similarity in the afternoon suggests that the vapor condensation and deposition result from the external forcing imposed in the model. The large difference in the morning suggests that the external forcing may not be responsible for the development of the nocturnal condensation. Thus, a scale analysis of the vapor condensation and deposition rates is conducted to explain the reason for the cause in the maximum in the condensation and deposition rates in the morning.

Following Tao et al. (1989), the sum of the vapor condensation (P_{CND}) and deposition (P_{DEP}) rates can be expressed by

$$P_{CND} + P_{DEP} = \frac{1}{\Delta t} \frac{q_v - (q_{ws} + q_{is})}{1 + QT}, \tag{9}$$

where

$$QT = \frac{A_1 q_c q_{ws} + A_2 q_i q_{is}}{q_c + q_i} \frac{L_v (T - T_{oo}) + L_s (T_o - T)}{c_p (T_o - T_{oo})}. \tag{9a}$$

The definitions of the variables and constants can be found in Tao et al. (1989). Since $QT << 1$, the sum of vapor condensation and deposition rates can be simplified as

$$P_{CND} + P_{DEP} = \frac{q_v - (q_{ws} + q_{is})}{\Delta t}. \tag{10}$$

Each variable can be decomposed into a daily mean (m) and a diurnal anomaly (d). $T = T_m + T_d$, and $q_v = q_{vm} + q_{vd}$. The diurnal anomaly of $P_{CND} + P_{DEP}$ then becomes

$$(P_{CND} + P_{DEP})_d = \frac{q_{vd} - q_{wisd}}{\Delta t}, \tag{11}$$

where

$$q_{wisd} = \frac{E_1 T_d}{(T_m - F_1)^2} q_{wsm} - \frac{E_2 T_d}{(T_m - F_2)^2} q_{ism}. \tag{11a}$$

Note that the relations $1/(1+x) = 1 - x$ and $e^x = 1 + x$ for $x << 1$ are used in the derivation of Eq. (11). Equation (11) indicates that the diurnal variation of vapor condensation and deposition is determined by the diurnal variation of temperature and specific humidity. The diurnal variation of temperature and specific humidity is therefore further analyzed.

The linear correlation coefficients between the diurnal anomalies of vapor condensation and deposition rates and mass-weighted mean temperature and between the diurnal anomalies of vapor condensation and deposition rates and precipitable water are –0.68 and 0.52, respectively. The correlation coefficient for 24 samples at the 99% confidence level is 0.5. Thus, the temperature correlation is well above, but the moisture correlation is only marginally above the 99% confidence level. Furthermore, taking into account that $<T_m>$ = 261.5 K, $<T_d>$ = 0.5 K, $[q_{vm}]$ = 54.7 mm, and $[q_{vd}]$ = 1.0 mm (<> and [ ] are mass-weighted mean and vertically integration respectively), $[q_{vd}]$ is about 1.8% of $[q_{vm}]$, and $[q_{wisd}]$ is about 4.5% of $[q_{wsm}] + [q_{ism}]$. These suggest that the diurnal variation of temperature is the primary factor in the diurnal variation of the vapor condensation and deposition rates and the surface rain rate. The diurnal variation of moisture is a secondary factor. The negative correlation between the diurnal vapor condensation and deposition anomalies and the mass-weighted mean temperature indicates that colder temperatures cause lower saturated mixing ratios making it easier for water vapor to be condensed and deposited into precipitation. Therefore, nocturnal radiative cooling leads to colder air temperatures that make it easier for clouds to develop and hence rainfall.

5. Dominant cloud microphysical processes

Cloud microphysical processes play an important role in determining the vertical distributions of cloud hydrometeors and their interaction with radiation. However, latent heat is largely balanced by vertical thermal advection over the tropics so that their residual determines the vertical temperature structure. Thus, accurate parameterization of cloud microphysical processes is of fundamental importance in numerical simulations of tropical climate. In their 7-day cloud-resolving simulations during TOGA COARE, Li et al. (1999) used the improved schemes for the growth of cloud ice by the Bergeron process and the conversion of cloud ice to snow of Krueger et al. (1995) and found that the increase in the mixing ratio of cloud ice led to better simulations of atmospheric thermodynamic states and surface fluxes. Grabowski et al. (1999) in their 7-day cloud-resolving simulations during Phase III of GATE found that cloud microphysics affects the temperature and moisture profiles in a way that retains relative humidity, responses of surface processes to cloud microphysics are important when the radiative tendencies are prescribed, and the temperature in the upper troposphere is modified by the effect of cloud microphysics on the anvil clouds when the radiation and clouds are fully interacted. In their 39-day cloud-resolving simulations during TOGA COARE, Wu ct al.

(1999) showed that simulations of cloud radiative properties are improved by modified ice microphysical parameterization schemes. They also found that radiative flux, cloud radiative forcing, and albedo are sensitive to the effective radius of ice particles. Liu et al. (1997) carried out a multiscale numerical study of Hurricane Andrew (1992), using an improved version of the Penn State-National Center for Atmospheric Research (NCAR) non-hydrostatic model (MM5), and found that including microphysical parameterization schemes in MM5 led to more realistic simulations of hurricane cloud structures when compared to the observations.

Analysis of cloud microphysical budgets by Li et al. (2002c) shows that vapor condensation enhances cloud water, the collection of cloud water by raindrops produces rain, and the riming of cloud water enhances precipitation ice in the early stage of development of tropical convection. In the later stages, the melting of graupel becomes a dominant rain-producing process. Vapor deposition and the riming process are equally important in the development of ice clouds. The vertical profile of upward motion affects the production of rain and surface rain rates through the collection of cloud water by raindrops and the riming of cloud water by graupel. Upward motion in the mid and lower troposphere causes large collection rates of cloud water by raindrops that lead to large growth rates of raindrops and surface rain rates, whereas upward motion in the upper troposphere produces large riming rates of cloud water by graupel that results in large growth rates of graupel. Thus, surface rain rates respond more to upward motion in the mid and lower troposphere than to upward motion in the upper troposphere in deep tropical convection.

Li et al. (2002c) found from their analysis of cloud microphysical budgets that in the deep tropical convective regime, the magnitudes of 12 terms out of total 29 cloud microphysical processes are negligibly small. Thus, they proposed a simplified set of cloud microphysical equations, which saves 30-40% of CPU time. The neglected terms in the simplified set include the accretion of cloud ice and snow by raindrops, the evaporation of melting snow, the accretion of cloud water and raindrops by snow, the accretion of raindrops and the homogeneous freezing of cloud water by cloud ice, the accretion and freezing of raindrops by graupel, the growth of cloud water by the melting of cloud ice, and the growth of cloud ice and snow by the deposition of cloud water. An experiment with the simplified set of cloud microphysical equations was conducted and compared to an experiment with the original set of cloud microphysical equations. Both experiments show similar time evolution and magnitudes of temperature and moisture profiles, surface rain rates including stratiform percentage and fractional coverage of convective, raining and non-raining stratiform clouds. This suggests that the original set of cloud microphysical equations could be replaced by the simplified set in simulations of tropical oceanic convection.

6. Precipitation efficiency

Precipitation efficiency is an important physical parameter for measuring the interaction between convection and its environment. Its definition may vary. For large-scale applications involving cumulus parameterization (e.g., Kuo 1965, 1974), the precipitation efficiency (PE) is defined as the ratio of the surface rain rate to the sum of the surface evaporation and the vertically-integrated horizontal and vertical advection of specific humidity (moisture convergence) and is referred to as large-scale precipitation efficiency (LSPE). The LSPE is similar to the precipitation efficiency defined by Braham (1952).

For smaller scale cloud-resolving models (e.g., Li et al. 1999), the PE is defined as the ratio of the surface rain rate to the sum of the vertically-integrated condensation and deposition rates. This is referred to as cloud-microphysics precipitation efficiency (CMPE). The CMPE is similar to the precipitation efficiency defined by Weisman and Klemp (1982) and Lipps and Hemler (1986).

Li et al. (2002a) found that the LSPE can exceed 100% for strong convection. This suggests that the surface rain rate could be larger than the total moisture convergence, which is contrary to the assumption in Kuo's scheme (1965, 1974) that a small portion of the surface evaporation and moisture convergence (say 5%) is used to moisten the atmosphere.

The PE may depend on the environmental conditions and strength of convection. Ferrier et al. (1996) showed that wind shear and updraft structure play a role in determining the PE. Li et al. (2002a) showed that the CMPE increases with increasing mass-weighted mean temperature and surface rain rate (Fig. 4). This suggests that precipitation processes are more efficient for heavy rain regime in a warm environment.

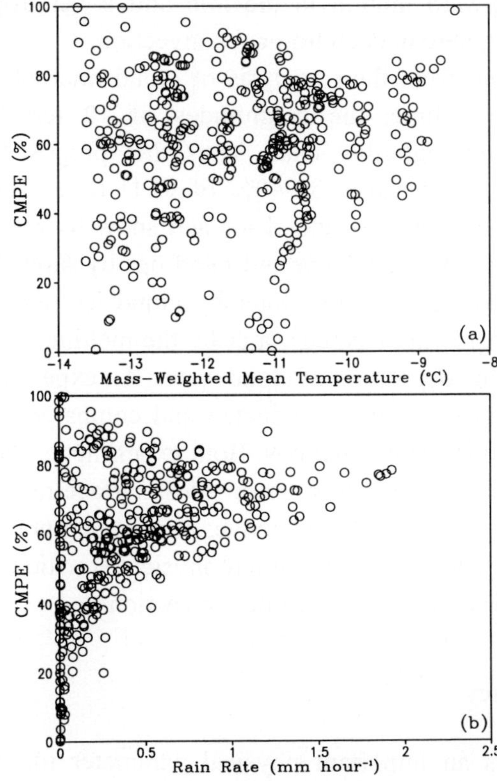

Fig. 4 (a) CMPE (%) versus mass-weighted mean temperature (°C), and (b) CMPE (%) versus surface rain rate (adapted from Li et al. 2002a).

7. A new look at the phase relationship between convection and its environment

Tropical convection occurs as a result of instability in environment. The large-scale environment provides favorable thermal and moisture conditions for the occurrence and development of convection. In return, it is affected by the vertical redistribution of temperature, moisture, and momentum as a result of the convection. Such an interaction allows us to use environmental conditions to estimate the properties of the convection such as the precipitation. Since the environmental time scales (a few days and longer) are much longer than the convective time scales (a few hours or less), the rate of production of available potential energy by the large-scale processes is nearly balanced by the rate of consumption of the available potential energy by the convection (Manabe and Strickler 1964). This quasi-equilibrium concept is the basic premise of the cumulus parameterization scheme proposed by Arakawa and Schubert (1974). A decrease in the convective available potential energy (CAPE), that measures the thermal and moisture conditions of the environment, often coincides with the development of convection so that the CAPE and rain rate are negatively correlated (e.g., Thompson et al. 1979; Cheng and Yanai 1989; Wang and Randall 1994; Xu and Randall 1998). The phase relation between the CAPE and rainfall is due to the coupling between the environmental dynamic and thermodynamic fields (Cheng and Yanai 1989).

The phases of CAPE and rainfall could be different because it takes time for clouds to develop. This phase difference can be included by relaxing the quasi-equilibrium assumption in cumulus parameterization (e.g., Betts and Miller 1986; Randall and Pan 1993). The minimum CAPE typically occurs a few hours after the maximum rainfall. Such a phase lag was also demonstrated by Xu and Randall (1998) in their 2-D cloud-resolving simulations. Xu and Randall (1998) interpreted the maximum phase lag as the adjustment time-scale from disequilibrium to equilibrium states in the presence of time-varying large-scale forcing. Since the CAPE is calculated in a Lagrangian framework and the relevant equations cannot be derived in that framework, the physical processes responsible for the phase difference between the CAPE and the surface rain rate cannot be examined. Potential and kinetic energy in an Eulerian framework represent the CAPE and surface rain rate in a Lagrangian framework, respectively.

Lorenz (1955) first introduced the concept of available potential energy for a dry atmosphere that represents the portion of the potential energy that can be transferred into kinetic energy. He defined the available potential energy for a dry atmosphere as the difference between the actual total enthalpy and the minimum total enthalpy that could be achieved by rearranging the mass under adiabatic flow. The dry enthalpy per unit mass is defined as the product of the temperature and the specific heat at constant pressure. In the absence of energy sources and sinks, the total kinetic energy and total enthalpy are conserved during adiabatic expansion. In a moist atmosphere, latent heat energy should be included in the energy conservation. The latent heat energy per unit mass is defined as the product of the specific humidity and the latent heat of vaporization at 0°C. In the absence of energy sources and sinks, the total kinetic energy, enthalpy and latent heat energy are conserved during dry and subsequent saturated adiabatic expansion. Therefore, the moist available potential energy is defined as the difference between the actual moist potential energy (sum of the enthalpy and latent heat energy) and the minimum moist potential energy that could be achieved by rearranging the mass under moist adiabatic processes. Li et al. (2002b) derived a set of equations for conversions between the moist available potential energy and kinetic energy in an Eulerain framework. Their equations were demonstrated to be the same as those derived by Lorenz (1955) in the absence of moisture.

Lag correlation analysis by Li et al. (2002b) showed that the maximum perturbation kinetic energy associated with the simulated convective events and its maximum growth rate lags and leads the maximum imposed large-scale upward motion by about 1-2 hours respectively, indicating that the convection is phase locked with the imposed large-scale forcing. Their imposed large-scale vertical velocity had time-scales longer than the diurnal cycle, whereas the simulated convective events had an average lifetime of about 9 hours. The imposed large-scale upward motion decreases the horizontal-mean moist available potential energy by the associated vertical advective cooling, providing a favorable environment for the development of convection.

They further show that the maximum latent heating and vertical heat transport by perturbation circulations cause maximum growth of perturbation kinetic energy to lead maximum loss of perturbation available potential energy by about 3 hours. The maximum vertical advective cooling, the horizontal-mean cloud-related heating, and perturbation radiative processes cause maximum loss of perturbation moist available potential energy to lead maximum loss of the horizontal-mean moist available potential energy by about one hour. Consequently, the maximum gain of perturbation kinetic energy leads the maximum loss of horizontal-mean moist available potential energy by about 4-5 hours (about half of the lifetime of the simulated convection).

8. A coupled ocean-cloud-resolving atmosphere model

Precipitation and the associated stratification of salinity affect both the sea surface temperature (SST) by changing the mixed-layer depth (e.g., Miller 1976; Li et al. 1998) and the upper ocean thermal structure by forming a barrier layer between the halocline and the thermocline (e.g., Godfrey and Linstrom 1989; Lukas and Linstrom 1991; Vialard and Delecluse 1998a,b). Miller (1976) found precipitation could induce shallower mixed-layers in numerical simulations. Since the effect of heating/cooling is inversely related to the depth of the mixed layer, shallower mixed layers will undergo larger temperature changes than deeper mixed layers with the same thermal forcing. Li et al. (1998) found that salinity stratification can cause instances when entrainment will result in warming the mixed layer as opposed to the normal cooling associated with entrainment. Cooper (1988) demonstrated the importance of salinity in numerical simulations of the Indian Ocean and found that salinity effects could account for as much as a $0.5^{\circ}C$ temperature bias and a $0.1\ ms^{-1}$ velocity bias near the surface after 110 days of integration. In their ocean modeling study, Murtugudde and Busalacchi (1998) found that the differences in annual mean SST between simulations with salinity and climatological precipitation and without could be as much as 0.5°C, indicating the inclusion of salinity effects is necessary to simulate realistic climatic systems. Yang et al. (1998) also found that in the western Pacific warm pool, SST would be 0.6°C lower if there were no salinity effect associated with precipitation. Sui et al. (1997b) found the mixed-layer temperature to be sensitive to the temporal scale of the atmospheric forcing in simulations of the mixed-layer during TOGA COARE. The model used a modified profile of solar radiation in the upper ocean (Sui et al. 1998b).

Due to the limitation of computational resources, the time step and horizontal grid lengths of an ocean general circulation model (OGCM) are usually much larger than the temporal- and spatial-scales of atmospheric convection. Thus, the effects of convective-scale ocean disturbances induced by atmospheric convection through surface heat and fresh water fluxes on ocean mixed-layer heat and salt budgets should be evaluated. A two-dimensional coupled ocean-cloud-resolving atmosphere model developed by Li et al. (2000) is a unique tool for such evaluation. The coupled model consists of a cloud-resolving model and an ocean circulation mixed-layer model developed by Adamec et al. (1981) with the mixing scheme of Niiler and Kraus (1977).

When the effects of fresh water flux and salinity were included in the coupled model, differences in the horizontal-mean mixed-layer temperature and salinity between 1-D and 2-D experiments were about 0.4°C and 0.3 PSU, respectively. The mean salinity difference was larger than the mean temperature difference in terms of their contributions to the mean density difference. In the 2-D experiment, the surface heat flux showed a significant diurnal signal with the dominance of downward solar radiation during daytime and upward flux (longwave, sensible and latent heat fluxes) during nighttime at each grid, although the amplitude was affected by precipitation. Thus, there was a strong thermal correlation between grids. Narrow cloudy areas were surrounded by broad cloud-free areas. Horizontal-mean precipitation could occur, whereas the precipitation may not occur in most of the integration period. Thus, there is very low correlation between horizontal-mean and grid value of the fresh water fluxes. Since the rain rates have significant spatial variations, the fresh water flux has much larger spatial fluctuations than the saline entrainment has. Therefore, the fresh water flux determines large spatial salinity fluctuations, which contributes to large mean salinity difference between the 1-D ocean model experiment and 2-D ocean model experiment.

9. Conclusion

In this paper, cloud modeling in the tropical deep convective regime is reviewed based on the author's research work in recent years. The cloud-resolving model has been demonstrated to simulate tropical thermodynamic states reasonably well. The model can simulate the genesis, development, and decay of convective systems and clouds so that convective-radiative responses to large-scale energy sources and sinks can be examined. Thus, the model can serve as a test bed for the improvement of cloud microphysical parameterization and precipitation prediction, and a research tool for the study of cloud processes, which is fundamentally important for understanding the role of cloud microphysics in atmospheric thermodynamics and energetics.

Acknowledgments. The author thanks Drs. C.-H. Sui and K.-M. Lau at NASA/GSFC for their support during the cloud modeling studies at Goddard, Prof. M. Zhang at the State University of New York at Stony Brook for providing us with his TOGA COARE forcing data, and Mr. N. Sun for his editorial assistance.

References

Adamec, D., R. L. Elsberry, R. W. Garwood, and R. L. Haney, 1981: An embedded mixed-layer-ocean circulation model, *Dyn. Atmos. Oceans*, **6**, 69-96.

Arakawa, A., and W. H. Schubert, 1974: Interaction of a cumulus cloud ensemble with the large-scale environment. Part I. *J. Atmos. Sci.*, **31**, 674-701.

Betts, A. K., and M. J. Miller, 1986: A new convective adjustment scheme. Part II: Single column tests using GATE wave, BOMEX, ATEX and arctic airmass data sets. *Quart. J. Roy. Meteor. Soc.*, **112**, 692-709.

Braham, R. R. Jr., 1952: The water and energy budgets of the thunderstorm and their relation to thunderstorm development. *J. Meteor.*, **9**, 227-242.

Cess, R. D., and Co-authors, 1990: Intercomparison and interpretation of climate feedback processes in 19 atmospheric general circulation models. *J. Geophys. Res.*, **95**, 16601-16615.

Cheng, M.-D., and M. Yanai, 1989: Effects of downdrafts and mesoscale convective organization on the heat and moisture budgets of tropical cloud cluster. Part III: Effects of mesoscale convective organization. *J. Atmos. Sci.*, **56**, 3028-3042.

Chou, M.-D., D. P. Kratz, and W. Ridgway, 1991: IR radiation parameterization in numerical climate studies. *J. Climate*, **4**, 424-437.

Chou, M.-D., and M. J. Suarez, 1994: An efficient thermal infrared radiation parameterization for use in General Circulation Model. NASA Technical Memorandum 104606, Vol. 3, 85pp.

Chou, M.-D., M. J. Suarez, C.-H. Ho, M. M.-H. Yan, and K.-T. Lee, 1997:Parameterizations for cloud overlapping and shortwave single-scattering properties for use in General Circulation and Cloud Ensemble Models. *J. Climate*, **11**, 202-214.

Cooper, N.S., 1988: The effect of salinity on tropical ocean models, J. Phys. Oceanogr., 18, 697-707.

Ferrier, B. S., J. Simpson, and W.-K. Tao, 1996: Factors responsible for different precipitation efficiencies between midlatitude and tropical squall simulations. *Mon. Wea. Rev.*, **124**, 2100-2125.

Gates, W. L., 1992: AMIP: The atmospheric model intercomparison project. *Bull. Amer. Met. Soc.*, **73**, 1962-1970.

Godfrey, J. S., and E. J. Lindstrom, 1989: The heat budget of the equatorial western Pacific surface mixed layer. *J. Geophys. Res.*, **94**, 8007-8017.

Grabowski, W. W., X. Wu, and M. W. Moncrieff, 1996: Cloud-resolving model of tropical cloud systems during Phase III of GATE. Part I: Two-dimensional experiments. *J. Atmos. Sci.*, **53**, 3684-3709.

Grabowski, W. W., X. Wu, and M. W. Moncrieff, 1999: Cloud-resolving model of tropical cloud systems during Phase III of GATE. Part III: Effects of cloud microphysics. *J. Atmos. Sci.*, **56**, 2384-2402.

Gray, W. M., and R. W. Jacobson, 1977: Diurnal variation of deep cumulus convection. *Mon. Wea. Rev.*, **105**, 1171-1188.

Kraus, E. B., 1963: The diurnal precipitation change over the sea. *J. Atmos. Sci.*, **20**, 546-551.

Krueger, S. K., Q. Fu, K. N. Liou and H.-N. S. Chin, 1995: Improvement of an ice-phase microphysics parameterization for use in numerical simulations of tropical convection. J. Appl. Meteor., 34. 281-287.

Kuo, H. L., 1965: On formation and intensification of tropical cyclones through latent heat release by cumulus convection. *J. Atmos. Sci.*, **22**, 40-63.

Kuo, H. L., 1974: Further studies of the parameterization of the influence of cumulus convection on large-scale flow. *J. Atmos. Sci.*, **31**, 1232-1240.

Li, X., C.-H. Sui, D. Adamec, and K.-M. Lau, 1998: Impacts of precipitation in the upper ocean in the western Pacific warm pool during TOGA COARE. J. Geophys. Res., 103, C3, 5347-5359.

Li, X., C.-H. Sui, K.-M. Lau, and M.-D. Chou, 1999: Large-scale forcing and cloud-radiation interaction in the tropical deep convective regime. *J. Atmos. Sci.*, **56**, 3028-3042.

Li, X., C.-H. Sui, K.-M. Lau, and D. Adamec, 2000: Effects of precipitation on ocean mixed-layer temperature and salinity as simulated in a 2-D coupled ocean-cloud resolving atmosphere model. *J. Meteor. Soc. Japan*, **78**, 647-659.

Li, X., C.-H. Sui, and K.-M. Lau, 2002a: Precipitation efficiency in the tropical deep convective regime: A 2-D cloud resolving modeling study. *J. Meteor. Soc. Japan*, **80**, 205-212.

Li, X., C.-H. Sui, and K.-M. Lau, 2002b: Interactions between tropical convection and its environment: An energetics analysis of a 2-D cloud resolving simulation. *J. Atmos. Sci.*, **59**, 1712-1722.

Li, X., C.-H. Sui, and K.-M. Lau, 2002c: Dominant cloud microphysical processes in a tropical oceanic convective system: A 2-D cloud resolving modeling study. *Mon. Wea. Rev.*, **130**, 2481-2491.

Lipps, F. B., and R. S. Hemler, 1986: Numerical simulation of deep tropical convection associated with large-scale convergence. *J. Atmos. Sci.*, **43**, 1796-1816.

Lin Y. L., R. D. Farley, and H. D. Orville, 1983: Bulk parameterization of the snow field in a cloud model. *J. Climate Appl. Meteor.*, **22**, 1065-1092.

Liu, C., and M. W. Moncrieff, 1998: A numerical study of the diurnal cycle of tropical oceanic convection. *J. Atmos. Sci.*, 55, 2329-2344.

Liu, Y., D.-L. Zhang, and M. K. Yau, 1997: A multiscale numerical study of Hurricane Andrew (1992). Part I: Explicit simulation and Verification. *Mon. Wea. Rev.*, **125**, 3073-3093.

Lorenz, E. N., 1955: Available potential energy and the maintenance of the general circulation. *Tellus*, **7**, 157-167.

Lukas, R., and E. Lindstrom, 1991: The mixed layer of the western equatorial Pacific Ocean. *J. Geophys. Res.*, 96, 3343-3457.

Manabe, S., and R. F. Strickler, 1964: Thermal equilibrium of the atmosphere with a convective adjustment. *J. Atmos. Sci.*, **21**, 361-385.

Miller, J.R., 1976: The salinity effect in a mixed-layer ocean model, *J. Phys. Oceanogr.*, 6, 29-35.

Murtugudde, R. and A. J. Busalacchi, 1998: Salinity effects in a tropical ocean model. *J. Geophys. Res.*, **103**, 3283-3300.

Niiler, P. P., and E. B. Kraus, 1977: One-dimensional models, in Modeling and Prediction of the Upper Layers of the Ocean, edited by E. B. Kraus, Pergamon, New York, 143-172.

Randall, D. A., and D.-M. Pan, 1993: Implementation of the Arakawa-Schubert cumulus parameterization with a prognostic closure. *The Representation of Cumulus Convection in Numerical Models of the Atmosphere*, K. A. Emanuel and D. J. Raymond, Eds. *Meteor.Monogr.*, No. 46, 37-144.

Rutledge, S. A., and R. V. Hobbs, 1983: The mesoscale and microscale structure and organization of clouds and precipitation in midlatitude cyclones. Part VIII: A model for the "seeder-feeder" process in warm-frontal rainbands. *J. Atmos. Sci.*, **40**, 1185-1206.

Rutledge, S. A., and R. V. Hobbs, 1984: The mesoscale and microscale structure and organization of clouds and precipitation in midlatitude clcones. Part XII: A diagnostic modeling study of precipitation development in narrow cold-frontal rainbands. *J. Atmos. Sci.*, **41**, 2949-2972.

Soong, S. T., and Y. Ogura, 1980: Response of tradewind cumuli to large-scale processes. *J. Atmos. Sci.*, **37**, 2035-2050.

Sui, C.-H., K.-M. Lau, W.-K. Tao, and J. Simpson, 1994: The tropical water and energy cycles in a cumulus ensemble model. Part I: Equilibrium climate. *J. Atmos. Sci.*, **51**, 711-728.

Sui, C.-H., K.-M. Lau, Y. Takayabu, and D. Short, 1997a: Diurnal variations in tropical oceanic cumulus ensemble during TOGA COARE. *J. Atmos. Sci.*, **54**, 639-655.

Sui, C.-H., X. Li, K.-M. Lau, and D. Adamec, 1997b: Multi-scale air-sea interactions during TOGA COARE. *Mon. Wea. Rev.*, **125**, 448-462.

Sui, C.-H., X. Li, and K.-M. Lau, 1998a: Radiative-convective processes in simulated diurnal variations of tropical oceanic convection. *J. Atmos. Sci.*, **55**, 2345-2359.

Sui, C.-H., X. Li, and K.-M. Lau, 1998b: Selective absorption of solar radiation and upper ocean temperature in the equatorial western Pacific. *J. Geophys. Res.*, **103**, C5, 10313-10321.

Takayabu, Y. N., K.-M. Lau, and C.-H. Sui, 1996: Observation of a quasi-2-day wave during TOGA COARE. *Mon. Wea. Rev.*, **124**, 1892-1913.

Tao, W.-K., J. Simpson, and M. McCumber, 1989: An ice-water saturation adjustment. *Mon. Wea. Rev.*, **117**, 231-235.

Tao, W.-K., and J. Simpson, 1993: The Goddard Cumulus Ensemble model. Part I: Model description. *Terr. Atmos. Oceanic Sci.*, **4**, 35-72.

Tao, W.-K., S. Lang, J. Simpson, C.-H. Sui, B. S. Ferrier, and M.-D. Chou, 1996: Mechanisms of cloud-radiation interaction in the Tropics and midlatitude. *J. Atmos. Sci.*, **53**, 2624-2651.

Thompson, R. M., Jr., S. W. Payne, E. E. Recker, and R. J. Reed, 1979: Structure and properties of synoptic-scale wave disturbances in the intertropical convergence zone of the eastern Atlantic. *J. Atmos. Sci.*,**36**, 53-72.

Vialard, J., and P. Delecluse, 1998a: An OGCM study for the TOGA decade. Part I: Role of salinity in the physics of the western Pacific fresh pool. *J. Phys. Oceanogr.*, **28**, 1071-1088.

Vialard, J., and P. Delecluse, 1998b: An OGCM study for the TOGA decade. Part II: Barrier layer formation and variability. *J. Phys. Oceanogr.*, **28**, 1089-1106.

Wang, J., and D. A. Randall, 1994: The moist available energy of a conditionally unstable atmosphere. Part II: Further analysis of GATE data. *J. Atmos. Sci.*, **51**, 703-710.

Weisman, M. L., and J. B. Klemp, 1982: The dependence of numerically simulated convective storms on vertical wind shear and buoyancy. *Mon. Wea. Rev.*, **110**, 504-520.

Weller, R. A., and S. P. Anderson, 1996: Surface meteorology and air-sea fluxes in the western equatorial Pacific warm pool during TOGA COARE. *J. Climate*, **9**,1959-1990.

Wu, X., W. W. Grabowski, and M. W. Moncrieff, 1998: Long-term evolution of cloud systems in TOGA COARE and their interactions with radiative and surface processes. Part I: Two-dimensional cloud-resolving model. *J. Atmos. Sci.*, **55**, 2693-2714.

Wu, X., W. D. Hall, W. W. Grabowski, M. W. Moncrieff, W. D. Collins, and J. T. Kiehl, 1999: Long-term evolution of cloud systems in TOGA COARE and their interactions with radiative and surface processes. Part II: Effects of ice microphysics on cloud-radiation interaction. *J. Atmos. Sci,,* **56**, 3177-3195.

Xu, K.-M., and D. A. Randall, 1996: Explicit simulation of cumulus ensembles with the GATE Phase III data: Comparison with observations. *J. Atmos. Sci.*, **53**, 3710-3736.

Xu, K.-M., and D. A. Randall, 1998: Influence of large-scale advective cooling and moistening effects on the quasi-equilibrium behavior of explicitly simulated cumulus ensembles. *J. Atmos. Sci.*, **55**, 896-909.

Yang, S., K.-M. Lau, and P. S. Schopf, 1998: Sensitivity of the tropical Pacific Ocean to precipitation induced freshwater flux. *Clim. Dyn.*, **15**, 737-750.

TURBULENCE-MICROPHYSICS INTERACTION IN STRATOCUMULUS CLOUDS: PROCESS AND PARAMETERIZATION

SHOUPING WANG

Naval Research Laboratory
7 Grace Hopper Ave., Monterey, CA 93943, USA
E-mail: wang@nrlmry.navy.mil

QING WANG

Naval Postgraduate School
589 Dyer Rd., Monterey, CA 93943, USA
E-mail: qwang@nps.navy.mil

(Manuscript received 24 January 2003)

The objective of this work is twofold: (1) to understand how the turbulence interacts with the microphysics to produce the turbulent liquid water flux; (2) to develop a new framework of turbulence-microphysics coupling parameterization for stratocumulus clouds. The approach is to analyze the turbulence and microphysics data produced by a coupled large-eddy simulation and bin microphysics model. The analysis demonstrates that the condensation time scale regulates the turbulence fields, because it affects the condensation fluctuation. The liquid water flux can be parameterized in terms of the down-gradient formulation, the fluxes of conservative thermodynamic variables, and both the condensation and the dominant turbulence time scales. The new parameterization explicitly parameterizes the turbulence-microphysics coupling based on a third-order turbulence closure model, and is shown to be able to simulate some key features of the coupling process.

1. Introduction

Marine stratocumulus clouds persist within the boundary layers over relatively cool ocean surfaces and cover extensive areas in the subtropics. These clouds drastically reduce the amount of solar radiation absorbed by the earth, but have little effect on the emitted infrared radiation to space. Therefore, it has been increasingly recognized that this low-level cloud system is an important component of the earth climate. Recent attention has been focused on the so-called "indirect aerosol effects" in which aerosols may substantially modify the cloud life time and its optical properties (e.g., Penner *et al.*, 2001). Furthermore, because of the turbulent nature of the cloud and the fine scale of the microphysics, it is inevitable that these clouds and relevant physical processes must be parameterized in the current meso-scale and general circulation models. It is in this context that the interaction between the turbulence dynamics and aerosol-cloud microphysics has become a focus of the research community.

This study is focused on the basic aspects of the interaction: the coupling of turbulence and condensation/evaporation (CE). It is well known that vertical motion strongly controls cloud droplet activation, condensation and liquid water transport. An excess of water vapor is produced at the cloud base by turbulent updrafts and initiates droplet activation. In the updrafts, the newly activated droplets are transported upward and the condensational latent heat release further enhances upward motion. A proper treatment of these processes in a coupled turbulence-microphysics parameterization, however, has proven to be difficult (e.g., Ackerman *et al.* 1996; Bott *et al.* 1996; Stevens *et al.* 1998; Svensson and Seinfeld, 2002). For example, the diffusion formulation has been frequently used for parameterization of the turbulent liquid water flux, even though it does not represent the nature of the turbulent liquid water transport. Many of the ensemble mean models use the mean supersaturation for the droplet growth and activation, which leads to major underestimates of peak supersaturations and severe errors in the microphysical structure as a consequence. The difficulty lies with the fact that the microphysical properties are almost entirely dependent on the turbulence flow, which can only be parameterized in an empirical manner in a turbulence closure model. The primary challenge is to obtain a clear understanding of the coupling, and then transfer it into a simple model in the current framework of the turbulence parameterization.

It remains unclear how the droplet population interacts with turbulent eddies to produce the upward liquid water transport. What is the role of the diffusion process in the turbulent transport? How should one parameterize the liquid water and the droplet number turbulent fluxes?

This paper summarizes the author's recent work that attempts to address these issues with two distinctively different approaches. One is coupled large-eddy simulation (LES) and bin microphysics (BM) modeling approach used to understand the physical processes. The other is a new framework of coupled turbulence-microphysics parameterization used to represent the coupling in a ensemble mean model. The results are presented in sections 2[a] and 3; and concluding remarks are given in section 4. In following sections, conventional meteorology notation is followed unless noted otherwise.

2. Physical processes – LES-BM results

Large-eddy-simulation is a numerical technique with which the contribution of large, energy-carrying stochastic eddies to turbulent transport is *explicitly resolved*. The bin microphysics technique explicitly resolves the statistical size distribution of cloud droplets given the dynamic and thermodynamic fields. The spectrum is defined by four microphysical processes: droplet activation, the condensation/evaporation, coalescence/collection, and gravitational settling. The advantage of a coupled LES-BM model is that it avoids major assumptions regarding the structures of the turbulence and microphysics, and thus provides a consistent and coherent evolution of both. The LES model used in this study is the same as that described by Stevens *et al.* (1999). The two-moment bin-microphysical (BM) model developed by Feingold *et al.* (1994) is adopted here, although only the condensation and activation are considered. The coupled LES-BM is used to provide detailed turbulence and microphysics data, which are used to perform budget analysis of the turbulent liquid water flux. The LES-BM model is used to simulate the case of stratocumulus clouds as described in Wang *et al.*

[a] This section is discussed in detail in Wang *et al.* (2003)

(2003). Fig. 1 gives a general view of the mean and turbulence profiles of the cloud variables. The droplet number concentration $\bar{N}$ is relatively constant within the cloud, while the liquid water content $\bar{q}_l$ increases linearly with height. The liquid water flux $\overline{w'q_l'}$ is weakly negative at the cloud base, increases with height, and becomes positive for most of the cloud layer. The droplet number flux $\overline{w'N'}$ increases rapidly from a negative value to the maximum, and then decreases with height. The negative values of the fluxes at the cloud base result from the downward turbulent transport of the cloud droplets, leading to larger q_l and N in the downdrafts than in the updrafts.

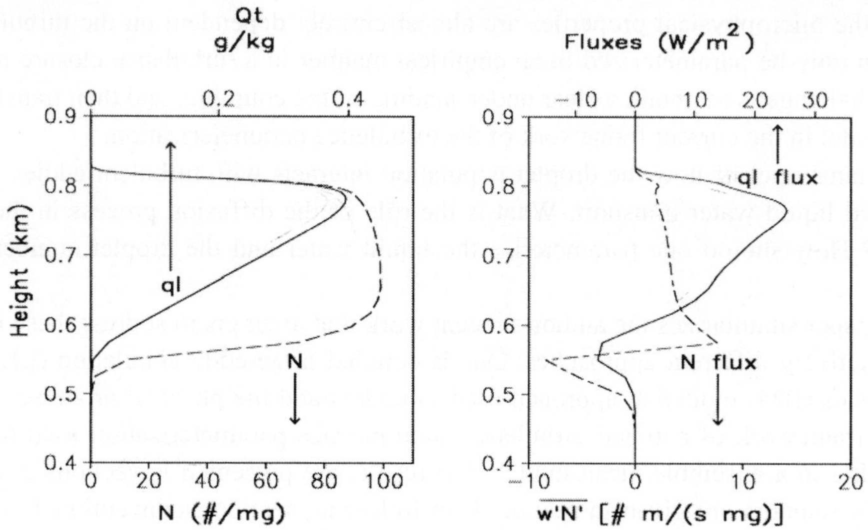

Figure 1. Cloud variables: left: liquid water content and the number concentration; right: The liquid water flux and the cloud droplet number flux. (Adapted from Wang *et al.* 2003)

2.1. *Liquid water flux budget*

One can derive following liquid water flux budget based on the conservation of cloud droplet mass and the vertical momentum equation [see Wang *et al.* (2003) for the details]

$$\frac{\partial \overline{w'q_l'}}{\partial t} = \underbrace{-\frac{\partial \overline{w'w'q_l'}}{\partial z}}_{T} \underbrace{- \overline{w'^2}\frac{\partial \bar{q}_l}{\partial z}}_{G} + \underbrace{\frac{g}{\theta_0}\overline{\theta_v' q_l'}}_{B} \underbrace{- \frac{1}{\rho_0}\overline{q_l' \frac{\partial p'}{\partial z}}}_{P} \tag{1}$$

$$+ \underbrace{4\pi \rho_l G(\bar{T},\bar{p}) \left[\underbrace{\bar{R}\,\overline{w'S'}}_{M1} + \underbrace{\bar{S}\,\overline{w'R'}}_{M2} + \underbrace{\overline{w'S'R'}}_{M3} \right]}_{M}$$

where S is the available water vapor surplus (i.e., the difference between the water vapor and saturated water vapor mixing ratio), R the spectrum-integrated radius and ρ_l the liquid water density, $G(\overline{T}, \overline{p})$ a thermodynamic function. The first tem on the right hand side is transport *T*; the second, the gradient production *G*; the third, the buoyant production *B*; the fourth, the pressure correlation P; the fifth, the microphysics *M*, with different components denoted by *M1*, *M2* and *M3*. The data produced by the LES-BM are used to evaluate the $\overline{w'q_l'}$ budget, as shown in Fig. 2. The most striking feature in the $\overline{w'q_l'}$ budget is a close balance below 700m between the gradient (*G*) and the microphysical (M) terms in (3). The mean gradient term represents the down-gradient transport (or the diffusive mixing). The large negative values are a direct result of the large gradient of $\overline{q}_l$. The positive values of the microphysics term are due to the fact that condensation mainly occurs in the updrafts while evaporation is likely to occur in the downdrafts, leading to a positive correlation between *w* and CE perturbations. The buoyancy and pressure correlation terms are small below 760 m.

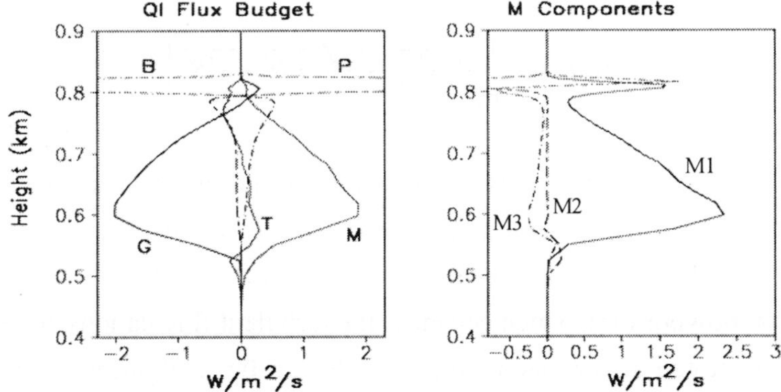

Figure 2. The $\overline{w'q_l'}$ budget defined by (1). Left: microphysical term (M), mean gradient (G), transport (T), buoyancy correlation (B, dot-dash), and pressure correlation (P, long-short); right: $\overline{w'S'}$ term (denoted by M1), ($\overline{w'R'}$) term (M2), and $\overline{w'S'R'}$ term (M3). (Adapted from Wang *et al.* 2003).

The contribution of the $\overline{w'S'}$ is dominant among the microphysical terms (Fig. 2b), and thus represents the fundamental element of the turbulence-microphysics coupling. The large positive value of $\overline{w'S'}$ results from the adiabatic cooling associated with the upward motion. One may further consider the steady-state *S* tendency equation to derive following approximation (Wang *et al.* 2003)

$$\overline{w'S'} \cong -\tau_{CE}(1+\gamma)\overline{w'^2}\frac{d\overline{q}_s}{dz}. \tag{2}$$

where $\gamma = (L/c_p)(\partial q_s/\partial t)$, q_s is the mean saturation water vapor mixing ratio, and τ_{CE} the CE time scale defined by

$$\tau_{CE} = \frac{1}{4\pi\rho_l(1+\gamma)\overline{G}\overline{R}} \tag{3}$$

Substituting (2) in (1), one can see that the gradient and the $\overline{w'S'}$ terms are balanced.

This analysis shows that the $\overline{w'S'}$ term balances the negative contribution of the gradient term to produce the positive $\overline{w'q_l'}$. Therefore, this term must be considered in the parameterization of the turbulent liquid water droplet fluxes when a supersaturation-based microphysics model is coupled to a turbulence parameterization.

2.2. *Dynamic feedback of τ_{CE}*

When considering the impact of condensation, one usually assumes that it occurs instantly because of its short time scale (~ 3-10 s) compared to that of the large turbulent eddies (~ 10 min). Despite the significant difference, however, the finite τ_{CE} has an important role in regulating the turbulence dynamics. Using the conserved variables of total water q_t, liquid water potential temperature θ_l, and the definition of S, one can derive following expression:

$$\overline{w'q_l'} = \frac{1}{1+\gamma}\left(\overline{w'q_t'} - \frac{c_p\gamma}{L}\overline{w'\theta_l'} - \overline{w'S'}\right) \tag{4}$$

where c_p and L are the specific heat capacity for air and latent heat for water, respectively. Because $\overline{w'S'}$ depends on τ_{CE} through (2), $\overline{w'q_l'}$ is regulated by τ_{CE}. A smaller τ_{CE} leads to a larger $\overline{w'q_l'}$. It results because a smaller τ_{CE} leads to more condensation (evaporation) in turbulent updrafts (downdrafts), which enhances the turbulent fluctuations w' and q_l'.

This argument can also be supported by a diagnostic relationship derived from (1). Assuming quasi-equilibrium condition ($\partial/\partial t = 0$) in (1), neglecting the turbulent transport and splitting the pressure correlation term into a buoyancy and a return-to-isotropy term i.e., $-1/\rho_0 \times \overline{q_l' \partial p'/\partial z} = -0.5\, g/\theta_0 \times \overline{\theta_v' q_l'} - \overline{w'q_l'}/\tau_R$ (Moeng and Wyngaard, 1986), where τ_R is return-to-isotropy time scale., one may obtain

$$\overline{w'q_l'} \cong \frac{\tau_R\left(-\overline{w'^2}\frac{\partial \overline{q}_l}{\partial z}\right) + \frac{\tau_R}{\tau_{CE}(1+\gamma)}\left(\overline{w'q_t'} - \frac{c_p\gamma}{L}\overline{w'\theta_l'}\right)}{1+\frac{\tau_R}{\tau_{CE}}\left(1-\frac{\overline{S}}{\overline{q}_l(1+\gamma)}\right)}. \tag{5}$$

Note that the first term in the numerator is the down-gradient diffusion flux, the second one represents the microphysical source term, and the buoyancy term is eventually dropped for a better comparison with LES results. A smaller τ_{CE} (or larger $\overline{R}$) would tend to reduce the effect of the diffusion flux, which can be explained as follows. The *effective* time scale over which the turbulence can affect $\overline{w'q_l'}$ is $(1/\tau_R + 1/\tau_{CE})^{-1}$. Therefore, when $\tau_R/\tau_{CE} >> 1$, the turbulence has a response time scale equivalent to τ_{CE}. Consequently, a smaller τ_{CE} would result in a smaller down-gradient diffusion contribution and a larger $\overline{w'q'_l}$. Furthermore, if $\tau_R / \tau_{CE} \to 0$, $\overline{w'q'_l}$ can be represented by the diffusion only. If $\tau_R / \tau_{CE} \to \infty$, $\overline{w'q'_l}$ is determined by the microphysical source contribution, which is the same as (4) with $\overline{w'S'}$ being zero.

To further evaluate these ideas, two more simulations are performed (Fig. 3). One (S2) uses the same LES-BM model, but with a background CCN number of 1000/mg. The second simulation (S3) uses the same LES dynamic model but with a saturation adjustment (SA) scheme, which diagnoses liquid water content at its equilibrium level (i.e., $\tau_{CE} \to 0$).

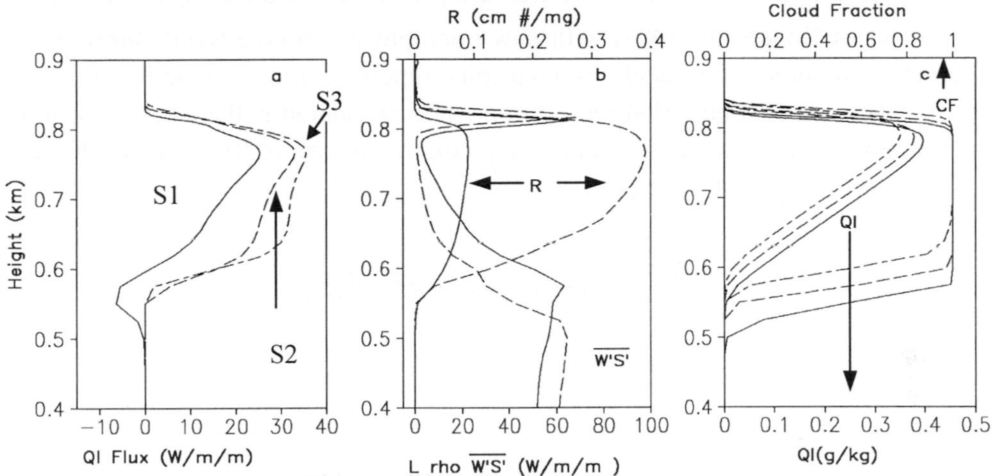

Figure 3. Impacts of condensation time scale. *a*: resolve liquid water fluxes ; *b*: integral radius (R) and super saturation flux; *c*: liquid water content and cloud fraction (CF). Solid lines denote S1 (CCN = 100/mg), long dashed S2 (CCN=1000/mg) and short-long dashed S3 (Saturation adjustment) (adapted from Wang *et al.* 2003).

Fig. 3 shows that both S2 and S3 result in significantly larger resolved $\overline{w'q'_l}$ than S1, as expected from the above discussion. Particularly, $\overline{w'q'_l}$ from S2 is closer to S3 than to S1 despite the fact that S3 is run with the LES-SA model. The integral droplet radius $\overline{R}$ in S2 is about 4 times that of S1, indicating a smaller τ_{CE} and leading to a reduction in $\overline{w'S'}$. This result is consistent with the earlier discussion in which we argued that an increase in $\overline{R}$ will decrease $\overline{w'S'}$ and thus increase $\overline{w'q'_l}$. Since local CE rates in turbulent eddies increase with $\overline{R}$ (or decreasing τ_{CE}), the cloud-top entrainment is enhanced. Consequently $\overline{q}_l$ and cloud fraction from S3 are reduced. Because LES-SA results in the largest instantaneous local CE rate among the three simulations, the entrainment from S3 is strongest and $\overline{q}_l$ is the smallest.

3. A coupled turbulence-closure-microphysics model

There are two major difficulties with a coupled turbulence-closure and BM models (Stevens *et al.*, 1998). They are the diffusion-like formulation used for $\overline{w'q'_l}$ and $\overline{w'N'}$, and the activation that depends on the ensemble mean supersaturation. In this section, a new framework of coupled turbulence-microphysics model is introduced to address these issues.

3.1. *Model description*

The proposed framework of the turbulence-microphysics coupling has following three basic elements. First, the droplet spectrum is assumed to be log-normal distribution with a fixed breadth parameter as in Feingold *et al.* (1998); second, all the major turbulence-microphysics moments (e.g., $\overline{w'q_l'}$, $\overline{w'N'}$, $\overline{S'R'}$, $\overline{S'^2}$, etc.) are predicted based on the 3rd-order turbulence closure model of Bougeault (1985) with new microphysics source terms; third, a statistical distribution is assumed for local instantaneous supersaturation. In addition, the coalescence/collection and sedimentation processes are not considered in the current version. Based on this framework, one may have a complete predictive system of 81 variables. For example, the source terms for $\overline{q}_l$ and $\overline{w'q_l'}$ are:

$$\left(\frac{\partial \overline{q}_l}{\partial t}\right)_{CE} = 4\pi G(T,p)\rho_l\left[\overline{R}\overline{S} + \overline{S'R'}\right] \tag{6}$$

and

$$\left(\frac{\partial \overline{w'q_l'}}{\partial z}\right)_{CE} = 4\pi G\rho_l\left[\overline{R}\overline{w'S'} + \overline{S}\overline{w'R'}\right] \tag{7}$$

where the 3rd-moments are neglected, S' is related to the fluctuations θ_l', q_t' and q_l' and R' can be expressed in terms of q_l' and N' based on the lognormal droplet number distribution (Wang *et al*, 2003). Using the modified Twomey relationship between activated *CCN* number and *S* (Khairoutdinov and Kogan 2000), we obtain the activation tendencies

$$\left(\frac{\partial \overline{N}}{\partial t}\right)_{Act} = \frac{1}{\Delta t}\int_{S_0}^{\infty} f(\overline{S},\sigma_s)\left(CCN\sqrt{\mathrm{Min}(S/S_{\max},1)} - \overline{N}\right)dS \tag{8}$$

and

$$\left(\frac{\partial \overline{w'N'}}{\partial t}\right)_{Act} = \sqrt{\frac{2}{\pi}}\sqrt{\overline{w'^2}}\left(\frac{\partial \overline{N}}{\partial t}\right)_{Act} \tag{9}$$

where Δt is the time step, $f(\overline{S},\sigma_s)$ the triangle probability density function, S_0 the minimum *S* above which the integrand of (8) is positive and the activation is assumed to occur only in the updrafts. Readers are referred to Bougeault (1985) and Wang and Wang (2000) for more details.

3.2. Results

The coupled turbulence-closure and microphysical model described above is run with the large-scale conditions of a stratocumulus cloud case (Moeng *et al.* 1996). As shown in Fig. 4, $\overline{q}_l$ increases linearly with height, while $\overline{N}$ is almost constant within the cloud layer, a feature consistent with the LES-BM result and observation data (Fig. 1). The liquid water flux increases with height from the cloud base and reaches the maximum near the cloud top due to the coupling term in (1). Particularly, the supersaturation flux is the main forcing term that significantly contributes to the positive liquid water flux. The negative values of $\overline{w'N'}$ are due to the large gradient of $\overline{N}$ at the cloud base. The N flux increases significantly from the minimum at the cloud base due to the activation of the droplets in the updrafts. Although the values are different, the profiles of the fluxes are similar to the LES-BM results (Fig. 1). The negative values of $\overline{w'N'}$ likely reflect the weak activation source shown in (9) where only averaged updrafts characteristics are considered.

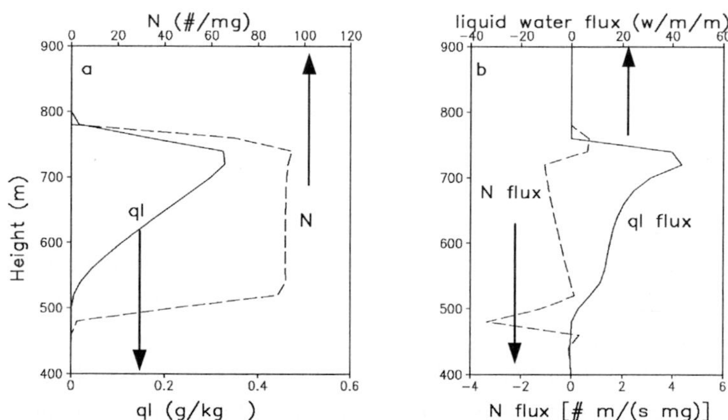

Figure 4. Simulated cloud variables. a: $\overline{q}_l$ (lower axis), $\overline{N}$ (upper); b: $L\overline{w'q_l'}$ (upper axis), $\overline{w'N'}$ (lower).

The mean and turbulence contributions to the net CE rate have similar magnitudes, but different signs, leading to a net condensation at the cloud base and near the top, and evaporation in the inversion as shown in Fig. 5. Note that the turbulence covariance $\overline{S'R'}$ in (6) is the major source of condensation, which is consistent with both the basic understanding and the LES-BM results (Wang *et al.* 2003). The mean CE profile is a direct result of the mean supersaturation, which is in general, negative (Fig. 5b). Integrating the assumed triangle PDF of supersaturation based on the mean and standard deviation gives the mean positive S, which controls the droplet activation rates. It is generally understood that the cloud droplets are activated in the updrafts and exit in the downdrafts at the cloud base. These results are consistent with this basic understanding. It should be emphasized that the microphysics related turbulence statistics (mainly S' related statistics) play essential roles in regulating the simulated cloud structure as shown in (6)-(9). For example, without the turbulence contribution terms to the activation and condensation, it would not be possible to obtain the constant $\overline{N}$ profile and positive $\overline{w'q_l'}$ in the cloud layer. This is also the main reason why the previous parameterizations failed in these aspects as pointed out in the introduction.

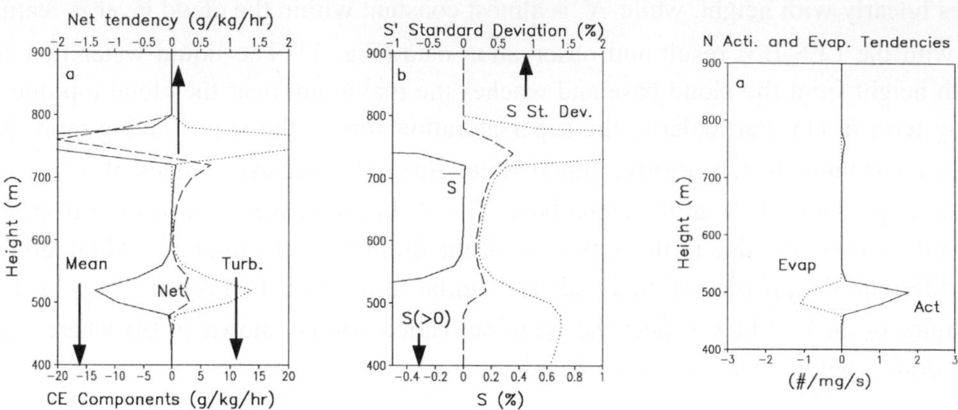

Fig. 5 Left: different CE terms in (6); center: the mean (solid), positive supersaturation (dashed), and the standard deviation (dotted); and right: tendency of the activation and the evaporation loss.

4. Concluding remarks

This work is intended to address two questions regarding the turbulence-microphysics coupling in stratocumulus clouds. How does the turbulence interacts with the microphysics to produce the liquid water fluxes? How should one parameterize this coupling within a framework of turbulence closure model.

The LES-BM results show that the condensation time scale may regulate the turbulence structure as it affects the condensation fluctuations in the turbulent updrafts and downdrafts. The diffusion process can only have limited effects on the liquid water flux because the "effective" turbulence time scale is overwhelmingly determined by the condensation time scale. It is also shown that the liquid water flux can be related to the diffusion and the conserved thermodynamic fluxes weighted by the condensation and main turbulence time scales.

A new framework of microphysics parameterization based on a third-order turbulence closure model is introduced. The essential element of the new model is the inclusion of the turbulence-microphysics covariance terms in all the liquid water mass and number related prognostic equations. Therefore, the turbulence-microphysics coupling is *explicitly parameterized.* The results show that the liquid water flux is mostly positive and the activation mainly occurs at the cloud base which leads to the constant droplet number density profile.

Two areas in the coupled parameterization need to be improved. First, it does not include the processes of droplet collection/coalescence and sedimentation. The drizzle parameterization of Khairoutdinov and Kogan (2000) may be included in the current framework. Second, the coupled parameterization is too complicated to be used in a large-scale model. One possibility is to modify the parameterization based on the 2nd-order closure technique with a simplified scheme of the coupling process. Currently, we are working on both areas toward a comprehensive and physically based turbulence-microphysics parameterization.

Acknowledgments. We acknowledge the contribution of Graham Feingold to section 2 of the paper. The comments of James Doyle are greatly appreciated. We also thank three anonymous reviewers for their constructive comments. S. Wang was supported by the Office of Naval Research under PE 062435N and Q. Wang was supported by NSF Grants ATM-9700845 and ATM-9900496.

References:

Ackerman, A .S., O. B. Toon, and P. V. Hobbs, 1995: A model for particle microphysics, turbulent mixing, and radiative transfer in the stratocumulus-topped marine boundary layer and comparison with measurements. *J. Atmos. Sci.*, **52,** 1204-1236.

Bott, A., T. Trautmann, W. Zdunkowki, 1996: A numerical model of cloud-topped planetary boundary-layer: Radiation, turbulence and spectral microphysics in marine stratus. *Quart. J. Roy. Meteor. Soc.* **122**, 635-667.

Bougeault, Ph., 1981: Modeling the trade-wind cumulus boundary layer. Part II: A higher-order one-dimensional model. *J. Atmos. Sci.*, 38, 2429-2439.

Feingold, G., B. Stevens, W. R. Cotton, and R. L. Walko, 1994: An explicit cloud microphysics/LES model designed to simulate the Twomey effect. *Atmos. Res.*, 33, 207-233.

Feingold, G., R. L. Walko, B. Stevens, and W. R Cotton, 1998: Simulations of marine stratocumulus using a new microphysical parameterization scheme. *Atmos. Res.*, **47-48**, 505-523.

Khairoutdinov, M., and Y. Kogan, 2000: A new cloud physics parameterization in a large-eddy simulation model of marine stratocumulus. *Mon. Wea. Rev.*, 128, 229-243.

Moeng, C.-H., and J. C. Wyngaard, 1986: An analysis of closures for pressure-scalar covariances in the convective boundary layer. *J. Atmos. Sci.*, **43**, 2499-2513.

Moeng, C.-H., and coauthors, 1996: Simulation of a stratocumulus-topped PBL: Intercomparisons among different numerical codes. *Bull. Amer. Meteor. Soc.*, **77**, 261-278.

Penner, J.E., and co-authors, 2001: Aerosols, their direct and indirect effects. In: Climate Change 2001: The Scientific Basis. *IPCC report.* Cambridge University Press, Cambridge, United Kingdom and New York, NY, USA, 881pp.

Stevens, B., W. R. Cotton, G. Feingold, 1998: A critique of one- and two-dimensional models of boundary layer clouds with a binned representations of drop microphysics. *Atmos. Res.*, 47-48, 529-553.

Stevens, B., C.-H. Moeng, and P. P. Sullivan, 1999: Large-eddy simulation of radiatively driven convection: Sensitivities to the representation of small scales. *J. Atmos. Sci.,* 56, 3963-3984.

Svensson, G., and J. H. Seinfeld, 2002: A numerical model of the cloud-topped marine boundary layer with explicit treatment of supersaturation-number concentration correlations. *Q. J. R. Meteorol. Soc.* **128**, 535-558.

Wang, S., and Q. Wang, 2000: A cloud microphysical parameterization for higher-order turbulence closure models. Proceedings, *13th International Conference on Clouds and Precipitation*, Reno, Nevada USA, ICCP, 557-560.

Wang, S., Q. Wang, and G. Feingold, 2003: Turbulence, condensation and liquid water transport in numerically simulated non-precipitating stratocumulus clouds. *J. Atmos. Sci.*, **60**, 262-278.

SOME CONCEPTS AND METHODS OF ATMOSPHERIC DATA ASSIMILATION — A TUTORIAL NOTE

XIAOLEI ZOU
Department of Meteorology, Florida State University
404 Love Building, Tallahassee, FL 32306, USA
E-mail: zou@met.fsu.edu

(Manuscript received 27 February 2003)

This paper covers topics of atmospheric data analysis and assimilation algorithms, ranging from function fitting methods to three-dimensional or four-dimensional variational approaches. The author has evaluated relevant materials in linear algebra, statistics, and optimization, and discussed major features of various methods covered. Finally, the challenges and research needed in assimilating a new type of atmospheric observation are briefly summarized.

1. Introduction

Numerical weather prediction (NWP) consists of a numerical integration of a set of model equations, which express the conservation of momentum, mass, energy, and water substances in various phases, in order to predict the future weather with given initial and/or boundary conditions (Pielke 1984, pages 3-19). The input initial and/or boundary conditions (called model inputs) are a set of values taken by independent model variables such as the three-dimensional (3D) wind vector, temperature, water vapor content, pressure, as well as values taken by various hydrometeor variables such as cloudwater, rainwater, snow ice, and graupel. The future weather (called model outputs) is usually described by the same set of model variables as those initial and boundary conditions in addition to quantities such as hourly surface rainfall amount. While encouraging progress has been made in NWP over the past decade (Kalnay 2003, pages 1-4), there is still room for further improvement in prediction of many high-impact severe weather systems, such as hurricanes and storms that strike regions of high population and property growth and result in large natural disasters.

How can NWP be further improved? Recently, the ever increasing performance of computers makes it possible to conduct NWP at very high resolutions. Extensive observational datasets, often consisting of multi-variate data with a huge number of sample locations or grid points and multiple time steps, provide a great opportunity to reduce errors in model formulation and model inputs, if they are carefully processed, analyzed, understood, and incorporated into NWP models. This is justified by the belief that errors made in model formulation and model inputs are usually greater than errors in observations. It is, therefore, desirable to examine and modify model formulation and model inputs by comparing with and making use of observations. Atmospheric data assimilation provides tools to do this. For example, data assimilation is carried out to satisfy various needs such as estimates of a field from sparse data of the same quantity, estimates of an unobserved field from observations of other quantities under a dynamic constraint or a set of physical laws, estimates of parameters in physical laws, designs for atmospheric observing systems, and tests of scientific hypotheses.

In general, atmospheric data assimilation deals with problems relating to merging observations with a prior knowledge of the atmosphere and its motion to provide more accurate model inputs and model formulation. Therefore, the emphasis for atmospheric data assimilation is on both the data and their assimilation. We need to know not only assimilation methods but also the properties of the data. Observations are often irregularly-distributed in space and time. Also, observations come from different observing systems varying from direct model variables (temperature, specific humidity, wind, and surface pressure) to indirect quantities (precipitable water, rainfall, reflectivity, radiance, ozone, and bending angle). The goal of data assimilation is to effectively extract useful information from (either limited or abundant) observations in order to maximally improve NWP and increase physical understandings of the evolution of weather systems. Quality control, physical consistency, spatial and temporal coherency, and noise suppression are four major concerns for atmospheric data assimilation no matter which data assimilation algorithm is used. Ideally, all of the following goals ought to be achieved: (i) outliers and redundant information are removed from observations; (ii) errors inherent in the original observations are adequately quantified and not carried over to the analysis values; (iii) interpolation algorithms are computationally efficient with low memory requirements and fast performance times; (iv) all spatial scales in the original sampled field needed by the forecast model are faithfully reproduced and undesired scales are filtered completely; and (v) the field estimates are consistent with all other atmospheric variables, i.e., together they satisfy the constraining relations of atmospheric motion. In reality, however, not all of these goals can be accomplished simultaneously, due to various constraints ranging from computing capability to human effort. In addition, these goals often conflict with one another. For example, the introduction of the geostrophic assumption ensures consistency between the height and wind analysis, but may also introduce additional errors. This occurs when the actual observational data contain scales of variation other than geostrophic flow features, and the error characteristics of these other variations are not properly quantified and utilized by the data assimilation algorithms.

This short tutorial note is part of author's lecture notes on atmospheric data assimilation taught in the Department of Meteorology, Florida State University. It is written with two motives: (i) simple enough to allow students and other readers with little or no background in atmospheric data assimilation to understand; and (ii) comprehensive enough to provide some advanced, state-of-the-art knowledge of atmospheric data assimilation. Materials that are important but omitted in this note include data sampling theory, spectral response and filtering, statistical structures of meteorological fields, specification of static and flow-dependent background (e.g., short-term forecast) error covariance matrices, parameter estimation, as well as estimates of model errors and their incorporation in data assimilation procedures.

In the following section, we discuss an early function fitting method and introduce simple ideas of atmospheric data assimilation. A statistical background of a least-squares fit problem is also included in section 2. Successive correction and optimal interpolation are presented in section 3. Kalman filter and ensemble Kalman filter are covered in section 4. Variational approaches are provided in section 5, which includes adjoint techniques, optimization theory, and the generalized inverse method. Space-born GPS (Global Positioning System) occultation measurements and some assimilation results of this data are briefly summarized in section 6. Section 7 concludes the paper.

2. Early methods — function fitting

2.1. Function fitting

The main sources of observations in the early days of meteorology were surface stations, radiosondes and pilot balloons. They were irregularly-distributed in space. On the other hand, estimates of the model fields (observed and unobserved) at regular grid points are required for NWP. The earliest data analysis was subjective. This method was used by Richardson (1922) to conduct the first NWP experiment. Subjective analysis was considered as being too slow and not reproducible; however, it was used as a benchmark for the development of various automated data analysis methods in the early days. After the Second World War, advancements in computer science made the dream of NWP a reality, and improvements in communication made assembly of ready and timely data possible. Both advances allowed development of an automated data analysis method (called objective analysis) to produce a gridded analysis for a finite-differenced NWP model.

The first major contribution to objective analysis was made by Panofsky (1949), who sought a horizontal 2D analysis field of the geopotential height $h(x, y)$ on a pressure surface from both height and wind observations. His main procedure is as follows: First, the entire model domain is divided into several sub-domains, each containing ten or more independent observations. Second, a local coordinate system is introduced and expresses the 2D height function by a Taylor expansion truncated with third-order accuracy:

$$\begin{aligned} h(x_0 + x, y_0 + y) &= h_0 + xh_{0x} + yh_{oy} + \frac{x^2}{2!}h_{0xx} + \frac{y^2}{2!}h_{oyy} + xyh_{0xy} \\ &\quad + \frac{x^3}{3!}h_{0xxx} + \frac{y^3}{3!}h_{oyyy} + \frac{x^2y}{2!}h_{0xxy} + \frac{xy^2}{2!}h_{oxyy}, \end{aligned} \tag{1}$$

where (x_0, y_0) is the center of a sub-domain, $h_0 = h(x_0, y_0)$, $h_{0x} = (\partial h/\partial x)|(x_0, y_0)$, etc. Equation (1) is a third degree polynomial approximation of the height field over the selected sub-domain. Although the location (x, y) and the point measurement $h(x_0 + x, y_0 + y)$ are known for each observation, $h_0 = h(x_0, y_0)$ and the derivatives of the field at (x_0, y_0) (h_{0x}, h_{oy}, h_{0xx}, h_{oyy}, h_{0xxx}, h_{oyyy}, h_{0xxy}, and h_{oxyy}) (a total of ten unknowns) are unknown. If h_0 and the derivatives can be estimated directly or indirectly based on point (station) measurements, then the values of the height field at any analysis grid within the sub-domain can be calculated by using eq. (1).

If there are ten observations whose locations and point measurements $h(x_0+x, y_0+y)$ are known, we can substitute each of them into (1) to obtain a linear system of ten equations:

$$\mathbf{Ax} = \mathbf{y}^{\text{obs}}, \tag{2}$$

where the coefficient matrix $\mathbf{A}$ consists of elements relating to the position of observational locations:

$$\mathbf{A} = \begin{pmatrix} 1 & x_1 & y_1 & \frac{x_1^2}{2} & \frac{y_1^2}{2} & x_1y_1 & \frac{x_1^3}{6} & \frac{y_1^3}{6} & \frac{x_1^2y_1}{2} & \frac{x_1y_1^2}{2} \\ 1 & x_2 & y_2 & \frac{x_2^2}{2} & \frac{y_2^2}{2} & x_2y_2 & \frac{x_2^3}{6} & \frac{y_2^3}{6} & \frac{x_2^2y_2}{2} & \frac{x_2y_2^2}{2} \\ \cdot & \cdot & \cdot & \cdot & \cdot & \cdot & \cdot & \cdot & \cdot & \cdot \\ 1 & x_{10} & y_{10} & \frac{x_{10}^2}{2} & \frac{y_{10}^2}{2} & x_{10}y_{10} & \frac{x_{10}^3}{6} & \frac{y_{10}^3}{6} & \frac{x_{10}^2y_{10}}{2} & \frac{x_{10}y_{10}^2}{2} \end{pmatrix}, \tag{3}$$

and (x_k, y_k) represents the coordinates of the kth observation relative to (x_0, y_0). The vector $\mathbf{x}$ represents a total of ten unknowns and the forcing vector $\mathbf{b}$ has the ten actual height observations as its components:

$$\begin{aligned} \mathbf{x} &= (h_0, h_{0x}, h_{0y}, h_{0xx}, h_{0yy}, h_{0xy}, h_{0xxx}, h_{0yyy}, h_{0xxy}, h_{0xyy})^T, \\ \mathbf{y}^{obs} &= \left(h_1^{\text{obs}}, h_2^{\text{obs}}, h_3^{\text{obs}}, h_4^{\text{obs}}, h_5^{\text{obs}}, h_6^{\text{obs}}, h_7^{\text{obs}}, h_8^{\text{obs}}, h_9^{\text{obs}}, h_{10}^{\text{obs}}\right)^T, \end{aligned} \tag{4}$$

where the superscript $(\ )^T$ stands for a transpose. A unique solution for $\mathbf{x}$ can be obtained if the coefficient matrix $\mathbf{A}$ is non-singular. In this case, the analysis problem is solved through a simple interpolation.

It is possible that the ten observations are located in such a way that $\mathbf{A}$ is singular. It is also possible that observations contain errors and observations from nearby stations are not in exact agreement. In addition, it is desirable to make use of wind observations if they are available. However, when $\mathbf{A}$ is singular (the system is under-determined) or there are more than ten observations (the system is over-determined), the analysis problem becomes ill-posed. A least-squares fit was used to solve these ill-posed problems in order to find an estimate of $\mathbf{h}$ which is a best fit to data from more than ten stations with wind observations included (Panofsky (1949) suggested using 12-14 height observations and 14-20 (no more than 20) wind reports to enhance error suppression).

For a total of M height and N wind observations, equation (2) is not solved exactly, but by minimizing the following mean-square total error:

$$J(\mathbf{x}) = \sum_{m=1}^{M} \alpha_m^2 (h_m - h_m^{\text{obs}})^2 + \sum_{n=1}^{N} \beta_n^2 \left[(u_n - u_n^{\text{obs}})^2 + (v_n - v_n^{\text{obs}})^2\right], \tag{5}$$

where h_m is the right-hand-side of (1) (e.g., (1) evaluated at the mth observational location), and u_n and v_n are calculated from h_n based on the geostrophic balance constraint. Panofsky (1949) chose constant weighting coefficients α_m and β_n whose values are specified as 30 feet (9.14 m) and 5 m s^{-1} based on the assumption that a height deviation of 30 feet corresponds to a 5 m s^{-1} wind deviation. Ideally, α_m^2 and β_n^2 should be inversely proportional to discrepancy variances including the effects of observation errors, the third-degree polynomial truncation errors, and errors associated with the geostrophic approximation to the actual winds. Noisy observations and observations over regions where the geostrophic assumption is less valid should be weighted less than the ones accurately determined. The weights also make each sum in (5) dimensionless such that wind and height observations can be combined in the same analysis.

The unknown variables $\mathbf{x}$ are obtained by solving a set of ten equations

$$\frac{\partial J}{\partial x_i} = 0, \quad i = 1, 2, \ldots, 10. \tag{6}$$

The coefficient matrix in (6) is different from (2) and depends not only on observational locations but also on height and wind observations. The rate of occurrence of the coefficient matrix in (6) being singular is smaller than that of $\mathbf{A}$.

With a polynomial fitted in each sub-domain, Panofsky (1949) then interpolated the coefficients across the boundaries of sub-domains, and finally evaluated the composite polynomial at each analysis grid point.

By examining Panofsky's approach, we find that analyses at the regular grid points are obtained from irregularly distributed observations through polynomial function fitting; observations are weighted according to their accuracy (in a rather simple way), a simple dynamic constraint is incorporated to combine height and wind observations in a single analysis, and a least-squares fit procedure is used to solve an ill-posed problem.

2.2. About a least-squares fit

Many examples in data assimilation, such as the one in section 2.1, invoke a least-squares fit at different stages of the data analysis procedure. Least-squares fitting is used in many other applications as well. What are the implications behind a least-squares fit from a statistical point of view?

Given a set of random functions $f_k(x_1, x_2, \ldots, x_N)$ $(k = 1, 2, \ldots, K)$, each obeying a distribution function (e.g., a probability density function):

$$f_k(\mathbf{x}; \boldsymbol{\theta}_{L_k}), \qquad k = 1, 2, \ldots, K, \tag{7}$$

where $\mathbf{x}$ are state variables to be estimated from a known data sample, and $\boldsymbol{\theta}_{L_k}$ are distribution parameters whose values are empirically given. If each function is sampled independently, their joint probability density is the product of the distribution (density) functions evaluated for each of the independent random functions:

$$L(\mathbf{x}; \boldsymbol{\theta}_{L_1}, \ldots, \boldsymbol{\theta}_{L_K}) = \Pi_{k=1}^{K} f_k(\mathbf{x}; \boldsymbol{\theta}_{L_k}). \tag{8}$$

The function L is called the likelihood function.

Based on the assumption that when the variable estimates are close to their true values, the likelihood function will be large (Menke 1984), the *method of maximum likelihood* provides an estimator that maximizes the likelihood function. In other words, the maximum likelihood estimator is the solution of following maximization problem:

$$\max_{\mathbf{x}} L(\mathbf{x}; \boldsymbol{\theta}_1, \ldots, \boldsymbol{\theta}_{L_K}). \tag{9}$$

Therefore, the likelihood function (8) can be used to provide estimators for state variables that are to be determined from data samples.

Now let's take a look at a linear system (2). In general, observations are not flawless so that the equalities suggested in (2) are not perfect. Rather, each observation is accompanied by an error which quantifies the difference between the actual observation (y_k^{obs}) and its true value (y_k^{true}), i.e.

$$\mathbf{y}^{\text{true}} = \mathbf{y}^{\text{obs}} + \boldsymbol{\epsilon}. \tag{10}$$

Therefore, eq. (2) must be written as

$$\mathbf{A}\mathbf{x} - \mathbf{y}^{obs} = \boldsymbol{\epsilon}. \tag{11}$$

It is also possible that in some instances the linear system that is to be solved contains (model) errors. In this case, ϵ_k will represent both observational and model errors, and together we will call it the discrepancy between the model and the observations or the residual of model.

When $\{\epsilon_k\}$ are not identically equal to zero, one way to solve problem (11) is to derive the maximum likelihood estimate. As will be shown below, under various assumptions about the distribution of ϵ_k, one obtains different minimization problems based on the maximum likelihood estimate.

If the observational errors (e.g., the random functions $\mathbf{f} = \mathbf{A}\mathbf{x} - \mathbf{b}$) are assumed to be random variables drawn from a set of Gaussian distributions:

$$f(\epsilon_k) = \frac{1}{\sqrt{2\pi E_k^2}} e^{-\frac{1}{2E_k^2}(\epsilon_k - \mu_k)^2}, \qquad k = 1, 2, \ldots, K, \tag{12}$$

where μ_k and E_k^2 are the mean and variance of the kth function, then the joint probability density function is the product of these Gaussian distributions:

$$\begin{aligned} &L(\epsilon_k; \mu_1, \mu_2, \ldots, \mu_K, E_1, E_2, \ldots, E_K) \\ &\quad = \Pi_{k=1}^K \frac{1}{\sqrt{2\pi E_k^2}} e^{-\frac{1}{2E_k^2}(\epsilon_k - \mu_k)^2} = \frac{1}{\sqrt{(2\pi)^K \Pi_{k=1}^K E_k^2}} e^{-\sum_{k=1}^K \frac{1}{2E_k^2}(\epsilon_k - \mu_k)^2} \end{aligned} \tag{13}$$

The maximum likelihood estimate (9) corresponds to finding the minimum in the argument of the exponential:

$$J(x_1, x_2, \ldots, x_N) \equiv \sum_{k=1}^{K} \frac{1}{E_k^2} (\epsilon_k - \mu_k)^2. \tag{14}$$

If the residuals are assumed to have zero mean ($\mu_k = 0$), the above cost function becomes

$$J(x_1, x_2, \ldots, x_N) = \sum_{k=1}^{K} \frac{1}{E_k^2} (\epsilon_k)^2 = \sum_{k=1}^{K} \frac{1}{E_k^2} (\mathbf{A}\mathbf{x} - \mathbf{b})_k^2. \tag{15}$$

Thus, the least-squares fit used in (5) to minimize the sum of squares of the discrepancy of (2) is equivalent to the maximum likelihood estimate under the assumptions that the errors are Gaussian and unbiased, and the weights (α_m and β_m in (5)) are the inverses of the observational error variances. If the errors are assumed to have the same variance, we obtain the following minimization problem

$$J(x_1, x_2, \ldots, x_N) = \frac{1}{E^2}\sum_{k=1}^{K}(\epsilon_k)^2 = \frac{1}{E^2}\sum_{k=1}^{K}(\mathbf{Ax}-\mathbf{b})_k^2\,, \tag{16}$$

which is the sum of the mean square values of the residuals corresponding to the L_2 norm fit.

What minimization problem do we obtain if observational errors are assumed to be random variables drawn from a distribution that is not Gaussian? Let's assume that observational errors (or residuals) are random variables drawn from a two-sided exponential distribution:

$$f(\epsilon_k) = \frac{1}{\sqrt{2}E_k}e^{-\sqrt{2}\frac{|x-\mu_k|}{E_k}}, \qquad k = 1, 2, \ldots, K. \tag{17}$$

Following the same procedure, we obtain that maximizing the likelihood (joint probability function) is equivalent to minimizing the following cost function

$$J(x_1, x_2, \ldots, x_N) = \sum_{k=1}^{K}\frac{1}{E_k}|\epsilon_k - \mu_k|. \tag{18}$$

If the residuals are assumed to have zero mean and the same variance, we obtain

$$J(x_1, x_2, \ldots, x_N) = \frac{1}{E}\sum_{k=1}^{K}|\epsilon_k|, \tag{19}$$

which is the sum of the absolute values of the residuals corresponding to the L_1 norm fit.

We therefore find that selecting a particular norm for each fitting of experimental data implies a choice about the distribution of residuals. A least-squares fit works best if residuals follow a Gaussian distribution with zero mean.

3. Successive correction and optimal interpolation — Minimum variance estimates

3.1. Successive correction method

An analysis is obtained directly from observations in the function fitting procedure in which background fields (obtained from climatology or short-range forecasts) are not introduced in the procedure. Successive correction (Cressman 1959; Barnes 1964) makes use of both observations and background fields for the analysis. Successive correction is a simple, computationally inexpensive and reasonably accurate scheme that is much more successful than function fitting and was the first scheme used in operational numerical weather forecasting. Two main features of the successive correction (SC) algorithm are: (i) observation increments within a specified influence radius R (say a total of K_i for the analysis at the ith point) are averaged to produce an analysis increment

$$x^a(\vec{r}_i) = x^b(\vec{r}_i) + \sum_{k=1}^{K_i} W_{ik}[x^{obs}(\vec{r}_k) - x^b(\vec{r}_k)], \tag{20}$$

where W_{ik} is called the *a posteriori* weights; and (ii) *a posteriori* weights are specified *a priori* to be monotonically decreasing functions of the distance between the observation station and the analysis point (Daley 1991, pages 66-68):

$$W_{ik} = \frac{E_o^{-2}(k)w(|\vec{r}_k - \vec{r}_i|)}{E_b^{-2} + \sum_{k=1}^{K_i} E_o^{-2}(\vec{r}_k)w(|\vec{r}_k - \vec{r}_i|)}, \tag{21}$$

where $w(|\vec{r}_k - \vec{r}_i|)$ is an empirical function of the distance between the observation location ($\vec{r}_k$) and the gridpoint ($\vec{r}_i$) whose value decreases from a unit value when $\vec{r}_k$ and $\vec{r}_i$ are collocated to 0 at R (the radius of the influence).

In the original schemes used by both Cressman (1959) and Barnes (1964) and in the absence of information about background and observation error variances, $E_o^{-2} = E_b^{-2} = 1$, and the SC formula is simply

$$x^a(\vec{r}_i) = x^b(\vec{r}_i) + \sum_{k=1}^{K_i} \underbrace{\frac{w(|\vec{r}_k - \vec{r}_i|)}{\sum_{k=1}^{K_i} w(|\vec{r}_k - \vec{r}_i|)}}_{W_{ik}} [x^{obs}(\vec{r}_k) - x^b(\vec{r}_k)]. \tag{22}$$

The SC algorithm expressed by (20)-(21) is obtained by a minimum variance estimate[1] of the following $K_i + 1$ estimates of x at the grid point $\vec{r}_i$:

$$x^b(\vec{r}_i), \ \underbrace{x^b(\vec{r}_i) + \left[x^{\text{obs}}(\vec{r}_k) - x^b(\vec{r}_k)\right]}_{x_k^c(\vec{r}_i)}, \quad k = 1, \cdots, K_i, \tag{23}$$

with specified error variances

$$E_b^2, \ \underbrace{\frac{E_o^2}{w(|\vec{r}_k - \vec{r}_i|)}}_{E_c^2(k)}, \qquad k = 1, \cdots, K_i, \tag{24}$$

where E_b^2 and E_o^2 are error variances of the background field and the observations, respectively. The analysis error variance is

$$E_a^2 = \frac{1}{E_b^{-2} + \sum_{k=1}^{K_i} E_o^{-2}(\vec{r}_k)w(|\vec{r}_k - \vec{r}_i|)}. \tag{25}$$

[1]the expected error variance of the minimum variance estimate (which is an unbiased linear estimate) is smaller than than any other unbiased linear estimate.

Examining (23)-(25) we find that the minimum variance estimate produces an analysis such that (i) each additional observation (or estimate in this case), no matter how inaccurate, always reduces the expected error variance of the analysis; and (ii) the expected error variance of the minimum variance estimate is no greater than the smallest expected error variance of the observations. In other words, the following two inequalities hold true:

$$E_a^2(\text{with } K_i \text{ observations}) \;<\; E_a^2(\text{with } (K_{i-1}) \text{ observations}), \tag{26}$$

$$E_a^2(\text{with } K_i \text{ observations}) \;\leq\; \text{the smallest of } \{E_b^2, E_c^2(1), \ldots, E_c^2(K_i)\}. \tag{27}$$

These results are desirable for atmospheric data assimilation.

Therefore, the SC algorithm is a statistical method with simplifying assumptions made to both the K_i+1 estimates (23) and to the error variances (24) of these estimates. Insights into the implications of these assumptions can be obtained by expanding the error variances for the 2nd through (K_i+1)th estimates in (23) as

$$E_c^2(k) \equiv \overline{(x_k^c(\vec{r}_i) - x^t(\vec{r}_i))^2} = E_O^2 + 2(E_b^2 - B_{ik}) = E_O^2 + 2(E_b^2 - E_b^2 \rho_{ik}^b), \tag{28}$$

where $x^t(\vec{r}_i)$ is the true value at the analysis gridpoint $\vec{r}_i$, the overline represents the expectation (mean) operator, and ρ_{ik}^b is the background error correlation between the ith analysis point and the kth observational location. The assumption made on the error variance of $x_k^c(\vec{r}_i)$ in (24) implies

$$w(|\vec{r}_i - \vec{r}_k|) = \frac{1}{1 + \frac{2E_b^2}{E_O^2}(1 - \rho_{ik}^b)}. \tag{29}$$

Therefore, the weight function should approximate the function defined in (29).

In the SC algorithm, (i) background errors are assumed homogeneous; (ii) both background and observational errors are assumed unbiased; (iii) background errors are assumed to be uncorrelated with observational errors; and (iv) the effect of the background error correlations in space on the analysis, reflected by the assumed error variances for the second through (K_i+1)th estimates, is accounted for through an empirical weighting function (29). The success of the SC method for the analysis of a particular field depends, therefore, on how much these assumptions are not violated. Bratseth (1986) showed that SC can be made to have some desirable properties of an optimal interpolation (OI) method, such as the use of observation error statistics and the ability to produce a multivariate analysis, through an iterative procedure.

3.2. Optimal interpolation

The analysis increment of OI takes the same form as that of the SC algorithm (20). However, unlike the SC method in which the weighting coefficients are specified empirically, OI uses the optimal weighting coefficients W_{ik}, $k = 1, \ldots, K_i$ that produce the minimum analysis error variance[2]. In other words, the weighting coefficients are

[2]The minimum variance estimate in SC is applied to the background and the K_i estimates which are obtained approximately. The minimum variance estimate in OI is applied directly to the analysis x_i^a.

obtained by minimizing the analysis error variance:

$$E_{ai}^2(W_{i1},\ldots,W_{iK_i}) = \overline{\{(x^a(\vec{r}_i) - x^t(\vec{r}_i))^2\}}. \tag{30}$$

Substituting (20) into (30), carrying out a similar manipulation as in (28), and assuming that there is no correlation between the background and observation error, one obtains (Daley 1991, pages 102-105)

$$E_{ai}^2 = E_{bi}^2 - 2\vec{W}_i^T\vec{B}_i + \vec{W}_i^T[\mathbf{B}+\mathbf{O}]\vec{W}_i, \tag{31}$$

where

$$\vec{W}_i = \begin{pmatrix} W_{i1} \\ W_{i2} \\ \cdot \\ \cdot \\ \cdot \\ W_{iK} \end{pmatrix}, \quad \vec{B}_i = \begin{pmatrix} E\{(x^b(\vec{r}_1) - x^t(\vec{r}_1))(x^b(\vec{r}_i) - x^t(\vec{r}_i))\} \\ E\{(x^b(\vec{r}_2) - x^t(\vec{r}_2))(x^b(\vec{r}_i) - x^t(\vec{r}_i))\} \\ \cdot \\ \cdot \\ \cdot \\ E\{(x^b(\vec{r}_K) - x^t(\vec{r}_K))(x^b(\vec{r}_i) - x^t(\vec{r}_i))\} \end{pmatrix}, \tag{32}$$

and

$$\begin{aligned} \mathbf{B} &= (B_{kl}) = \Big(E\{(x^b(\vec{r}_k) - x^t(\vec{r}_k))(x^b(\vec{r}_l) - x^t(\vec{r}_l))\}\Big)_{K_i\times K_i} \\ &= E\{(\vec{x}^b(\vec{r}_k) - \vec{x}^t(\vec{r}_k))(\vec{x}^b(\vec{r}_l) - \vec{x}^t(\vec{r}_l))^T\}, \\ \mathbf{O} &= (O_{kl}) = \Big(E\{(x^{\text{obs}}(\vec{r}_k) - x^t(\vec{r}_k))(x^{\text{obs}}(\vec{r}_l) - x^t(\vec{r}_l))\}\Big)_{K_i\times K_i} \\ &= E\{(\vec{x}^{\text{obs}}(\vec{r}_k) - \vec{x}^t(\vec{r}_k))(\vec{x}^{\text{obs}}(\vec{r}_l) - \vec{x}^t(\vec{r}_l))^T\}. \end{aligned} \tag{33}$$

The necessary condition for the optimal weighting to produce the minimum analysis error variance is that $W_{i1},\ldots,W_{iK_i}$ are the stationary points of E_{ai}^2, which leads to the following set of K_i linear equations:

$$\frac{\partial E_{ai}^2}{\partial W_{ik}} = 0, \qquad k = 1,\ldots,K_i. \tag{34}$$

Substituting (31) into (34) we obtain

$$[\mathbf{B}+\mathbf{O}]\vec{W}_i = \vec{B}_i. \tag{35}$$

Substituting the optimal weights (35) into (31) we obtain the minimum analysis error variance:

$$E_{ai}^2 = E_{bi}^2 - \vec{B}_i^T[\mathbf{B}+\mathbf{O}]^{-1}\vec{B}_i. \tag{36}$$

Obviously, the analysis error variance is smaller than the background error variance. By definition, the OI analysis error is the least of all expected analysis error variances produced by a linear estimate under the assumption that observation and background errors are uncorrelated, meaning that it is also smaller than the analysis error variance of any SC algorithm.

OI consists of (35)-(36). In reality, correct values of the background ($\mathbf{B}$ and $\vec{B}_i$) and observation ($\mathbf{O}$) error covariances are not known precisely since the truth is not known. Their values are either estimated from observations (Daley 1991) and model forecasts (Parrish and Derber 1992), or prespecified (Buell 1972). Therefore, the actual OI systems implemented at various operational centers are only suboptimal.

The role of the background error covariances can be illustrated more clearly by substituting the weight equation (35) into the analysis equation (20), through which we obtain

$$x^a(\vec{r}_i) - x^b(\vec{r}_i) = \underbrace{\left\{\vec{B}_i^T\mathbf{B}^{-1}\right\}}_{\text{interpolation}} \underbrace{\left\{\mathbf{B}[\mathbf{B}+\mathbf{O}]^{-1}\right\}}_{\text{filter}}[\mathbf{x}^{\text{obs}} - \mathbf{x}^b]. \tag{37}$$

Here $\vec{B}_i^T\mathbf{B}^{-1}$ is a vector that interpolates the filtered observation increments to the analysis gridpoint to generate the analysis increment. $\mathbf{B}[\mathbf{B}+\mathbf{O}]^{-1}$, whose spectrum determines the filtering property of OI, is independent of the analysis point and is sometimes called the response matrix. The vector $\mathbf{d} = \mathbf{B}[\mathbf{B}+\mathbf{O}]^{-1}[\mathbf{x}^{\text{obs}} - \mathbf{x}^b]$ represents filtered observation increments at the K observation locations.

Analysis increments at a total of I analysis gridpoints can be written as

$$\underbrace{\begin{pmatrix} x^a(\vec{r}_1) - x^b(\vec{r}_1) \\ x^a(\vec{r}_2) - x^b(\vec{r}_2) \\ \cdots \\ x^a(\vec{r}_I) - x^b(\vec{r}_I) \end{pmatrix}}_{\mathbf{x}^a - \mathbf{x}^b} = \underbrace{\left\{\begin{pmatrix} B_1^T \\ B_2^T \\ \cdots \\ B_I^T \end{pmatrix} \mathbf{B}^{-1}\right\}}_{\text{interpolation}} \underbrace{\left\{\mathbf{B}[\mathbf{B}+\mathbf{O}]^{-1}\right\}[\mathbf{x}^{\text{obs}} - \mathbf{x}^b]}_{\text{filter}}. \tag{38}$$

Some insight can be obtained with regard to taking $\vec{B}_i^T\mathbf{B}^{-1}$ as an interpolation operator. Consider a special case in which the analysis gridpoints coincide with the observation locations. Then the OI interpolation operator is

$$\underbrace{\left\{\begin{pmatrix} B_1^T \\ B_2^T \\ \cdots \\ B_I^T \end{pmatrix} \mathbf{B}^{-1}\right\}}_{\text{interpolation}} = \mathbf{I}. \tag{39}$$

No interpolation is done as it is not needed in this case.

As seen in (35), the matrix $\mathbf{B}+\mathbf{O}$ of order $K_i \times K_i$ must be inverted to produce the OI analysis weights. In practice, the size of a matrix that can be inverted is limited by computer capacity while the number of observations can reach millions. Therefore, each grid point is analyzed separately by using only those observations that are within a certain distance (the radius of influence) of the analysis gridpoint. A type of data thinning is also necessary for observations that are of high resolution (such as radar and some satellite measurements) to reduce the dimension of $\mathbf{B}+\mathbf{O}$ and to avoid ill conditioning of $\mathbf{B}+\mathbf{O}$.

The primary differences in OI algorithms implemented at various operational centers rest with (i) the observations that are included in the analysis, (ii) the rule by which the data selection is conducted, (iii) the construction of the error matrices $\mathbf{B}_{K_i \times K_i}$, $\mathbf{O}_{K_i \times K_i}$, and $\vec{B}_{K_i \times 1}$, and (iv) the method for solving for the weights in (35). An excellent source of information about OI's theory and implementation can be found in Gandin (1965) and Daley (1991).

3.3. Simple examples

We take several extremely simple cases (Daley 1991, pages 131-133) to illustrate some interesting features of SC and OI. In all cases considered below, it is assumed that observation errors are uncorrelated and background errors are homogeneous. Such assumptions imply that

$$B_{kl} = E_b^2 \rho_{kl}(|\mathbf{r}_k - \mathbf{r}_l|), \ O_{kl} = E_o^2(k), \ B_{ik} = E_b^2 \rho_{ik}(|\mathbf{r}_i - \mathbf{r}_k|), \tag{40}$$

where $\rho_{kl}(r)$ and $\rho_{ik}(r)$ are the background error correlations depending only on distance. The OI formulae (20), (35) and (36) under these assumptions can be written as

$$x^a = x^b + \sum_{k=1}^{K} W_{0k}[x^{\mathrm{obs}}(\vec{r}_k) - x^b(\vec{r}_k)], \ \sum_{l=1}^{K} W_{0l}[\rho_{kl} + \epsilon_k^2] = \rho_{k0}, \ \epsilon_a^2 = 1 - \sum_{k=1}^{K} \rho_{0k} W_{0k}, \tag{41}$$

where we have dropped the index "i" for analysis point, $\epsilon_k^2 = E_o^2(k)/E_b^2$, and $\epsilon_a^2 = E_a^2/E_b^2$.

Let's first consider a single observation located at $\vec{r}_1$, $x^{\mathrm{obs}}(\vec{r}_1)$, to be incorporated into the analysis at $\vec{r}_0$. The SC analysis produces the following results

$$x^a = x^b + W_{01}[x^{\mathrm{obs}}(\vec{r}_1) - x^b(\vec{r}_1)], \ W_{01} = \frac{w_{10}}{w_{10} + \epsilon_1^2}, \ \epsilon_a^2 = \frac{\epsilon_1^2}{w_{10} + \epsilon_1^2}, \tag{42}$$

where $w_{10} = w(|\vec{r}_1 - \vec{r}_0|)$ is the empirical weighting function.

The OI analysis gives

$$x^a = x^b + W_{01}[x^{\mathrm{obs}}(\vec{r}_1) - x^b(\vec{r}_1)], \ W_{01} = \frac{\rho_{10}}{1 + \epsilon_1^2}, \ \epsilon_a^2 = \frac{1 + \epsilon_1^2 - \rho_{10}^2}{1 + \epsilon_1^2}. \tag{43}$$

From (42)-(43), it is found that the *a posteriori* weight depends on observation and background error variances as well as the empirical weight function (in SC) or the background error correlation (in OI). The analysis error variance from both analyses is smaller than the background error variance ($\epsilon_a^2 < 1 \longrightarrow E_a^2 < E_b^2$). Since the weight function w_{10} in SC and the background error correlation ρ_{10} in OI decrease as the distance between the observation location and analysis grid ($d_{01} = |\vec{r}_1 - \vec{r}_0|$) increases, the *a posteriori* weights determined by the weight equations in (42) and (43) in the two algorithms decrease as the observation is located further away from the analysis point. Since the maximum values of w_{10} and ρ_{10} are both equal to the unit value, the *a posteriori* weight in SC is greater than that in OI if w_{10} varies with the distance d_{01} between the analysis point and the observation location in a similar manner as that of ρ_{10}. In other words, the SC algorithm tends to give a larger weight to the same observation increment than the OI algorithm. This may suggest the use of an empirical weighting function which decreases with the distance faster than the background error correlation function.

If the observation location coincides with the analysis point, $w_{10} = \rho_{10} = 1$, the SC and OI algorithms produce the same analysis:

$$x^a = x^b + W_{01}[x^{\text{obs}} - x^b],\ W_{01} = \frac{1}{1+\epsilon_1^2},\ \epsilon_a^2 = \frac{\epsilon_1^2}{1+\epsilon_1^2}. \tag{44}$$

From (44), it is concluded that when a single observation at the analysis point is available, a weighted observational increment is added to the background field to obtain the analysis at the same point in both SC and OI analysis algorithms. The weight is inversely proportional to the observational error variance. Observations with larger errors have smaller impacts on the analysis. If an observation is known to have a smaller error, then the weight is larger and the analysis is more accurate. These are desirable features for any data assimilation algorithm to have.

The second example incorporates two observations into the SC and OI analyses. Assuming that the observation error variances are equal ($\epsilon_1^2 = \epsilon_2^2 = \epsilon^2$), the distances from the two observations to the analysis grid is equal ($\rho_{10} = \rho_{20} \equiv \tilde{\rho}$), and the observations are on either side of the grid point or collocated, the OI formulae in (41) become

$$x^a = x^b + W_{01}[x^{\text{obs}}(\vec{r}_1) - x^b(\vec{r}_1)] + W_{02}[x^{\text{obs}}(\vec{r}_2) - x^b(\vec{r}_2)], \tag{45}$$

$$W_{01} = \frac{\tilde{\rho}(1+\epsilon^2) - \tilde{\rho}\rho_{12}}{(1+\epsilon^2)^2 - \rho_{12}^2},\ W_{02} = \frac{\tilde{\rho}(1+\epsilon^2) - \tilde{\rho}\rho_{12}}{(1+\epsilon^2)^2 - \rho_{12}^2}, \tag{46}$$

$$\epsilon_a^2 = 1 - \frac{2\tilde{\rho}^2(1+\epsilon^2) - 2\tilde{\rho}^2\rho_{12}}{(1+\epsilon^2)^2 - \rho_{12}^2}. \tag{47}$$

Two special cases to be considered are: (i) The two observation locations are sufficiently separated so that $\rho_{12} \approx 0$; (ii) The two observation locations are collocated so that $\rho_{12} \approx 1$. If the two observation locations are sufficiently separated, (46)-(47) simplify to

$$W_{01} = W_{02} = \frac{\tilde{\rho}}{1+\epsilon^2},\ \epsilon_a^2 = 1 - \frac{2\tilde{\rho}^2}{1+\epsilon^2}. \tag{48}$$

If the two observations are collocated, we have

$$W_{01} = W_{02} = \frac{\tilde{\rho}}{2+\epsilon^2},\ \epsilon_a^2 = 1 - \frac{2\tilde{\rho}^2}{2+\epsilon^2}. \tag{49}$$

Comparing the results in (49) with (48), we find that when two observations are sufficiently separated, the weights are greater and the analysis error is smaller. This implies that evenly distributed observations produce a better analysis than clustered observations. Therefore, if the same number of observations is deployed, their spatial distribution will affect the analysis quality.

By comparing the results in (49) with (43), it is found that when two observations are available at a single location (which does not have to coincide with the analysis point), the weights are smaller and the analysis error is smaller than if only a single observation is used. This implies that more observations help improve the analysis.

The two special cases for the analysis considered above that had two observations equidistance from the analysis point are not distinguishable in the SC algorithm because the correlations of background errors among observation locations are not taken into account. The SC analysis for two observations which have the same distance from the analysis grid (either separated or collocated) is always

$$W_{01} = W_{02} = \frac{\tilde{w}}{2\tilde{w} + \epsilon^2}, \ \epsilon_a^2 = \frac{\epsilon^2}{\epsilon^2 + 2\tilde{w}}, \tag{50}$$

where $\tilde{w}$ is the value of the empirical weight function at the distance the two observations are from the analysis grid.

Comparing the OI analysis with two observations that are separated with equal distance to the analysis point (48), the OI analysis with two observations that are collocated (49), and the SC analysis for both cases (50), we find that the SC gives weights that are too large (small) when the two observations are separated (collocated).

In spite of being a more efficient method, the SC algorithm produces a less accurate analysis than OI in general. It is typical in atmospheric data assimilation that one scheme is less accurate but more efficient than another scheme. It is often difficult to have both the maximum accuracy and the greatest efficiency although they are both the most desirable qualities of an assimilation scheme.

4. Kalman filter

The Kalman filter incorporates the data assimilation procedure into the forward model integration at every time step whenever there are observations. It not only produces an optimal linear unbiased estimate of the state of a dynamical system, but also predicts and propagates optimally the error in the estimate due to observational and system errors.

To briefly describe the discrete-time Kalman filtering algorithm, we consider a sequence of state vectors, $\{\mathbf{x}_n\}$, satisfying a forward-time-stepping model in discrete time

$$\mathbf{x}_n^f = \mathbf{M}(t_n, t_{n-1})\mathbf{x}_{n-1}^a + \boldsymbol{\epsilon}_n^f, \quad n = 1, 2, \cdots, \tag{51}$$

with an initial condition $\mathbf{x}_0$, and a sequence of measurement data (or observations) $\{\mathbf{y}_n^{\text{obs}}\}$:

$$\mathbf{y}_n^{\text{obs}} = \mathbf{H}_n\mathbf{x}_n + \boldsymbol{\epsilon}_n^o, \quad n = 1, 2, \cdots, \tag{52}$$

where $\{\mathbf{M}(t_n, t_{n-1})\}$ and $\{\mathbf{H}_n\}$ are two sequences of the time-varying system and measurement matrices, $\mathbf{x}_n = \mathbf{x}(t_n)$ represents all model variables at time $t_n = t_0 + n\Delta t$ (Δt is the time step), $\{\boldsymbol{\epsilon}_n\}$ and $\{\boldsymbol{\epsilon}_b^o\}$ are two noise sequences. The dynamic equation (51) and the measurement equation (52) are both linear.

At time t_n, once we obtain the model forecast $\mathbf{x}_n^f$ from the analysis $\mathbf{x}_{n-1}^a$ at the previous time step t_{n-1}, a linear unbiased data estimate is carried out based on new observations $\mathbf{y}_n^o$:

$$\mathbf{x}_n^a = \mathbf{x}_n^f + \mathbf{K}_n(\mathbf{y}_n^{\text{obs}} - \mathbf{H}_n\mathbf{x}_n^f), \tag{53}$$

where $\mathbf{K}_n$ is a weighting matrix (or the gain matrix as it is called in engineering literature) obtained by minimizing the expected analysis error variance:

$$\min J \equiv E\{(\mathbf{x}_n^a - \mathbf{x}_n^t)^T(\mathbf{x}_n^a - \mathbf{x}_n^t)\}. \tag{54}$$

It turns out that such a solution exists under the following conditions (Chen, 1993):

1. Errors in the initial state vector is Gaussian with zero mean and covariance $\mathbf{B}_0$ being given; and
2. the two noise sequences $\{\epsilon^f\}$ and $\{\epsilon^o\}$ are Gaussian, stationary, mutually independent, and mutually independent of $\mathbf{x}_0$, with zero mean and known covariances $\mathbf{Q}_n = E(\epsilon_n^f(\epsilon_n^f)^T)$ and $\mathbf{O}_n = E(\epsilon_n^o(\epsilon_n^o)^T)$.

The weight matrix $\mathbf{K}_n$ in (52) at each time step can be obtained by setting the derivative of J with respect to each element of $\mathbf{K}_n$ equal to zero. A unique matrix for $\mathbf{K}_n$ is attained for

$$\mathbf{K}_n = \mathbf{K}_n^* \equiv \mathbf{B}_n\mathbf{H}_n^T(\mathbf{H}_n\mathbf{B}_n\mathbf{H}_n^T + \mathbf{O}_n)^{-1}, \tag{55}$$

where $\mathbf{B}_n$ is the error covariance matrix of the forecast $\mathbf{x}_n^f$:

$$\mathbf{B}_n = E\{(\mathbf{x}_n^f - \mathbf{x}_n^t)(\mathbf{x}_n^f - \mathbf{x}_n^t)^T\}, \tag{56}$$

which is governed (can be predicted) by

$$\mathbf{B}_n = \mathbf{M}(t_n, t_{n-1})\mathbf{A}_{n-1}\mathbf{M}^T(t_n, t_{n-1}) + \mathbf{Q}_{n-1}. \tag{57}$$

Here, the matrix $\mathbf{A}_n$ is the error covariance matrix of $\mathbf{x}_n^a$:

$$\mathbf{A}_n = E\{(\mathbf{x}_n^a - \mathbf{x}_n^t)(\mathbf{x}_n^a - \mathbf{x}_n^t)^T\}, \tag{58}$$

and can be derived by the following equation:

$$\mathbf{A}_n = (\mathbf{I} - \mathbf{K}_n\mathbf{H}_n)\mathbf{B}_n(\mathbf{I} - \mathbf{K}_n\mathbf{H}_n)^T + \mathbf{K}_n\mathbf{O}\mathbf{K}_n^T. \tag{59}$$

When $\mathbf{K}_n = \mathbf{K}_n^*$, the formula for $\mathbf{A}_n$ (59) can be simplified as follows

$$\mathbf{A}_n = (\mathbf{I} - \mathbf{K}_n^*\mathbf{H}_n)\mathbf{B}_n. \tag{60}$$

We thus see that if at time t_{n-1}, $\mathbf{x}_{n-1}^a$, $\mathbf{B}_{n-1}$, $\mathbf{A}_{n-1}$, and $\mathbf{Q}_{n-1}$ are produced from the previous analysis step, given $\mathbf{y}_n^o$, $\mathbf{H}_n$, $\mathbf{O}_n$, and $\mathbf{Q}_n$, the next new analysis $\mathbf{x}_n^f$, $\mathbf{B}_n$, $\mathbf{K}_n^*$, $\mathbf{x}_n^a$, $\mathbf{A}_n$ can be calculated as follows:

Forecast step :

$$\mathbf{x}_n^f = \mathbf{M}(t_n, t_{n-1})\mathbf{x}_{n-1}^a + \epsilon_n^f, \tag{61}$$

$$\mathbf{B}_n = \mathbf{M}(t_n, t_{n-1})\mathbf{A}_{n-1}\mathbf{M}^T(t_n, t_{n-1}) + \mathbf{Q}_{n-1}, \tag{62}$$

Analysis step :

$$\mathbf{K}_n^* \equiv \mathbf{B}_n\mathbf{H}_n^T(\mathbf{H}_n\mathbf{B}_n\mathbf{H}_n^T + \mathbf{O}_n)^{-1}, \tag{63}$$

$$\mathbf{x}_n^a = \mathbf{x}_n^f + \mathbf{K}_n^*(\mathbf{y}_n^{\text{obs}} - \mathbf{H}_n\mathbf{x}_n^f), \tag{64}$$

$$\mathbf{A}_n = (\mathbf{I} - \mathbf{K}_n^*\mathbf{H}_n)\mathbf{B}_n. \tag{65}$$

We notice that the analysis step in the Kalman filter is the same as that of OI.

Some remarkable advantageous features of the Kalman filter are that (i) it can be implemented recursively; (ii) it is the optimal estimator of all possible estimators giving linear unbiased estimates of the unknown vectors under the specified conditions; and (iii) it propagates the optimally predicted and estimated error covariances in the optimal estimator of the state of a dynamical system for the subsequent analysis step.

Often, the full implementation of a Kalman filter algorithm is impossible due to (i) the nonlinear nature of the dynamic and measurement equations of the system; (ii) the insufficient computational power for calculating the forecast error covariance ($\mathbf{B}_n$) and the weighting matrix ($\mathbf{K}_n$) (to store and invert large-dimension matrices); as well as (iii) the lack of complete knowledge of required statistical inputs (the mean and the covariance matrix of the Gaussian initial state vector and the two noise sequences in the dynamic and measurement equations). As a result, various modified versions of the Kalman filter, called *approximate Kalman filters*, have been developed (Chen, 1993). An extended Kalman filter is an extension of the standard Kalman filtering theory to estimating the unknown states of a nonlinear system. An adaptive Kalman filter is an adaptation of the Kalman filter or extended Kalman filter to either unknown noise statistics or non-Gaussian noise (but is probably a sum of many Gaussian noises). For example, the ensemble Kalman filter is a type of adaptive Kalman filter.

In the extended Kalman filter, (61)-(62) are replaced with

$$\mathbf{x}_n^f = M(t_n, t_{n-1})(\mathbf{x}_{n-1}^a) + \boldsymbol{\epsilon}_n^f, \quad i = 1, 2, \cdots, \tag{66}$$

$$\mathbf{B}_n = \mathbf{M}(t_n, t_{n-1})\mathbf{A}_{n-1}\mathbf{M}^T(t_n, t_{n-1}) + \mathbf{Q}_{n-1}, \tag{67}$$

where $M(t_n, t_{n-1})$ represents the nonlinear model forecast from the previous analysis time t_{n-1} to the current time t_n, and $\mathbf{M}(t_n, t_{n-1})$ is a linear operator that predicts the background error covariance matrix at t_n from the analysis error covariance at time t_{n-1}. Various simplifications can be made with regard to $\mathbf{M}$, which is used to evolve the forecast error covariance matrix in the extended Kalman filter. A straightforward choice of $\mathbf{M}(t_n, t_{n-1})$ is the tangent linear model (see section 5.3). Other simplifications include the use of approximated dynamics, reduced resolution, and/or simplified physics. Any simplification in the extended Kalman filter algorithm will result in some loss of optimality.

The ensemble Kalman filter (EnKF) combines state estimation with probabilistic ensemble forecasting. It makes use of an ensemble forecast to approximately calculate the error covariances required in the gain matrix ($\mathbf{K}_n$). First, an ensemble of II initial conditions $\mathbf{x}_0^a(i)$ is generated to represent uncertainties in the analysis at initial time $\mathbf{x}_0^a$. The formulas in (61), (62) and (64) are modified as:

$$\mathbf{x}_n^f(m) = M(t_n, t_{n-1})(\mathbf{x}_{n-1}^a(m)) + \boldsymbol{\epsilon}_n^f, \tag{68}$$

$$\mathbf{B}_n\mathbf{H}_n^T \approx \frac{1}{MM-1}\sum_{m=1}^{MM}\left(\mathbf{x}_n^f(m) - \overline{\mathbf{x}_n^f}\right)\left[H(\mathbf{x}_n^f(m) - \overline{\mathbf{x}_n^f(m)})\right]^T, \tag{69}$$

$$\mathbf{H}_n\mathbf{B}_n\mathbf{H}_n^T \approx \frac{1}{MM-1}\sum_{m=1}^{MM} H(\mathbf{x}_n^f(m) - \overline{\mathbf{x}_n^f})\left[H(\mathbf{x}_n^f(m) - \overline{\mathbf{x}_n^f(m)})\right]^T, \tag{70}$$

$$\mathbf{K}_n^* \equiv \mathbf{B}_n\mathbf{H}_n^T(\mathbf{H}_n\mathbf{B}_n\mathbf{H}_n^T + \mathbf{O}_n)^{-1}, \tag{71}$$

$$\mathbf{x}_n^a(m) = \mathbf{x}_n^f(m) + \mathbf{K}_n^*(\mathbf{y}_n^{\text{obs}} + \mathbf{v}_n(m) - \mathbf{H}_n\mathbf{x}_n^f(m)), \tag{72}$$

where "i" represents the ith ensemble member, and $\mathbf{y}_n^o + \mathbf{v}_n(i)$ is the ensemble of observations produced by the real observation $\mathbf{y}_n^o$ plus a noise ensemble $\mathbf{v}_n(i)$ based on specified observational error statistics. The EnKF approximates the extended Kalman filter.

In practice, the performance of the EnKF is limited by sampling errors in its covariance estimates. Various techniques, such as the covariance localization methods of Houtekamer and Mitchell (2001) and the deterministic covariance update of Whitaker and Hamill (2002), can be used to ameliorate the influence of the sampling errors. EnKF is appealing because of the minimal effort required in implementing one. For examples, a forecast model used in EnKF may be revised with minimal changes to an EnKF system; and the EnKF requires no linear or adjoint models of the forecast model whose development is intensive.

Experimentations implementing an ensemble Kalman filter have mostly been conducted using simulated observations and/or simple models (Houtekamer and Michell 2001; Anderson 2001). The performance of an ensemble Kalman filter for atmospheric data assimilation incorporating real observations and using a primitive equation model has yet to be fully demonstrated.

5. Variational approaches

5.1. *A variational problem*

The variational method was first attempted by Sasaki (1958, 1970), and further advanced by Le Dimet and Talagrand (1986). The concepts of strong and weak constraint were introduced by Sasaki and the practical implementation of the variational approach by using a forecast model as a strong constraint was made feasible by Le Dimet and Talagrand. For simplicity, we use the following simple problem to illustrate some basic concepts of variational approaches.

Given a prognostic equation

$$\frac{\partial \phi(x,t)}{\partial t} + c\frac{\partial \phi(x,t)}{\partial x} = F, \; \phi(x,0) = \phi_0(x), \tag{73}$$

where c and F are constants and $\phi(x,t)$ is a periodic function defined over an interval $a \le x \le b$, a set of observations

$$\phi_{mn}^{\text{obs}} = \phi^{\text{obs}}(x_m, t_n), \quad m = 1,2,\ldots,M; n = 1,2,\ldots,N, \tag{74}$$

and a background field of ϕ at $t = 0$, $\phi_b(x)$, find an analysis ϕ that minimizes a prescribed cost function:

$$J(\phi(x,t)) = \frac{1}{2}\sum_{n=1}^{N}\sum_{m=1}^{M} R_\phi(\phi(x_m,t_n) - \phi_{mn}^{\text{obs}})^2 + \frac{1}{2}\int_x B_\phi(\phi_0(x) - \phi_b(x))^2 dx, \tag{75}$$

while the constraint (73) is satisfied exactly or approximately. Here R_ϕ and B_ϕ are prespecified weighting coefficients.

One way to solve the strong constrained minimization problem (73) and (75) is to find the stationary points of the following Lagrange function:

$$\begin{aligned} L(\phi(x,t), \lambda(x,t)) &= \frac{1}{2}\sum_{n=1}^{N}\sum_{m=1}^{M} R_\phi(\phi(x_m,t_n) - \phi_{mn}^{\text{obs}})^2 + \frac{1}{2}\int_x B_\phi(\phi_0(x) - \phi_b(x))^2 dx \\ &+ \int_t\int_x \left\{\lambda\left(\frac{\partial\phi}{\partial t} + c\frac{\partial\phi}{\partial x} - F\right)\right\} dxdt, \end{aligned} \tag{76}$$

where λ is a Lagrange multiplier.

The first-order variation δL resulting from perturbation $\delta\phi$ and $\delta\lambda$ is equal to

$$\begin{aligned} \delta L &= \int_t\int_x \left\{R_\phi \sum_{n=1}^{N}\sum_{m=1}^{M}(\phi - \phi^{\text{obs}})\delta(x - x_m)\delta(t - t_n) - \frac{\partial\lambda}{\partial t} - c\frac{\partial\lambda}{\partial x}\right\}\delta\phi dxdt \\ &+ \int_t\int_x \left\{\delta\lambda\left(\frac{\partial\phi}{\partial t} + c\frac{\partial\phi}{\partial x} - F\right)\right\} dxdt \\ &+ \int_x B_\phi\left\{\phi_0(x) - \phi_b(x) - \lambda(x,0)\right\}\delta\phi_0(x)dx, \end{aligned} \tag{77}$$

where $\delta(x - x_m)$ and $\delta(t - t_n)$ are delta functions, and $\delta\phi(a,t) = \delta\phi(b,t) = 0$ (fixed boundary) and $\lambda(x,T) = 0$ are assumed based on variational calculus. The point $(\phi,\ \lambda)$ is an extremum of L_1 if the coefficients of $\delta\phi(x,t)$, $\delta\lambda(x,t)$ and $\delta\phi(x,0)$ in δL all vanish. These conditions require ϕ and λ to satisfy the following so-called Euler-Lagrange equation

$$-\frac{\partial\lambda}{\partial t} - c\frac{\partial\lambda}{\partial x} = -R_\phi\sum_{n=1}^{N}\sum_{m=1}^{M}(\phi - \phi^{\text{obs}})\delta(x - x_m)\delta(t - t_n), \tag{78}$$

$$\lambda(x,T) = 0, \tag{79}$$

$$\frac{\partial\phi}{\partial t} + c\frac{\partial\phi}{\partial x} = F, \tag{80}$$

$$\phi(x,0) = \phi_b(x,0) + B_b^{-1}\lambda(x,0). \tag{81}$$

Therefore, minimization of L_1 under the constraint of (73) is converted to solving the above Euler-Lagrange equations. The constraining relation are introduced with a Lagrange multiplier. The stationary point is found by solving a set of two equations.

Examining (78)-(81) we find that λ appears in the initial condition for ϕ and ϕ appears in the forecast model for λ. Numerical methods for approximately solving Euler-Lagrange equations can be found in Bennett (2002). However, deriving and solving a coupled system (78)-(81) could be very difficult, especially for nonlinear constraints such as the primitive equation NWP models. An alternative is to use an optimal control method which is described in the following subsection.

5.2. Optimal control method

In the variational analysis utilizing Lagrange multipliers, the variation is performed with respect to the full unknown field $\phi(x,t)$ on both a spatial and time domain Σ, instead of only those ϕ's along part of the boundary Γ of Σ. The entire field of ϕ on Σ serves as the control variables which are varied to minimize J. However, from the constraint of (73), a solution of the model variable $\phi(x,t)$ will be uniquely defined by the specification of ϕ at the boundary (initial time) $t = t_0$, $\phi_0(x)$. The constrained minimization problem (75) can then be stated equivalently in another form: find the initial condition of $\phi_0(x)$ such that the corresponding solution of (73) minimizes J defined in (75). The problem thus becomes an unconstrained minimization problem since no particular condition is imposed on the control variable $\phi_0(x)$. Only the field of $\phi_0(x)$ is varied to minimize J and will serve as the control variable of an optimization process.

There are two direct consequences of such a transformation from a constrained minimization problem to an unconstrained minimization problem: (i) the dimension of the control variable is greatly reduced; and (ii) how J varies with respect to the initial condition must be determined. The reduction of the control variable is crucial given the large dimension of most NWP models which already taxes the capability of the largest available computers. Although standard differentiation techniques will not in general lead to a practically usable form for the gradient of J, ∇J, the introduction of the adjoint concept allows an accurate gradient vector to be calculated with a single backward time integration of the adjoint model (see section 5.3.3).

With a way to calculate the values of J and ∇J, a minimization algorithm is employed to compute a sequence of approximate solutions which are closer and closer to a local minimizer. Suppose we are concerned with the problem of calculating the smallest value of a given function:

$$\text{minimize} J(\mathbf{x}), \qquad \mathbf{x} = (x_1, x_2, \ldots, x_N)^T, \tag{82}$$

in the case that only $J(\mathbf{x})$ and its first-order derivatives

$$g_i(\mathbf{x}) = \frac{\partial J(\mathbf{x})}{\partial x_i}, i = 1, 2, \ldots, N, \tag{83}$$

are known. A very basic iterative process for finding the minimizer of the problem (82) can be written as (Nash and Sofer 1996)

$$\mathbf{x}_{k+1} = \mathbf{x}_k + \alpha_k \mathbf{d}_k, \quad k = 0, 1, \ldots, \tag{84}$$

where k is the iteration number and the variables $\alpha_k (> 0)$ and $\mathbf{d}_k$ are called step size and search direction, respectively. The search direction $\mathbf{d}_k$ is a descent direction, that is,

$$\mathbf{d}_k^T \nabla J(\mathbf{x}_k) < 0. \tag{85}$$

The inequality equation (85) implies that the angle between the search direction $\mathbf{d}_k$ and the gradient $\nabla J(\mathbf{x}_k)$ (along which J increases most rapidly) at the point $\mathbf{x}_k$ is greater than 90°. The step size α_k is the minimizer of the following problem

$$\text{minimize}_\alpha J(\mathbf{x}_k + \alpha \mathbf{d}_k). \tag{86}$$

In other words, from the point $\mathbf{x}_k$ we search for an approximate minimum point along the direction $\mathbf{d}_k$. This point is taken to be $\mathbf{x}_{k+1}$ for which

$$J(\mathbf{x}_{k+1}) \equiv J(\mathbf{x}_k + \alpha_k \mathbf{d}) < J(\mathbf{x}_k). \tag{87}$$

Equations (85) and (87) are minimum requirements for convergence. Additional requirements on $\{\mathbf{d}_k\}$ and $\{\alpha_k\}$ are needed to guarantee convergence of the solutions $\{\mathbf{x}_k, k = 0, 1, \ldots\}$ (see section 5.4).

Equation (84) summarizes an underlying structure for almost all the descent algorithms of nonlinear programming. One starts at an initial guess point, determines, according to a fixed rule, a direction of movement (e.g., the descent direction $\mathbf{d}_k$), and then moves in that direction a distance to reach a relative minimum of J on that line (completed by a line search procedure which determines the step size α_k). At the new point, a new search direction and a new step size are determined and the process is repeated until some prespecified convergence is obtained. The primary differences between various optimization algorithms rest with (i) the rule by which successive directions of movement are selected and (ii) how the movement to the minimum point along the search direction was done. We observe that higher-dimensional problems (5) can be solved by executing a sequence of successive line searches minimizing a set of single-variable problems (86).

5.3. Numerical models and adjoint concepts

One major purpose of atmospheric data assimilation is to provide initial conditions for NWP models. A better initial condition is expected to produce an improved forecast. In this section, a general brief description of an NWP model and its tangent linear and adjoint models are given, along with the concept of gradient of a forecast aspect and its calculation using adjoint techniques. The tangent linear and adjoint models are additional tools for understanding corresponding NWP model and model predictions.

5.3.1. Nonlinear model

NWP models become major tools for studying the atmosphere and ocean, both in its practical aspects in weather forecasting by using primitive equation models, and its theoretical aspects in understanding atmospheric dynamics using idealized models. A model is constructed on the basis of physical laws that govern the temporal evolution of the flow of interest. No matter how simple or complicated it is, an NWP model is a computer program that provides a temporal evolution of the atmospheric flow depicted by specified values of input parameters. The model state inputs include initial conditions, lateral boundary conditions (for limited area models only), and parameters that define the physical and numerical schemes of the model.

Any NWP model can be written as (neglecting the time discretization)

$$\frac{\partial \mathbf{x}(t)}{\partial t} = F(\mathbf{x}(t)), \tag{88}$$

where $\mathbf{x}(t)$ is an N-dimensional vector describing the atmospheric state at time t, and $F(\mathbf{x}(t))$ is a vector of nonlinear functions of the model state $\mathbf{x}$. F represents all the operations consisting of dynamical terms, physical parameterizations, and model numerical schemes such as diffusion and filtering. Given an initial condition (we assume no lateral boundaries) of the model state: $\mathbf{x}|_{t=t_0} = \mathbf{x}_0$, (88) determines a unique solution of the model state at any future time t ($t > t_0$). The solution of nonlinear model can also be written as

$$\mathbf{x}(t_r) = M(\mathbf{x}_0), \tag{89}$$

where $M(t_r, t_0)$ represents the operations performed in the nonlinear model to obtain the model forecast at time t_r from an initial condition at time t_0, $\mathbf{x}_0$.

In many situations, solutions of a problem requires an explicit (quantitative) determination of the sensitivities of a certain forecast aspect (at a future time) to model inputs. In other words, information in model inputs must be inferred from future information of the atmosphere in model outputs. If we say that the nonlinear forecast model is a forward process, solving problems listed above requires a backward process. In order to solve these problems, the tangent linear model and the adjoint model are needed.

5.3.2. Tangent linear model

Assume that $\mathbf{x}(t)$ is a solution of the nonlinear model (88) corresponding to the initial condition $\mathbf{x}_0$. A perturbation $\delta\mathbf{x}_0$ added to the initial condition $\mathbf{x}_0$ will result in a different model solution $\mathbf{x}^{ptb}(t)$. Since both $\mathbf{x}(t)$ and $\mathbf{x}^{ptb}(t)$ satisfy (88), the differences between these two forecasts, $\mathbf{x}^{ptb}(t) - \mathbf{x}(t)$, satisfy

$$\frac{\partial}{\partial t}\left(\mathbf{x}^{ptb}(t) - \mathbf{x}(t)\right) = \frac{\partial F(\mathbf{x}(t))}{\partial \mathbf{x}}\left(\mathbf{x}^{ptb}(t) - \mathbf{x}(t)\right) + \underbrace{O\left(\|\mathbf{x}^{ptb}(t) - \mathbf{x}(t)\|^2\right)}_{\text{second and higher order terms}}, \tag{90}$$

where the *Jacobian operator* $\partial F(\mathbf{x}(t))/\partial \mathbf{x} = \{F_{ij}\}$ is an $N \times N$ matrix made up of the derivatives of F taken at the model state $\mathbf{x}(t)$: $F_{ij} = \partial F_i/\partial x_j$.

The first-order variation of the model forecasts resulted from the perturbation $\delta\mathbf{x}_0$ in the initial condition $\mathbf{x}_0$, $\delta\mathbf{x}$, thus satisfy the following so-called *tangent linear model*:

$$\frac{\partial \delta\mathbf{x}(t)}{\partial t} = \frac{\partial F(\mathbf{x}(t))}{\partial \mathbf{x}}\delta\mathbf{x}(t), \quad \delta\mathbf{x}|_{t=t_0} = \delta\mathbf{x}_0. \tag{91}$$

The solution of the linear model (91) can be written as

$$\delta\mathbf{x}(t) = \mathbf{M}(t, t_0)\delta\mathbf{x}_0, \tag{92}$$

where $\mathbf{M}(t, t_0)$ is called the *resolvent* between times t_0 and t ($t > t_0$). (92) is equivalent to (91) and is also called the tangent linear model of (88).

5.3.3. Adjoint model

There are two definitions of adjoint in linear algebra: the adjoint of the linear transform and the adjoint of the linear differential equation (Nering 1970). Corresponding to the two forms of the tangent linear model, (91) and (92), the adjoint model can be defined as either the adjoint of the linear differential equation (91)

$$-\frac{\partial \hat{\mathbf{x}}(t)}{\partial t} = \left(\frac{\partial F(\mathbf{x}(t))}{\partial \mathbf{x}}\right)^T \hat{\mathbf{x}}(t), \tag{93}$$

or the transpose of the tangent linear operator (resolvent) $\mathbf{M}(t, t_0)$ in (92):

$$\hat{\mathbf{x}}_0 = \mathbf{M}^*(t, t_0)\hat{\mathbf{x}}(t), \tag{94}$$

where $\hat{\mathbf{x}}(t)$ is the adjoint variable[3].

Accordingly, there are two ways of obtaining a discretized adjoint model for a given forward nonlinear model. One way is to obtain a computer program of the adjoint model with little concern for what the analytic form of the adjoint model equations looks like, that is to develop (94). The other is to derive an analytic form of adjoint model equations (93) and then to discretize and turn it into a computer program of the adjoint model. Here, we adopt the names of adjoint of finite-difference (AFD) for the first type and the finite-difference of adjoint (FDA) for the second, as was used by Sirkes and Tziperman (1997).

The AFD method starts with a discretized numerical model which is a sequence of computer codes solving equations (88). It consists of two procedures: (i) linearizing the discretized nonlinear model to obtain the discretized tangent linear model and (ii) transposing the discretized tangent linear model to obtain the discretized adjoint model. Linearization of the forward nonlinear model is carried out with respect to the nonlinear model trajectory and is conducted at the coding level. The same discretization schemes are automatically kept in the tangent linear model as in the original nonlinear model. The resulting sequence of computer codes of the tangent linear model will produce future perturbations for any given initial perturbation, i.e., it completes the operation expressed by (92).

Once the tangent linear model is developed, the adjoint model development follows. However, it is worth noting that although we have a tangent linear model, we don't have the matrix $\mathbf{M}$ itself. Theoretically speaking, a single integration of the tangent linear model with an initial perturbation of which the ith component equals a unit value and the rest of the components equal zero, will produce the ith column of $\mathbf{M}$. However, the dimension of the model variables, L, is usually too large (more than 10^6) to allow integrating the tangent linear model L times and to store an $L \times L$ matrix on the best computers available for use today.

An alternative for the adjoint model development is not to generate the matrix $\mathbf{M}^T$ itself, but to obtain a sequence of computer codes which produce the results of the

[3]The "initial" condition for the adjoint variable is not specified yet. It will be determined later when we have a purpose for integrating the adjoint model.

matrix $\mathbf{M}^T$ multiplied by a vector $\hat{\mathbf{x}}(t)$. This can be done by taking the resolvent of the tangent linear model, $\mathbf{M}(t, t_0)$, as a product of multiple matrices:

$$\mathbf{M}(t, t_0) = \mathbf{M}_N \cdots \mathbf{M}_1, \tag{95}$$

where $\mathbf{M}_n$ $(n = 1, \ldots, N)$ can be a single DO loop, or a subroutine, or a combination of both. Corresponding to each $\mathbf{M}_n$, the computer codes which realize the operation of $\mathbf{M}_n^T$ must be developed. Then, all the operations $\mathbf{M}_n$, $n = 1, \ldots, N$ must be linked together to obtain the sequence of computer codes which consist of the resolvent of the adjoint model (94):

$$\mathbf{M}^T(t, t_0) = \mathbf{M}_1^T \cdots \mathbf{M}_N^T. \tag{96}$$

We notice that the order of adjoint operations is reversed with respect to operations in the nonlinear and tangent linear models. Since the coefficients in each pair of operators ($\mathbf{M}_n$, $\mathbf{M}_n^T$) are the same functions of nonlinear model states (called basic states), the adjoint coding has to deal with two reverse orders: the basic state goes forward and the adjoint variable is calculated backward. In other words, the adjoint model operator (96) can be written more explicitly as

$$\mathbf{M}^T(t, t_0) = \mathbf{M}_1^T \cdots \underbrace{\mathbf{M}_{N-1}^T M_{N-2}}_{\text{stage 2}} \underbrace{\mathbf{M}_N^T M_{N-1}}_{\text{stage 1}}. \tag{97}$$

We notice that operations in M_{N-2} are included in M_{N-1} and are therefore conducted twice in (97). If all the input basic state variables for M_{N-2} are saved during stage 1, the calculations in the operator M_{N-2} do not need to be repeated for the adjoint code of $\mathbf{M}_{N-1}^T$. Therefore, depending on the choice of saving computer time or reducing memory usage, an adjoint model can be made faster if more memory is used to avoid re-calculations of basic state variables. Usually, the CPU time for a tangent linear model integration is slightly less than 2 times that of the nonlinear model integration. It is ideal to have an adjoint model integration which is about 3-6 times more expensive than the nonlinear model.

Therefore, in the entire procedure of tangent linear and adjoint model development using the AFD method, we never store the full matrices $\mathbf{M}$ and $\mathbf{M}^T$. We are only concerned with the following task: Given an input vector $\delta\mathbf{x}$ to $\mathbf{M}$ or an input vector $\hat{\mathbf{y}}$ to $\mathbf{M}^T$, we will be able to obtain an output vector of $\mathbf{M}\delta\mathbf{x}$ or $\mathbf{M}^T\hat{\mathbf{y}}$.

The FDA method, on the other hand, derives an analytic adjoint equation for a given set of forward model equations using the Lagrangian multiplier method (see section 5.1), followed by a discretization procedure. To illustrate this, let's consider a cost function

$$J(\mathbf{x}(t)) = \sum_{n=0}^{N} J_n(\mathbf{x}(t_n)), \tag{98}$$

where J_n is a functional of the model state $\mathbf{x}(t_n)$).

Minimizing J under the constraint (88) is equivalent to finding the stationary points of the following Lagrange function:

$$L(\mathbf{x}(t), \boldsymbol{\lambda}(t)) = J + \int_t < \boldsymbol{\lambda}, \frac{\partial \mathbf{x}}{\partial t} - F(\mathbf{x}) > dt, \tag{99}$$

where $\boldsymbol{\lambda}$ is a Lagrange multiplier. Finding the stationary points of the Lagrange function is equivalent to the determination of $\mathbf{x}$ and $\boldsymbol{\lambda}$ subject to the condition that the gradient of the Lagrange function vanishes. It results in the following equations:

$$\frac{\partial L(\mathbf{x}, \boldsymbol{\lambda})}{\partial \boldsymbol{\lambda}} = 0, \quad \frac{\partial L(\mathbf{x}, \boldsymbol{\lambda})}{\partial \mathbf{x}} = 0. \tag{100}$$

The first equation in (100) recovers the original nonlinear model equation (88). As shown below, the second equation (100) results in the adjoint differential equation (93).

Substituting (98) and (99) into the second equation in (100) we obtain

$$\int_t \sum_{n=0}^{N} \left(\nabla_{\mathbf{x}(t_n)} J_n\right) \delta(t - t_n) dt + \frac{\partial}{\partial \mathbf{x}} \int_t \left\{ < \boldsymbol{\lambda}, \frac{\partial \mathbf{x}}{\partial t} - F(\mathbf{x}) > \right\} dt = 0. \tag{101}$$

where $\delta(t - t_n)$ is a delta function. Since

$$< \boldsymbol{\lambda}, \frac{\partial \mathbf{x}}{\partial t} >= \frac{\partial}{\partial t} < \boldsymbol{\lambda}, \mathbf{x} > - < \frac{\partial \boldsymbol{\lambda}}{\partial t}, \mathbf{x} >, \tag{102}$$

and assuming $\boldsymbol{\lambda}(t_R) = 0$, we obtain from (101) the following

$$\int_t \sum_{n=0}^{N} \left(\nabla_{\mathbf{x}(t_n)} J_n\right) \delta(t - t_n) dt - \int_t \left\{ \frac{\partial \boldsymbol{\lambda}}{\partial t} + \left(\frac{\partial F}{\partial \mathbf{x}}\right)^T \boldsymbol{\lambda} \right\} dt = 0. \tag{103}$$

The above equation is valid for any length of assimilation window $[t_0, t_R]$. Thus the integrated function should be zero, which results in the adjoint equation:

$$-\frac{\partial \boldsymbol{\lambda}}{\partial t} = \left(\frac{\partial F}{\partial \mathbf{x}}\right)^T \boldsymbol{\lambda} - \underbrace{\sum_{n=0}^{N} \left(\nabla_{\mathbf{x}(t_n)} J_n\right) \delta(t - t_n)}_{\text{forcing term}}. \tag{104}$$

Eliminating the second term on the right-hand side of (104), which is a forcing term relating to a specific response function J, results in the same equation as (93) — the adjoint equation. Therefore, the Lagrange multiplier method can be used to derive an analytic adjoint differential equation, and the Lagrange multipliers are equivalent to the adjoint variables.

The nonlinear model makes a prediction from a given initial condition. The tangent linear model predicts a future perturbation from a known initial perturbation. What does the adjoint model do? A particular application of the adjoint model is to calculate the gradient of a scalar function of the nonlinear model forecast at a future time with respect to the initial model state, which is required by the optimal control method (see section 5.2). Let's first introduce the gradient definition. Consider a cost functional defined as in (98). We can view it as a function of the initial condition (a vector of control variables) $\mathbf{x}_0$, i.e.,

$$J(\mathbf{x}_0) = \sum_{n=0}^{N} J_n(\mathbf{x}(t_n)). \tag{105}$$

The gradient $\nabla_{\mathbf{x}_0} J$ is defined as a vector in the same space $\mathcal{R}$ as $\mathbf{x}_0$, and is such that the first-order variation δJ resulting from any perturbation $\delta \mathbf{x}_0$ to $\mathbf{x}_0$ is equal to the inner product

$$\delta J \equiv < \nabla_{\mathbf{x}_0} J, \delta \mathbf{x}_0 > . \tag{106}$$

Based on the observation that if we set all the components of $\delta \mathbf{x}_0$ to zero except the ith component of $\delta \mathbf{x}_0$ as $\delta x_0^{(i)}$, we can derive the ith component of the gradient $\nabla_{\mathbf{x}_0} J$ through the relation

$$(\nabla_{\mathbf{x}_0} J)_i = \frac{\delta J_i}{\delta x_0^{(i)}}, \tag{107}$$

where δJ_i is the difference between the value of $J(\mathbf{x}_0 + \delta \mathbf{x}_0^{(i)})$ and $J(\mathbf{x}_0)$, where $\delta \mathbf{x}_0^{(i)} = (0, \ldots, \delta x_0^{(i)}, 0, \ldots)^T$. One may think of using the forward finite-difference method to get an estimate for the gradient. However, this may turn out to be too expensive for large-dimensional problems since the forward model has to be integrated as many times as the dimension of the control variable.

According to the techniques of *optimal control* (Le Dimet and Talagrand 1986), it is shown below that the gradient of J with respect to the initial condition can be explicitly computed at a cost comparable to the nonlinear model integration. The first-order variation δJ resulting from a perturbation $\delta \mathbf{x}_0$ in the control variable vector $\mathbf{x}_0$ can now be derived as follows:

$$\begin{aligned} \delta J &= \sum_{n=0}^{N} < \nabla_{\mathbf{x}(t_n)} J_n, \delta \mathbf{x}(t_n) >= \sum_{n=0}^{N} < \nabla_{\mathbf{x}(t_n)} J_n, \mathbf{M}(t_n, t_0) \delta \mathbf{x}_0 > \\ &= < \sum_{n=0}^{N} \mathbf{M}^T(t_n, t_0) \nabla_{\mathbf{x}(t_n)} J_n, \delta \mathbf{x}_0 > . \end{aligned} \tag{108}$$

Comparing (108) with (106) we obtain

$$< \nabla_{\mathbf{x}_0} J, \delta \mathbf{x}_0 >=< \sum_{n=0}^{N} \mathbf{M}^T(t_n, t_0) \nabla_{\mathbf{x}(t_n)} J_n, \delta \mathbf{x}_0 > . \tag{109}$$

Equation (109) is valid for any $\delta \mathbf{x}_0$, which leads to the following expression for the gradient calculation:

$$\nabla_{\mathbf{x}_0} J = \sum_{n=0}^{N} \mathbf{M}^T(t_n, t_0) \nabla_{\mathbf{x}(t_n)} J_n, \tag{110}$$

i.e., the gradient of J (defined by the forecast) with respect to the initial condition $\mathbf{x}_0$ can be exactly obtained by (110), which consists of the following operations:

1. Integrate the nonlinear model forward from t_0 to t_N, with $\mathbf{x}_0$ as the initial condition of the model state at t_0;

2. Store the model forecast at every time step, which is used to make up the coefficients of the adjoint operator $\mathbf{M}^T$;

3. Calculate and store the forcing term $\nabla_{\mathbf{x}(t_n)} J_n$ $(n = 0, 1, \ldots, N)$ which will be taken as input to the adjoint model integration;

4. Integrate the adjoint model backward in time from t_N to t_0, with $\nabla_{\mathbf{x}(t_N)} J_N$ as the value of the adjoint variable at the time t_N; and

5. Add the saved forcing term $\nabla_{\mathbf{x}(t_n)} J_n$ to the currently computed adjoint model variable at time t_n $(n = N-1, \ldots, 0)$.

The final result of the adjoint model integration, i.e., the value of the adjoint variable at time t_0, is the gradient $\nabla_{\mathbf{x}_0} J$. This describes the entire procedure of obtaining the gradient of the cost function (98) by using the AFD method, i.e.,

$$\hat{\mathbf{x}}_0^{\mathrm{AFD}} = \nabla_{\mathbf{x}_0} J. \tag{111}$$

The same gradient can be obtained using the FDA method by integrating (104) which gives

$$\boldsymbol{\lambda}_0^{\mathrm{FDA}} = \nabla_{\mathbf{x}_0} J. \tag{112}$$

Although the AFD and FDA model integrations produce the same result at t_0, the time evolutions of the two adjoint solutions are different:

$$\hat{\mathbf{x}}^{\mathrm{AFD}}(t) \neq \boldsymbol{\lambda}^{\mathrm{FDA}}(t) \qquad \text{when } t \neq t_0. \tag{113}$$

For $\boldsymbol{\lambda}^{\mathrm{FDA}}(t)$, we have

$$\boldsymbol{\lambda}^{\mathrm{FDA}}(t) = \nabla_{\mathbf{x}(t)} J \qquad \text{when } \; t_0 \leq t \leq t_N. \tag{114}$$

(114) is a useful feature for any adjoint application in which the gradient of J with respect to the entire model trajectory is required (e.g., continuous sensitivity study). However, the two-time-step computational mode that exists in forward finite-difference models which use a leapfrog time integration scheme makes the time evolution of the AFD adjoint variables to be characterized by a non-physical oscillation with finite-amplitude (Sirkes and Tziperman 1997). As a result,

$$\hat{\mathbf{x}}^{\mathrm{AFD}}(t) \neq \nabla_{\mathbf{x}(t)} J \qquad \text{when } \; t_0 < t < t_N. \tag{115}$$

Realizing that differences in the AFD and FDA are due to the differences in the way the backward time integration of the adjoint model is carried out, Zou et al. (2001) proposed simple modifications that eliminate the computational mode from the AFD model solution. Therefore, an accurate time evolution of the gradient over the entire time period can also be obtained by using the AFD method.

The capability that the adjoint model offers us for calculating the exact gradient of any forecast aspect allows many NWP questions mentioned at the end of section 5.3.1 to be sought quantitatively. For atmospheric data assimilation, it has made four-dimensional variational data assimilation (4D-Var, see section 5.6) become a reality for large-scale models.

5.4. Selected introductory minimization concepts

We are interested in finding a *local minimizer* of $J(\mathbf{x})$ which is a point $\mathbf{x}^*$ that satisfies the condition

$$J(\mathbf{x}^*) \leq J(\mathbf{x}) \quad \text{for all } \mathbf{x} \text{ such that } \|\mathbf{x} - \mathbf{x}^*\| < \epsilon, \tag{116}$$

where ϵ is some positive number. When the cost function J is convex, a local minimizer is a global minimizer.

The minimization of a given cost function described in section 5.2 involves an iterative procedure in which a search direction and a step size need to be determined. These two procedures are briefly described in the following.

5.4.1. Line search

The process of (approximately) determining the minimum point on a given line is called a *line search*. This process is actually accomplished by searching, in an intelligent manner, along the line for the minimum point. It is a procedure for solving a single-scalar-variable minimization problem. The line search techniques form the backbone of nonlinear programming algorithms.

During minimization, some *nontrivial* reduction in the function value (J) must be obtained in each iteration. Nontrivial is measured in terms of the Taylor series. A linear approximation to $J(\mathbf{x}_k + \alpha\mathbf{d}_k)$ is obtained from

$$J(\mathbf{x}_k + \alpha_k\mathbf{d}_k) = J(\mathbf{x}_k) + \alpha_k\mathbf{g}_k^T\mathbf{d}_k, \tag{117}$$

where $\mathbf{g}_k = \nabla J|_{\mathbf{x}=\mathbf{x}_k}$ and $\mathbf{d}_k$ is a specified search direction vector. In the line search the step length α_k is required to produce a decrease in the function value that is at least some fraction $\beta' \in (0,1)$ of the decrease predicted by the linear approximation, i.e.,

$$J(\mathbf{x}_k + \alpha_k\mathbf{d}_k) \leq J(\mathbf{x}_k) + \beta'\alpha_k\mathbf{g}_k^T\mathbf{d}_k. \tag{118}$$

Equation (118) is called the Armijo condition (Nash and Sofer 1996). If α is small, the linear approximation will be good, and the Armijo condition will be satisfied. If α is large, the decrease predicted by the linear approximation may differ greatly from the actual decrease in J. The Armijo condition prevents α from being "too large".

In addition to requiring a decrease in the function value, a decrease in the gradient value is guaranteed during a linear search. Let

$$F(\alpha) \equiv J(\mathbf{x}_k + \alpha\mathbf{d}_k). \tag{119}$$

If α_k approximately minimizes F, then

$$F'(\alpha_k) \approx 0. \tag{120}$$

In the linear search, the step size α_k is required to produce a decrease in the gradient value (Nash and Sofer 1996; Nocedal and Wright 1999), i.e.,

$$|F'(\alpha_k)| \leq \beta|F'(0)|, \quad \beta \in [0,1), \tag{121}$$

which can be written as

$$\left| \frac{\mathbf{g}_{k+1}^T \mathbf{d}_k}{\mathbf{g}_k^T \mathbf{d}_k} \right| \leq \beta. \tag{122}$$

Equation (122) is called the Wolfe condition.

It is usually too expensive to solve the minimization problem along the search direction *exactly*, so in practice an *approximate* minimizer is accepted instead. The minimum point of $F(\alpha)$ is to be determined approximately by calculating the values of F and F' at a certain number of points and check if the two conditions on the step length α_k, (118) and (122), are satisfied to guarantee convergence in searching for an approximate minimum along a given line. Keep in mind that each calculation of F (and F') is *costly* since each calculation of F involves a forward model integration and each calculation of F' requires a backward adjoint model integration.

There are several line search methods. One of those is the *bracketing method.* The bracketing method searches for an interval $[\underline{\alpha}, \overline{\alpha}]$ with

$$F'(\underline{\alpha}) < 0, \quad F'(\overline{\alpha}) > 0. \tag{123}$$

At some point in the interval there must be an α satisfying $F'(\alpha) = 0$. This α, or an approximation to it, will be chosen as α_k. In other words, the interval $[\underline{\alpha}, \overline{\alpha}]$ brackets a minimizer of $F(\alpha)$.

Therefore, the bracketing method consists of two steps: (i) finding the interval $[\underline{\alpha}, \overline{\alpha}]$ for which (123) holds true; and (ii) finding the step size within that interval that is a minimizer of $F(\alpha)$ and for which both the Armijo and Wolfe conditions are satisfied.

Since $F'(0) = \mathbf{g}_k^T \mathbf{d}_k < 0$, $\underline{\alpha} = 0$ provides an initial lower bound on the step size interval we are looking for. To obtain an upper bound $\overline{\alpha}$, an increasing sequence of values of α are examined until one is found that satisfies $F(\overline{\alpha}) > 0$.

Once an interval $[\underline{\alpha}, \overline{\alpha}]$ is chosen, a quadratic or cubic polynomial approximation $P(\alpha)$ to the function $F(\alpha)$ can be applied which meets the following conditions:

$$P(\underline{\alpha}) = F(\underline{\alpha}), \; P'(\underline{\alpha}) = F'(\underline{\alpha}), \; P(\overline{\alpha}) = F(\overline{\alpha}), \; P'(\overline{\alpha}) = F'(\overline{\alpha}). \tag{124}$$

The quadratic or cubic polynomial must have a local minimizer $\hat{\alpha} \in (\underline{\alpha}, \overline{\alpha})$. For instance, if a quadratic polynomial is used, it is possible to fit the quadratic

$$P(\alpha) = F(\underline{\alpha}) + F'(\underline{\alpha})(\alpha - \underline{\alpha}) + \frac{F'(\overline{\alpha}) - F'(\underline{\alpha})}{\overline{\alpha} - \underline{\alpha}} \frac{(\alpha - \underline{\alpha})^2}{2}, \tag{125}$$

which has the same corresponding values:

$$P(\underline{\alpha}) = F(\underline{\alpha}), P'(\underline{\alpha}) = F'(\underline{\alpha}), P'(\overline{\alpha}) = F'(\overline{\alpha}). \tag{126}$$

We can then calculate an estimate $\hat{\alpha}$ of the minimum point of F by finding the minimum point of $P(\alpha)$, which can be obtained by setting

$$0 = P'(\alpha) = F'(\underline{\alpha}) + \frac{F'(\overline{\alpha}) - F'(\underline{\alpha})}{\overline{\alpha} - \underline{\alpha}} (\hat{\alpha} - \underline{\alpha}). \tag{127}$$

Therefore,

$$\hat{\alpha} = \underline{\alpha} - F'(\underline{\alpha})\left[\frac{\overline{\alpha} - \underline{\alpha}}{F'(\overline{\alpha}) - F'(\underline{\alpha})}\right]. \tag{128}$$

If $\hat{\alpha}$ satisfies the Armijo and Wolfe conditions, then $\alpha_k = \hat{\alpha}$. The line search at the kth iteration finishes. Otherwise, one of $\underline{\alpha}$ or $\overline{\alpha}$ is replaced by $\hat{\alpha}$ depending on the sign of $F'(\hat{\alpha})$. A new cubic polynomial fit is carried out and the process is repeated. During the bracketing step, the Armijo and Wolfe conditions are always checked for any trial value of α. If an α is found to satisfy the Armijo and Wolfe conditions, the line search at the kth iteration is also determined with that trial value as the step length α_k. The minimization then moves to the next iteration.

5.4.2. Search direction

We will examine first the search directions used by two traditional optimization methods: the steepest descent method and Newton's method. We will then discuss how the search direction is obtained for minimization algorithms used in atmospheric data assimilation. The steepest descent method and Newton's method provide two benchmarks for the calculation of the search direction of other algorithms which are always expected to perform better than the steepest descent method and nearly as good as Newton's method.

The steepest descent method is one of the oldest and most widely known methods and provides a clear picture of how the iterative procedure of minimization is carried out. It is defined by the iterative algorithm

$$\mathbf{x}_{k+1} = \mathbf{x}_k - \alpha_k \mathbf{g}_k, \tag{129}$$

i.e., the search direction in the steepest descent method is defined as the negative of the gradient:

$$\mathbf{d}_k = -\mathbf{g}_k. \tag{130}$$

Obviously, this search direction is a descent direction since

$$\mathbf{d}_k^T \mathbf{g}_k = -(\mathbf{g}_k)^T \mathbf{g}_k < 0, \tag{131}$$

i.e., (85) is verified.

If the linear search is exact, the steepest descent method converges linearly. Although the costs of the steepest descent method per iteration are low, its linear convergence can be so slow that $\mathbf{x}_{k+1} - \mathbf{x}_k$ is below the precision of computer arithmetic and the method may fail. As a result, the overall costs of solving an optimization problem could be high.

Newton's method is defined by the following iterative algorithm

$$\mathbf{x}_{k+1} = \mathbf{x}_k - \nabla^{-2} J(\mathbf{x_k})\mathbf{g}_k \equiv \mathbf{x}_k + \mathbf{d}_k, \tag{132}$$

i.e., the search direction in Newton's method is defined by the negative of the inverse Hessian matrix multiplied by the gradient

$$\mathbf{d}_k = -\nabla^{-2} J(\mathbf{x_k})\mathbf{g}_k \tag{133}$$

The step size in Newton's method is always $\alpha_k = 1$.

The new estimate $\mathbf{x}_{k+1}$ in Newton's method is obtained as the minimum point of the following quadratic function approximating the cost function J locally at $\mathbf{x}_k$:

$$q(\mathbf{x}) = J(\mathbf{x}_k) + \mathbf{g}_k(\mathbf{x} - \mathbf{x}_k) + \frac{1}{2}(\mathbf{x} - \mathbf{x}_k)\nabla^2 J(\mathbf{x}_k)(\mathbf{x} - \mathbf{x}_k). \tag{134}$$

Setting $\partial q/\partial \mathbf{x} = 0$ we obtain (132).

The search direction in Newton's method is a descent method if

$$\mathbf{d}_k^T \mathbf{g}_k = -\left(\mathbf{g}_k\right)^T \nabla^{-2} J(\mathbf{x}_k)\mathbf{g}_k < 0, \tag{135}$$

which can be satisfied if $\nabla^2 J(\mathbf{x}_k)$ is positive definite. But the positive definiteness of the Hessian matrix is not the necessary condition for (135) to be true. Methods to guarantee that the search directions are descent directions can be found in Nash and Sofer (1996).

A principal advantage of Newton's method is that it usually converges quadratically. However, Newton's method is not used in atmospheric data assimilation because it requires second derivatives in the Hessian matrix to be evaluated and the Hessian matrix to be inverted, which is too computationally expensive.

Therefore, the steepest descent method has a slow convergence, and Newton's method is computationally too expensive. Hence, the minimization software which is employed in atmospheric data assimilation is a more practical method lying somewhere between the steepest descent and Newton's method. These methods include the quasi-Newton methods and the limited-memory quasi-Newton methods. The idea underlying these methods is to replace the inverse Hessian, $\nabla^{-2} J(\mathbf{x}_k)$, by an approximate matrix $\mathbf{H}_k$ for the search direction:

$$\mathbf{d}_k = -\mathbf{H}_k \mathbf{g}_k. \tag{136}$$

The approximations of the inverse Hessian are built up from gradient information obtained at previous points. Assuming that J has continuous second partial derivatives, we have

$$\mathbf{g}_{k+1} - \mathbf{g}_k = \mathbf{F}(\mathbf{x}_{k+1})(\mathbf{x}_{k+1} - \mathbf{x}_k), \tag{137}$$

where $\mathbf{F}(\mathbf{x}_k) = \nabla^2 J(\mathbf{x}_k)$ is the Hessian matrix. Equation (137) can be written in a more compact form as

$$\mathbf{q}_k = \mathbf{F}(\mathbf{x}_{k+1})\mathbf{p}_k, \tag{138}$$

where $\mathbf{q}_k = \mathbf{g}_{k+1} - \mathbf{g}_k$ and $\mathbf{p}_k = \mathbf{x}_{k+1} - \mathbf{x}_k$.

From (138) we obtain the condition used to define the quasi-Newton approximation:

$$\mathbf{H}_{k+1}\mathbf{q}_k = \mathbf{p}_k, \tag{139}$$

where $\mathbf{H}_{k+1}$ denotes the $(k+1)$th approximation to the inverse Hessian $\mathbf{F}^{-1}$.

Now how is the sequence of $\mathbf{H}_k$ generated? The fundamental idea behind the quasi-Newton methods is to construct an approximation of the inverse Hessian, with the current approximation $\mathbf{H}_k$ being updated to form the next approximation of the inverse

Hessian $\mathbf{H}_{k+1}$. Ideally, the approximations converges to the inverse Hessian at the solution point.

A simple and efficient rank two correction can be applied to update $\mathbf{H}_k$:

$$\mathbf{H}_{k+1} = \mathbf{H}_k + a\mathbf{z}_1\mathbf{z}_1^T + b\mathbf{z}_2\mathbf{z}_2^T, \tag{140}$$

where the two constant parameters a and b as well as the two vectors $\mathbf{z}_1$ and $\mathbf{z}_2$ are to be determined. Substituting (140) into (139), we obtain

$$\mathbf{H}_k\mathbf{q}_k + a\mathbf{z}_1\mathbf{z}_1^T\mathbf{q}_k + b\mathbf{z}_2\mathbf{z}_2^T\mathbf{q}_k = \mathbf{p}_k. \tag{141}$$

Since both $\mathbf{z}_1^T\mathbf{q}_k$ and $\mathbf{z}_2^T\mathbf{q}_k$ are scalar, we observe that $\mathbf{z}_1$ and $\mathbf{z}_2$ can be expanded by the vectors $\mathbf{p}_k$ and $\mathbf{H}_k\mathbf{q}_k$. In order for (141) (and therefore (139)) to be valid, a convenient choice for the vectors and scalars in the second and third terms in (141) would be (Fletcher 1987):

$$\mathbf{z}_1 = \mathbf{p}_k, \qquad \mathbf{z}_2 = \mathbf{H}_k\mathbf{q}_k, \tag{142}$$

$$a\mathbf{z}_1^T\mathbf{q}_k = 1, \qquad b\mathbf{z}_2^T\mathbf{q}_k = -1. \tag{143}$$

It follows from (143) that

$$a = \frac{1}{\mathbf{p}_k^T\mathbf{q}_k}, \qquad b = -\frac{1}{(\mathbf{H}_k\mathbf{q}_k)^T\mathbf{q}_k}. \tag{144}$$

Substituting (142) and (144) into (140) we obtain the DFP rank two update formula:

$$\mathbf{H}_{k+1} = \mathbf{H}_k + \frac{\mathbf{p}_k\mathbf{p}_k^T}{\mathbf{p}_k^T\mathbf{q}_k} - \frac{\mathbf{H}_k\mathbf{q}_k\mathbf{q}_k^T\mathbf{H}_k}{\mathbf{q}_k^T\mathbf{H}_k\mathbf{q}_k}. \tag{145}$$

It is considered the most effective rank two correction. The DFP formula ensures that if $\mathbf{H}_k$ is symmetric and positive definite, $\mathbf{H}_{k+1}$ is symmetric and positive definite. If J is a quadratic functional with constant Hessian $\mathbf{F}$, then $\mathbf{H}_N = \mathbf{F}^{-1}$ after the DFP update formula is carried out N steps.

The updating formulae for the inverse Hessian $\mathbf{F}^{-1}$ considered in the previous rank one or two corrections are based on (139) which is derived from (138). It is also possible to update approximations to the Hessian $\mathbf{F}$ itself. Denoting the kth approximation of $\mathbf{F}$ by $\mathbf{B}_k$, we seek the update of $\mathbf{B}_k$ which satisfies

$$\mathbf{q}_k = \mathbf{B}_{k+1}\mathbf{p}_k. \tag{146}$$

Equation (146) is similar to (139) except that $\mathbf{q}_k$ and $\mathbf{p}_k$ are interchanged and $\mathbf{H}$ is replaced by $\mathbf{B}$. This implies that any update formula for $\mathbf{H}$ derived to satisfy (139) can be transformed into a corresponding formula for $\mathbf{B}$. For example, from the DFP formula (145) we can write its corresponding complementary formula for updating $\mathbf{B}$:

$$\mathbf{B}_{k+1} = \mathbf{B}_k + \frac{\mathbf{q}_k\mathbf{q}_k^T}{\mathbf{q}_k^T\mathbf{p}_k} - \frac{\mathbf{B}_k\mathbf{p}_k\mathbf{p}_k^T\mathbf{B}_k}{\mathbf{p}_k^T\mathbf{B}_k\mathbf{p}_k}. \tag{147}$$

Taking the inverse of (147) using the Sherman-Morrison formula:

$$[\mathbf{A} + \mathbf{a}\mathbf{b}^T]^{-1} = \mathbf{A}^{-1} - \frac{\mathbf{A}^{-1}\mathbf{a}\mathbf{b}^T\mathbf{A}^{-1}}{1 + \mathbf{b}^T\mathbf{A}^{-1}\mathbf{a}}, \tag{148}$$

we obtain the following formula

$$\mathbf{H}_{k+1}^{\mathrm{BFGS}} = \mathbf{H}_k + \left(\frac{1 + \mathbf{q}_k^T\mathbf{H}_k\mathbf{q}_k}{\mathbf{q}_k^T\mathbf{p}_k}\right)\frac{\mathbf{p}_k\mathbf{p}_k^T}{\mathbf{p}_k^T\mathbf{q}_k} - \frac{\mathbf{p}_k\mathbf{q}_k^T\mathbf{H}_k + \mathbf{H}_k\mathbf{q}_k\mathbf{p}_k^T}{\mathbf{q}_k^T\mathbf{p}_k}, \tag{149}$$

which is called the BFGS formula following the initials of the names of the four people who developed it, Broyden, Fletcher, Goldfard, and Shannon (Nash and Sofer, 1996). Numerical experiments have repeatedly indicated that its performance is superior to that of the DFP formula (145) and it is therefore generally preferred.

Methods using the DFP or BFGS formula are called quasi-Newton methods. The new approximation $\mathbf{H}_{k+1}$ is obtained from $\mathbf{H}_k$ by using $O(n^2)$ arithmetic operations since the difference $\mathbf{H}_{k+1} - \mathbf{H}_k$ only involves products of vectors. The search direction can also be calculated by using $O(n^2)$ arithmetic operations. Compared with the usual cost of solving a system of linear equation $(O(n^3))$, the quasi-Newton method represents a significant saving in computational time. In addition, the approximation $\mathbf{H}_k$ can be found by using only first-derivative information. The method does not converge quadratically, but can converge superlinearly. However, quasi-Newton methods still require matrix storage. Modifications to the quasi-Newton methods that do not use matrix storage are the limited-memory quasi-Newton methods that follow BFGS (L-BFGS), which are used in atmospheric data assimilation. The L-BFGS method considers a simplification of the BFGS quasi-Newton method. It applies the BFGS update to a unit matrix $\mathbf{I}$ at the initial m steps $(k = 0, 1, \ldots, m)$, and rather than to $\mathbf{H}_k$, i.e.,

$$\mathbf{H}_{k+1}^{\mathrm{L-BFGS}} = \mathbf{I}_k + \left(\frac{1 + \mathbf{q}_k^T\mathbf{q}_k}{\mathbf{q}_k^T\mathbf{p}_k}\right)\frac{\mathbf{p}_k\mathbf{p}_k^T}{\mathbf{p}_k^T\mathbf{q}_k} - \frac{\mathbf{p}_k\mathbf{q}_k^T + \mathbf{q}_k\mathbf{p}_k^T}{\mathbf{q}_k^T\mathbf{p}_k} \equiv \mathbf{I} + f(\mathbf{p}_k, \mathbf{q}_k), \tag{150}$$

and then applies the BFGS update (149) to $\mathbf{H}_l$ for the following m steps to obtain $\mathbf{H}_{l+1}$, $l = k - m + 1, k - m + 2, \ldots, k$. From (149) and (150) we find that $\mathbf{H}_{l+1}$ becomes a function of $(\mathbf{p}_l, \mathbf{q}_l)$, $l = k - m, k - m + 1, \ldots, k$ without explicit reference to the matrix $\mathbf{H}_l$. Hence the memory required by the L-BFGS is $O(n)$, and is much less than that of a full matrix $(\mathbf{H}_k\ (O(n^2))$.

We can now write the L-BFGS minimization procedure as follows:

1. Start at an initial guess $\mathbf{x}_0$, compute the gradient $\mathbf{g}_0 = \nabla J(\mathbf{x}_0)$.

2. Set $\mathbf{H}_0 = \mathbf{I}$ and $k = 0$.

3. Set $\mathbf{d}_k = -\mathbf{H}_k\mathbf{g}_k$.

4. Carry out a line search to find the step size α_k:

$$J(\mathbf{x}_k + \alpha_k\mathbf{d}_k) = min_\alpha J(\mathbf{x}_k + \alpha\mathbf{d}_k) \tag{151}$$

5. Update the variable: $\mathbf{x}_{k+1} = \mathbf{x}_k + \alpha_k \mathbf{d}_k$.

6. Compute the new gradient value: $\mathbf{g}_{k+1} = \nabla J(\mathbf{x}_{k+1})$.

7. Check for convergence: If $\|\mathbf{g}_{k+1}\| \leq \epsilon \ \max\{1, \|\mathbf{x}_{k+1}\|\}$, stop, where $\epsilon = 10^{-5}$.

8. Otherwise, compute the new search direction $\mathbf{d}_{k+1} = -\mathbf{H}_{k+1}\mathbf{g}_{k+1}$, where

$$\mathbf{H}_{k+1} = \begin{cases} \mathbf{I} + \sum_{l=1}^{k} \left(\left(\frac{1+\mathbf{q}_l^T \mathbf{H}_l^0 \mathbf{q}_l}{\mathbf{q}_l^T \mathbf{p}_l} \right) \frac{\mathbf{p}_l \mathbf{p}_l^T}{\mathbf{p}_l^T \mathbf{q}_l} - \frac{\mathbf{p}_l \mathbf{q}_l^T \mathbf{H}_l^0 + \mathbf{H}_l^0 \mathbf{q}_l \mathbf{p}_l^T}{\mathbf{q}_l^T \mathbf{p}_l} \right), & \text{if } 0 < k \leq m; \\ \mathbf{I} + \sum_{l=k-m}^{k} \left(\left(\frac{1+\mathbf{q}_l^T \mathbf{H}_l^0 \mathbf{q}_l}{\mathbf{q}_l^T \mathbf{p}_l} \right) \frac{\mathbf{p}_l \mathbf{p}_l^T}{\mathbf{p}_l^T \mathbf{q}_l} - \frac{\mathbf{p}_l \mathbf{q}_l^T \mathbf{H}_l^0 + \mathbf{H}_l^0 \mathbf{q}_l \mathbf{p}_l^T}{\mathbf{q}_l^T \mathbf{p}_l} \right), & \text{otherwise.} \end{cases} \tag{152}$$

where m is a constant. Data assimilation experiments by Zou et al. (1993) show that the best value for m is between 5 to 11.

9. Set $k = k + 1$ and return to the line search step (4).

Besides having a good line search scheme and efficient search directions, the performance of the minimization algorithms depends also on the particular choice of variables $\mathbf{x}$ used to define the cost function J, $J(\mathbf{x})$. A new choice of variables — scaling — may substantially alter the convergence characteristics (Navon et al. 1992).

5.5. A brief description of 3D-Var/4D-Var problems

The variational approaches discussed in this section, up to now, mainly address issues related to finding the solution of a minimization problem under a strong constraint. Not much discussions have been provided on the definition of cost function itself, which is, if not more, as important as solving minimization problems. How is the cost function defined for atmospheric data assimilation? Factors that must be considered in the definition of the cost function include assimilating all direct and indirect observations, introducing background information, accounting for background and observational errors, incorporating dynamic and physical constraints, suppressing noises, etc.

A general form of the cost function that a three-dimensional or four-dimensional variational data assimilation (3D-Var/4D-Var) system minimizes can be written as:

$$J(\mathbf{x}_0) = \underbrace{\frac{1}{2}(\mathbf{x}_0 - \mathbf{x}_b)^T \mathbf{B}^{-1} (\mathbf{x}_0 - \mathbf{x}_b)}_{J_b} + \underbrace{\frac{1}{2}(\mathbf{y}^{\text{obs}} - H(\mathbf{x}_0))^T \mathbf{R}^{-1} (\mathbf{y}^{\text{obs}} - H(\mathbf{x}_0))}_{J_o}, \tag{153}$$

where $\mathbf{x}_0$ is the analysis vector on the forecast-model grid, $\mathbf{x}_b$ is the forecast background vector, $\mathbf{B}$ is the background error covariance matrix, $\mathbf{y}^{\text{obs}}$ is the vector of observations, $\mathbf{R}$ is the sum of the observation error covariance matrix $\mathbf{O}$ and the forward model error covariance matrix $\mathbf{F}$, and H in 3D-Var is the observation operator which transforms the model variables to the observational quantities. In 4D-Var, H transforms the model initial condition $\mathbf{x}_0$ into observed quantities. In other words, H in 4D-Var includes a model prediction from the initial condition to the observation time. We must keep in mind that H in (153) actually represents the operator $H(M(\mathbf{x}_0))$ for 4D-Var, where M is the nonlinear model operator. 4D-Var allows information in a time series of observations to be fully utilized. A third term J_c may be added to suppress additional noises that are not sufficiently filtered by the background error covariances.

Unlike OI in which data are incorporated locally, all observations are used simultaneously to perform the analysis globally in 3D-Var/4D-Var. Indirect observations, such as measurements of microwave radiance and bending angle from radio occultation techniques (see section 5), can be assimilated as long as they can be expressed as a function of the model variables. *A priori* retrieval, which is often under-determined and introduces ad hoc assumptions, is not required for the use of indirect observations in 3D-Var/4D-Var. Since the retrieval products may be contaminated by errors inherent in the retrieval procedure and the characteristics of these errors are difficult to quantify and often not available, the use of the "raw" form of data, instead of retrieval products, is always preferred.

As discussed in section 5.4, standard minimization software, such as the L-BFGS method that is used to find the minimum of (153), requires users to provide the values of the cost function J and its gradient ∇J at every iteration for any given values of the control variables $\mathbf{x}_0$. New adjoint operators, $\mathbf{H}^T$, for the calculation of the gradient of J:

$$\nabla J = \mathbf{B}^{-1}(\mathbf{x}_0 - \mathbf{x}_b) + \mathbf{H}^T\mathbf{R}^{-1}\left(H(\mathbf{x}_0) - \mathbf{y}^{\text{obs}}\right), \tag{154}$$

must be developed.

Therefore, defining and solving a 3D-Var/4D-Var problem requires not only a solid mathematical background of linear algebra, statistics, variational calculus, programming and optimization, but also meteorological knowledge of NWP models, dynamic and physical constraints of the atmospheric motion applicable at various scales, atmospheric observations, and estimates of the background and observation errors.

The cost function defined in (153) consists of two terms. The first term in (153), denoted by J_b, measures the misfit between the analysis and the background field. It contains all the available information prior to the analysis. The second term in (153), denoted by J_o, measures the distance between the analysis and the observations. For both the background field $\mathbf{x}_b$ and observations $\mathbf{d}^{\text{obs}}$, their bias errors are assumed to be removed and their covariances enter the formulation of J as inverse weighting matrices.

Large residuals occur if the model does not match the observations. It may imply either a need to construct a new and better model, or the existence of outliers in the observations. Erroneous data must be either removed from the assimilation scheme by quality control or assigned lower weights to deemphasizes their effect. If the cause of large residuals is due to the model, then the model needs to be fixed. For example, if observed microwave brightness temperatures are included in the assimilation, large residuals could occur when an NWP model produces an erroneous snow-ice prediction or a radiative transfer model does not separate snow ice and graupel. The brightness temperatures calculated from these erroneous models could be largely different from the observations, which leads to the large residuals. The models will have to be improved if brightness temperatures are to be kept in the assimilation. Therefore, data assimilation may help identifying problems in models.

5.6. 3D-Var/4D-Var as a general discrete inverse problem

Atmospheric data assimilation uses both partial or complete information contained in physical laws, measurements of the observable variables, and *a priori* background information to infer the values of model inputs, which are expected to improve the forward model prediction. Is it legitimate to ask why 3D-Var/4D-Var solves the problem defined by (153)? What are the implications behind the least-squares estimate defined by (153) from a statistical point of view? To answer this question, we must consider a general discrete inverse problem.

It was argued in Tarantola (1987) that the most general way of describing a state of information over a parameter set is by defining a probability density function over the corresponding parameter space. Assuming that measurements of the observable quantities (data), *a priori* information on model variables, and information on the physical system can all be described using probability density functions, a general inverse problem can then be defined as finding the probability density function of the *a posteriori* state of information which "combines" all this information under some simplifying assumptions. The purpose of data assimilation is to simply extract some features from this *a posteriori* density function such as the maximum likelihood estimator (the analysis) and the covariances (the analysis error covariances). We will briefly describe such a process in order to provide a statistical background for 3D-Var/4D-Var data assimilation approaches.

Let $p_o(\mathbf{y}^{\mathrm{obs}}, \mathbf{y})$ represent the probability density for the observed value $\mathbf{y}^{\mathrm{obs}}$, where $\mathbf{y}$ denotes the coordinates of a point in data space $\mathcal{R}_{ND}$, $p_m(\mathbf{y}^{\mathrm{model}}, \mathbf{x}_0)$ be the probability density describing model errors, where $\mathbf{y}^{\mathrm{model}} = H(\mathbf{x})$ and $\mathbf{x}_0$ are the atmospheric state variables in model space $\mathcal{R}_N$, and $p_b(\mathbf{x}_b, \mathbf{x}_0)$ be the probability density for the background value to be $\mathbf{x}_b$. Combination of these three states of information can be defined as their joint probability density. If these three types of information are assumed independent, the joint probability density is the product of these three density functions:

$$\sigma(\mathbf{y}, \mathbf{x}_0) = p_o(\mathbf{y}^{\mathrm{obs}}, \mathbf{y}) p_m(\mathbf{y}^{\mathrm{model}}, \mathbf{x}_0) p_b(\mathbf{x}_b, \mathbf{x}_0), \tag{155}$$

which will be called the probability density of the *a posteriori* state of information and is defined on both observation and model spaces $\mathcal{R}_{ND} \cup \mathcal{R}_N$.

The *a posteriori* information in the model space is the marginal density function

$$\sigma(\mathbf{x}_0) = \int_D d\mathbf{y} \sigma(\mathbf{y}, \mathbf{x}_0). \tag{156}$$

The function $\sigma(\mathbf{x}_0)$ is called the *a posteriori* probability density function of the state variable $\mathbf{x}_0$ and is the solution of the general discrete inverse problem. From (156), we may derive the maximum likelihood estimate which can be used as the new analysis of the inverse method. The analysis error covariances can also be derived from (156).

Let's assume that

1. Observations contain the observed output, $\mathbf{y}^{\mathrm{obs}}$, from a measuring instrument with Gaussian statistics:

$$p_o(\mathbf{y}^{\mathrm{obs}}, \mathbf{y}) = ((2\pi)^{ND} det\mathbf{O})^{-1/2} exp\left(-\frac{1}{2}(\mathbf{y} - \mathbf{y}^{\mathrm{obs}})^T \mathbf{O}^{-1}(\mathbf{y} - \mathbf{y}^{\mathrm{obs}}),\right) \tag{157}$$

where ND is the dimension of the data space D (number of data parameters) and $\mathbf{O}$ is the covariance operator describing experimental uncertainties.

2. The forward model errors are also Gaussian, i.e.,

$$p_m(\mathbf{y}^{model}, \mathbf{x}_0) = ((2\pi)^{ND} det\mathbf{F})^{-1/2} exp\left(-\frac{1}{2}(\mathbf{y} - H(\mathbf{x}_0))^T \mathbf{F}^{-1}(\mathbf{y} - H(\mathbf{x}_0))\right), \tag{158}$$

where $\mathbf{F}$ is the covariance operator describing forward model uncertainties.

3. The probability density representing the *a priori* background information is also Gaussian, i.e.,

$$p_b(\mathbf{x}_b, \mathbf{x}_0) = ((2\pi)^N det\mathbf{B})^{-1/2} exp\left(-\frac{1}{2}(\mathbf{x}_0 - \mathbf{x}_b)^T \mathbf{B}^{-1}(\mathbf{x}_0 - \mathbf{x}_b)\right), \tag{159}$$

where N is the dimension of the model space, $\mathbf{x}_b$ is the *a priori* background field, and $\mathbf{B}$ is the covariance operator describing the estimated uncertainties in $\mathbf{x}_b$.

Substituting (157)-(159) into (156) we obtain (see Tarantola 1987)

$$\frac{\sigma(\mathbf{x}_0)}{p_b(\mathbf{x}_b, \mathbf{x}_0)} = C \times \exp\left[-\frac{1}{2}\left(\mathbf{c} - \mathbf{b}^T \mathbf{A}^{-1} \mathbf{b}\right)\right], \tag{160}$$

where C is a constant and the following notations have been used:

$$\begin{aligned} \mathbf{R} &= \mathbf{O} + \mathbf{F}, &(161)\\ \mathbf{b}^T &= (\mathbf{y}^{\text{obs}})^T \mathbf{O}^{-1} + H^T(\mathbf{x}_0)\mathbf{F}^{-1}, &(162)\\ \mathbf{c} &= (\mathbf{y}^{\text{obs}})^T \mathbf{F}^{-1} \mathbf{y}^{\text{obs}} + H^T(\mathbf{x}_0)\mathbf{F}^{-1} H(\mathbf{x}_0). &(163) \end{aligned}$$

Substituting (161)-(163) into (160), the exponential expression in (160) can be written as

$$\begin{aligned} \mathbf{c} - \mathbf{b}^T\mathbf{R}^{-1}\mathbf{b} = {} & (\mathbf{y}^{\text{obs}})^T \left[\mathbf{O}^{-1} - \mathbf{O}^{-1}\left(\mathbf{O}^{-1} + \mathbf{F}^{-1}\right)^{-1}\mathbf{O}^{-1}\right]\mathbf{y}^{\text{obs}} \\ & + H^T(\mathbf{x}_0)\left[\mathbf{F}^{-1} - \mathbf{F}^{-1}\left(\mathbf{O}^{-1} + \mathbf{F}^{-1}\right)^{-1}\mathbf{F}^{-1}\right]H(\mathbf{x}_0) \\ & - 2H^T(\mathbf{x}_0)\mathbf{F}^{-1}\left(\mathbf{O}^{-1} + \mathbf{F}^{-1}\right)^{-1}\mathbf{O}^{-1}\mathbf{y}^{\text{obs}}. \end{aligned} \tag{164}$$

The above expressions can be simplified by using the following matrix identities:

$$\begin{aligned} \mathbf{O}^{-1}(\mathbf{O} + \mathbf{F}) &= \left(\mathbf{O}^{-1} + \mathbf{F}^{-1}\right)\mathbf{F}, &(165)\\ \left(\mathbf{O}^{-1} + \mathbf{F}^{-1}\right)^{-1}\mathbf{O}^{-1} &= \mathbf{F}(\mathbf{O} + \mathbf{F})^{-1}, &(166)\\ \mathbf{F}^{-1}\left(\mathbf{O}^{-1} + \mathbf{F}^{-1}\right)^{-1}\mathbf{O}^{-1} &= (\mathbf{O} + \mathbf{F})^{-1}, &(167)\\ \mathbf{O}^{-1} - \mathbf{O}^{-1}\left(\mathbf{O}^{-1} + \mathbf{F}^{-1}\right)^{-1}\mathbf{O}^{-1} &= (\mathbf{O} + \mathbf{F})^{-1}, &(168)\\ \mathbf{F}^{-1} - \mathbf{F}^{-1}\left(\mathbf{O}^{-1} + \mathbf{F}^{-1}\right)^{-1}\mathbf{F}^{-1} &= (\mathbf{O} + \mathbf{F})^{-1}. &(169) \end{aligned}$$

Substituting (166)-(169) into (164) we obtain

$$\mathbf{c} - \mathbf{b}^T\mathbf{R}^{-1}\mathbf{b} = \left(H(\mathbf{x}_0) - \mathbf{y}^{\text{obs}}\right)^T (\mathbf{O} + \mathbf{F})^{-1} \left(H(\mathbf{x}_0) - \mathbf{y}^{\text{obs}}\right). \tag{170}$$

Substituting (159) and (170) into (160) we obtain

$$\sigma(\mathbf{x}_0) = C \times \exp\left(-\frac{1}{2}\left((\mathbf{x}_0 - \mathbf{x}_b)^T\mathbf{B}^{-1}(\mathbf{x}_0 - \mathbf{x}_b) + (H(\mathbf{x}_0) - \mathbf{y}^{\text{obs}})^T\mathbf{R}^{-1}(H(\mathbf{x}_0) - \mathbf{y}^{\text{obs}})\right)\right). \tag{171}$$

The probability density function (163) is the solution of a general discrete inverse problem under the assumptions of (157)-(159).

Let's consider two special cases before we link (162) to the 3D-Var/4D-Var problem (153). The first case is that the equation solving the forward model is linear:

$$\mathbf{y}^{model} = \mathbf{H}\mathbf{x}_0. \tag{172}$$

In this case, the *a posteriori* probability density is *Gaussian*, and can be written as (in normalized form)

$$\sigma(\mathbf{x}_0) = C \times exp\left(-\frac{1}{2}(\mathbf{x}_0 - \overline{\mathbf{x}}_0)^T\mathbf{C}_a^{-1}(\mathbf{x}_0 - \overline{\mathbf{x}}_0)\right), \tag{173}$$

where $\overline{\mathbf{x}}_0$ and $\mathbf{C}_a$ are the mean and covariance operator of the *a posteriori* information. They can be calculated from the following formula

$$\begin{aligned} \overline{\mathbf{x}}_0 &= \mathbf{x}_b + \left(\mathbf{H}^T\mathbf{R}^{-1}\mathbf{H} + \mathbf{B}^{-1}\right)^{-1} \mathbf{H}^T\mathbf{R}^{-1}(\mathbf{y}^{\text{obs}} - \mathbf{H}\mathbf{x}_b) \\ &= \mathbf{x}_b + \mathbf{B}\mathbf{H}^T\left(\mathbf{H}\mathbf{B}\mathbf{H}^T + \mathbf{R}\right)^{-1}(\mathbf{y}^{\text{obs}} - \mathbf{H}\mathbf{x}_b), \end{aligned} \tag{174}$$

and

$$\mathbf{C}_a = \left(\mathbf{H}^T\mathbf{R}^{-1}\mathbf{H} + \mathbf{B}^{-1}\right)^{-1} = \mathbf{B} - \mathbf{B}\mathbf{H}^T\left(\mathbf{H}\mathbf{B}\mathbf{H}^T + \mathbf{R}\right)^{-1}\mathbf{H}\mathbf{B}. \tag{175}$$

The mean $\overline{\mathbf{x}}_0$, being the center of the Gaussian and the maximum likelihood point, is referred to as the solution of the inverse problem.

The second case is that $H(\mathbf{x}_0)$ is not linear, but can be linearized around the *a priori* background field $\mathbf{x}_b$, i.e.,

$$H(\mathbf{x}_0) \approx H(\mathbf{x}_b) + \mathbf{H}_b(\mathbf{x}_0 - \mathbf{x}_b), \qquad \mathbf{H}_b = \left.\frac{\partial H(\mathbf{x}_0)}{\partial \mathbf{x}_0}\right|_{\mathbf{x}_0 = \mathbf{x}_b}. \tag{176}$$

The *a posteriori* probability density is approximately Gaussian, with the mean and covariance operator being given by

$$\begin{aligned} \overline{\mathbf{x}}_0 &\approx \mathbf{x}_b + \left(\mathbf{H}_0^T\mathbf{R}^{-1}\mathbf{H}_b + \mathbf{B}^{-1}\right)^{-1} \mathbf{H}_b^T\mathbf{R}^{-1}(\mathbf{y}^{\text{obs}} - H(\mathbf{x}_b)) \\ &= \mathbf{x}_b + \mathbf{B}\mathbf{H}_b^T(\mathbf{H}_b\mathbf{B}\mathbf{H}_b^T + \mathbf{R})^{-1}(\mathbf{y}^{\text{obs}} - H(\mathbf{x}_b)), \end{aligned} \tag{177}$$

and

$$\mathbf{C}_a \approx \left(\mathbf{H}_b^T\mathbf{R}^{-1}\mathbf{H}_b + \mathbf{B}^{-1}\right)^{-1} = \mathbf{B} - \mathbf{B}\mathbf{H}_b^T(\mathbf{H}_b\mathbf{B}\mathbf{H}_b^T + \mathbf{R})^{-1}\mathbf{H}_b\mathbf{B}. \tag{178}$$

It is clear that if $H(\mathbf{x}_0)$ is not a linear function of $\mathbf{x}_0$, $\sigma(\mathbf{x}_0)$ is not Gaussian. The more nonlinear $H(\mathbf{x}_0)$ is, the further $\sigma(\mathbf{x}_0)$ is from a Gaussian function. For many indirect observations such as space-borne GPS bending angle measurements, the forward model $H(\mathbf{x}_0)$ is nonlinear. For a 4D-Var problem, $H(\mathbf{x}_0)$ includes the forecasting model and is therefore nonlinear even if the observation operator is linear. Eq. (171) cannot be further simplified, as has been done for the two special cases just considered. Obtaining an analytic solution of the mean and covariance operators of the *a posteriori* information is not straightforward.

One of the easiest ways to solve such a nonlinear problem is to obtain a central estimator of $\sigma(\mathbf{x}_0)$ — the maximum likelihood point $\mathbf{x}_0^*$:

$$\sigma(\mathbf{x}_0) \;\; \text{MAXIMUM for} \quad \mathbf{x}_0 = \mathbf{x}_0^*, \tag{179}$$

which is equivalent to

$$J(\mathbf{x}_0) \;\; \text{MINIMUM for} \quad \mathbf{x}_0 = \mathbf{x}_0^*, \tag{180}$$

where $J(\mathbf{x}_0)$ is the argument of the exponential of $\sigma(\mathbf{x}_0)$ in (171) and is exactly the 3D-Var/4D-Var problem introduced in (153). Therefore, 3D-Var/4D-Var is designed to find the maximum likelihood estimator of the *a posteriori* probability density function in model space which combines information from three sources: observations, physical models, and background fields. Errors from all these sources are assumed to be Gaussian.

If $H(\mathbf{x}_0)$ can be linearized around the true maximum likelihood point $\mathbf{x}_0^*$, linearization of $H(\mathbf{x}_0)$ around $\mathbf{x}_0^*$ provides an estimation of the *a posteriori* covariance operator:

$$\mathbf{A} \approx \left(\mathbf{H}_*^T\mathbf{O}^{-1}\mathbf{H}_* + \mathbf{B}^{-1}\right)^{-1} = \mathbf{B} - \mathbf{B}\mathbf{H}_*^T(\mathbf{H}_*\mathbf{B}\mathbf{H}_*^T + \mathbf{O})^{-1}\mathbf{H}_*\mathbf{B}, \tag{181}$$

where

$$\mathbf{H}_* = \left.\frac{\partial H(\mathbf{x}_0)}{\partial \mathbf{x}_0}\right|_{\mathbf{x}_0=\mathbf{x}_0^*}. \tag{182}$$

For uncorrelated errors,

$$R_{ij} = E_{oK_i}^2 \delta_{K_iK_j}, \qquad B_{ij} = E_{bi}^2 \delta_{ij}, \tag{183}$$

where $E_{oK_i}^2$ and E_{bi}^2 are observational and background error variances, respectively, and the indices K_i and K_j $(1 \le K_i, K_j \le ND)$ indicate components of $\mathbf{y}^{\text{obs}}$ and the indices i and j $(1 \le i, j \le N)$ represent components of $\mathbf{x}_0$.

The cost function (153) in this case becomes

$$J(\mathbf{x}_0) = \frac{1}{2}\left(\sum_{K_i=1}^{ND} \frac{\left(H_{K_i}(\mathbf{x}_0) - y_{K_i}^{\text{obs}}\right)^2}{E_{oK_i}^2} + \sum_{i=1}^{N} \frac{\left(x_i - x_i^b\right)^2}{E_{bi}^2}\right)^2. \tag{184}$$

Eq. (184) explains why 3D-Var/4D-Var is referred to as a least-squares fit problem.

As was shown in the above derivation, the fitting criterion defined by 3D-Var/4D-Var in (153) is intimately related with the Gaussian probability assumption on all errors (model errors, observational errors, errors in the *a priori* information). Therefore, we shall limit the 3D-Var/4D-Var techniques to cases where this assumption is not strongly violated.

5.7. The 3D-Var/4D-Var analysis increments and the role of background error covariances and the forecast model constraint in 3D-Var/4D-Var analysis

3D-Var/4D-Var obtains an analysis in the model space based on given observations by minimizing a cost function (153) defined in terms of the deviation of the desired analysis from the first guess (background) field and observations. The background field and the observations are weighted by the inverse of the background and observation error covariance matrices respectively. The forecast model serves as a strong constraint to assimilate observations at their exact times in 4D-Var. The role of the background error covariances can be examined by expressing the analysis increment $(\mathbf{x}_a - \mathbf{x}_b)$ in terms of the observational increment $(\mathbf{y}^{\text{obs}} - H(\mathbf{x}_b))$. Assuming that the background field $\mathbf{x}_b$ is not too far from the optimal solution $\mathbf{x}_a$ (e.g., the final analysis field), then $H(\mathbf{x}_a)$ can be expressed by the first two terms of the Taylor expansion around $\mathbf{x} = \mathbf{x}_b$:

$$H(\mathbf{x}_a) = H(\mathbf{x}_b) + \mathbf{H}(\mathbf{x}_a - \mathbf{x}_b), \tag{185}$$

where $\mathbf{H} = \partial H(\mathbf{x})/\partial \mathbf{x}|_{\mathbf{x}=\mathbf{x}_b}$.

At the minimum of J, $\nabla J = 0$, i.e.,

$$\mathbf{B}^{-1}(\mathbf{x}_a - \mathbf{x}_b) + \mathbf{H}^T\mathbf{R}^{-1}\left(H(\mathbf{x}_a) - \mathbf{y}^{\text{obs}}\right) = 0. \tag{186}$$

Substituting (185) into (186) we obtain the 3D-Var/4D-Var analysis increment

$$\mathbf{x}_a - \mathbf{x}_b = \left(\mathbf{H}^T\mathbf{R}^{-1}\mathbf{H} + \mathbf{B}^{-1}\right)^{-1}\mathbf{H}^T\mathbf{R}^{-1}(\mathbf{y}^{\text{obs}} - H(\mathbf{x}_b)). \tag{187}$$

It is interesting to notice an obvious equivalence between the 3D-Var/4D-Var analysis increment and the OI analysis increment if $H = \mathbf{I}$ ($\mathbf{I}$ is the unit matrix). In this case, $\mathbf{R} = \mathbf{O}$, $\mathbf{y}^{\text{obs}} = \mathbf{x}^{\text{obs}}$, and the observation locations coincide with the analysis points. The 3D-Var/4D-Var analysis increment in (187) can be written as

$$\mathbf{x}_a - \mathbf{x}_b = (\mathbf{O}^{-1} + \mathbf{B}^{-1})^{-1}\mathbf{O}^{-1}(\mathbf{x}^{\text{obs}} - \mathbf{x}_b) \equiv \mathbf{B}(\mathbf{O} + \mathbf{B})^{-1}(\mathbf{x}^{\text{obs}} - \mathbf{x}_b), \tag{188}$$

where the matrix identity (161) is used. On the other hand, the OI analysis increment is (see (20) and (35))

$$\mathbf{x}_a - \mathbf{x}_b = (\ldots, \vec{W}_i, \ldots)^T(\mathbf{B} + \mathbf{O})^{-1}(\mathbf{x}^{\text{obs}} - \mathbf{x}_b) \equiv (\ldots, \vec{B}_i, \ldots)^T(\mathbf{B} + \mathbf{O})^{-1}(\mathbf{x}^{\text{obs}} - \mathbf{x}_b). \tag{189}$$

If observation locations coincide with analysis points, (189) becomes identical to (188).

If the observation error covariance matrix $\mathbf{R}$ is diagonal, the background error statistics $\mathbf{B}$, along with the tangent linear and adjoint operators $\mathbf{H}$ and $\mathbf{H}^T$, solely determines the spread of the increments brought by each observation to its nearby regions and to other model variables not directly associated with the observed quantity (see (187)). The relationship between analysis increments and background error covariances is often examined by conducting a single-observation experiment (Parrish and Derber 1992; Courtier *et al.* 1998). In the following, we express $\mathbf{x}_a - \mathbf{x}_b$ in terms of $\mathbf{B}$ instead of $\mathbf{B}^{-1}$ (see (187)).

Assume $\mathbf{R}$ is diagonal with error variances E_o^2. For a single observation (y_{obs}), $\mathbf{H}$ contains only one row and $\mathbf{H}^T$ is a vector. Using the Sherman-Morrison formula (148), the analysis increment in (187) can be written as

$$\mathbf{x}_a - \mathbf{x}_b = \frac{\mathbf{BH}^T}{E_o^2 + E_b'^2}(y^{\text{obs}} - H(\mathbf{x}_b)), \tag{190}$$

where $E_b'^2 = \mathbf{HBH}^T$ represents the implicitly specified error variance of the background.

The influence of background error variances on the analysis can be illustrated by examining the analysis increment at the observational point, which can be obtained by applying $\mathbf{H}$ to (190):

$$\mathbf{H}(\mathbf{x}_a - \mathbf{x}_b) = \frac{E_b'^2}{E_o^2 + E_b'^2}(y^{\text{obo}} - H(\mathbf{x}_b)). \tag{191}$$

Equation (191) shows that a single observation will affect the analysis with varying magnitudes depending on background and observation error variances, or more precisely the ratio of the two variances. If the background is far more accurate than the observation, then $\mathbf{x}_a = \mathbf{x}_b$. If the observation is more accurate than the background, then the observation with a smaller error variance will have a larger impact on the analysis. For a fixed observation error variance E_o^2, an under-estimate of the background error variance $E_b'^2$ (i.e., when the estimated background error variance is smaller than the true variance) will reduce the impact of the observations.

The influence of background error covariances to analysis increments near the observational point can be illustrated by assuming that $\mathbf{H}^T$ is a unit vector with its ith element equal to 1 and the rest of the elements being equal to 0, i.e., the single observation y_{obs} is a dircct measurement of the model variable at the jth analysis grid. In this case, (190) becomes

$$x_a(i) - x_b(i) = \frac{1}{E_o^2 + E_b'^2} B_{ij}(y^{\text{obs}} - x_b(j)), \tag{192}$$

where $(B_{1j}, B_{2j}, \ldots, B_{Ij})^T$ is the jth column of the background error covariance matrix $\mathbf{B}$. Therefore, for a non-diagonal background error covariance matrix, the one-point observation will not only modify the observed analysis variable at the observation location, but also at the surrounding areas as well as other variables not directly related to the observed variable. In other words, $\mathbf{B}$ serves as an interpolation vehicle and defines a structure for the observed information to be spread in the analysis variable space.

When the observation location coincides with the analysis point ($i = j$), (192) becomes

$$x_a(i) - x_b(i) = \frac{E_b^2}{E_o^2 + E_b^2}(y^{\text{obs}} - x_b(j)), \tag{193}$$

which is the same as (44) that is obtained by SC and OI algorithms.

The influence of the forecast model constraint on an analysis increments near an observational point can be illustrated by assuming **B** is a diagonal matrix. Equation (190) becomes

$$x_a(i) - x_b(i) = \frac{1}{E_o^2 + E_b'^2} H_{ji}(y^{\mathrm{obs}} - x_b(j)), \tag{194}$$

where **H** is the Jacobian matrix of the forecast model depending on the basic state.

We realize that generation of a multivariate analysis is solely dependent on background error covariances in 3D-Var, while the forecast model also contributes to the multivariate analysis in 4D-Var. Operational implementations of 4D-Var require that covariances of background errors are estimated and that the inverse of the background error covariance matrix can be calculated.

6. An example of assimilating a new type of atmospheric observation

Observations fuel the need for data assimilation. Without observations, analysis and assimilation are not needed. It is extremely important to know what is observed, how much error exists in the observation, why the observation is used, what improvements are expected in the analysis and forecast, and how they can be assimilated into the model inputs and/or model formulation with maximum accuracy and efficiency. An example reflecting how these considerations are taken into account in the assimilation of a new type of observation is provided by a series of papers concerning the assimilation of space-borne GPS occultation measurements (Zou et al. 1999; Zou et al. 2000; Liu et al. 2001; Zou et al. 2002; Shao and Zou 2002).

GPS radio occultation technology provides measurements of phase delay of microwave radio signatures (at two frequencies: $1227.6MHz$ and $1575.42MHz$) passing through the Earth's atmosphere. The magnitudes of the phase delay are determined by the thermodynamic state of the atmosphere that radio waves experience. The time derivative of phase excess (the Doppler shift excess) characterizes the atmospheric and ionospheric effect on the Doppler frequency shift and can be treated as the basic measurement data. These space-borne GPS occultation measurements are unique in a sense that they have very high vertical resolutions, are not affected by weather, and verify exceptionally well with radiosonde measurements. The basic physics and an observation operator of GPS occultation were described in Zou *et al.* (1999). Several modifications are recently made to improve the efficiency and accuracy of the raytracing model.

The GPS operators were linked to the NCEP Spectral Statistical Interpolation (SSI) analysis system (Zou *et al.* 2000). Actual bending angle occultation measurements were incorporated into the GPS 3D-Var system. The average magnitudes of the adjustments in temperature and specific humidity fields were calculated. The greatest adjustments to the temperature were found in the middle and upper troposphere and stratosphere, while the major changes in specific humidity occurred in the lower troposphere. Zou et al. (2002) conducted a statistical estimate of errors in the forward modeling and the assumption of spherical symmetry of the atmosphere. Impacts of GPS/MET bending angle observations on assimilation and forecasts were examined in Liu et al. (2001) in which a total of 837 GPS/MET bending angle profiles during the period June 20-30, 1995, were incorporated. Collocated radiosonde observations were used as independent data sources for the numerical evaluation of GPS/MET data assimilation results. The sensitivity of GPS/MET bending angle assimilation to the specifications of observational weightings was studied by Shao and Zou (2002). This is an ongoing data assimilation project. Besides the work mentioned above, at the time of writing, analysis and forecast results from incorporating CHAMP (Challenging Minisatellite Payload, a new German GPS occultation mission) occultations into the NCEP global 3D-Var system are being analyzed, and an accurate and efficient scheme for the assimilation of GPS refractivity is under development.

7. Concluding Remarks

Due to the voluminous material covered in the literature in this area, any attempt to include all of the topics in atmospheric data assimilation in one single paper would be a daunting task and would probably invite reader discontent. Therefore, in this paper, relevant topics in data assimilation are examined only in a limited scope.

Acknowledgments

This effort was supported by the National Science Foundation under project No. ATM-9908939.

References

Anderson, J. L., 2001: An ensemble adjustment Kalman filter for data assimilation. *Mon. Wea. Rev.*, **129**, 2884-2903.

Barnes, S. L., 1964: A technique for maximizing details in numerical weather map analysis. *J. Appl. Meteor.*, **3**, 396-409.

Bennett, A. F., 2002: *Inverse Modeling of the Ocean and Atmosphere.* Cambridge University Press. 234pp.

Bratseth, A. M., 1986: Statistical interpolation by means of successive corrections. *Tellus*, **38A**, 439-447.

Buell, C., 1972: Correlation functions for wind and geopotential on isobaric surface. *J. Appl. Meteor.*, **11**, 51-59.

Chen, G., 1993: *Approximate Kalman Filtering. Series in Approximations and Decompositions*, **2**, World Scientific, Singapore, 226pp.

Courtier, P., E. Andersson, W. Heckley, J. Pailleux, D. Vasiljević, M. Hamrud, A. Hollingsworth, F. Rabier and M. Fisher, 1998: The ECMWF implementation of three-dimensional variational assimilation (3D-Var). I: Formulation, *Q. J. R. Meteorol. Soc.*, **124**, 1783-1808.

Cressman, G. P., 1959: An operational objective analysis system. *Mon. Wea. Rev.*, **87**, 367-374.

Daley, R., 1991: *Atmospheric Data Analysis*, Cambridge, New York, 457pp.

Flecher, F., 1987: *Practical methods of optimization.* Second edition, John Willey & Sons. 436pp.

Gandin, L. S., 1965: *Objective analysis of meteorological fields.* U. S. Dept. of Commerce, 242pp.

Houtekamer, P. L., and H. L. Mitchell, 2001: A sequential ensemble Kalman filter for atmospheric data assimilation. *Mon. Wea. Rev.*, **129**, 796-811.

Kalnay, E., 2003: *Atmospheric Modeling, Data Assimilation and Predictability.* Cambridge University Press. 341pp.

Le Dimet, F. X., and O. Talagrand, 1986: Variational algorithms for analysis and assimilation of meteorological observations: Theoretical aspects. *Tellus*, **38A**, 97-110.

Liu, H., X. Zou, R. A. Anthes, J. C. Chang, J.-H. Tseng, and B. Wang, 2001: The Impact of 837 GPS/MET bending angle profiles on assimilation and forecasts for the period June 20-30, 1995, *J. Geoph. Res.*,, **106**, 31771-31786.

Menke, W., 1984: *Geophysical Data Analysis: Discrete Inverse Theory*, Academic Press, Orlando, 260pp.

Nash, S. G., and A. Sofer, 1996: *Linear and Nonlinear Programming.* The McGraw-Hill Companies, Inc. New York, 692pp.

Navon, I. M., X. Zou, J. Derber and J. Sela, 1992: Variational data assimilation with an adiabatic version of the NMC spectral model. *Mon. Wea. Rev.*, **120**, 1433-1446.

Nering, E. D., 1970: *Linear Algebra and Matrix Theory.* Second Edition. New York. 352pp.

Nocedal, J., and S. J. Wright, 1999: *Numerical Optimization.* Springer. 636pp.

Panofsky, H. A., 1949: Objective weather-map analysis. *J. Meteor.*, **6**, 386-392.

Parrish, D. F. and J. Derber,1992: The National Meteorological Center's spectral and statistical-interpolation analysis system. *Mon. Wea. Rev.*, **120**, 1747-1763.

Pielke, R. A., 1984: *Mesoscale Meteorological Modeling.* Academic Press. 612pp.

Richardson, L. F., 1922: *Weather Prediction by Numerical Process.* Cambridge University Press, Cambridge.

Sasaki, Y., 1958: An objective analysis based on the variational method. *J. Meteor. Soc. Japan*, **36**, 77-88.

Sasaki, Y., 1970: Some basic formalisms in numerical variational analysis. *Mon. Wea. Rev.*, **98**, 875-883.

Shao Hui, and X. Zou, 2002: On the observational weighting and its impact on GPS/MET bending angle assimilation. *J. Geoph. Res.*,, (accepted)

Sirkes, Z. and E. Tziperman, 1997: Finite difference of adjoint or adjoint of finite difference? *Mon. Wea. Rev.*, **125**, 3373-3378.

Tarantola, A., 1987: *Inverse Problem Theory: Method for Data Fitting and Model Parameter Estimation.* New York, 613pp.

Whitaker, J. S., and T. M. Hamill, 2002: Ensemble data assimilation without perturbed observations. *Mon. Wea. Rev*, **130**, 1913-1924.

Zou, X., H. Liu, and R. A. Anthes, 2002: A statistical estimate of errors in the calculation of radio occultation bending angles caused by a 2D approximation of raytracing and the assumption of spherical symmetry of the atmosphere. *JTECH.*, **19**, 51-64.

Zou, X., I. M. Navon, M. Berger, Paul K. H. Phua, T. Schlick, and F. X. LeDimet, 1993: Numerical experience with limited-memory quasi-Newton and truncated-Newton methods. *SIAM Journal on Optimization*, **3**, 582-608.

Zou, X., K. Sriskandarajah, Q.-N. Xiao, W. Yu, and S.-Q. Zhang, 2001: A note on eliminating finite-amplitude non-physical oscillations from the solution of an adjoint of finite-difference model. *Tellus*, **53A**, 578-584.

Zou, X., B. Wang, H. Liu, R. A. Anthes, T. Matsumura, and Y.-J. Zhu, 2000: Use of GPS/MET refraction angles in 3D variational analysis. *Quart. J. Roy. Meteor. Soc.*, **126**, 3013-3040.

Zou, X., F. Vandenberghe, B. Wang, M. E. Gorbunov, Y.-H. Kuo, S. Sokolovskiy, J. C. Chang, J. G. Sela, and R. Anthes, 1999: A raytracing operator and its adjoint for the use of GPS/MET refraction angle measurements. *J. Geoph. Res.*,, **104**, 22,301-22,318.

PROBABILISTIC FORECASTS AND EVALUATIONS BASED ON A GLOBAL ENSEMBLE PREDICTION SYSTEM

YUEJIAN ZHU

Environmental Modeling Center
NOAA/NCEP/National Weather Service, Washington, DC 20233, USA
E-mail: Yuejian.Zhu@noaa.gov

(Manuscript received 2 December 2002)

In the past decade, ensemble forecasting has developed into a major component of numerical weather prediction. With increases in computing resources, it is becoming more realistic to produce operational ensemble forecasts for compatible members with comparable resolutions in many numerical weather prediction centers around the world. Probabilistic forecasts based on a global ensemble prediction system, especially flow-dependent forecast probability distribution, which can be readily generated from an ensemble, allow for the identification of weather systems with high and low uncertainties. The potential economic benefit achieved by using ensemble probabilistic forecasts is significant when compared to that of a deterministic forecast. Among NCEP global ensemble-based applications, the relative measure of predictability is one of the excellent prediction tools used to estimate forecast uncertainty. Probabilistic quantitative precipitation forecasts can supplement short-/medium-range forecasts. However, the ensemble forecasts, like any numerical weather prediction system, are biased. The bias of ensemble forecasts is very similar to those of a deterministic forecast and comes from the imperfections of a numerical model such as initial conditions, physical parameterizations, numerical schemes, etc. The bias of ensemble forecasts can be removed, however, by applying a statistical calibration method. With the use of such a method, calibrated ensemble forecasts and ensemble-based, calibrated probabilistic forecasts can offer the possibility of bias-free products to the meteorological community and other users.

1. Introduction

One of the goals of the United States National Weather Service (US/NWS) for 2000-2005 is to provide weather, water and climate forecasts in probabilistic terms by the year 2005 (NWS 1999). In the past decade, global ensemble forecasting has been implemented into a major component of numerical weather prediction, such as the National Centers for Environmental Prediction (NCEP) ensemble prediction system (EPS), which is based on the breeding of growing mode method (Toth and Kalnay 1993; Tracton and Kalnay 1993; Toth and Kalnay 1997), and the European Center for Medium-Range Weather Forecasts (ECMWF) EPS, which is based on a singular vector method (Palmer et al 1992; Molteni et al 1996). Both of the NCEP and ECMWF ensemble forecast systems add a small perturbation to the initial state of the model analysis. The Meteorological Service of Canada also uses an EPS which is based on the system simulation experiment, but uses different methods to generate initial perturbations (Houtekamer and Derome 1996), including the use of two different numerical models, different physical parameterization packages which are designed to simulate observation errors, model errors, imperfect boundary conditions, etc. The operational applications of EPS in major centers around the world have been offering dramatic information in addition to deterministic forecasts. There are many advantages of using ensemble forecasts, such as potentially providing case-dependent estimates of forecast uncertainty. The initial uncertainty and model uncertainty are two main sources that limit the skills of single/deterministic forecasts in a highly flow-dependent forecasting system. From a long term objective evaluation, there is a gain in skill of about 24 hours when comparing the ensemble mean (Fig. 1a and 1b, ENS, closed circle marks) to higher resolution deterministic forecast (MRF, cross marks) for root means square (RMS) errors and pattern anomaly correlation (not shown here) for an analysis based on Northern Hemisphere extra-tropical (20°N-80°N) 500hPa geopotential height 6-day forecasts (Zhu et al 1996; Fig. 1a and 1b). The season statistics of the ensemble spread (SPR, open square marks, where SPR is defined as standard deviation of the ensemble members from the ensemble mean). RMS errors between the NCEP analysis and climatology (CLM, closed square marks) are also shown in Fig. 1a and 1b. Note that the usage of computer resource for low resolution ensemble forecasts is less than or equal to the higher resolution deterministic forecast, but the ensemble probabilistic forecasts have more skill than the higher resolution deterministic forecast even for extreme events (Zhu and Toth 2001), based on Brier Skill Score statistics. In fact, the potential economic value of an ensemble-based weather forecast (Zhu et al 2002) has indicated many advantages of using ensemble probabilistic forecast information. In the next section, I will briefly describe the current NCEP configuration, public data access and experiment setup. In section 3, selected current probabilistic forecast products would be introduced. The two probabilistic evaluation methods will be discussed in section 4. Finally, in section 5, conclusions and further discussions will be presented.

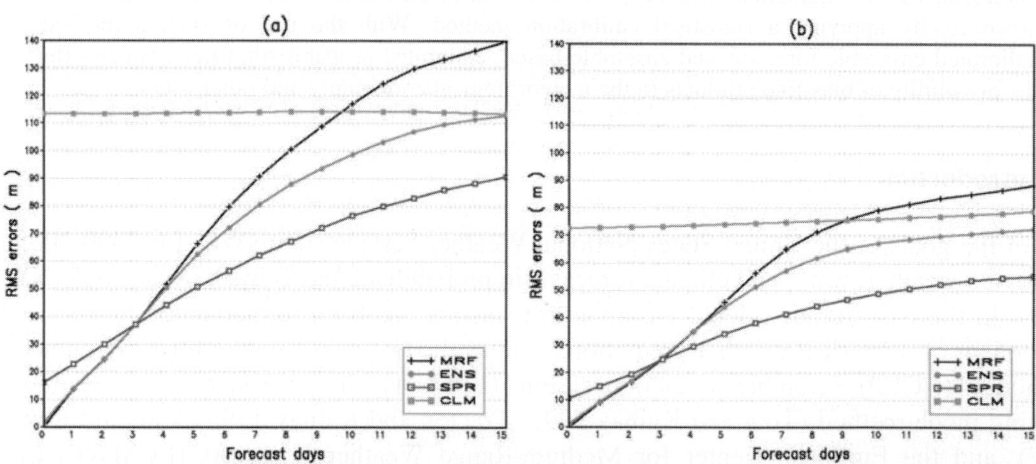

Figure 1. RMS errors for Medium-Range forecasts (MRF) and ensemble mean of northern hemisphere extra-tropical 500 hPa geopotential height: (a) for winter season (December 2001, January and February 2002), (b) for summer season (June, July and August 2002).

2. Ensemble Forecast

2.1. *NCEP Global Ensemble Configuration*

Currently, by adding small initial perturbations to an operational global analysis, the NCEP global ensemble forecasts are generated at 0000 UTC and 1200 UTC every day (Zoltan and Kalnay 1993; Tracton and Kalnay 1993; Zoltan and Kalnay 1997). The NCEP global ensemble forecasts consist each day of 25 individual independent forecasts run out to 16 days lead time, of which 5 members are control forecasts started from unperturbed analyses at 0000 UTC (global forecast system (GFS), which merges the former medium-range forecast (MRF) and aviation forecast after October 2002 and ensemble control), 0600 UTC (GFS), 1200 UTC (GFS) and 1800 UTC (GFS), and 20 members are perturbed forecasts at 0000 UTC (10) and 1200 UTC (10) from initial conditions where bred perturbations of the size of estimated analysis uncertainty are added. Four GFS forecasts are integrated at T170/L42 resolution out to 180 hours, and then reduced to T62/L28 resolution out to 16 days. All perturbed members and an ensemble control forecast are run at T128/L28 resolution out to 84 hours, and then reduced to T62/L28 resolution out to 16 days. The breeding cycle is 24 hours in the current NCEP global EPS.

2.2. *NCEP Ensemble Data*

Two public access ftp sites are available to users for NCEP global ensemble forecast data:
1) ftp://tgftp.nws.noaa.gov
2) ftp://ftpprd.ncep.noaa.gov/pub/data/nccf/com/mrf/prod.
All available files are updated daily on these NCEP public access ftp sites.

2.3. *Experimental Data*

The data for this study are the NCEP global ensemble and deterministic forecasts (MRF) of 500hPa geopotential height (period from 01 DEC 2001 to 28 FEB 2002 for a winter season, period from 01 JUN 2002 to 31 AUG 2002 for a summer season), 850hPa temperature (period from 01 DEC 2001 to 28 FEB 2002) and gage 24 hour accumulated total precipitation (period from 01 DEC 2001 to 28 FEB 2002). The gage 24-hour accumulated precipitation data are used as observed precipitation. The climatology of 500hPa geopotential height and 850hPa temperature is from NCEP/NCAR re-analysis. All the model analysis and forecast data are calculated at 2.5 by 2.5 degree resolutions globally except of the precipitation analysis, which is based on 80km ETA grids (Baldwin and Mitchell 1996).

2.4. *Best Ensemble Forecast*

What is the best ensemble forecast? There are many methods to measure ensemble forecasts. One of them is evaluating the difference between RMS errors of ensemble mean and ensemble spread. A perfect ensemble model assumes that an initial perturbation represents growing errors from an analysis, and also assumes that the forecasting model is perfect (Toth and Kalnay 1993; Toth and Kalnay 1997). Therefore, the spread of the ensemble forecasts should be fully equal to the size of the RMS errors of the ensemble mean when compared to observations. However, in a real application, the forecast model is not perfect and the initial perturbations do not fully represent the analysis growing errors, and so differences appear between RMS errors and the spread. In Figs. 1a and 1b, the differences of the ensemble mean RMS errors (ENS, closed circles) and spread (SPR, open squares) indicate the imperfection levels of NCEP ensemble forecasts relative to the analysis. The measurement would be more accurate if the observations, rather than analyses, could be used directly in the verification. In general, when a spread is smaller than RMS error, it means the ensemble forecast is under representing the model errors and forecast uncertainties; otherwise it is over representing them.

3. Probabilistic Forecast

It is well known that all environmental forecasts are associated with uncertainty and that the amount of uncertainty can be situation dependent. The use of probabilistic forecasts helps in estimating this uncertainty. By considering a wide range of forecasting information, forecasters could subjectively generate probabilistic forecasts by using different methods. For example, probabilistic forecasts could be generated from a set of deterministic forecasts valid at the same time, such as a lag forecast, which was very often used for climate prediction. Meanwhile, probabilistic forecasts could be made by using historical forecasts and observations, or by statistical methods such as the model output statistics (MOS) forecast, or by a multi-model super-ensemble which is based on a set of numerical models that use statistical regression to weight each ensemble member (Krishnamurti et al 1999). In this study, probabilistic forecasts that are based on initial perturbed NCEP global ensemble forecasts from only the same numerical model will be discussed.

There are many applications of probabilistic forecasts for the short/medium range. A "spaghetti diagram" is one such product. Two other major applications that will be described below include the probabilistic quantitative precipitation forecast (PQPF) and the relative measure of predictability (RMOP). PQPF is a product to help with a deterministic quantitative precipitation forecast (QPF). RMOP is a predictability measure for large and small uncertainty.

3.1. *PQPF Forecasts*

The PQPF forecasts have been produced operationally since 1997 in NCEP and are based on global ensemble forecasts using T62/L28 (T126/L28 out to 60 hours after 26 June 2000, T126/L28 out to 84 hours after 9 Jan. 2001) model resolution. The product includes nine threat amounts (0.254, 1.0, 2.54, 5.0, 6.35, 10.0, 12.7, 25.4, 50.8 mm) of 24-hour accumulation precipitation. To generate PQPF forecasts, the number of ensemble members that exceed a given precipitation threat level is divided by the total number of ensemble members at each grid point. For example, if the 24-hour forecasted precipitation amount exceeds 2.54 mm in 7 out of 10 total ensemble members at a particular grid point, then a 70%

probability of rainfall exceeding 2.54 mm (0.1 inches) for a 24-hour period (Zhu and Toth 1998) is assigned to that grid point. The synoptic PQPF maps shown in Fig. 2 show a strong fall/winter storm affecting most of the U. S. East Coast, with another system off the West Coast of Washington State. The four panels show 60-84 hour lead-time PQPF forecasts with four different 24-hour threat amounts of 2.54, 6.35, 12.7 and 25.4 mm. This is a highly predictable storm system (Predictability will be discussed in section 3.2). Additional sets of PQPF maps (not shown here) are very useful as well. For example, we could use a 3-parameter Gamma distribution or L-moment method to generate continued PQPF forecasts. By using continued PQPF forecasts, we may create a different precipitation forecast map for a specified probability (for example 30% probability, 75% probability, etc., Zhu and Toth 2003).

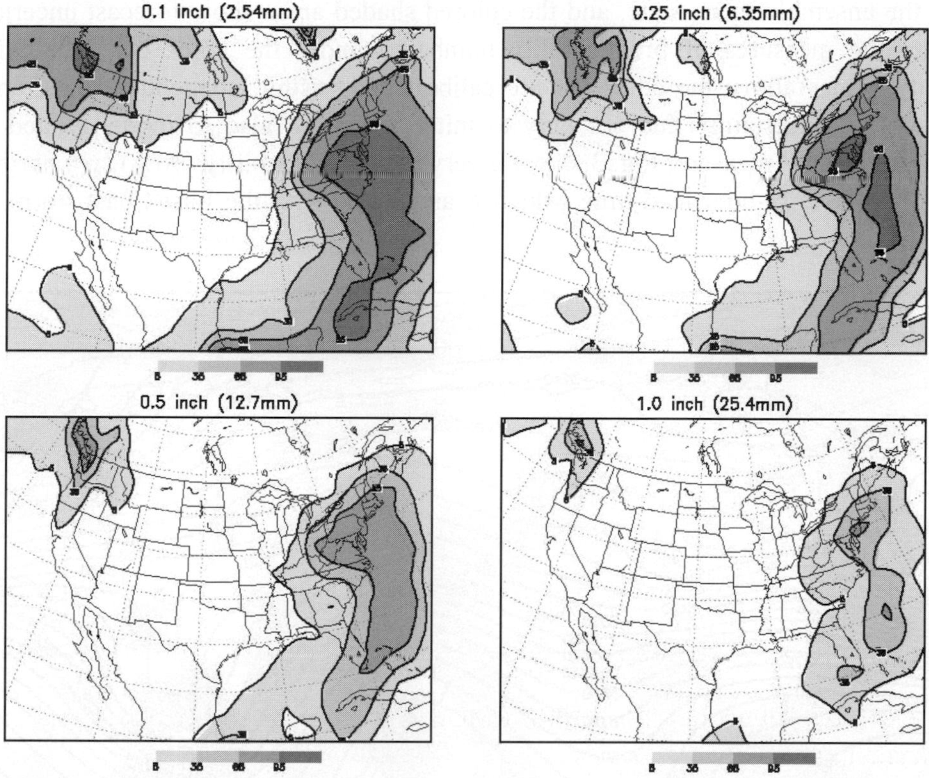

Figure 2. The 3-day lead-time probabilistic quantitative precipitation forecasts for 0.1, 0.25, 0.5 and 1.0-inch thresholds of the 24-hour period ending at 1200 UTC 17 Nov 2002, based on 23-member NCEP global ensemble forecasts. Contour lines are drawn at 5%, 35%, 65% and 95% probability levels.

As we know from long terms statistical objective evaluations, model forecasts are biased in most cases, and ensemble forecasts are biased as well. Especially for precipitation (Zhu and Toth, 1999, Zhu and Toth, 2003), most numerical model forecasts tend to over-forecast small precipitation amounts and under-forecast extreme amounts. In order to make reliable PQPF forecasts, it is necessary to remove the bias (or, first moments) and adjust the spread (or, second moments) if possible. Real time ensemble based calibrated PQPF and GFS (or MRF) based calibrated QPF forecast maps can be found on the NCEP website at http://wwwt.emc.ncep.noaa.gov/gmb/ens/enshome.html.

3.2. *RMOP Forecasts*

RMOP is a probabilistic measure to assess the flow-dependent uncertainty in a single forecast. Statistics indicate that, for certain cases, 10-13 day lead time forecast skill of the low uncertainty (high predictability, top 10%-15%) cases could be as good as one day forecast skill of the high uncertainty (low predictability, top 10%-15%) cases (Toth et al. 2001). Continuing with the same synoptic case from section 2, Fig. 3 shows the 3-day RMOP forecast map for 500hPa geopotential height, valid at 0000 UTC 17 Nov. 2002. The contours indicate the ensemble mean state, and the colored shaded areas show forecast uncertainties, which are the measures of predictability numbered under the color bar. The reference probability values above the color bar are calibrated by using independent data, and they reflect forecast uncertainty due not only to initial error but also to model related errors. Based on the RMOP measure, Fig. 3 shows a very high predictability (90%) area north of the Gulf of Mexico at 72-hour lead-time, which is associated with the East Coast storm system (see Fig. 2).

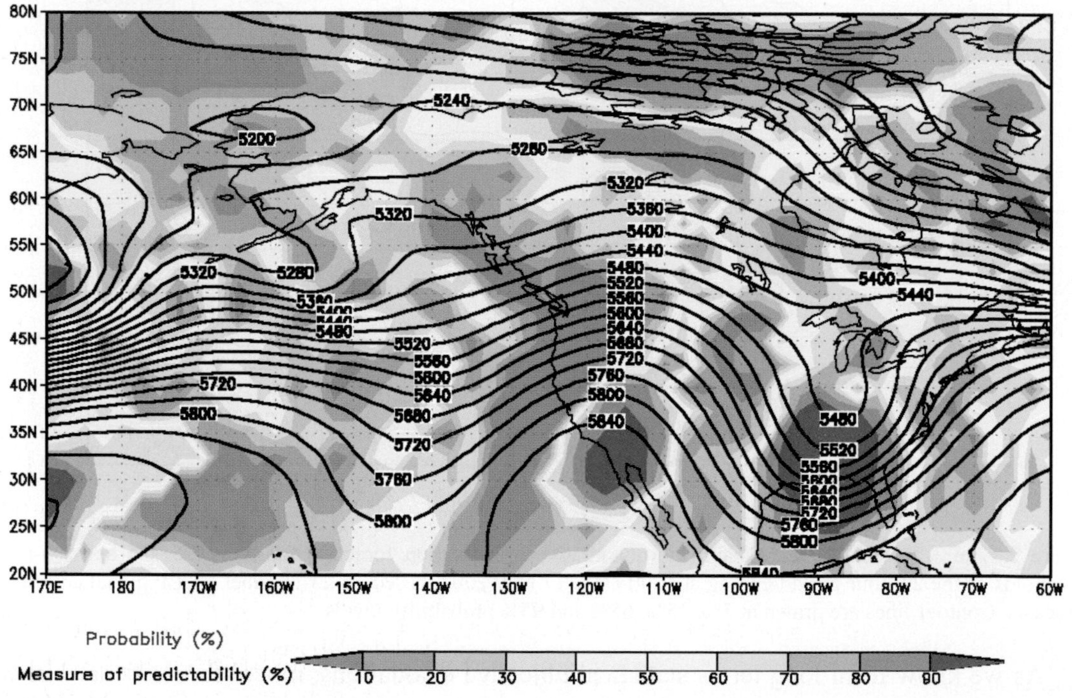

Figure 3 (color): The 3-day lead-time 500hPa geopotential height 10-member NCEP global ensemble mean forecast (contour lines) and associated relative measure of predictability (shaded), valid at 0000 UTC 17 Nov. 2002.

4. Probabilistic evaluation

An operational global ensemble forecast at NCEP and ECMWF is made by using an initial perturbation method, which is based on the assumption of a perfect forecast model. All errors are then considered to result from observations, first guess error and an imperfect data assimilation scheme. From this assumption, clearly the imperfections of the forecast model are not considered to be a source of errors. The forecast uncertainty is fully represented by model initial uncertainty. However, the forecast model is not perfect due to physical parameterization, boundary forces and other factors. Therefore, the forecast uncertainties are truly from both the initial condition (analysis) and the use of an imperfect numerical model. It is very difficult to separate initial errors and model errors quantitatively by using probabilistic evaluations of a model forecast. However, the probabilistic evaluations are the basic tools to measure model forecast uncertainties, which represent model resolution and reliability (Toth et al. 2002). There are many probabilistic evaluation methods that can be used, such as Talagrand distributions, rank probability scores, relative operating characteristics area, information contents, Brier scores, outlier maps, etc. Two methods will be described below which can assess the statistical reliability and resolution of a numerical model (Wilks 1995).

4.1. *Reliability Diagram*

For a given set of forecast probabilities of an event, one can compare with observations and determine the relative frequency at which an event with that forecast probability is observed. Ideally, one would like to see that the observed frequency is close to the forecast frequency, which would indicate perfect reliability (Fig. 4, diagonal line), in fact, the result is a curve indicating a model that is less than perfectly reliable (Fig. 4, solid line with closed circles). However, the reliability curve could be adjusted to near diagonal (dash line with open circles) by simple calibration by using past forecast and analysis information that is independent of current forecast data. This reliability diagram is based on Northern Hemisphere extratropical 500hPa geopotential height forecasts at day 5 (120-hour). The forecasts and verifying analysis data are separated at each grid point into 10 climatological equally likely intervals (bins). These intervals are defined uniquely for each grid point and each month of the year using NCEP/NCAR re-analysis (Kalnay et al, 1996). From this study (Fig. 4), the forecasts are still not perfect after a simple calibration (dash line, open circles), but they are more reliable when compared to the uncalibrated (raw) forecasts.

4.2. *Economic Value*

A decision maker becomes a user of a weather forecast if he/she alters his/her actions based on forecast information. In 2002, Zhu et al introduced the concept of economic value (EV) of a weather forecast by using cost-loss analysis method and considering user reaction. Different from the reliability measurement that was discussed in section 4.1, the EV is an estimation of forecast resolution, which is the ability of a forecast system to discern sub-sample forecast periods with different relative frequencies of an event. In this study, we evaluate the potential EV associated with the use of ensemble forecasts (T126/L28 out to 84 hours, then reduced to T62/L28), in terms of equivalent costs, as compared to a higher resolution deterministic forecast (MRF, T170 /L42 out to 7 days, then reduced to T62/L28). For this objective comparison, 850hPa temperatures in the Northern Hemisphere extratropics (20°N-80°N) have been used. By comparing lower resolution ensemble forecasts and higher resolution MRF forecasts at day 5 (Fig. 5), ensemble forecasts (solid line, closed circles) have better EV's than MRF forecasts (dash line, open circles) over all reasonable and selected cost-loss ratios from 0.01 to 1.0. Based on 3-month statistics, the highest EV is apparently around 1:10 cost-loss range for both the ensemble and deterministic forecasts, which means that the probabilistic forecasts from numerical models are the best fit for user groups who are most comfortable with around a 1:10 cost-loss ratio. Therefore, when considering 1:10 cost-loss ratio only, Fig. 6 shows the EV's for lead time out to 15-day. The ensemble forecasts (solid line, closed circles) have 20% more economic value than the MRF forecasts (dash line, open circles) at early lead times. The EV's of ensemble forecasts are almost double those of the MRF forecasts at longer lead time, but the EV's are already very small and close to zero.

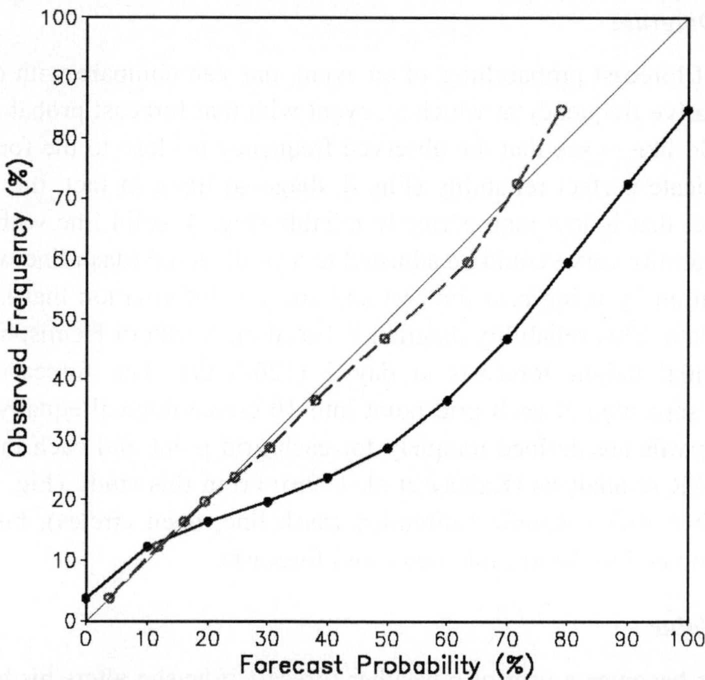

Figure 4. The 5-day lead-time ensemble-based reliability diagram for January 2002 for NH extra-tropical 500hPa geopotential height. Calibrated forecast probabilities (dash line, open circles) are based on observed frequencies associated with the same number of ensemble members falling in a climatologically equally likely bin during 1-20 December 2001. The uncalibrated (raw) forecast is shown on the solid line (closed circles). The vertical axis shows observed relative frequencies.

5. Summary and Conclusions

Ensemble-based probabilistic forecast products can provide dramatic information to users. With the use of such probabilistic forecasts, users have access to significantly more information than is available through a single, deterministic forecast, and this information can be used to make more accurate forecasts and allow users to alter their decisions, especially for extreme events that can be associated with higher and lower uncertainty weather systems. PQPF forecasts and RMOP forecasts are two examples of probabilistic products that were discussed in this paper. The PQPF predicts the future possibilities for many precipitation amount thresholds at all lead times and physical locations, while the RMOP describes the degree of uncertainty associated with the forecasting of future weather systems.

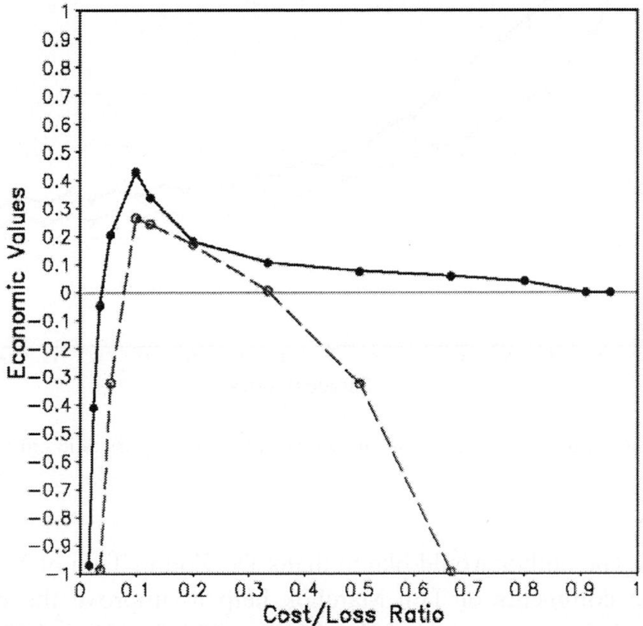

Figure 5. *The economic values for NH extratropical 850hPa temperatures, 5-day lead 10-member NCEP ensemble forecasts (solid line, closed circles) and NCEP deterministic forecast MRF (dash line, open circles) for a winter season (Dec. 2001, Jan. and Feb. 2002). The vertical axis is economic values.*

Reliability and resolution are two major attributes that are considered in the evaluation of these ensemble-based probabilistic forecasts. The reliability indicates how statistically consistent these probabilistic forecasts are with observations, such as through the use of Talagrand distributions, reliability diagrams, outlier maps, etc. The resolution summarizes how much more information the forecasts have with respect to climatology, such as through the use of EV's, relative operating characteristics areas, information contents, etc.

Finally, recent studies indicate that statistical calibration can improve model forecast reliability, although it does not have much effect on the resolution. Since forecast reliability is mostly due to system bias, that bias can be removed by applying historical information from previous forecasts with the same model, if there has been no significant change in the forecasting model in the recent past.

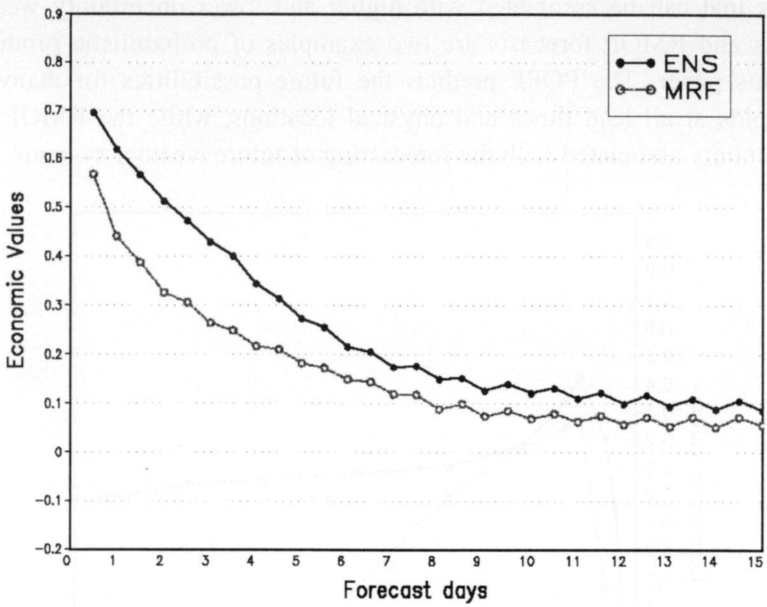

Figure 6: The same as Fig. 5 except for all lead-time forecasts and 1:10 cost-loss ratio only.

Acknowledgments. The author would like to thank Dr. Zoltan Toth of NOAA/NCEP for his encouragement. The comments of Tim Marchok help to improve the presentation of the manuscript. I acknowledge the support of Hua-Lu Pan, Chief of Global Climate and Weather Modeling Branch, EMC and Stephen Lord, Director of EMC.

References:

Baldwin, M.E., and K. E. Mitchell, 1996: The NCEP hourly multi-sensor U. S. precipitation analysis. Preprints, *11th AMS Conf. on Numerical Weather Prediction*, Norfolk, VA. Amer. meteor. Soc. J95-96.

Houtekamer, P. L., and J. Derome, 1995: Methods for ensemble prediction. *Mon. Wea. Rev.*, **123**, 2181-2196.

Kanamitsu, M., and Coauthors, 1991: Recent changes in the global forecasting system at NMC. *Weather and Forecasting*, **6**, 425-435.

Krishnamurti, T. N., C. M. Kishtawal, T.LaRow, D. Bachiochi, Z. Zhang, C. E. Williford, S. Gadgil, and S. Surendran, 1999: Improved weather and seasonal climate forecasts from multimodel superensemble. *Science*, **285**, 1548-1550.

Molteni, F., R. Buizza, T. N. Palmer, and T. Petroliagis, 1996: The ECMWF ensemble prediction system: Methodology and validation. *Quart. J. Roy. Meteor. Soc.*, **122**, 73-119.

NWS, 1999: *Vision 2005: National Weather Service Strategic Plan for Weather, Water, and Climate Services 2000-2005.* 24 pp. [Available from NWS, 1315 East-West Highway, Silver Spring, MD 20910]

Palmer, T. N., F. Molteni, R. Mureau, R. Buizza, P. Chapelet, and J. Tribbia, 1992: Ensemble prediction. *ECMWF Research Department Tech. Memo.* **188**, 45pp.

Toth, Z., and E. Kalnay, 1993: Ensemble forecasting at NMC: The generation of perturbations. *Bull. Amer. Meteor. Soc.*, **74**, 2317-2330.

Toth, Z., and E. Kalnay, 1997: Ensemble forecasting at NCEP and the breeding method. *Mon. Wea. Rev.*, 125, 3297-3319.

Toth. Z., Y. Zhu, and T. Marchok, 2001: On the ability of ensembles to distinguish between forecasts with small and large uncertainty. *Weather and Forecasting*, 16, 436-477.

Toth, Z., O. Talagrand, G. Candille, and Y. Zhu, 2002: Probability and ensemble forecasts. In: *Forecast Verification: A Practitioner's Guide in Atmospheric Science.* Editors.: I. T. Jolliffe and D. B. Stephenson. Wiley, 137-164.

Tracton, M. S., and E. Kalnay, 1993: Ensemble forecasting at NMC: Operational implementation. *Wea. Forecasting*, **8**, 379-398.

Wilks, D. S., 1995: *Statistical Methods in the Atmospheric Sciences.* Academic Press, New York, 467pp.

Zhu, Y., G. Iyengar, Z. Toth, S. Tracton, and T. Marchok, 1996: Objective evaluation of the NCEP global ensemble forecasting system. *15th AMS Conf. on Weather Analysis and Forecasting.* Norfolk, VA. Amer. Meteor. Soc. J79-82.

Zhu, Y., Z. Toth, E. Kalnay, and S. Tracton 1998: Probabilistic Quantitative Precipitation Forecasts based on the NCEP global ensemble. *Special Symposium on Hydrology*, Phoenix, AZ. Amer. Meteor. Soc. J8-11.

Zhu, Y., and Z. Toth, 1999: Objective Evaluation of QPF and PQPF Forecasts Based on NCEP Ensemble, Preprints, *Third International Scientific Conference on the Global Energy and Water Cycle*, 16-19 June 1999 Beijing China, 47-48.

Zhu, Y., and Z. Toth, 2001: Extreme weather events and their probabilistic prediction by the NCEP ensemble forecast system. Preprints, *Symposium on Precipitation Extremes: Prediction, Impact, and Responses*, Albuquerque, NM. Amer. Meteor. Soc., 82-85.

Zhu, Y., Z. Toth, R. Wobus, D. Richardson, and K. Mylne 2002: On the economic value of ensemble based weather forecasts. *Bull. of Amer. Meteor. Soc.*, **83**, 73-83.

Zhu, Y., and Z. Toth, 2003: A synoptic evaluation of ensemble based probabilistic quantitative precipitation forecasts. Submit to *J. Hydrometeorology*.

MODELING LAND SURFACE PROCESSES IN SHORT-TERM WEATHER AND CLIMATE STUDIES

ZONG-LIANG YANG

Department of Geological Sciences
University of Texas at Austin, Austin, Texas 78712, USA
E-mail: liang@mail.utexas.edu
Website: www.geo.utexas.edu/climate

(Manuscript received 31 January 2003)

Land exchanges momentum, energy, water, aerosols, carbon dioxide and other trace gases with its overlying atmosphere. The land surface influences climate on local, regional and global scales across a wide range of timescales. This review concentrates on the rapid (i.e., seconds to seasons) biophysical and hydrological aspects of land surface processes. This paper provides the historical development of land surface models designed for short-term weather and climate studies, ranging from the early, simple "bucket" models to recent sophisticated soil-vegetation-atmosphere transfer schemes. Major research issues are reviewed by grouping into datasets, coupling to atmospheric models, component processes, and sub-grid-scale variability and scaling. Significant problems remain to be addressed, including the difficulties in parameterizing hillslope runoff, fractional snow cover, stomatal resistance, evapotranspiration, and sub-grid-scale variability and scaling. However, further progress is expected as the results of large-scale field experiments and satellite datasets are exploited.

1. Introduction

Land covers about 30% of the Earth's surface. The land surface consists of soil, vegetation, snow, glaciers, inland water, mountains, animals, human beings, their shelters, and much more. Land surface processes, in principal, refer to the exchanges of heat, water, CO_2, and other trace constituents among these components. In particular, land-atmosphere interaction refers to the exchanges of momentum, energy and mass (water, aerosols, and other important chemical constituents) between land surfaces and the overlying atmosphere. Land and oceans are coupled through river runoff and breezes that develop near the coast.

Unlike other surfaces of the Earth, land plays a distinctive role in weather and climate. Land has considerable heterogeneity such that bare soil, rock, short grass, tall grass, trees and snow patches can all coexist in one small area (Figure 1). This surface variability not only determines the microclimate but also affects the mesoscale atmospheric circulation (Giorgi and Avissar 1997; Pielke 2001). Land provides humans with a habitable place, which explains why over land there has been more historical documentation of routine observations of climatic variables such as temperature, clouds, and precipitation. Due to much less storage and negligible horizontal transport of heat over land, the diurnal, synoptic and seasonal variations of temperature are greater than those over water surfaces. The land orography not only exerts thermal and dynamic influences on weather and climate, but also has a direct effect on the surface hydrologic cycle (e.g., hillslope evaporation and runoff, Figure 1). The surface roughness over land is much greater than that over the oceans and affects greatly the heat and water transfer between the land surface and the overlying atmosphere. Land surfaces are more changeable than ocean surfaces. Such examples include urbanization, agricultural use, irrigation, deforestation, desertification, and bush fire as well as large natural seasonality in vegetation greenness and snow cover. Land surfaces store large quantities of carbon (more than twice that in the atmosphere). About half of the current anthropogenic emissions of CO_2 are being absorbed by the ocean and by land ecosystems, but there is a concern over how the terrestrial carbon sink is sensitive to climate and whether the sink will become a source due to global warming.

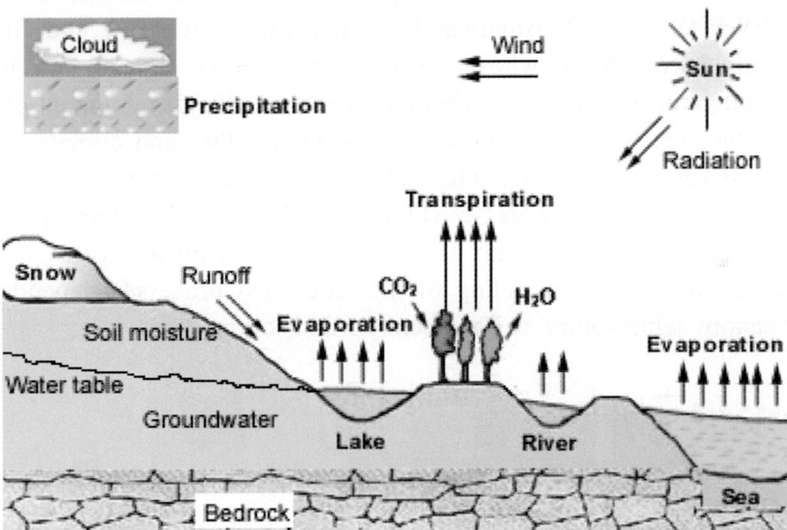

Figure 1. Schematic representation of the coupling between land, the atmosphere and oceans through the hydrologic cycle.

Not only should the land surface play an important role in weather and climate from the above intuitive reasoning, the evidence is increasing that the influence of the land surface is significant on local, regional and global climate on timescales from seconds to millions of years. On timescales of seconds to hours, land-atmosphere interaction is dominated by the rapid biophysical and biogeochemical processes that exchange momentum, energy, water, carbon dioxide and other chemical constituents between the land surface and the atmosphere (Jarvis 1976; Farquhar and Sharkey 1982; Dickinson 1983, 1992; Sellers 1992; Garratt 1993; Giorgi and Avissar 1997; Pielke 2001; Pitman 2003). On timescales of days to seasons, land-atmosphere interaction occurs through changes in the store of soil moisture (Yang 1995; Chen et al. 2001; Koster and Suarez, 2001), changes in snowpack (Robock et al. 2003), changes in carbon allocation, and vegetation phenology (e.g., budburst, leaf-out, senescence, dormancy) (Foley et al. 1996; Lu et al. 2001; Dickinson et al. 1998, 2002). Liu (2003) reviews monthly and seasonal variability of the land-atmosphere system. On timescales of years to centuries, vegetation structure and function (e.g., disturbance, land use, stand growth) is strongly determined by climate influences, primarily through temperature ranges and water availability (Foley et al. 1996). During the early to mid-Holocene period (6,000 years ago), important interactions and feedbacks between climate, vegetation, soil, snow, and lake may have taken place (Pielke et al. 1998). During the late Pleistocene period, glacial-interglacial cycles probably involve changes in the geographical distribution of vegetation, soils, surface albedo, and biogeochemical cycling in response to orbitally-induced insolation variations. On even longer geological timescales spanning hundreds of millions of years, the Earth's climate has been tightly coupled to atmospheric CO_2 levels through the carbonate-silicate cycle and/or the organic carbon cycle (Pielke et al. 1998).

This paper concentrates on land surface processes on timescales of a year and less. These timescales also are focused by international programs of Global Energy and Water cycle Experiment (GEWEX), GEWEX Americas Prediction Project (GAPP), and GEWEX Asian Monsoon Experiment (GAME), whose goals are to predict climate and hydrology on intraseasonal to interannual timescales. This article reviews the development of land surface models (LSMs) designed for use in three-dimensional weather and climate models with a focus on the rapid (biophysical) and intermediate (out to around a year; biogeochemical, hydrological, and phenological) processes (Figure 2). Understanding these processes is also central to an effective coupling with long-term biogeochemical cycles and vegetation dynamics because an accurate modeling of the latter two processes depends on credible simulations of canopy temperature, soil temperature and soil moisture.

2. Early Land Surface Models

2.1. *Bucket Model*

While biophysically realistic land surface processes are clearly important in weather and climate modeling, they were not included in atmospheric general circulation models until the late 1980s (Dickinson et al. 1986; Sellers et al. 1986; Abramopoulos et al. 1988).

In early GCMs, the land surface parameterization schemes are represented in a very crude way. Global soil was assumed to have a fixed water holding capacity of 15 cm (Manabe 1969). At each land grid square and each time step, the "bucket" is filled with precipitation and emptied by evaporation. The excess above its field capacity or a critical value is termed runoff. The evaporation rate is a product of the coefficient (the β function) and the potential evaporation. The coefficient, commonly called "soil wetness", or "moisture availability", is assumed to be a linear function of soil moisture content.

The empirical basis for the bucket parameterization is diurnally averaged data. It is most useful in GCMs that use diurnally averaged solar heating and therefore may not be appropriate for climate models with a diurnal cycle of solar radiation.

Sato ct al. (1989) found that the constant moisture availability in the bucket model can make the estimate of evapotranspiration several times too large compared to the more realistic

Penman-Monteith equation (Monteith 1981). Sellers and Dorman (1987) found that evapotranspiration changes nonlinearly with the soil moisture especially when the soil is drying and when soil wetness drops below about 0.5.

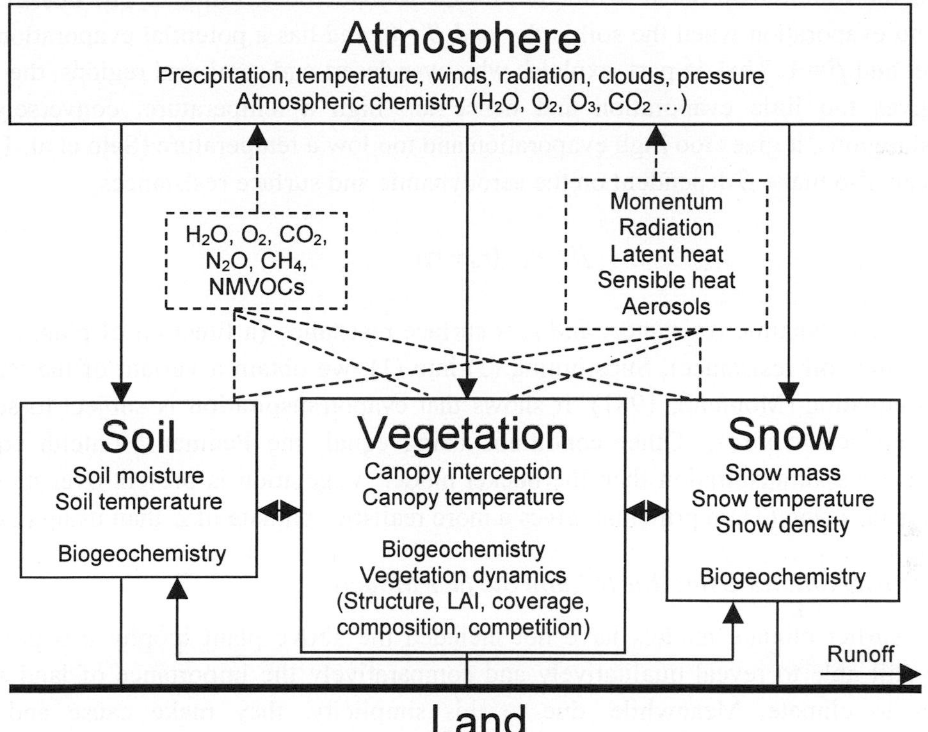

Figure 2. Schematic representation of the components and processes in a land-surface model. The framework may be used for studies of weather, climate, air quality, and water quality. The arrows indicate interactions between different components (solid boxes). NMVOCs are nonmethane volatile organic compounds (e.g., isoprene (C_5H_8) monoterpenes ($C_{10}H_{16}$)). More components may be added to represent glaciers, lakes, wetland, and urban areas.

The bucket model ignores the complex processes of soil water movement (e.g., capillary and gravitational processes) and the uptake of water by roots in the presence of vegetation (Yang 1995). In addition, the bucket model fails to represent the β function in terms its dependence on the soil moisture content through aerodynamic, stomatal and soil surface resistances (Yang, 1995).

Evapotranspiration may be expressed in the electrical analogue form by using the resistance formulation, whereby

$$E = \beta\rho[(q_s(T_g) - q_r]/r_a\,, \tag{1}$$

where ρ = air density, $q_s(T_g)$ = saturation mixing ratio at ground surface temperature T_g, q_r = mixing ratio at a reference height, r_a = aerodynamic resistance, β = soil wetness. The β function in (1) is defined as

$$\beta = W / W_k , \tag{2}$$

where W and W_k are soil moisture content and its critical value, respectively. The above relation is the well-known linear β-function for the Manabe bucket model. The bucket model predicts no evaporation when the soil is dry and $\beta = 0$, and has a potential evaporation when soil is wet and $\beta = 1$. This, in part, explains why over desert and semi-arid regions, the bucket model gives too little evaporation and hence too high a temperature; conversely over rainforest regions, it gives too high evaporation and too low a temperature (Sato et al. 1989).

We can also make β dependent on the aerodynamic and surface resistances,

$$\beta = r_a / (r_a + r_s), \tag{3}$$

where r_a = aerodynamic resistance, and r_s = surface resistance (a function of plant stomatal resistance and soil resistance). Substituting (3) into (1), we obtain a variant of the Penman-Monteith equation (Monteith, 1981). It shows that evapotranspiration is subject to soil and plant biophysical controls. Other conditions being equal, the Penman-Monteith equation gives lower evapotranspiration than the bucket model. Vegetation is present over most land surfaces. Thus using (3), in principal, gives a more realistic estimate of E than using (2).

2.2. *Sensitivity Studies Using Early Land Surface Models*

Although earlier climate models have not included the above plant biophysical processes, they are still able to reveal qualitatively and comparatively the importance of land surface processes to climate. Meanwhile, due to this simplicity, they make cause and effect relationships easier to describe.

Charney et al. (1977) first used a GCM to study the albedo effects on the initiation and maintenance of drought in the semi-arid regions. Other GCM studies have examined the modeled sensitivities to surface roughness, soil wetness and emissivity (Pielke et al., 1998). The surface hydrological variables, temperature, heat fluxes and circulations are shown to be very sensitive to the prescribed anomalies of the surface components. The experiments reveal that the deforestation issue may be better studied using GCMs with more elaborate land surface models that treat soil and canopy effects in a more realistic way.

3. Advanced Land Surface Models

Deardorff (1978) first proposed an advanced land surface model suitable for use in GCMs. In his model, Deardorff considered one-layer vegetation in addition to a two-layer soil. The evaporation from the soil layer and the wet canopy, the interception, and the transpiration from the dry parts of the canopy were considered. This formulation opened the way for the following developments toward constructing advanced land surface models in GCMs.

Over the past two decades, active researches on land-atmosphere interactions have led to more than two dozen LSMs constructed for use in GCMs (e.g., Henderson-Sellers et al. 1993;

Slater et al. 2001). There is much similarity and overlap among these models (Figure 2). Two well known land surface parameterizations are discussed below. The emphasis is placed on the structure of the models, requirements of GCMs, and sensitivity studies. The structure part is discussed in the order of vegetation, soil, and snow sub-models.

3.1. *Biosphere-Atmosphere Transfer Scheme (BATS)*

The development of the Biosphere-Atmosphere Transfer Scheme (BATS) is well documented in Dickinson et al. (1986, 1993, 1998, 2002). Originally, BATS was designed for use in the National Center for Atmospheric Research (NCAR) Community Climate Model (CCM). But its philosophy, as well as many aspects of its physical parameterizations, have been adopted in other LSMs (e.g., Noilhan and Planton 1989; Dai et al. 2003).

BATS has three soil layers and one vegetation layer. There are eight prognostic variables: leaf temperature, surface soil/snow temperature, subsurface soil/snow temperature, surface soil water, root-zone soil moisture, total soil water, snow cover amount measured in terms of liquid water content, and canopy water store. There are 18 surface types, with 15 types for vegetation which are based on Matthews (1983) and Wilson and Henderson-Sellers (1985). The soil type data are based on Wilson and Henderson-Sellers (1985). For each vegetation type, there are about 27 derived parameters which determine the morphological, physical and physiological properties of vegetation and soil.

Vegetation in BATS is assumed to be a flat, porous and uniform layer. It may cover the whole grid square of a GCM. The foliage is assumed to have zero heat capacity, and photosynthetic and respiratory energy transformations are neglected. The vegetation temperature is obtained by solving a vegetation energy balance equation,

$$R_n(T_f) = L\, E_f(T_f) + H_f(T_f), \tag{4}$$

where R_n, E_f, H_f are net radiation, canopy evapotranspiration, and sensible heat flux, respectively. T_f is the leaf temperature, and L is the latent heat of vaporization.

The rate of change of water store per unit land-surface area is calculated as

$$dW_{dew} / dt = \sigma_f P - E_f + E_{tr}, \tag{5}$$

where W_{dew} = water store on canopy surface, σ_f = fractional vegetation cover, P = precipitation rate, E_f = evapotranspiration rate, E_{tr} = transpiration rate. An accurate simulation of E_{tr} largely depends on the way stomatal functioning is described. The stomatal resistance takes a form of

$$r_s = r_{smin} f_R f_S f_{VPD} f_M, \tag{6}$$

where r_s = stomatal resistance, r_{smin} = minimum stomatal resistance, f_R = a factor dependent on solar radiation (visible), f_S = a seasonality factor dependent on leaf temperature, f_{VPD} = a factor dependent on VPD, f_M = a factor dependent on soil moisture potential and distribution of roots. In (6), r_{smin} is defined to be average value for the whole canopy in earlier versions of BATS, whereas it is the minimum stomatal resistance at the top of the canopy in Dickinson et al. (1998).

When calculating f_R, a four-level canopy is simply considered to take account of different amounts of radiation received by leaves. Earlier version of BATS (Dickinson et al. 1993) calculated the leaf area index (LAI) as a function of temperature between prescribed maximum and minimum values, while Dickinson et al. (1998) simulates the growth and loss of the green foliage by describing leaf CO_2 assimilation in addition to leaf water use. The linkage between carbon assimilation and the reciprocal of stomatal resistance (i.e., stomatal conductance) is described by a derivative of that given by Ball et al. (1987), hereafter referred to as the Ball-Berry equation, that is

$$g_s = m(A_n/C_s)F_eP + g_0, \tag{7}$$

where g_s is stomatal conductance for water vapor transfer, g_0 is a prescribed minimum stomatal conductance, m is a slope parameter (equal to 9 for C_3 plants), A_n is the net carbon assimilation, C_s is the carbon dioxide partial pressure adjacent to the leaf, P is atmospheric pressure, and F_e is a humidity-dependency stress factor. Dickinson et al. (2002) include the effects of nitrogen cycling in (7).

A bulk canopy stomatal resistance is given by r_s/LAI, where LAI is the effective leaf area index used to account for the attenuation of radiation as light passes through the canopy and the coincident decrease in plant surface which is actively transpiring.

The soil and subsurface temperatures are obtained using the force-restore method (Dickinson 1988). Soil moisture contents are computed for three overlapping soil layers: the upper layer, the root zone and the total active layer. All share the same incident precipitation, drip from foliage, snowmelt, surface evaporation and surface runoff, while they have different transpiration rates because of different distribution of roots. Fluxes between soil layers are parameterized.

Snow and soil share the same thermal balance equations while the effects of snow on modifying thermal properties, albedo and surface fluxes are described by a simple roughness and snow masking relation. The effects of refreezing of melt water are ignored.

The requirement of BATS for inclusion in GCMs is that the host GCM contains diurnal variation of insolation. The input atmospheric variables are incident shortwave and longwave radiation at the surface, precipitation rate, temperature, water vapour mixing ratio and wind velocity at the lowest model level.

There are numerous sensitivity studies using BATS carried out in standalone (off-line) mode and coupled mode. The off-line studies include (1) the model's overall performance against the observations for different land types and under different forcing conditions (e.g., Unland et al. 1996; Yang et al. 1997, 1999; Sen et al. 2000), (2) sensitivity to the inevitable errors or uncertainties of input parameters for a given biome (Gupta et al. 1999), (3) sensitivity to the initialization of its prognostic variables (Yang et al. 1995), and (4) sensitivity to the aggregation methods employed to convert from finer resolution to the coarser grids of GCMs (Shuttleworth et al. 1997; Burke et al. 2000). BATS has been used in global climate models to explore the regional climatic impacts of tropical deforestation (e.g., Dickinson and Henderson-Sellers 1988), and the impacts of doubled stomatal resistance on the water resources of the American Southwest (Martin et al. 1999). BATS also is used in high-resolution regional climate models (e.g., Giorgi et al. 1999).

3.2. *Simple Biosphere (SiB) Model*

The development of the Simple Biosphere Model (SiB) is well documented in Sellers et al. (1986, 1996a,b). The philosophy of the model design is to model the vegetation itself and let the vegetation determine the ways in which the land surface interacts with the atmosphere. Specifically, when vegetation is present, it plays an important role in radiation absorption (i.e., high absorptivity in the visible wavelength interval and moderate reflectivity in the near-infrared region), biophysical control of evapotranspiration (e.g., through stomatal resistance), momentum transfer (e.g., by roughness length), soil moisture availability (i.e., by the depth and density of the vegetation roots), and insulation (i.e., by vegetation shelter).

The original SiB model has 3 soil layers and 2 vegetation layers. There are 8 prognostic physical-state variables, 3 temperatures (one for the canopy vegetation, one for both the ground cover and the soil surface, and one for the deep soil layer; 2 interception water stores (one for the canopy, and one for the ground cover); and three soil moisture stores. The revised SiB model has changed the original two-layer vegetation canopy structure to a single layer and incorporated a patchy snowmelt treatment (Sellers et al. 1996a).

In SiB, the global vegetation is classified into 12 ecotypes. They are originally based on Kuchler (1983) that recognizes 32 natural vegetation communities, and the land use data base of Matthews (1983). The 12 ecotypes are stored at $1°\times1°$. For each vegetation type, there are about 54 parameters for both canopy and ground cover, which determine the morphological, physical and physiological properties of the biome. All these parameters, together with prognostic variables of SiB and the atmospheric boundary conditions, are used to determine the fluxes between the surface and the atmosphere. The vegetation phenology is described by use of satellite data (Sellers et al. 1996b).

In the original SiB model (Sellers et al. 1986), there are two vegetation layers representing two morphological groups. The top layer consists of trees or shrubs while the ground layer is for grasses and other herbaceous plants. Either or both or neither may exist in one grid square of model land. Unlike BATS which neglects canopy heat storage, SiB assumes it is a function of leaf area index and intercepted water on the canopy.

In the revised SiB model (Sellers et al. 1996a), there is only one canopy layer. Some of the original vegetation classes are combined to reduce the number of distinct vegetation classes from 12 to 9. A canopy photosynthesis submodel is incorporated. This submodel makes explicit calculation of the photosynthetic CO_2 flux between the atmosphere and the land surface. The leaf photosynthesis-conductance model used in the new model is similar to that used in Dickinson et al. (1998), with, however, somewhat different implementation. SiB includes description of C_4 photosynthesis in addition to C_3 photosynthesis. The photosynthetic rate of the canopy as a whole is estimated from that of the uppermost leaves by multiplying by a factor that allows for the absorption of photosynthetically active radiation through the canopy. The canopy conductance then is estimated using the Ball-Berry equation (7), with the humidity stress factor set equal to relative humidity. Canopy transpiration thus is related directly to the whole-canopy carbon assimilation via the canopy conductance, but transpiration itself may feed back on the canopy conductance by influencing the canopy environment. The net CO_2 flux is assumed to be the difference between the soil respiration and the net carbon assimilation rate.

The soil model has three soil layers: an upper thin soil layer, a root zone and an underlying recharge layer. The topsoil layer is chosen to be thin to ensure there can be a significant rate of withdrawal of water by direct evaporation into the air when the pores of the soil are at or near saturation. The root zone layer contains all the roots for the two vegetation layers. The recharge layer is included to account for water transfer by gravitational drainage and hydraulic diffusion.

Although there are three governing equations for the three moisture stores in the three soil layers, there is only one equation for the soil temperature: deep soil temperature. This equation, together with the equation for the ground surface temperature, is formulated in the framework of the force-restore method (Deardorff 1977).

In SiB, the snow model is very simple compared to the sophisticated vegetation and soil models. The snow depth is explicitly predicted though it is very crude. There is no explicit treatment of snow temperature; rather it is included in ground surface temperature for a combination of three surface types: ground vegetation, soil and snow. Simple modifications of ground albedo, roughness length, the heat capacity of the canopy or ground, latent heat flux, and runoff are proposed to include snow effects (Sellers et al. 1986, 1996a).

SiB requires a GCM with diurnal variation of insolation. The atmospheric forcings include the incident radiative flux, the precipitation, air temperature, vapor pressure and wind components at the lowest model level. The incident radiative flux must be partitioned into five components: visible or PAR (< 0.72 μm) direct beam radiation, visible or PAR (< 0.72 μm) diffuse radiation, near infrared (0.72–4.0 μm) direct beam radiation, near infrared (0.72 – 4.0 μm) diffuse radiation, thermal infrared (8.0–12.0 μm) diffuse radiation, The precipitation includes large scale and convective parts.

Numerous studies have been conducted using SiB in off-line mode and coupled mode. The off-line studies include (1) the model's overall performance against the observations for different land types and under different forcing conditions (e.g., Sellers and Dorman 1987; Sellers et al. 1989; Sen et al. 2000), and (2) production of global biophysical land-surface dataset (Dorman and Sellers 1989; Los et al. 2000). SiB has been linked with GCMs to study the implementation strategy (Sato et al. 1989; Randall et al. 1996), explore the impacts of tropical deforestation on regional climate (Nobre et al. 1991), compare radiative and physiological effects of doubled atmospheric CO_2 on climate (Bounoua et al. 1999), simulate terrestrial surface CO_2 concentrations and fluxes (Denning et al. 1996), assess δO^{18} in atmospheric CO_2 (Ciais et al. 1997), assess sensitivity of climate to changes in vegetation growth as measured by NDVI (Bounoua et al. 2000), and examine the effects of vegetation on the diurnal temperature range (Collatz et al. 2000).

Xue et al. (1991) described a simplified version of SiB model (SSiB) by reducing the number of input parameters to 26 from the original 54. Among those 28 eliminated parameters, 17 are due to change of model structure (i.e., reducing the vegetation to one layer from the original two storeys) and the remaining 11 are from the simplification of parameterizations (e.g., in radiation fluxes, stomatal resistance and aerodynamic resistances). The simplified version is shown to reproduce the original results quite closely but with the computational cost reduced by about 55%.

4. Major Issues of Land-Surface Modeling

From a brief review of BATS and SiB as two examples of current LSMs, several general conclusions can be drawn. *First*, these models by design may be regarded as one-dimensional models, i.e., only layers in the vertical (*z*) direction are considered. While they are to be used ultimately in 3-dimensional atmospheric models, they ignore the horizontal interactions of land surface processes between adjacent grid squares. Processes that occur in the soil-vegetation-snow-atmosphere system of one grid square are not affected by what happen in systems of the neighboring grid squares. For example, the runoff is not modeled to take this into account. *Secondly*, vegetation is treated as a "big leaf", i.e., (linearly) scaling from a size of a normal leaf up to a grid square of 10 km × 10 km to 100 km × 100 km so that a single stoma is associated with the big leaf. *Thirdly*, only three land components (soil, snow and vegetation) are treated explicitly whereas land ice and lakes are neglected. *Fourthly*, while carbon fluxes and plant growth are included, the number and areal coverage of vegetation types within a grid are prescribed.

While some progress has been made to address some of the above issues, e.g. topographic-based runoff (Famiglietti and Wood 1994), vegetation dynamics (Foley et al. 1996), and effects of lakes in LSMs (Bonan 1995), it is likely that they will remain active research topics in next five to ten years. Meanwhile, further research is required to resolve uncertainties in several aspects discussed below.

4.1. *Datasets*

The dataset of vegetation and soil is an indispensable and important part of LSMs. However, the chosen number of vegetation and soil types, the assignment of the dominant type for each grid cell, and the derived secondary parameters for each type are still in "trial-and-error" stage. The sensitivity of LSMs to the chosen procedure has not yet been adequately explored. In particular, the sensitivity of the schemes to the aggregation procedure used to transform the data from the original high resolution (say, 1km×1km) to the required host model resolution has just begun to receive attention (e.g., Shuttleworth et al. 1997; Bonan et al. 2002b).

A wide variety of approaches has been used to specify vegetation and soil data. Matthews' (1983) vegetation dataset has been previously used in generating the data for BATS and SiB. However, the application of this dataset varies from scheme to scheme. For example, BATS also utilizes the Wilson and Henderson-Sellers vegetation dataset. The SiB model mainly uses the Kuchler (1983) vegetation dataset. Nowadays, LSMs have used satellite-derived land cover classifications and vegetation seasonality (Los et al. 2000; Bonan et al. 2002a,b).

The soil datasets are different in all the schemes although FAO/UNESCO (1974) is the main data source for most of the schemes. These digital archives of global vegetation and soil types are generally obtained by digitizing published global maps of vegetation and soil (Matthews 1983; Wilson and Henderson-Sellers 1985). Their approach was to choose one or a few main maps as main references for the globe, while the other maps were used to complement these for particular regions. Therefore, the data quality is heterogeneous across the globe. In addition, since their sources are generally based on past publications, their data cannot reflect recent rapid changes of land surface features (e.g., tropical deforestation).

As for the secondary parameters (e.g., roughness length, zero-plane displace height, leaf area index, canopy height, minimum stomatal resistance), most of them can be estimated by numerical and field experiments (Sellers and Dorman 1987; Noilhan and Planton 1989; Sellers et al. 1989; Sen et al. 2001), while some of them can be inferred from satellite observations (Sellers et al. 1996b; Zeng et al. 2000; Schaudt and Dickinson 2000). However, many have to be inferred by "intelligent guessing" as guided by the literature (Dickinson et al. 1993; 2002). Dorman and Sellers (1989) undertook an interesting study to provide a mutually consistent global climatology of surface albedo, surface roughness and the minimum stomatal resistance on a $1°\times1°$ grid by running their SiB submodels with prescribed PAR and wind velocity. This global dataset has been updated recently by Los et al. (2000). Their dataset may be used in GCMs which do not have biophysically based LSMs.

Satellite imaging has been used to obtain integrated and consistent information about land-surface conditions. There are an increasing number of studies that have used satellite-derived vegetation types and/or seasonality in the coupled LSMs and GCMs (Bounoua et al. 2000; Buermann et al. 2001; Bonan et al. 2002b). Wei et al. (2001) used AVHRR-derived surface albedos to evaluate the coupled LSMs and GCM simulations. Tsvetsinskaya et al. (2002) related MODIS-derived surface albedo to soils and rock types over arid regions.

The past twenty years have also seen a number of important field experiments, e.g., ARME (Shuttleworth et al. 1984), BOREAS (Sellers et al. 1997), FIFE (Sellers et al. 1992), and HAPEX-MOBILHY (Andre et al. 1996) (see Appendix A for the meanings of these acronyms). These field observations need to be integrated with remotely sensed data to produce a new generation of global land surface parameters.

4.2. *LSM and GCM Compatibility*

When developing a land surface model for a GCM, one has to make certain compromises between its realism and its suitability for use in a climate model. Scientists in plant and soil sciences have developed their own complex framework which may be comparable to the climate models in terms of complexity. However, their framework must be somewhat simplified before being used in climate models. On the other hand, land surface processes have to be adequately treated to capture their importance to climate.

As discussed earlier, the minimum resolution requirements of LSMs are at least two soil layers and one canopy layer. Physically, the two soil layers are necessary since the top thin layer responds faster with the diurnal cycle while the deeper layer controls the seasonal changes (Deardorff 1977, 1978). More layers of vegetation are desirable for realistic calculations of radiation transfer and momentum exchange, but a larger number of parameters must be prescribed and calibrated (Sellers et al. 1989; Gupta et al. 1999). Most LSMs have only one vegetation layer.

Owing to the diurnal variations of incident radiation flux and its partition of sensible and latent fluxes, LSMs require GCMs to include a diurnal cycle. Most LSMs need the incident shortwave radiation in two parts split at 0.7μm. Compared to the other LSMs, SiB requires a more detailed description of the incident radiation flux from GCMs. The five components of the fluxes must be provided for use in the radiative transfer calculations in the canopy. In LSMs, the radiation transfer is generally calculated every time step while in GCMs radiation is computed every 3 hours for shortwave radiation and every hour for longwave radiation to save CPU time.

Generally, GCMs provide the atmospheric forcings for the LSMs such as incident radiation at the surface, precipitation, temperature, water vapour mixing ratio and the wind components. The modeled variables such as surface solar radiation, mixing ratio and precipitation may be questionable at present because of possibly inadequate treatments in cloud, aerosol properties and convection. As the rainfall intensity strongly affects the rainfall interception and re-evaporation by the canopy, the nature of the precipitation (frontal or convective) and its sub-scale variability determine the mutual requirements of GCMs and LSMs (Pitman et al. 1990).

The height of the lowest GCM model level is an important factor for LSMs. In calculating the turbulent transfers of momentum, sensible and latent fluxes between the surface and the lowest model level, the fluxes are assumed to be independent on the height above the surface. This approximation can hold only over flat and homogeneous surfaces (below 100m). Most GCMs to date treat the PBL in a very simple fashion. Only one or two model layers are located within the PBL, and the PBL processes are heavily parameterized with large-scale prognostic variables. On the other hand, most mesoscale models pay greater attention to computation of PBL processes. A lowest model level < 100 m seems necessary, and improved resolution PBL models may be needed in order for the GCM to be fully compatible with some LSMs.

4.3. *Surface Temperature*

The thermal balance at the land surface is important to the near ground climate. Two schemes are commonly used: the slab model and the force-restore method. A slab model assumes a thermal uniform layer with fixed thermal properties and heat conduction equations are solved for the temperature using a finite difference method or finite element method. Its accuracy is increased when the number of layers is increased. Due to the computational cost, the number of layers cannot be very high. The Community Land Model (CLM: Dai et al. 2003, Bonan et al. 2002a) has ten soil layers, while many LSMs have three layers.

The force-restore method is formulated from an analytical solution of the soil heat conduction equation under assumptions of periodic forcings and homogeneous medium

(Deardorff 1977, 1978; Dickinson 1988). One of the two prognostic variables interacts (rapidly) with the forcing term, and the other responds (slowly) with the storage term. This efficient scheme has been used in many LSMs including BATS and SiB.

The above two assumptions in the force-restore method are not always valid. The periodic solar forcing is often disturbed by other factors (e.g., clouds and trees), and heterogeneity rather than homogeneity is common (snow, soil, soil moisture, roots, etc) within a soil layer. Therefore, a number of approaches have been adopted to modify the original force-restore method. Dickinson (1988) takes account of the contribution from snow and soil moisture to the heterogeneity, and proposes a generalized force-restore method with a seasonal cycle included, which has been added in an updated version of BATS. CLM solves the heat diffusion equation explicitly for soil temperatures.

For canopy temperature calculations, the degree of complexity of schemes used is much lower than that adopted for calculating canopy albedo. For example, in SiB, a reasonably realistic two-stream approximation has been used to calculated albedo and five components of solar flux are considered. In contrast, a single temperature is assigned to the whole canopy with a single prescribed canopy heat capacity. This limitation will influence the calculation of stomatal resistance since it is a function of vapor pressure deficit that is in turn dependent on leaf temperature. The ground has one temperature for ground cover and bare soil, and each soil layer is assumed to be isothermal with a single temperature.

Similar approaches are also used in BATS for calculating vegetation temperature except that the heat capacity is assumed to be zero. The scheme used in Abramopoulos et al. (1988) in calculating canopy temperature is slightly more elaborate. It differentiates canopy temperature for wet and dry surfaces. It calculates surface temperatures and soil temperatures for shielded and bare portions of land, separately. Wang and Leuning (1998) developed an efficient one-layered, two-leaf canopy model which calculates the fluxes of sensible heat, latent heat and CO_2 fluxes separately for sunlit and shaded leaves.

4.4. *Soil Moisture and Canopy Interception*

Prognostic variables for describing soil water movements vary. For example, in BATS, they are soil moisture content measured in terms of water equivalent depth d_w, in meters. In Abramopoulos et al. (1988), they are mass of liquid water per unit lateral area in the soil layer, which is a product of water density and the depth, i.e., $d_w\ \rho_w$, (kg m^{-2}). In ISBA (Noilhan and Planton 1989) and CLASS (Verseghy 1991), they are measured with actual volumetric soil moisture X, (m^{3}m^{-3}). In SiB, they are soil moisture wetness W which is a ratio of actual volumetric soil moisture (X) in a layer to its value at saturation (X_s) or porosity. W is dimensionless. In the UKMO LSM (Warrilow et al. 1986), they are products of water density and actual volumetric soil moisture, and are also called soil moisture concentration, in units of kg m^{-3}. All these are essentially the same and related through

$$W = X/X_s = (X\rho_w)/(X_s\rho_w) = (d_w/D)/X_s = d_w/D_w = (d_w\rho_w)/(D_w\rho_w), \tag{8}$$

where D is the depth of soil layer (m) and D_w is the soil moisture capacity (m).

Under a series of assumptions, e.g., a spatially homogeneous soil layer with no horizontal water movement and no melting or freezing within it, vertical movement of water soil will follow Darcy's law (noting that the actual form of the equation is dependent upon the direction of z and that of the water flux). This flux-gradient has been used in all the LSMs discussed in previous sections. Therefore, all of them treat more or less the same processes when calculating soil water. They include, for instance, the forcing (throughfall, canopy drip and snowmelt), surface soil evaporation, surface runoff, capillary and gravitational drainage, and transpiration by the canopy. Only in Abramopoulos et al. (1988), are the surface slope and the interstream distances taken into account in order to calculate the underground runoff and soil moisture content. Some LSMs have included explicit treatment of the frozen soil water budget (Verseghy 1991; Dai et al. 2003).

LSMs calculate the canopy vegetation water store with similar governing equations. BATS uses the same method proposed by Deardorff (1978) to calculate the wet fraction of a canopy, i.e., the 2/3 power law,

$$L_w = (W_{dew}/W_{max})^{2/3}, \tag{9}$$

where L_w = fractional area of leaves covered by water, W_{dew} = water store on the surface of canopy, W_{max} = maximum water the canopy can hold .

In SiB, it is prescribed as

$$L_w = W_{dew} / W_{max} \tag{10}$$

under the condition of the saturation vapor pressure at canopy temperature which is less than the vapor pressure in canopy air space. L_w = 1 otherwise. In earlier version of BATS, W_{max} = $0.2\sigma_f$LAI (mm); in SiB (Sellers et al. 1989), $W_{max} = 0.1\sigma_f$LAI (mm)

4.5. *Evapotranspiration*

In general, surface evapotranspiration may be categorized into three components: soil evaporation, canopy evaporation and transpiration. They all look similar at first sight but are in fact quite different. They can, however, all be written as a product of potential evaporation and a coefficient in the form of (1). The fact that they are dissimilar is mostly because of the coefficient and partly due to the potential evaporation. In the case of soil, the coefficient is controlled by the soil wetness and the soil properties (Abramopoulos et al. 1988). In the case of canopy evaporation, the coefficient is determined by the amount of liquid water on surfaces of the canopy and by the plant morphology, e.g., vegetation cover and leaf area index (Rutter and Morton 1977; Sellers et al. 1986). In the case of canopy transpiration, the coefficient depends upon plant physiology, morphology and the environmental conditions (Sellers et al. 1986). The potential evaporation from the canopy surface is different from that from a soil surface partly because the aerodynamic resistances are different and partly because the temperatures at the soil surface and canopy surfaces differ. In the following we will discuss how they are parameterized in LSMs.

Soil evaporation

Soil evaporation is the water flux from the soil surface to its overlying atmosphere. It is limited by diffusion. Therefore in BATS, evaporation from the soil surface is given by the minimum of potential evaporation and maximum moisture flux through the wet surface that the soil can sustain. This diffusion-limiting maximum moisture flux is parameterized from a multi-layer soil model. However, in SiB, a different approach is used. A soil surface resistance is calculated explicitly and then added to the aerodynamic resistance. The resistance formulation of bare soil evaporation takes the form similar to (3).

Canopy evaporation

Canopy evaporation or wet-canopy evaporation is sometimes called interception loss since it is that portion of the rainfall that is held on canopy surfaces as liquid water and which then evaporates to the atmosphere without reaching the soil moisture store (Rutter and Morton 1977). Researches on interception loss from forests can be represented by two directions: one is guided by Rutter and his co-workers who used a complex numerical model; the second is represented by Gash (1979) who used an analytical model, although conceptually similar to the Rutter model.

In both groups of models, the evaporation from a saturated canopy during rainfall is estimated from the Penman-Monteith equation, i.e., a product of a wetting function and the potential evapotranspiration (Deardorff 1978). The wetting function has been discussed before, and the potential evapotranspiration is the evaporation which would occur from a totally wet canopy assuming the water flux comes freely from the water at the surface of canopy (i.e., canopy surface resistance is zero). Both groups of models were tested against measurements (Rutter and Morton 1977).

Many LSMs simplified Rutter's approach to calculate the evaporation from the canopy surface water store. For example, these LSMs use one prognostic variable only for the whole canopy surfaces (including leaves and stems/trunks). In the resistance formulation of evaporation, the only resistance term is aerodynamic resistance. Therefore, it is essentially equal to potential evaporation rate as for a wet soil surface, except where the resistance is replaced by a bulk boundary layer resistance for the canopy leaves, which is different from the common approach of micrometeorologists and forest scientists who have typically an aerodynamic resistance above canopy under neutral conditions (Shuttleworth et al. 1984).

Canopy transpiration

Canopy transpiration, sometimes called dry-canopy evaporation, is a physiological process associated with water transfer from the soil through roots, stems, branches and leaves (Sellers 1992). Most LSMs use the same formulation of transpiration, i.e., Penman-Monteith's combination equation (Monteith 1981), but more closely related to a soil-plant-atmosphere model for transpiration proposed by Federer (1979).

In general, the resistance term is a sum of the bulk stomatal resistance for the canopy and the bulk boundary layer resistance for the canopy leaves, although in Federer's original model, the aerodynamic resistance was approximated by an equation for a thermally neutral atmosphere above the canopy, which is similar to that used by forest scientists for calculation of interception loss. The transpiration is then limited by the supply of water from the roots and atmospheric conditions of demand. For example, the canopy cannot transpire when there is dew forming onto its surface, nor can it when the soil moisture potential drops below the wilting point. However, the parameterizations of stomatal resistance, the dry fraction of transpiring canopy and roots-limiting factor are different in various LSMs. In Federer's approach, the roots-limiting factor was formulated according to Cowan (1965), but this is somewhat simplified in all the LSMs. Recently, a database of vertical root profiles was developed from the literature with 475 profiles from 209 geographic locations (Schenk and Jackson, 2002), and this root database was incorporated in CLM (Dai et al., 2003).

4.6. *Stomatal Resistance*

Stomatal resistance is a biophysical sub-grid scale process which is difficult to parameterize while retaining sufficient generality for use in climate models. Although numerous factors determine leaf stomatal resistance, LSMs consider these environmental conditions: solar radiation, temperature, vapour pressure deficit, leaf water potential (soil water potential), ambient carbon dioxide, and nutrient level. Various schemes exist in the literature (e.g., Jarvis 1976; Farquhar and Sharkey 1982; Sellers et al. 1996a; Dickinson et al. 1998, 2002). In LSMs, a simple strategy is used to scale stomatal resistance from a leaf to a grid. After the stomatal resistance (r_s) of a leaf is calculated, the bulk stomatal resistance (r_c) is derived by assuming all the leaves of the canopy are parallel and then applying Ohm's law. A quantity over a grid square is thus obtained by multiplying vegetation cover fraction (σ_f) to scale up from a canopy to a grid square. In general, there are some uncertainties in the above formulations for the bulk stomatal resistance. For example, the coefficients and parameters that determine the stress factors are not readily available. They must be determined from complex physiological experiments and more advanced theories need to be established.

4.7. *Canopy Drip*

In all LSMs, the calculation of retained water on the canopy surface (including leaves and stems/trunks) is analogous to the bucket model for soil moisture content, whereas a more general drip formulation is developed by Massman (1980). A universal water holding capacity on the canopy surface is equivalent to the soil field capacity. The water level falls because of canopy evaporation and rises because of intercepted rainfall or dew formed onto the surface. The excess amount of water beyond the maximum water storage is called canopy drip, which is equivalent to the soil surface runoff.

In most LSMs, the intercepted rainfall is simply a proportion of incident precipitation above the canopy according to the calculated vegetation cover fraction. In SiB, this is calculated in a way similar to the exponential attenuation of radiation through the canopy (Sellers et al. 1986), and the drip is formulated to be dependent upon the sub-grid scale precipitation (Sato et al. 1989). A similar approach was described by Warrilow et al. (1986).

How the interception and canopy drip is calculated affects the infiltration, surface runoff and evapotranspiration, as demonstrated by Pitman et al. (1990).

4.8. *Runoff*

Runoff, as a part of the hydrological cycle, is important in hydrology and climatology though it is viewed differently in both sciences. It is often perceived in hydrology as a direct response to precipitation (i.e., overland flows in the form of sheets, rivulets, streams and rivers, and/or near-surface flows guided by underlying impermeable layers or subground air channels) with evapotranspiration being a residual; conversely, it is often treated in climatology as the residual after evapotranspiration requirements have been satisfied (Dickinson 1992). In the latter case, the determination of surface evapotranspiration is based on well-established theories by micro-meteorologists and plant scientists.

Hydrologists and climatologists also have different interests in the scales of runoff. Hydrologists are primarily concerned with the local or catchment scale runoff, while climatologists consider runoff at a much broader scale (Shuttleworth 1988; Dickinson 1992). Recently, both views have begun to converge and there is an emergence of the so-called macrohydrology or global-scale hydrology (Shuttleworth 1988).

Warrilow et al. (1986) have given a useful introduction to runoff generating mechanisms. These are summarized as follows: (1) Horton runoff, (2) Dunne runoff, (3) saturation through flow, occurring at less permeable levels in the soil which having become saturated enable moisture to move parallel to the soil horizons in a downslope direction rather than vertically and (4) deep percolation to the water table and thus to surface drainage.

Horton runoff or infiltration excess overland flow, is generated due to the excess of precipitation intensity over soil infiltration capacity at a point. It accounts for only a small fraction of the surface runoff contribution to streamflow. Dunne runoff or saturation excess overland flow is caused due to the occurrence of precipitation over saturated and impermeable surfaces. It is largely responsible for the rapid response of streams to precipitation.

The runoff process is difficult to model due to its complex nature. Nevertheless, there is still a rich coverage in the literature on modelling runoff process over a small scale (e.g., Maidment 1993). L'vovich (1979) provided global maps of annually-averaged surface and ground-water runoff. In contrast, runoff in early generation of climate models simply is overland flow. In more advanced LSMs, it also includes the gravitational drainage. In BATS, surface runoff is parameterized as

$$R_s = (\rho_w / \rho_{wsat})^4 G \text{, if } T_g > 0°\text{C} \tag{11a}$$
$$R_s = (\rho_w / \rho_{wsat}) G \text{, if } T_g < 0°\text{C} \tag{11b}$$

where ρ_w is the averaged soil density toward the top layer, ρ_{wsat} is the saturated soil water density, G is the throughfall plus snowmelt and canopy drip minus soil evaporation. The total runoff is equal to the sum of the surface runoff and the subsoil drainage. In SiB, runoff is defined as precipitation excess plus gravitational outflow from the lowest soil moisture store.

Considerable progress in runoff research has been made recently that includes hillslope processes in LSMs. In particular, a topography-related parameterization of runoff (Beven and Kirkby 1979) has been incorporated in LSMs (Famiglietti and Wood 1994; Stieglitz et al. 1997; Koster et al. 2000; Yang and Niu 2003; Niu and Yang 2003). More research is needed to quantify the role of microtopography at ~2 m horizontal grid and macropores in soil moisture and runoff simulations (Liu and Dickinson 2003).

Runoff is treated as a diagnostic variable in all the LSMs and generally regarded as the excess of water in the soil reservoir. This excess amount plays no further part in the model's hydrological cycle, though in CLASS, an attempt has been made to save surface runoff (overland flow) as ponded water between time steps (Verseghy 1991). Since there are no rivers or lake levels allowed explicitly in climate models, runoff cannot be used to increase lake levels or strengthen the river flows. Input of land runoff water has just begun to be used to alter the salinity of the ocean in the oceanic GCMs. In addition, those simplifications may be in accordance with omitted treatment of sub-grid scale distribution of topography and horizontal water flow processes, though surface slope and interstream distance has only been considered in the GISS LSM (Abramopoulos et al. 1988).

4.9. *Snow*

Of all the large-scale surface features, snow cover exhibits the largest spatial and temporal fluctuations, ranging from 7% to 40% in the Northern Hemisphere during the annual cycle. Associated with these fluctuations are variations in the surface albedo and radiation balance as well as water vapor input to the atmosphere through sublimation and evaporation and water input to the soil and river systems through melt. Therefore, snow cover represents an important component in the Earth's climate system (Yang et al., 1999).

The snow energy balance equations in BATS and SiB are solved following the force-restore approach, which uses a composite snow-soil energy budget and a single snow mass layer. Recent studies have shown that such approach is inadequate to simulate snow mass and snow temperature during melt and that the results are improved with a multi-layer snowpack model (e.g., Lynch-Stieglitz 1994; Jin et al 1999; Boone and Etchevers 2001; Yang and Niu, 2003).

LSMs simulate co-existence of snow-covered canopy, snow-covered bare ground, snow cover underneath the canopy, snow-free canopy, and snow-free bare ground (e.g., Verseghy 1991; Dickinson et al. 1993). The parameterization of snow cover fraction has been shown to play an important role in modulating surface radiation balance and hence snow mass balance through surface albedo (Yang et al. 1997; Slater et al. 2001; Wei et al. 2001). Further research is needed to provide an accurate parameterization of snow masking over the vegetated surface and in the mountainous regions.

4.10. *Sub-grid Scale Variability*

Sub-grid parameterization within climate models "will always remain the Achilles' heel of numerical climate simulation" no matter how fine the model's resolution is (Entekhabi and

Eagleson 1989). Currently there are three main approaches to account for the sub-grid scale variability.

Component approach

As in BATS and SiB, each land grid is treated as a combination of basic components such as soil, vegetation and snow (Figure 2). Temperature and water content equations are established explicitly for each component. Linked with the common overlying atmosphere, these components may be coupled with one another through the radiative, aerodynamic and physiological feedbacks. More components may be included to increase the realism.

Tile approach

In a grid square which consists of one single component, there still exists heterogeneity. For example, Wetzel and Chang (1987) reported that for a grid square with a spacing of 100 km or greater and comprising a soil surface only, the expected sub-grid scale variability of soil moisture may be as large as the total amount of potentially available water in the soil.

Avissar and Pielke (1989) have discussed an approach to account for the non-uniform surfaces of a grid-square. In this approach, the grid-square consists of numerous patches and each patch can be assumed to be homogeneous. It is further assumed that there are no horizontal interactions among these patches, and only processes occurring between the surfaces and the overlying atmosphere are treated. A total flux of the grid is obtained by adding up the area-weighted individual fluxes.

This approach has been incorporated into regional meteorological models to study relationships between meso-scale circulations and land surface heterogeneities (e.g., random distribution of topographical variations, land use and land cover change patterns) (Avissar and Pielke 1989). They found that strong contrasts in sensible heat fluxes due to land surface heterogeneities could generate atmospheric circulation as strong as sea breezes. The tile approach is recently incorporated into global climate models (e.g., Bonan et al. 2002a,b).

Statistical approach

In this approach (Giorgi and Avissar, 1997), some key surface variables such as the grid-square area-means are assumed to have subgrid variance and follow a probability density function (pdf). Precipitation, especially convective, was the first such variable to be studied seriously. The grid size in a climate model is large enough to include a number of convective storms. Consequently there must be great spatial variability of distribution of precipitation over a grid square. However, GCMs assume precipitation uniformly covers the grid square. Precipitation intensity is thus underestimated and this in turn leads to an unrealistic estimation of canopy drip, soil moisture, infiltration, runoff and evapotranspiration. Eagleson et al. (1987) conducted theoretical and observational studies on the relationship between convective storms and the spatial distribution of surface wetting for a catchment area. Warrilow et al. (1986) assumed that precipitation falls on a proportion, μ, of the grid area, and within μ, the local precipitation rate P_1 is represented by a pdf of an exponential form as

$$f_p(P_1) = (\mu/P)\exp(-\mu P_1/P), \tag{12}$$

where P_1 = local point precipitation rate, P = grid square area-averaged precipitation, a prognostic variable from climate models, $f_p(P_l)$ = probability distribution function, μ = fraction of grid-square receiving precipitation.

Assuming a point runoff rate is

$$R_1 = \max(P_1 - F, 0), \tag{13}$$

one has

$$R = P\exp(-\mu F/P),\ \ F_e = P\,[1 - \exp(-\mu F/P)], \tag{14a, b}$$

where R_1 , R = local point and area-averaged runoff rate, respectively. F_e, F = effective and maximum surface infiltration rates. A similar approach was adopted to simulate canopy interception of precipitation by Shuttleworth (1988).

The above schemes have been incorporated into BATS for sensitivity tests over the Amazon rain forest region by Pitman et al. (1990). They found that surface variables in terms of evaporation and runoff are very sensitive to precipitation regimes. Sato et al. (1989) have used a similar exponential pdf for precipitation and canopy interception in SiB and made GCM studies with this version of SiB. Further research is required to compare these two methods at regional scales using observational data.

Entekhabi and Eagleson (1989) further generalized Warrilow et al.'s approach, using a gamma pdf for soil moisture

$$F_s(s) = [\lambda^{\alpha}/\Gamma(\alpha)]s^{\alpha-1}\exp(-\lambda s)\,,\ \ \lambda, \alpha, s > 0\,, \tag{15}$$

where s = soil wetness, λ, α = parameters determining the variance of the mean $E(s)$.

For precipitation, an exponential form of pdf is used, which is a simplification of Γ-pdf when α = 1. Based on both P_l and s and the deterministic equations describing basic soil moisture physics, they then derived a number of grid-square averaged dimensionless quantities including surface runoff ratio (surface runoff to grid square mean precipitation), infiltration rate, bare soil evaporation efficiency (ratio of actual to potential evaporation) and transpiration efficiency.

In their derivations, the Horton and Dunne components of surface runoff are considered. The derived dimensionless quantities are useful to determine the key variables such as runoff, infiltration, bare soil evaporation and transpiration for a grid-square. For example, given a grid-square mean precipitation which is supplied from a climate model, the actual runoff of a grid square is obtained as a product of the runoff ratio and the grid-square mean precipitation. Entekhabi and Eagleson (1989) found that by having such formulations, runoff and evapotranspiration rates are very sensitive to the fraction of surface wetting and the spatial variability of the soil moisture. Liang et al. (1994) incorporated the sub-grid variable infiltration capacity in a land-surface model. However, whether these approaches are valid in GCMs has not yet been investigated fully.

5. Summary

A comprehensive review is given to the history, development, current status and future of biophysical and hydrological aspects of LSMs. It is pointed out that an improved understanding of these processes is critical for an accurate prediction of climate variability on intraseasonal to interannual timescales. The improved understanding also is essential for an effective coupling of these schemes with biogeochemical cycles and vegetation dynamics in GCMs to simulate the impacts of land-cover change and the impacts of increasing CO_2 on climate. Significant problems remain to be addressed, including the difficulties in parameterizing hillslope runoff, fractional snow cover, stomatal resistance, evapotranspiration, and sub-grid-scale variability and scaling. However, further progress is expected as the results of large-scale field experiments and satellite datasets are exploited.

Acknowledgments. It would not have been possible to complete this review without the encouragement of Dr. Xun Zhu. I thank Dr. Mark Cloos, Dr. Robert Scott and two anonymous reviewers for their useful comments, and Qianru Zeng for assistance in drawing Figure 1. This manuscript evolved from the author's dissertation (Yang, 1992). The work reported here is supported by NASA under grant NAG5-12577, NAG5-10209 and by NOAA under grant NA03OAR4310076.

References:

Abramopoulos, F., C. Rosenzweig, and B. Choudhury, 1988: Improved ground hydrology calculations for global climate models (GCMs): soil water movement and evapotranspiration, *J. Climate*, **1**, 921-941.

Andre, J. C., J. P. Goutorbe, and A. Perrier, 1986: HAPEX-MOBILHY: a hydrologic atmospheric experiment for the study of water budget and evaporation flux at the climatic scale, *Bull. Amer. Meteorol. Soc.*, **67**, 138-144.

Avissar, R. and R. A. Pielke, 1989: A parameterization of heterogeneous land surfaces for atmospheric numerical models and its impact on regional meteorology, *Mon. Wea. Rev.*, **117**, 2113-2136.

Ball, J. T., I.E. Woodrow, and J.A. Berry, 1987: A model predicting stomatal conductance and its contribution to the control of photosynthesis under different environmental conditions, *Progress in Photosynthesis Research*, Vol. 1, J. Biggins, Ed., Martinus Nijhof, 221-234.

Beven, K.J. and M. J. Kirkby, 1979. A physically based, variable contributing model of basin hydrology. *Hydrol. Sci. Bull.*, **24**: 43-69.

Bonan, G. B., 1995: Sensitivity of a GCM simulation to inclusion of inland water surfaces, *J. Climate*, **8**, 2691-2704.

Bonan, G. B., K. W. Oleson, et al., 2002a: The land surface climatology of the community land model coupled to the NCAR community climate model, *J. Climate*, **15**, 3123-3149.

Bonan, G. B., S. Levis, et al., 2002b: Landscapes as patches of plant functional types: An integrating concept for climate and ecosystem models, *Global Biogeochem. Cy.,* **16**, art. no. 1021.

Boone, A., and P. Etchevers, 2001: An intercomparison of three snow schemes of varying complexity coupled to the same land surface model: Local-scale evaluation at an Alpine site, *J. Hydrometeorol.*, **2**, 374-394.

Bounoua, L., G. J. Collatz, et al., 2000: Sensitivity of climate to changes in NDVI, *J. Climate*, **13**, 2277-2292.

Bounoua, L., G. J. Collatz, et al., 1999: Interactions between vegetation and climate: Radiative and physiological effects of doubled atmospheric CO_2, *J. Climate*, **12**, 309-324.

Buermann,W., J. R. Dong, et al., 2001: Evaluation of the utility of satellite-based vegetation leaf area index data for climate simulations, *J. Climate*, **14**, 3536-3550.

Burke, E.J., W.J. Shuttleworth, et al., 2000: The impact of the parameterization of heterogeneous vegetation on the modeled large-scale circulation in CCM3-BATS, *Geophys. Res. Lett.*, **27 (3)**, 397-400.

Charney, J. G., W.J. Quirk, et al., 1977: A comparative study of the effects of albedo change on drought in semi-arid regions, *J. Atmos. Sci.*, **34**, 1366-1385.

Ciais, P, A. S., Denning, et al., 1997: A three-dimensional synthesis study of δO^{18} in atmospheric CO_2. 1. Surface fluxes, *J. Geophys. Res.*, **102** (D5), 5857-5872.

Collatz, G. J., L. Bounoua, et al., 2000: A mechanism for the influence of vegetation on the response of the diurnal temperature range to changing climate, *Geophys. Res. Lett.*, **27**, 3381-3384.

Cowan. I.R., 1965: Transport of water in the soil-plant-atmosphere system, *J. Appl. Ecol.*, **2**, 221-239.

Chen, F., R.A. Pielke, Sr. and K. Mitchell, 2001: Development and application of land-surface models for mesoscale atmospheric models: Problems and promises. In Land-Surface Hydrology, Meteorology, and Climate: Observations and Modeling Water Science and Application, V. Lakshmi, J. Alberston, and J. Schaake, Eds., American Geophysical Union, Volume 3, 107-135.

Dai, Y., X. Zeng, et al., 2003: The Common Land Model (CLM), *Bull. Amer. Meteor. Soc.*, **84** (8), 1013-1023.

Deardorff, J. W., 1977: Parameterization of ground surface moisture content for use in atmospheric prediction models, *J. Appl. Meteor.*, **16** (11), 1182-1185.

Deardorff, J. W., 1978: Efficient prediction of ground surface temperature and moisture, with inclusion of a layer of vegetation, *J. Geophys. Res.*, **83**, 1889-1903.

Denning, A. S., G. J. Collatz, et al., 1996: Simulations of terrestrial carbon metabolism and atmospheric CO_2 in a general circulation model. 1. Surface carbon fluxes, *Tellus*, **B48**, 521-542.

Dickinson, R.E., 1983: Land surface processes and climate – surface albedos and energy balance, *Adv. Geophys.*, **25**, 305-353.

Dickinson, R.E., 1988: The force-restore model for surface temperatures and its generalizations. *J. Climate*, **1**, 1086-1097.

Dickinson, R.E., 1992: Land surface, In *Climate System Modeling*, K.E. Trenberth (Ed.), Cambridge University Press, 149-171.

Dickinson, R.E., J.A. Berry, et al., 2002: Nitrogen controls on climate model evapotranspiration, *J. Climate*, **15**(3), 278-295.

Dickinson, R.E. and A. Henderson-Sellers, 1988: Modeling tropical deforestation – A study of GCM land surface parameterizations, *Q. J. Roy. Meteor. Soc.*, **114**, 439-462.

Dickinson, R.E., A. Henderson-Sellers, et al., 1986: Biosphere Atmosphere Transfer Scheme (BATS) for the NCAR Community Climate Model. *NCAR Technical Note*, NCAR/TN-275 + STR.

Dickinson, R.E., A. Henderson-Sellers, and P.J. Kennedy, 1993: Biosphere Atmosphere Transfer Scheme (BATS) Version 1e as Coupled to the NCAR Community Climate Model. *NCAR Technical Note*, NCAR/TN-387 + STR.

Dickinson, R.E., M. Shaikh, et al., 1998: Interactive canopies for a climate model, *J. Climate*, **11**, 2823-2836.

Dorman, J. L., and P. J. Sellers, 1989: A global climatology of albedo, roughness length and stomatal resistance for atmospheric general circulation models as represented by the simple biosphere model (SiB), *J. Appl. Meteorol.*, **28**, 833-855.

Eagleson, P.S., N.M. Fennessy, et al., 1987: Application of spatial Poisson models to air mass thunderstorm rainfall, *J. Geophys. Res.*, **92**, 9661-9678.

Entekhabi, D., and P. S. Eagleson, 1989: Land surface hydrology parameterization for atmospheric general circulation models including subgrid scale spatial variability, *J. Climate*, **2**, 816-831.

Famiglietti, J.S., and E.F. Wood, 1994: Multiscale modeling of spatially variable water and energy-balance processes, *Water Resour. Res.*, **30**(11), 3061-3078.

FAO/UNESCO, 1974: Soil Map of the World, 1:5,000,000, FAO, Paris.

Farquhar, D. D., and T. D. Sharkey, 1982: Stomatal conductance and photosynthesis, *Ann. Rev. Plant Physiol.*, **33**, 317-345.

Federer, C. A., 1979: A soil-plant-atmosphere model for transpiration and availability of soil water, *Water Resour. Res.*, **15**, 555-562.

Foley, J. A., I. C. Prentice, et al., 1996: An integrated biosphere model of land surface processes, terrestrial carbon balance, and vegetation dynamics, *Global Biogeochem. Cy.*, **10**, 603-628.

Garratt, J. R., 1993: Sensitivity of climate simulations to land-surface and atmospheric boundary layer treatments – A review, *J. Climate*, **6**, 419-449.

Giorgi, F., and R. Avissar, 1997: Representation of heterogeneity effects in earth system modeling: Experience from land surface modeling, *Rev. Geophys.*, **35**(4), 413-437.

Giorgi, F., Y. Huang Y, K., et al., 1999: A seasonal cycle simulation over eastern Asia and its sensitivity to radiative transfer and surface processes, *J. Geophys. Res.*, **104**, 6403-6423.

Gupta, H.V., Bastidas, L.A., et al., 1999: Parameter estimation of a land surface scheme using multicriteria methods, *J. Geophys. Res.*, **104**, 19491-19503.

Henderson-Sellers, A., Z.-L. Yang and R.E. Dickinson, 1993: The Project for Intercomparison of Land-surface Parameterization Schemes, *Bull. Amer. Meteor. Soc.*, **74**, 1335-1349.

Jarvis, P.G., 1976: The interpretations of the variation in leaf water potential and stmatal conductance found in canopies in the field, *Phil. Tran. Roy. Soc. London*, **273**, 593-610.

Jin, J.M., X. Gao, et al., 1999: Comparative analyses of physically based snowmelt models for climate simulations. *J. Climate,*, **12**, 2643-2657.

Koster, R.D. and M.J. Suarez, 2001: Soil moisture memory in climate models, *J. Hydrometeor.*, **2**, 558-570.

Koster, R. D., M. J. Suarez, et al., 2000: A catchment-based approach to modeling land surface processes in a general circulation model 1. Model structure, *J. Geophys. Res.*, **105**, 24,809-24,822.

Kuchler, A. W., 1983: World map of natural vegetation, *Goode's World Atlas* (16th ed.), 16-17, Rand McNally.

Liang, X., D. P. Lettenmaier, et al., 1994: A simple hydrologically based model of land surface water and energy fluxes for general circulation models, *J. Geophys. Res.*, **99**, 14,415-14,428.

Liu, Q. and R. E. Dickinson, 2003: Use of a two-mode soil pore size distribution to estimate soil water transport in a land surface model, *Geophys. Res. Lett.*, **30** (6), Art. No. 1331.

Liu, Y.-Q., 2003: Monthly and seasonal variability of the land-atmosphere system, this issue.

Los, S. O., G. J. Collatz, et al., 2000: A global 9-yr biophysical land surface dataset from NOAA AVHRR data, *J. Hydrometeorol.*, **1**, 183-199.

Lu, L., R.A. Pielke, et al., 2001: Implementation of a two-way interactive atmospheric and ecological model and its application to the central United States, *J. Climate*, **14**, 900-919.

L'vovich, M. I., 1979: *World Water Resources and Their Future*, American Geophysical Union, Washington, D.C., 415 pp.

Lynch-Stieglitz, M., 1994: The development and validation of a simple snow model for the GISS GCM, *J. Climate,* **7**, 1842-1855.

Maidment, D. R., 1993: *Handbook of Hydrology*, McGraw-Hill, Inc.

Manabe, S., 1969: Climate and the ocean circulation I. The atmospheric circulation and the hydrology of the Earth's surface, *Mon. Wea. Rev.*, **97**, 739-774.

Martin, M., R.E. Dickinson, and Z.-L. Yang, 1999: Use of a coupled land surface general circulation model to examine the impacts of doubled stomatal resistance on the water resources of the American Southwest, *J. Climate*, **12**, 3359-3375.

Massman, W.J., 1980: Water storage on forest foliage: A general model, *Water Resour. Res.*, **16**, 210-216.

Matthews, E., 1983: Global vegetation and land use: new high-resolution data base for climate studies, *J. Clim. Appl. Meteor.*, **22**, 474-487.

Monteith, J. L., 1981: Evaporation and surface temperature, *Q. J. Roy., Meteor. Soc.*, **107**, 1-27.

Niu, G.-Y. and Z.-L. Yang, 2003: The Versatile Integrator of Surface and Atmosphere processes (VISA) Part II: Evaluation of three topography-based runoff schemes, *Global and Planetary Change*, **38**, 191-208.

Nobre, C. A., P. J. Sellers, and J. Shukla, 1991: Amazonian deforestation and regional climate change, *J. Climate*, **4**, 957-988.

Noilhan, J. and S. Planton, 1989: A simple parameterization of land surface processes for meteorological models, *Mon. Wea. Rev.*, **117**, 536-549.

Pielke, R.A., R. Avissar, et al., 1998: Interactions between the atmosphere and terrestrial ecosystems: influence on weather and climate, *Global Change Biology*, **4**, 461-475.

Pielke, R.A. , 2001: Influence of the spatial distribution of vegetation and soils on the prediction of cumulus convective rainfall, *Rev. Geophys.*, **39**, 151-177.

Pitman, A.J., A. Henderson-Sellers and Z.-L. Yang, 1990: Sensitivity of regional climates to localised precipitation in global models, *Nature,* **346**, 374-737.

Pitman, A.J., 2003: The evolution of, and revolution in, land surface schemes designed for climate models, *Int. J. Climatol.*, **23 (5)**, 479-510.

Randall, D. A., D. A. Dazlich, et al., 1996: A revised land surface parameterization (SiB2) for GCMs. 3. The greening of the Colorado State University general circulation model, *J. Climate*, **9**, 738-763.

Robock, A., M.Q. Mu, et al., 2003: Land surface conditions over Eurasia and Indian summer monsoon rainfall, *J. Geophys. Res.*, **108 (D4)**, art no. 4131.

Rutter, A.J. and A.J. Morton, 1977: A predictive model of rainfall interception in forests. III. Sensitivity of the model to stand parameters and meteorological variables, *J. Appl. Ecol.*, **14**, 567-588.

Sato, N., P. J. Sellers, et al., 1989: Effects of implementing the simple biosphere model in a general circulation model, *J. Atmos. Sci.*, **46** (18), 2757-2782.

Schaudt, K. J., and R. E. Dickinson, 2000: An approach to deriving roughness length and zero-plane displacement height from satellite data, prototyped with BOREAS data, *Agr. Forest. Meteorol.*, **104**, 143-155.

Schenk, H.J. and R.B. Jackson, 2002: The global biogeography of roots, *Ecol. Monog.*, **72**, 311-328.

Sellers, P.J., 1992: Biophysical models of land surface processes, In *Climate System Modeling*, K.E. Trenberth (Ed.), Cambridge University Press, 451-490.

Sellers, P.J., and J. L. Dorman, 1987: Testing the simple biosphere model (SiB) using point micrometeorological and biophysical data, *J. Clim. Appl. Meteorol.*, **26**, 622-651.

Sellers, P.J., Y. Mintz, et al., 1986: A simple biosphere model (SiB) for use within general circulation models, *J. Atmos. Sci.*, **43**, 305-331.

Sellers, P. J., W. J. Shuttleworth, et al., 1989: Calibrating the simple biosphere model for Amazonian tropical forest using field and remote-sensing data. 1. Average calibration with field data, *J. Appl. Meteorol.*, **28**, 727-759.

Sellers, P. J., F. G. Hall, et al., 1992: An overview of the 1st international satellite land surface climatology project (ISLSCP) field experiment (FIFE), *J. Geophys. Res.*, **97**, 18345-18371.

Sellers, P.J., D. A. Randall, et al., 1996a: A revised land surface parameterization (SiB2) for atmospheric GCMs. 1, Model formulation. *J. Climate*, **9**, 676-705.

Sellers, P. J., S. O. Los, et al., 1996b: A revised land surface parameterization (SiB2) for atmospheric GCMs. 2. The generation of global fields of terrestrial biophysical parameters from satellite data, *J Climate*, **9**, 706-737.

Sellers, P.J., F. G. Hall, et al., 1997: BOREAS in 1997: Experiment overview, scientific results, and future directions, *J. Geophys. Res.*, **102**, 28731-28769.

Sen, O.L., L.A. Bastidas, et al., 2001: Impact of field-calibrated vegetation parameters on GCM climate simulations, *Q. J. Roy. Meteor. Soc.*, **127**, Part B, 1199-1223.

Sen, O.L., W.J. Shuttleworth and Z.-L. Yang, 2000: Comparative evaluation of BATS2, BATS and SiB2 with Amazon data, *Journal of Hydrometeorology*, **1** (2), 135-153.

Shuttleworth, W. J., 1988: Macrohydrology – The new challenge for process hydrology, *J. Hydrol.*, **100**, 31-56.

Shuttleworth, W.J., J.H.C. Gash, et al., 1984: Eddy-correlation measurements of energy partition for Amazonian forest, *Q. J. Roy. Meteor. Soc.*, **110**, 1143-1162.

Shuttleworth, W.J., Z.-L. Yang and A.M. Arain, 1997: Aggregation rules for surface parameters in global models, *Hydrology and Earth System Sciences*, **2**, 217-226.

Slater, A., C.A. Schlosser, et al., 2001: The representation of snow in land-surface schemes: results from PILPS 2(d), *Journal of Hydrometeorology*, **2**, 7-25.

Stieglitz, M., D. Rind, et al., 1997: An efficient approach to modeling the topographic control of surface hydrology for regional and global climate modeling, *J. Climate*, **10**, 118-137.

Tsvetsinskaya, E. A., C. B. Schaaf, et al., 2002: Relating MODIS-derived surface albedo to soils and rock types over Northern Africa and the Arabian peninsula, *Geophys. Res. Lett.*, **29**, art. no. 1353.

Unland, H., P. Houser, et al., 1996: Surface flux measurement and modeling at a semi-arid Sonoran desert site, *Agricultural and Forest Meteorology,* **82**, 119-153.

Verseghy, D.L., 1991: CLASS: A Canadian land surface scheme for GCMs, I. Soil model, *Int. J. Climatol.*, **11**, 111-133.

Wang, Y.P. and R. Leuning, 1998: A two-leaf model for canopy conductance, photosynthesis and partitioning of availability energy I: model description and comparison with a multi-layered model, *Agric. Forest Meteor.*, **91 (1-2)**, 89-111.

Warrilow, D.A., A.B. Sangster and A. Slingo, 1986: Modeling of land surface processes and their influence on European climate, *Dynamic Climatology Tech. Note* No. 38, Met. Office, UK, 94 pp.

Wei, X., R.E. Dickinson, et al., 2001: Comparison of the albedo computed by land surface models and their evaluation against remotely sensed data, *J. Geophys. Res.,* **106,** 20687-20699.

Wetzel, P. and J.T. Chang, 1987: Concerning the relationship between evapotranspiration and soil moisture, *J. Clim. Appl. Meteor*., **26**, 18-27.

Wilson, M. F., and A. Henderson-Sellers, 1985: A global archive of land cover and soil data for use in general circulation models, *J. Climatol*., **5**, 119-143.

Xue, Y., P. J. Sellers, et al., 1991: A simplified biosphere model for global climate studies, *J. Climate*, **4**, 345-364.

Yang, Z.-L., 1992: Land Surface Processes in 3-Dimensional Climate Models, Ph.D. thesis, Macquarie University, 437 pp.

Yang, Z.-L., 1995: Investigating impacts of anomalous land-surface conditions on Australian climate with an advanced land-surface model coupled with the BMRC AGCM, *Int. J. Climatol*., **15**, 137-174.

Yang, Z.-L. and G.-Y. Niu, 2003: The Versatile Integrator of Surface and Atmosphere Processes (VISA) Part 1: Model description, *Global and Planetary Change*, **38**, 175-189.

Yang, Z.-L., R.E. Dickinson, et al. 1999: Simulation of snow mass and extent in global climate models, *Hydrological Processes*, **13 (12-13)**, 2097-2113.

Yang, Z.-L., R.E. Dickinson, et al., 1995: Preliminary study of spin-up processes in landsurface models with the first stage data of Project for Intercomparison of Land Surface Parameterization Schemes Phase 1(a), *J. Geophys. Res*., **100**, 16,553-16,578.

Yang, Z.-L., R.E. Dickinson, et al., 1997: Validation of the snow sub-model of the Biosphere-Atmosphere Transfer Scheme with Russian snow cover and meteorological observational data, *J. Climate*, **10**, 353-373.

Zeng, X., R.E. Dickinson, et al., 2000: Derivation and evaluation of global 1-km fractional vegetation cover data for land modeling, *J. Appl. Meteorol*., **39**, 826-839.

AIR-SEA INTERACTIONS ASSOCIATED WITH MESOSCALE WEATHER SYSTEMS

LIAN XIE
North Carolina State University
Raleigh, NC 27695-8208, USA
E-mail: lian_xie@ncsu.edu

(Manuscript received 4 October 2002)

The benefit of coupled models has been well documented for seasonal to interannual climate predictions and global climate change assessments, whereas the advantage of mesoscale coupled models in weather and ocean predictions has only recently been exploited. Observational evidence supports the notion that mesoscale atmospheric processes over the tropical ocean and the coastal waters are coupled to the underlying ocean. Experimental forecasts of coastal ocean circulation and winter weather using coupled models confirmed the improvement in forecast skills over uncoupled models. Recent studies also showed that remarkable improvement in hurricane intensity forecasts could be achieved by using coupled hurricane-ocean models. In this paper, we provide a brief review on the subject of air-sea interactions associated with mesoscale weather systems and its application in operational weather and ocean predictions. Possible directions and opportunities for future studies are discussed.

1. Introduction

The subject of mesoscale air-sea interaction discussed in this article deals with the causal and feedback impacts between the atmosphere and the ocean at spatial and temporal scales encompassed by mesoscale weather systems. The atmospheric mesoscale can be divided into three sub-scales: meso-γ (2-20 km), meso-β (20-200 km) and meso-α (200-2000 km) with temporal periods ranging from O (10) minutes to O (10) hours (Orlanski 1975). On the other hand, oceanic processes at the scale of 10^0-10^2 km are considered as mesoscale, which coincides with the meso-β and γ scales of the atmosphere. Atmospheric meso-α generally corresponds to the synoptic length scale in the ocean. Thus, studies of air-sea interactions associated with mesoscale weather systems must deal with oceanic processes not only at mesoscale but also at synoptic scale (10^2-10^3 km) with temporal extents of up to two weeks.

The wide spectrum of spatial and temporal scales and the mismatch of these scales between the atmosphere and the ocean provide complex but challenging problems. Among them, there are two basic forms of coupling issues: 1) concurrent coupling, and 2) sequential coupling (Xie et al., 1999a). Concurrent coupling refers to air-sea feedback processes occurring between specific atmospheric and oceanic systems within their respective life cycles, such as the mutual impact of an atmospheric cyclone and a warm-core eddy in the ocean. Since any significant mutual responses in the coupled atmosphere-ocean system require the presence of a mechanism to create large enough changes in both media of the coupled system within the period during which they are co-located, a tight concurrent coupling can only occur when the mesoscale weather system has sufficient time to interact with the underlying ocean feature. Thus, a tight concurrent coupling is more likely to occur when the atmospheric system moves at low speed and when there exists a mechanism to accelerate the response between the ocean and the atmosphere than otherwise. For example, a hurricane moving at a slower speed and parallel to the Gulf Stream is more likely to intensify than one that moves at a faster speed and across the Gulf Stream (Bright et al. 2002). Sequential coupling refers to the air-sea feedback processes occurring during two or more sequential time periods, and often encompassing more than one event. For example, interactions between an atmospheric frontal system and the warm western boundary current, such as the Gulf Stream, often produce conditions favorable for subsequent coastal frontogenesis and cyclogenesis.

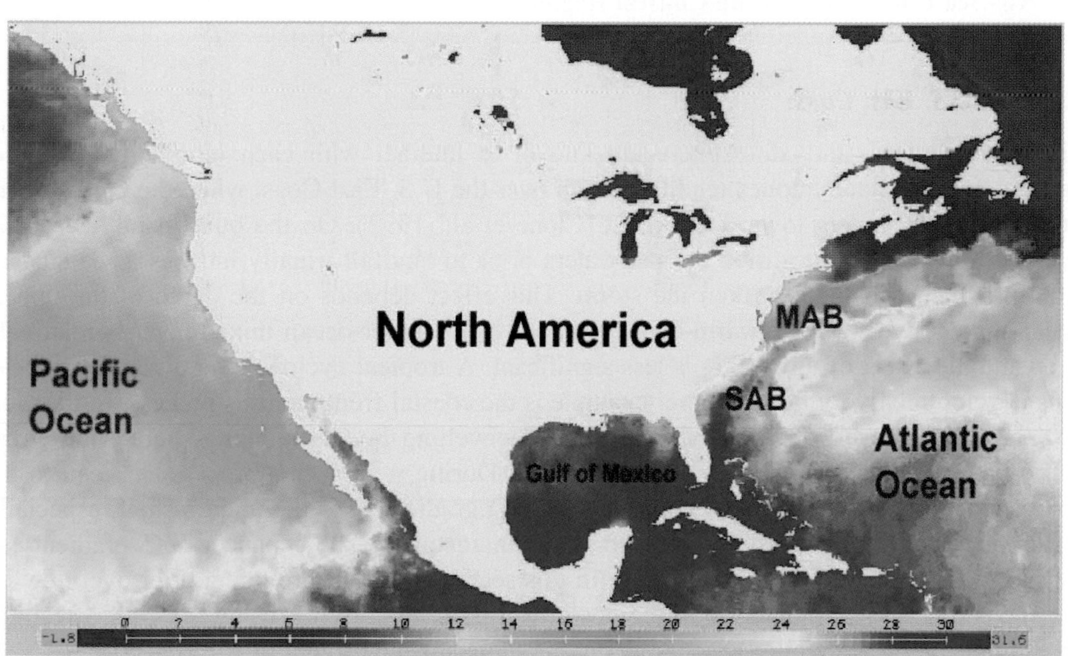

Figure 1. Regions of mesoscale air-sea interactions referred to in the text. The sea surface temperature image is obtained from NOAA NCDC through its website: http://www.ncdc.noaa.gov.

In this paper, we will discuss mesoscale air-sea interactions in two geographical regions, namely, the coastal zone and the tropics (Figure 1). The preference of mid-latitude cyclones to develop off the east coast of North America and Asia in the vicinity of the warm western boundary currents (the Gulf Stream and the Kuroshio Current) points to the importance of air-sea interaction in coastal cyclogenesis. On the other hand, these western boundary currents and the circulation on adjacent continental shelves are likely strongly influenced by wind stress and surface buoyancy flux. Thus, the atmosphere and the ocean are inextricably coupled in coastal region of western boundaries of major ocean basins. Coupled processes have also been observed along the U.S. West Coast where a very stable marine atmospheric boundary layer (MABL) forms over cold coastal currents, which allows the coastal atmospheric disturbances to propagate along the interface, and stationary flow patterns governed by hydraulics to develop within the layer (Dorman 1985). These atmospheric flows can further modify the coastal ocean circulation locally, or even remotely by generating coastal-trapped waves. In the tropics, the impact of air-sea interaction on the generation and intensification of tropical cyclones has been exploited recently. The effect of mesoscale air-sea exchange on intraseasonal to interannual climate variability has also received attention.

In the sections to follow, we will review recent advances in mesoscale air-sea interaction studies in the coastal region, using wintertime air-sea coupling off the Southeast U.S. coast as an illustration, and in the tropics, with a focus on tropical cyclone-ocean interaction. References on the broad topic of air-sea interactions can be found in Csanady (1999) and Kraus and Businger (1994).

2. Air-Sea Interaction in the Coastal Region

2.1. *The U.S. East Coast*

The coastal ocean and atmosphere are known to interact with each other. For example, extratropical coastal cyclones tend to deepen over the U.S. East Coast when the Gulf Stream Front (GSF) meanders toward the coast (Cione et al., 1993). On the other hand, a tropical cyclone moving over stratified coastal waters prior to landfall usually induces a sea surface cooling which can then weaken the storm. This effect depends on the depth of the upper-ocean mixed layer. Over a warm-core ocean eddy, the upper-ocean mixed layer is relatively deep and the effect of upwelling is less significant. A tropical cyclone over a warm eddy can thus develop relatively fast. Another example is the coastal frontogenesis process. In the Gulf Stream region, onshore transport driven by downwelling-favorable along-shore wind stress can cause advective warming on the outer shelf. During winter, this process can lead to the enhancement of the sea surface temperature (SST) gradient over the middle shelf, especially when near-shore water is cooled by cold air simultaneously. The enhanced SST gradient can further modify the surface wind pattern. In this section, we will provide a brief overview on the subject of air-sea interaction in regions of warm ocean currents, using the Gulf Stream as an example. A more detailed discussion on this subject can be found in Xie et al. (1999a).

2.1.1 Response of Coastal Ocean Circulation to Atmospheric Forcing

The atmosphere plays a central role in determining the coastal ocean circulation through either local or remote wind and buoyancy forcing. Generally speaking, the atmosphere influences the coastal ocean in two ways: through the mechanic forcing of the wind stress; and through the buoyancy forcing related to surface heat and salinity fluxes. The upper ocean receives kinetic energy directly from the atmosphere through the first mechanism and potential energy by way of the second mechanism.

a) Effect of Buoyancy Forcing

Buoyancy forcing is determined by heat flux, evaporation and fresh water from run off and precipitation. Intense heat and moisture fluxes are frequently observed over the Gulf

Stream and the adjacent ocean during cold-air outbreaks and coastal storm development in winter. Based on observations made along a cross Gulf Stream section obtained during the Genesis of Atlantic Lows Experiment (GALE), there was a total heat loss of 3.2×10^{13} J (per meter along-stream) from the upper layer of the Gulf Stream during the period January 26-30, 1986. This heat loss resulted in a 1.5°C decrease of Gulf Stream SST and an average mixed-layer deepening of 35 m in the Gulf Stream. The observed mixed-layer deepening and SST reduction in the Gulf Stream have been reproduced in numerical model simulations (Xue et al. 1995).

Surface heat fluxes can also significantly modify the ocean circulation on the continental shelf. In an analysis of the shelf circulation in the Carolina Capes (off the U.S. Carolinas), Pietrafesa and Janowitz (1979) indicated that, in the absence of a wind stress, a positive surface buoyancy flux, could produce along-shore velocities on the order of 0.2 m s^{-1} and an upwelling-like cross-shelf circulation. When combined with downwelling favorable winds, this could lead to oceanic frontogenesis on the shelf. The combined effects of wind and buoyancy forcing on oceanic frontogenesis in shelf water have also been confirmed numerically by Xie et al. (1995).

b) Effect of Wind Forcing

Wind stress was known to affect the cross-stream circulation in the vicinity of the Gulf Stream and on the continental shelf. Adamec and Elsberry (1985) showed that by way of the exchange of momentum in the vertical, currents at the surface decrease in magnitude, whereas the currents immediately below the surface layer increase in magnitude. This results in a thermally-direct cross-stream circulation which flows down the pressure gradient, from warm to cold, at the surface and flows up the pressure gradient, from cold to warm, below the surface layer. This cross-stream circulation tends to weaken the horizontal temperature gradient at the front.

In the presence of bottom friction, the response of coastal ocean currents to downwelling-favorable along-shore wind stress consists of a barotropic, quasi-geostrophic, along-shore current near the coast, onshore Ekman drift currents near the surface, and offshore bottom Ekman drift currents near the bottom. This response, termed Ekman Frictional Equilibrium Response (Beardsley and Butman 1974), has been observed on continental shelves around the United States.

The importance of cross-shelf Ekman transport in the onshore intrusion of upper Gulf Stream water in the South Atlantic Bight (SAB) region, and particularly the Carolina Capes, is supported by the fact that SST fronts often form on the Carolina Capes' shelf during persistent southward winds accompanying cold air outbreaks. Oey (1986) proposed the possibility that during winter in the SAB, onshore Ekman transport driven by strong southward wind stress can cause shoreward intrusion of upper Gulf Stream water onto the continental shelf, and that continental shelf fronts form when the warm Gulf Stream water mixes with cooler shelf water. Oey's working hypotheses which describe these intrusion and mixing processes are: 1) Strong wintertime wind stress (~2 dyn cm^{-2}) and cold, dry continental air produce an upper mixed-layer depth of O (100 m). As a result, shelf water is vertically well mixed and the "west wall" of the Gulf Stream Front (GSF) surfaces to form a front at the shelf break. The cold wind induces heat and mass losses from the ocean's surface and continuously cools the shelf waters. The temperature of Gulf Stream water, however, remains essentially unchanged because of a continuous replenishment of warm water from the south. Thus, the shelf-break front separates warm upper Gulf Stream water from cooler shelf water. 2) Strong southward wind impulses of about 10 dyn cm^{-2} can break down the shelf-break front. Warm Gulf Stream water then spills onto the continental shelf where it mixes with cooler shelf water to produce frontal zones. Finally, 3) Continental shelf fronts, once formed, must be maintained by southward wind stresses, which continuously drive warmer offshore water towards a convergence at the front. The wind also induces downward turbulent diffusion of heat, which balances the downwelling (weakening), seaward advection of cold water at the foot of the front. These hypotheses have been subsequently examined with a two-dimensional numerical model and with the GALE observations, and generally consistent features have been found.

In addition to cross-shelf transport by Ekman drift, alongshore advection near the coast may also contribute to mid-shelf frontogenesis. Pietrafesa et al. (2002) showed that the Chesapeake Bay plume and Middle Atlantic Bight waters can breach Diamond Shoals and invade the South Atlantic Bight during the passage of extratropical cyclones, contrary to the notion that Virginia Coastal Water (VCW) is entirely entrained into the Gulf Stream. In fact the cooler and fresher VCW can appear as far south as Frying Pan Shoals, covering the entire North Carolina Capes inner to mid shelf, facilitating the formation of mid-shelf fronts in the Carolina Capes.

2.1.2 Effect of Gulf Stream on Coastal Weather Systems

It is well known that mid-latitude cyclones are generated most frequently off the east coast of North America and Asia in winter (e.g., Sanders and Gyakum, 1980; Anderson and Gyakum, 1989). These cyclones generally track northward and intensify along the western and northern boundaries of the Gulf Stream and Kuroshio Current. This suggests that large surface heat and moisture fluxes in the vicinity of strong SST gradient have important effects on winter storm development. There is some evidence indicating that the contribution of these heat and moisture fluxes to the development of winter storms is most significant during the incipient and early coastal development stages and diminishes during the rapid intensifying stage (Kuo et al. 1991). Thus, air-sea heat and moisture fluxes are important on the genesis of mesoscale atmospheric systems, including coastal fronts, mesoscale vortices, and low-level jets, and on the preconditioning of the coastal atmosphere for later development of coastal cyclones (Bosart et al. 1995; Xie and Lin 1996; Xie et al., 1996). Observational evidence, numerical model results and theoretical analysis of such coastal developments are reviewed below.

a) Coastal frontogenesis influenced by SST distribution

The coastal front is a well-known mesoscale weather system along the east coast of the United States. The general characteristics of a coastal front has been documented for both the New England coastal region (Ballentine 1980) and the North Carolina coastal region (Riordan 1990). The importance of coastal front is underscored when its role in east coast cyclogenesis is considered. Thus, the significance of the coastal front reaches beyond a contributor to local surface weather conditions, to one of regional importance (Riordan 1990).

Coastal fronts off the Carolina Coast often form on the shelf during easterly ambient winds after a major cold-air outbreak. The effect of a cross-shelf SST gradient (or mid-shelf front) on the coastal front was suggested by Riordan (1990). He showed that while confluence and diabatic heating processes associated with cross-shelf SST gradient nearly compensate, their combined effect at the surface is frontogenetic with maxima in concentrated zones along the existing confluent axis. Based on his calculation, Riordan (1990) suggests that direct thermal circulation forced by differing surface thermal properties across the shelf and the Gulf Stream may promote the establishment of convergence associated with frontogenesis. Such a mechanism has been found previously to be important for New England coastal frontogenesis cases as well (Ballentine 1980). Thus, in order to accurately predict the formation and evolution of coastal fronts, it is necessary to have information on the cross-shelf and cross-Gulf Stream SST distribution.

b) Carolina Coastal Low-Level Jet (LLJ) Induced by Cross-Shelf SST Gradient

Bonner's (1968) LLJ climatology for the United States indicates a prominent area of LLJ occurrence over the Great Plains and a secondary frequency maximum along the North Carolina coast. The Carolina coastal LLJ is a mesoscale phenomenon that frequently develops during the autumn and winter months within the baroclinic zones along the Carolina coast and the GSF with the preferred wind direction being from the north or east at the jet stream level within the marine atmospheric boundary layer. The structure of these LLJs have been analyzed by Doyle and Warner (1991) using meteorological observations obtained during GALE IOP 2. They have identified three separate northeasterly boundary-layer LLJs located in the Carolina region. One low-level wind maximum is positioned just to the east of the Appalachians in the Piedmont region. The second is confined to the coastal region associated with land-sea thermal contrast, and the third located to the rear of an offshore cold front over the Gulf Stream.

Several mechanisms have been proposed in the literature as responsible for LLJ formation in the Great Plains of the United States. Blackadar (1957) attributed the nocturnal LLJ development in the Great Plains to the supergeostrophic flow that occurs when the ageostrophic wind component undergoes an inertial oscillation, as the flow at sunset becomes decoupled from the surface frictional stress. Wexler (1961) concluded that the great Plains LLJ develops as a result of the northward deflection of the easterly flow by the Rocky mountain chain. Other studies have emphasized the importance of the development of a nocturnal low level geostrophic wind maximum as a result of the diurnal heating and cooling over sloping terrain (Holton 1967). Upper-level jet streak forcing has also been hypothesized as responsible for the development of Great Plains LLJs. Unlike the LLJs in the Great Plains, the formation mechanism for the LLJs in the Carolina coastal region is not as well understood. The shallow baroclinicity between the cold and dry air east of the Appalachian Mountains and the warm and moist air over the Gulf Stream is strengthened during episodes of cold-air damming (CAD) events that typically occur when northeasterly winds associated with a surface high pressure centered near the Great Lakes funnel cold air against the eastern slope of the Appalachian Mountains during the cold season. Such shallow baroclinicity has been shown to contribute to the development of LLJs that typically have a direction parallel to the coastline and are confined to the boundary layer. The SST distribution in coastal waters has been found to significantly affect the Carolina coastal LLJs (Doyle and Warner 1993).

c) Effects of Gulf Stream Frontal Events

Gulf Stream frontal events manifest themselves as "meanders" and "filaments", and account for much of the current and hydrographic variability on the outer shelf. Pietrafesa (1983) provided a detailed description of a three-dimensional conceptual view of currents and density structure of a Gulf Stream filament. It is defined by a warm, folded-back tongue of anticyclonically flowing near surface water colder, cyclonically flowing near bottom water. Offshore of the warm water tongue is a cyclonically spinning cold dome of upwelled water. A Gulf Stream filament event similar to that described above was observed from 9-11 February 1986 during the GALE 1986. Using data from King Air and Electra aircraft, ships, buoys, Cross-Chain Loran Atmospheric Sounding System (CLASS) and satellite imagery during this period, Reddy and Raman (1994) showed that somewhat quiescent conditions existed on 9 February but a mesoscale circulation formed over the filament on 10 February. This low intensified into a cyclone on 11 February. In addition to the formation of a meso-low over the filament, the analysis also indicated a significant increase in wind speed and a change in wind direction across the filament up to a height of about 300m. Northeasterly ambient winds changed to northwesterly, thus with winds almost at right angles to the filament. Vertical profiles of wind speeds over the filament indicated low-level acceleration of winds due to a mesoscale convergence. This, in combination with the effects of the horizontal temperature gradients between the cold core adjacent to the filament and the Gulf Stream, appears to have induced a closed mesoscale circulation leading to the generation of a meso-low.

d) Effect of Gulf Stream Offshore Position on Coastal Storm Development

Besides filaments, shoreward excursions of Gulf Stream meanders also appear to contribute to the intensity of offshore storms (Halliwell and Mooers 1983). Such a hypothesis has been substantiated by a statistical analysis of Cione et al. (1993). In this study, a climatological analysis was undertaken to test the hypothesis that storm intensification is influenced by the Gulf Stream's relative location with respect to the North Carolina coast. One hundred and sixteen storm events off the Carolina-Virginia coast from 1982-1990 were surveyed and analyzed regarding their relationships with the offshore distance between the GSF and the coast at Wilmington and at Cape Hatteras, NC. Among the 116 storms, 19 (group A) experienced pressure drops (dP/dt) > 12 mb/12hrs while the other (group B), was entirely composed of disturbances with pressure decreases less than or equal to 12mb/12hrs. For the 19 group-A storms, the mean distance to the Gulf Stream's edge was 56.6 km, with a standard deviation of 17.1 km. For the 97 group B storms, the mean and standard deviation were 71.4 km and 26.8 km, respectively. Using a two sample T-test that assumed unequal variance, a true difference between means was found well beyond the 0.5% significance level. These results suggest that the Gulf Stream position may indeed affect initial deepening rates since the Gulf Stream to Hatteras distance is clearly less during periods of deep offshore intensification.

e) Modeling Studies

Observational evidence indicates that the effects of SST gradients on the continental shelf and across the Gulf Stream may facilitate coastal developments of meso-β scale coastal fronts, LLJs, meso-cyclones and occasionally, explosive cyclogenesis off the east coast of the United States during late fall, winter and early spring. However, numerical model studies showed that a broad spectrum of cyclone life cycles and their sensitivity to SST variations are possible (Kuo et al. 1991; Bosart et al. 1995). Petterssen et al. (1962), in a study of characteristic Atlantic cyclones, concluded that oceanic surface sensible and latent heat fluxes had little impact on cyclogenesis because these fluxes were maximized in the cold air behind the cold front tailing the surface cyclone. The physical basis for their argument was that this distribution of oceanic sensible and latent heat fluxes acted to dampen the thermal wave associated with the cyclone, thereby reducing the eddy available potential energy of the large-scale flow. In several other studies using the Penn State-NCAR mesoscale mode, oceanic sensible and latent heat fluxes were found to be secondary, but still important, in the development of coastal cyclones (e.g., Kuo et al. 1991). Bosart and Lin (1984) demonstrated that the oceanic sensible and latent heat fluxes in advance of the Presidents' Day storm of February 1979 along the east coast of the United States played an important role in contributing to the warming, moistening and destabilizing of the air mass flowing westward toward the developing cyclone center to the south of a massive anticyclone. A similar situation was reported by Bosart et al. (1995) for the February 1974 cyclone. Furthermore, Bosart et al. (1995) suggested that the development of the February 1974 cyclone was favored by the presence of a Gulf Stream warm core ring and an unusual penetration of the Gulf Stream much farther to the north and west than is typical (Cione et al. 1993).

2.2. *The U.S. West Coast*

One of the most noticeable mesoscale phenomena in the coastal waters of the eastern Pacific is the upwelling off the California coast. It occurs in the spring and summer when the Northeast Pacific high-pressure system is established over the ocean and the low-pressure systems over land supporting upwelling favorable southward winds in the coastal zone. Wind-driven coastal ocean upwelling provides nutrient-rich waters to the coastal environment, which supports approximately 50% of the world's annual seafood production (Packard, 1981).

Coastal upwelling is primarily wind-driven. It produces much cooler SST than that of the ambient surface water. As a result, sharp temperature fronts often form between the cooler water near shore and the warmer water offshore. Such SST fronts are observed in many coastal upwelling regions and can produce complex responses in overlying MABL, which has a mean depth of about 200 m along the central California coast and 400 m in the southern California coastal region. When warm air flows over cold coastal waters in the upwelling region, a large air temperature inversion develops. In addition to advection, inversions in the trade wind region can also be caused by the subsidence associated with the subtropical high-pressure system and the cooling of surface air due to turbulent mixing over cold surface water. The temperature difference across the inversion in the California upwelling region is typically on the order of 10°C. Such a strong inversion separates the near-surface marine air from the air above it. The inversion layer may support Kelvin waves, gravity currents, or both (Dorman 1985). The characteristics of the waves are determined by the Froude number (*Fr*) defined as the ratio of the wind speed to the long gravity wave speed. If *Fr* is greater than 1, the flow is referred to as supercritical and long gravity waves cannot propagate upstream. The MABL off the coast of California has been observed to be frequently under high *Fr* region (supercritical) in the inner 100 km (Dorman et al. 2000).

The effect of coastal inversion can be quite different in regions of complex coastal topography. When the inversion height is lower than the immediate topography, the boundary layer flow is constrained by the coastline. When the coastal topography changes shape, such as in the vicinity of capes, the boundary layer flow is disrupted. If the topography turns away from the flow, the boundary layer height decreases and wind speed increases downstream. When the coastal topography bends into the flow, the boundary layer height increases and wind speed decreases. In this case, a hydraulic jump occurs. Changes in the inversion height can be detected from the changes of hydrostatic pressure at the surface. Numerical model simulations show compression bulges with increased MABL depth and slowing of the boundary layer air on the upwind side of capes (Koracin and Dorman 2001), but such features have not been detected in observations.

The presence of stable MABL has significant implications in coastal air-sea interaction. Turbulence is suppressed and often weak or intermittent within the shallow stable MABL. As a result, developing a similarity theory for the stable MABL is a challenging task. The interaction between the gravity waves and the turbulence further complicates the problem. Recent attempts to develop a local similarity theory for offshore flow in stable MABL showed some success, but developing a universal similarity theory for stable MABL is still an illusive goal (Nappo and Johansson, 1999).

Considerable effects of air-sea interaction on coastal weather are observed in the California coastal upwelling region. These include strong low-level jets and large spatial variations in surface wind stress and wind stress curl. However, evidence of coupled feedback between the stable MABL and the coastal upwelling is yet to be established. Enriquez and Friehe (1994) speculate that the changes in sea surface temperature due to upwelling are not large enough to cause significant feedback.

2.3. *The Gulf Coast*

Mesoscale air-sea interactions off the Gulf Coast range from the coupling of the land-sea breeze circulations and the SST on the continental shelf of the northern Gulf of Mexico, the interaction of winter cyclones with the Loop Current, and the interaction between the Loop Current and warm-core rings with tropical cyclones. Interactions of the Loop Current with extra-tropical storms during the winter and with tropical cyclones during summer are of particular interest because of the associated large heat fluxes and the potential threat of these storms to the coastal communities along the Gulf Coast. Air-sea interaction associated with tropical cyclones will be discussed in detail in Section 5.

Winter season cyclogenesis often occurs in the northwestern Gulf of Mexico (Johnson et al. 1984). Johnson et al. (1984) suggest that the bend of the coastline and the Loop Current in the northwestern Gulf of Mexico provide an initial basis for cyclogenesis there. The combination of the coastal shape and the SST gradient on the shelf of the northwestern Gulf of Mexico can induce a gradual spatial change in MABL height, surface air temperature and surface pressure and the isobars tend to parallel to the shape of the coast. Thus, as a cold front moves offshore, the modification of MABL by shelf waters and the Loop Current produces a pressure field that decreases offshore with isobars parallel to the coastline. This causes cyclonic vorticity and even a low-level convergence line, both of which are favorable for coastal cyclogenesis.

3. Air-Sea Interaction in Tropical Cyclones

3.1. *Ocean Response to Hurricanes*

The response of the upper ocean to hurricanes has been studied extensively in the past. One of the first comprehensive observational studies was made by Leipper (1967). Leipper's observation in the Gulf of Mexico following the passage of Hurricane Hilda in 1964 indicated that upwelling of water from a depth of 60 m occurred along the path of the hurricane. Coincident with the upwelling near the storm center was an outward transport of warm surface water and a downwelling of this water around the periphery of the storm. The net result of upwelling and mixing was a decrease in sea surface temperature (SST) of about 5°C along the wake of the hurricane. Black and Withee (1976) analyzed buoy data acquired during hurricane Eloise of 1975 and found a decrease of sea surface temperature of 2.5°C in the 24 hour period following the passage of the storm. Their data showed that a longer residence time of a hurricane over a given point resulted in greater cooling.

After the passage of a hurricane, internal inertia-gravity waves could be induced in the ocean. Hurricane-induced internal inertia-gravity waves have been observed in the Gulf of Mexico (Brooks, 1983), the Mid Atlantic Bight (MAB) (Mayer et al., 1981), and the South Atlantic Bight (SAB) (Xie et al., 1998). The observation of the wake of Hurricane Allen 1980 in the Gulf of Mexico (Brooks, 1983) showed that near-inertial wake oscillations commenced as the eye of the storm passed. The period of the fully-developed wake was 22-23 h, or about 85% of the local inertial period (IP). The early part (1~2 IP) of the wake response was nearly in-phase vertically, indicating a very large vertical length scale in the thermocline. By the time of the most energetic wake stage (~ 3 IP), the phase lead of the velocity and temperature oscillations clearly increased with depth, indicating downward radiation of energy. The maximum phase difference between the current oscillations at 200 and 575 m depths was about one-third of a cycle, indicating the vertical scale in the thermocline was about 1 km. The large vertical scale of the wake resulted in an estimate of vertical energy transport velocity of approximately 60 m/day, which provides a radiation energy flux sufficient to account for the observed depletion of the mid-thermocline wake kinetic energy. Energy reflection from the shelf edge or the bottom may have distorted the wake after about 4 IP (Brooks, 1983).

Over the continental shelf, the oceanic response is more complicated than that over the deep ocean due to the influence of the complex coastline and large variations in water depth. Using mooring records of Hurricane Belle passing over the outer shelf in the MAB, Mayer et al. (1981) found that the response of the coastal ocean to the hurricane depended strongly on water depth. They found that inertial motion is much weaker at shallower stations due to a lack of inertial resonance to impulsive forcing and large frictional effects and hence a more heavily damped response.

In the SAB, the characteristics of the ocean response to hurricanes are complicated by the presence of the Gulf Stream and its frontal features. In a study of the ocean response to Hurricane Fran of 1996, Xie et al. (1998) suggested that pre-existing Gulf Stream frontal features, such as the oceanic trough located on the outer shelf and upper slope downstream of Charleston, South Carolina, could enhance near-inertial-gravity waves induced by the storm. In a more recent study, Xie et al. (1999) noted the existence of both inertial and subinertial (synoptic) responses in the wake of hurricane Fran. Synoptic variability has long been known to exist in the Gulf Stream system in the vicinity and downstream of the Charleston bump (Pietrafesa et al., 1978; Brooks and Bane, 1978) and trough (Piefrafesa, 1983). The current understanding (eg. Sun and Pietrafesa, 1996) is that synoptic variability of the Gulf Stream system is primarily a result of its internal instability, with meteorological forcing not exerting a dominant influence. Albeit, in this study, we will show that, although synoptic variability of the Gulf Stream is principally a manifestation of Gulf Stream internal dynamic instability, interaction between storm-induced oscillations and intrinsic Gulf Stream variability may result in a more pronounced response in both inertial and subinertial (synoptic) frequencies.

3.2. *Effect of Ocean on Hurricane Intensity Change*

The importance of ocean as an energy source for tropical cyclogenesis is demonstrated by the requirement of a threshold SST (26.5°C) below which no tropical cyclones form. Merrill (1987) examined the relationship between tropical cyclone intensity and the climatological SST over a 12-year period and concluded that the occurrence of the most intense hurricanes in the Atlantic Ocean coincided with the regions of the warmest SST. The warm water, although not sufficient to form tropical cyclones, provided an upper limit of tropical cyclone intensity. Emanual (1988) referred to the upper limit of tropical cyclone intensity as the maximum potential intensity (MPI), which is a function of SST, relative humidity of the atmospheric boundary layer and the temperature of the outflow air at the top of the storm. MPI increases with the increase of SST. The difference between the observed tropical cyclone intensity and its MPI can be used as a predictor for tropical cyclone intensity change (DeMaria and J. Kaplan, 1994).

3.3. *Coupled Hurricane-Ocean Models*

The importance of air-sea interaction in tropical cyclone (TC) intensity change has been well-documented. Hurricanes induce a surface cooling in the ocean as they pass and the reduced SST has a negative impact on the storm's intensity. Gallacher et al. (1989) pointed out that a mere 2.5°C decrease in SST near the core of the storm would suffice to shut down the energy production entirely. Schade (2000) have shown in a coupled hurricane-ocean model that the negative feedback from the storm-induced sea surface cooling can reduce the intensity of the hurricane by over 50%, depending on the translation speed of the storm. The depth of the warm water in the upper ocean mixed layer also plays an important role in the oceanic feedback to tropical cyclone intensity change. Deep ocean mixed layer associated with warm western boundary currents and eddies tend to reduce the negative feedback (Shay et al. 2000, Hong et al., 2000). Hurricane Opal, 1995 was a good example of reduced negative feedback and accelerated storm intensification. Hurricane Opal traveled over a warm-core ocean eddy in the Gulf of Mexico at a moderate speed on October 3 1995 which coincided with a rapid intensification of the storm from 965 hpa to 916 hPa within a 14 hour period. Shay et al. (2000) argued that the relatively large upper ocean heat content associated with warm-core eddies in the ocean contributed to the rapid intensification of Opal. Similarly, warm ocean current features associated with the Gulf Stream can reduce the negative feedback of hurricane-ocean interaction. Recently, Bright et al. (2002) carried out a statistical study on the effect of the Gulf Stream on the intensity change of landfalling TCs off the Southeast coast of the United States based on historical storm data and satellite imagery. Their results confirm the notion that the Gulf Stream is conducive to TC intensification. Conversely, cold-core eddies in the ocean may limit the growth of TCs (Xie et al. 1998). There is growing evidence suggesting that the use of coupled hurricane-ocean models can significantly improve the skill of hurricane intensity forecasts (Bender and Ginnis 2000; Emanuel 1999).

The interaction between tropical storms and the warm western boundary currents and warm-core rings can be explained by the response of the deep ocean mixed layer to moving tropical storms and the feedback of the reduced surface cooling in the deep mixed layer on the further development of the storm. In the absence of a deep mixed layer, such as in the stratified open ocean or in cold core eddies, as a storm moves over the ocean, upward Ekman pumping produces a ridge of cold water along the storm's path (Xie et al. 1999b). The cold-water ridge can persist for several days. The evolution of the cold wake can be explained by analytical solutions of linear inertia-gravity waves (Geisler 1970) as well as numerical solutions of nonlinear internal inertia-gravity waves (Xie et al. 1998). The reduction of SST beneath the storm generally causes weakening in the intensity of the storm. However, since the mixed layer along the western boundary current and in the warm-core ring is deeper than that in the open ocean, surface cooling caused by storms passing over these warm current features are significantly less than it would be in other areas. Therefore, a tropical cyclone over the western boundary current and its warm-core eddies can extract greater amount of energy from the ocean and achieve stronger intensity than it would be over the open ocean.

There is also indication that there exists some level of correlation between high tropical storm frequency and the Loop Current and its warm-core rings in the Gulf of Mexico (Lewis and Hsu 1992). If considering only those tropical storms that originate within the Gulf, significant influences from the Loop Current and its warm-core eddies can be found on the tropical storms.

4. Summary and Concluding Remarks

The subject of air-sea interaction encompasses cause and effect feedbacks between the atmosphere and the ocean. In this article, a brief review is given on the meteorological aspects of mesoscale air-sea interactions. In this context, the coupling between the ocean and mesoscale weather systems occurs over scales ranging from kilometers to hundreds of kilometers and minutes to days. The wide spectra of coupled phenomena cover scales which are both disparate and overlapping, and herein lies the great challenge of not only understanding these couplings conceptually but also establishing both diagnostic and prognostic predictive capabilities. Recent efforts in developing and applying coupled models in hurricane prediction have already shown promise (Bender and Ginnis 2000; Emanuel 1999).

Although the role of oceanic sensible and latent heat fluxes in the offshore development of winter storms has been the subject of much conjecture, observations indicate that large surface heat and moisture fluxes in the vicinity of the Gulf Stream and the Kuroshio Current have significant effects on the development of mesoscale atmospheric systems near the coast. These atmospheric phenomena in turn drive energetic ocean currents along and across the shelf bringing disparate water masses into close confluence, affecting ocean frontogenesis, and the feedback continues. This suggests the necessity of utilizing coupled atmosphere-ocean models in coastal ocean prediction. However, in order to further advance the subject of mesoscale air-sea interaction, several roadblocks must be overcome. First, mesoscale atmospheric and oceanic systems are far more energetic than large-scale systems. Estimating air-sea momentum, heat and moisture fluxes under high-energy, or high wind speed, weather events remains a challenge. How do ocean surface waves develop and break, and modify air-sea exchange of momentum, heat and moisture under high wind events? How do winds, ocean surface waves, and ocean currents interact under various weather conditions and in different parts of the ocean (i.e. deep vs. coastal and near-shore ocean)? Xie et al. (2001) revealed that surface waves and ocean currents at all depths can interact strongly in shallow coastal waters and thus the effect of wind stress on coastal ocean circulation depends not only on wind-wave interaction, but also on wave-stress interaction near the sea bed. However, the 3-way interaction among wind, wave and currents is still hard to quantify both observationally and numerically, particularly because the ocean circulation models are three-dimensional whereas only two-dimensional ocean wave models are available. Second, the mechanisms for a tight coupling between mesoscale atmospheric and oceanic systems are still poorly understood, particularly for short-lived, fast moving systems. Does there exist a mechanism that accelerates the exchange of properties between the ocean and the atmosphere under certain conditions? Under light wind conditions, ocean currents could play a significant role in air-sea exchange of momentum, heat and moisture (Rooth and Xie, 1992). Alternatively, in a numerical study of air-sea interaction associated with a tropical squall-line, Bao et al. (2002) revealed that precipitation can create a shallow stable layer in the water near the sea surface, which acts to isolate the surface water from the ocean interior, and allows the skin temperature to respond more quickly to atmospheric forcing. However, Bao et al. (2002) simulated only one single squall-line system. The effect of precipitation on air-sea exchange can be more convincingly addressed statistically using a large number of cases. Finally, high-resolution atmosphere and ocean monitoring systems are needed to provide sufficient data for model initialization and validation if any coupled mesoscale models are to be used in operational weather and ocean forecasting. Such monitoring systems are just beginning to be established. With increased availability of coupled mesoscale atmosphere-ocean observations and high-performance computing machines, advances on the subject of mesoscale air-sea interactions are expected to accelerate in the 21 century.

Acknowledgments. The National Oceanic and Atmospheric Administration is acknowledged for supporting this work under Grant NA060C0 with Waterstone Group Inc. and North Carolina State University. The constructive comments from three anonymous reviewers are gratefully acknowledged.

References:

Adamec, D. and R. L. Elsberry, 1985: The response of intense ocean current systems entering regions of strong cooling. *J. Phys. Oceanogr.*, **15**, 1284-1295.

Anderson, J.R. and J.R. Gyakum, 1989: A diagnostic study of Pacific basin circulation regimes as determined from extropical cyclone tracks. Mon. Wea. Rev., **117**, 2672-2686.

Ballentine, R. J., 1980: A numerical investigation of New England coastal frontogenesis. *Mon. Wea. Rev.*, **108**, 1479-1497.

Bao, S., S. Raman, and L. Xie, 2002: A numerical simulation of the oceanic response to precipitation over the western Pacific warm pool during TOGA COARE. *Pure and Applied geophysics* (in press).

Beardsley, R. C. and B. Butman, 1974: Circulation the New England continental shelf: response to strong winter storms. *Geophys. Res. Lett.*, **1**, 181-184.

Bender, M. A., and I. Ginis, 2000: Real-case simulations of hurricane-ocean intensification using a high-resolution coupled model. Effects on hurricane intensity. *Mon. Wea. Rev.*, **128**, 917-945.

Black, and G. Withee, 1976: The effect of Hurricane Eloise on the Gulf of Mexico. Proc. Second Conf. Ocean-Atmosphere Interactions, Seattle, Amer. Meteor. Soc., Abstract in Bull. Amer. Meteor. Soc., 57, 139.

Blackadar, A. K., 1957: Boundary layer wind maxima and their significance for the growth of nocturnal inversions. *Bull. Amer. Meteor. Soc.*, **38**, 283-290.

Bonner, W.D., 1968: Climatology of the low level jet. *Mon. Wea. Rev.*, **96**, 833-850.

Bosart, L. F., and S. C. Lin, 1984: A diagnostic study of the President's Day Storm of February 1979. *Mon. Wea. Rev.*, **112** , 2148-2177.

Bosart, L. F., C.-C. Lai, and E. Rogers, 1995: Incipient explosive marine cyclogenesis: Coastal development. *Tellus*, **47A**, 1-29.

Bright, R., L. Xie, and L. J. Pietrafesa, 2002: Evidence of the Gulf Stream's influence on tropical cyclone intensity. *Geophy. Res. Lett.*, **29**, 1801-1804.

Brooks, D. A., 1983: The wake of Hurricane Allen in the Gulf of Mexico. J. Phys. Oceanogr., 13, 117-129.

Cione, J. J., S. Raman, and L. J. Pietrafesa 1993: The effect of Gulf Stream induced-baroclinicity on U.S. east coast winter cyclones. *Mon. Wea. Rev.*, **121** , 421-430.

Csanady, G.T, 1999: Air-Sea Interaction: Laws & Mechanisms. Cambridge University Press, 248p.

DeMaria, M., J. Kaplan, 1994: Sea surface temperature and the maximum intensity of Atlantic tropical cyclones. J. Climate, 7, 1324-1334.

Dorman, C. E., 1985: Evidence of Kevin waves in California's marine layer and related eddy generation. *Monthly Weather Review*, **113**, 827-839.

Dorman , C. E., T. holt, D. P. Rogers and K. Edwards, 2000: Large-scale structure of the June-July 1996 marine layer along California and Oregon. *Monthly Weather Review*, **128**, 1632-1652.

Doyle, J. D. and T. T. Warner, 1991: A Carolina coastal low-level jet during GALE IOP 2. *Mon. Wea. Rev.*, **119**, 2414-2428.

Doyle, J. D. and T. T. Warner, 1993: The impact of the sea surface temperature resolution on mesoscale coastal processes during GALE IOP 2. *Mon. Wea. Rev.*, **121**, 313-334.

Emanuel, K. A., 1999: Thermodynamic control of hurricane intensity. *Nature*, **401**, 665-669.

Enriquez, A. G. and C. A. Friehe, 1995: Effects of wind stress and wind stress curl variability on coastal upwelling. *J. Phys. Oceanogr.*, **25**, 1651-1671.

Gallacher, P. C., Rotunno, R., and K.A. Emanuel, 1989: Tropical cyclogenesis in a coupled ocean-atmosphre model. *Preprints of the 18th Conf. On Hurricanes and Tropical Meteorology*, American Meteorological Soc., Boston, Mass.

Geisler, J. E., 1970: Linear theory on the response of a two layer ocean to a moving hurricane. *Geophys. Fluid Dyn.*, **1**, 249-272.

Halliwell, G. R. and C. N. K. Mooers, 1983: Meanders of the Gulf Stream downstream from Cape Hatteras, 1975-1978. *J. Phys. Oceanogr.*, **13**, 1275-1292.

Holton, J.R., 1967: The diurnal boundary layer wind oscillation above sloping terrain. *Tellus*, **19**, 199-205.

Hong, X., S.W. Chang, S. Raman, L.K. Shay and R. Hodur, 2000: The interaction between Hurricane Opal (1995) and a warm-core ring in the Gulf of Mexico. Mon. Wea. Rev., 128, 1347-1365.

Johnson, G. A., E. A. Meindl, E. B. Mortimer and J. S. Lynch, 1984: Features associated with repeated strong cyclogenesis in the western Gulf of Mexico during the winter of 1982-83. *Third Conference on Meteorology of the Coastal Zone*, American meteorological Society, Boston, Mass.

Kraus, E.B. and J.A. Businger, 1994: Atmosphere-ocean interaction. Oxford Monograph on Geology and geophysics, No. 27. 363p.

Koracin, D. and C. E. Dorman, 2001: Marine atmospheric boundary layer divergence and clouds along California in June 1996. *Mon. Wea. Rev.*, **129**, 2040-2056.

Kuo, Y.-H., R. J. Reed, and S. Low-Nam , 1991: Effects of surface energy fluxes during the early development and rapid intensification stages of seven explosive cyclones in the western Atlantic. *Mon. Wea. Rev.*, **119**, 457-476.

Leipper, D. F., 1967: Observed ocean conditions and Hurricane Hilda ,1964. J. Atoms. Sci., 24, 182-196.

Lewis, J. K. and S. A. Hsu, 1992: Mesoscale air-sea interactions related to tropical and extratropical storms in the Gulf of Mexico. *J. Geophys. Res.*, **97**, 2215-2228.

Mayer, D. H., H. O. Mofjeld and K. D. Leaman, 1981: Near-internal waves on the outer shelf in the middle Atlantic bight in the wake of Hurricane Belle. J. Phys. Oceanogr., 11, 86-106.

Merill, R.T., 1988: environmental influence on hurricane intensification. J. Atmos. Sci., 45, 1678-1687.

Nappo, C.J. and P.E. Johnasson, 1999: Summary of the Lovanger international workshop on turbulence and diffusion in the stable planetary boundary layer. Bound.-Layer Meteorol., 90, 345-374.

Oey, L.-Y., 1986: The formation and maintenance of density fronts on U.S. southeastern continental shelf during winter. *J. Phys. Oceanogr.*, **16**, 1121-1135.

Orlanski, I, 1975: A rational subdivision of scales for atmospheric processes. *Bull. Amer. Meteor. Soc.*, **56**, 527-530.

Packard, T.T, 1981: Organiser's remarks in Coastal Upwelling. Edited by F.A. Richards. American Geophysical Union, Coastal and Estuarine Sciences Series 1. 529p.

Petterssen, S., D. L. Bradbury, and K. Pedersen, 1962: The Norwegian cyclone models in relation to heat and cold sources. *Geofys. Publ.*, **24**, 243-280.

Pietrafesa, L. J., 1983: Survey of a Gulf Stream frontal filament. *Geophys. Res. Lett.*, **10**, 203-206.

Pietrafesa, L. J. and G. S. Janowitz, 1979: On the effects of buoyancy flux on Continental Shelf circulation. *J. Phys. Oceanogr.*, **9**, 911-918.

Pietrafesa, L. J., C.N. Flagg, L. Xie, G.L. Weatherly, J. M. Morrison, 2002: The winter/spring 1996 OMP current, meteorological, sea state and coastal sea level fields. *Deep-Sea Res. II*, **49**, 4331-4354.

Reddy, N. C. and S. Raman, 1994: Observations of a mesoscale circulation over the Gulf Stream. *The Global Atmosphere-Ocean Systems*, **2**, 21-39.

Riordan, A. J., 1990: Examination of the mesoscale features of the GALE coastal front of 24-25 January 1986. *Mon. Wea. Rev.*, **118**, 258-282.

Rooth, C. and L. Xie, 1992: Air-sea boundary layer dynamics in the presence of mesoscale surface currents. *J. Geophys. Res.*, **97**, 14,431-14,438.

Sanders, F. and J. R. Gyakum, 1980: Synoptic-dynamic climatology of the "Bomb". *Mon. Wea. Rev.*, **108**, 1589-1606.

Schade, L.R., 2000: tropical cyclone intensity and sea surface temperature. J. Atmos. Sci., 57, 3122-3130.

Shay, L. K., G. J. Goni, and P. G. Black, 2000: Effects of a warm oceanic feature on Hurricane Opal. *Mon. Wea. Rev.*, **128**, 1366-1383.

Wexler, H., 1961: A boundary layer interpretation of the low-level jet. *Tellus*, **13**, 368-378.

Xie, L., K. Wu, L. J. Pietrafesa, and C. Zhang, 2001: A numerical study of wave-current interaction through surface and bottom stresses: Part I: Wind-driven circulation in the South Atlantic Bight under uniform winds. *J. Geophysical Research*, **106**, 16,841-16,855.

Xie, L., L. J. Pietrafesa, and S. Raman, 1999a: Coastal ocean-atmosphere coupling. In *Coastal ocean Prediction*, C.N.K Mooers ed. American Geophysical Union, Coastal and Estuarine Studies Series, 523 pp.

Xie, L., L. J. Pietrafesa, and C. Zhang, 1999b: Subinertial response of the Gulf Stream system to Hurricane Fran of 1996. *Geophysical Research Letters*, **26**, 3457-3460.

Xie, L., L. J. Pietrafesa E. Bohn, C. Zhang, and X. Li, 1998: Evidence and mechanism of hurricane Fran-induced ocean cooling in the Charleston trough. *Geophysical Res. Letters*, **25**, 769-772.

Xie, L. and Y.-L. Lin, 1996: A numerical study of stratified flow over an elliptical mesoscale heat source with application to Carolina frontogenesis. *Monthly Weather Review*, **124**, 2807-2827.

Xie, L., L. J. Pietrafesa and Sethu Raman, 1996: Mesoscale air-sea interaction off the Carolina coast. *Global Ocean and Atmosphere Systems*, **4**, 65-88.

Xie, L. and L. J. Pietrafesa, 1995: Shoreward intrusion of upper-layer warm water induced by prescribed shelf-break temperature perturbation and surface wind stress. *Geophysical Research Letters*, **22**, 2585-2588.

Xue, H., J.M. Bane, Jr. and L. M. Goodman, 1995: Modification of the Gulf Stream through strong air-sea interaction in winter: Observations and numerical simulations. *J. Phys. Oceanogr.*, **25** (4), 533-557.

SIMULATION OF SHORT-TERM VARIABILITY OF THE ASIAN SUMMER MONSOON USING THE NCEP REGIONAL ETA MODEL

YIMIN JI

School of Computational Science, George Mason University, Fairfax, VA 22030-4444, USA
E-mail: yji@tsdis.gsfc.nasa.gov

(Manuscript received 13 November 2002)

To improve summer monsoon simulation, the regional high-resolution Eta model of the National Centers for Environmental Prediction is modified and nested in the Center for Ocean-Land-Atmosphere Studies spectral global general circulation model (GCM). A traditional one-way nesting scheme is developed and used such that the lateral boundary conditions of the Eta model are derived from GCM simulations every six hours. The Eta model domain (30°–140°E, 30°S–50°N) covers the whole Asian monsoon region, which includes the Indian, East Asian, and Southeast Asian monsoons. Both the GCM and the Eta model are integrated continuously from mid-April to the end of September to simulate the monsoon rainfall of 1987, an El Niño year, and 1988, a La Niña year, with the prescribed seasonally varying sea surface temperature. Three separate runs are made for each year with atmospheric conditions for April 14, 15, and 16. The ensemble rainfall means of the three simulations for both years are calculated for the GCM and Eta model and compared to *in situ* observations. Overall, the Eta model simulation shows much-improved seasonal precipitation mean and patterns, as well as intraseasonal and interannual variability as compared to its parental GCM simulations.

1. Introduction

The Asian summer monsoon is one of the most complex circulation systems in the world. Its interannual variability has a profound influence on more than 50% of the world's population. The importance of the Asian summer monsoon as a global forcing has also received considerable interest for many years. During the Northern Hemisphere summer, the Asian summer monsoon has been found to be the major source of energy for the global general circulation. The Asian summer monsoon affects a vast area. The spatial variability within the region is so complicated that it is often difficult to treat it as a single unit. Traditionally, the Asian summer monsoon is divided into three major components: the Indian monsoon, Southeast Asian monsoon, and East Asian monsoon.

The summer monsoon is a regular event in that it normally arrives and retreats in the same time period each year. However, all of its major components, including the total amount of seasonal rainfall, the onset date, and the number and duration of breaks, are subject to large year-to-year variations. This variability sometimes determines whether drought or flooding is experienced in the region. The interannual variability of the monsoon circulation is largely determined by the slowly varying surface boundaries, such as sea surface temperature (SST), surface albedo, and soil moisture. The large amplitude of the SST anomaly over a large area occurs mainly in the central and eastern equatorial Pacific Ocean associated with the El Niño-Southern Oscillation (ENSO) cycle. The large SST anomaly shifts the normal precipitation patterns, creates an anomalous heat source in the atmosphere, and modulates the normal roles of both the Hadley and Walker circulations. Rasmusson and Carpenter (1983) explored the relationship between El Niño events and the annual rainfall over India and Sri Lanka. They noted a strong tendency of below-normal rainfall occurrence during the episodes of warm SST anomalies. The deviation of Indian monsoon rainfall after 1980 was studied by Krishnamurti et al. (1990). They showed below-normal Indian monsoon rainfall in 1982 and 1987 (El Niño years) and above-normal rainfall in 1983 and 1988 (La Niña years). The Asian summer monsoon also undergoes a rich spectrum of intraseasonal variability. The three major observed modes of oscillation of the Asian summer monsoons are 4-6 days, 10-20 days, and 30-50 days. The 4-6 day oscillation is observed mainly in the monsoon trough region. The 10-20 day oscillation is associated with the northward propagation of the convection bands. The 30-50 day oscillation is the global low-frequency mode (Madden and Julian 1971).

Global general circulation models (GCMs) have traditionally been the primary tool for simulating monsoon circulations, in spite of the fact that the horizontal resolution of these models allows for only a highly smoothed representation of topographic features and coast lines. Some of the initial attempts in simulating the summer monsoon with GCMs were able to reproduce the gross features of the seasonal mean circulation and pressure system but not those of precipitation. State-of-the-art GCMs have shown some ability to simulate the interannual variability, especially the large-scale characteristics of the observed anomalies in circulation (Palmer et al. 1992; WMO 1994). However, it has been found that most GCMs have serious deficiencies in simulating Asian summer monsoon precipitation patterns, especially over the India and East China regions (WMO 1994). The model deficiencies of GCMs in the representation of physical and dynamical processes undoubtedly contribute to these results. However, GCMs have other significant problems in simulating the Asian summer monsoon. One of these problems is the horizontal and vertical resolution. Most GCMs have been run at resolutions varying in a range of 2.5° to 10° for both latitude and longitude. These resolutions are too coarse to adequately describe mesoscale forcing and yield accurate regional climate details in the Asian summer monsoon region. Trying to diagnose physical and dynamical processes in GCMs at these scales would introduce uncertainties that often preclude a clear analysis of cause and effect.

Although the use of the sigma coordinate system, which has been widely accepted in GCMs as well as other models, precludes the intersection of predictive surfaces and the ground, the steep slopes of sigma surfaces over mountainous terrain may present potential errors in calculating pressure gradient forces. In such regions, They are often equal to the sum of two large opposing terms (Mesinger 1984; Mesinger et al. 1988). The Eta model adopted at the National Centers for Environmental Prediction (NCEP) uses a step mountain coordinate in which the coordinate surfaces are quasi-horizontal while retaining the simplicity of the sigma system (Mesinger et al. 1990; Janjic 1990). The relatively realistic features other than the resolutions of the Eta model compared to other models include orographic effects, convective processes, and planetary boundary layer processes.

In order to simulate realistic features, especially precipitation patterns and variability, of the Asian summer monsoon, we developed a procedure in which the Eta model with 80-km horizontal resolution and 38 vertical layers is nested in the Center for Ocean-Land-Atmosphere Studies (COLA) spectral GCM with rhomboidal truncation of 40 waves and 18 vertical sigma levels. The purpose of this study is to improve the summer monsoon rainfall simulation and address the nature of the relationship between the Asian summer monsoon and SST anomalies associated with the ENSO cycle. We focus our study on seasonal monsoon rainfall and the onset and break of the Indian summer monsoon. All of these components are subject to large interannual variability associated with the ENSO cycle. However, the physics and dynamics of these components themselves are not fully understood.

2. Description of Models

Regional models have been used in climate studies for several decades (Anthes, 1977; Dickinson et al. 1989; Giorgi 1990). The domain sizes of these regional climate models are about 3,000 km × 3,000 km or smaller. Most of these models use a traditional one-way nesting scheme, in which large-scale information is provided to the regional model through lateral boundaries by a GCM or global analysis. In this study, the size of the domain of interest is about 12,000 km × 8,000 km, an order of magnitude larger than that in previous studies. At this scale, the ability of the regional model to simulate realistic large-scale circulation system is crucial. The regional high-resolution model used in this study is adopted from the NCEP 1996 version operational weather forecast Eta model. The horizontal domain extends from 30° to 140° E and 30° S to 50° N, with the center at 10° N, 85° E. The 80-km Eta model has been used in weather forecasting in Northern America which has a similar size to that of the summer monsoon area. The Eta coordinate is normalized with respect to mean sea-level pressure instead of surface pressure as in the sigma system; thus, the Eta coordinate surface is quasi-horizontal.

The GCM used in this study is an improved version of the COLA GCM (Fennessy et al. 1994). The COLA GCM is one of the few GCMs that have shown efficacy in simulating the observed interannual variability of monsoon circulations. Physical characteristics of the Eta model and the COLA GCM are summarized in Table 1. Soil moisture and snow cover are initialized with climatological values but are predicted by both models. Daily variations of sea surface temperature are prescribed, as is the seasonal variation of vegetation cover. Both models include diurnal cycle and cloud radiation interactions. The initial data for the GCM are from the NCEP reanalysis in the R40 L18 spectral form. They are also interpolated to the E-grid form as the initial conditions for the Eta model. The SSTs used in the GCM are optimally interpolated weekly 1° x 1° data in 1987 an 1988. They are linearly interpolated to daily time interval and input into the Eta model.

Table 1 Physical characteristics of the Eta model and GCM

	Eta Model	COLA GCM
Horizontal Space	E grid (80 km (0.75°), Arakawa and Lamb, 1977)	R40 (Model's Gaussian grid at 1.76° lat 2.81° lon)
Vertical Space	38 Eta Levels	18 Sigma levels
Convection	Betts-Miller (Betts and Miller 1986)	Kuo's scheme (Kuo 1965), modified by Anthes (1977)
Land Surface Parameterization	OSU land surface model (Chen et al. 1996)	SSiB (Xue et al. 1990)
Radiation	Fels-Schwarzkopf (1975)	Harshvardhan et al. (1987) and Lasis and Hansen (1974)
Vertical Exchange	Miyakoda et al. (1986), Mellor-Yamada (1974)	Mellor-Yamada (1974) level 2
Lateral Diffusion	Non-linear second order scheme	

3. Nesting Scheme and Experimental Design

The 1993 version of the Eta model uses a bucket model for ground hydrology and land surface parameterizations. There are also deficiencies in the evaporation scheme over ocean. This version of the Eta model shows a drying-out tendency for long-term integration (Ji and Vernekar 1997). Under these circumstances, a specific procedure has been used for the GCM/Eta interactions, that is; the GCM not only provides lateral boundary conditions for the Eta model every 6 hours, but also provides initial conditions every 48 hours to re-initialize the Eta model. The re-initialization nesting procedure for the Eta model rainfall simulation is valid because the model spin-up time period is relatively short (3~6 hours). The details of the nesting procedure and model spin-up of that version of the model system were given by Ji and Vernekar (1997). The re-initialization procedure has been widely used for Eta model climate sensitivity studies using reanalyses as driving forcing (e.g., Xue et al. 1998; Nickovic and Dobricic 1996.).

The 1993 version of the GCM/Eta system has been used to simulate the 1987 and 1988 summer monsoons. Despite the nesting scheme, the Eta model has shown an ability to simulate regional climates that are closer to the observations than the parental GCM simulations. Such simulations were used to study monsoon onset and intraseasonal variability (Vernekar and Ji 1999).

The 1996 version of the Eta model was improved by coupling a land surface model (Chen et al. 1996) to parameterize the land surface processes. The evaporation scheme was also improved. The buffer zone at the lateral boundary was increased from two-line to four-line. In the 1996 version of the GCM/Eta system, the Eta model is integrated continuously without re-initialization. The 1996 version of the model system was also re-configured such that users are able to choose either GCM or NCEP reanalyses to drive the Eta model. Both re-initialization and non-stop integration nesting schemes were tested for the 1987/1988 monsoon simulations. The re-initialization nesting scheme showed the best simulations provided reanalyses were used as the drive forcing. Large-scale circulation systems generated by the two GCM/Eta nesting schemes do not differ substantially although some noticeable improvements on local details are observed from the1996 version of the GCM/Eta system.

In this study, the Eta model is integrated continuously for six months, being provided lateral boundary conditions every six hours from the GCM simulations. We choose 1987 as a typical El Niño case and 1988 as a typical La Niña case. Three simulations were made using the same surface boundary conditions but different initial conditions for three consecutive days each year: April 14, 15, and 16. The differences between the 1988 and 1987 experiments were only the SSTs and initial conditions.

During the 1987 and 1988 years, the ENSO cycle swung from a warm episode to a cold episode. The SST differences between 1988 and 1987 revealed clear evidence of the interannual SST variability associated with the ENSO cycle, with a positive anomaly in the eastern equatorial Pacific in 1987 and negative anomalies in the same region in 1988.

The *in situ* observations (Singh et al. 1992) from 1987 and 1988 showed contrasting behavior in the Indian summer monsoons. In particular, 1987 was a severe drought year while 1988 was an above-normal rainfall monsoon year over most parts of India. The most striking contrast occurred during July and August. The total rainfall in July 1987 was only about 60% of the climatological value. The monsoon break lasted for almost one month. In July 1988, the total rainfall was about 30% above normal.

4. Results

It is speculated that a horizontal resolution of about 50 km may be a necessary to realistically simulate the Asian monsoon precipitation and mesoscale circulation features due to the complex topography in the area. The JJA mean Indian rainfall from *in situ* observations (Singh et al. 1992); ensemble means from the Eta model and the parental GCM simulations are shown in Fig. 1. As compared to the observations, the Eta model simulation shows a much better pattern and more accurate rain amount than the GCM simulations across the country. The precipitation patterns are clearly related to the model topographic features described by Ji and Vernekar (1997). The major deficiency of GCM simulations is the dry biases along the west coast area. The correlation between the Eta ensemble JJA mean rainfall and observations is about 0.92 and between GCM ensemble JJA mean rainfall and observations is about 0.55.

The Eta model showed an ability to reproduce the observed interannual variability of rainfall in Asia. Tables 2 and 3 show the comparisons of seasonal mean rainfall amount derived from Eta, GCM simulations and *in situ* observations in India and Southeast China in 1987 and 1988, respectively. The Eta model showed a more improved rainfall simulation than the GCM simulations in both regions. In both India and Southeast China, the GCM simulations show dry biases. The fact that the seasonal rainfall amount from the Eta model simulation in all cases is much closer to the observations than the GCM simulations reflects the positive influence of higher resolution, better topographic representation, and more sophisticated skills in the precipitation calculation of the Eta model. The predictive value of the Eta model is also higher than that of the GCM. In India, the variance due to the SST anomaly is about eight times larger than the variance due to the initial conditions in the Eta model simulation. This ratio is about four in GCM simulations. The significance level due to the SST anomaly is 99% for the Eta model and 95% for the GCM. In Southeast China, the GCM does not effectively simulate the influence of the SST anomaly.

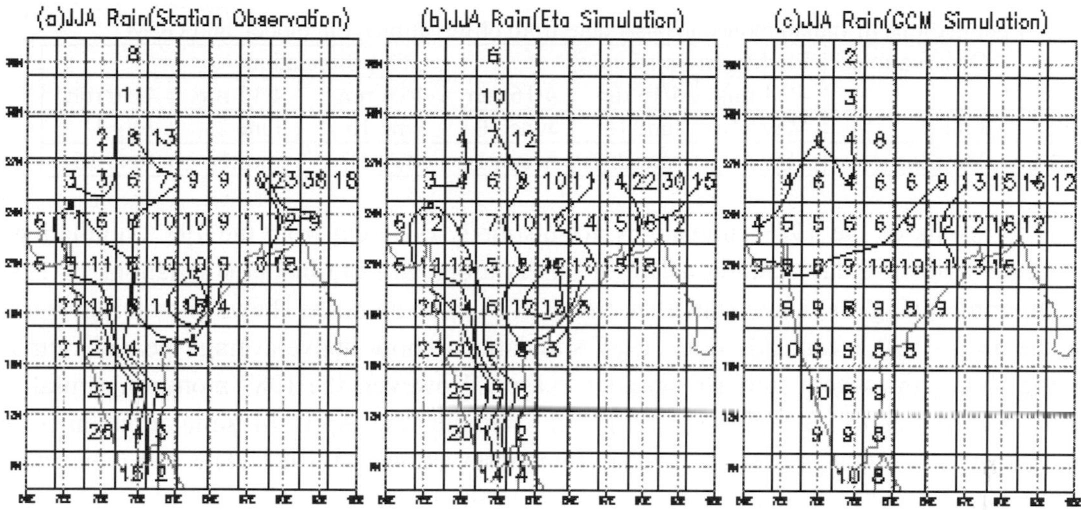

Figure 1. JJA mean rainfall in 1988 (mm/day). (a) Observations; (b) Eta simulation; (c) GCM simulation

Table 2 Comparison of model simulated JJA mean rainfall with *in situ* observations in India

JJA mean rainfall	Eta JJA ensemble mean	GCM JJA ensemble mean	In situ observations
1987 El Nino	695 mm	450 mm	739 mm
1988 La Nina	958 mm	702 mm	972 mm
1971-1990 mean			820 mm
Statistical F test	Significant at 99% level	Significant at 95% level	

Table 3 Comparison of model simulated July mean rainfall with *in situ* observations in Southeast China

July mean rain	Eta July ensemble mean	GCM July ensemble mean	In situ observations
1987 El Nino	202 mm	137 mm	206 mm
1988 La Nina	115 mm	106 mm	120 mm
Statistical F test	Significant at 99% level	Not significant	

The Eta regional model also improves the simulation of abrupt transitions of the atmosphere during the onset period of the summer monsoon. The onset of the summer monsoon in a particular region is characterized by an abrupt increase of rainfall for a sustained period. The mean onset date at the southern tip of the Indian peninsula is around the first day of June. The onset line of the Indian monsoon gradually propagates northward, and the monsoon rains cover the whole of the Indian subcontinent by the middle of July. The average onset date along the South China coast is about May 10. The onset line of the monsoon migrates to the north slowly from May to early June. During the middle of June, it advances into the Yangtze Valley, which marks the beginning of the Mei Yu season.

The estimated onset date based on pentad precipitation from observations and each of these simulations are listed in Table 4. The Eta model results for different initial conditions are closer to each other than the GCM simulations.

Table 4 Estimations of Indian monsoon onset date from observations, Eta model, and GCM

Year	Observations	Eta Model Simulation			GCM Simulation		
		4/14 run	4/15 run	4/16 run	4/14 run	4/15 run	4/16 run
1987	June 3	May 31	June 10	June 10	June 16	June 21	July 1
1988	May 26	May 26	May 26	May 31	May 21	June 5	June 16

The results of the 1988 simulations are distinctly different from the 1987 simulations. The time series (30-day running mean) of observations and Eta model simulations of 1988 Indian rainfall (Fig. 2a) show consistent trends. The thick bold line shows climatology based on the 1971-1990 mean. All three 1988 simulations show above-average pentad rainfall during almost the whole summer monsoon season. However, the 1987 monsoon rainfall is more challenging to simulate (Fig. 2b). For example, the April 14 simulation failed to reproduce the August monsoon revival from observations of 1987. The 1987 ensemble simulations also show a wider spread than the 1988 simulations. All 1987 Eta simulations show below-average pentad rainfall as compared to the climatology. However, observations show more dramatic oscillations. The time series from GCM simulations are not comparable to the observations due to the dry biases and significant deviation between ensemble simulations.

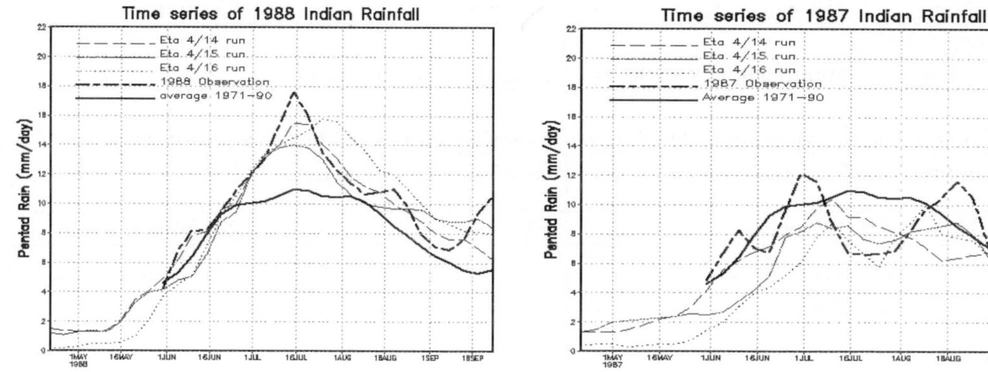

Figure 2. Time series of mean precipitation averaged over India: a) 1988 monsoon season; b) 1987 monsoon season.

The average onset of the East China monsoon occurred in the middle of May 1987 and in early May 1988 in Eta model simulations. These results are similar to the climate data.

The observed active and break monsoons are also better simulated by the Eta model. The latitude-time evolution of Indian monsoon rainfall in 1987 (averaged from 70° – 80° E) from observation (Fig. 3, left panel, top) shows a strong break from mid-July to early August. The active and break monsoon evolution from Eta model simulation (Fig. 3, left panel, middle) is similar to the observations except that the break period in northern India is relatively shorter. The rainfall amount from the Eta simulation is also much closer to the station observations than the GCM simulations. The GCM simulation shows dry biases throughout the season as compared to the observations (Fig. 3, left panel, bottom). The observations and Eta simulation indicate that the drought in 1987 was dominated by the prolonged break in July while the GCM simulation indicates a drought throughout the whole season.

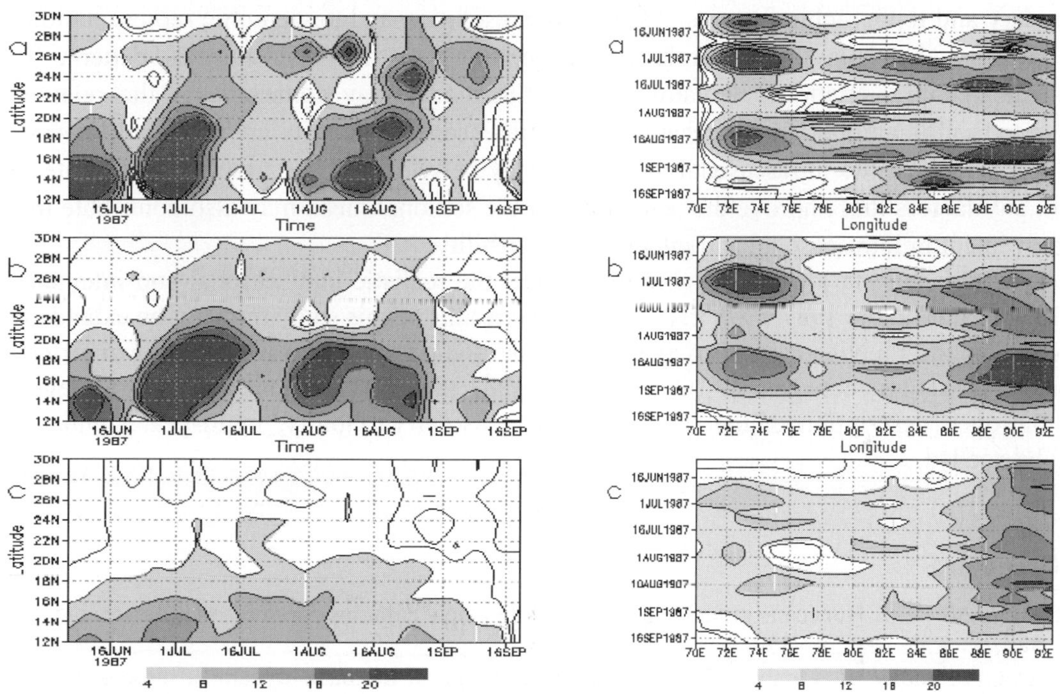

Figure 3. Latitude-time (left panel) and longitude-time (right panel) evolutions of Indian monsoon rainfall (a) Observation, (b) Eta model simulation , (c) GCM simulation.

The longitude-time evolution (Fig. 3, right panel) of 1987 monsoon rainfall (averaged from 15° – 25° N) also indicates the improvements of the Eta model in simulating the intraseasonal variability of the Indian monsoon rainfall.

5. Summary and Discussion

Numerical simulations of the Asian summer monsoon were performed with the high-resolution regional Eta model nested in the COLA GCM. Much-improved simulations of seasonal mean rainfall and interannual variability were made by the regional model system as compared to its parental GCM. The precipitation pattern and value simulated by the regional model in both India and East China are much closer to the observations than the GCM simulations. The correlation between Eta simulation and ground observation is about 0.90

that is significantly higher than the average correlation of about 0.50 between GCM simulations and observations. The Eta model also showed the ability to simulate realistic abrupt monsoon events such as monsoon onset and break. The simulations with the same SST anomaly but different initial conditions showed consistent results in both 1987 and 1988, reflecting the positive predictive value of the regional model system. Ensemble simulations exclude the possibility that these Eta simulations of seasonal mean, monsoon onset, active and break monsoons, and interannual variability, are marching the observations by chance. Our speculation is that the monsoon variability is determined by interactions between surface boundary and topographic features. As such, the resolution and topographic representation of the models shall play a key role in determining the monsoon rainfall pattern. The sophisticated precipitation calculation of the Eta model may contribute to the relatively accurate rainfall amount. The mechanisms of intraseasonal variability are not clear. However, we speculate that the intraseasonal variability may be related to the base seasonal state. As such, any factors that improve the simulation of the seasonal mean may also contribute to an improvement in intraseasonal and interannual variability.

Although the Eta model was able to simulate the realistic intraseasonal and interannual variability during the 1987/1988 ENSO cycle, the results of this study may not be extended to all El Niño and La Niña cases. For example, Indian station observations have shown different monsoon behaviors between the 1982/1983 and 1987/1988 ENSO events. In order to develop an improved description of the ENSO-related monsoon variability, multi-year simulations will be needed.

References:

Anthes. R. A., 1977: Hurricane model experiments with a new cumulus parameterization scheme. *Mon Weather Rev*, **105**, 287-300.

Arakawa, A., and V. R. Lamb, 1977: Computational design of the basic dynamical processes of the UCLA general circulation model. Methods in Computational Physics, *Academic Press*, **17**, 173-265.

Betts, A. K., and M. J. Miller, 1986: A new convective adjustment scheme. Part II: Single column tests using GATE wave, BOMEX, ATEX and Arctic air mass data sets. *Quart. J. Roy. Meteor. Soc.* **112**, 693-709.

Chen, F., K. Mitchell, J. Schaake, Y. Xue, H. L. Oan, V. Koren, Q. Y. Duan, M. Ek, and A. Betts, 1996: Modeling of land surface evaporation by four schemes and comparison with FIFE observations. *J. Geophys. Res.*, **102**, 7251-7268.

Dickinson, R. E., R. M. Errico, F. Giorgi, and G. T. Bates, 1989: A regional climate model for the western U.S.. *Climate Change*, **15**, 383-422.

Fels. S. B., and M. D. Schwarzkopf, 1975: The simplified exchange approximation : Anew method for radiative transfer calculations. J. Atmos. Sci, **32**, 1474-1488.

Fennessy, M. J., J.L. Kinter III, B. Kirtman, L. Marx, S. Nigam, E. Schneider, J. Shukla, D. Straus, A. Vernekar, Y. Xue and J. Zhou, 1994: The simulated Indian monsoon; A GCM sensitivity study. J. Climate, **7**, 33-43

Giorgi, F., 1990: On the simulation of regional climate using a limited area model nested in a general circulation model. *J. Climate*, **3**, 941-963.

Harshvardhan, R. Davies, and T. G. Corsetti, 1984: Long wave radiation parameterization for UCLA/GLAS climate model. *NASA Tech. Memo*, 86072, pp. 51.

Janjic, Z. I., 1990: The step-mountain coordinate model: Physical package. *Mon Weather Rev.*, **118**, 1429-1443.

Ji, Y., and A. D. Vernekar, 1997: Simulation of the Asia summer monsoon of 1987 and 1988 with a regional model nested in a global GCM. *J. Climate*, **10**, 1965-1979.

Krishnamurti, T. N., H. S. Bedi and M. Subramaniam, 1990: The summer monsoon of 1988. Meteorol. *Atmos. Phys.*, **42**, 19-37.

Kuo, H. L., 1965: On the formation and intensification of tropical cyclones through latent heat release by cumulus convection. *J. Atmos. Sci.*, **22**, 40-63.

Lasis, A. A., and J. E. Hansen, 1974: A parameterization of the absorption of solar radiation in the earth's atmosphere. *J. Atmos. Sci.*, **31**, 118-133.

Madden, R., and P. Julian, 1971: Detection of a 40-50 day oscillation in the zonal wind. *J. Atmos. Sci.*, **28**, 702-708.

Mellor, G. L., and T. Yamada, 1974: A hierarchy of turbulence closure models for planetary boundary layers. *J. Atmos. Sci.*, **31**, 1791-1806.

Mesinger, F., 1984: A blocking technique for representation of mountains in atmospheric models. *Rev. Meteor. Aeronautica*, **44**, 195-202.

Mesinger, F., T. L. Black, D. W. Plummer and J. H. Ward, 1990: Eta model precipitation forecasts for a period including tropical storm Allison. *Wea. Forecasting*, **5**, 483-493.

Mesinger, F., Z. I. Janjic, S. Nickovic, D. Gavrilov and D. G. Deaven, 1988: The step-mountain coordinate: Model description and performance for cases of Alpine lee cyclogenesis and for a case of an Appalachian redevelopment. *Mon Weather Rev.*, **116**, 1493-1518.

Miyakoda, K., J. Sirutis and J. Ploshay, 1986: One-month forecast experiments without anomaly boundary forcing. *Mon Weather Rev.*, **114**, 2363-2401.

Nickovic, S., and S. Dobricic, 1996: A model for long-range transport of desert dust. *Mon. Wea. Rev.*, **124**, 2537-2544.

Palmer, T. N., C. Brankovic, P. Viterbo and M. J. Miller, 1992: Modeling interannual variations of summer monsoons. *J. Climate.*, **5**, 399-417.

Rasmusson, E. M., and T. H. Carpenter, 1983: The relationship between eastern equatorial Pacific sea surface temperature and rainfall over India and Sri Lanka, *Mon Weather Rev.*, **111**, 517-528.

Singh, S. V., R. H. Kripalani, and D. R. Sikka, 1992: Interannual variability of the Madden-Julian Oscillations in India summer monsoon rainfall, *J. Climate.*, **5**, 973-978.

Vernekar, A. D., and Y. Ji, 1999: Simulation of the onset and intraseasonal variability of two contrasting summer monsoons, *J. Climate.*, **12**, 1707-1725.

WMO, 1994: International conference on monsoon variability and prediction. Triete, Italy, May 9-13, 1994. WMO/TD - No. 619, WCRP-84,

Xue, Y., P. J. Sellers, J. Kinter and J. Shukla, 1991: A simplified biosphere model for global climate studies. *J. Climate.*, **4**, 345-364.

Xue, Y., F.J. Zeng, Y. Ji, K. Mitchell, and Z. Janjic, 1998: The impact of land surface processes and reanalysis data on the U.S. weather prediction. Proceedings of the First WCRP International Conference on Reanalyses, WMO/TD-NO. 876, 243-246.

EFFECT OF FORCING DATA ERRORS ON CALIBRATION AND UNCERTAINTY ESTIMATES OF THE CHASM MODEL: A MULTI-DATASET STUDY

YOULONG XIA, PAUL L. STOFFA, CHARLES JACKSON AND MRINAL K. SEN
Institute for Geophysics, the John A. and Katherine G. Jackson School of Geosciences
University of Texas at Austin, 4412 Spicewood Spring Road, Austin, TX, USA
E-mail: youlong@utig.ig.utexas.edu

(Manuscript received 15 November 2002)

We investigate the effects of forcing data errors on the calibration and uncertainty estimates of the Chameleon Surface Model (CHASM) land surface model. We use Bayesian Stochastic Inversion (BSI) that is based on Bayes theorem, importance sampling, and Very Fast Simulated Annealing to search for optimal parameters and estimate the posterior probability density function (PPD) for three experiments with varying levels of forcing uncertainty. Since the complete representation of the PPD is impossible in a highly multidimensional parameter space, marginal PPDs and their maximum values are used to describe uncertainties and sensitivities in the parameters of the CHASM land surface model for five biomes representative of tropical forests, tropical pastures, midlatitude grasslands, midlatitude forests, and semiarid grasslands. The results show that forcing data errors have little effect on the calibration of the most important parameters and the estimation of their PPDs, although forcing data errors have significant effect on the selection of non-dominant optimal parameters and estimation of their PPDs. The CHASM land surface model predictions are insensitive to reasonable forcing errors for all sites except Loobos site. The Loobos example suggests that the effect of forcing errors is at least vegetation and climate dependent.

1. Introduction

The results from the Project for Intercomparison of Land-surface Parameterization Schemes (PIPLS) reveal that there are large differences among land surface schemes (Henderson-Sellers et al., 1995) that come from model parameters and model structures (Henderson-Sellers, 1996; Chen et al., 1997; Qu et al., 1998; Desborough, 1999; Schlosser et al., 2000; Slater et al., 2001) when assuming the observations for forcing are 'perfect' for these models. One may use global optimization techniques such as the multi-criteria (MC) approach to search for the optimal parameter values that minimize the differences between observations and model predictions. The results of Gupta et al. (1999) show that the Biosphere-Atmosphere Transfer Scheme (BATS) performs better when its parameters are optimized using the MC method. Xia et al. (2002) show that complex models perform better than simple models when optimal model parameters are used. Recently, Bayesian Stochastic Inversion (BSI) was used to search for optimal parameters along with their uncertainties for the CHASM model (Jackson et al., 2002b; Pitman et al., 2002; Xia et al., 2003a). The results show that the performance of BSI and MC consistently identify similar optimal parameter sets for the seven PILPS sites. Furthermore, BSI is able to quantify the uncertainties of the CHASM model parameters and surface energy flux predictions (e.g. sensible heat fluxes, latent heat fluxes, Xia et al., 2003b) that stem from observational uncertainties in the 'target' observations (but not the forcing data.

However, all these studies assume that the forcing data are accurate and do not contain observational and other artificial errors (e.g. interpolation errors). In fact, both errors indeed exist in forcing data. Observations typically involve 5m to 40m observational towers and may have gaps due to instrument damage. These gaps are usually filled by numerical methods, such as the closest station method (Beljaars and Bosveld, 1997) or interpolation techniques (Xia et al., 1999a; Xia et al., 1999b). This filling will introduce artificial errors into the forcing data. On the other hand, synthetic meteorological data (see Pitman et al., 1999), four-dimensional assimilation data (e.g., NCEP (National Center for Environmental Prediction) data base), the data interpolated from observational stations (see Xia et al., 1999a, Xia et al., 1999b), or a combination of these approaches (Bowling et al., 2002) were often employed as the forcing data for PILPS phase 1 (see Pitman, 1993) and phase 2 experiments (Wood et al., 1998; Bowling, et al., 2002). Therefore, both observations from towers and other sources (see Henderson-Sellers et al., 2002) may contain errors. It is not clear whether and/or how these errors affect the calibration and uncertainty estimates of land surface models.

In this paper we use seven field observation sites (two tropical forest sites, two mid-latitude grassland sites, one mid-latitude forest site, one tropical pasture site, and one semi-arid site) to investigate the impact of forcing data errors (hereafter called forcing errors) on calibration and uncertainty estimates of the CHASM land surface model. To do this, we design three experiments: one experiment with fixed forcing data; a second experiment with forcing data consistent with small errors as estimated in Beljaars and Bosveld (1997) for Cabauw; and a third experiment with forcing data containing larger errors. We consider how forcing errors affect optimal parameter values, parameter sensitivity analyses, estimates of model parameter uncertainties, and predictions of annual mean energy fluxes.

2. Sites, Model, Optimization Method and Experiment Design

2.1. *Sites*

Observations of model forcing and surface energy fluxes collected at seven sites were used in this study. These sites were chosen based upon data availability, different climate and vegetation characteristics representative of grasslands, tropical pastures, tropical forests, and mid-latitude forests. As suggested by Sen et al. (2001), these land surface biomes cover over 50% of the world's land area. Table 1 shows site locations, observational period and data intervals, and vegetation types. All forcing data include either downward short-wave radiation or net radiation, as well as downward long-wave radiation, precipitation, air temperature, wind speed, and specific humidity. The surface energy fluxes used to calibrate the model include sensible heat flux (SH) and latent heat flux (LH). More details can be found in Xia et al. (2003a).

Table 1. Descriptions of seven PILPS sites

Site name	Site location Lat Lon Elevation (m)	Observational period	Observational intervals	Vegetation type
Abracof	10°5'S 61°55'W 120 m	May 1992 – December 1993	60 minutes	Tropical forest
Abracop	10°45'S 62°22'W 220 m	May 1992 – December 1993	60 minutes	Tropical pasture
Amazon	2°57'S 59°57'W 80 m	January 1997 – December 1998	30 minutes	Tropical forest
Armcart	36°36'N 97°29'W 318 m	April 1995 – August 1995	30 minutes	Mid-latitude grassland
Cabauw	51°58'N 4°56'E -0.7 m	January 1987 – December 1987	30 minutes	Mid-latitude grassland
Loobos	52°10'N 5°44'E 52 m	January 1997 – December 1998	30 minutes	Mid-latitude Scott pine
Tucson	32°13'N 111°5'W 730 m	May 1993 – May 1994	20 minutes	Semi-arid grass and shrubs

2.2. *CHASM model*

CHASM (CHAmelon Surface Model) land surface model (Desborough, 1999; Pitman et al., 2002; Xia et al., 2002) has been used within offline experiments of the PILPS phase 2d and 2e (Project for Intercomparison of Land-surface Parameterization Schemes, Schlosser et al., 2000; Slater et al., 2001, Bowling et al., 2002), and experiments using a coupled global climate model (Deborough et al., 2001) and a regional climate model (Zhang et al., 2001). It was designed to explore the general aspects of land-surface energy balance representation within a common modeling framework (Desborough, 1999) that can be run in a variety of formulations ranging from a complex mosaic type structure (see Koster and Suarez, 1992) all the way to the most simple Manabe type formulation (Manabe, 1969). Here we use the complex mosaic-type representation. Within this representation the land-atmosphere interface is divided into two tiles. The first tile is a combination of bare ground and exposed snow with the second tile consisting of dense vegetation. The tiles may be of different sizes and the energy fluxes of each tile are area-weighted. Because separate surface balance is calculated for each tile, temperature variations may exist across the land-atmosphere interface. A prognostic bulk temperature for the storage of energy and a diagnostic skin temperature for the computation of surface energy fluxes are calculated for each tile. Snow cover fractions for both ground and foliage surfaces are calculated as functions of the snowpack depth, density, and the vegetation roughness length. The vegetation fraction is further divided into wet and dry fractions if canopy interception is considered. This model has explicit parameterizations for canopy resistance, canopy interception, vegetation transpiration and bare ground evaporation, but has no explicit canopy-air space (see Pitman et al., 2002).

CHASM uses the formulation of Manabe (1969) for the hydrologic component of the land surface in which the root zone is treated as a bucket with finite water holding capacity. Any water accumulation beyond this capacity is assumed to be runoff. In addition to moisture in the root zone, water can be stored as snow on the ground or on the canopy. Soil temperature is calculated within four soil layers using a finite difference method and zero-flux boundary condition. Each tile has four evaporation sources for canopy evaporation, transpiration, bare ground evaporation, and snow sublimation.

2.3. *Bayesian Stochastic Inversion (BSI)*

The Bayesian formulation of an inverse problem has been widely used in geophysics (Tarantola, 1987; Sen and Stoffa, 1995; Sen and Stoffa, 1996) and in climate-change detection and attribution (Hasselmann, 1998). The Bayesian method accounts for uncertainties in the observed data and model simulations. Sen and Stoffa (1996) and Jackson et al. (2002a) provide a detailed review of parameter and uncertainty estimation of geophysical models and climate models based on Bayesian statistics. The basic idea of the BSI is to identify parameter sets that are consistent with observational uncertainty. We have applied this method to evaluate optimal parameters and estimate parameter uncertainties at each of the seven PILPS sites considered here using the CHASM land surface model with fixed forcing (Xia et al., 2003a, 2003b).

2.4. *Experiment Design and Analysis of Parameters*

Three experiments are designed in this study. The first experiment uses fixed forcing to calibrate 12 CHASM model parameters. The second experiment uses forcing that includes small but reasonably sized forcing errors to calibrate 12 CHASM land surface parameters and 6 forcing parameters. The third experiment is the same as the second experiment but uses larger forcing errors. In all experiments, 12 CHASM parameters and their ranges can be obtained from Xia et al. (2002) and Leplastrier et al. (2002). In the second experiment, selection of error ranges for six forcing variables is based upon the analysis in Beljaars and Bosveld (1997) for Cabauw. Accuracies of wind and temperature are suggested to be 1% or 0.1 K, respectively. The wind effect results in an underestimation varying between 2% and 11% for precipitation. Specific humidity has a typical error of 0.3 g Kg^{-1}, and downward solar and long-wave radiation is believed to be accurate to within a few percent. We use this information for the other sites for the second experiment because no information on forcing uncertainties has been provided at these sites. In the third experiment, error ranges are increased to two to ten times as those in the second experiment following Qu et al. (1998) and Beringer et al. (2002). Even the enlarged ranges are believed to be reasonable according to the previous estimates of interpolation errors (Xia et al., 1999a; 1999b). The description and default values of 18 parameters are shown in Table 2. The forcing data have been modified in the following manner: the solar radiation, long-wave radiation, precipitation and wind speed have been modified by multipliers csol, clrad, cpre, and cv, respectively. The specific humidity and air temperature now include offsets cq and ct, respectively. The ranges of 18 calibration parameters are shown in Table 3. Sensible and latent heat fluxes at all seven sites were used as 'target' observational data for all three experiments. Ratio of the variance of the errors to the variance of observation (RVE) is used as the error function (cost function) for the BSI calibration. RVE is defined as

$$RVE = \sum_{n=1}^{N}(obs_n - sim_n)^2 \Big/ \sum_{n=1}^{N}(obs_n - \overline{obs_n})^2 \,, \tag{1}$$

where N is the number of observational data, obs_n is the observed data, sim_n is the simulated data, and $\overline{obs_n}$ is mean values of observed data.

Table 2. The parameter description and default values of CHASM parameters for seven PILPS sites (adapted from Chen et al.(1997) and Desborough (1999), ABF=Abracof, ABP=Abracop, AMA=Amazon, ARM=Armcart, CAB=Cabauw, LOB=Loobos, TUC=Tucson)

Para-meters	ABF AMA	ABP	ARM	CAB	LOB	TUC	Description
albg	0.20	0.20	0.20	0.20	0.20	0.30	Bare ground albedo
albn	0.75	0.75	0.75	0.75	0.75	0.75	Snow albedo
albv	0.14	0.14	0.23	0.23	0.18	0.30	Vegetation Albedo
lefm	6.00	4.00	4.00	4.00	4.00	4.00	Maximum LAI
veg	0.90	0.90	0.90	0.95	0.80	0.70	Maximum fractional vegetation cover
vegs	0.25	0.25	0.25	0.25	0.25	0.25	Fractional vegetation cover seasonality
rcmn	50.0	50.0	40.0	40.0	40.0	40.0	Minimum canopy resistance (s/m)
wrmax	234.	234.	141.	141.	122.	122.	Available water holding capacity (mm)
z0g	0.01	0.01	0.01	0.01	0.01	0.01	Ground roughness length (m)
z0n	4.0	4.0	4.0	4.0	4.0	4.0	Snow roughness length (10^{-4}m)
z0v	2.00	0.20	0.15	0.15	2.00	0.02	Vegetation roughness length (m)
ts	300.	300.	279.	279.	279.	284.	Initial surface temperature (K)
csol	1.0	1.0	1.0	1.0	1.0	1.0	Downward short-wave solar radiation
clrad	1.0	1.0	1.0	1.0	1.0	1.0	Downward long-wave radiation
cpre	1.0	1.0	1.0	1.0	1.0	1.0	Precipitation
cq	0.0	0.0	0.0	0.0	0.0	0.0	Specific humidity
cv	1.0	1.0	1.0	1.0	1.0	1.0	Wind speed
ct	0.0	0.0	0.0	0.0	0.0	0.0	Air temperature

3. Parameter Sensitivities and Correlations

Parameters that strongly influence model predictions tend to also have strongly peaked posterior probability density functions (PPDs) as there tends to only be a limited range of parameter values that are consistent with observational uncertainty. We use the peak value of the PPD for each parameter to give a bulk sense of parameter sensitivities. As reported in Xia et al. (2003a), the PPD is proportional to exp(-E) value (here E is the error function). A minimum E generates a maximum PPD and a maximum E generates a minimum PPD. As minimum PPD values are close to zero, differences between maximum and minimum PPD values can be approximately represented using maximum PPD values. Therefore, the maximum PPD values which mean strong sensitivity on these parameters are used to analyze the relative importance of the model parameters.

Table 3. Ranges for 12 CHASM model parameters and 6 forcing parameters for all three experiments

Parameters	Experiment 1		Experiment 2		Experiment 3	
	minimum	Maximum	minimum	maximum	minimum	maximum
albg	0.05	0.40	0.05	0.40	0.05	0.40
albn	0.70	1.00	0.70	1.00	0.70	1.00
albv	0.05	0.40	0.05	0.40	0.05	0.40
lefm	4.00	6.00	4.00	6.00	4.00	6.00
veg	0.20	1.00	0.20	1.00	0.20	1.00
vegs	0.23	0.26	0.23	0.26	0.23	0.26
rcmn	0.00	300.00	0.00	300.00	0.00	300.00
wrmax	10.00	600.00	10.00	600.00	10.00	600.00
z0g	0.00	0.01	0.00	0.01	0.00	0.01
z0n	$1.0\text{x}10^{-4}$	$6.0\text{x}10^{-4}$	$1.0\text{x}10^{-4}$	$6.0\text{x}10^{-4}$	$1.0\text{x}10^{-4}$	$6.0\text{x}10^{-4}$
Grassland Forest (z0v)	0.0 0.8	0.4 2.5	0.0 0.8	0.4 2.5	0.0 0.8	0.4 2.5
ts	275.0	305.00	275.0	305.00	275.0	305.00
csol	1.0	1.0	0.9	1.1	0.8	1.2
clrad	1.0	1.0	0.9	1.1	0.8	1.2
cpre	1.0	1.0	1.0	1.15	1.0	1.3
cq	0.0	0.0	$-3.0\text{x}10^{-4}$	$3.0\text{x}10^{-4}$	$-1.0\text{x}10^{-3}$	$1.00\text{x}10^{-3}$
cv	1.0	1.0	0.99	1.01	0.8	1.2
ct	0.0	0.0	-0.1	0.1	-1.0	1.0

Figure 1 shows the maximum marginal PPD values for the three experiments and the seven sites. At the mid-latitude grassland sites (i.e. Armcart, Cabauw), minimum stomatal resistance (rcm), ground roughness length (z0g) and vegetation roughness length (z0v) are important for model predictions (Figures 1d and 1e). At the semi-arid Tucson site, vegetation roughness length (z0v) dominates the predictions of sensible heat fluxes and latent heat fluxes although ground roughness length (z0g) and minimum stomatal resistance are also important (Figure 1g). At the mid-latitude Scott pine forest site (i.e. Loobos), minimum stomatal resistance (rcm) is a dominant parameter although forcing errors are important for experiments two and three (see Figure 1f). At tropical forest and pasture sites (i.e. Abracof and Abracop), minimum stomatal resistance (rcm), vegetation roughness length (z0v) and ground roughness length (z0g) are important although maximum water holding capacity shows its importance (figures 1a-1c).

Our results show that the parameters that are the most important remain the same at six of the seven PILPS sites (Loobos is the only exception) for the three experiments. At Loobos, parameters associated with the forcing are the most important of all parameters (Figure 1f). The small PPD values for the parameters modifying solar radiation (csol), downward longwave radiation (clrad), precipitation (cpre), specific humidity (cq), wind speed (cv), and air temperature (ct) that occur at all sites except Loobos mean that the forcing errors should not be critical for CHASM model predictions. This suggests that the forcing data with reasonable errors may be used for the sensitivity analyses as we do here. This result is not consistent with other sensitivity studies that explore the effects of variable forcing (Qu et al., 1998; Lynch et al., 2001).

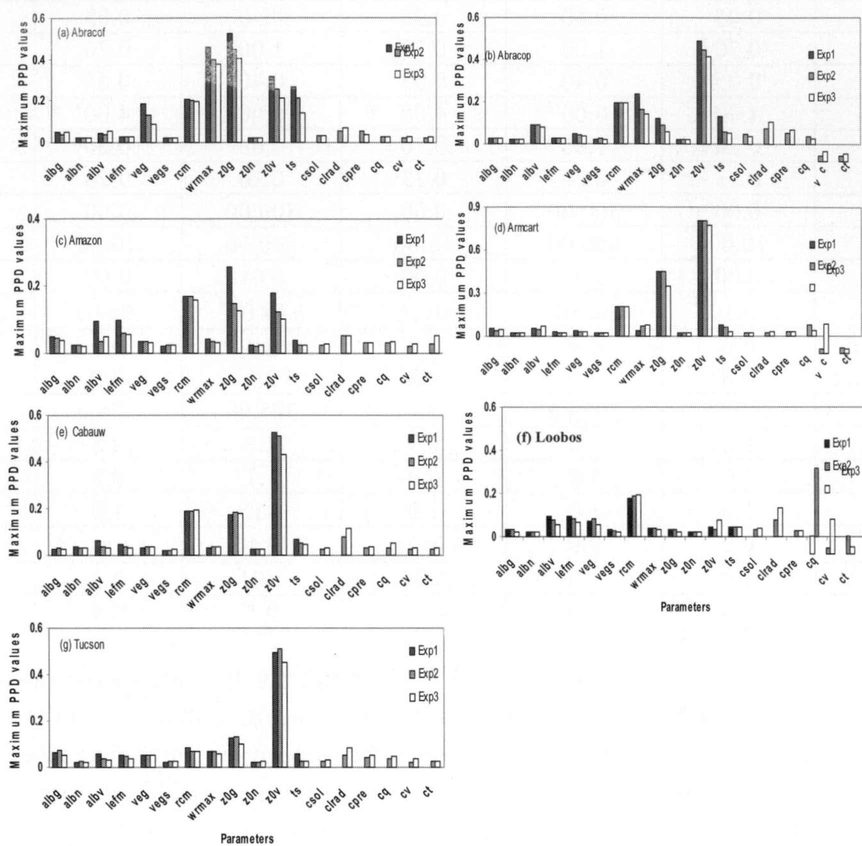

Figure 1. Maximum PPD values for 18 parameters and three experiments at the seven PILPS sites

There are several reasons that may explain this inconsistency. First, the purpose and experiment design of our work are very different from those of Qu et al. (1998) and Lynch et al. (2001). Qu et al. (1998) artificially adds ±2 K to air temperature while all other forcing data remained as observed, and then ran 23 land surface schemes in order to obtain a first-order estimate of the sensitivity of the PILPS schemes to changed air temperatures. Lynch et al. (2001) use a reduced form model to discuss the sensitivity of land surface models to climate change by varying all forcing. Our purpose is to investigate the effect of reasonable forcing errors on calibration and the uncertainty estimations for land surface models. For this reason the amplitude of the forcing errors that we use are necessarily smaller than has been investigated previously. Second, the relative importance of parameters is likely to depend on the type of vegetation (e.g. the CHASM model is sensitive to the forcing errors at mid-latitude Scott pine forest, see Figure 1f) and climate (Lynch et al., 2001). Third, it is possible that certain combinations of selected parameters may result in unnatural model predictions. As reported in Lynch et al. (2001), a combination of both large increases in downward longwave and incoming shortwave radiation, which is unlikely in a realistic climate change scenario, may lead to sensitivities that have no physical meaning (Beringer et al., 2002). In contrast, in our study, BSI uses Very Fast Simulated Annealing to search for the global minimum of an error function by progressively selecting parameter combinations and weighting their relative likelihood in an effort to identify the optimal parameter set. The constraint of making errors between observed and simulated fluxes (e.g., sensible heat fluxes, latent heat fluxes) minimum excludes many unreasonable parameter combinations. There is a negative correlation (-0.4 to -0.2, parameters 13 and 14 in Figure 2) between parameters modifying incoming shortwave radiation (csol) and downward longwave radiation (clrad) for all sites. These negative correlation coefficients mean that combinations of downward longwave radiation and incoming shortwave radiation contain two cases: either increased downward longwave radiation and decreased incoming short wave radiation or decreased downward longwave radiation and increased incoming shortwave radiation.

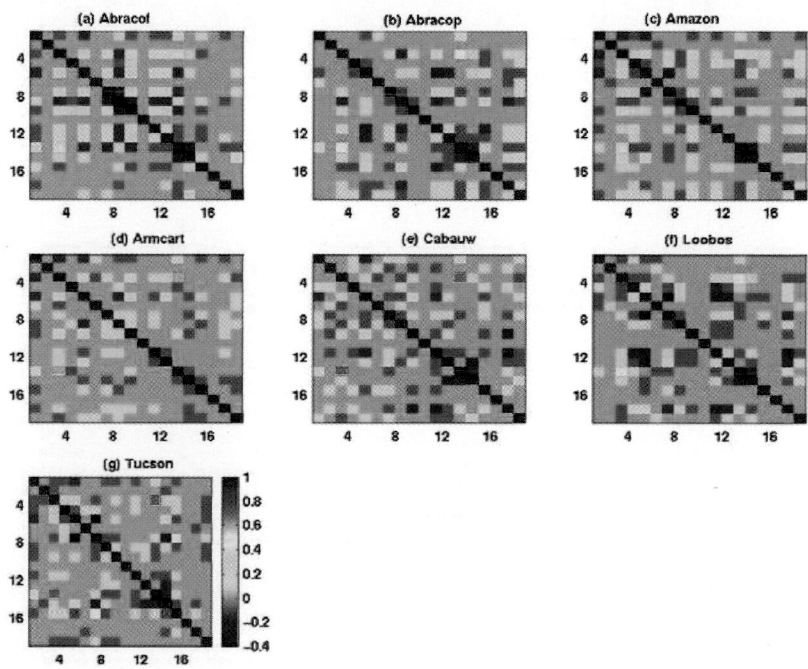

Figure 2. Correlation matrices between 18 parameters for the seven PILPS sites (parameters 1 to 18 are consistent with parameters ablg to ct in Table 3.

Analysis of the posterior correlation matrices for the seven PILPS sites gives a measure of the interdependences that exist between model parameters and/or forcing errors. From Figure 2 we see that some relationships (see parameters 13 and 3 in Figure 2) exist at all sites while other relationships may vary from site to site (see parameters 15 and 7 in Figure 2). For example, there is a positive correlation (0.40 to 0.68) between the parameter affecting incoming shortwave radiation (clrad, parameter 13) and vegetation albedo (albv, parameter 3) for all sites, while a positive correlation between the parameters affecting precipitation (cpre, parameter 15) and minimum stomatal resistance (rcm, parameter 7) can only be seen at the Tucson site. Therefore, it is possible for some model parameters and forcing errors to interact although albv and clrad do not happen to be the most important parameters. We observe that the correlation coefficients between forcing errors and the most important parameters are very small for all sites, indicating that the forcing errors and the most important parameters are nearly independent (see Figure 2).

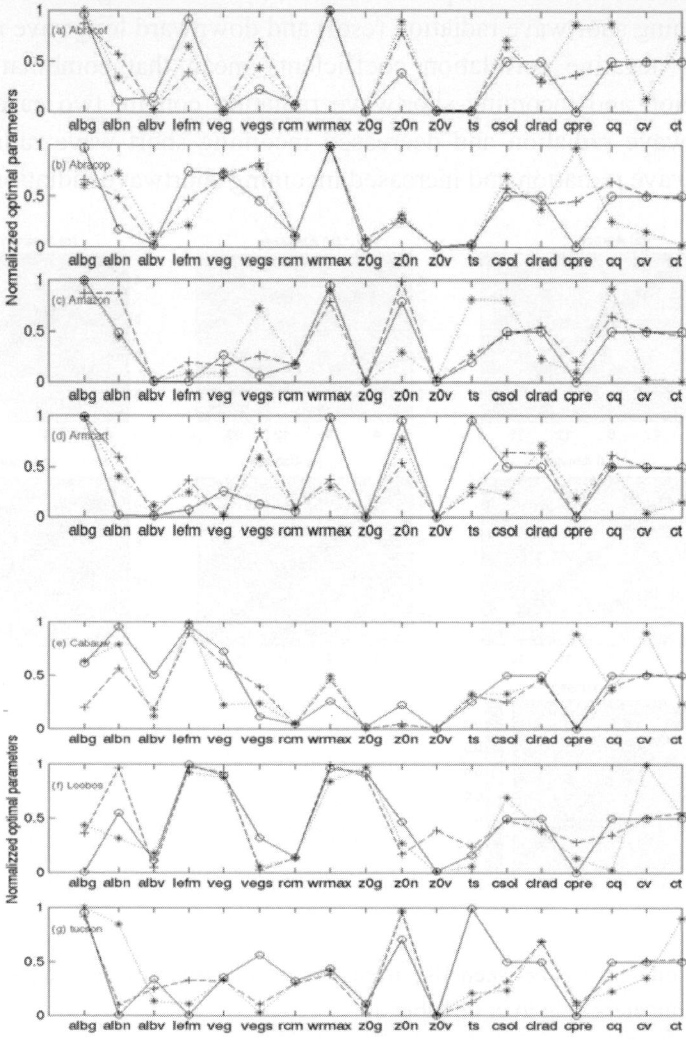

Figure 3. 18 normalized optimal parameter values identified using the BSI approach for the seven PILPS sites (experiment one is represented in solid lines with circles, experiment two is represented in dashed lines with pluses, and experiment is represented in dotted lines with stars). Each parameter is given as a fraction between its minimum and maximum values.

4. Effect of forcing errors on calibrations of the CHASM model

The purpose of calibrating a land surface model is to identify a set of parameters that optimize the model's surface energy flux predictions. The calibrated parameters are shown in Figure 3 for the three experiments and the seven PILPS sites. Experiment one is represented by solid lines with circles, experiment two is represented by dashed lines with pluses, and experiment three is represented by dotted lines with asterisks. These parameters are normalized by this method, that is, for each parameter and each experiment, difference between optimized parameter value and minimum parameter value (see Table 3) is divided by the difference between maximum parameter value and minimum parameter value shown in Table 3. The results presented in Figure 3 show that there are differences among the three experiments, but comparisons of Figure 1 and Figure 3 show that the values of the important parameters are similar for the three experiments. For example, nearly identical parameter values occur for veg, rcm, wrmax, z0g, z0v and ts at Abracof (Figure 3a), albv, rcm, wrmax, z0g, z0v, and ts at Abracop (Figure 3b), rcm, z0g, and z0v at Amazon (Figure 3c), rcm, z0g, and z0v at Armcart (Figure 3d), rcm, z0g, and z0v at Cabauw (Figure 3e); and z0v at the Tucson (Figure 3g). This means that forcing errors have little effect on the ability to identify optimal values for the important parameters at these six sites. At Loobos (Figure 3f), optimal values of the important parameters are different for cq, clrad and cv for experiments two and three, indicating the effect of forcing errors.

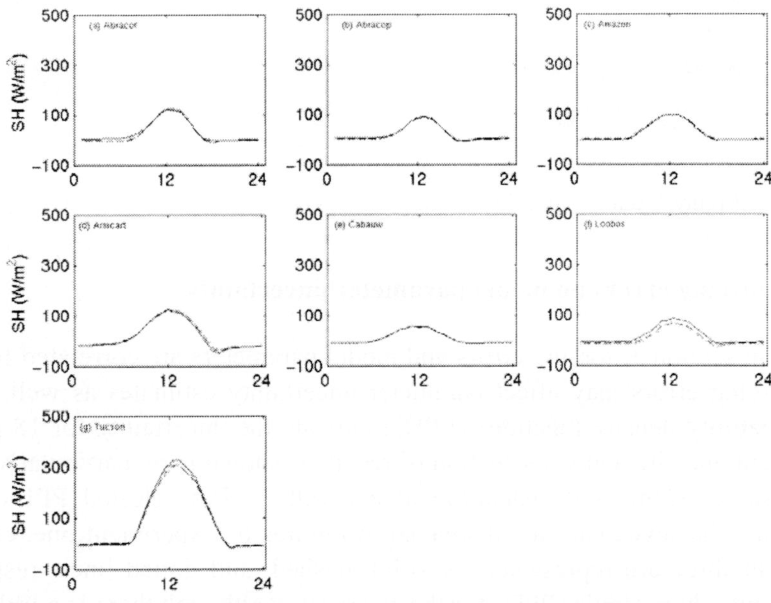

Figure 4. June to August averaged diurnal cycles of simulated sensible heat fluxes for seven PILPS sites (experiment 1 in solid lines, experiment 2 in dashed lines, experiment 3 in dotted lines.

Using the optimal parameters obtained from the three experiments, we use the CHASM model to predict seasonal variations and diurnal cycles of sensible heat fluxes for the seven sites (Figures 4 and 5). The results show that the three experiments give consistent and similar predictions for all seven sites except for Loobos. These conclusions also hold for latent heat flux simulations. Therefore, similar optimal parameters give similar predictions of SH and LH for different time scales. Inconsistent simulations at the Loobos site are due to the effect of forcing errors. Experiments two and three underestimate sensible heat fluxes for different time scales (see Figures 4f and 5f). This underestimation is due to prediction of the small net radiation (see Figure 8) caused by the presence of forcing errors.

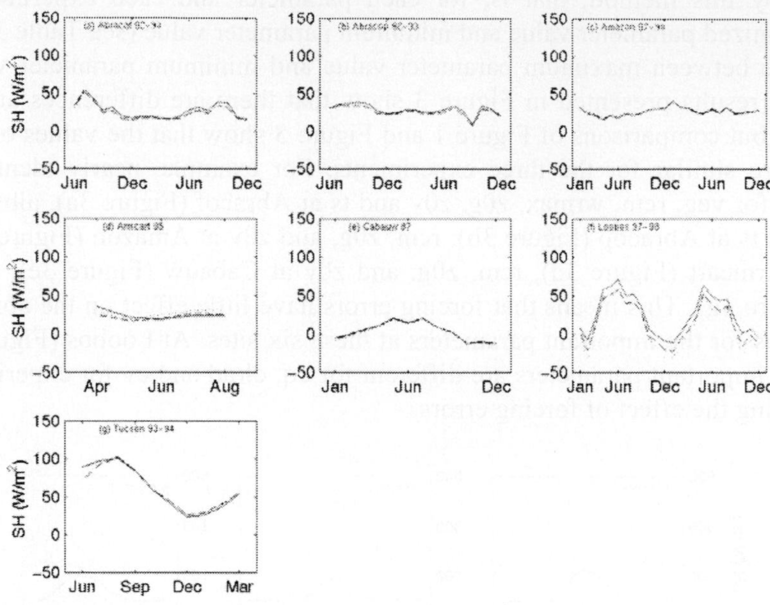

Figure 5. Same as Figure 3 but for monthly means.

5. Effect of forcing errors on model parameter uncertainty

As discussed in section 3, forcing errors and model parameters are correlated for some cases. Therefore, forcing errors may affect parameter uncertainty estimates as well. The marginal posterior probability density functions (PPDs) indicate the uncertainty for 18 parameters for each experiment and site. Only a selection of results is shown here, particularly for minimum stomatal resistance (rcm) and vegetation albedo (albv). The selected PPDs for the three experiments and the seven sites are displayed in Figures 6. Experiment one, experiment two and experiment three are represented by solid, dashed, and dotted lines, respectively. The three experiments show similar PPDs for the seven sites, although there is a little difference at Armcart, Cabauw, Loobos, and Tucson (Figure 6). A similar result occurs for all important parameters as can be expected when forcing errors and parameters are not correlated. So when the correlation between forcing errors and parameters is small, we do not expect the parameter PPDs to be affected.

Figure 7 shows PPDs of vegetation albedo (albv) for the three experiments and the seven sites. The PPDs are now different for the three experiments. Forcing errors either change the shapes of the PPDs for the Abracof, Abracop, Amazon, Artmcart, and Loobos sites or increase the widths of the PPDs at the Cabauw and Tucson sites. because there are positive correlations between albedo vegetation and downward longwave radiation (see Figure 2).

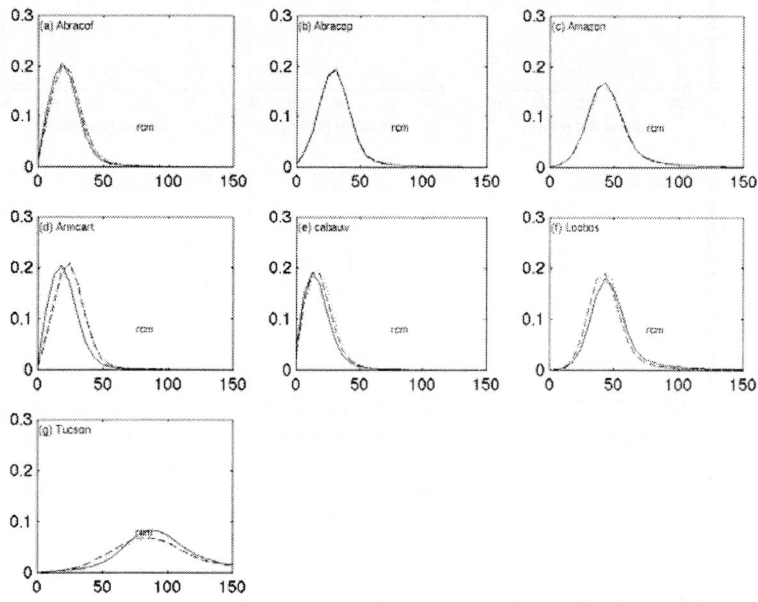

Figure 6. Mrgirginal posterior probability density functions (PPDs) of minimum stomatal resistance (rcm) for three experiments and the seven PILPS sites (experiment 1 in solid lines, experiment 2 in dashed lines, experiment 2 in dotted lines).

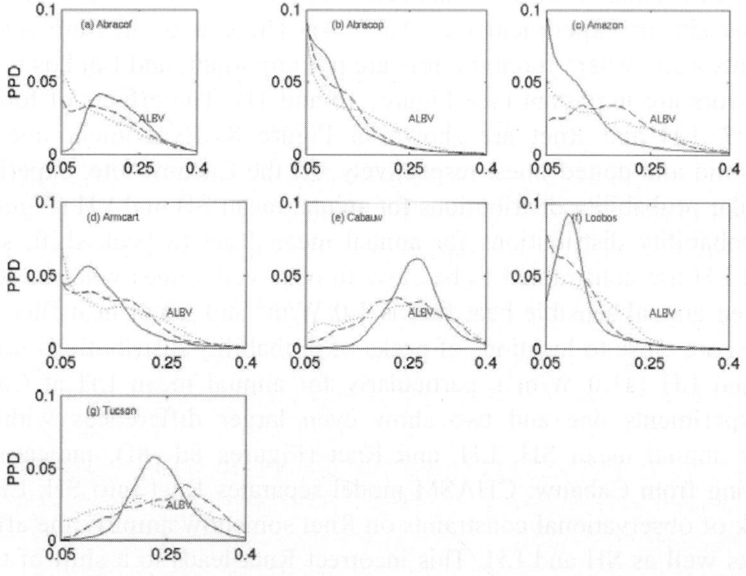

Figure 7. Same as Figure 6 but for vegetation albedo (albv).

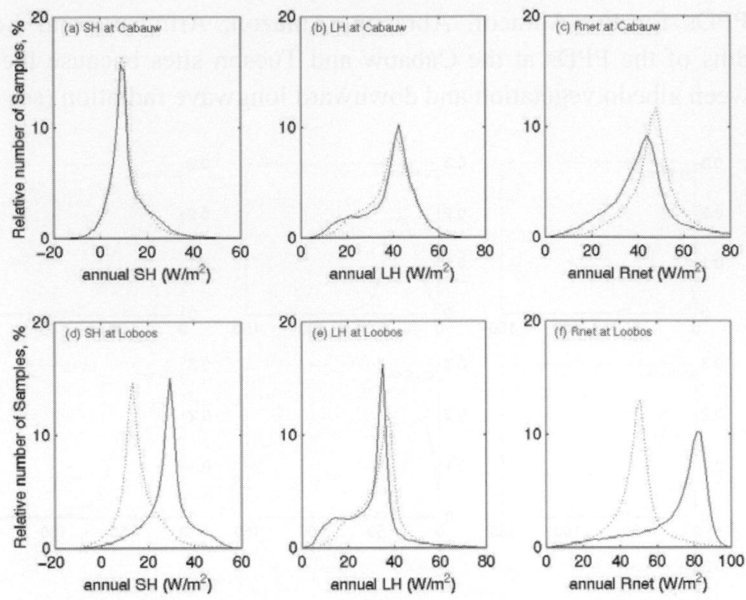

Figure 8. Probability distributions of simulated annual sensible heat fluxes, latent heat fluxes and net radiation at Cabauw and Loobos sites (experiment 1 in solid lines, experiment 2 in dotted lines.

6. Effect of forcing errors on annual mean energy fluxes

In order to study the effect of forcing errors on annual mean sensible heat fluxes (SH), latent heat fluxes (LH), and net radiation (Rnet), we compare probability distributions for experiment one and experiment two at the Cabauw and Loobos sites. These probability distributions are calculated from a normalized histogram based on 65,344 to 86,033 sets depending on the site or experiment (see Table 4). These two sites are selected because Cabauw represents a site where forcing errors are not important, and Loobos represents a site where forcing errors are important (see Figures 1e and 1f). The effects of forcing errors on annual mean SH, LH and Rnet are shown in Figure 8. Experiment one and two are represented by solid and dotted lines, respectively. At the Cabauw site, experiment one and two display similar probability distributions for annual mean SH and LH (Figures 8a and 8b) and different probability distributions for annual mean Rnet (a peak shift, see Figure 8c) because SH and LH are constrained to be close to observed values whereas Rnet is not. At Cabauw, observed annual sensible heat flux is 1.0 W/m^2 and latent heat flux is 41.0 W/m^2. These two values are close to locations of peaks of probability distributions for annual mean SH (5 W/m^2) and LH (41.0 W/m^2), particularly for annual mean LH at Cabauw. At the Loobos site, experiments one and two show even larger differences within probability distributions for annual mean SH, LH, and Rnet (Figures 8d -8f), indicating a different energy partitioning from Cabauw. CHASM model separates Rnet into SH, LH, and ground heat flux. A lack of observational constraints on Rnet somehow amplify the affect of forcing errors on Rnet as well as SH and LH. This incorrect Rnet leads to a shift of the probability distribution to low values for annual SH and shift of the probability distributions to high values for annual LH. The low sensible heat fluxes are consistent with the results in section 4 where simulated sensible heat fluxes in the experiment two are lower than those in experiment one for seasonally averaged diurnal cycles (Figure 4f) and monthly variations (Figure 5f). Therefore, the forcing errors have a considerable effect on the predictions of annual mean SH and LH, particularly for SH. In contrast, at the Cabauw site changes of annual Rnet do not lead to significant shifts in the probability distributions for annual mean SH and LH because predictions of SH and LH are not as sensitive to forcing errors. We do not have a hypothesis for why Loobos is so sensitive to forcing errors.

7. Summary and Conclusions

Forcing errors were not found to significantly affect parameter calibration and uncertainty estimation of a land surface model. Forcing errors do not strongly affect parameter sensitivities at most of the PILPS sites except for Loobos. Analysis of correlation matrixes between 18 parameters shows that forcing errors and/or model parameters are correlated. For example, there is a negative correlation between solar radiation and downward longwave radiation, and positive correlation between solar radiation and vegetation albedo. Forcing errors have little effect on the ability to identify optimal parameters for important parameters or their marginal PPDs for all sites, although forcing errors have considerable effect on the optimization of non-dominant parameters and their associated PPDs. Forcing errors have little effect on the predicted annual mean sensible heat and latent heat fluxes at Cabauw but show significant effects on simulated energy fluxes at Loobos. This difference is due to the sensitivity of land surface models to forcing errors at different sites, although we were not able to identify the factors that control this sensitivity. However, the Loobos example suggests that the effect of forcing errors is at least vegetation and climate dependent. Like other optimized approaches, such as multicriteria method (Gupta et al., 1999) and GLUE method (Frank and Beven, 1997), BSI also contains subjective choice of parameter ranges. In order to understand effect of parameter ranges, we used the ranges of parameters which are smaller than those listed in Table 3 to redo experiments 1 and 3. The experiment results show that forcing errors indeed have little effect on the identification of the relatively important parameters and selection of their optimal values. Therefore, selection of parameter ranges has little effects on our results presented in this study.

Acknowledgments. The G. Unger Vetlesen Foundation and the UTIG provide financial support for this study. YX would like to thank the discussions with Dr. indrajit Roy, Dr. Qiaozhen Mu and Mr. Chengshu Wang. Our special thanks go to Dr. Andy Pitman for providing the CHASM land surface model. We acknowledge the Royal Netherlands Meteorological Institute for providing the Cabauw data set.

References:

Beljaars, A. C. M., and F. Bosveld, 1997: Cabauw data for the validation of land surface parameterization schemes. *J. Climate,* **10**, 1172-1193.

Beringer, J., S. McIlwaine, A. H. Lynch, F .S. Chapin, and G. B. Bonan, 2002: The use of a reduced form model to assess the sensitivity of a land surface model to biotic surface parameters. *Climate Dynamics*, **19**, 455-466.

Bowling, L. C., D .P. Lettenmaier, and Co-authors, 2002: Simulation of high latitude hydrological processes in the Torne-Kalix basin: PILPS Phase 2(e) 1. Experiment description and summary intercomparisons. *Global Planetary Change*, in press.

Chen, T. H., A. Henderson-Sellers, and Co-authors, 1997: Cabauw experimental results from the Project for Intercomparison of Land-surface Parameterization Schemes. *J. Climate,* **10**, 1194-1215.

Desborough, C .E., 1999: Surface energy balance complexity in GCM land surface models, *Climate Dynamics*, **15**, 389-403.

Desborough, C. E., A. J. Pitman, and B. McAvaney, 2001: Surface energy balance complexity in GCM land surface models, Part II: Coupled simulations. *Climate Dynamics*, **17**, 615-626.

Frank, S. W., and K.J. Beven, 1997: Bayesian estimation of uncertainty in land surface-atmosphere flux predictions. *J. Geophys. Res.*, **102**(D20), 23,991-23,999.

Gupta, H. V., L. A. Bastidas, S. Sorooshian,S., W. J. Shuttleworth, and Z.-L. Yang, 1999: Parameter estimation of a land surface scheme using multicriteria methods. *J. Geophys. Res.*, **104**, 19,491-19,503.

Hasselmann, K., 1998: Conventional and Bayesian approach to climate-change detection and attribution. *Quart. J. Roy. Meteorol. Soc*., 124, 2541-2564.

Henderson-Sellers, A., A. J. Pitman, P.Love, P. Irannejad, and T. Chen, 1995: The Project for Intercomparison of Land-surface Parameterization Schemes (PILPS): Phases 2 and 3. *Bull. Amer. Meteor. Soc.*, **76**, 489-503.

Henderson-Sellers, A., K. McGuffie, and A. J. Pitman, 1996: The Project for Intercomparison of Land-surface Parameterization Schemes (PILPS): 1992-1995. *Climate Dynamics*, **12**, 849-859.

Henderson-Sellers, A., A. J. Pitman, P. Irannejad, and K. McGuffie, 2002: Land –surface simulations improve atmospheric modeling. *EOS*, **83**, 145-152.

Jackson, C., M. K. Sen, and P. L. Stoffa, 2002a:An efficient stochastic Bayesian approach to optimal parameter and uncertainty estimation for climate model predictions. *Mon. Wea. Rev*., submitted.

Jackson, C., Y. Xia, M. K. Sen, and P.L. Stoffa, 2002b: Optimal parameter estimation and uncertainty analysis of a land surface model: A case study using data from Cabauw, Netherlands. *J. Geophys. Res.*, submitted.

Koster, R. D., and M. J. Suarez, 1992: Modeling the land surface boundary in climate models as a composite of independent vegetation stands. *J. Geophys. Res.*, **97**, 2697-2715.

Leplastrier, M., A.J. Pitman, H. Gupta, and Y. Xia, 2002: Exploring the relationship between complexity and performance in a land surface model using the multicriteria method. *J. Geophys. Res.*, **107**, No.D20, 4443, doi: 10.1029/2001JD000931.

Lynch, A. H., S. McIlwaine, J. Beringer, and G. B. Bonan, 2001: An investigation of the sensitivity of a land surface model to climate change using a reduced form model. *Climate Dynamics*, **17**, 643-652.

Manabe, S., 1969: Climate and the ocean circulation: 1, The atmospheric circulation and the hydrology of the earth's surface. *Mon. Wea. Rev*., **97**, 739-805.

Pitman, A. J, 1993: Assessing the sensitivity of a land-surfacc scheme to the parameter values using a single column model. *J. Climate*, **7**, 1856-1869.

Pitman A. J., A. Henderson-Sellers, and Co-authors, 1999: Results from the off-line control simulation phase 1c of the Project for Intercomparison of Land-surface Parameterization Schemes (PILPS). *Climate Dynamics*, **15**, 673-684.

Pitman, A .J., Y. Xia, M. Leplastrier, and A. Henderson-Sellers, 2002: The CHAmeleon Surface Model (CHASM): description and use with the PILPS Phase 2e forcing data. *Global and Planetary Change*, in press.

Qu, W., A. Henderson-Sellers, and Co-authors, 1998: Sensitivity of latent heat flux from PILPS land surface schemes to perturbations of surface air temperature. *J. Atmos. Sci.*, **55**, 1909-1927.

Schlosser, C. A., A. G. Slater, and Co-authors, 2000: Simulation of a boreal grassland hydrology at Valdai, Russia: PILPS phase 2(d). *Mon. Wea. Rev.*, **128**, 301-321.

Sen, O. L., L. A. Bastidas, and Co-authors, 2001: Impact of field-calibrated vegetation parameters on GCM climate simulations. *Quart. J. Roy. Meteorol. Soc.*, **127**, part B, 1199-1223.

Sen, M. K., and P. L. Stoffa, 1995: *Global Optimization Methods in Geophysical Inversion*, Elsevier, Amsterdam, 281pp.

Sen, M. K., and P. L. Stoffa, 1996: Bayesian inference, Gibbs' sampler and uncertainty estimation in geophysical inversion. *Geophysical Prospecting*, **44**, 313-350.

Slater, A. G. and Co-authors, 2001: The representation of snow in land-surface scheme: results from PILPS 2(d). *J. Hydrometeorology*, **2**, 7-25.

Tarantola, A., 1987: *Inverse problem theory, Methods of Data Fitting and Model Parameter Estimation*, Elsevier Science Publishing Company, 613pp.

Wood, E. F., D. Lettenmaier, and Co-authors, 1998: The prohect for intercomparison of land-surface parameterization scheme (PILPS) phase 2e, Red-Arkansas river experiment: I. Experiment description and summary intercomparisons. *Global Planetary Change*, **19**, 115-136.

Xia, Y., P. Fabian, A. Stohl, and M. Winterhalter, 1999a: Forest Climatology: Estimation of missing values for Bavaria, Germany. *Agricultural and Forest Meteorology*, **96**, 103-116.

Xia, Y., M. Winterhalter, and P. Fabian, 1999b: A model to interpolate monthly mean climatological data at Bavarian forest climate stations. *Theoretical and Applied Climatology*, **64**, 27-38.

Xia, Y., A. J. Pitman, and Co-authors, 2002: Calibrating a land surface model of varying complexity using multi-criteria methods and the Cabauw data set. *J. Hydrometeorology*, **3**, 181-194.

Xia, Y., M. K. Sen, C. Jackson, and P. L. Stoffa, 2003a: Analysis for the CHASM land surface model: Part I. Estimation of optimal parameters. *J. Appl. Meteor.*, submitted.

Xia, Y., M. K. Sen, C. Jackson, and P. L. Stoffa, 2003b: Analysis for the CHASM land surface model: Part II. Uncertainty estimation. *J. Appl. Meteor.*, submitted.

Zhang, H., A. Henderson-Sellers, and Co-authors, 2001: Limited-area model sensitivity to the complexity of representation of the land surface energy balance. *J. Climate,* **14**, 3965-3986.

Part III

Radiative Transfer and Remote Sensing

RADIATIVE TRANSFER IN THE MIDDLE ATMOSPHERE AND PLANETARY ATMOSPHERES*

XUN ZHU

The Johns Hopkins University Applied Physics Laboratory
11100 Johns Hopkins Road, Laurel, MD 20723-6099, USA
E-mail: xun.zhu@jhuapl.edu

(Manuscript received 5 September 2002)

Radiative heat exchange is the major energy source in the Earth's middle atmosphere and most models of planetary atmospheres. In this article, the author will start with the basic radiative transfer equation and its formal solution from the perspective of radiative heating and cooling rate calculations in numerical models for the middle atmosphere and planetary atmospheres. The exposition of the radiative transfer theory will then be presented along three major lines that lead to the goal of heating and cooling rate calculations in a clear-sky atmosphere. First, the basic quantum concepts of the microscopic interactions between photons and particles and the physical mechanisms of absorption and emission line profiles will be summarized. Second, the traditional approaches of band models associated with frequency integration will be briefly reviewed with a special emphasis on the correlated-*k* distribution method. Third, the physical nature of the source function associated with non-local thermodynamic equilibrium processes will be discussed. Finally, a few selected applications of radiative transfer theory in the middle atmosphere and planetary atmospheres will be presented in the context of how the objective of high accuracy and efficiency is accomplished while solving various types of radiative transfer problems.

1. Introduction

It is often said that solar radiation is the ultimate source of energy for atmospheric and oceanic circulations because the Earth has no significant internal source of its own energy. However, to understand the physical basis of atmospheric and oceanic circulations and their variabilities, one must understand the direct energy or momentum sources that drive the circulations. For example, many of the observed characteristics of large-scale oceanic

* Professor Jijia Zhang taught many of the authors in this collection of papers a course "Atmospheric General Circulations" at Nanjing Institute of Meteorology in the early 1980s. Part of the lecture notes (Zhang 1981) introduced a simple radiative transfer theory of a gray atmosphere in which the theory of the vertical thermal structure of the Earth's atmosphere was developed following the historical works by Emden (1913), Kibel (1943), et al. The modern theory of radiative transfer has introduced many fundamental advances in the understanding of the physical nature of the emission/absorption processes and in techniques for calculating radiative heat exchange in the atmosphere. However, the basic ideas of radiative heat exchange, energy balance, radiative-convective adjustment, etc., embedded in those lectures remain valid and also inspired us to pursue our careers in different fields of atmospheric sciences. Furthermore, some studies in the planetary atmospheres are following a path similar to that of the early history of meteorology due to a similar situation of having only limited sets of observations. The author presents this paper in memory of Professor Jijia Zhang.

circulations can be entirely attributed to forcing by the surface wind stress (e.g., Pedlosky 1987, Ch. 5). Three well-known types of energy that drive the atmospheric circulation are net radiative heating, sensible heating, and latent heating. Calculations of the energy budget in the atmosphere on these three types of energy show comparable contributions from all the sources (Peixoto and Oort 1992, p. 366). Since the troposphere consists of 90% of the whole atmosphere, the energy budget reflects mainly the atmospheric structure in the troposphere.

Sensible and latent heat sources mostly result from the unstable convection and the formation or dissipation of clouds, which often need to be parameterized in large-scale numerical models (e.g., Zhang 2004). However, few clouds exist in the Earth's middle atmosphere (altitude: $z \sim$ 10-100 km) due to its relatively stable vertical stratification. Hence we expect that radiative heat exchange is the dominant diabatic energy source in the middle atmosphere. Furthermore, because of the scarcity of clouds in the middle atmosphere, radiative transfer can often be studied without considering multiple scattering processes. Emission and absorption take place in the middle atmosphere under "clear-sky" conditions, where the absorbers' radiation properties are slowly varying with time and space. As a result, the net radiative heating rate (H_{r_net}) is often the most accurately evaluated forcing term in middle atmosphere models with limited computational resources. Since the contributions from the sensible and latent heating are negligibly small in magnitude with greater uncertainties H_{r_net} is the only thermal forcing term in many middle atmosphere models.

In this article, the ideas and techniques of middle atmosphere radiative transfer theory and its applications will be explored in an informal and hierarchical style from which readers with different backgrounds could all benefit. This article is tutorial in nature, and most of the material can serve as an introduction to the field. It also contains a brief review on the recent developments in the field and several expository views by the author from different perspectives. Section 2 gives a descriptive introduction to the basic quantum concept of discrete energy levels and some consequences in applying the idea to atmospheric radiation. Section 3 establishes the macroscopic radiative transfer equation (RTE) and its formal solution for heating and cooling rate calculations. The basic ideas of two remote sounding techniques (occultation and limb emission-absorption) together with their pros and cons are also briefly described based on the formal solution of the RTE. Section 4 briefly shows how the atmospheric spectra are formed. Section 5 reviews the two most widely used tools in the integration of radiance over frequency. Section 6 establishes the microscopic RTE and its relation to the macroscopic one. The rate equation and a few simple solutions to the source function are also described. Readers who are newcomers to the field of atmospheric radiation or who are interested only in stratospheric radiation theory may skip this section. Section 7 presents several applications of the radiative transfer theory to planetary atmospheres and the Earth's middle atmosphere, based on author's own research experience. Special emphasis is placed on how to reach an appropriate balance between accuracy and efficiency for different types of problems. Finally, Section 8 provides concluding remarks and additional references.

2. The Basic Quantum Concepts of Interaction Between Radiation and Matter

In the late 1800s, physicists were convinced that light, heat radiation, and radio waves were all electromagnetic waves differing only in frequency and wavelength. However, the spectral distribution of radiation from a heated cavity could not be explained by classical (Maxwell) wave theory. Max Planck's solution to the problem was to introduce the new concept that radiation could only take discrete or quantized energies (joule = kg m^2 s^{-2}) given by

$$E_{light} = nh\nu, \tag{1}$$

where n is an integer, h (= 6.626×10^{-34} J s) is the Planck constant, and ν (s^{-1}) is the wave frequency. Therefore, light consists of a stream of discrete photons. A photon can be thought of as a wave packet that carries an electromagnetic field of energy $E = h\nu$. For example, each photon emitted by a radio station of frequency, say, 630 kHz (= 6.3×10^5 s^{-1}) carries 4.17×10^{-28} joules of energy. Each photon emitted from a red traffic light ($\nu \approx 4 \times 10^{14}$ s^{-1}) carries 2.7×10^{-19} joules of energy.

A photon has no rest mass. However, based on Einstein's well-known formula, $E = mc^2$, where c (= 2.9979×10^8 m s^{-1}) is the speed of light, it does possess a moving mass (m: kg) of $h\nu/c^2$. Since h is a constant, a photon's energy ($h\nu$) can also be expressed by its frequency: $\nu = E/h$. Furthermore, wavelength (λ: m = 100 cm = 10^6 μm = 10^9 nm = 10^{10} Å), wavenumber ($\tilde{\nu}$: m^{-1} = 0.01 cm^{-1}), and wave frequency (ν: s^{-1}) of a traveling photon with a fixed speed of light are related: $c = \lambda\nu$ and $\tilde{\nu} = 1/\lambda$. Therefore, one can also express a photon's energy in terms of its wavenumber: $\tilde{\nu} = E/hc$. When one says a photon of energy 667 cm^{-1} or 15 μm what that really means is that the photon has energy of 667 cm^{-1} $\times 6.626 \times 10^{-34}$ J s $\times 2.9979 \times 10^{10}$ cm s^{-1} = 1.3×10^{-20} J.

A particle, or matter (such as a molecule or an electron), has both rest mass (kg) and energy (J). The quantum view of matter as advanced by Niels Bohr to Max Planck's quantum hypothesis is that it too can only exist at discrete energy levels. For example, a molecule can rotate and vibrate only at certain discrete rates of frequency. A particle makes a transition from one of its energy states (E_i) to another (E_j) by absorbing or emitting a photon that has energy ($h\nu_0$) identical to the energy difference ($E_j - E_i$) of those two states. Since both the particle and photon can only have discrete energy levels, it should be quite clear at this point that a transition is impossible if $E_j - E_i \neq h\nu_0$. Furthermore, the energy levels of all substances are determined intrinsically by their masses and the internal structure of more elemental particles (such as electronic or atomic configurations in an atom or molecule). Hence, the absorption and emission spectra of various substances can be used as fingerprints of the matter. We have gained almost all our knowledge of the physical and chemical structure of stars and other planets by analyzing the emission and absorption spectra collected either from ground telescopes or from spacecrafts.

The microscopic interactions between a particle and a photon or among particles are usually studied on a one-to-one basis. The natural phenomena we encounter in daily life can be considered ensemble averages (statistical means) of various interactions between individual particles and discrete photons (radiative transitions) or among the individual particles (non-radiative transitions). Based on the concept of discrete energy levels one may ask: How could we have all those photons whose energies happen to match the exact differences of discrete energy levels of those matters? The simple answer is: those photons may have been originally released by particles when they made transitions from the higher energy levels to the lower ones.

Constant interactions between matter and photons (such as absorption and emission) are the nature of the existing world. An energy difference (~ 667 cm^{-1} or ~ 15 μm) between two vibrational states of a carbon dioxide (CO_2) molecule in the atmosphere happens to match that of many photons emitted from the Earth's surface. A photon has to undergo repeated absorption and emission with CO_2 molecules before it finally escapes to outer space (a probability of less than 0.001) carrying energy or heat with it. The increase of CO_2 in the Earth's atmosphere further reduces the escape probability of the photons to space and may cause global warming. The very thick CO_2 atmosphere of Venus makes it almost impossible for its photons (at 15 μm) to escape to space. This leads to a very high surface temperature (~750 K) for Venus. One may wonder that if there are hardly any photons escaping from Venus surface then how do we know and verify its surface temperature? The answer to this question rests on the fact that there are lots of photons at other wavelengths (such as microwaves with wavelengths of millimeters and centimeters) that can escape directly from the surface of Venus to outer space and provide us with information.

Even if the condition $E_j - E_i = h\nu_0$ holds, it does not necessarily mean that the transition between E_i and E_j is possible. For example, the increase or decrease of the rate of vibration for a molecule can only occur one step at a time between two neighboring energy levels. Whether a particular transition is possible or not is determined by detailed analytical and numerical calculations of quantum mechanics and spectroscopy theory. The results can be summarized into sets of selection rules for different types of transitions (rotational, vibrational, and electronic) with some exceptions. For more detailed descriptions of the physical nature of absorption and emission processes readers are referred to McCartney (1983), Houghton and Smith (1966), and Levine (1975).

3. Macroscopic Radiative Transfer Equation and Its Formal Solution

The basic radiation quantity is the monochromatic specific intensity or the monochromatic radiance, $L_\nu(\mathbf{r}, \mathbf{\Omega})$, which can be considered a macroscopic ensemble of a stream of photons in a unit solid angle $d\Omega$, passing a spatial point $\mathbf{r}$, traveling in direction $\mathbf{\Omega}$, with frequency ν and speed c. The definition of $L_\nu(\mathbf{r}, \mathbf{\Omega})$ is also reflected in its unit: J m^{-2} steradian^{-1} = W m^{-2} Sr^{-1} Hz^{-1}. The macroscopic radiative transfer equation (RTE) is a continuity equation that describes the energy conservation of the photon stream as it interacts with matter along the path:

$$\frac{\partial L_\nu}{\partial t} + c\frac{\partial L_\nu}{\partial s} = c(\text{source} - \text{sink}), \tag{2}$$

where t is time and s the distance along the slant path of the photon stream. For most applications, the first term, the local time derivative in (2), is negligible because of the assumption of statistical equilibrium. The standard macroscopic RTE describes a spatial variation of a photon stream for an infinitesimal element ds (Fig. 1):

$$\frac{dL_\nu(s,\mathbf{\Omega})}{ds} = -k_\nu \rho_a L_\nu(s,\mathbf{\Omega}) + k_\nu \rho_a J_\nu \,, \tag{3}$$

where ρ_a (kg m^{-3}) is the density of radiatively active gas, k_ν (cm^2 kg^{-1}) is the extinction coefficient that linearly parameterizes the "sink" term in (2), and the source function J_ν in the "source" term ($k_\nu \rho_a J_\nu$) describes the rate of increase of photon energy along the path. It is noted from (3) that the extinction of monochromatic radiance is linearly dependent on both ρ_a and L_ν. If we instead choose number density (n_a: m^{-3}) as our measure of the absorber amount in (3) then the extinction coefficient is referred to as a cross section (σ_ν: m^2) with $\sigma_\nu n_a = k_\nu \rho_a$. There are two physical processes, absorption and scattering, that remove photons from the photon stream along the path. In this article, we focus our quantitative discussion mainly on the absorption process in the middle atmosphere and planetary atmospheres.

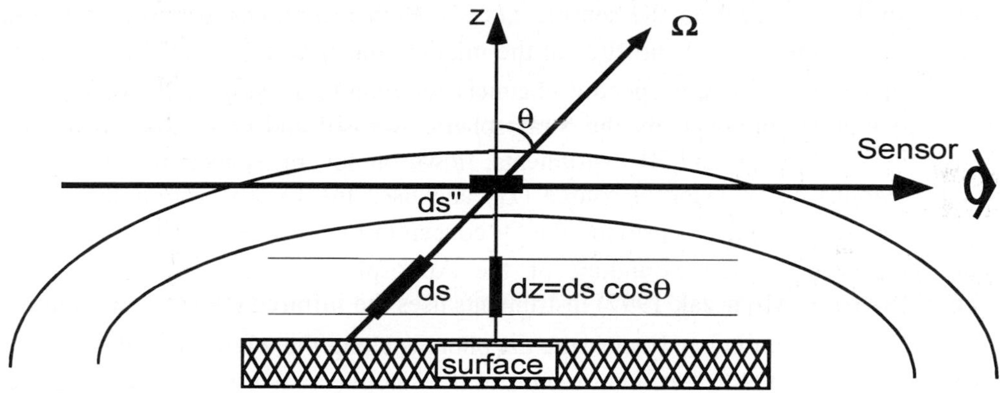

Figure 1. Infinitesimal element of a slant path (*ds*) and schematic illustration of limb (tangentially viewing) measurements of radiance.

Equation (3) can be easily integrated to yield the formal solution of monochromatic radiance,

$$L_\nu(s,\mathbf{\Omega}) = L_\nu(s_0,\mathbf{\Omega})\tau_\nu(s_0,s) + \int_{s_0}^{s} [k_\nu(s')\rho_a(s')J_\nu(s',\mathbf{\Omega};L_\nu)]\tau_\nu(s,s')ds' \,, \tag{4}$$

where the *monochromatic transmission function,*

$$\tau_\nu(s,s') = \exp\left[-\left|\int_{s}^{s'} k_\nu(s'')\rho_a(s'')ds''\right|\right] \equiv \exp[-u_{\nu,\Omega}(s,s')] \,, \tag{5}$$

is a basic building block in calculating any radiation quantity. The above solution also defines the *optical path* ($u_{\nu,\Omega}$) between s and s'. The solution (4) describes the fact that $L_\nu(s, \mathbf{\Omega})$ is composed of contributions from the boundary radiance $L_\nu(s_0, \mathbf{\Omega}) \equiv L_0$ exponentially attenuated by the optical path between s_0 and s plus the sum of the infinitesimal source function emission contributions from the internal atmospheric source term ($k_\nu \rho_a J_\nu ds' \equiv L_{atm}$) in the volume elements at positions s' along the path.

To illustrate the physical significance of the formal solution (4) and to also briefly review our knowledge of the interactions between radiation and matter described in the last section, let us consider two applications of (4) in retrievals of ozone (O_3) concentration (n_a) in the middle atmosphere. Depending on whether the largest contribution to $L_\nu(s, \mathbf{\Omega})$ in (4) comes from L_0 or from L_{atm}, the formal solution of (4) is the basis of two remote sounding techniques: occultation and emission-absorption (Fig. 1). The occultation technique neglects any contribution from L_{atm} and measures the ratio of the attenuated to the un-attenuated radiances, L_ν / L_0, from which the physical parameters, such as $\rho_a(s'')$, contained in $\tau_\nu(s_0, s)$ is derived. On the other hand, the emission-absorption technique measures the photons that are emitted within the atmosphere, L_{atm}, that have been attenuated by the rest of the atmosphere along the path as they reach to the sensor, $\tau_\nu(s, s')$. Both techniques are currently used for continuous measurements of O_3 profiles in the middle atmosphere by two different satellite instruments that select different spectral channels for remote sensing. The solar (or lunar) occultation technique employed by the Stratospheric Aerosol and Gas Experiment (SAGE, Chu and Veiga 1998) I, II, and III instruments, flown on several spacecraft platforms, uses the Sun (or Moon) as an external source (L_0) and uses the O_3 absorption feature in the ultraviolet (UV) wavelengths to provide the O_3 concentration. The limb emission-absorption technique employed by the Sounding of the Atmosphere using Broadband Emission Radiometry (SABER, Mlynczak 1997) instruments uses the infrared (IR) O_3 emission feature at the 9.6-μm band to continuously measure the photons along the whole satellite orbit.

Since the Sun will get to the line-of-sight of the satellite sensor only twice (sunrise and sunset) in each orbit cycle, the spatial and temporal coverage of the occultation technique is usually very limited. Furthermore, nearly half of the section along the limb line-of-sight is in dayside whereas the rest is in nightside. The retrieval by the solar occultation is mathematically underdetermined if there exists a significant difference in O_3 concentration across the narrow twilight zone. Additional constraints are needed for a meaningful retrieval or interpretation in the retrieval algorithm. On the other hand, global coverage is much wider and more frequent when using the emission-absorption technique because the instrument is able to continuously collect photons along the whole satellite orbit. One disadvantage of the emission-absorption technique is that its measurements only provide indirect information about n_a. As an O_3 molecule in the atmosphere absorbs an UV photon in a Sun ray it makes a transition from its lower energy state (E_i: ground state) to an upper energy state (E_j). Therefore, the measured attenuation using the occultation technique $\tau_\nu(s_0, s)$ reflects the ozone population at its ground state (n_i). On the other hand, the photons collected by the emission-absorption technique represent the ozone population at its upper energy state (n_j: excited state) because the strength of the emission is nearly proportional to the relative population at its excited state (n_j / n_i). When $n_j \ll n_i$, the ozone population at its ground state as measured by the occultation technique can be considered the actual ozone concentration ($n_i \approx n_a$) whereas the emission-absorption technique relies on additional physical modeling that can accurately relate $n_j \ll n_i$ to n_a (e.g., Mlynczak and Drayson 1990).

The above two examples also illustrate a particular aspect of spectral channel selection in remote sensing, i.e., solar UV channel (0.29 μm) for absorption and IR 9.6-μm band for emission of O_3 in the mesosphere. Often, a retrieval problem can be better solved if a physical variable is dominantly sensitive to a particular radiation process of absorption, emission, or scattering (e.g., Chen 2002; G. Liu 2004).

Once the monochromatic radiance is known we can construct other radiation quantities that often involve integrating $L_\nu(s, \mathbf{\Omega})$ over a solid angle ($\mathbf{\Omega}$) to yield irradiance and over frequency (ν) to yield energy flux. A solid angle is determined by a zenith angle ($0 \le \theta < \pi$) and an azimuthal angle ($0 \le \varphi < 2\pi$). For middle atmospheric heating and cooling rate calculations it is often a good approximation to assume a plane-parallel atmosphere (Fig. 1) and an isotropic source function:

$$ds = \frac{dz}{\cos\theta} \equiv \frac{dz}{\mu}, \quad J_\nu(s', \mathbf{\Omega}; L_\nu) = J_\nu(s'; \overline{L}_\nu), \tag{6a,b}$$

where $\overline{L}_\nu$ denotes the mean radiance averaged over the 4π solid angle (see Section 6). The basic building block between pressure levels p and p' can then be written as

$$\tau_\nu(p, p'; \mu) = \exp[-u_\nu(p, p') / \mu], \tag{7}$$

where the optical thickness between p and p' is

$$u_\nu(p, p') = \left| \int_p^{p'} k_\nu(p'')[r_a(p'')/g]dp'' \right|. \tag{8}$$

In (8), g (m s^{-2}) is the gravitational constant, and $r_a(p'') = \rho_a / \rho_{air}$ is the mass mixing ratio at pressure p'' with ρ_{air} defined as air density. We have also used the hydrostatic relation ($dp = -g\rho_{air}dz$) while converting (5) to (7). Note that the radiance is independent of φ except for the contribution from the boundary, which can be easily separated as shown in (4). The upward/downward energy flux (W m^{-2}) of the photon stream over a frequency band ($\Delta\nu$) can be calculated by

$$F^{\pm}(p) = 2\pi \int_{\Delta\nu} d\nu \int_0^{\pm 1} \mu L_\nu(p, \mu) d\mu. \tag{9}$$

Note, from (9), that both F^+ and F^- are positive. The net energy flux is defined by the difference $F_n(p) = F^+(p) - F^-(p)$. Its convergence defines the radiative heating rate per unit mass (H_r: K s^{-1}) and is given by

$$H_r = \frac{-1}{c_p \rho_{air}} \frac{\partial F_n}{\partial z} = \frac{g}{c_p} \frac{\partial F_n}{\partial p}. \tag{10}$$

According to (4)-(9), calculations of energy flux and radiative heating rate involve multiple integrations over p, μ, and ν, respectively. Two basic quantities often used in the energy flux and radiative heating calculations that contain these multiple integrations are the flux transmission function (τ_f) and flux escape function (Γ_f) (Andrews et al. 1987) defined as

$$\tau_f(p,p') = \frac{2}{\Delta\nu}\int_{\Delta\nu}\int_0^1 \mu \exp\left[-\left|\int_p^{p'} k_\nu(p'')[r_a(p'')/g]dp''\right|\Big/\mu\right] d\mu d\nu\,, \tag{11}$$

$$\Gamma_f(p,p') = \frac{1}{S[T(p)]}\int_{\Delta\nu}\int_0^1 k_\nu[T(p)] \exp\left[-\left|\int_p^{p'} k_\nu(p'')[r_a(p'')/g]dp''\right|\Big/\mu\right] d\mu d\nu\,, \tag{12}$$

respectively, where $S[T(p)]$ is the total band strength that is a function of temperature, T, at a given pressure level (p). We have also explicitly denoted that both the absorption coefficient k_ν and the mass mixing ratio of absorber r_a are functions of spatial position in terms of pressure.

The flux escape function $\Gamma_f(p, p')$ represents a radiative heat exchange of a frequency band ($\Delta\nu$) between p and p', which is the focus of this article. The net effect of such an exchange between one level and the rest of the atmosphere is the net radiative heating rate. By careful algebraic manipulations the radiative heating rate can be expressed as (e.g., Andrews et al. 1987; Zhu 1994)

$$H_r(p) = \frac{2\pi r_a(p)S[T(p)]}{c_p}\{-J_\nu(p)\Gamma_f(p,p_\infty) + [L_\nu(p_{00}) - J_\nu(p)]\Gamma_f(p,p_{00})$$

$$+\int_{\Gamma_f(p,p_\infty)}^1 [J_\nu(p') - J_\nu(p)]d\Gamma_f(p,p') + \int_{\Gamma_f(p,p_{00})}^1 [J_\nu(p') - J_\nu(p)]d\Gamma_f(p,p')\}\,. \tag{13}$$

The four terms within the curly brackets in the above equation represent the radiative heat exchange between the photon energy at level p, $J_\nu(p)$, with (i) space ($L_\nu(p_\infty) = 0$); (ii) the lower boundary emission ($L_\nu(p_{00})$); and emissions from (iii) overlying and (iv) underlying layers, respectively. Equations similar to (13) can also be derived for the energy flux and flux transmission function. The fact that the heating rate is the vertical derivative of energy flux as defined in (10) is also reflected by the following relationship between τ_f and Γ_f,

$$\Gamma_f(p,p') = \frac{g\Delta\nu}{2r_a S[T(p)]}\left|\frac{\partial\tau_f(p,p')}{\partial p}\right|. \tag{14}$$

Note that the crucial quantity $\Gamma_f(p,p')$ for the radiative heating or cooling rate calculation can be determined numerically either from its integral expression (12) or from the differential expression (14) once $\tau_f(p,p')$ is given.

Since the integrands in (11) and (12) are smoothly varying functions of p and μ, the integrations over p and μ are straightforward. Furthermore, the integrations over μ can be expressed in closed forms by exponential integrals (e.g., Abramowitz and Stegun 1965):

$$\tau_f(p,p')=\frac{2}{\Delta\nu}\int_{\Delta\nu}\mathrm{E}_3[u_\nu(p,p')]d\nu\,, \tag{15}$$

$$\Gamma_f(p,p')=\frac{1}{S[T(p)]}\int_{\Delta\nu}k_\nu[T(p)]\mathrm{E}_2[u_\nu(p,p')]d\nu\,, \tag{16}$$

where $\mathrm{E}_n(x)=\int_1^\infty t^{-n}e^{-xt}dt$ and $u_\nu(p,p')$ is as defined in (8). However, the integration over frequency poses a cumbersome problem in radiative transfer theory because the integrands that contain the absorption coefficient k_ν are composed of many molecular absorption lines and therefore are rapidly varying functions of ν. The traditional method of random band models explores the multiplication property of the exponential function $e^{x+y}=e^x e^y$ (see Section 5). Since neither exponential integrals nor the form of integrand in (12) possesses this property, i.e., $E_n(x+y)\neq E_n(x)E_n(y)$ and $(x+y)e^{x+y}\neq(xe^x)(ye^y)$, the random band models can only be applied to (11) by first integration over frequency, yielding the slant path transmission function. We will sketch the basic idea behind the traditional random band models and also present a recent approach of the correlated-k distribution (CKD) method in Section 5. But let us first briefly review some background on how line profiles are formed.

4. Quantum Concepts of Absorption/Emission Coefficients and Line Profiles

When the energy difference (E_j-E_i) between two discrete states (n_j and n_i) is identical to that of a photon ($h\nu_0$), then there exists a possibility that the two will interact in various ways, for example:

Absorption:	$n_i+h\nu_0\rightarrow n_j\,,$	(17a)
Emission:	$n_j\rightarrow n_i+h\nu_0\,,$	(17b)
Stimulated emission:	$n_j+h\nu_0\rightarrow n_i+2h\nu_0\,,$	(17c)
Scattering:	$n_i+h\nu_0\rightarrow n_j^*\rightarrow n_i+h\nu_0\,,$	(17d)
Fluorescence:	$n_i+h\nu_0\rightarrow n_j^*\rightarrow n_i'+h\nu'\,,$	(17e)

where n^*_j is the so-called "excited" state that can only exist for a very short time. In quantum mechanics, the energy dependence of particles (n_j versus n_i) is viewed as each excited state being a different particle. Hence, the above equations look just like chemical reaction equations in their appearance and contents. Note that the scattered photon on the right-hand-side of (17d) travels in a different direction from the incoming photon whereas the induced photon on the right-hand-side of (17c) travels in the same direction and with the phase as the incoming photon. In the case of fluorescence, both the state n'_i and the scattered photon energy $h\nu'$ could be different from the original ones of n_i and $h\nu_0$.

Up to this point, we have only presented the simple view that radiative transitions occur only when the photon energy coincides exactly with the difference between two transition states $E_j - E_i = h\nu_0$. This corresponds to a resonant interaction when both states in transition are assumed to be stationary, which correspond to pure sinusoidal waves with fixed amplitudes. To quantitatively illustrate such a transition and relate the process to a δ-function distribution in frequency space we note that quantum mechanics expresses the state of a system in the form of wave functions. The wave functions for states i and j of energies E_i and E_j can be expressed as ($\hbar = h/2\pi$)

$$\Psi_i \sim \exp(-iE_i t/\hbar)\,,\ \Psi_j \sim \exp(-iE_j t/\hbar)\,. \tag{18}$$

The intensity of a radiative transition (a_t) between states i and j is proportional to ($\omega_0 = 2\pi\nu_0$)

$$a_t = \Psi_i^* \Psi_j \sim \exp(-i(E_j - E_i)t/\hbar) = \exp(-i\omega_0 t)\,. \tag{19}$$

The photon distribution function in frequency ($f_\omega \sim f_\nu$) is proportional to $|a_\omega|^2$ from which we have

$$f_\omega = |a_\omega|^2 \sim \delta(\omega - \omega_0), \tag{20}$$

where $a_\omega = \int_{-\infty}^{\infty} a_t \exp(i\omega t)dt$ is the Fourier transform of a_t, and we have also normalized the frequency distribution function $\int_{-\infty}^{\infty} f_\omega d\omega = 1$.

Both laboratory measurements and more detailed theoretical analyses suggest that emission and absorption lines have finite widths due to several mechanisms. One mechanism that broadens the frequency distribution function is the finite lifetime of an emitting particle (atom or molecule) staying at a particular state. To illustrate this process, let us consider the effect of finite lifetime of the state j when a radiative transition from state j to state i occurs. The probability P_{decay} for an emitting particle to remain in its initial state j before making a transition by emitting a photon is $P_{decay}(t) \sim \exp(-t/\tau_d)$, where τ_d is the mean lifetime of the state j. Since an observed emission line represents an ensemble average of many photons with various lifetimes, the Fourier component of the transition intensity can be written as

$$\bar{a}_\omega = \int_0^{\infty} a_t \exp(i\omega t)\exp(-t/\tau_d)dt = \frac{1}{\tau_d^{-1} + i(\omega - \omega_0)}\,. \tag{21}$$

Substituting (21) into (20) yields the following normalized Lorentz line profile

$$f_\omega^L(\xi) = |\bar{a}_\omega|^2 \sim \frac{\alpha_L}{\pi}\frac{1}{\alpha_L{}^2 + (\omega - \omega_0)^2} = \frac{\alpha_L}{\pi}\frac{1}{\alpha_L{}^2 + \xi^2}\,, \tag{22}$$

where $\alpha_L = 1/\tau_d$ is the Lorentz half-width and $\xi = \omega - \omega_0$. Note that δ-function can be expressed in many different limiting forms, specifically (e.g., Arfken 1985),

$$\delta(y) = \frac{1}{\pi} \lim_{x\to 0} \frac{x}{x^2 + y^2}. \tag{23}$$

Therefore, the idealized line profile (20) is recovered, $\lim_{\alpha_L \to 0} f_\omega^L = \delta(\xi) = \delta(\omega - \omega_0)$, as $\alpha_L \to 0$ or $\tau_d \to \infty$, which corresponds to the case of a stationary state.

Another broadening mechanism arises from the Doppler effect of thermal motion of the emitting particles. A photon wave packet of frequency ω traveling at the speed of light c emitted by a moving particle with a velocity u ($<< c$) in the direction of propagation of the photon is perceived by a stationary observer as having the emission frequency ω_0 replaced by a Doppler-shifted frequency $\omega' = \omega_0(1 + u/c)$. Again, the bulk effect of many moving particles on an emission line can be derived by averaging the Doppler shifts to the profiles by all the emitting particles,

$$f_\omega^V(\xi) = \frac{\alpha_L}{\pi} \int_{-\infty}^{\infty} \frac{P_D(u)du}{\alpha_L^2 + (\xi - u\omega_0/c)^2}, \tag{24}$$

where $P_D(u)$ is the Maxwellian distribution of the thermal velocity,

$$P_D(u) = \left(\frac{m_p}{2\pi k_B T}\right)^{1/2} \exp\left(-\frac{m_p u^2}{2k_B T}\right). \tag{25}$$

In (25), m_P (kg) is the mass of an emitting particle, T (K) is the temperature, and k_B ($= 1.38 \times 10^{-23}$ J K^{-1}) is the Boltzmann constant. $f_\omega^V(\xi)$, defined in (24), includes both Lorentz and Doppler effects and is called the Voigt profile. Based on the above derivation, the Voigt profile includes statistical means over two physical processes that can be considered to be independent: radiative decay and the thermal motion of moving particles. That an individual wave packet can carry both properties and be averaged is similar to a case that a physical quantity (F) can be a function of x, y, z, and t over which one can perform averages over more than one variables.

Substituting (25) into (24) and changing the integral variable, we have

$$f_\omega^V(\xi) = \int_{-\infty}^{\infty} \frac{\alpha_L}{\pi^{3/2}\alpha_D} \frac{\exp\left(-\eta^2/\alpha_D^2\right)d\eta}{\alpha_L^2 + (\xi - \eta)^2}. \tag{26}$$

Letting $\alpha_L \to 0$ and applying (23) to (26), we obtain the pure Doppler line profile

$$f_\omega^D(\xi) = \left(\pi^{1/2}\alpha_D\right)^{-1} \exp\left[-\xi^2 / \alpha_D{}^2\right], \quad (27)$$

where $\alpha_D = (\omega_0 / c)(2k_BT / m_p)^{1/2}$ and $(\ln 2)^{1/2}\alpha_D$ is the Doppler half-width. Equation (26) indicates that, mathematically, the Voigt line profile has been expressed as a convolution of the Doppler and Lorenz line profiles

$$f_\omega^V(\xi) = \int_{-\infty}^{\infty} f_\omega^D(\eta) f_\omega^L(\xi - \eta) d\eta . \quad (28)$$

According to the convolution theorem for a pair of functions related by an integral transform the convolution relation between $f_\omega^L(\xi)$ and $f_\omega^D(\xi)$ implies a product relation between their inverse-transformed functions in the t-variable (e.g., Arfken 1985)

$$f_t^V(t) = f_t^L(t) f_t^D(t), \quad (29)$$

where f_t^L and f_t^D are the inverse Fourier transforms of $f_\omega^L(\xi)$ and $f_\omega^D(\xi)$:

$$f_t^L(t) = (2\pi)^{-1} \exp(-\alpha_L |t| - i\omega_0 t), \; f_t^D(t) = (2\pi)^{-1} \exp[-(\alpha_D^2 t^2 / 4) - i\omega_0 t] . \quad (30)$$

Our derivation of the Voigt line profile (26) demonstrates a clear physical significance of $f_\omega^L(\xi)$, $f_\omega^D(\xi)$, and the convolution relation (28). Specifically, $f_\omega^D(\xi)$ has been derived *after* $f_t^V(t)$ was derived. Also, note that an approximation of a linear relationship between the Doppler-shifted frequency and the velocity has been used in the derivation to yield the convolution relation (28). As a result, the physical significance of f_t^L and f_t^D in the product relation (29) in t-variable is not obvious at all. This is a common situation in a convolution-product relationship associated with an integral transform where only one side of the equation shows a clear physical significance. Two other examples in the field of radiative transfer are the frequency-dependent coefficients of phenomenological conductivity and permeability (Bohren and Huffman 1983, p. 15) and the multiplication property of the transmission function (Zhu 1995). In the first example, electric polarization and magnetization of the electromagnetic waves will be related to the electric and magnetic fields through convolution integrals of Fourier transforms. Only the frequency-dependent phenomenological coefficients that define product relations carry a clear physical meaning in this case. In the second example, the composite k-coefficient of a gas mixture can be related to those of individual gases through convolution integrals of Laplace transforms (Zhu 1995). The physical basis of such a convolution relationship is the multiplication property of the transmission function. Although only one side of the convolution relationship has a clear physical significance, the correct usage of its mathematical relationship will help avoid unnecessary confusions in applications (Goody et al. 1989; Zhu 1995).

5. Integration of Radiation Functions Over Frequency

The absorption coefficient k_ν is composed of all (N) the allowed transitions among different states with each transition having its line strength, s_i, and line profile, $f_\nu^i = f^i(\nu - \nu_i)$

$$k_\nu = \sum_{i=1}^{N} s_i f_\nu^i , \tag{31}$$

where we have substituted ω with $\nu = \omega / 2\pi$. Since the line profile has been normalized, the total band strength over a frequency band that covers many lines is given by $S \equiv \int_{\Delta\nu} k_\nu d\nu = \sum_i s_i$.

Generally speaking, s_i is a function of temperature whereas f_ν^i depends on both temperature and pressure (e.g., McCartney 1983, Section 6.1). Fig. 2a (Section 5.2) shows a synthesized k_ν at three different pressures ($p_1 < p_2 < p_3$) and temperatures. From (27), we know that the Doppler half-width $\alpha_D \propto \sqrt{T}$. On the other hand, the collision-induced Lorentz half-width goes as $\alpha_L (= \tau_L^{-1}) \propto \rho_{air} \propto p$ because τ_L is proportional to the mean free path of particles, which is inversely proportional to ρ_{air}. Note that pressure varies by several orders of magnitude in the middle atmosphere whereas temperature usually varies by no more than a factor of 2. Therefore, differences in absorption/emission spectra among different levels in the Earth's middle atmosphere are mostly caused by the pressure variation.

One example that illustrates the strong effect of α_D on line profiles due to temperature variations is the hydrogen Lyman-α emission and absorption at 121.6 nm. The Lyman-α emission line from the Sun is the strongest line in the UV wavelengths, and it plays an important role in mesospheric O_3 photochemistry. A very high temperature in the solar hydrogen (H) atmosphere produces the emission line broadened by the Doppler shifting according to (27). On the other hand, the absorption by terrestrial H in the upper thermosphere has a narrower profile because of a much lower temperature than that of the Sun. As a result, the Lyman-α radiation has a local minimum at its line center when reaching the lower thermosphere where the most absorption is by O_2 continuum (e.g., Nicolet 1985).

Note from Fig. 2a, that k_ν rapidly varies with ν when the spectrum consists of many lines. Furthermore, a typical vibration-rotation band of a simple molecule, such as H_2O, CO_2, or O_3, will often consist of hundreds or thousands of significant lines. Therefore, a straightforward evaluation of integrals over frequency in (11) and (12), the so-called line-by-line (LBL) integration, is extremely time-consuming and impractical in many applications. Various techniques have been developed to efficiently and accurately approximate the integration over frequency (Goody and Yung 1989). In this section, we briefly review two such techniques: random band models and correlated-k distribution method.

5.1. *Classic Random Band Models*

The simplest situation for integrating (11) and (12) over frequency is the weak absorption for a homogeneous medium when line overlap is negligible. Let us consider the following integral that defines slant path transmittance in (11) or escape function in (12), respectively

$$\bar{\Phi} \equiv \frac{1}{\Delta v}\int_{\Delta v} \Phi_v dv \equiv \frac{1}{\Delta v}\int_{\Delta v} [h_v \exp(-k_v m/\mu)] dv , \tag{32}$$

where $m = \left|\int_p^{p'} [r(p'')/g] dp''\right|$ is the absorber amount and $h_v = 1$ for (11) and $h_v = k_v$ for (12). The homogeneous assumption has been used in (32) when assuming k_v is independent of the path (p''). Substituting (31) into (32) and making a few approximations, we obtain

$$\begin{aligned}
\bar{\Phi} &= \frac{1}{\Delta v}\sum_{i=1}^{N}\int_{v_i-(v_i-v_{i-1})/2}^{v_i+(v_{i+1}-v_i)/2} h_v \exp(-k_v m/\mu) \\
&\approx \frac{1}{\Delta v}\sum_{i=1}^{N}\int_{v_i-(v_i-v_{i-1})/2}^{v_i+(v_{i+1}-v_i)/2} \left(\frac{1}{s_i f_v^i}\right) \exp(-s_i f_v^i m/\mu) dv \\
&\approx \frac{1}{\Delta v}\sum_{i=1}^{N}\int_{-\infty}^{\infty} \left(\frac{1}{s_i f_v^i}\right) \exp(-s_i f_v^i m/\mu) dv \\
&= \frac{1}{\Delta v}\int_0^{\infty} p(s) ds \int_{-\infty}^{\infty} \left(\frac{1}{s f_v^i}\right) \exp(-s f_v^i m/\mu) dv \\
&= \frac{1}{\Delta v}\int_{-\infty}^{\infty} dv \int_0^{\infty} \left(\frac{1}{s f_v^i}\right) \exp(-s f_v^i m/\mu) p(s) ds .
\end{aligned} \tag{33}$$

The first line of the above expression replaces the integral over a narrow band (Δv) by a summation of integrals over sub-intervals containing individual spectral lines ($\int_{\Delta v} \to \sum_i \int_{\Delta v_i}$). The second line neglects contributions of line strengths from those outside the individual frequency interval, i.e., neglects the line overlap ($k_v \to s_i f_v^i$) when lines are well separated. The third line approximates the finite integrals by infinite ones in order to derive closed expressions for certain f_v^i profiles ($\int_a^b \to \int_{-\infty}^{\infty}$). The fourth line replaces the summation over individual lines of different line strengths by an infinite integral with a line strength distribution function ($\sum_i X(s_i) \to \int_0^{\infty} p(s) X(s) ds$), where $p(s)$ represents the fractional numbers of lines with line strengths lying between s and $s + ds$. The last line exchanges the order of the two integrals since the integration over s can often be easily evaluated (e.g., Zhu 1989a).

The line strength distribution $p(s)$ is chosen in such a way that the double integral can be evaluated analytically so that $\bar{\Phi}$ becomes a closed expression in terms of a few parameters. Those parameters describe the characteristics of line profile (e.g., $\bar{\alpha}_L$ and $\bar{\alpha}_D$ for the mean

Lorentz and Doppler half-widths, respectively) and the distribution function $p(s)$ (e.g., δ and β for the mean wavenumber spacing and relative population of weak lines, respectively) (e.g., Zhu 1989a; Zhu et al. 1991)

$$\bar{\Phi} = \bar{\Phi}(\bar{\alpha}_L, \bar{\alpha}_D, \delta, \beta, ...)\,. \tag{34}$$

Various analytic expressions of (34) have been derived for either a pure Lorentz profile or a pure Doppler profile (e.g., Goody 1952; Malkmus 1967; Zhu 1989a, 1990). To derive closed expressions for the most general Voigt profile one has to introduce piecewise approximations for the Voigt profile (Fels 1979; Zhu 1988). Here, it needs to be emphasized that we have assumed a homogeneous path while driving (33). Therefore, the optical thickness has been replaced by a product between a constant k_ν and the absorber amount along the path: $u_\nu = k_\nu m$. The mean Lorentz and Doppler half-widths along the path, $\bar{\alpha}_L$ and $\bar{\alpha}_D$, reflect this approximation when the medium is inhomogeneous. Another important approximation used in deriving (33) is the neglect of the overlap effect, i.e., that lines are well separated. This approximation usually holds only when the absorption is weak ($u_\nu / \mu << 1$). Finally, note that (33) applies to both the transmission and escape functions.

To include overlap effects among the different absorption lines one has to use the multiplication property for the transmission function ($h_\nu = 1$ and $\bar{\Phi} \equiv \bar{\tau}$); this is based on the assumption that the transmission functions of any two lines are uncorrelated

$$\bar{\tau}_{12} \equiv \frac{1}{\Delta\nu}\int_{\Delta\nu} \tau_{12} d\nu = \frac{1}{\Delta\nu}\int_{\Delta\nu} \exp[-(k_{\nu 1} + k_{\nu 2})m/\mu]d\nu = \bar{\tau}_1\bar{\tau}_2 + \overline{\tau'_{\nu 1}\tau'_{\nu 2}} \approx \bar{\tau}_1\bar{\tau}_2\,, \tag{35}$$

where $k_{\nu 1} + k_{\nu 2} = s_1 f_\nu^1 + s_2 f_\nu^2$ is the composite absorption coefficient of line profiles, and ($\bar{\tau}_1, \bar{\tau}_2$) and ($\tau'_{\nu 1}, \tau'_{\nu 2}$) are the mean and perturbed transmission functions, respectively, e.g.,

$$\bar{\tau}_1 = \frac{1}{\Delta\nu}\int_{\Delta\nu} \exp(-k_{\nu 1} m/\mu) d\nu\,,\ \ \tau'_{\nu 1} = \exp(-k_{\nu 1} m/\mu) - \bar{\tau}_1 \equiv \tau_{\nu 1} - \bar{\tau}_1\,. \tag{36}$$

Note that the multiplication property strictly holds for the monochromatic transmission function ($\tau_{\nu 12} = \tau_{\nu 1}\tau_{\nu 2}$). It is also expected to be a good approximation for a narrow-band slant-path transmission function ($\bar{\tau}_{12} \approx \bar{\tau}_1\bar{\tau}_2$) when the lines are randomly distributed (Goody 1952; Goody and Yung 1989), which means that two group of lines are positioned randomly in frequency. However, one should expect the multiplication property to be a bad approximation for the flux transmission function of (11) ($\bar{\tau}_{f12} \neq \bar{\tau}_{f1}\bar{\tau}_{f2}$) since the exponential integral, $E_3(x)$, does not possess the multiplication property. In practice, applications of random band models are often associated with the introduction of a diffusivity factor, $\eta \equiv 1 / \mu_0 \approx 1.66$, that approximates the integration with respect to μ in (11) for calculating flux by a mean value of μ_0. Most importantly, the one-term approximate flux transmission function still possesses the multiplication property that is crucial for the random band models to include the overlap effect as shown below. Apruzese (1980) found different values of the optimum diffusivity for the heating rate calculations.

Now, let us extend the multiplication property to many lines and define the band transmission function (Goody and Yung 1989)

$$\tilde{\tau} = \prod_{i=1}^{N} \overline{\tau}_i = \prod_{i=1}^{N} (1 - \frac{W_i}{N\delta}), \tag{37}$$

where $\delta = \Delta\nu / N$ is the mean spacing between two lines and W_i is the equivalent width of a single line that characterizes the absorption

$$W_i \equiv \int_{-\infty}^{\infty} [1 - \tau_{\nu i}] d\nu = \int_{-\infty}^{\infty} [1 - \exp(-s_i f_\nu^i m / \mu)] d\nu . \tag{38}$$

Equation (37) can be rewritten as

$$\tilde{\tau} = \exp\{\ln \prod_{i=1}^{N} (1 - \frac{W_i}{N\delta})\} = \exp\{\frac{1}{N} \sum_{i=1}^{N} \ln(1 - \frac{W_i}{N\delta})^N\} . \tag{39}$$

Noting that $\lim_{N \to \infty} (1 - W_i / N\delta)^N = \exp(-W_i / \delta)$, we finally get the band transmission function including the line overlap effect

$$\tilde{\tau} \approx \exp\{\frac{-1}{N} \sum_{i=1}^{N} (W_i / \delta)\} = \exp\{\frac{-1}{\Delta\nu} \sum_{i=1}^{N} W_i\} = \exp(-\overline{A}) . \tag{40}$$

In (40), $\overline{A}$ $(= 1 - \overline{K}$ for $h_\nu = 1)$ is the band absorption for weak absorption as defined in (32). On the other hand, $\tilde{\tau}$ is also related to the band absorption $\tilde{A}$ by $\tilde{\tau} = 1 - \tilde{A}$. When $\overline{A} << 1$, which corresponds to a case of negligible overlap effect, we recover the weak approximation (34) for the band transmission function ($h_\nu = 1$): $\tilde{A} \approx \overline{A}$ or $\tilde{\tau} \approx \overline{\tau} = 1 - \overline{A}$.

5.2. *Correlated-k Distribution Method*

In the last decade or so, it has become clear that the correlated-*k* distribution (CKD) method is a powerful tool for calculating radiation functions in inhomogeneous atmospheres. The *k*-distribution and its relation to band models were first introduced into atmospheric applications by Arking and Grossman (1972). The classical description of CKD emphasized the frequency distribution function *f*(*k*) (Domoto 1974). Shi (1981) suggested that the integration over frequency could be greatly accelerated by sorting an absorption spectrum into a monotonic function. Lacis and Oinas (1991) applied the *k*-distribution to radiative transfer problems in a vertically inhomogeneous atmosphere, formally defined CKD, and, more importantly, established a close connection between the cumulative frequency distribution, *g*(*k*), and the sorted absorption spectrum. West et al. (1990) suggested dispensing with $f(k)$ conceptually and only considering the cumulative distribution function $g(k)$ in radiation calculations. Although one needs $f(k)$ to efficiently derive $g(k)$, in numerical calculations the physical insights of radiative transfer in an inhomogeneous atmosphere can be mostly gained by only considering $g(k)$. By introducing a perfect random band model in analyses, Zhu (1994) established a universal relationship among the exact LBL integration, CKD, and traditional random band models.

The k-distribution can be introduced from two different perspectives. One is to consider the k-distribution as a generalization of the traditional exponential-sum-fitting method where the slant-path transmission function for a homogeneous path can be expressed as (e.g., Wiscombe and Evans 1977)

$$\tau(m) = \sum_i a_i \exp(-k_i m) = \int_0^\infty f(k)\exp(-km)dk \equiv \mathcal{L}\{f(k)\}, \tag{41}$$

where $L\{\ \}$ denotes the Laplace transform. In (41), the first equality can be considered a discretization of frequency integration whereas the second equality applies the same tactic as the fourth line in (33) to k. Based on the second equality, $f(k)dk$ ($\approx a_i$) is the fraction of the frequency domain occupied by the absorption coefficient between k and $k+dk$. The second perspective of the k-distribution method is to first map the absorption coefficient k_ν in frequency ν into a monotonic function k_g in a normalized variable $g \in (0, 1)$. Then, the integration over any radiation function such as (15) or (16) can be carried out with respect to g,

$$\frac{1}{\Delta\nu}\int_{\Delta\nu} R(k_\nu, m, \ldots)d\nu = \int_0^1 R(k_g, m, \ldots)dg\,. \tag{42}$$

Since k_ν, as defined by (31), is a rapidly varying function of ν, whereas k_g is a monotonic function of g expected to slowly vary, the right-hand-side of (42) can be evaluated far more efficiently than the left-hand-side.

The connection between the two perspectives can be established by introducing the cumulative distribution

$$g(k) = \int_0^k f(k')dk'\,. \tag{43}$$

Note that $dg = f(k)dk$. Furthermore, $g(k)$ is monotonic with $g(\infty) = 1$, i.e., the fraction of frequency that includes all values of the absorption coefficient is unity. From (41) and (42) we conclude that the integral variable g in (42) is the cumulative distribution function defined in (43). Based on the (41) and (43) we have

$$f(k) = \mathcal{L}^{-1}\{\tau(m)\}\,,\ g(k) = \mathcal{L}^{-1}\{\tau(m)/m\}\,. \tag{44}$$

The Laplace transform relationship between $\tau(m)$ and $f(k)$ (or $g(k)$) as indicated by (41) and (44) implies a convolution relation in $f(k)$ (or $g(k)$) when the multiplication property of $\tau(m)$ holds (Zhu 1995). The convolution formulas can then be used to derive the composite k-distributions for a gas mixture when the k-distributions for the individual gases are known.

Equation (41) suggests a one-to-one correspondence between the transmission function, $\tau(m)$, and the k-distribution, say, $f(k)$. However, $\tau(m)$ and $f(k)$ are completely different in terms of how they can be used to calculate other radiation quantities. The traditional $\tau(m)$ is an integral quantity that depends on the absorber amount (m) along the path. On the other hand, the k-distribution, $f(k)$, or k_g, is a localized quantity. In other words, even though the k-distribution as introduced by (41) and (42) has assumed a homogeneous path the derived $f(k)$ or k_g can be considered a parameter that may vary along the path. This crucial property of the k-distribution, first pointed out explicitly by Zhu (1995), is the reason that the k-distribution defined by (42) can be used to calculate any radiation function including those that have multiple scattering processes. The difference between a localized quantity (k_g) and an integral quantity ($\tau(m)$) is similar to the difference between an intensive property and an extensive property in thermodynamics (e.g., Pathria 1996). The former (e.g., p and T) is independent of the size of the system whereas the latter (e.g., volume and mass) will dependent on the system size. Radiative heat exchange usually involves inhomogeneous atmospheres where the emission and absorption occur at different altitudes with significantly different spectra (Fig. 2a). Lacis and Oinas (1991) were the first to relate the validity or the accuracy of (42) to the correlation of spectra among different levels of an inhomogeneous atmosphere. They found the errors to be small (~1%) for the calculations of cooling rates by H_2O, CO_2, and O_3, and suggested CKD. Similar results of about 1 to 2% error were also obtained while applying CKD to calculating other radiation quantities and with other gases (e.g., Fu and Liou 1992).

To illustrate the relation between the correlation spectra and the validity of the CKD method we first map the synthesized k_ν at level 2 of (p_2, T_2) into a monotonic k_g as shown in Figs. 2b and 2c. However, since the radiative heat exchange is a non-localized process, the spectra at other altitudes should also be sorted accordingly when calculating the radiative heat exchange among those levels. Assuming that the absorption and emission among different levels are strictly correlated (e.g., the maximum k_ν's at levels 1 and 3 occur at the same ν where k_ν at level 2 reaches its maximum), we can map the rest of the spectra at levels 1 and 3 into monotonic k_g spectra as shown in Fig. 2b. Immediately, we find that the c-k coefficients are much smoother than the original absorption coefficients. As a result, the integration over frequency in radiation calculations is greatly simplified and computationally comparable to that of a random band model such as (34) or (40).

The assumption of strict correlation in spectra among different levels (i.e., CKD) means that the absorption coefficients in Fig. 2a having the same ν at different levels will have the same g when they are mapped to the g-variable. There are few cases where this condition is strictly satisfied (Goody et al. 1989; Fu and Liou 1992). In general, the spectra among different levels are not strictly correlated. In other words, if we sort the spectra in all the other levels according to a reference level at which the absorption coefficient is mapped to a monotonically increasing k-distribution, then the sorted spectra in those levels will not increase monotonically. Figure 2c shows the k-coefficients at levels 1 and 3 sorted according to the mapping at level 2. As a result of such a simultaneous mapping, the calculated radiation quantities among these 3 levels will be identical for absorption coefficients in either Fig. 2a or Fig. 2c. Clearly, the difference between Fig. 2b and Fig. 2c determines the validity and accuracy of CKD.

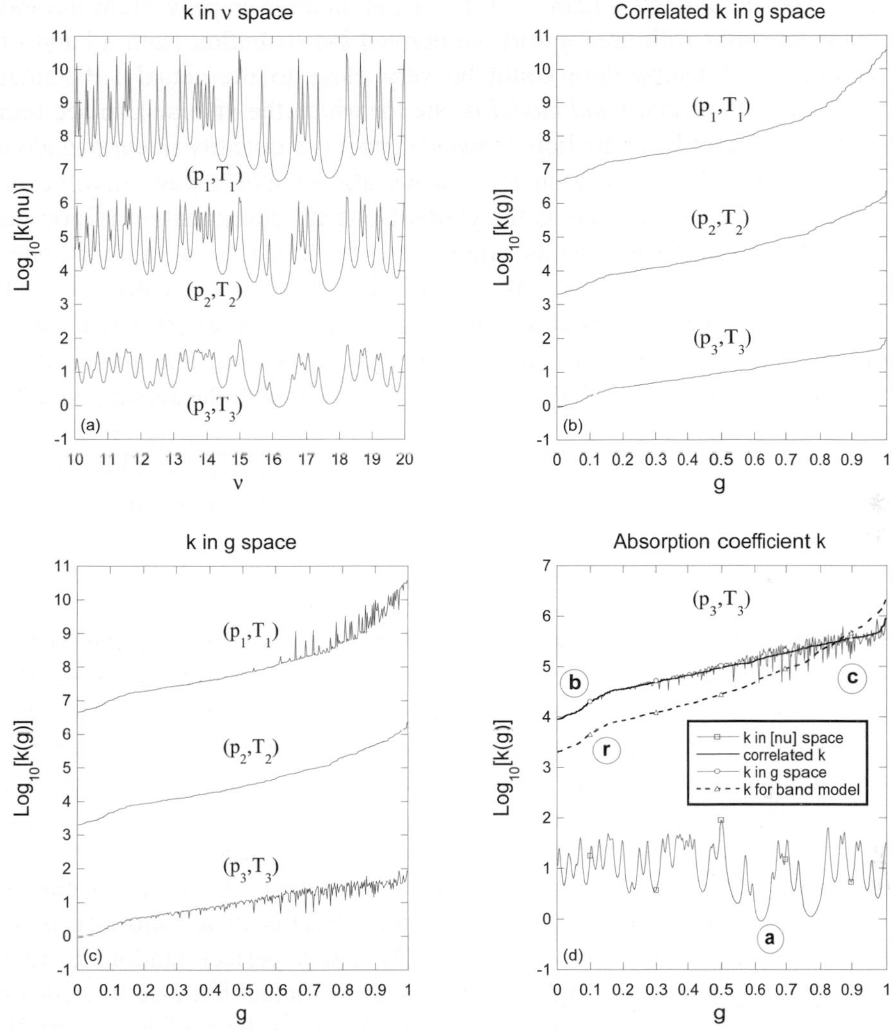

Figure 2. (a) Three absorption spectra in frequency variable synthesized from 60 Voigt lines with random positions and line strengths. (b) CKD derived directly by mapping each of the absorption spectra in (a) to a monotonically increasing *k*-distribution in a normalized *g*-variable. (c) When the spectrum at level 2 is mapped to a monotonically increasing *k*-distribution, the spectra at levels 1 and 3 are sorted accordingly. (d) All the absorption coefficients defined in (a-c) plus a *k*-distribution for a perfect random band model (denoted by **r**) and for level 3. In panels (a-c) the coefficients at level 1 have been moved up 4 units and those at level 3 have been moved down 4 units. In panel (d), only the original absorption coefficient in frequency space has been moved down 4 units (adapted from Zhu 1994).

To provide a universal assessment of validity for CKD without going over countless real examples of specific atmospheric and spectral specifications let us compare CKD with the traditional random band models, as described in the previous section, that are still widely used. For a random band model with a given $\tau(m)$, Eq. (44) defines the corresponding k-distribution for the band model. With the introduction of additional free parameters in (34) there will always be certain areas such as line shapes (Zhu 1988), line intensity distributions (Zhu 1989a), or band parameters (Zhu 1991) that can undergo improvement in random band models. This means that with great effort the derived k-distribution from a band model at a reference pressure and temperature could be very close to the actual LBL integrated k-distribution. A *perfect random band model* is one for which the inverse Laplace transform of its $\tau(m)$ matches the exact k-distribution. However, $\tau(m)$ is a quantity integrated along a finite path. Applications of $\tau(m)$ to a nonhomogeneous atmosphere always involve the use of certain scaling approximations such as the Curtis-Godson approximation (Goody and Yung 1989) such that the nonhomogeneous atmosphere is scaled to a uniform one with an equivalent pressure and temperature, say, (p_2, T_2). In other words, a perfect random band model will have its k-distributions at all levels identical to that of its reference level (Zhu 1994). Since the k-distribution is a localized quantity, the spatial inhomogeneity along the path can be easily incorporated by letting k-distribution vary with the actual p and T along the path, as shown in Fig. 2b.

A comparison of CKD with the LBL integration and a perfect random band model is shown by placing all three absorption coefficients defined in Figs. 2a-c for level 3 in the same figure plus the k-coefficient for the perfect random band model as in Fig. 2d. The latter is the k-coefficient at reference level 2 and is denoted by (**r**) in Fig. 2d. First, we note that the sorted curve (**c**) varies more rapidly than the original spectrum (**a**). Therefore, if we do not use the "correlated" assumption, then mapping spectra alone does not simplify the calculations of radiative heat exchange. The figure shows that the CKD shown by curve (**b**) can be considered a smoothed spectrum of the exact curve (**c**). The most important fact from Fig. 2d is the comparison between curves (**b**) and (**r**), which illustrates the improvement of the CKD over a perfect random band model in accuracy. Since the major advantage of using a random band model is its computational speed, our analyses here demonstrate the additional superiority of CKD to the random band model of accurately incorporating the inhomogeneous effects. In other words, CKD takes advantage of both the accuracy of LBL integration and the efficiency of random band models. CKD is always more accurate because there exists a noticeable difference between CKD and a perfect random band model as indicated in Fig. 2d by curves (**b**) and (**r**). CKD is also efficient because it needs only a few points in g to integrate any radiation quantity by (42) regardless of how complicated the spectra are in ν.

To summarize the merits of the CKD method in comparison to the traditional random band models, we first note that the accuracy of the CKD method for a specific spectrum can be easily improved by increasing the quadrature points in the g-variable. On the other hand, a perfect random band model is only an ideal case that can hardly be realized to an arbitrary spectrum. In other words, even the accuracy of the frequency integration of a radiation quantity along a homogeneous path can be more systematically improved by the CKD method. The second merit of the CKD method is its more accurate treatment of the atmospheric inhomogeneity. This includes the elimination of both the path-length weighted means of the band parameters (e.g., Curtis-Godson approximation) and the diffusivity factor

(η). The third merit is that any radiation quantity, such as $\Gamma_f(p,p')$, can be computed directly based on its well-defined formula, such as (12), by the CKD method. The last two merits have clearly been established on the basis that the k-distribution is a localized quantity. Radiation algorithms can easily take all these merits while adopting the CKD method. For example, Lacis and Oinas (1991) showed noticeable errors in $\tau_f(p, p')$ when the one-point diffusivity approximation is used in (11) for the integration over μ. Therefore, algorithms adopting the CKD method while still using the 1.66-diffusivity approximation have not taken full advantage of the method.

6. Microscopic RTE and Source Function

Equipped with the powerful computational tools for doing integrations over frequency as described in the last section, we can easily evaluate the radiative heating rate (13) once the source function (J_ν) is given. Under thermodynamic equilibrium (TE), J_ν depends only on T and is given by the well-known *Planck function*, $B_\nu(T)$ (e.g., Goody and Yung 1989)

$$J_\nu = B_\nu(T) \equiv \frac{2h\nu^3}{c^2}[\exp(h\nu / k_B T) - 1]^{-1}. \tag{45}$$

The basic physics of (45) is that the emission depends only on the relative population of the excited state n_j / n_i. Furthermore, n_j / n_i depends only on a well-defined T in a system under TE through thermal contact (i.e., through collision processes) with a very large heat reservoir. When other processes such as radiative heat exchange become important, J_ν will no longer be determined by (45) but by a more general rate equation that includes both collisional and radiation processes. To derive J_ν under the breakdown of TE let us first derive the microscopic RTE that relates J_ν in (3) to the detailed interactions between radiation and matter. Assume that the lower (i) and upper (j) states have number densities n_i and n_j, and the degeneracies g_i and g_j, respectively. There are three fundamental radiation processes responsible for the change of radiance: absorption, spontaneous emission, and induced emission; see (17a-c).

The first process is the direct absorption of radiation (17a) that leads to an upward transition from i to j. The rate at which this process occurs for radiation of radiance L_ν can be written in terms of the *Einstein coefficient* B_{ij} as $(n_i f_\nu)B_{ij}L_\nu(d\Omega/4\pi)$, where $(n_i f_\nu)$ is the number density in state i that can absorb radiation at frequencies on the range (ν, $\nu+d\nu$). Here, f_ν is the normalized line (or band) profile. In making the transition from i to j, particles absorb photons of energy $h\nu_0 = E_j - E_i$. Thus, the rate at which energy is removed from an incident beam of radiation is given by

$$k_\nu^* \rho_a L_\nu \equiv s_i^* f_\nu L_\nu = n_i B_{ij}(h\nu_0 / 4\pi) f_\nu L_\nu, \tag{46}$$

where $s_i^* f_\nu = k_\nu^* \rho_a$ denotes a macroscopic absorption coefficient per unit length, uncorrected for stimulated emission. For particles returning from j to i, the two processes of spontaneous and stimulated emissions (17b-c) are possible. The rates of energy emission are given by $n_j(A_{ji}h\nu_0/4\pi)f_\nu$ and $n_j(B_{ji}h\nu_0/4\pi)f_\nu L_\nu$, respectively, where A_{ji} and B_{ji} are the Einstein coefficients for the spontaneous and stimulated emissions (e.g., Goody and Yung 1989).

Applying the continuity equation (2) or (3) of the energy conservation microscopically to a photon stream, we can write the RTE in terms of the microscopic quantities:

$$\frac{dL_\nu}{ds} = [-n_i B_{ij} L_\nu + n_j (A_{ji} + B_{ji} L_\nu)](h\nu_0/4\pi) f_\nu , \tag{47}$$

or a form similar to (3) in terms of the source function (L_ν) and line strength (s_i):

$$\frac{dL_\nu}{ds} = -s_i f_\nu (L_\nu - J_\nu) , \tag{48}$$

where

$$J_\nu = \frac{n_j A_{ji}}{n_i B_{ij} - n_j B_{ji}} , \quad s_i = \frac{h\nu_0}{4\pi}\left(n_i B_{ij} - n_j B_{ji}\right) . \tag{49}$$

Equation (47) indicates that the change of radiance depends on the number densities (n_i, $i = 0, 1, 2, \ldots$) of particles at different energy levels. Interactions between matter and radiation (radiative transition processes) as indicated by (17a-c) will change the distribution of n_i at various energy levels. Furthermore, interactions among the particles will also change the number distributions at different energy levels. For example, the following forward and backward collisional reactions will change the number densities:

De-excitation: $$n_j + M \rightarrow n_i + M' , \quad k_{ji} \tag{50a}$$

Excitation: $$n_i + M' \rightarrow n_j + M , \quad k_{ij} \tag{50b}$$

where k_{ji} and k_{ij} (m^{-3} s^{-1}) denote rate coefficients for the reactions.

Equation (50a) denotes that a particle (say, a CO_2 molecule of ~360 part per million in volume = 3.6×10^{-4} in the Earth's atmosphere) at state i collides with a background medium M (a "heat bath" or a heat reservoir; for instance, an air molecule energetically coupled with all the other air molecules) to give out a portion of energy $\Delta E = E_j - E_i$ and to transit into state j. Equation (50b) denotes a similar reaction for the collision except in the reverse direction in which the particle acquires a portion of energy ΔE from the background medium. Here, the background medium is so large that it can absorb or give out almost any amount of energy without changing its own physical state ($M' = M$).

Considering a system that consists of many particles (say, $3.6 \times 10^{-4} \times 2.687 \times 10^{19}$ molecules $cm^{-3} \times 1\ cm^3 \approx 10^{16}$ molecules of CO_2) that is in contact with a heat bath, we ask the question of how the probability P_k (say, $= \prod_{j=1}^{1e16} p_{j,k}$) of the kth quantum state of the system depends on the value of its energy E_k (say, $=\sum_{j=1}^{1e16}\varepsilon_{j,k}$). According to statistical mechanics (e.g., Pathria 1996), in a long enough time (depending on the collision frequency and the efficiency of the energy transfer per collision) the system will reach a state of TE with P_k given by the following canonical distribution

$$P_k = \frac{g_k \exp(-E_k / k_b T)}{\sum_k g_k \exp(-E_k / k_b T)} \equiv \frac{g_k \exp(-E_k / k_b T)}{Q}, \tag{51}$$

where g_k is the degeneracy of the kth state and the normalization factor Q is called the partition function. The above equation suggests that under TE the system will have a greater probability of staying at a lower energy state than at a higher energy one.

The word "thermodynamic" contains the following two implications about the state of equilibrium. First, the equilibrium is established under a fixed thermal temperature T (or one could also consider that (51) leads to the definition of temperature). Second, it is a dynamical (rather than static) balance. In other words, it is the continuing energy exchange through (50) that brings about a balanced probability distribution (51). Based on this argument, the collisional rate coefficients k_{ji} and k_{ij} are related through $k_{ji}n_iM = k_{ij}n_jM$, or

$$k_{ij} / k_{ji} = (n_j / n_i)_{\text{equil}} = (g_j / g_i)\exp(-\Delta E / k_b T), \tag{52}$$

where (51) has been used to derive the second relation.

TE also implies that the radiance is homogeneous ($dL_\nu / ds = 0$) and isotropic ($L_\nu = J_\nu = B_\nu$). From (47), we have

$$-n_i B_{ij} L_\nu + n_j (A_{ji} + B_{ji} L_\nu) = 0, \tag{53}$$

or

$$L_\nu = \frac{(n_j / n_i) A_{ji}}{B_{ij} - (n_j / n_i) B_{ji}} = \frac{(A_{ji} / B_{ji})}{\dfrac{g_i B_{ij}}{g_j B_{ji}} \exp(h\nu_0 / k_b T) - 1} = B_\nu(T). \tag{54}$$

Comparing (54) with (45), we obtain the relationships among different Einstein coefficients

$$A_{ji} = (2h\nu_0^3 / c^2) B_{ji}, \quad g_i B_{ij} = g_j B_{ji}. \tag{55}$$

The absorption and emission coefficients are derived from the steady state measurements (and spectroscopy theory). However, microscopically, the absorption and emission coefficients are the intrinsic properties that describe the interactions between the photons and matter. This remains invariant even if the system deviates away from TE. The same idea also applies to the relation (52) between forward and backward collisional or chemical reaction rate coefficients k_{ji} and k_{ij} that have also been derived based on the assumption that the system is in TE. Microscopically, k_{ji} and k_{ij} are related to the intrinsic physical properties of those collisions among the particles and should be independent of whether the macro-system is in TE or not. Actually, our analysis on how the system arrives at TE through (50) has implicitly assumed that k_{ji} and k_{ij} would remain constant so that the system can reach a state of dynamical equilibrium.

Substituting (55) into (49) we finally relate the source function and the absorption coefficient to the particle populations n_j

$$J_\nu = \frac{2h\nu^3}{c^2}\left[\frac{g_j n_i}{g_i n_j} - 1\right]^{-1}, \; s_j = A_{ji}\frac{c^2 n_i}{8\pi\nu^2}\frac{g_j}{g_i}\left[1 - \frac{g_i n_j}{g_j n_i}\right] \equiv s_j^*\left[1 - \frac{g_i n_j}{g_j n_i}\right] \tag{56}$$

subject to the constraint that the sum of n_i at all states gives the total number density of absorber: $\sum_i n_i = n_a$. In (56), s_j^* is the line strength without correction for stimulated emission. Laboratory measurements of absorption coefficients assume the TE condition and are given by

$$s_{TE} = s_i^*\left[1 - \exp(-h\nu_0 / k_b T)\right], \tag{57}$$

where we have applied (52) to the last identity in (56).

One still needs a set of equations that relate n_j to a known physical field or radiance to bring the system to a closure. In the most general approach, one has to again start from a conservation law of the particle stream similar to (2) that describes how n_i varies with time and space (e.g., Lifshitz and Pitaevskii 1981). The steady state solution (52) shows that n_i is solely determined by T under the TE condition. Equation (52) has been derived from the statistical physics that studies the special laws governing the behavior and properties of macroscopic systems. It reduces the total number of freedoms of the microscopic system (say, 10^{16} molecules) into a very small number of the corresponding macroscopic system (say two variables: T and p) as shown in (51) and (52). However, most useful results in statistical physics have been derived based on the fundamental premise that the (closed) system is under TE. This assumption eliminates both temporal and spatial changes.

To account for the temporal-spatial changes of a system while still preserving the key idea of macroscopic TE we review the concept of a continuum or fluid. Any small volume element in a fluid is always supposed to be large enough (greater than the mean free path of the particles) that it contains a great number of particles and photons. This assumption makes TE a very good approximation in the volume. In addition, the volume element is also relatively small so that its macro-physical properties can be considered a (slowly varying) continuous function of time and space. For example, instead of a fixed temperature T, we have the temperature of a fluid varying continuously with time and space: $T(t,x,y,z)$. Furthermore, the concept and the term "thermodynamic equilibrium" are expanded into "local thermodynamic equilibrium" (LTE) to reflect the fact that the canonical distribution (51) locally holds even though T is not a constant. The equilibrium n_j distribution at two energy levels (52) in a continuum becomes

$$[n_j(t,x,y,z)/n_i(t,x,y,z)]_{\text{equil}} = (g_j/g_i)\exp[-\Delta E/k_bT(t,x,y,z)] \qquad (58)$$

and the corresponding LTE source function is given by (45). Specifically, if $n_i = n_0 \approx n_a$ ($\Delta E \approx E_j$) is the ground state, then (58) provides the relative populations of the excited states at various energy levels E_j ($j = 1, 2, 3, \ldots$).

To illustrate some implications of LTE, we note that (58) has been derived based on the collisional processes as to (50) by assuming a statistical equilibrium in a temporal-spatial domain over which an averaged T can be well defined according to (51). Since both relationships (51) and (58) depend on the energy levels, it is only natural to argue that whether a system can reach an equilibrium state through various collisional processes depends on ΔE ($\approx E_j$) and how efficiently the energy is exchanged among different particles through collisions. Hence, the concept of LTE or the definition of $T(t, x, y, z)$ based on (51) under LTE is also related to a specific frequency range ($\Delta E = h\nu_0$) over which the LTE relationship holds. From this perspective, the term "local" in LTE can also be understood as a portion of the whole energy or frequency coverage. The energies of different categories of electronic, vibrational, rotational, and translational transitions vary by many orders of magnitude. Furthermore, TE is attained by exchanging energy between the sub-system and a heat reservoir (through thermal contact, i.e., (50)). Therefore, a sub-system reaches LTE more easily at a state corresponding to a translational or rotational transition with a smaller ΔE than at a vibrational or electronic transition with a greater ΔE.

LTE also depends on the size of the region over which one allows the collisional processes to occur to reach an equilibrium state. In the Earth's exosphere ($z > 450$ km), particles become near collisionless due to very low air densities ($<5 \times 10^7$ cm^{-3}). A moving particle may easily escape to space or be subject to the influence of Earth's gravitational potential and travel in a ballistic orbit. As a result, the Maxwellian distribution (25) can no longer be used to describe particle's kinetic energy that yields a well-defined translational T (e.g., Chamberlain and Hunten 1987; Strobel 2002) in the Earth's exosphere. On the other hand, many astronomical phenomena derived from ground-based observations can be explained by a well-defined (say, rotational) T (e.g., McKee et al. 1982) even when the particle densities in those objects (say, nebulae) are much lower (say, $< 10^6$ cm^{-3}). The implication of these two contrasting examples is that the measured nebula T has been averaged over a much greater volume in which particles experience collisions of many times to maintain an equilibrium distribution. In other words, the so-called intensive variables, such as n_i ($= N_i / V$) and T, are derived not by assuming the total number of particles at state i (N_i) and the volume (V) approaching to zero but by assuming that they are approaching infinity (e.g., Pathria 1996).

If additional processes such as radiative transitions and chemical reactions are needed and play non-negligible roles in terms of their magnitudes in order to derive n_i, then, we say that the system is in a state of non-LTE. The so-called rate equations illustrate the balance of populations when additional processes are included in considering a system in a statistical equilibrium. Here, we consider a radiative two-level model of the two vibration-rotation states j and i (e.g., Lopez-Puertas and Taylor 2001; Zhu 2003)

$$n_j\left(A_{ji}+B_{ji}\bar{L}_\nu+k_{ji}[M]\right)+n_jD_j=n_i\left(B_{ij}\bar{L}_\nu+k_{ij}[M]\right)+\sum\nolimits_\ell P_{\ell\to j}+P_{cj}, j=1,2,3,\ldots \quad (59)$$

where $k_{ji}n_jM$ and $k_{ij}n_iM$ represent the collisional processes (50), and $\bar{L}_\nu = (4\pi S)^{-1} \times\int_{\Delta\nu}\int_{4\pi}k_\nu L_\nu d\Omega d\nu$ is the mean radiance. The terms in the parentheses on both sides represent the rates of loss and production, respectively, for the state j. n_jD_j and $\sum_\ell P_{\ell\to j}$ denote the rates of loss and production, respectively, by collisional transitions other than (50) (Zhu 1990). The last term P_{cj} represents a net chemical production generated by photochemical reactions for the state j. Many applications of IR radiative transfer in the middle atmosphere and planetary atmospheres use the vibration-rotation spectra of gas molecules. In this case, it can be shown that non-LTE has little effect (< 5%) on s_i. On the other hand, J_ν can significantly deviate from its LTE value given by (45). Therefore, many non-LTE studies in the middle atmosphere and planetary atmospheres focus only on the non-LTE source functions. The rate equations (59) describe a localized statistical equilibrium at given states of molecules that correspond to certain energy levels. In principle, they are solved together with RTEs that relate the non-localized radiative energy exchanges among different layers of atmosphere to the changes of populations at those states.

It is also worth pointing out the difference between non-LTE and "not in TE." We note that neither the standard RTE (3) nor (59) contains any terms of time derivative in the closed system. The non-LTE rate equations (59) still describe a state of dynamical balance between the production and the loss of the state population n_j except that additional physical processes have been added to TE. On the other hand, "not in TE" that refers to evolving L_ν or n_j often needs time-dependent continuity equations, such as (2), to describe the adjustment processes as the system deviates away from TE (e.g., Lifshitz and Pitaevskii 1981).

In general, the coupled equations are nonlinear and have to be solved through iteration (e.g., Zhu and Strobel 1990; Kutepov et al. 1998; Lopez-Puertas and Taylor 2001). One often requires calculations of broadband radiation quantities such as cooling rates or spectrally integrated limb radiances of radiatively active species (e.g., Goody and Yung 1989) for various modeling and retrieval applications. Under these circumstances, various approximations can be used to linearize the coupled equations to yield a closed expression for J_ν. Here, we briefly review and summarize a few approximations.

A vibration-rotation transition typically will have an energy difference ranging from 10^2 cm^{-1} to 10^4 cm^{-1}. In this case, n_j decreases rapidly with increasing j. The simplest non-LTE model is the classic two-level model that includes only the fundamental band (0↔1) corresponding to the transition between $j = 0$ and $j = 1$ (e.g., Houghton 1986). The non-LTE system is linearized and reduced into one rate equation and one RTE. Specifically, J_ν is the weighted average of the LTE blackbody source function B_ν and the mean radiance $\bar{L}_\nu$ whereas H_r is proportional to the difference between $\bar{L}_\nu$ and J_ν (Houghton 1986):

$$J_\nu = \frac{\phi_{10} B_\nu + \overline{L}_\nu}{\phi_{10} + 1}, \quad H_r = \frac{4\pi Sr}{c_p}(\overline{L}_\nu - J_\nu), \tag{60}$$

where $\phi_{10} = k_{10}M / A_{10}$ is the ratio of collisional de-excitation to the Einstein coefficient. In the limit of LTE ($\phi_{10} \to \infty$) we have $J_\nu = B_\nu$ and $H_r \propto (\overline{L}_\nu - B_\nu)$. The latter relationship indicates a technical difficulty of evaluating H_r, for it usually involves a small difference between two large quantities. In the limit of strong non-LTE, when $\phi_{10} \to 0$, we have $J_\nu = \overline{L}_\nu$ and $H_r = 0$. This limit corresponds to an isotropic and conservative scattering. J_ν and H_r can further be expressed in closed matrix forms.

When the fundamental band becomes extremely thick in exchanging radiative heat (say, $\Gamma_f << 0.01$), contributions from the weak hot bands (say, 1↔2 and 2↔3) and minor isotope bands become comparable to or more important than those from the fundamental band. One has to consider a generally nonlinear multi-level model to derive accurate source functions for different sub-bands. A major source of the nonlinearity comes from the so-called vibration-vibration (V-V) transitions, such as ($j = 1$ and 2),

$$n_{j-1} + n_{j+1} \leftrightarrow n_j + n_j, \quad k_{vvf}, k_{vvb} \tag{61}$$

where the forward and backward transition rates are given by the nonlinear terms $k_{vvf} n_{j-1} n_{j+1}$ and $k_{vvb} n_j n_j$, respectively (e.g., Zhu and Strobel 1990).

To reduce the generally nonlinear multi-level system into a linear two-level system Zhu (1990) and Zhu and Strobel (1990) explicitly developed and tested an equivalent two-level model that neglects V-V transitions for the middle atmosphere broadband cooling rate calculations, with J_ν and H_r being similar to (60) for individual sub-bands. Since the broadband cooling rate of an equivalent two-level model using a broadband Curtis matrix is efficient and accurate, it has been used in some radiation algorithms for middle atmosphere models (e.g., Zhu 1994). An improvement that partially includes the V-V transitions by parameterizing J_ν as a linear combination of two limiting J_ν's with an adjustable parameter was also suggested in Zhu (1989b). In Zhu (2003), the equivalent two-level model was further extended to include an additional source term of chemical production generated by photochemical reactions for the upper-level energy states of gas molecules. The closed expressions of J_ν and H_r are similar to those of the equivalent two-level model except that B_ν is modified by a factor $(1 + \phi_c)$, with ϕ_c $(= P_{cj} / k_{ij} n_i [M])$ characterizing the ratio of chemical production (P_{cj}) to collisional excitation ($k_{ij} n_i [M]$).

Non-LTE of CO_2 has been investigated extensively while studying the remote sensing and radiative energy budget in the atmospheres of three inner planets: Venus, Earth, and Mars (e.g., Lopez-Puertas and Taylor 2001). On the other hand, CH_4 is the most abundant radiatively active species in the outer planetary system (e.g., Yung and DeMore 1999). Many atmospheric models for the outer planetary atmospheres often include CH_4 non-LTE processes (e.g., Appleby 1990; Yelle 1991; Strobel et al. 1994).

7. Applications of Radiative Transfer to Planetary Atmospheres and the Earth's Middle Atmosphere

The last two sections described the frequency integration ($\int_{\Delta\nu} L_\nu(\mathbf{r},\mathbf{\Omega})d\nu$) and the source function (J_ν) determination. These are two essential parts of the theory and applications of radiative transfer in a clear-sky atmosphere. One may look at the problems of radiative transfer in a scattering medium in a similar way where the integration over the solid angle ($\int_{\Delta\Omega} L_\nu(\mathbf{r},\mathbf{\Omega})d\Omega$) becomes critical and cumbersome because of the rapidly varying phase function (e.g., Bohren and Huffman 1983; Liou 2002). Different techniques such as (33) and (42) can be considered different ways to sum up the contributions from individual building blocks. From this perspective, solutions to RTEs can be considered bookkeeping exercises because the RTE (3) and its solution (4), including more specific ones such as (13), are simple. Most importantly, as we have already pointed out that, the coupled system of (3) and (59) does not contain any time derivative terms even though the concept of continuum (i.e., fluid) was used while deriving those equations. Note that many interesting phenomena and difficult theories in fluid mechanics (e.g., waves, instabilities, solitons, chaos, etc.) are associated with the evolving nature of the system that corresponds to the existence of time derivative terms (e.g., Andrews et al. 1987; Craik 1985). From this perspective, the radiative transfer theory as described by the steady state system of (3) and (59) is often considered a technique or a tool to be used in a broader context of scientific researches. Such a view toward radiation theory naturally leads us to propose an empirical rule of "optimizing computational accuracy and efficiency" for solving specific radiation problems.

We have shown in Section 5 that the CKD method takes advantage of the accuracy of LBL integration and the efficiency of random band models. Therefore, the CKD method should be considered a starting point for calculating any radiation quantity. With such an empirical principle in mind, the author developed radiative heating and cooling modules for modeling the atmospheres of Pluto and Titan (Saturn's largest moon) where the radiative cooling by the methane (CH_4), ethane (C_2H_6), and acetylene (C_2H_2) vibration-rotation bands and the radiative heating by CH_4 and a haze layer were treated uniformly by the CKD method (Strobel et al. 1996; Zhu 2002). Previously, a radiation module of cooling and heating rates by sulfur dioxide (SO_2) with similar spectral complexities was developed for modeling Io's (Jupiter's closest moon discovered by Galileo) atmosphere in which several forms of random band models were adopted (Strobel et al. 1994). Although the analytical expressions such as (34) could be simple and fast, in some cases the evaluation of the mean parameters ($\bar{\alpha}_L, \bar{\alpha}_D, \delta, \beta, \ldots$) was generally computationally time-consuming for an inhomogeneous atmosphere. One possible advantage of using random models could be one's research heritage where the preexisting programs and experience may lead to savings in developing a new set of programs for different sets of spectral parameters and atmospheric conditions. However, the author also found that the CKD method could handle different specifications (e.g., T, p, frequency coverage, absorbers, path geometry) more easily than random band models since the k-coefficients can be derived by a general LBL integration whereas analytical expressions such as (34) exist only under certain special conditions.

Another significant advantage of using the CKD method, especially for cooling rate calculations, is to avoid calculating a small difference between two large quantities. We have already shown (Section 5.1) that only $\tilde{\tau}$ and $\tau_f(p,p')$ are available from random models for an overlap band though the cooling calculations require $\Gamma_f(p,p')$ that is related to $\tau_f(p,p')$ by the derivative relationship (14). We also note from (15) and (16) that $\tau_f(p,p) = \Gamma_f(p,p) = 1$. It is common in many applications that $\Gamma_f(p,p') << 0.1$ when $|p-p'|$ only corresponds to a few layers of atmosphere in a model. This suggests that the computation of $\Gamma_f(p,p')$ based on (14) will involve a small difference between two nearly equal values of $\tau_f(p,p')$. As a result, the cooling rate calculation in the Earth's middle atmosphere by the CO_2 15-μm band often requires $\tau_f(p, p')$ to be accurate up to five significant digits. Since the *k*-coefficient is a localized quantity that can be used to compute any radiation function according to (42), we can derive $\Gamma_f(p,p')$ directly from its integral definition (16) to avoid finite differencing $\tau_f(p, p')$. In addition to the radiative heating and cooling rates, photolysis rate calculations in modeling the photochemistry of the middle atmosphere and planetary atmospheres sometimes also involve integrations over the frequency with complex spectra. Under these circumstances, the CKD method should also be considered a preference (e.g., Minschwaner and Siskind 1993; Zhu et al. 1999).

We listed several advantages for the CKD method for solving general radiative transfer problems regarding accurately and efficiently handling the radiative heat exchange among different layers. There exist also some special cases of radiation problems where techniques other than CKD may be better in terms of an overall optimization with respect to computational accuracy and speed. We have shown in Section 5 (Fig. 2) that the improvement of the CKD method over the random band models in terms of accuracy is its more accurate treatment of inhomogeneity when the radiative energy is exchanged among different layers with different spectra. In some applications of planetary atmospheres, we have only limited information regarding the bulk structure of an atmosphere. The bulk property will be determined by the integral quantity such as the absorption or transmission of a whole atmospheric layer. In this case, the random band model is a better approach due to its simplicity. Here, we show an application of retrieving Io's SO_2 atmosphere by the Malkmus model that corresponds to the specification of $\alpha_D = 0$ and $p(s) = s^{-1}e^{-s/\sigma}$ in (33) (Malkmus 1967; Zhu 1989a).

In Section 3 we described two techniques, occultation and limb emission-absorption, for atmospheric remote sounding. A third technique of analyzing the absorption-reflection spectra is illustrated in the retrieval of Io's SO_2 atmosphere and its physical properties. The solar photons reflected from Io were collected by a detector (Faint Object Spectrograph) on the Hubble Space Telescope (HST). The spectra of the reflected solar radiation depend on the atmospheric transmission and surface reflectance. Since gas and solid spectra are fundamentally different even for the same molecule (e.g., Houghton and Smith 1966), the spectral analyses of the so-called geometric albedo from Io provide us with information about the chemical composition and physical properties of Io's SO_2 atmosphere (Ballester et al. 1994). The geometric albedo is a function of the atmospheric transmission function ($\tilde{\tau}$) that is sensitive to SO_2 temperature (T) and its column density (N). Figure 3 shows one example of the measured and modeled Io geometric albedos over an SO_2 spectral region 1980-2300 Å for a hemispheric SO_2 atmosphere at 250 K and with $N = 4.7\times10^{15}$ cm^{-2}. In this case, we are not interested in details of photon exchanges within the atmosphere. The needed transmission function at each sub-band (i) for the whole atmosphere can be best derived by the simple Malkmus model (Ballester et al. 1994)

$$\tilde{\tau}_i(N) = \exp\left(-\frac{\pi y_i}{2}\left[\left(1+\frac{4\sigma_i(T)}{\pi y_i}N\right)^{1/2} - 1\right]\right) \quad \text{for each sub-band } \Delta\nu_i, \tag{62}$$

where the band parameters y_i and $\sigma_i(T)$ (i = 1, 2, ..., 219) can be derived from laboratory measurements of high-resolution SO_2 spectra. A reasonably good fit between the measured and modeled geometric albedos, as shown in Fig. 3, justifies the inferred atmospheric T and N. Recently, using the improved solar UV spectrum and a better database for the SO_2 spectrum, the differences in the measured and modeled geometric albedos with additional constraints have been further reduced (personal communications, K. L. Jessup).

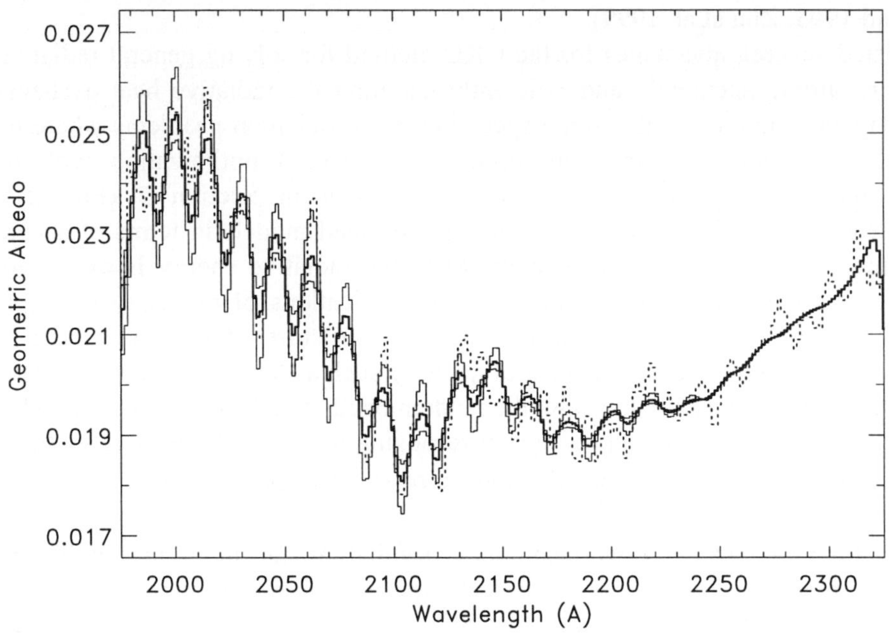

Figure 3. Measured and modeled geometric albedos over an SO_2 spectral region 1980-2300 Å for a hemispheric SO_2 atmosphere at 250 K. Short-dashed line: HST observed data; thick solid line: best fit by least-squares method with column density of 4.7×10^{15} cm^{-2}; thin solid lines: model albedos for the same atmosphere with the column density scaled up (down) by a factor of 2 that show greater (smaller) spectral contrast (after Ballester et al. 1994).

Another special method that is superior to the CKD method in terms of accuracy and efficiency is the table look-up method. When the physical property of the atmospheric state, such as $T(p)$ or mass mixing ratio $r_d(p)$, varies only slightly with respect to a reference state, the induced changes of the radiation quantities such as τ_f and Γ_f associated with the flux and cooling rate calculations are also expected to be small. In this case, the most efficient way to calculate the perturbed radiation quantity is the table look-up method. Chou and Arking (1978) first put forward such a radiation algorithm for the transmission function $\tau_f(p, p')$ to use in atmospheric numerical models. According to (11) and our discussions in Section 4, $\tau_f(p,p')$ is a functional of the atmospheric parameters of $T(p'')$, p'', and $r_a(p'')$ between levels p and p'. Chou's algorithm (Chou and Kouvaris 1991; Chou et al. 1995) stores the precomputed three-dimensional absorptance, A_{ijk}, as look-up tables

$$A_{ijk} = 1 - \tau_f(p_{eff}^i, w^j, T_{eff}^k), \quad i=1,2,\ldots,I; \ j=1,2,\ldots,J; \ k=1,2,\ldots,K, \tag{63}$$

where p_{eff} and T_{eff} are the effective pressure and temperature, respectively, averaged by the Curtis-Godson approximation, and w is the absorber amount between p and p'. Often, both p_{eff} and w vary by several orders of magnitude whereas T_{eff} varies no more than a factor of 2 in atmospheric models. For example, for the CO_2 15-μm band in the middle atmosphere, parameters I, J, and K can be set at 26, 21, and 3, respectively (Chou and Kouvaris 1991). Given $T(p'')$ and $r_a(p'')$ in a numerical model, $\tau_f(p, p')$ can be derived by interpolation formulas according to the stored A_{ijk} and the calculated (p_{eff}, w, T_{eff}) between p and p'.

Note that in addition to storing accurate tables A_{ijk} that can be derived by the LBL integration, the accuracy of Chou's interpolation algorithm also depends on the validity of the Curtis-Godson approximation. Fels and Schwarzkopf (1981) developed an interpolation algorithm that removed this limitation by a direct Taylor expansion of the transmission function between two model grids of p_j and p_k, $\tau_{jk} \equiv \tau_f(p_j, p_k)$ $(j,k=1,2,\ldots,N)$

$$\tau_{jk} \approx \tau_{jk}^0 + \sum_{i=j}^{k} \frac{\partial \tau_{jk}}{\partial T_i^0}(T_i - T_i^0) + \sum_{i=j}^{k} \frac{\partial \tau_{jk}}{\partial r_{ai}^0}(r_{ai} - r_{ai}^0) \equiv \tau_{jk}^0 + \sum_i B_{ijk}^0 (\Delta T)_i + \sum_i D_{ijk}^0 (\Delta r_a)_i, \tag{64}$$

where τ_{jk}^0 is the transmission function calculated at the reference state of $T^0(p'')$ and $r_a^0(p'')$, and three-dimensional arrays $B_{ijk}^0 \equiv \partial \tau_{jk} / \partial T_i^0$ and $D_{ijk}^0 \equiv \partial \tau_{jk} / \partial r_{ai}^0$ are the stored coefficients used for the interpolation when the model $T(p'')$ and $r_a(p'')$ deviate from the reference ones. For the CO_2 15-μm band, $r_a(p'')$ is a constant in most numerical models. Hence, Fels and Schwarzkopf (1981) only considered the case $D_{ijk}^0 = 0$. The size $(=N^3)$ of the stored look-up table B_{ijk}^0 that needs to be precomputed depends on the number (N) of model grids in vertical. Fels and Schwarzkopf (1981) factored the three-dimensional look-up table into a two-dimensional one plus a vector that greatly saved off-line storage $(= N^2 + N)$ and computational efforts for deriving the look-up tables

$$B_{ijk}^0 = F_{jk} G_i, \quad (i,j,k=1,2,\ldots,N), \tag{65}$$

from which the interpolation algorithm for the CO_2 15-μm band becomes

$$\tau_{jk} \approx \tau_{jk}^0 + F_{jk}\sum_i G_i(T_i - T_i^0),\ (i, j, k = 1, 2, \ldots, N). \tag{66}$$

Most solutions to the radiative transfer problems in planetary atmospheres (as contrasted to stellar atmospheres in astrophysics) start with the transmission function, for instance, (63) and (64). One possible reason for this is that, historically, the most powerful tool of the random band models can only be developed for the transmission function, using the multiplication property to include the overlap effect (Section 5.1). However, (15) and (16) suggest that there is little difference in computational efforts if the LBL integration or CKD method is used to perform the frequency integration over different radiation functions. Furthermore, heating or cooling rate calculations are better formulated by $\Gamma_f(p, p')$ than by $\tau_f(p, p')$, as shown in (13). Applying the similar idea of factoring to $\Gamma_f(p, p')$, Zhu (1990, 1994) developed a table look-up algorithm for calculating the middle atmosphere cooling rate by the CO_2 15-μm band and O_3 9.6-μm band

$$\Gamma_{jk} \approx \Gamma_{jk}^0 + \alpha_j(T_j - T_j^0) + \beta_{jk}\sum_i G_i(T_i - T_i^0) + \gamma_j(r_{aj} - r_{aj}^0) + \delta_{jk}\sum_i G_i'(r_{ai} - r_{ai}^0), \tag{67}$$

where Γ_{jk}^0 is the flux escape function at the reference temperature (T_i^0) and mixing ratio (r_{ai}^0) profiles, β_{jk}, δ_{jk}, α_j, γ_j, G_i, and G_i' are the stored arrays and vectors all calculated off-line by the CKD method. Direct interpolation of (67) to obtain Γ_{jk} is intrinsically more accurate than that of (66) for the cooling rate calculations because it eliminates the unnecessary error introduced while taking the finite difference of τ_{jk} according to (14).

It is also interesting to make a comparison between Chou's algorithm based on the look-up tables (63) and those based on (66) and (67) from another perspective. Chou's algorithm has greater flexibility since the look-up tables are independent of individual models. On the other hand, the look-up tables in (66) and (67) are calculated based on the fixed model grids of p_k (k = 1, 2, ..., N). However, the interpolations in (66) and (67) are explicit and straightforward whereas the multi-dimensional interpolation in Chou's algorithm requires the intermediate input of p_{eff} and T_{eff} that costs additional computational resources. This is one example illustrating a common trade-off between efficiency and flexibility one often encounters in the field of optimization. The so-called "No Free Lunch" theorems establish that an algorithm that is effective on one class of problems will always be offset by ineffectiveness on another class (Wolpert and Macready 1997). Therefore, selection of a particular algorithm or approach to a specific radiative transfer problem in planetary atmospheres often involves considering an appropriate balance among accuracy, efficiency, and available resources.

The author's view of radiative transfer as a tool or a technique is based on the fact that given an input of atmospheric parameters (e.g., T and r_a) and a spectral database (e.g., Rothman et al. 1998), solution of the RTE (3) can be considered a well-defined and solved problem. The output could be various radiation quantities such as radiative energy flux or heating rate of the atmosphere. Note that the RTE (2) or (3) represents the basic law of energy conservation. Most science issues involving the Earth's middle atmosphere need to be resolved by incorporating additional conservation laws for momentum and mass, meaning that radiative transfer is coupled with dynamical, chemical, and transport processes in modeling and theories (e.g., Andrews et al. 1987; Brasseur and Solomon 1984; Zhu et al. 2000). This is mostly due to the advances in our acquired knowledge of the atmospheric state and the associated physical processes as a result of modern technology that provided us with tremendous amounts of measurements of both temporal and spatial coverage. Compared to the Earth's atmosphere, information derived from the observations of planetary atmospheres is still very limited. Often, the observed phenomena can only be investigated quantitatively by a particular conservation law, such as the RTE corresponding to the radiative energy conservation, even though processes associated with dynamics or other energy sources would certainly play roles. In this case, the solutions of the RTE associated with the energy conservation often involve specific science problems.

One application of using the RTE alone to solve a science problem is to determine and explain the vertical temperature profiles of planetary atmospheres. A good example of this type of work is the seminal paper by Hulburt (1931), who investigated the vertical temperature of the Earth's atmosphere based on limited observations and a clear-sky radiative transfer model. Using the measured absorbers and dividing the absorption spectrum into three sub-bands in solving the RTE, Hulburt (1931) was able to derive the vertical temperature profile of the Earth's atmosphere below 12 km that was consistent with the measured average temperature profile. We mentioned in the introduction that clouds play a major role in determining the energy budget in the troposphere. Even though the clouds were excluded in a clear-sky RTE, their effects were implicitly included in Hulburt's model for he assumed "... the incoming solar radiation of which 32 percent is reflected out to space by the surface of the earth and the clouds." Introducing an effective albedo to solar radiation while adopting a clear-sky RTE for cloudy atmospheres is an approximation often used even in modern research (e.g., Kasting 1988).

From the perspective of limited observations, the current studies of planetary atmospheres, especially the atmospheres of outer planets and their moons, are similar to that of the studies of the Earth's atmosphere in earlier history. Furthermore, our modern understanding of the problem allows us to make a retrospective assessment of earlier works. Therefore, it is worthwhile to list the following significant contributions and interesting points of the paper by Hulbert (1931):

- Specific absorbers and the measured absorption coefficients of H_2O and CO_2 were used. These are still the major absorbers in current modeling studies of Earth's atmosphere.
- A convective adjustment scheme was presented that redistributes the radiative energy in the vertical.
- Sensitivity studies of doubling and tripling CO_2 in the atmosphere were performed.
- Several feedback mechanisms among the surface temperature, H_2O, vegetation, snow covers, etc. associated with the CO_2-induced global warming were suggested.

- A major feature of an isothermal lower stratosphere above 12 km as observed at the time was not reproduced in the model since Hulburt (1931) did not include a UV spectrum in his model. This lack of reproducing observations without a correct mechanism makes the above-mentioned contributions even more valuable.

For readers' convenience and entertainment, we show below three excerpts from Hulbert (1931) regarding the ideas of energy balance and convective adjustment, the doubling of CO_2 induced global warming, and the feedback mechanisms, respectively:

"If the convective region extended only to, say, 5 km we find that the atmosphere again is dynamically unstable and in addition that the total radiation emitted from the earth and the atmosphere is less than the energy received from the sun. When the convective region extends to 10 or 12 km (as is observed) the atmosphere is found to be stable, the calculated sea level temperature is about 290° (close to the observed value 287°) and the total radiation emitted from the earth and the atmosphere is equal to the received solar energy. If the convective region extended to a level greater than 12 or 15 km the outgoing radiation is less than the incoming solar radiation."

"Calculation shows that doubling or tripling the amount of the carbon dioxide of the atmosphere increases the average sea level temperature by about 4° and 7°K, respectively; halving or reducing to zero the carbon dioxide decreases the temperature by similar amounts."

"Further, an increase or decrease in the world-wide average atmospheric temperatures of a few degrees would give rise to other changes. The water vapor in the atmosphere would be increased or decreased, this would accentuate the temperature changes. At the same time changes in the areas covered by vegetation and snow fields would take place, thus changing the optical properties, that is, the emissive power *a*, section 5, of the surface of the earth."

Note that the 4°K increase in global mean surface temperature for doubling atmospheric CO_2 derived more than 70 years ago conforms to the current consensus derived by much more comprehensive three-dimensional general circulation models. However, the modern theory of global warming attributes major contributions to the feedback mechanisms (such as that due to water vapor) that were absent in Hulburt's model. Many physical numbers (not physical constants) undergo continuous revisions and fine tunings as new influencing factors, better measurements, and more efficient computational tools become available. It is often the fresh and fundamental ideas presented in original papers such the one by Hulburt (1931) that will last for a quite long time.

8. Concluding Remarks

In this article, the author presents a hierarchy of basic ideas and techniques of middle atmosphere radiative transfer. Though equations of various complexities are presented throughout the article, the main focus is a step-by-step exposition of how to design a physically based algorithm for calculating any radiation quantity such as heating rates in planetary atmospheres. The establishment of RTE (2) or (3) is based on the energy conservation law. For clear-sky problems involving only absorption and emission processes, understanding of the spectral formation (Sections 2 and 4) and our two basic methods of integration over frequency (Section 5) will allow modelers to quickly design an effective algorithm for a specific problem. The rationales of several accurate and efficient approaches were explained through real examples, most of which the author had directly worked on (Section 7). A non-LTE parameterization for the source function (Section 6) often needs to be included in an algorithm for low-pressure conditions (say, < 0.1 mb = 10 Pascal for O_3 9.6 μm band). The radiative transfer theory has been presented mostly based on a viewpoint that radiation is a stream of photon particles that interact with matter particles. A theory based on a contrary viewpoint that considers radiation as electromagnetic waves is presented by another overview paper in this volume (Q. Liu 2004). The materials presented in this article have been mostly based on a part (~ 8 hours) of a graduate course (Planetary Atmospheres) taught regularly by Darrell F. Strobel and the author at the Johns Hopkins University for the past decade. In this article, the author's expository review is limited to the theory of clear-sky radiative transfer. For more detailed theory and applications including the scattering, students are referred to other textbooks such as those by Houghton (1986), Andrews et al. (1987), Goody and Yung (1989), and Liou (2002). A modern textbook by Thomas and Stamnes (1999) that appropriately balances the depth and breadth of the field of radiative transfer has also been added to the reference list in recent years.

Acknowledgments. This research was supported by NASA grant NAG5-11962 and in part by NSF grant ATM-0091514 to The Johns Hopkins University Applied Physics Laboratory. The author thanks Dr. Elsayed Talaat for editorial assistance. Comments by Dr. Ming-Dah Chou and Professor Darrell F. Strobel on the original manuscript are greatly appreciated.

References:

Abramowitz, M. J., and I. A. Stegun, 1965: *Handbook of Mathematical Functions.* Dover, New York, 1046 pp.

Andrews, D. G., J. R. Holton, and C. B. Leovy, 1987: *Middle Atmosphere Dynamics.* Academic Press, New York, 489 pp.

Appleby, J. F., 1990: CH_4 nonlocal thermodynamic equilibrium in the atmospheres of giant planets. *Icarus*, **85**, 355-379.

Apruzese, J. P., 1980: The diffusivity factor reexamined. *J. Quant Spectrosc. Radiat. Transfer.* **24**, 461-470.

Arfken, G., 1985: *Mathematical Methods for Physicists.* 3rd Edition. Academic Press, Inc., New York, 985 pp.

Arking, A., and K. Grossman, 1972: The influence of line shape and band structure on temperature in planetary atmospheres. *J. Atmos. Sci.*, **29**, 937-949.

Ballester, G. E., M. A. McGrath, D. F. Strobel, X. Zhu, P. D. Feldman, and H. W. Moos, 1994: Detection of the SO_2 atmosphere on Io with the Hubble space telescope. *Icarus*, **111**, 2-17.

Bohren, C. F., and D. R. Huffman, 1983: *Absorption and Scattering of Light by Small Particles.* John Wiley & Sons, Inc., 530 pp.

Brasseur, G., and S. Solomon, 1984: *Aeronomy of the Middle Atmosphere: Chemistry and Physics of the Stratosphere and Mesosphere.* D. Reidel Publishing Company, Boston, 441 pp.

Chamberlain, J. W., and D. M. Hunten, 1987: *Theory of Planetary Atmospheres: An Introduction to Their Physics and Chemistry.* 2nd Edition. Academic Press, 481 pp.

Chen, H.-B., 2002: A concept for measuring liquid water path from microwave attenuation along the satellite-Earth path. *Chinese J. Atmos. Sci.*, **26**, 401-408.

Chou, M.-D., and A. Arking, 1978: An infrared radiation routine for use in numerical atmospheric models. Paper presented at *Third Conference on Atmospheric Radiation*, Am. Meteorol. Soc., Davis, Calif..

Chou, M.-D., and L. Kouvaris, 1991: Calculations of transmission functions in the infrared CO_2 and O_3 bands. *J. Geophys. Res.*, **96**(D5), 9003-9012.

Chou, M.-D., W. L. Ridgway, and M.-H Yan, 1995: Parameterizations for water vapor IR radiative transfer in both the middle and lower atmospheres. *J. Atmos. Sci.*, **52**, 1160-1167.

Chu, W. P., and R. Veiga, 1998: SAGE III/ EOS, *SPIC*, **3501**, pp. 52-60.

Craik, A. D., 1985: *Wave Interactions and Fluid Flows.* Cambridge Univ. Press, Cambridge, 322 pp.

Domoto, G. A., 1974: Frequency integration for radiative transfer problems involving homogeneous non-gray gases: The inverse transmission function. *J. Quant. Spectrosc. Radiat. Transfer*, **14**, 935-942.

Emden, R., 1913: Uber Strahlungsgleichgewicht und atmospharische Strahlung. Ein Beitrag zur Theorie der oberen Inversion. Sitzungsberichte, Adademie der Wissenschaften, Munchen, No. 1, 55-142.

Fels, S. B., 1979: Simple strategies for inclusion of Voigt effects in infrared cooling rate calculations. *Appl. Opt.*, **18**, 2634-2637.

Fels, S. B., and M. D. Schwarzkopf, 1981: An efficient, accurate algorithm for calculating CO_2 15 μm band cooling rates. *J. Geophys. Res.*, **86**, 1205-1232.

Fu, Q., and K. N. Liou, 1992: On the correlated *k*-distribution method for radiative transfer in nonhomogeneous atmospheres. *J. Atmos. Sci.*, **49**, 2139-2156.

Goody, R. M., 1952: A statistical model for water-vapour absorption. *Quant. J. R. Meteorol. Roc.*, **78**, 165-169.

Goody, R., R. West, L. Chen, and D. Crisp, 1989: The correlated-*k* method for radiation calculations in nonhomogeneous atmospheres. *J. Quant. Spectrosc. Radiat. Transfer*, **42**, 539-550.

Goody, R. M., and Y. L. Yung, 1989: *Atmospheric Radiation: Theoretical Basis.* 2nd Edition. Oxford Univ. Press, New York and Oxford, 519 pp.

Houghton, J. T., 1986: *The Physics of Atmospheres.* 2nd Edition. Cambridge Univ. Press, New York, 271 pp.

Houghton, J. T., and S. D. Smith, 1966: *Infra-red Physics.* Oxford Univ. Press (Clarendon), London and New York, 319 pp.

Hulburt, E. O., 1931: The temperature of the lower atmosphere of the Earth. *Phys. Rev.*, **38**, 1876-1890.

Kasting, J. F., 1988: Runaway and moist greenhouse atmospheres and the evolution of Earth and Venus. *Icarus*, **74**, 472-494.

Kibel', I. A., 1943: The temperature distribution in the Earth's atmosphere. *Dokl. Akad. Nauk SSSR*, **39** (1), 18-22 (in Russian).

Kutepov, A. A., O. A. Gusev, and V. P. Ogibalov, 1998: Solution of the non-LTE problem for molecular gas in planetary atmospheres: Superiority of accelerated lambda iteration. *J. Quant. Spectrosc. Radiat. Transfer*, **60**, 199-220.

Lacis, A. A., and V. Oinas, 1991: A description of the correlated *k* distribution method for modeling nongray gaseous absorption, thermal emission, and multiple scattering in vertically inhomogeneous atmospheres. *J. Geophys. Res.*, **96**(D5), 9027-9063.

Levine, I. N., 1975: *Molecular Spectroscopy*. John Wiley & Sons, New York, 491 pp.

Lifshitz, E. M., and L. P. Pitaevskii, 1981: *Course of Theoretical Physics. Vol. 10: Physical Kinetics.* Butterworth-Heinenann, Oxford, 452 pp.

Liou, K. N., 2002: *An Introduction to Atmospheric Radiation.* 2nd Edition. Academic Press, New York, 583 pp.

Liu, G., 2004: Satellite microwave remote sensing of clouds and precipitation. *This volume.*

Liu, Q., 2004: Polarimetric radiative transfer theory and its applications: An overview. *This volume.*

Lopez-Puertas, M., and F. W. Taylor, 2001: *Non-LTE Radiative Transfer in the Atmosphere.* World Scientific Pub. Co., Singapore, 487 pp.

Malkmus, W., 1967: Random Lorentz band model with exponential-tailed S^{-1} line intensity distribution function. *J. Opt. Soc. Am.*, **57**, 323-329.

McCartney, E. J., 1983: *Absorption and Emission by Atmospheric Gases: The Physical Processes.* John Wiley & Sons, New York, 320 pp.

McKee, C. F., J. W. V. Storey, D. M. Watson, and S. Green, 1982: Far-infrared rotational emission by carbon monoxide. *Astrophys. J.*, **259**, 647-656.

Minschwaner, K., and D. E. Siskind, 1993: A new calculation of nitric oxide photolysis in the stratosphere, mesosphere, and lower thermosphere. *J. Geophys. Res.*, **98**, 20,401-20,412.

Mlynczak, M. G., 1997: Energetics of the mesosphere and lower thermosphere and the SABER experiment. *Adv Space Res.*, **20**(6), 1177-1183.

Mlynczak, M. G., and S. R. Drayson, 1990: Calculation of infrared limb emission by ozone in the terrestrial middle atmosphere. 1. Source function. *J. Geophys. Res.,* **95**(D10), 16,497-16,511.

Nicolet, M., 1985: Aeronomical aspects of mesospheric photodissociation: processes resulting from the solar H Lyman-alpha line. *Planet. Space Sci.*, **33**, 69-80.

Pathria, R. K., 1996: *Statistical Mechanics.* 2nd Edition. Butterworth Heinemann, Oxford, 529 pp.

Pedlosky, J., 1987: *Geophysical Fluid Dynamics.* 2nd Edition. Springer-Verlag, New York, 710 pp.

Peixoto, J. P., and A. H. Oort, 1992: *Physics of Climate.* American Institute of Physics, New York, 520 pp.

Rothman, L. S., C. P. Rinsland, and Coauthors, 1998: The HITRAN molecular spectroscopic database and HAWKS (HITRAN Atmospheric Workstation): 1996 edition. *J. Quant. Spectrosc. Radiat. Transfer,* **60**, 665-710.

Shi, G.-Y., 1981: An accurate calculation and representation of the infrared transmission function of the atmospheric constituents. *Ph. D. thesis*, Dept. of Science, Tohoku Univ. of Japan, 191 pp.

Strobel, D. F., 2002: Aeronomic systems on Planets, Moons, and Comets. In *Atmospheres in the Solar System: Comparative Aeronomy*, edited by M. Mendillo, A. Nagy, and J. H. Waite, pp. 7-22, AGU, Washington, DC.

Strobel, D. F., X. Zhu, and M. E. Summers, 1994: On the vertical thermal structure of Io's atmosphere. *Icarus*, **110**, 18-30.

Strobel, D. F., X. Zhu, M. E. Summers, and M. H. Stevens, 1996: On the vertical thermal structure of Pluto's atmosphere. *Icarus*, **120**, 266-289.

Thomas, G. E., and K. Stamnes, 1999: *Radiative Transfer in the Atmosphere and Ocean.* Cambridge Univ. Press, Cambridge, 517 pp.

West, R., D. Crisp, and L. Chen, 1990: Mapping transformations for broadband atmospheric radiation calculations. *J. Quant. Spectrosc. Radiat. Transfer,* **43**, 191-199.

Wiscombe, W. J., and J. W. Evans, 1977: Exponential-sum fitting of radiative transmission functions. *J. Comp. Phys*., **24**, 416-444.

Wolpert, D. H., and Macready, W. G., 1997: No Free Lunch Theorems for optimization. *IEEE Trans. Evolutionary Computation*, **1**, 67-82.

Yelle, R. V., 1991: Non-LTE models of Titan's upper atmosphere. *Astrophy. J.*, **383**, 380-400.

Yung, Y. L., and W. B. DeMore, 1999: *Photochemistry of Planetary Atmospheres*. Oxford Univ. Press, Oxford, 456 pp.

Zhang, G. J., 2004: Parameterization of convection in global climate models. *This volume.*

Zhang, J., 1981: *Lecture Notes on Atmospheric General Circulation (in Chinese).* Nanjing Institute of Meteorology Press, Nanjing, China, 313 pp.

Zhu, X., 1988: An improved Voigt line approximation for the calculations of equivalent width and transmission. *J. Quant. Spectros. Radiat. Transfer,* **39**, 421-427.

Zhu, X., 1989a: Radiative cooling calculated by random band models with $S^{-1-\beta}$ tailed distribution. *J. Atmos. Sci.,* **46**, 511-520.

Zhu, X., 1989b: A parameterization of cooling rate calculation under the non-LTE condition: Multi-level model. *Adv. Atmos. Sci.*, **6**, 403-413.

Zhu, X., 1990: Carbon dioxide 15-μm band cooling rates in the upper middle atmosphere calculated by Curtis matrix interpolation. *J. Atmos. Sci.,* **47**, 755-774.

Zhu, X., 1991: Spectral parameters in band models with distributed line intensity. *J. Quant. Spectrosc. Radiat. Transfer,* **45**, 33-46.

Zhu, X., 1994: An accurate and efficient radiation algorithm for middle atmosphere models. *J. Atmos. Sci.*, **51**, 3593-3614.

Zhu, X., 1995: On overlapping absorption of a gas mixture. *Theor. Appl. Climatol.*, **52**, 135-142.

Zhu, X., 2002: Numerical modeling of Titan's stratosphere by a two-dimensional coupled model. *Bull. Amer. Astron. Soc.,* **34**, 901-902.

Zhu, X., 2003: Parameterization of non-local thermodynamic equilibrium source function with chemical production by an equivalent two-level model. *Adv. Atmos. Sci.*, **20**, 487-495.

Zhu, X., and D. F. Strobel, 1990: On the role of vibration-vibration transitions in radiative cooling of the CO_2 15 μm band around the mesopause. *J. Geophys. Res.,* **95**(D4), 3571-3577.

Zhu, X., M. E. Summers, and D. F. Strobel, 1991: Analytic models for the ozone radiative absorption rate at 9.6 μm in the mesosphere. *J. Geophys. Res.,* **96**(D10), 18,551-18,559.

Zhu, X., J.-H. Yee, S. A. Lloyd, and D. F. Strobel, 1999: Numerical modeling of chemical-dynamical coupling in the upper stratosphere and mesosphere. *J. Geophys. Res.,* **104**(D19), 23,995-24,011.

Zhu, X., J.-H. Yee, and D. F. Strobel, 2000: Coupled models of photochemistry and dynamics in the mesosphere and lower thermosphere. In *Atmospheric Science Across the Stratopause*, edited by D. E. Siskind, M. E. Summers, and S. D. Eckermann, pp. 337-342, AGU, Washington, DC.

SATELLITE MICROWAVE REMOTE SENSING OF CLOUDS AND PRECIPITATION

GUOSHENG LIU
Department of Meteorology, Florida State University
Tallahassee, FL 32306, USA
E-mail: liug@met.fsu.edu

This article provides a tutorial for remote sensing of cloud liquid/ice water path and precipitation with satellite passive microwave measurements. First, the emission and scattering signatures in microwave measurements, and their connection with atmospheric hydrological variables are explained, followed by a description of the theoretical bases for retrieving cloud and precipitation properties with microwave observations. The retrieval methods are then briefly reviewed, and some details are provided for those algorithms developed by the author. The products derived from the microwave measurements are utilized to determine the global distribution of cloud and precipitation, assess the variability of these variables, and understand physical and radiative processes. Some challenging issues in satellite microwave remote sensing are also discussed.

1. Introduction

Correctly parameterizing processes related to cloud and precipitation in numerical weather prediction and climate models is one of the most challenging tasks in today's atmospheric research (Del Genio, et al. 1996). The major obstacle for developing and validating these parameterizations arises from the lack of reliable observations of cloud water and precipitation intensity over the time and space scales suitable to the scales of the models (Rasch and Kristjansson 1998). First of all, there are few surface-based observations over the oceanic areas, which cover about two thirds of the Earth's surface. Additionally, clouds and precipitation are highly variable in both time and space compared to other atmospheric variables such as temperature and pressure. For example, rainfall rate measured by a raingauge at a given location can be significantly different from that measured just a couple of hundred meters away. Similarly, rainfall measured at a given time can be significantly different from that just minutes earlier or later. As a result, a high-density raingauge network is required in order to reasonably measure rainfall over an area comparable to the grid size of numerical weather prediction and climate models, if one attempts to derive the area rain total from raingauge observations alone. Such a high-density raingauge network is generally not available even in well-developed countries and regions. These problems related to surface-based measurements make satellite remote sensing of clouds and precipitation indispensable.

Quantitative retrievals of cloud water path and precipitation intensity using satellite remote sensing have been advanced during the last two decades with the advance of space technology (Rossow and Schiffer 1991; Adler et al. 2001). Depending on the wavelength used by the remote sensors, cloud water path and precipitation intensity may be derived from the following four types of signatures: visible reflectance, infrared cloud top temperature, low frequency microwave emission, and high frequency microwave scattering signatures. Visible and infrared measurements have a longer history, and therefore provide longer data records for climatological studies (Arkin et al. 1994), while microwave measurements provide more accurate instantaneous rainfall retrievals due to the direct physical relationship between microwave radiation and column rainwater (Ebert and Manton 1998). In this paper, we particularly focus on the techniques and applications that utilize microwave measurements. Unlike visible and infrared radiations that characterize cloud properties around the top portion of cloud layers, microwave radiation possesses much stronger penetration ability so that the observed radiance reflects the integrated radiative property of the entire atmosphere.

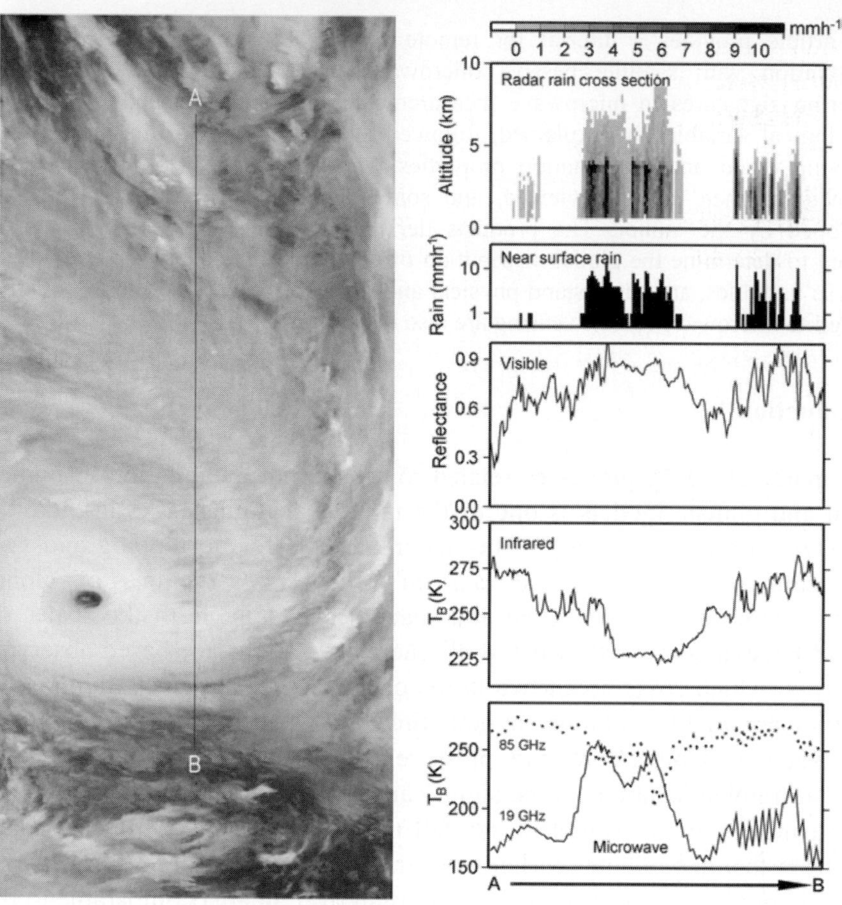

Figure 1. Left: Satellite infrared image of a hurricane over the southeastern Pacific Ocean. The imagery covers an area approximately 720 km wide by 3050 km long on 5 January 1998. Right: Radiative properties of hurricane clouds along line A-B shown in the satellite imagery on left. From top down, distance-height cross-section of rainfall rate from radar, near-surface rainfall rate, reflectance in visible, brightness temperature in thermal infrared, and brightness temperatures at 19 and 85 GHz microwave frequencies (After Liu 2002).

To better understand the differences of the cloud and precipitation signatures received by various satellite sensors, we show in Figure 1 the radiative signatures from a hurricane's clouds, simultaneously observed by several instruments on the Tropical Rainfall Measuring

Mission (TRMM) satellite (Liu 2002). The satellite imagery on the left of the figure shows the clouds associated with a hurricane in the South Pacific Ocean. On the right, we show the observed radiative properties at satellite nadir along the line A-B, which crosses the outer cloud band of the hurricane. The parameters shown here include the space-radar-derived rainfall rate distance-height cross-section, the near-surface rainfall rate (also derived from the space-radar), the reflectance at visible (0.63 μm) wavelength, the thermal infrared (11 μm) brightness temperature and the passive microwave brightness temperatures at 19 and 85 GHz frequencies. Compared to those in the cloud-free area near point B, radiometric properties for rainy areas show the following features: high reflectance in the visible, low brightness temperatures in the thermal infrared, high brightness temperatures at 19 GHz, and low brightness temperatures at 85 GHz. More than half the areas along the line A-B are actually not associated with rain, although clouds in those areas have low infrared brightness temperature and high visible reflectance. It is the microwave brightness temperatures that most closely follow the radar-observed rainfall variation. That is, corresponding to the increase of surface rainfall, the brightness temperature at low microwave frequencies (e.g., 19 GHz) increases, and the brightness temperature at high microwave frequencies (e.g., 85 GHz) decreases. Because of this physical directness between microwave signature and cloud/rain, microwave satellite measurements have been gaining attention in quantitative remote sensing of clouds and precipitation during the last two decades.

While this article emphasizes on remote sensing in the microwave spectrum, it is worth mentioning that the combined use of microwave with visible and infrared measurements provides many unique features for cloud remote sensing. Examples of such combinations include performing cloud classification (Liu et al. 1995), determining cloud temperature and layering (Lin et al 1998a; 1998b), retrieving droplet size and assessing aerosols' indirect radiative effect (Greenwald et al. 1995; Zuidema and Hartmann 1995; Liu et al. 2003), and inferring cloud ice water path (Lin and Rossow 1996).

2. Theoretical Basis of Microwave Remote Sensing

2.1. *Microwave Emission and Scattering Signatures*

Microwave brightness temperatures measured from a satellite-borne radiometer result from the integrated effects of surface emission and reflection, absorption and emission by atmospheric gases, and the absorption, emission and scattering of cloud and precipitation particles. To accurately describe the microwave signatures, a radiative transfer model with full inclusion of the aforementioned effects, particularly the multiple scattering by precipitating particles, is required. The methods to numerically solve such a model equation are discussed by other authors in this book (e.g., Q. Liu, X. Zhu) and are not repeated here. Rather, in the following, we will use a simplified radiative transfer equation to illustrate the basic physical principles of microwave remote sensing of cloud and precipitation.

Let us consider an idealized atmosphere that contains liquid water drops in the lower level and ice water particles on the top level. Also, let's ignore the emission and absorption by atmospheric gases for the moment for simplicity, although they should be considered in actual retrieval algorithms. The microwave radiation reaching to the satellite may then be expressed by

$$T_B = \left[\varepsilon_s T_s e^{-\tau_W} + \int_0^{\tau_W} T(t) e^{-t} dt + (1-\varepsilon_s) e^{-\tau_W} \int_0^{\tau_W} T(t) e^{-(\tau_W - t)} dt \right] \times e^{-\tau_i}, \tag{1}$$

where T_B is the brightness temperature received by a radiometer on the satellite, T_s and ε_s are, respectively, the surface temperature and the surface emissivity, τ_W and τ_i are, respectively, the optical depths (optical paths for slant viewing geometry) for the liquid water drops and ice water particles, and *T(t)* is the temperature at the altitude where optical depth equals *t*. To derive the above equation, we assumed that the ice particle layer does not emit microwave radiation, and the contribution of scattered radiation to the radiometer's field-of-view is also ignored. As a result, the ice layer acts only as an attenuator to the upwelling radiation, which is expressed by the transmission function $e^{-\tau_i}$. It should be noted that this simplification is not valid for optically thick ice layers, such as precipitating ice layers with large graupels or hailstones.

In the bracket of the right-hand-side of (1), the first term is the surface emission term; the second term defines the integrated liquid water emission; and the third term corresponds to the downwelling radiation emitted by the liquid water drops, reflected at the surface and then transmitted to the top of atmosphere. Here, a Fresnel surface is assumed. A simplification of (1) can be further made if we assume that the emitting temperature of the liquid water drops is constant and the same as sea surface temperature. Applying this approximation, we obtain

$$T_B = T_s[1 - e^{-2\tau_W}(1-\varepsilon_s)]e^{-\tau_i}. \tag{2}$$

Depending on the radiation frequency and surface type, we may further simplify (2) as shown in Table 1, in which the following assumptions are introduced:

(1). Land surface emissivity is close to unity, i.e., $\varepsilon_s \sim 1$ (Ulaby et al. 1986, Chapter 19);

(2). At low microwave frequencies (15 to 40 GHz), ice clouds are mostly transparent, i.e., $\tau_i \sim 0$ (Wilheit et al. 1977);

(3). At high microwave frequencies (>80 GHz), liquid water layer (cloud and rains) are nearly opaque, i.e., $\tau_W \to \infty$ (Spencer et al. 1989).

These assumptions are generally valid for low and mid-latitude conditions. At high latitudes, particularly during winter, the liquid layer becomes very shallow, or even nonexistent; the last assumption then becomes invalid.

Table 1. Simplified form of terms in the right-hand side of Eq. (2).

	Over Land ($\varepsilon_s \sim 1$)	Over Ocean
Low Frequency	T_s	$T_s[1 - e^{-2\tau_W}(1-\varepsilon_s)]$
High Frequency	$T_s e^{-\tau_i}$	$T_s e^{-\tau_i}$

From Table 1, it becomes immediately clear that no cloud and precipitation information may be detected at low microwave frequencies over land. Over ocean where surface emissivity is significantly smaller than one, the brightness temperature increases with τ_W, the optical depth resulted from liquid water drops, at low microwave frequencies. Since τ_W is a measure of column integrated liquid water, or liquid water path (LWP), the more liquid water drops exist in the atmosphere, the higher the brightness temperature will be. This signature indicates the intensity of microwave emission (and scattering when raindrops are large) by liquid water drops; therefore, it is called the emission signature. At high frequencies, on the other hand, the brightness temperature decreases with the increase of τ_i, the optical depth resulted from ice particles. A higher concentration of ice particles leads to a lower brightness temperature. Since the extinction of the upwelling microwave intensity is caused by the scattering of ice particles, the signature at high frequency microwaves is called the scattering signature.

Although the above discussion on microwave signatures lacks many details, it explains, on the first order, the responses of microwave brightness temperatures to rainfall rate as shown in Figure 1. That is, over ocean, brightness temperature increases at 19 GHz (low frequency) but decreases at 85 GHz (high frequency) when rainfall rate increases.

2.2. *Emission-Based Sensing*

As shown in Table 1, the satellite-received brightness temperatures at low frequency (below 40 GHz) microwaves respond to the emission of cloud liquid droplets and raindrops over ocean surface. The optical depth resulted from liquid water drops may be written as

$$\tau_W = \int_{\Delta Z} \sigma_{ext}(\lambda, z)dz\,, \tag{3}$$

where ΔZ is the depth of the liquid water layer, $\sigma_{ext}(\lambda, z)$ is the volume extinction cross-section at wavelength λ and height z.

For liquid cloud droplets, the radius (~10 μm) is much smaller than the wavelength of microwave (mm to cm). Consequently, scattering becomes negligibly small and the extinction reduces to absorption. At the small particle limit, the absorption cross-section, $\sigma_{abs}(\lambda,z)$, may be expressed as

$$\sigma_{ext}(\lambda, z) \approx \sigma_{abs}(\lambda, z) \approx \frac{6\pi}{\lambda}\mathrm{Im}(-K)\int_0^\infty \frac{4}{3}\pi r^3 n(r)dr = \frac{6\pi}{\lambda\rho_w}\mathrm{Im}(-K)LWC(z)\,, \tag{4}$$

where $n(r)$ is the number concentration of drops with radius of r; ρ_w is the density of liquid water, $LWC(z)$ is liquid water content, K is a function of the complex refractive index of water, m, $K = (m^2 - 1)/(m^2 + 2)$. Thus, (3) may be further expressed by

$$\tau_W = \frac{6\pi}{\lambda\rho_w}\mathrm{Im}(-K)\int_{\Delta Z} LWC(z)dz \equiv \frac{6\pi}{\lambda\rho_w}\mathrm{Im}(-K)LWP\,, \tag{5}$$

where LWP is liquid water path, i.e., vertically integrated liquid water content. From (5), we find that the optical depth of cloud droplets at microwave frequencies is proportional to liquid water path regardless the specific droplet size distribution. It also depends on complex refractive index through K. This equation, together with Table 1, lays the basic foundation for remote sensing of cloud liquid water path.

It is worth mentioning that ice clouds nearly have no effect on radiation in the low frequency microwaves. The imaginary part of the refractive index m for ice is about 3 orders smaller than that for water in the low microwave frequencies. Consequently, the Im($-K$) term for ice is close to 0. Therefore, the ice emission/absorption is generally negligible. This property allows liquid water alone to be inferred in a mixed-phase cloud, or when liquid water cloud being covered by cirrus ice cloud.

For rainfall remote sensing, the problem becomes more complicated than sensing of liquid water path for nonprecipitating clouds because of the following three factors: First, raindrops are comparable to microwave wavelength and the scattering by raindrops cannot be ignored. Thus, the brightness temperature varies with not only liquid water path but also size distribution of raindrops. The τ_W-LWP relationship can no longer be expressed by an analytic equation as is shown in (5). Second, rainfall rate is a measure of water flux at the ground level, while the satellite-received radiation reflects the vertically integrated property of the column, resulting in that the brightness temperature is a function of the vertical profile of rain as well, not just rainfall rate at the surface. The third complication arises from the fact that rainfall field often has a very high horizontal variability and the Field-of-View (FOV) of a microwave sensor is large, typically several tens of kilometers, so that rainfall is not uniformly distributed within a satellite pixel. Together with the highly nonlinear relation between rainfall rate and brightness temperature (Wilheit et al., 1977; also see the nonlinear relation between T_B and τ_W in Table 1), it follows that the pixel-averaged brightness temperature depends not only on the pixel-averaged rainfall rate, but also on the horizontal distribution of the rain within the pixel. Because of these many contributors to upwelling radiation, the retrieval of rainfall rate is an ill-conditioned inversion problem, i.e., more unknowns than the information content provided by measurements. Therefore, retrieval algorithms often have to depend more or less on statistical relations on the horizontal and vertical distributions of rain fields. These statistical relations can be either derived from cloud resolving models or based on limited observations.

2.3. *Scattering-Based Sensing*

At high microwave frequencies (>80 GHz), ice scattering is the dominant signature in satellite received brightness temperatures. The optical depth τ_i in the equations shown in Table 1 is the vertical integration of volume scattering cross-section due to ice particles, which, under Rayleigh approximation, may be expressed as

$$\sigma_{sca}(\lambda,z)=\frac{128\pi^5}{3\lambda^4}|K|^2\int_0^\infty r^6 n(r)dr=\frac{24\pi^3}{\lambda^4\rho_i^2}|K|^2\int_0^\infty n(r)M^2(r)dr\,, \tag{6}$$

where $M(r)$ is the mass of ice particle with radius of r, and ρ_i is the density of ice particles. Unlike liquid water absorption that is proportional to the mass of liquid water content, the ice scattering cross-section is proportional to the 6th power of r. Therefore, the ice optical depth, which is the vertical integration of $\sigma_{scat}(\lambda, z)$, is not only related to ice water path (IWP), but also strongly depends on particle size distribution. Increasing either the amount or the size of particles will lead to an increase of ice optical depth and, in turn, a decrease of brightness temperature (cf. Table 1). To conduct ice water retrieval requires either a prior knowledge of particle size distribution or to perform simultaneous retrieval of both size distribution and IWP. In addition, the ice water path retrieval problem is further complicated by the bulk density of ice particles, which may vary from less than 0.1 g cm^{-3} to larger than 0.9 g cm^{-3} depending on ice particle type (Pruppacher and Klett 1997).

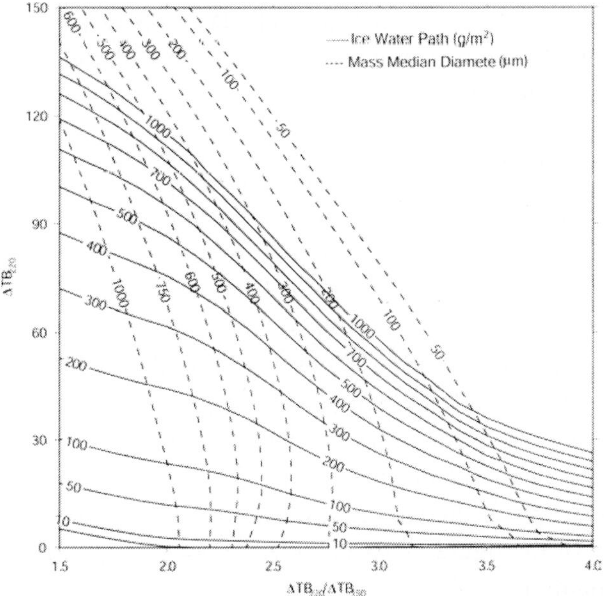

Figure 2. Schematic diagram for simultaneously retrieving ice water path and mass median diameter (After Liu and Curry 2000).

The simultaneous retrieval of IWP and ice particle size distribution may be achieved by observing radiances at two high microwave frequencies (Liu and Curry 2000; Zhao and Weng 2002). Figure 2 illustrates such an approach, in which brightness temperature depressions at 150 and 220 GHz are utilized. The brightness temperature depression is defined by $\Delta T_B = T_{Bclr} - T_B$, where T_{Bclr} is the clear-sky brightness temperature. The mass median diameter, D_{mm}, is defined as the diameter by which ice particles in the entire spectrum are divided into two groups with equal mass. It is indicated from the figure that ΔT_B at 220 GHz varies largely with IWP while the ratio, $\Delta T_{B220}/\Delta T_{B150}$, is largely determined by D_{mm}. Therefore, once brightness temperature depressions at the two frequencies are derived from observations, both IWP and D_{mm} may be determined from a lookup table like Figure 2.

A similar argument may be made for retrieving snowfall from scattering signatures, only that snowfall rate is a quantity of water flux at ground level while the scattering signature represents the vertically integrated effect of ice scattering. Accordingly, the vertical profile of ice particles becomes additionally needed information for snowfall retrieval, just as rain profile is needed for rainfall rate retrieval. Therefore, a statistical relationship between surface snowfall rate and vertical distribution of snow particles is required for developing snowfall retrieval algorithms. This type of information has not been available so far; future investigations through field experiments are highly desirable.

3. Microwave Remote Sensing Methods

The basic principles for retrieving cloud liquid and ice, rainfall and snowfall have been discussed in the previous section. Many retrieval algorithms based on these physical principles have been developed in the past two decades, especially with the success of Special Sensor Microwave/Imager (SSM/I) and TRMM Microwave Imager (TMI). Here, we review some of the methods published in the literature although the main focus will be on the methods developed by the author.

3.1. *Cloud Liquid Water Path*

As pointed out in the previous section, cloud liquid water path may be retrieved over ocean from microwave satellite observations, without requiring the knowledge of drop size distribution. Several retrieval algorithms have been published in the literature, such as Greenwald et al. (1993), Liu and Curry (1993), Petty (1990) and Weng and Grody (1994). All of these algorithms are based on SSM/I observations, although they may be modified for other satellite sensors. Additionally, Chen (2002) proposed a method to derive LWP using the attenuation along satellite-Earth path to signals transmitted by communication satellites. Here, we introduce the method of Liu and Curry (1993).

Considering an atmosphere containing a nonscattering liquid cloud layer and a Fresnel surface, the emissivity of the liquid water cloud (ε_c) may be approximately expressed by the following quadratic equation:

$$a\varepsilon_c^2 + b\varepsilon_c + c = 0, \tag{7}$$

and

$$a = \chi T_c, \quad b = T_{B0} - (1+\chi)T_c, \quad c = T_B - T_{B0}, \quad \chi = 1 - \frac{T_{B0}}{2}\left(\frac{1}{T_a} - \frac{1}{T_s}\right), \tag{8}$$

where T_B is the observed brightness temperature by satellite, T_{B0} is the brightness temperature under clear-sky conditions, and T_c, T_a, and T_s are the temperatures of cloud (mean), lower atmosphere (weighted by absorption coefficients), and sea surface, respectively. T_{B0} may be derived from actual satellite observations at clear-sky areas adjacent to the cloud, or calculated using radiative transfer models, given the atmospheric temperature and moisture profiles are known. T_a, T_c and T_s may be calculated using climatological data. For a nonscattering cloud layer, the sum of emissivity and transmission ($e^{-\tau W}$) equals unity. From (5), LWP may be derived by

$$LWP = \frac{\lambda \rho_w}{6\pi \operatorname{Im}(-K)} \ln\left(\frac{1}{1-\varepsilon_c}\right) \cos\theta , \tag{9}$$

where θ is the sensor viewing zenith angle. Utilizing Eqs.(7)-(9), LWP can be calculated for every microwave channel. The final retrieval may be determined by a combination of several channels to reduce uncertainties.

Results of error analysis indicate that among all the contributing variables, T_{B0} and T_c are the most important ones to the retrieval accuracy. T_{B0} is the background radiation; an accurate value for T_{B0} will eliminate retrieval bias. The importance of T_c arises because refractive indices of water for microwave strongly depend on temperature; so does K in (9). Therefore, an accurate retrieval of LWP requires good estimates of the mean temperature or the altitude of the liquid water cloud layer, which is a rather nontrivial task, particularly for mixed phase and multilayer clouds.

3.2. *Rainfall*

Many microwave rainfall retrieval algorithms have been developed in the last two decades. For detailed reviews, readers are referred to Smith et al. (1998) and Adler et al. (2001). Here, we introduce the algorithms in terms of categories. Except for a few pure regression-type algorithms, a characteristic of the microwave methods is that they rely on radiative transfer models either at the algorithm development stage or during the retrieval computation. Through a radiative transfer model, microwave brightness temperatures are directly connected to the amount and distribution of precipitating hydrometeors. The microwave rainfall algorithms may be grouped into the categories of emission-based, scattering-based, combined emission and scattering, and radiative transfer model-based profiling techniques.

The emission signature as discussed in the previous section provides the most direct physical relation between rainfall and brightness temperature. In emission-based algorithms, the relation between brightness temperature and rainfall rate is usually determined using radiative transfer models by specifying surface properties and the profiles of the following atmospheric quantities: temperature, humidity, cloud water, cloud ice, rain water and precipitating ice. Figure 3 shows such a relationship for 19.35 GHz calculated using a radiative transfer model (Liu 1998) for atmospheres with freezing level heights at 1, 3, and 5 km. In these calculations, no raindrops are assumed to exist above freezing level and vertically constant rainrate profiles are assumed below freezing level, although the pattern of rainrate profiles has a great variety in actual rainclouds (Liu and Fu 2001; Fu and Liu 2001). Because of the polarized nature of water surface emission (reflection), vertically-polarized brightness temperatures are larger than horizontally-polarized ones at 53° viewing angle. As rainfall rate increases, brightness temperature increases at first, but reaches saturation eventually and starts to decrease. The rainfall rate at which saturation starts depends on the depth of the rain layer; a deeper rain layer causes saturation to start at a lower rainrate.

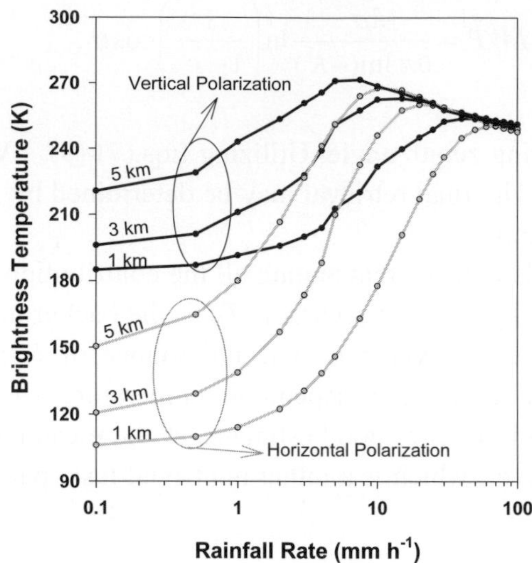

Figure 3. Radiative transfer model simulations of brightness temperatures at 19.35 GHz for rainfall over ocean. A viewing angle of 53° is assumed. Dark curves are for vertical, and light curves are for horizontal polarizations. The values in km indicated on the curves are freezing level heights.

There are several problems associated with emission-based algorithms. (1) They may only be applied over ocean; the high land surface emissivity prevents emission signatures from being detected by low-frequency microwave radiometers (cf. Table 1). (2) Brightness temperature saturates for heavy rain (Fig. 3). This problem is particularly serious for tropical regions where the rain-layer is deep. (3) Nonuniform rainrate across the satellite field of view causes underestimation of rainfall rate. As shown in Figure 3, the brightness temperature versus rainfall rate relation is highly nonlinear. The spatial resolution of a satellite pixel for microwave radiometers is on the order of several tens of kilometers. The rain field within one satellite pixel is generally inhomogeneous. If $R = R(T_B)$ is the theoretical relation between brightness temperature T_B and rainfall rate R for a homogeneous rain field, the retrieval resulting from the field-of-view-averaged brightness temperature, $R(\overline{T}_B)$, does not equal the field-of-view-averaged rainfall rate, $\overline{R(T_B)}$. Instead, in the case of microwave emission, it is always true that $R(\overline{T}_B) < \overline{R(T_B)}$, i.e., the technique underestimates rainfall rate.

Rainfall rate retrievals have also been conducted based on the scattering signature at high frequency microwaves (Ferraro and Marks 1995). The scattering signature is less directly related to precipitation than the emission signature because it is an indication of the ice amount above freezing level. Frequencies higher than 80 GHz are primarily used for scattering-based algorithms. Figure 4 shows the radiative transfer model simulated relation between vertically-polarized brightness temperatures at 85 GHz and surface rainfall rate assuming a viewing angle of 53°. In these calculations, it is assumed that rainfall rates are vertically constant below freezing level (5 km). Above freezing level, all precipitating particles are ice with a density of 0.3 g cm^{-3}. Ice precipitation rate is assumed to decrease linearly from freezing level to cloud top. The cloud top heights are 6, 8, and 10 km for the three curves shown in the figure, resulting in the ice layer depth being 1, 3, and 5 km, respectively.

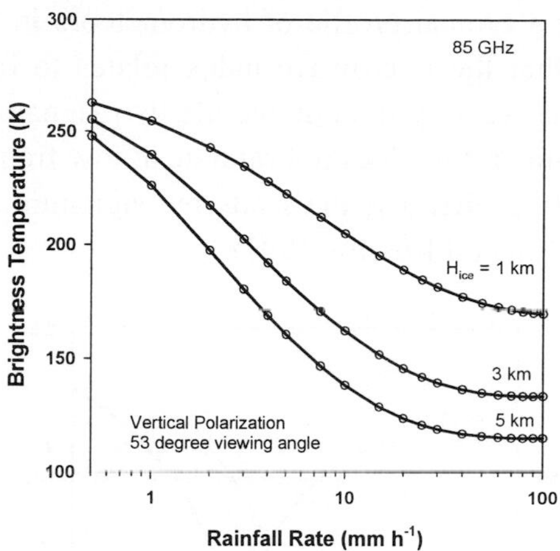

Figure 4. Radiative transfer model simulated relation between brightness temperature at 85 GHz and surface rainfall rate for a viewing angle of 53°. Only vertical polarization is shown. H_{ice} is the depth of ice layer above freezing level.

The advantage of scattering-based algorithms is that they can be applied over both ocean and land. However, although statistically it is true that heavier surface rainfalls usually correspond to greater amounts of ice particles aloft, this corresponding relation is neither unique nor well understood. An obvious exception is warm rain, in which surface rainfall does not associate with ice particles at all. On the other hand, anvil clouds contain ice particles while they seldom produce rainfall.

For rain associated with a shallow rain-layer, scattering-based techniques fail to work because of the lack of ice scattering. For heavy rainfall with deep rain-layers, emission-based techniques cannot correctly determine rainfall rate because brightness temperature saturates (cf. Fig.3). A natural solution is to take advantage of both emission and scattering signatures by combining them in a single algorithm. One such combination for SSM/I observations is proposed by Liu and Curry (1992; 1996), in which a parameter called "microwave index" is defined as follows:

$$MWI = (1 - D / D_0) + 2(1 - PCT / PCT_0), \qquad (10)$$

where $D = T_{B19V} - T_{B19H}$ is the polarization difference at 19 GHz, $PCT = (1 + \alpha)T_{B85V} - \alpha T_{B85H}$ is the polarization-corrected temperature at 85 GHz as defined by Spencer et al. (1989) and $\alpha = 0.818$; T_{B19V}, T_{B19H}, T_{B85V} and T_{B85H} are brightness temperatures at 19 GHz vertical, 19 GHz horizontal, 85 GHz vertical and 85 GHz horizontal polarizations, respectively; D_0 and

PCT_0 are D and PCT at the onset of rain. The first and the second term in the right-hand side of (10) represent the emission by raindrops and the scattering by ice particles and, therefore are emission and scattering signatures, respectively. Both terms are normalized so that *MWI* is zero at the onset of rain. To illustrate the advantage of the combined signature over either emission or scattering signature, we show in Figure 5 the relations of rainfall rate versus horizontally-polarized brightness temperatures at 19 and 85 GHz, and *MWI*, calculated for a 53° viewing angle assuming a typical profile of hydrometeors in tropical deep convections (Liu and Fu 2001). Notice that the microwave index relates to rainfall rate monotonically without saturation. An alternative way to combine the two signatures is to use the emission signature when brightness temperature does not saturate at low frequency, or where cloud top temperature is warmer than 0°C, then use the scattering signature at higher rainfall rates for other situations (Petty 1994; Lin and Rossow 1997).

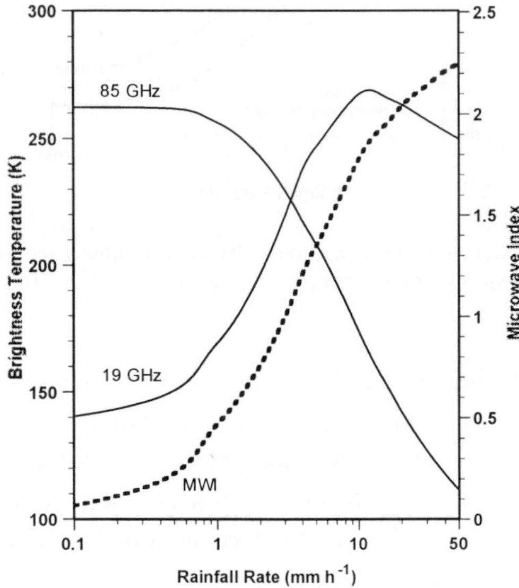

Figure 5. Radiative transfer model calculated brightness temperatures at 19 and 85 GHz, and microwave index (MWI) over ocean as a function of rainfall rate for tropical convective rains. A viewing angle of 53° is assumed and the brightness temperatures shown are for horizontal polarization (After Liu 2002).

If the surface emissivity and vertical distributions of atmospheric temperature, humidity, and hydrometeors are known, brightness temperatures from any given set of frequencies can be calculated with a radiative transfer model. Inversely, if brightness temperatures observed at several frequencies match well with those calculated by a radiative transfer model, it is then possible that the profiles assumed in the model are the same as those in the actual rain clouds. Based on this logic, radiative transfer model-based retrieval algorithms have been constructed (Kummerow et al. 1996; Evans et al. 1995; Smith et al. 1994). While they differ in details, this type of algorithm generally consists of the following retrieval procedures: First, a large database of vertical profiles of hydrometeors must be prepared. This database should possess the characteristic feature of hydrometeor profiles that occur in nature. Because of the lack of observational data, this database has so far been constructed with simulated results from numerical cloud-resolving models. Radiative transfer model calculations are performed using the hydrometeor profiles in the database, which result in many sets of calculated brightness temperatures. The set that best matches the satellite-observed brightness temperatures is selected, and the hydrometeor profile used to produce the best match is determined to be the retrieval. The retrieval gives not only rainfall rate at surface but also its vertical distribution. Model-based algorithms have the advantage of fully using physical relations between cloud microphysics and microwave radiation. With more observational data becoming available in the future to build the database of hydrometeor profiles, this approach is expected to play a more significant role in satellite remote sensing. However, there are two major problems associated with this technique. First, the retrieval depends heavily on the pre-constructed database, which, at present, relies on numerical cloud models because observational data are insufficient. Any cloud model deficiency could directly affect the quality of the rainfall retrieval. The second problem arises from the ill-posed problem in finding the best match between the observed and the calculated brightness temperatures. The number of unknowns in the retrieval problem (all components that interacts with microwave radiation) is far greater than the information content (number of independent information in brightness temperatures). Several totally different hydrometeor profiles may result in a similar "good" match, causing non-uniqueness for the solution. This is usually dealt with by averaging the hydrometeor profiles of the closest brightness temperature matches. Interested readers are referred to the work by S. Yang in this book for further discussion on this method.

3.3. *Ice water Path and Snowfall*

Ice scattering causes a reduction of upwelling radiative energy at high frequency microwaves, which is the primary signature to be used for sensing ice water and snowfall. Retrieving ice water path (IWP) and snowfall by high frequency microwave observations is a very new area; algorithms mentioned in this article are still in an early developing stage. Several ice water path algorithms have been published, for example, by Deeter and Evans (2000), Liu and Curry (2000), Weng and Grody (2000). A snowfall retrieval algorithm has been published by Liu and Curry (1997). All these algorithms use observations at frequencies higher than 85 GHz.

The ice water path algorithm of Liu and Curry (2000) was based on the two-frequency approach as described in Figure 2. In this approach, the depression (relative to clear-sky) of brightness temperatures at 220 GHz (i.e., ΔT_{B220}) is the most sensitive to IWP where the ratio of the brightness temperature depressions at 220 GHz and 150 GHz is the most sensitive to the mass median diameter, a parameter that characterizes the ice particle size distribution. The actual retrieval starts with generating a look-up table by a radiative transfer model. The inputs of the look-up table are ΔT_{B220} and $\Delta T_{B220}/\Delta T_{B150}$, and the outputs are IWP and mass median diameter. In generating this look-up table, ancillary data of atmospheric temperature and humidity profiles, ice cloud height and depth, and cloud liquid water path are needed. These data may be obtained from numerical model analysis and satellite retrievals based on other low-frequency microwave sensors. Liu and Curry (2000) found that ice water path and mass median diameter may be reasonably retrieved for those non-precipitating clouds with IWP > 200 g m^{-2} and mass median diameter greater than 200 μm. For thinner clouds, higher frequency (e.g., 340 GHz) data are needed. For very thick, especially precipitating clouds, the combination of 90 and 150 GHz is preferable, such as the algorithm proposed by Weng and Grody (2000).

The snowfall algorithm proposed by Liu and Curry (1997) uses the scattering signature at 150 GHz of SSM/T2 (Special Sensor Microwave Water Vapor Sounder) data. First, a scattering parameter, β, is defined as follows:

$$\beta = \frac{T_{B0} - T_B}{T_{B0} - T_{BA}}, \tag{11}$$

where T_B and T_{B0} are, respectively, the brightness temperatures at 150 GHz under the conditions with and without ice clouds. T_{B0} is determined from the scatter-plot of brightness temperature at 92 GHz versus that at 150 GHz of SSM/T2 and is expressed as a function of brightness temperature at 92 GHz. T_{BA} is the typical value of 150 GHz brightness temperature under clear-sky conditions. In order to relate β to snowfall rate, radiative transfer simulations are performed for high-latitude winter conditions. Figure 6 shows the simulated results of the relationship between β and snowfall rate (R_s), assuming a 45°N standard atmosphere during winter. The ice particle sizes are assumed to range from 100 to 3000 μm in radius with size distribution given by Sekhon and Srivastava (1970). The cloud top changes from 1 to 6 km as snowfall rate increases from 0 to 1.5 mm h^{-1} (water equivalent). A constant cloud liquid water path (50 g m^{-2}) is assumed in the cloud. Five ice densities (0.1, 0.2, 0.3, 0.4, 0.5 g cm^{-3}) are used to represent different ice/snow types. We see that there is a clear relationship between snowfall rate and β although there exists significant scatter due to the variations in ice density and cloud vertical structure. Also shown in this figure is the best fit of the model results, which was used by Liu and Curry (1997) to determine snowfall rate from SSM/T2 observations. It must be noted that the assumptions used above in snow cloud structures and cloud liquid water path in the cloud have not been confirmed (or denied) by observations. Future field experiments are greatly needed in constructing a more realistic snow cloud model.

4. Selected Applications of Microwave Remote Sensing

While there may be various ways to group microwave satellite remote sensing applications, we elect to present the cloud and precipitation applications according to the following three categories: determining the global mean distribution of cloud and precipitation variables, assessing the variability of these variables, and understanding physical and radiative processes.

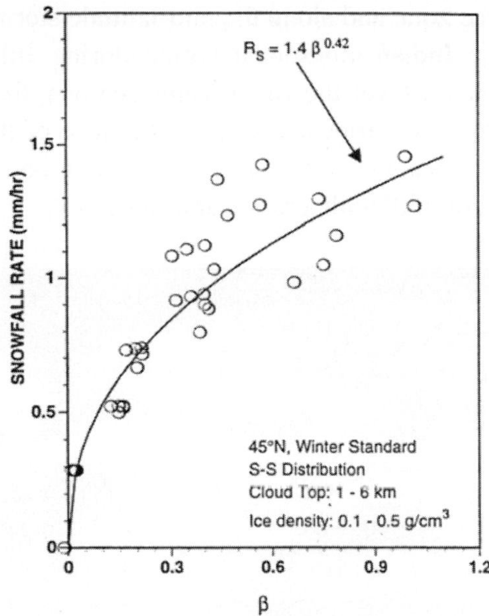

Figure 6. Relationship between snowfall rate and the scattering parameter β simulated by a radiative transfer model. The best-fit curve is also shown (After Liu and Curry 1997).

4.1. *Global Climatology of Cloud Liquid Water and Precipitation*

Probably the most significant development in the microwave remote sensing for atmospheric sciences is the successful determination of global distribution of rainfall (e.g., Xie and Arkin, 1998). Before satellite microwave data became available, rainfall estimates over oceanic areas were extrapolated from observations near the coasts, over scattered atolls and from infrequent ship reports (e.g. Legates and Wilmott 1990), while in the tropical regions estimates based on infrared satellite measurements were available (Arkin and Meisner 1987). The uncertainties associated with those estimates are largely unknown. Beginning in 1987, the launch of the first of a series of SSM/I satellites, several rainfall retrieval algorithm intercomparison projects (Ebert and Manton 1998; Adler et al. 2001) were conducted, in which over 50 algorithms were inter-compared and evaluated. The comparison results clearly showed that microwave retrieval algorithms are superior to other algorithms in deriving instantaneous rainfall rate over oceanic regions. By conducting these intercomparison studies, as well as further efforts made by NASA through TRMM project (Kummerow et al. 2000), global distribution of rainfall, particularly in the tropical regions, is much better estimated than a decade ago. One of the milestones in this progress is the generation of global monthly rainfall dataset under the Global Precipitation Climatology Project (GPCP) (Huffman et al. 1997), which is widely used by scientists in model validation and climate diagnostic studies. Figure 7 shows the rainfall rate maps for January and July averaged over 20 years (1979-1999) generated from GPCP merged satellite and raingauge data. Rain bands are shown in the inter-tropical convergence zone and along the mid-latitude storm tracks. Rain maximums are also very clear over the Indian monsoonal region during July and over the Amazon rainforest region during January. Over the vast oceanic regions, the main input for this rain product is the satellite microwave retrievals. As such, because of the availability of satellite microwave measurements, it becomes possible to document both the magnitude and the distribution pattern of global rainfall with a reasonable accuracy.

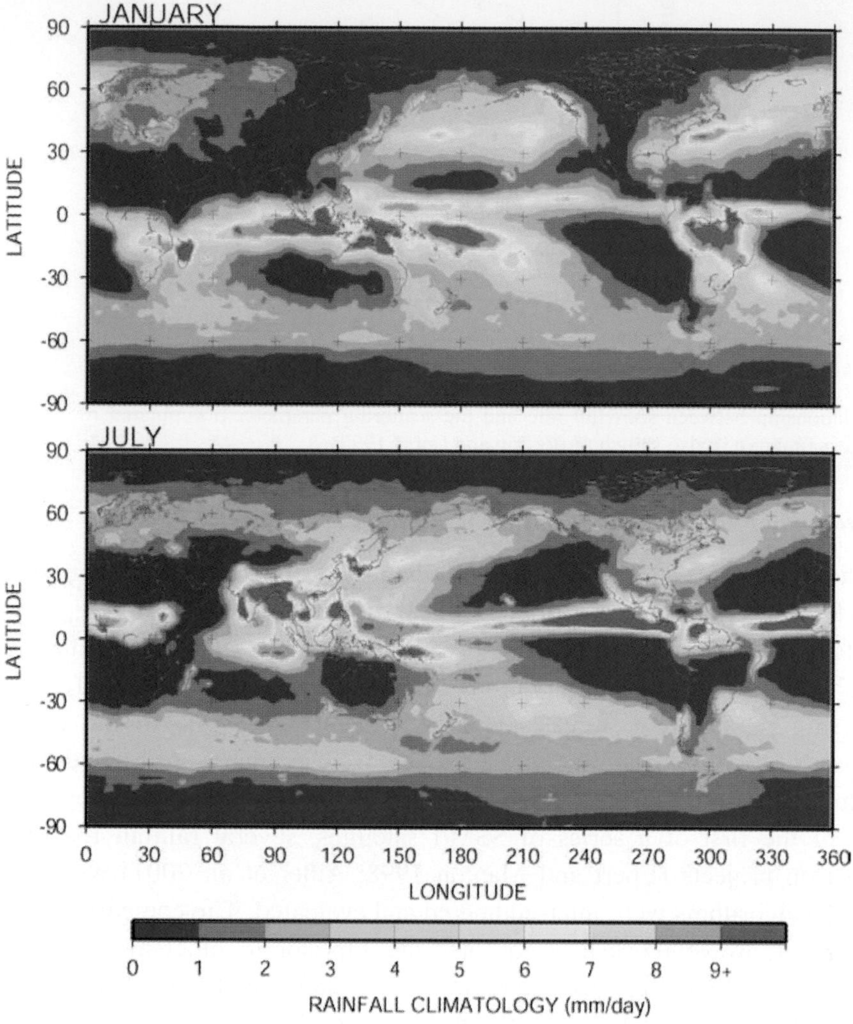

Figure 7. Global distribution of rainfall during January and July. This climatology is derived from a 20-year (1979-1999) GPCP dataset.

The magnitude and distribution of cloud liquid and ice water paths over the globe are important for assessing global radiative energy budget, as well as for evaluating general circulation models (GCMs), and yet, they are among the least known atmospheric variables today. Rasch and Kristjansson (1998) summarized the globally averaged values of cloud liquid and ice water paths resulted from several GCMs and satellite microwave retrievals, which are shown in Table 2 with slight modification by the author to include the work of Lin and Rossow (1996) and to exclude results from a GCM control run. The three LWP retrievals are all based on SSM/I data. The difference among the various LWP "climatologies" is about a factor of 2~3. The only published observational results of global IWP are from Lin and Rossow (1996). The models cannot even agree whether there is more liquid or ice globally. Clearly, more vigorous research in the direction of obtaining more reliable observational data on LWP, and any kind of global observation of ice water path, is needed.

Table 2 Comparison of globally averaged cloud liquid and ice water path resulted from three GCMs and three satellite microwave retrievals. GSVJ93 (Greenwald et al. 1993), WG94 (Weng and Grody 1994), LR96 (Lin and Rossow 1996), RK98 (Rasch and Kristjansson 1998), FRR96 (Fowler et al. 1996), DYKL96 (Del Genio et al. 1996).

	Observations			Models		
	GSVJ93	WG94	LR96	RK98	FRR96	DYKL96
LWP (g m^{-2})	81	44	50~60	32	45	90
IWP (g m^{-2})	-	-	70	20	19	150

4.2. *Rainfall Variation in East Asia*

With years of observational rainfall data available, it may be possible to assess its long-term variability. Figure 8 is such an example that shows a 23-year time series of rainfall anomaly in the East Asian region. The bottom diagram shows the latitude-time cross-section of rainfall anomaly that is averaged over the longitudinal range of 70-140°E. The largest anomaly occurs near 5 ~ 20°N, the region characterized by the Asian summer monsoon.

In the middle diagram of Figure 8, the Japanese Meteorological Agency's (JMA) ENSO (El Nino Southern Oscillation) index (sea surface temperature anomaly over the area of 4°S-4°N, 150°W-90°W) is shown. A warm episode (El Nino) of ENSO clearly corresponds to a decrease of rainfall near the tropical region in this longitude range. To further illustrate the rainfall variation in the East Asian region associated with ENSO events, time series of rainfall anomalies in the following 5 regions are also shown (top diagram): Indian monsoon region (5-25°N,70-100°E), Southeast Asia (5°S-20°N, 100-130°E), Southern China: (20-30°N, 100-120°E), Northern China (30-40°N, 100-120°E) and Japan (30-45°N, 120-140°E). Note that the naming of the regions is somewhat arbitrary and just shows a proxy of the locations. The correlation coefficients (r) between rainfall anomaly in each region and ENSO index is also indicated in the diagrams. There is a significant correlation between ENSO and rainfall in the Indian monsoon region and Southeast Asian region. These two regions happen to have large inter-annual rainfall variability. From the rainfall data, it is not clear whether ENSO has a clear correlation with rainfalls in the other three regions.

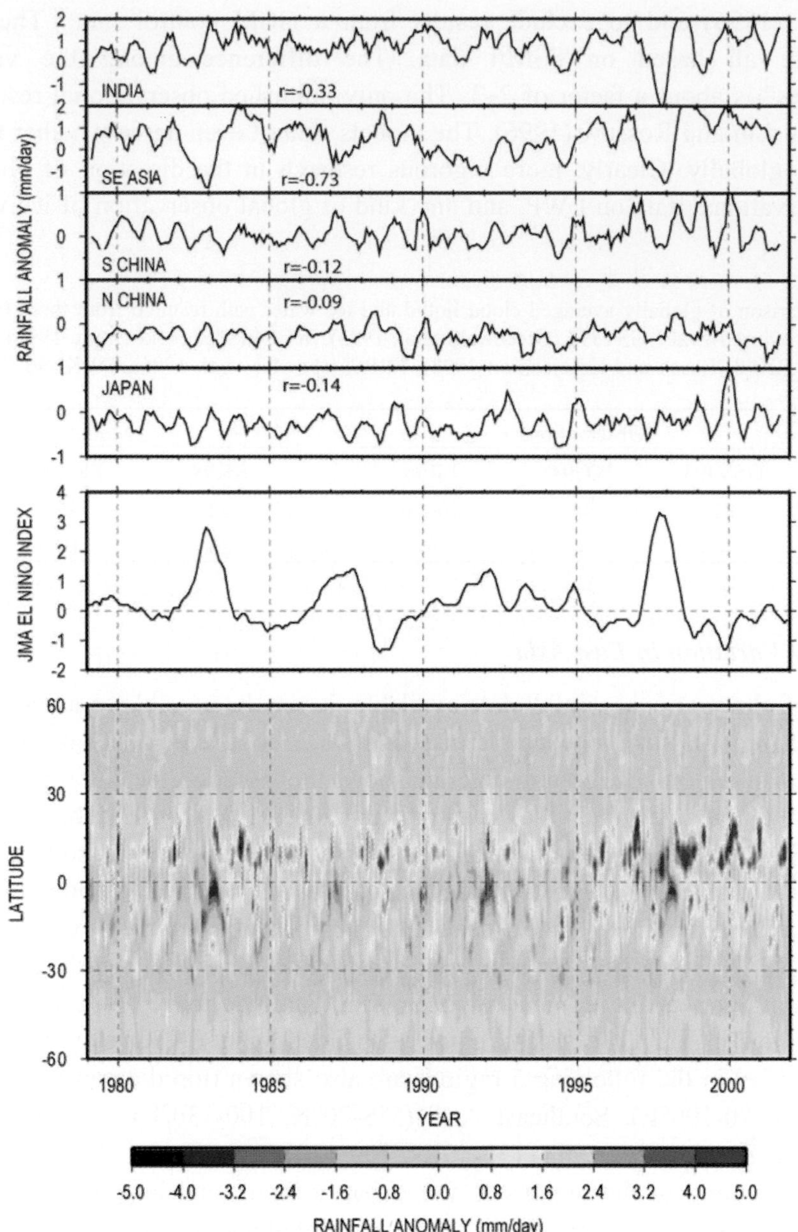

Figure 8. Bottom: Time-latitude cross-section of rainfall anomalies averaged between 70°E and 140°E. Middle: ENSO index compiled by Japanese Meteorological Agency. Large positive values correspond to El Nino events. Top: Rainfall anomalies averaged in 5 regions. INDIA:5-25°N, 70-100°E; SE ASIA: 5°S-20°N, 100-130°E; S CHINA: 20-30°N, 100-120°E; N CHINA: 30-40°N, 100-120°E; JAPAN: 30-45°N, 120-140°E.

4.3. *Physical Processes*

One important issue in the interaction between radiation and clouds is the so-called aerosols' indirect radiative effect, for which it is hypothesized that increasing concentration of anthropogenic aerosols would increase the number of cloud condensation nuclei and the number of cloud droplets, resulting in reducing cloud particle size, and consequently increasing cloud albedo (Twomey, 1991). To study this effect, microwave and visible remote sensing data are jointly used to retrieve both liquid water path and effective radius of cloud drops in the Indian Ocean during wintertime when the prevailing north wind carries polluted airmass from the Indian subcontinent to the ocean to form a clear contrast between clean air to the south of equator, and polluted air in the north near Indian coastal regions. Figure 9 shows the effective radius (r_e) frequency distribution derived for the northern (5-10°N) and southern (10-5°S) regions at 4 constant LWP values. It is shown that given the same cloud liquid water path the cloud particles are smaller in the northern polluted region than in the southern clean region; the difference is particularly evident for clouds with large value of liquid water paths.

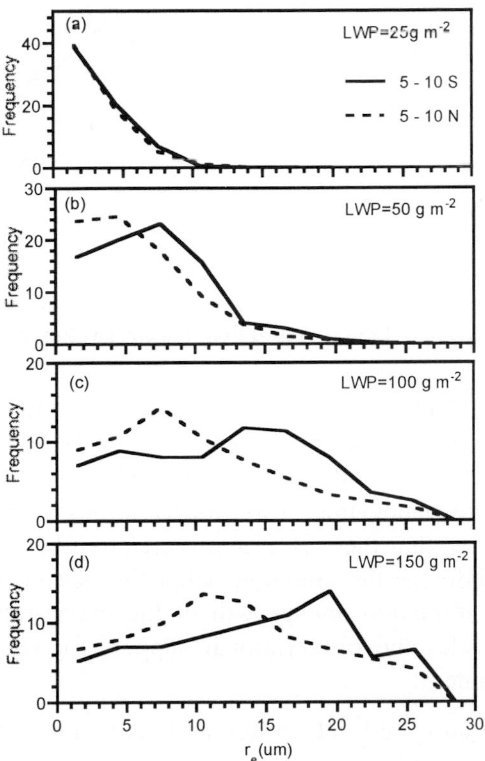

Figure 9. Frequency distribution of effective radius (r_e) at 4 constant liquid water path values for northern polluted and southern clean regions in the Indian Ocean approximately along 75°E longitude during wintertime (After Liu et al. 2003).

The ice fraction in mixed phase clouds, i.e., IWP/(IWP+LWP), is studied by Lin and Rossow (1996) using the combination of SSM/I retrievals and International Satellite Cloud Climatology Project (ISCCP) analysis. In their study, Lin and Rossow used SSM/I data to retrieve LWP and visible data to retrieve the total water path (LWP+IWP). IWP is then calculated by the difference between total water path and LWP. Figure 10 shows the relation between this fraction and cloud top temperature for clouds over global oceans. While there are some differences seasonally, the dominant feature is that clouds with colder tops have greater ice fractions. At cloud top temperature of −10°C, the fraction is ~50%. Then it gradually increases to about 80% as cloud top temperature decreases to −50°C. The reason why ice fraction decreases when cloud top temperature becomes colder than −50°C is not clear, although this may be artificially due to the low datum number in this temperature range. Also, it is noted that multilayer clouds may be another explanation.

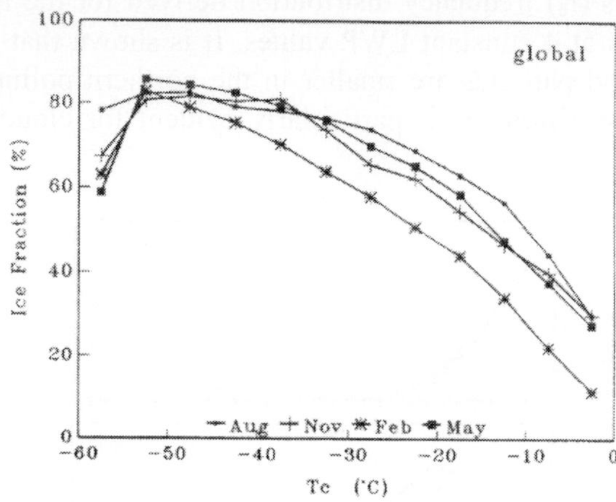

Figure 10. Relation between ice fraction and cloud top temperature for global oceans during August 1987, November 1987, February 1988 and May 1988 (Adapted from Lin and Rossow 1996).

Figure 11 shows the probability density function of ice water path and precipitation by Liu and Curry (1999). The isolines are for $\ln(n_{ij})$, where n_{ij} is the number of pixels that fall within the *i*th 50 gm^{-2} IWP-bin and *j*th 1 mm h^{-1} rainrate-bin. By comparing the ice water path retrieved from high-frequency microwave observations with the surface rainfall rate retrieved from SSM/I, we see that there is a high correlation between the two for clouds in the tropics. This result reinforces the common belief that ice microphysical process is the major mechanism for the formation and growth of the precipitation particles even in the tropics. Moreover, it provides the observational support for using microwave scattering signatures to infer rainfall rate.

In Figure 12, we show how the LWPs observed in cumulus cells are compared to values calculated by assuming a reversible saturated adiabatic process (Liu et al. 2001). Here, we use cloud top temperature at an indicator of cloud top height. All clouds in this case are water clouds with top temperature warmer than 0°C. A dot in the figure represents an averaged value over the area of a cumulus cell. The dashed curves show the relations equivalent to 1, 1/2, 1/4, and 1/8 of the adiabatic LWP values. The observations were made in the Indian Ocean by airborne remote sensing measurements. It is seen that actual LWPs are far less than the theoretical adiabatic values, and the two have a mean ratio of ~1/3.

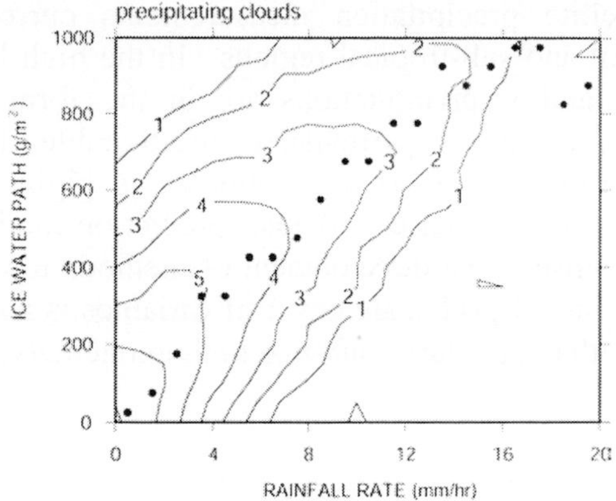

Figure 11. Two-dimension probability density function of ice water path and rainfall rate in the tropics derived from SSM/T-2 and SSM/I observations. The isolines are for $\ln(n_{ij})$, where n_{ij} is the number of pixels that fall within the ith 50 g m^{-2} IWP-bin and jth 1 mm h^{-1} rainrate-bin. Dots show the averaged values of IWP at each 1-mm h^{-1} rainfall rate bin (After Liu and Curry 1999).

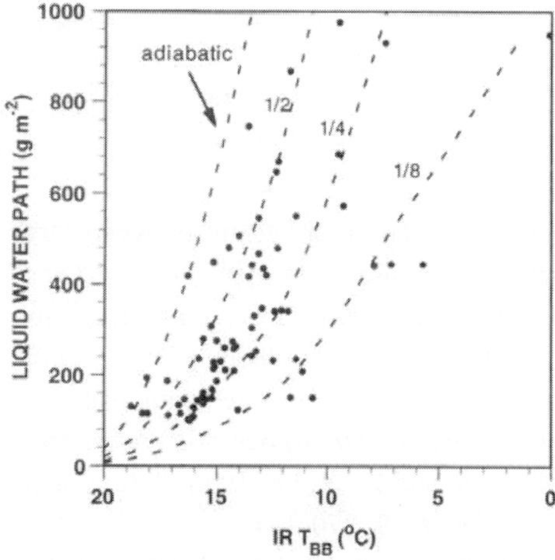

Figure 12. Scatterplot of cloud top temperature (IR T_{BB}) from a nadir-looking infrared radiometer versus retrieved LWP for convective cells. The dashed lines indicate the relations when assuming 1, 1/2, 1/4, and 1/8 times of adiabatic LWPs, respectively (After Liu et al. 2001).

5. Concluding Remarks

Although satellite microwave remote sensing has shown many promises, there are still many problems and issues in current retrieval algorithms. The most stunning issue might be the lack of global measurements of cloud ice water and the lack of consensus on the global averaged value of cloud liquid water path as pointed in the previous section. To solve this problem, the likely approaches are to put sensors with multiple high microwave frequencies (above 150 GHz) on satellites and conduct reliable surface validation experiments. So far, there have been no experiments to adequately validate satellite cloud liquid water retrievals. Future satellite programs using active sensing by cloud radars, such as CloudSat, will also be greatly beneficial. Satellite precipitation measurements currently have relatively high confidence in the tropical and sub-tropical regions. In the high latitudes, particularly where rain-layers are shallow and/or precipitations are in the form of snowfall, the retrieval accuracy of current algorithms is particularly questionable. Further research migrating precipitation retrieval from tropics to high-latitudes is highly desired. In recent years, assimilating satellite data into numerical weather prediction models has become a hot topic and has shown great promises. The development of methods to convert pixel-scale radiance observed by satellites to model grid-scale physical variables is another area in need of study, particularly for variables that have large sub-pixel inhomogeneity, such as rainfall.

Acknowledgements

During the preparation of this article, my research program has been benefited from support by the following US funding agencies: the National Science Foundation, the Department of Energy and the National Aeronautics and Space Administration. I thank Jim Lamm for editing the manuscript.

References

Adler, R. F., C. Kidd, G. Petty, M. Morrisey, and H. M. Goodman, 2001: Intercomparison of global precipitation products: The Third Precipitation Intercomparison Project (PIP-3). *Bull. Amer. Meteor. Soc.*, **82**, 1377-1396.

Arkin, P. A., and B. N. Meisner,1987: The relationship between large-scale convective rainfall and cold cloud over the Western Hemisphere during 1982-1984. *Mon. Wea. Rev.*, **115**, 51-74.

Arkin, P. A., R. Joyce, and J. E. Janowiak, 1994: The estimation of global monthly mean rainfall using infrared satellite data: The GOES precipitation index (GPI). *Remote Sensing Review*, **11**, 107-124.

Chen, H., 2002: A concept for measuring liquid water path from microwave attenuation along the satellite-Earth path. *Chinese Journal of Atmospheric Sciences*, **26**, 401-408.

Del Genio, A., M.-S. Yao, W. Kovari, and L. W. Lo, 1996: A prognostic cloud water parameterization for global climate models. *J. Climate*, **9**, 270-304.

Deeter, M. N., and K. F. Evans, 2000: A novel ice-cloud retrieval algorithm based on the Millimeter-wave Imaging Radiometer (MIR) 150- and 220-GHz channels. *J. Appl. Meteor.*, **39**, 623-633.

Ebert, E. E., and J. Manton, 1998: Performance of satellite rainfall estimation algorithms during TOGA COARE. *J. Atmos. Sci.*, **55**, 1537-1557.

Evans, K. F., J. Turk, T. Wong, and G. L. Stephens, 1995: A Bayesian approach to microwave precipitation profile retrieval. *J. Appl. Meteor.*, **34**, 260-279.

Ferraro, R. R., and G. F. Marks, 1995: The development of SSM/I rain rate retrieval algorithms using ground based radar measurements. *J. Atmos. Oceanic. Technol.*, **12**, 755-770.

Fowler, L., D. A. Randall, and S. A. Rutledge, 1996: Liquid and ice cloud microphysics in the CSU general circulation model. Part I: Model description and simulated microphysical processes. *J. Climate*, **9**, 489-529.

Fu, Y., and G. Liu, 2001: The variability of tropical precipitation profiles and its impact on microwave brightness temperatures as inferred from TRMM data. *J. Appl. Meteor.*, **40**, 2130-2143.

Greenwald, T. J., G. L. Stephens, T. H. Vonder Haar, and D. L. Jackson, 1993: a physical retrieval of cloud liquid water over the global oceans using Special Sensor Microwave/Imager (SSM/I) observations. *J. Geophys. Res.*, **98**, 18471-18488.

Greenwald, T. J., G. L. Stephens, S. A. Christopher, and T. H. Vonder Haar, 1995: Observations of the global characteristics and regional radiative effects of marine cloud liquid water. *J. Climate*, **8**, 2928-2946.

Huffman, G. J., and Coauthors, 1997: The Global Precipitation Climatology Project (GPCP) combined precipitation data set. *Bull. Amer. Meteor. Soc.*, **78**, 5-20.

Kummerow, C., W. S. Olson, and L. Giglo, 1996: A simplified scheme for obtaining precipitation and vertical hydrometeor profiles from passive microwave sensors. *IEEE Trans. Geosci. Remote Sens.*, **34**, 1213-1232.

Kummerow, C., and Coauthors, 2000: The status of the Tropical Rainfall Measuring Mission (TRMM) after two years in orbit. *J. Appl. Meteor.*, **39**, 1965-1982.

Legates, D., and C. J. Wilmott, 1990: Mean seasonal and spatial variability in gauge-corrected, global precipitation. *Int. J. Climatol.*, **10**, 111-127.

Lin, B., and W. B. Rossow, 1996: Seasonal variation of liquid and ice water path in nonprecipitating clouds over oceans. *J. Climate*, **9**, 2890-2902.

Lin, B., and W. B. Rossow, 1997: Precipitation water path and rainfall rate estimates over oceans using special sensor microwave imager and International Satellite Cloud Climatology Project data. *J. Geophys. Res.*, **102**, 9359-9374.

Lin, B., B. Wielicki, P. Minnis, and, W. B. Rossow, 1998a: Estimation of water cloud properties from satellite microwave, infrared and visible measurements in oceanic environment. I: Microwave brightness temperature simulations. *J. Geophys. Res.*, **103**, 3873-3886.

Lin, B., P. Minnis, B. Wielicki, D. Doelling, R. Palikonda, D. Young, and Uttal, 1998b: Estimation of water cloud properties from satellite microwave, infrared and visible measurements in oceanic environment. II: Results. *J. Geophys. Res.*, **103**, 3887-3905.

Liu, G., 1998: A fast and accurate model for microwave radiance calculations. *J. Meteor. Soc. Japan*, **76**, 335-343.

Liu, G., 2002: Satellite remote sensing: Precipitation. In "*Encyclopedia of Atmospheric Sciences*", J. Holton, J. Pyle, and J. Curry, ed., Academic Press. London, UK.

Liu, G., and J. A. Curry, 1992: Retrieval of precipitation from satellite microwave measurement using both emission and scattering. *J. Geophys. Res.*, **97**, 9959-9974.

Liu, G., and J. A. Curry, 1993: Determination of characteristic features of cloud liquid water from satellite microwave measurements. *J. Geophys. Res.*, **98**, 5069-5092.

Liu, G., and J. A. Curry, 1996: Large-scale cloud features during January 1993 in the North Atlantic Ocean as determined from SSM/I and SSM/T2 observations. *J. Geophys. Res.*, **101**, 7019-7032.

Liu, G., and J. A. Curry, 1997: Precipitation characteristics in Greenland-Iceland-Norwegian Seas determined by using satellite microwave data. *J. Geophys. Res.*, **102**, 13987-13997.

Liu, G., and J. A. Curry, 1999: Tropical ice water amount and its relations to other atmospheric hydrological parameters as inferred from satellite data. *J. Appl. Meteor.*, **38**, 1182-1194.

Liu, G., and J. A. Curry, 2000: Determination of ice water path and mass median particle size using multichannel microwave measurements. *J. Appl. Meteor.*, **39**, 1318-1329.

Liu, G., and Y. Fu, 2001: The characteristics of tropical precipitation profiles as inferred from satellite radar measurements. *J. Meteor. Soc. Japan*, **79**, 131-143.

Liu, G., J. A. Curry, and R.-S. Sheu, 1995: Classification of clouds over the western equatorial Pacific Ocean using combined infrared and microwave satellite data. *J. Geophys. Res.*, **100**, 13811-13826.

Liu, G., J. A. Curry, J. A. Haggerty, and Y. Fu, 2001: Retrieval and characterization of cloud liquid water path using airborne passive microwave data during INDOEX. *J. Geophys. Res.*, **106**, 28719-28730.

Liu, G., H. Shao, J. A. Coakley, Jr., J. A. Curry, J. A. Haggerty and M. A. Tschudi, 2003: Retrieval of cloud droplet size from visible and microwave radiometric measurements during INDOX: Implication to aerosols' indirect radiative effect. *J. Geophys. Res.* **108**, D1, 10.1029/2001JD001395.

Petty, G. W., 1990: On the response of the Special Sensor Microwave/Imager to the marine environment – Implications for atmospheric parameter retrievals. Ph. D. dissertation, Univ. of Wash., Seattle, 291pp.

Petty, G. W., 1994: Physical retrievals of over-ocean rain rate from multichannel microwave imagery. II: Algorithm implementation. *Meteor. Atmos. Phys.*, **54**, 101-122.

Pruppacher H. R., and J. D. Klett, 1997: Microphysics of Clouds and Precipitation. Kluwer Academic, 954pp.

Rasch, P. J., and J. E. Kristjansson, 1998: A comparison of the CCM3 model climate using diagnosed and predicted condensate parameterizations. *J. Climate*, **11**, 1587-1612.

Rossow, W. B., and R. A. Schiffer, 1991: ISCCP cloud data products. *Bull. Amer. Meteor. Soc.,***72**, 2-20.

Sekhon, R. S., and R. C. Srivastava, 1970: Snow size spectra and radar reflectivity. *J. Atmos. Sci.*, **27**, 299-307.

Smith, E. A., X. Xiang, A. Mugnai, and G. J. Tripoli, 1994: Design of an inversion-based precipitation profile retrieval algorithm using an explicit cloud model for initial guess microphysics. *Meteor. Atmos. Phys.*, **54**, 53-78.

Smith, E. A., and Coauthors, 1998: Results of WetNet PIP-2 project. *J. Atmos. Sci.*, **55**, 1483-1536.

Spencer, R. W., H. M. Goodman, and R. E. Hood, 1989: Precipitation retrieval over land and ocean with the SSM/I: Identification and characteristics of the scattering signal. *J. Atmos. Oceanic Technol.*, **6**, 254-273.

Twomey, S., 1991: Aerosols, clouds and climate. *Atmos. Env.*, **25A**, 2435-2442.

Ulaby, F. T., R. K. Moore, A. K. Fung, 1986: Microwave Remote Sensing – Active and Passive. Vol. III, Artech House, 2162pp.

Weng, F., and N. C. Grody, 1994: Retrieval of cloud liquid water using the Special Sensor Microwave Imager (SSM/I). *J. Geophys. Res.*, **99**, 25535-25551.

Weng, F., and N. C. Grody, 2000: Retrieval of ice cloud parameters using a microwave imaging radiometer. *J. Atmos. Sci.*, **57**, 1069-1081.

Wilheit, T. T., A. T. C. Chang, M. S. V. Rao, E. B. Rodgers, and J. S. Theon, 1977: A satellite technique for quantitatively mapping rainfall rates over the ocean. *J. Appl. Meteor.*, **16**, 551-560.

Xie, P., and P. A. Arkin, 1998: Global monthly precipitation estimates from satellite-observed outgoing longwave radiation. *J. Climate*, **11**, 137-164.

Zhao, L., and F. Weng, 2002: Retrieval of ice cloud parameters using the Advanced Microwave Sounding Unit. *J. Appl. Meteor.*, **41**, 384-395.

Zuidema, P., and D. L. Hartmann, 1995: Satellite determination of stratus cloud microphysical properties. *J. Climate*, **8**, 1638-1657.

POLARIMETRIC RADIATIVE TRANSFER THEORY AND ITS APPLICATIONS: AN OVERVIEW

QUANHUA LIU

Colorado State University
West Laporte Avenue, Fort Collins, CO 80523, USA
E-mail: Quanhua.liu@noaa.gov

(Manuscript received 8 October 2002)

This overview outlines the theoretical basis of polarimetric radiative transfer and its applications in the field of Earth science. The generic radiative transfer equation described here is applicable for the polarization and intensity in the microwave, infrared and visible portions of the spectrum. The equation is solved analytically to calculate the radiance and derivative of the radiance with respect to geophysical parameters. Both the radiance and its derivative are essential in data assimilation and in the application of remote sensing to numerical forecast models. This overview surveys the numerous applications of polarimetric signatures, such as the derivation of sea surface wind vector, sea ice parameters, long-wave radiation at the surface, and aerosol properties.

1. Introduction

Radiation is the energy transfer of electromagnetic waves. It can be characterized by the brightness (intensity), color (wavelength), and polarization. Human eyes are sensitive to the brightness and color, but insensitive to polarization. The brightness (from dark to bright) of daylight was known in ancient times. However, the color of natural light could not be physically understood until Sir Isaac Newton dispersed natural light into purple, violet, blue, green, yellow, orange, and red colors in the late seventeen century. Although a set of four parameters was introduced by Sir George Stokes in 1852 to represent the polarized light theoretically, the polarization nature of sunlight was unknown until Lord Rayleigh performed investigations in 1871 on the illumination and polarization of the sunlit sky. In 1901, Max Planck developed a theory for black body thermal radiation. The Planck function is one of the most applied functions used in the infrared and microwave remote sensing. Radiative transfer theory had been widely used by physicists for studying neutron transport and fusion in the late nineteen and early twenty centuries. Today, much attention is paid to radiative transfer theory and its applications because of the rapid development of remote sensing technology, and in particular satellite measurements. In this paper, we focus on radiative transfer theory and its applications in the Earth's atmosphere. Basic radiative transfer theory and its applications to atmospheric sciences can be found in the book of Liou (1980). A treatise on radiative transfer theory is given in the book by Chandrasekhar (1950). For his theories on stellar structures and dynamics, Chandrasekhar was awarded the Nobel Prize in physics in 1983.

Radiative transfer theory has been widely applied for studying the Earth's atmosphere from space (Lenoble 1985; Liou 1980). Radiative transfer theory provides the physical foundation for understanding the radiation budget at the Earth's surface and at the top of the atmosphere, climate change, and radiative cooling and heating rates of the atmosphere. In radiative transfer codes, one has to deal with huge absorption lines, wing and continuous absorptions. Effective methods of integration over frequency were developed by Zhu (1989), who also designed an effective algorithm for the middle atmosphere radiative heating and cooling rate calculations (Zhu 1994). Recently, polarization signatures have received great attention (Ulaby et al. 1982; Tsang et al. 1985; Liu 2000). Un-polarized natural light can be polarized by scattering from molecules, aerosols and clouds, and by the reflection and scattering from surfaces (Stephens 1994). The polarization of light in the atmosphere-surface system contains important signals that have led to major scientific breakthroughs that could not have been achieved by only studying the radiance. The polarization measured by microwave sensors provides unique information to calculate the sea surface wind speed (Goodberlet et al. 1990), sea surface wind vector (Wentz 1992), and sea ice (Miao et al, 2000). The judicious choice of polarization can also provide accurate measurement of water vapor (Schluessel and Emery 1990), cloud liquid water (Weng and Grody 1994), ice cloud parameters (Weng and Grody 2000), rainfall (Ferraro et al. 1996; Petty 1994), and snow cover (Grody 1991). The French satellite-based POLarization and Directionality of the Earth Reflectance (POLDER) instrument has been used to obtain the particle shape of ice clouds (Chepfer et al. 2001). Recently, Miao et al. (2002) have proposed a new millimeter/sub-millimeter technology to study cirrus cloud parameters. In this paper, we discuss the applications of polarized measurements for studying sea ice (Liu et al. 1998), retrieving long-wave radiation at the surface (Liu et al. 1997), and the applications of polarimetric sensors on the sea surface wind vector (Liu and Weng 2002) and the shape of cloud particles.

Polarized radiative transfer calculations are very important when studying the marine light field. Some very interesting biological applications of polarization have been investigated by scientists who demonstrated that many marine organisms exhibit a polarization axis. The Stokes vector radiation calculation is also used for atmospheric corrections of ocean color sensors. The atmospheric correction algorithm derives the water-leaving radiance by removing the radiation contribution from atmosphere and the air-sea interface, where water-leaving radiance is only about a few percent of the radiance at the top of the atmosphere. While various advanced forward models (Haferman 2000) had been developed with high enough accuracy long before the beginning of satellite data assimilation, the uncertainty of the optical properties in the operational retrieval is still a matter of concern. For sensors that do not explicitly measure polarization, radiometric response can be simulated using scalar radiative transfer. An example of these models is the well developed multi-layer discrete-ordinate radiative transfer scheme (DISORT) (Stamnes et al. 1988). Recently, DISORT was also expanded to include the full Stokes vector radiative transfer and is referred to as VDISORT (Weng 1992; Schulz et al. 1999). In fact, VDISORT can be best utilized for simulating the scattering and polarization properties of atmospheres. In addition to VDISORT, other schemes were developed, including the doubling and adding model (Evans and Stephens 1991) and the matrix operator method (Liu and Ruprecht 1996). Fast radiative transfer models (Kummerow 1993; Liu and Weng 2002a) were developed to provide operational retrievals of atmospheric parameters. An adjoint radiative transfer model having the capability of efficiently computing the radiance gradient (or Jacobian) for operational retrieval algorithms is under development (Weng and Liu 2003).

2. Basic Radiometric Quantities and Definition

Different definitions and terminologies (Liu and Simmer 1996) are used in the radiative transfer theory by physicists, astronomers, and atmospheric scientists. In this paper, we adopt the most accepted radiative quantities and definitions used in the atmospheric science community. The most important quantities are the radiation flux density, radiance, and phase function/matrix. The radiant flux density is the radiant energy across any planar element and can be measured in watts per square meter. Solar irradiance is the radiant flux intensity from sun. The radiance is defined as the radiation energy across a solid angle and crossing a surface perpendicular to the axis of propagation of the radiation beam, with unit of watts per square meter per steradian.

In the infrared and microwave regimes, the brightness temperature is converted from radiance using the Planck function. Stokes vectors are used to represent the polarimetric radiation and can be defined as: $\mathbf{S}=[I,Q,U,V]^t$ or $\mathbf{S}=[I_v,I_h,U,V]^t$, where the superscript t denotes the transpose. The scalar radiance I is defined as:

$$I = I_v + I_h, \tag{1}$$

and the polarization difference Q is defined as

$$Q = I_v - I_h, \tag{2}$$

where I_v and I_h the vertically and horizontally polarized radiances, respectively. The U and V components denote the plane of polarization and the ellipticity of the electromagnetic wave, respectively. The electric field vector of horizontally polarized radiation is normal to the *meridian plane* (which contains the z-axis and the direction of the electromagnetic wave), while the vertically polarized radiation is parallel to the plane.

For spherical and randomly oriented non-spherical particles the *scattering matrix* $\mathbf{P}$ can be written in the form (Mishchenko et al. 2000):

$$\mathbf{P}=\begin{bmatrix} a_1(\Theta) & b_1(\Theta) & 0 & 0 \\ b_1(\Theta) & a_2(\Theta) & 0 & 0 \\ 0 & 0 & a_{33}(\Theta) & b_2(\Theta) \\ 0 & 0 & -b_2(\Theta) & a_{44}(\Theta) \end{bmatrix}. \tag{3}$$

The scattering matrix **P** is defined in the *scattering plane*, which contains the incident and outgoing directions. Since the meridian plane is generally different from the scattering plane, the Stokes vector needs to be rotated from the meridian plane into the scattering plane by a rotation matrix (see Figure 1). After multiplication by the scattering matrix, the scattered Stokes vector needs to be rotated back from the scattering plane to the meridian plane by another rotation matrix. The product of the scattering matrix multiplied by the two rotation matrices may be called the *phase matrix*.

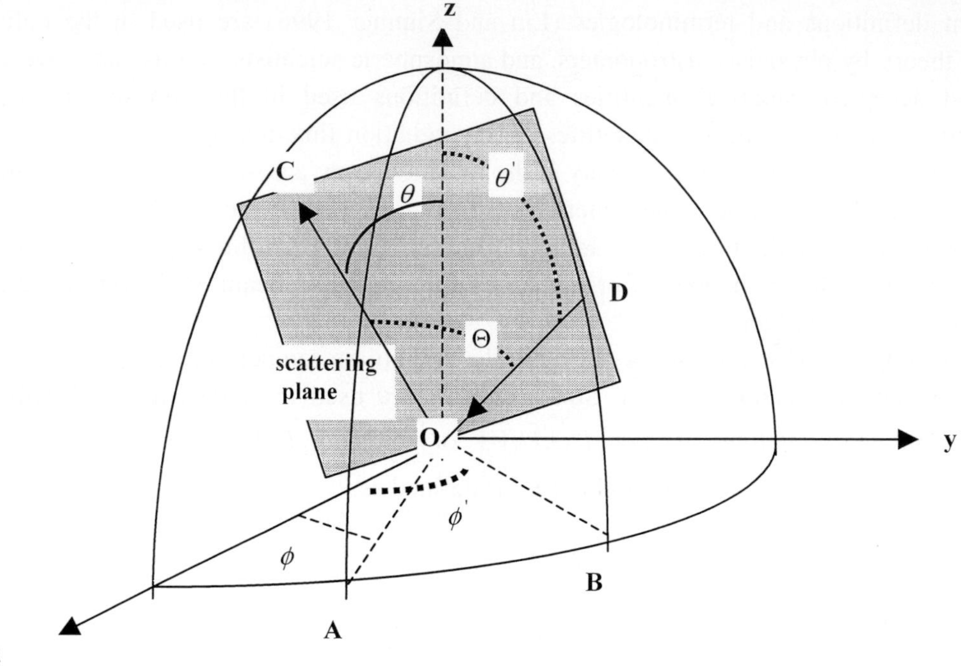

Figure 1. Illustration of the meridian (reference) plane and the scattering plane. The scattering plane contains the incoming direction OD and the outgoing direction OC. The plane Doz is the meridian plane for the incoming ray (θ', ϕ') and the plane Coz is the meridian plane for the outgoing ray (θ, ϕ).

There exists a useful relationship for Stokes vector as follows:

$$I^2 \geq Q^2 + U^2 + V^2 . \tag{4}$$

The equality can occur, when and only when the light is elliptically polarized (constant phase and amplitude). However, the actual light consists of many simple waves in very rapid succession. One then introduces a new parameter, degree of polarization, *P,* defined as:

$$P = \frac{(Q^2 + U^2 + V^2)^{1/2}}{I} . \tag{5}$$

Many applications only require the I_v and I_h components. These components are often expressed by the degree of linear polarization:

$$LP = \frac{-Q}{I} = -\frac{I_h - I_v}{I_h + I_v}. \tag{6}$$

For the Rayleigh scattering, the degree of linear polarization has an analytical function of the scattering angle Θ as:

$$LP = \frac{-Q}{I} = -\frac{\sin^2\Theta}{1+\cos^2\Theta}. \tag{7}$$

3. Measurement of Stokes Vector

There are various techniques to measure the Stokes vector. For visible, infrared and microwave radiometers, a polarizer is often used to change the angle Ψ to the reference direction and a wave plate is used to introduce a phase retardation ε (Liou 1980). For such a radiometer, the intensity can be written as (Stephens 1994):

$$R(\Psi,\varepsilon) = \frac{1}{2}[I + Q\cos 2\Psi + (U\cos\varepsilon - V\sin\varepsilon)\sin 2\Psi]. \tag{8}$$

The Stokes components can be obtained from Eq. (8):

$$I = R(0^\circ,0) + R(90^\circ,0), \tag{9a}$$

$$Q = R(0^\circ,0) - R(90^\circ,0), \tag{9b}$$

$$U = R(45^\circ,0) + R(135^\circ,0), \tag{9c}$$

$$V = R(135^\circ,90^\circ) - R(45^\circ,90^\circ). \tag{9d}$$

The six measurements on the right side of Eq. (9) complete the full Stokes vector. The technique is simple, but it is complicated by radiometric noise for the third and fourth components of the Stokes vector. Correlation radiometry can also be used to measure the Stokes vector. The technique is increasingly applied recently.

4. Vector Radiative Transfer Model

Various vector radiative transfer models deal with radiation transport. Among the one-dimensional radiative transfer models, the Eddington approximation (Kummerow 1993), two-stream model (Liu and Weng 2002a), the successive orders of scattering method (Simmer 1994), doubling and adding method (Hansen 1971; Evans and Stephens 1991), vector discrete ordinate method (VDISORT) (Weng 1992; Schultz et al. 1999), and matrix operator method (Liu and Ruprecht 1996) are commonly applied in the atmospheric sciences. The Monte Carlo method (Liu et al. 1996) is often applied for the three-dimensional radiative transfer problem. The method of the successive orders of scattering (Simmer 1994) computes the scattering effects by orders, which allows one to analyze the emission effect, single scattering effect and multiple scattering effects. The concept of the doubling and adding method is straightforward: in the adding method, any homogeneous layer can be built by successive addition of very thin layers of equal optical depth. The reflection and transmission for the thin layers can be easily found in the single scattering approximation. The reflection and transmission for the doubled layers can be obtained by computing the successive reflection back and forth between the two layers. VDISORT gives the explicit solution of the radiative transfer equation for the homogeneous layer in terms of the solved eigenvalues and eigenvectors. An inhomogeneous atmosphere can be divided into a set of homogeneous layers. VDISORT is an elegant method to solve for the inhomogeneous atmosphere from the solutions of homogeneous atmospheric layers based on the continuity of internal and external boundary conditions at the surface and the top of the atmosphere. The solution of the matrix operator method is an exponential function in a matrix form. The analytic solution has all analytic and derivable properties of the exponential function. The analytic derivative is very valuable in the retrieval of the geophysical parameters from the satellite measurements. Monte Carlo method is flexible for boundaries, so that it can be used to solve three-dimensional radiative transfer problems such as radiative transfer for finite clouds. In this section, the vector radiative transfer equation is solved using the analytic expression of the matrix operator method for the homogeneous layer. The elegant technique from VDISORT is applied to construct the solution for the inhomogeneous atmosphere from the solutions for the homogeneous layers (Weng and Liu 2003).

The equation describing vector radiative transfer in a plane-parallel medium (Liou 1980; Lenoble 1985) can be expressed as:

$$\mu \frac{d\mathbf{I}(\tau,\mu,\phi)}{d\tau} = -\mathbf{I}(\tau,\mu,\phi) + \frac{\varpi}{4\pi}\int_0^{2\pi}\int_{-1}^{1}\mathbf{M}(\tau;\mu,\phi;\mu',\phi')\mathbf{I}(\tau,\mu',\phi')d\mu' d\phi' + \mathbf{S}(\tau,\mu,\phi;\mu_0,\phi_0) \tag{10}$$

and

$$\mathbf{S} = (1-\varpi)B(T(\tau))\begin{bmatrix}1\\0\\0\\0\end{bmatrix} + \frac{\varpi F_0}{4\pi}\exp(-\tau/\mu_0)\begin{bmatrix}M_{11}(\mu,\phi;\mu_0,\phi_0)\\M_{12}(\mu,\phi;\mu_0,\phi_0)\\M_{13}(\mu,\phi;\mu_0,\phi_0)\\M_{14}(\mu,\phi;\mu_0,\phi_0)\end{bmatrix}, \tag{11}$$

where **M** is the phase matrix; $\mathbf{I}=[I,Q,U,V]^t$; $B(T)$ the Planck function of a temperature T; F_0 the solar spectral constant; μ_0 the cosine of sun zenith angle; ϖ the single scattering albedo; and τ the optical thickness.

In general, Eq. (10) can only be evaluated numerically. The azimuth-dependence of the radiance can be separated by use of the Fourier transformation for the azimuth angle. The phase matrix, the source function, and Stokes vector can be expanded as a series of sine and cosine function as follows:

$$\mathbf{M}(\tau,\mu,\phi;\mu',\phi')=\sum_{m=0}^{N}[\frac{\mathbf{M}_m^c(\tau,\mu,\mu')}{1+\delta_{0m}}\cos m(\phi'-\phi)+\mathbf{M}_m^s(\tau,\mu,\mu')\sin m(\phi'-\phi)], \tag{12}$$

$$\mathbf{I}(\tau,\mu,\phi)=\sum_{m=0}^{N}[\mathbf{I}_m^c(\tau,\mu)\cos m(\phi_0-\phi')+\mathbf{I}_m^s(\tau,\mu)\sin m(\phi_0-\phi')], \tag{13}$$

$$\mathbf{S}(\tau,\mu,\phi)=\sum_{m=0}^{N}[\mathbf{S}_m^c(\tau,\mu)\cos m(\phi_0-\phi')+\mathbf{S}_m^s(\tau,\mu)\sin m(\phi_0-\phi')], \tag{14}$$

where δ_{0m} is the Kronecker delta.

The phase matrix **M** has some special properties. For spherical particles or randomly oriented non-spherical particles, the sub-matrix off diagonal has only a non-zero sine part, whereas the sub-matrix in diagonal has only non-zero cosine part (Schultz et al. 1999). Substituting Eqs. (12)-(14) into Eqs. (10)-(11) and comparing the cosine and sine terms of equal order for the first and last two elements of the Stokes vector respectively, we have

$$\begin{aligned}\mu\frac{d\mathbf{I}_m(\tau,\mu)}{d\tau}&=-\mathbf{I}_m(\tau,\mu)+\varpi\int_0^1\mathbf{M}_m(\mu,\mu')\mathbf{I}_m(\tau,\mu')d\mu'\\&+\varpi\int_0^1\mathbf{M}_m(\mu,-\mu')\mathbf{I}_m(\tau,-\mu')d\mu'+\mathbf{S}_m,\end{aligned} \tag{15}$$

where

$$\mathbf{I}_m=[I_m^c,Q_m^c,U_m^s,V_m^s]^T, \tag{16}$$

$$\mathbf{S}_m=[S_m^c,S_m^c,S_m^s,S_m^s]^T, \tag{17}$$

$$\mathbf{M}_m = \begin{bmatrix} M_{11m}^c & M_{12m}^c & M_{13m}^s & M_{14m}^s \\ M_{21m}^c & M_{22m}^c & M_{23m}^s & M_{24m}^s \\ -M_{31m}^s & -M_{32m}^s & M_{33m}^c & M_{34m}^c \\ -M_{41m}^s & -M_{42m}^s & M_{43m}^c & M_{44m}^c \end{bmatrix}. \quad (18)$$

The harmonic mode of the phase matrix (Hovenier and van der Mee 1983) can also be represented as

$$\mathbf{M}_m(\mu,\mu') = \frac{1}{2\pi}\int_0^{2\pi}[\mathbf{E}\cos m(\phi-\phi') - \mathbf{D}\sin m(\phi-\phi')]\mathbf{M}(\mu,\phi;\mu',\phi')d\phi', \quad (19)$$

with the unit matrix $\mathbf{E} = \mathrm{diag}(1, 1, 1, 1)$ and the diagonal matrix $\mathbf{D} = \mathrm{diag}(1, 1, -1, -1)$.

Theoretically, the form for radiance containing the sine part as the first two components of Stokes vector and the cosine part as the rest components of Stokes vector is the same. However, the solution is null, since the solar and the thermal sources are even functions and unpolarized. Even for the rough ocean surface, the emitted thermal source has only the cosine part for the first two Stokes components and the sine part for the rest Stokes components (Yueh 1997).

The remaining integration over μ for each Fourier component or harmonic component in Eq. (15) can be replaced by a discrete sum (Stamnes et al. 1988) using Gaussian quadrature points μ_i and corresponding weights w_i. Although the solutions of the discrete ordinate method are given at the Gaussian quadrature points, the solution at a given desired angle, for example the satellite viewing angle, can be easily obtained by selecting additional quadrature points for the desired angle with an integration weight of zero. Thus, Eq. (15) for a set of quadrature points can be written as:

$$\begin{aligned} \mu_i \frac{d}{d\tau}\begin{bmatrix} \mathbf{I}_m(\tau,\mu_i) \\ -\mathbf{I}_m(\tau,\mu_{-i}) \end{bmatrix} &= \begin{bmatrix} \mathbf{I}_m(\tau,\mu_i) \\ \mathbf{I}_m(\tau,\mu_{-i}) \end{bmatrix} - \\ &\varpi\sum_{j=1}^{N}\begin{bmatrix} \mathbf{M}_m(\mu_i,\mu_j) & \mathbf{M}_m(\mu_i,\mu_{-j}) \\ \mathbf{M}_m(\mu_{-i},\mu_j) & \mathbf{M}_m(\mu_{-i},\mu_{-j}) \end{bmatrix}\begin{bmatrix} \mathbf{I}_m(\tau,\mu_j) \\ \mathbf{I}_m(\tau,\mu_{-j}) \end{bmatrix} w_j - \begin{bmatrix} \mathbf{S}_m(\tau,\mu_i,\mu_0) \\ \mathbf{S}_m(\tau,\mu_{-i},\mu_0) \end{bmatrix}. \end{aligned} \quad (20)$$

There are numerous techniques to solve Eq. (20). The solution in a matrix form at any atmospheric layer (by omitting the index m for the *m*-th Fourier component) can be written as:

$$\mathbf{I}_l(\tau) = \exp[\mathbf{A}_l(\tau-\tau_{l-1})]\mathbf{c}_l + \mathbf{s}_l(\tau). \quad (21)$$

The coefficients vector $\mathbf{c}_l$ can be determined from the continuity internal boundary conditions and the downward boundary condition at the top of the atmosphere, as well as the upward boundary condition at the surface. From the viewpoint of an earth-viewing satellite, it is important to determine the radiance at the top of the atmosphere. This radiance can be written as:

$$\mathbf{I}_1 = \mathbf{c}_1 + \mathbf{s}_1 . \tag{22}$$

Thus, the radiance gradient at the top of the atmosphere corresponding to any optical parameter (x_j) at the atmospheric layer j can be expressed as:

$$\frac{\partial \mathbf{I}_1}{\partial x_j} = \frac{\partial \mathbf{c}_1}{\partial x_j} + \frac{\partial \mathbf{s}_1}{\partial x_j} \delta_{1j} . \tag{23}$$

The radiance gradient to the geophysical parameters is just a linear combination of the radiance gradient to the optical parameters. The detailed solution can be found in Weng and Liu (2003).

5. Applications

Active and passive polarimetric sensors are widely applied for the retrieval of surface and atmospheric parameters. The Synthetic Aperture Radar (SAR, active sensor) of the European Earth Resource Satellite (ERS) has a high spatial resolution of 12.5 m at the surface. ERS SAR sends the vertically polarized signal and receives the backscatter vertically polarized signal in the *C*-band. The first ERS was launched in 1991 and the second ERS was launched in 1995. An advance SAR has been deployed on European Environment Satellite (ENVISAT), which was launched in March 2002. The SAR measurements have been successfully used in soil moisture studies, surface water body morphology, snow extent and condition, and polar sea ice (Drinkwater and Lytle 1997). As an example of the applications, the SAR measurements are examined to study the characteristics of a sea-ice pressure ridge in the Antarctic.

Sea-ice pressure ridges are generated during ice deformation events when ice floes are pressed against each other in a convergent drift regime, and finally break. Along the former floe edges, the resulting ice blocks are piled up above and below the water level, thus forming extended ridges or ridge zones. The amount of deformation is determined by atmospheric and oceanic forcing. The Synthetic Aperture Radar onboard the European Space Agency satellites ERS-1 and ERS-2 can resolve the ice characteristics to fine scales. The ice contained in pressure ridge sails has a lower density and higher porosity than the surrounding level ice. Also, many surfaces of the tilted ice blocks are oriented normal or at a high incident angle towards the incident radar waves, which have incidence angles ranging from 20 to 26° to the local zenith angle at the surface. This results in a higher backscatter compared to that from level ice (Drinkwater and Lytle 1997). On the summer scene from the Weddell Sea (see Figure 2), ridges and floes can be observed, and leads and ice are clearly distinguishable.

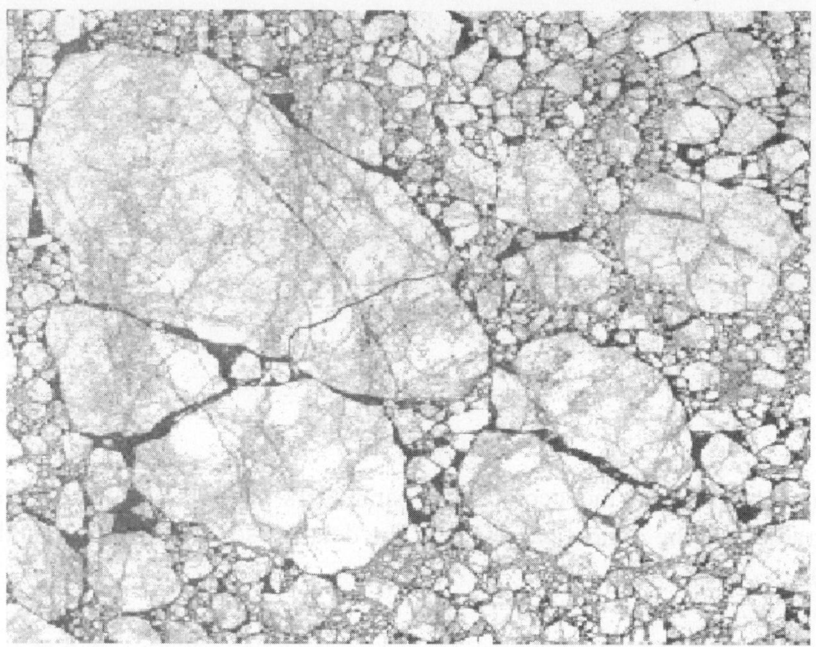

Figure 2. ERS-2 SAR image (20 km by 20 km) on the summer scene from the Weddell Sea.

The SAR measurements reveal useful characteristics of sea ice (Haas et al. 1999). The distributions of SAR backscatter coefficients in the Bellingshausen and Amundsen Seas (Figure 3) represent different sea-ice regimes. In the Amundsen Sea, the sea-ice appears relatively homogeneous, which leads to a narrow distribution of the SAR backscatter coefficients. In contrast, the distribution of the backscatter coefficients in the Bellingshausen Sea is multi-modal, resulting from alternating level floes and floes with ridges. Close to the Antarctic Peninsula, the distribution is very wide due to the occurrence of a mixture of water, thin ice, and level and deformed floes. The backscatter distributions of all processed images could be assigned to one of these classes, namely narrow peak, multi-mode, and wide peak (Figure 3).

Passive microwave radiometers have successfully been applied in the retrieval of geophysical parameters over oceans. Using the Special Sensor Microwave Imager (SSM/I) and collocated buoy data, Goodberlet et al. (1990) derived a simple regression algorithm to calculate the wind speed from SSM/I measurements. Future passive microwave sensors, such as U.S. Navy WindSAT/Coriolis and U.S. National Polar-orbiting Environmental Satellite System (NPOESS) Conical Microwave Imager Sounder (CMIS), are all developed with polarimetric sensors for global remote sensing of surface wind vectors. However, the variation of Stokes vector to the wind direction is generally less than 3 K (Figure 4). The amplitudes of the variation for the vertically, T_v, and horizontally, T_h, polarized components are about 2 K. For a wind speed of 10 $\mathrm{m\,s^{-1}}$, the amplitudes of the variation are about 3 K and 0.5 K for the third, U, and the fourth, V, components of Stokes vector, respectively. The surface emissivities for the vertical and horizontal polarization can be expanded into harmonic cosine series (Yueh 1997). The third and the fourth components of the emissivities can be expanded into harmonic sine series. For the retrieval of wind direction, the algorithm deals with these small signatures.

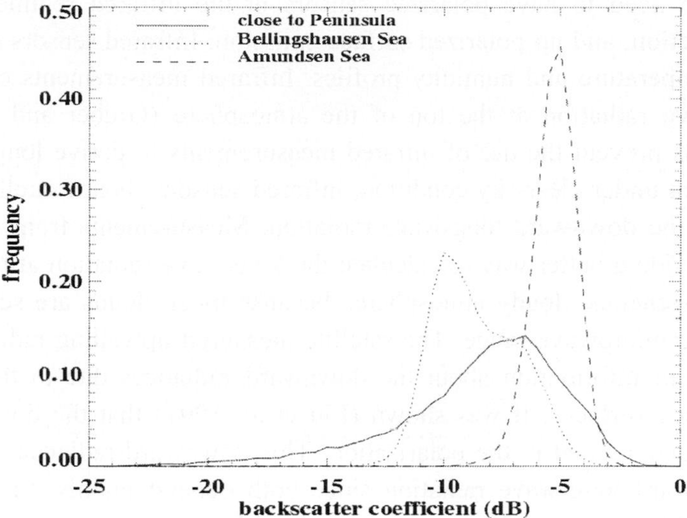

Figure 3. Backscatter coefficients from SAR images at three ice regimes.

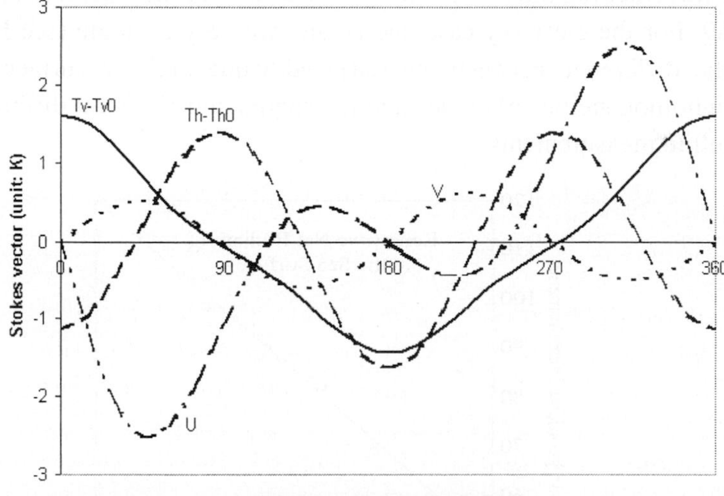

Figure 4. Variation of Stokes vector at 37 GHz with the relative azimuth angle for a wind speed of 10 m s^{-1} and the sea surface temperature of 300 K. The averaged values for the horizontally and vertically polarized brightness temperatures are extracted. The dashed, solid, dash-dotted, and dotted lines denote the horizontally polarized, vertically polarized, the third and the fourth Stokes components, respectively. The viewing zenith angle is 53°.

Piepmeier and Gasiewski (2001) studied the retrieval of the wind vector from multi-frequency and multi-looking angles measurements. They retrieved wind-speed and direction separately. Liu and Weng (2002b) applied a physical retrieval algorithm to calculate the wind speed and direction simultaneously. A study of simulated polarimetric signatures with radiometric noise from the radiometer specification document showed that the retrieval error is about 10 degrees and 0.6 m s^{-1} for the wind direction and speed, respectively.

There is no need to have polarized sensors in the infrared regime due to the weak scattering, reflection, and up-polarized surface emission. Infrared sensors are used to retrieve atmospheric temperature and humidity profiles. Infrared measurements can also be used to derive long-wave radiation at the top of the atmosphere (Gruber and Jacobowitz 1985). However, clouds prevent the use of infrared measurements to derive long-wave radiation at the surface. Even under clear-sky condition, infrared sensors aboard satellites do not provide information on the downward long-wave radiation. Measurements from passive microwave radiometers provide a better way to calculate the long-wave radiation at the surface for both clear and homogeneous cloudy atmospheres because most clouds are semi-transparent and detectable in the microwave range. The satellite-measured upwelling radiances over oceanic areas also contain information about the downward radiances due to the high microwave reflectivity of sea surfaces. It was shown (Liu et al. 1997) that the downward microwave radiance is linearly related to the polarization. This downward radiance is highly correlated with the downward long-wave radiation since both depend mainly on temperature, water vapor, and cloud liquid water. However, the retrieval is a high non-linear problem. Due to the high non-linearity, we applied a layered perceptron type artificial neural network (Liu et al. 1997). The artificial neural network was applied for the collocated SSM/I data and the measured long-wave net radiation over the North Sea during the International Ice Experiment in October 1989. For the clear sky case, the results are very accurate (see Figure 5). For the cloudy case, the difference between the retrieved value and the surface measurement is largely due to inhomogeneities of clouds and the spatial and temporal difference between the surface and satellite measurements.

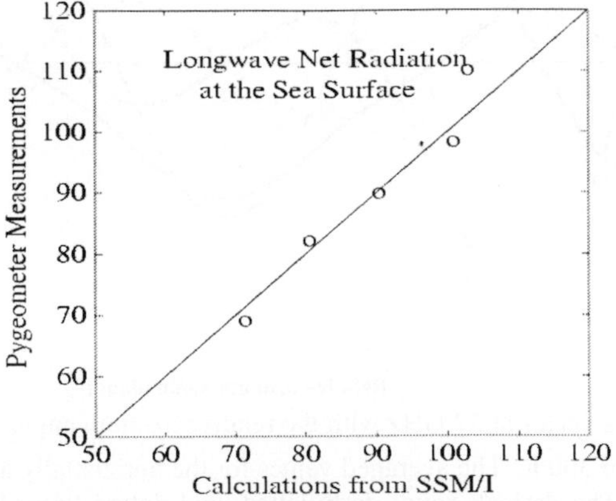

Figure 5. Comparison of the long-wave net radiation (clear sky cases) between the calculation from SSM/I data and the surface measurements during ICE Experiment in 1989.

In the visible spectral range, a polarimetric sensor like POLDER has demonstrated the capability to derive optical properties of aerosols and clouds. Recently, cloud microphysical parameters have been assimilated into numerical prediction models. Liou and Ou (1989) studied the shape of ice cloud crystals and its effect on climate change. Using a one-dimensional cloud and climate model, Liou and Ou (1989) found that the surface temperature would increase 0.4 K if one assumes spherical rather than the column/plate particles for the ice cloud. An uncertainty of 0.4 K in surface temperature is comparable in magnitude to the effect of greenhouse gases on climate change. In is increasingly important to study the sensitivity of cloud particle size and shape on the Stokes vectors and to utilize this sensitivity in the retrievals. In particular, the upcoming sensor, Aerosol Polarimetry Sensor (APS), of U.S. National polar-orbiting Environmental Satellite System (NPOESS), will have Stokes vector measurements from the blue to near infrared bands for multiple viewing directions. The multi-angle and multi-spectral Stokes vectors provide unique signatures for studying the particle size and shape of clouds. For example, the polarization reflectance obtained near 140° scattering angle in stratocumulus is quite different from that in cirrus clouds. A lot of attention has been paid to aerosols because of their importance in the environmental influence of climate change. Over oceans, the remote sensing of aerosols is relatively easy since the ocean reflectance is very low and the visible measurements are mostly due to the atmospheric parameters. However, land surfaces are usually very bright, which prevents an accurate retrieval of aerosols. The polarimetric sensor provides a new way to look at the aerosols since the polarization mainly results from atmospheric scattering. The polarization signature should therefore allow one to separate the aerosol features from the land surface. Furthermore, molecular scattering does not contribute to the V component (the fourth Stokes component) while the aerosol does produce the V component. Therefore, the V component can be used to separate molecular and aerosol scattering.

6. Discussion

This paper provides an overview of polarimetric radiative transfer theory and its applications. This overview is introductive rather than complete. Nowadays, polarization measurements can provide unique signatures for the determination of hydrometeors and surface parameters. However, more efforts are required for both measurements and theory. Calibration accuracy (Schmetz 1989) and the signal-to-noise ratio of polarization measurements can limit the accuracy of remote sensing products. Liu et al. (1998) found negative microwave brightness temperatures differences emitted from fresh ice. This microwave brightness temperature anomaly is a result of the coherence of the electromagnetic waves from air-ice and ice-water interfaces. Both TRMM TMI and SSM/I measurements also revealed the anomaly from the clouds. Figure 6 showed the distribution of the polarization difference from both TRMM TMI and SSM/I for Hurricane BONNIE in August 1998. It shows the negative polarization difference from both sensors where TRMM has a much better spatial resolution.

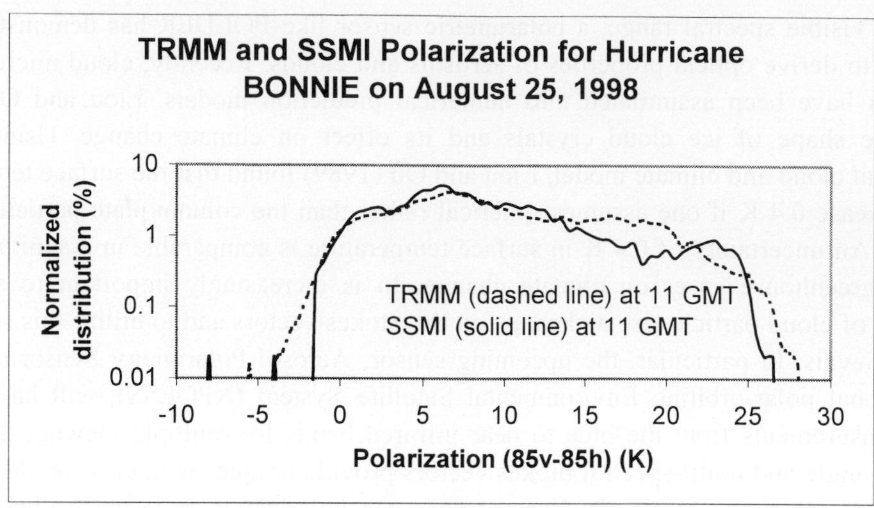

Figure 6. Histogram of the polarization at 85.5 GHz for Hurricane Bonnie at 11 GMT August 25, 1998 from the TRMM (dashed line) and SSM/I (solid line) measurements.

The polarimetric emission and the reflection (Haggerty and Curry 2001; Plokhenko and Menzel 2000) of the surface are a challenging topic. The emissivity models of snow, ice, and vegetation land are crucial in the retrieval and the forward model simulations. The three-dimensional radiation leakage problem in the case of inhomogeneous clouds is also important (Haferman 1993; Liu et al. 1996). Overall, the future of polarimetric measurements and their applications is very bright, but also very challenging. It should, however, be pointed out that the successful use of polarization depends on the particular geophysical variable and on the judicious choice of polarization for each variable.

Acknowledgments. The author thanks Dr. Steven A. Lloyd for editorial assistance.

References:

Chandrasekhar, 1960: *Radiative Transfer*, Dover Publications, INC, New York, 393p.

Evans, K. F., and G. L. Stephens, 1991: A new polarized atmospheric radiative transfer model. *J. Quant. Spectrosc. Radiat. Trans.*, **46**, 413-423.

Chepfer, H., P. Goloub, J. Riedi, J. F. De Haan, J. W. Hovenier, and P. H. Flamant, 2000: Ice crystal shapes in cirrus clouds derived from POLDER/ADEOS-1. *J. Geophys. Res.*, **106**, 7955-7966.

Drinkwater, M. R., and V. I. Lytle, 1997: ERS-1 SAR and field-observed characteristics of austral fall freeze-up in the Weddel Sea, Antarctica, *J. Geophys. Res.*, **102**, 12593-12608.

Goodberlet, M., M. C. T. Swift, J. Wilkerson, 1990: Ocean surface wind speed measurements of the Special Sensor Microwave/Imager (SSM/I). *IEEE Trans. Geosci. Remote Sensing*, ***28***, 823-828.

Grody, N., 1991: Classification of snow cover and precipitation using Special Sensor Microwave/Imager (SSM/I). *J. Geophys. Res.*, ***96***, 7423-7435

Gruber, A., and H. Jacobowitz, 1985: The longwave radiation emitted from NOAA polar orbiting satellite: An update and comparison with NIMBUS-7 ERB results. *Ad. Space Res.*, ***5***, 111-120.

Ferraro, R., F. Weng, N. Grody, and A. Basist, 1996: An eight year (1987-1994) time series of rainfall, clouds, water vapor, snow and sea ice derived from SSM/I measurements. *Bull. Amer. Meteor. Soc., 77, 891-905.*

Haas, C., Q. Liu, and T. Martin, 1999: Retrieval of Antarctic sea-ice pressure ridge frequencies from ERS SAR imagery by means of in situ laser profiling and usage of a neural network. *Int. J. Remote Sensing*, **20**, 3111-3123.

Haferman, J. L., W. F. Krajewski, T. F. Smith, and Sanchez, 1993: Radiative transfer for a three-dimensional raining cloud. *Appl. Opt.*, **32**, 2795-2802.

Haferman, J. L., 2000: Microwave Scattering by Precipitation, Chapter17 in Light scattering by nonspherical particles: Theory, measurements, and geophysical applications. Edited by Michael I. Mishchenko, Joachim W. Hovenier, and Larry D. Travis, Academic Press, San Diego.

Haggerty, J. A., and J. A. Curry, 2001: Variability of sea ice emissivity estimated from airborne passive microwave measurements during FIRE SHEBA. *J. Geophys. Res.*, **106**, 15265-15277.

Hansen, J. E., 1971: Multiple Scattering of Polarized Light in Planetary Atmospheres. Part I. The Doubling Method. *J. Atmos. Sci.*, **28**,120-125.

Hovenier, J. W., and C. V. M. van der Mee, 1983: Fundamental relationships relevant to the transfer of polarized light in a scattering atmosphere. *Astronomy and Astrophysics*, **128**, 1-16.

Kummerow, C., 1993: On the accuracy of the Eddington approximation for radiative transfer in the microwave frequencies. *J. Geophy. Res.*, **98**, 2757-2765.

Lenoble, J., 1985: Radiative transfer in scattering and absorbing atmospheres: standard computational procedures. A Deepak Publishing, 300p.

Liou, K. N., 1980: An introduction to atmospheric radiation. Academic Press, New York, 392p.

Liou, K. N., and S. Ou, 1989: The role of cloud microphysical processes in climate: An assessment from a one-dimensional perspective. *J. Geophys. Res.*, **94**,8599-8607.

Liu, Q., and E. Ruprecht, 1996: A radiative transfer model: matrix operator method. *Appli. Opt.*, **35**, 4229-4237.

Liu, Q., C. Simmer, and E. Ruprecht, 1996: 3-D radiative transfer effects of clouds in the microwave spectral range. *J. Geophy. Res.*, **101**, 4289-4298.

Liu, Q., and C. Simmer, 1996: Polarization and intensity in microwave radiative transfer. *Contri. Atmos. Phys.*, **69**, 535-545.

Liu, Q., C. Simmer, and E. Ruprecht, 1997: Estimating longwave net radiation at sea surface from the Special Sensor Microwave/Imager (SSM/I). *J. Appl. Meteorol.*, **36**, 7, 919-930.

Liu, Q., E. Augstein, and A. Darovskikh, 1998: Polarization anomaly of the microwave brightness temperature from ice, *Appl. Opt.*, **37**, 2228-2230.

Liu, Q., 2000: An improved look-up table technique for geophysical parameters from SSM/I. *Int. J. Remote Sensing*. **21**, 1571-1582.

Liu, Q., and F. Weng, 2002a: A Microwave Polarimetric Two-Stream Radiative Transfer Model. *J. Atmos. Sci.*, **59**, 2396-2402.

Liu, Q., and F. Weng, 2002b: Retrieval of Sea Surface Wind Vector from Simulated Satellite Microwave Polarimetric Measurements. *Radio Science*, in press.

Miao, J., K.-P. Johnsen, S. Kern, G. Heygster, and K. Kunzi, 2000: Signature of Clouds over Antarctic Sea Ice Detected by the Special Sensor Microwave/Imager. *IEEE Trans. Geosci. Remote Sensing*, **38**, 2333-2344.

Miao, J., T. Rose, K. Kunzi, G. and P. Zimmerman, 2002: A Future Millimeter/Sub-Millimeter Radiometer for Satellite Observation of Ice Clouds. *Int. J. Infrared Millimeter Waves.*, **23**, 1159-1170.

Mishchenko, M., J. W. Hovenier, and L. D. Travis, 2000: Light Scattering by Nonspherical Particles. Edited by Michael I. Mishchenko, Joachim W. Hovenier, and Larry D. Travis, Academic Press, San Diego.

Petty, G., 1994: Physical retrievals of over-ocean rain rate from multichannel microwave imagery. Part I: Theoretical characteristics of normalized polarization and scattering indices. *Meteorol. Atmos. Phys.*, **54**, 79-99.

Piepmeier, J. R., and A. J. Gasiewski, 2001: High-resolution passive Polarimetric microwave mapping of ocean surface wind vector fields. *IEEE Trans. Geosci. Remote Sens,* **39**, 606-622.

Plokhenko, Y., and P. Menzel, 2000: The effects of surface reflection on estimating the vertical temperature-humidity distribution from spectral infrared measurements. *J. Appl. Meteorol.*, **39**, 3-14.

Schluessel, P., and W. J. Emery, (1990). Atmospheric water vapor over oceans from SSM/I measurements. *Int. J. Remote Sensing*, **11**, 753-766.

Schmetz, J., 1989: Operational calibration of the METEOSAT water vapor channel by calculated radiances. *Appl. Opt.*, 28, **15**, 3030-3038.

Schultz, F. M., K. Stamnes, and F. Weng, 1999: An improved and generalized discrete ordinate radiative transfer model for polarized (vector) radiative transfer computations. *J. Quant. Spectrosc. Radiat. Trans.*, **61**, 105-122.

Simmer, C., 1994: Satellitenfernerkundung hydrologischer Parameter der Atmosphaere mit Mikrowellen. Verlag Dr. Kovac, Hamburg, 313p.

Stamnes, K., S-C. Tsay, W. Wiscobe, and K. Jayaweera, 1988: Numerical stable algorithm for discrete-ordinate-method radiative transfer in multiple scattering and emitting layered media. *Appli. Opt.*, **27**, 2502-2509.

Stephens, G. L., 1994: Remote Sensing of the Lower Atmosphere, An Introduction. Oxford University Press, 523p.

Tsang, L., J. A. Kong, and R. T. Shin, 1985: Theory of microwave remote sensing. John Wiley & Sons, New York, 613p.

Ulaby, F. T., R. K. Moore, and A. K. Fung, 1982: Microwave Remote Sensing, Active and Passive. Addison-Wesley Publication Company, 257-1064p.

Wentz, F. J., 1992, Measurement of the oceanic wind vector using satellite microwave radiometers, *IEEE Trans. Geosci. Remote Sens.,* **30**, 960-972.

Weng, F., 1992: A multi-layer discrete-ordinate method for vector radiative transfer in a vertically-inhomogeneous, emitting and scattering atmosphere I: Theory. *J. Quant. Spectrosc. Radiat. Trans.*, **47**, 19-33.

Weng, F., and N. C. Grody, 1994: Retrieval of cloud liquid water using the Special Sensor Microwave/Imager. *J. Geophys. Res.*, **99**, 25535-25551.

Weng, F., and N. C. Grody, 2000: Retrieval of ice cloud parameters using a microwave imaging radiometer. *J. Atmos. Sci.*, **57**, 1069-1081.

Weng, F., and Q. Liu, 2003: Toward Direct Uses of Satellite Cloudy Radiances in NWP Models. Part I: Forward and Adjoint Models, 12th Conference on satellite meteorology and oceanography, American Meteorological Society.

Yueh, S., 1997: Modeling of wind direction signals in polarimetric sea surface brightness temperatures. *IEEE Trans. Geosci. Remote Sensing*, **35**, 1400-1418.

Zhu, X., 1988: Radiative cooling calculated by random band models with $S^{-1-\beta}$ tailed distribution. *J. Atmos. Sci.*, **46**, 511-520.

Zhu, X., 1994: An accurate and efficient radiation algorithm for middle atmosphere models. *J. Atmos. Sci.*, **51**, 3593-3614.

ON THE SOLAR RADIATION BUDGET AND THE CLOUD ABSORPTION ANOMALY DEBATE

ZHANQING LI
Department of Meteorology and ESSIC
University of Maryland, College Park, MD 20742, USA
E-mail: zli@atmos.umd.edu, http://www.atmos.umd.edu/~zli/

(Manuscript received 29 January 2003)

This paper reviews the current state of knowledge, advances and challenges in short-wave earth radiation budget (ERB) studies and the cloud absorption anomaly (CAA) debate. The ERB issues deal exclusively with the solar energy disposition between the atmosphere, clouds and the surface. The ERB and its disposition have been derived from surface observations, satellite remote sensing and general circulation modeling. Major sources of uncertainties and discrepancies between observations and modeling are highlighted. Reported discrepancies between the ERB data sets obtained by various means are discussed, especially within the context of the recent debate concerning the CAA. This debate is documented thoroughly and critically in four stages: before and after the mid-1990s, and following two dedicated field experiments: the Atmospheric Radiation Measurement Enhanced Shortwave Experiment (ARESE I and II). An attempt is made to shed light on the causes of some controversial findings. It is now clear that the CAA is largely an artifact which does not emerge from a carefully designed closure test using the state-of-the-art radiative transfer models.

1. Introduction

Solar radiation is the ultimate source of energy for the planet Earth. Variable radiative heating/cooling in the atmospheric column drives vertical atmospheric convection, and global general atmospheric and oceanic circulations are the primary responses to the uneven distribution of the radiation budget. Modeling Earth's weather and climate and changes incurred by any external or internal forcing thus requires a good knowledge and understanding of the disposition and distribution of solar radiation, which is unfortunately still fraught with large uncertainties (Wild et al. 1995; Li et al. 1997; Arking 1999). Radiative transfer models founded on classical electromagnetic and quantum mechanics theories can compute the breakdown of solar energy in the atmosphere and at the surface. While the fundamentals of radiative transfer theories were well established about a century ago, it is still a daunting task to apply these theories to the complex real world in order to accurately compute the ERB. Complications stem from intricate and fickle cloud morphologies (commonly known as 3-D effects), odd ice crystal shapes, and multi-layer cloud structures, among other factors. Radiation processes are governed by a large number of radiatively sensitive variables that act on a wide range of scales (Wielicki et al. 1995).

Solar energy reaching our planet is partly reflected to space, partly absorbed in the atmosphere, and partly absorbed at the Earth's surface. This partitioning of the solar energy incident at the top of the atmosphere (TOA), hereafter called the solar energy disposition (SED), is determined by the optical properties of the atmospheric column that, in turn, is influenced by the SED. The key variables that control the SED include the amount, vertical distribution, and optical properties of clouds, aerosol and radiation sensitive gases, as well as surface properties. Feedbacks involving these variables and the SED are important in modeling the climate system and its response to external perturbations, such as changes in the concentrations of CO_2 and other greenhouse gases. At this point, cloud feedback is the principal contributor to the large uncertainty in the climate system response (Cess et al. 1989, 1996; Arking 1991). Not only does the SED play an active role in the energetics of the climate system, it is also closely linked to the hydrological cycle via dynamic and thermodynamic processes (Stephens and Greenwald 1991). About half the solar energy absorbed at the surface is used to evaporate water, which eventually forms clouds. Latent heat released during cloud formation is a major source of energy driving the atmospheric circulation, especially in the tropics, and is comparable in magnitude to the solar radiation directly absorbed by the atmosphere. A sensitivity study with a general circulation model (GCM) showed that modifying the partitioning of solar energy between the atmosphere and surface could substantially alter the modeled fields of cloud cover, temperature, precipitation, humidity, and the atmospheric circulation pattern (Kiehl et al. 1995). Understanding the Earth's climate and modeling it, therefore, requires an accurate representation of the radiation energy budget at the TOA (Hartmann et al. 1986; Ramanathan 1987) and at the surface (Suttles and Ohring 1986; Wielicki et al. 1996). Together they determine how much of the solar energy is absorbed in the atmosphere.

The importance of the earth radiation budget (ERB) has been underscored in numerous international programs. The Global Energy Water Cycle Experiment (GEWEX) has 5 projects directly related to the ERB including the Surface Radiation Budget SRB), the International Satellite Cloud Climatology Project (ISCCP), the Global Water Vapor Project (GVaP), the Global Aerosol Climatology Project (GACP), and the Baseline Surface Radiation Network (BSRN). Under the auspices of these projects, various ERB-related issues have been tackled. Among the largest undertakings that have flourished during the past decade are the Earth Observation System (EOS) of the National Aeronautics and Space Administration (NASA) and the Atmospheric Radiation Measurement (ARM) program of the Department of Energy (DOE) (Ackerman and Stokes 2003)

In light of the importance and quick development of these major earth science enterprises, this paper reviews some major advances made in the past few decades and challenges that still confront us. The review is focused on two intimately related and yet fast-developing topics, namely the solar radiation budget and its disposition, and the cloud absorption anomaly.

2. The Solar Radiation Budget

The ERB in the atmosphere-surface system has been monitored from space for more than two decades, while the surface radiation budget (SRB) has been observed at various sites, unevenly distributed over the globe, for more than a century. Both ERB and SRB observations have limitations on their accuracy, making it difficult to obtain a reliable estimate of the energy absorbed in the atmosphere since the latter is often determined as the difference between two large quantities. Consequently, fervent debate has been waged over the past 10 years or so concerning the amount of solar energy absorbed inside the atmospheric column as discussed in the following section.

2.1. *Surface Observations*

Prior to the space-borne Earth observation era inaugurated in the 1960s, radiation estimates were based primarily on surface measurements, although simple models of radiative transfer in the atmosphere were used to infer TOA fluxes from the surface measurements. Surface radiation is among the few meteorological variables that have been observed since the last century (Hunt et al. 1986). On the basis of very limited observations at different latitudes, Abbot and Fowle (1908) obtained the first estimate of the global annual mean planetary albedo (0.37) and near-surface (below 1800m) absorption (0.42). All numbers are normalized to the incoming solar flux at the TOA. Other investigators obtained similar estimates of the SED in the 1920s and 1930s (c.f. Table 3.2 of Budyko 1982). Spatial and temporal variations in the SED were first addressed by Simpson (1929). More extensive analyses were made in the middle of the 20th century (Houghton 1954; Budyko 1956; London 1957), based on increased surface observations, more sophisticated radiative transfer theory, and the beginnings of laboratory studies. Table 1 summarizes observation-based estimates of SED.

Table 1. Comparison of historical estimates of the solar energy disposition estimated from ground and/or satellite (*) observations (After Li et al. 1997).

Sources	Coverage	TOA	Surface	Atmosphere
Abbot & Fowle (1908)	NH	0.37	0.42	0.21
Houghton (1954)	NH	0.34	0.47	0.19
London (1957)	NH	0.35	0.475	0.175
Sasamori et al. (1972)	SH	0.35	0.45	0.20
Budyko (1982)	Global	0.30	0.46	0.24
Pinker & Laszlo (1992)	Global	0.29	0.50	0.21
Ohmura & Gilgen (1993)	Global	0.30	0.42	0.28
Li & Lighton (1993)*	Global	0.30	0.46	0.24
Rossow & Zhang (1995)*	Global	0.32	0.48	0.19

Note that the most extensive and complete compilation of the global surface energy balance (SEB) was carried out by Budyko (1982) and his colleagues. They generated several versions of the SEB atlas depicting the monthly-mean global distribution of various SEB components, including the SRB. Empirical relationships involving conventionally measured meteorological variables (e.g., cloud amount, sunshine duration, etc) were adopted. With improving techniques and a growing set of observations, their estimates of solar flux absorbed at the surface increased (Budyko 1982). Their latest estimates of the SED are nearly identical with the satellite-based estimates of Li and Leighton (1993). Interestingly, a recent ground-based estimate of surface absorption by Ohmura and Gilgen (1993) is similar to an earlier estimate by Abbot and Fowle (1908). The fact that a discrepancy exists between the two contemporary estimates of surface absorption, encompassing a range as much as reported in history, is a good testimony to the complexity of the problem. The problem is also attributed to meager ground-truth information available to characterize the SRB. Given a long history of ground-based radiometric observations, we are still confronted with large uncertainties despite the improved quality of radiometers (Michalsky et al. 1999; Dutton et al. 2001) employed in dedicated radiation campaigns such as the ARM program. Moreover, surface SRB measurements suffer from an inability to maintain uniform deployment standards and proper calibration, in addition to the problem of inadequate spatial sampling. Most of all, the surface measurements do not provide any information on radiation budget at the TOA which is best quantified by satellite observations.

2.2. *Satellite Remote Sensing*

Since 1960, meteorological satellites have radically advanced our knowledge of the ERB (House et al. 1986). In contrast to ground-based observations, space-borne observations have the advantage of a global coverage. From the space-borne radiometers of the first (TIROS-type) and second generations (Nimbus 3, ESSA and NOAA series), a global mean planetary albedo was found to be around 0.30 (Vonder Haar and Suomi 1971; Stephens et al. 1981). This estimate is significantly lower than the pre-satellite estimates but agrees fairly closely with the later observations by much more advanced sensors (c.f. Table 1) (Hartmann et al. 1986; Ramanathan 1987; Barkstrom et al. 1989). The geographical distribution of the TOA albedo for the four seasons was obtained by Raschke et al. (1973). These early estimates of regional radiative fluxes contain large uncertainties due in part to the crude treatment of the dependence of satellite radiance measurements on viewing geometry (Arking and Levine 1967; Raschke et al. 1973).

More meticulous monitoring of the spatial and temporal variations in the TOA albedo was accomplished by the third-generation radiometers, including the Earth Radiation Budget (ERB) sensors aboard Nimbus-7 (Jacobowitz et al. 1984) and the Earth Radiation Budget Experiment (ERBE) sensors aboard three satellites (Barkstrom et al. 1989). One of the major advances was the development of improved angular dependence models (ADMs) (Taylor and Stowe 1984; Suttles et al. 1988). Nevertheless, angular correction has been the primary source of uncertainty in ERB measurements (Arking and Vemury 1984; Stuhlmann and Raschke 1987; Suttles et al. 1992; Wielicki et al. 1995; Li 1996). The uncertainty results from difficulties in accounting for the dependence of the ADM on many variables such as surface and cloud properties. Chang et al. (2000) demonstrated a strong dependence of the ADM on cloud optical depth and cloud phase and derived a set of ADMs as a function of these variables using the French ScaRaB satellite data. The most extensive ADMs are being developed for the fourth generation of the ERB mission (Wielicki et al. 1996), which is expected to reduce the ADM-related uncertainty to less than 1% (Loeb et al. 2002). While well-calibrated broadband radiation measurements are most desirable, such data are only available from non-continuous experimental missions such as the Nimbus series, ERBE and CERES. Between these missions, there exist significant gaps, e.g. 8 years between ERBE and CERES. To fill such gaps, non-calibrated narrowband imager data were often utilized to estimate broadband fluxes, thanks to the strong correlation between the two quantities (Minnis and Harrison 1984; Li and Leighton 1992; Li and Trishchenko 1999). Lack of on-board radiometric calibration is a major source of error in these estimates (Trishchenko and Li 1998), which can be overcome/reduced by comparing the estimates against those from calibrated imaging sensors that were launched in recent years, e.g. the Visible Infrared Scanner (VIRS) and Moderate Resolution Imaging Spectroradiometer (MODIS) (Minnis et al. 2002).

Since a scanning radiometer only measure radiances exiting from the entire atmosphere-surface system, surface and atmospheric radiation budgets cannot be directly determined from satellites. Considerable success has been achieved in the retrieval of the solar SRB from TOA reflected flux or albedo measurements, as reviewed by Schmetz (1989) and Pinker et al. (1995). The retrieving algorithms generally fall under three categories: empirical relationships (Fritz et al. 1964; Tarpley 1979; etc.), parameterized schemes (Gautier et al. 1980; Chou 1989; Li et al. 1993) and radiative transfer models (Möser and Raschke 1983; Pinker and Ewing 1985; Stuhlmann et al. 1990; Rossow and Zhang 1995; etc). The first satellite-based estimation of the SED was made by Hanson et al. (1967) over the United States for the spring of 1962. Global SED data sets of multiple years are now available from both operational meteorological satellites (Pinker and Laszlo 1992; Rossow and Zhang 1995) and radiation research satellites (Li and Leighton 1993). The global mean surface absorptance estimated from these satellite observations ranges from 0.46 to 0.50 (Table 1). For a planetary albedo of 0.30, the global mean atmospheric absorptance therefore varies from 0.20 to 0.24. To understand why such a discrepancy in atmospheric absorptance exists, Li (1995) examined the differences between two global SRB datasets derived from ISCCP using the algorithm of Pinker and Laszlo (1992) and from ERBE using the algorithm of Li et al. (1993). It was found that use of different input datasets leads to random discrepancies, while systematic discrepancies (global mean being 0.2 versus 0.24) were traced to the use of an outdated water vapor absorption scheme (Lacis and Hansen 1974) in Pinker and Laszlo (1992). Replacement of this old scheme with a new water vapor absorption scheme essentially removed the systematic difference. This finding is corroborated with an independent validation study (Li et al. 1995) that shows no systematic difference relative to global ground-based radiation measurement. Figure 1 presents a sound estimate of the SED according to ERBE satellite measurements of the ERB and satellite-based retrievals of the

SRB (Li and Leighton 1993) (Wielicki et al. 1995). It should be pointed out, however, that the SRB is likely subject to large errors for regional and instantaneous values due in part to a lack of corrections for some secondary factors that could become dominant, such as aerosols, in heavy biomass burning regions (Li 1998; Li and Kou 1998). Modifications have been made to improve the SRB retrievals by accounting more explicitly for the effects of aerosol, ozone, cloud droplet size (Masuda et al. 1995) and ice cloud variables (Zhang et al. 2002).

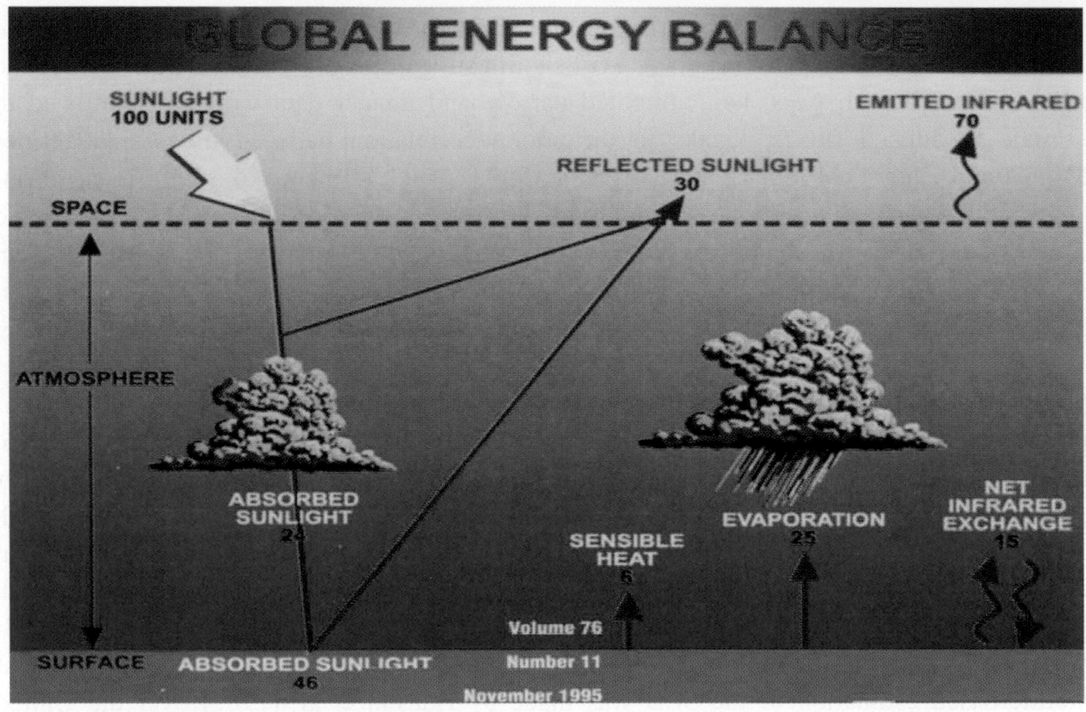

Figure1. The revised global energy balance per ERBE satellite observations of the radiation budget at the TOA and inference of the surface solar radiation budget by Li and Leighton (1993) (the cover of the *Bulletin of American Meteorological Society* from an article by Wielicki et al. (1995)).

2.3. *GCM Modeling*

In general circulation models (GCMs), the solar energy disposition is computed by fast radiative transfer codes with input parameters provided by the GCM. Since GCMs generally do not reproduce cloud properties well, barring more sound cloud amount and distribution, and since clouds are the most important factor in determining the SED, the SED from a GCM usually incurs larger uncertainties. However, the majority of GCMs may have been tuned to produce "reasonable" values for such highly averaged quantities as global and annual mean SED. In any event, a comparison of the SED from a model and from observations can help evaluate and improve the performance of the GCM. To evaluate the performance of a GCM, we need not only reliable observations of the SED, but also dependable estimates of the variables that influence the SED. The SED is mainly modified by clouds (fractional cover, thickness, height, microphysical parameters), water vapor (amount and vertical distribution), aerosols (amount, vertical profile, size distribution and optical properties), and surface albedo (including its spectral and angular dependencies). To date, many of these variables can be derived from satellite observations such as cloud from ISCCP (Rossow and Schiffer 1991) and surface albedo from ERBE (Li and Garand 1994). With these values, one is able to interpret the difference between modeled and observed SED in terms of the treatment of various physical processes and radiative transfer algorithms (Barker et al. 1994; Barker and Li 1995; Kiehl et al. 1994; Wild et al. 1995; Ward 1995; among others).

2.4. *Comparison of Observations, Inversions and Modeling*

Li et al. (1997) compared eight global radiation data sets: one based on surface observations that are extended globally using empirical relationships, three based on estimates from satellite measurements, and four from GCM simulations. Comparisons were made for global and annual means and zonal and monthly means, under both clear and all-sky conditions. Overall, the agreement at the top of the atmosphere is much better than at the surface and in the atmosphere. Global and annual mean TOA albedos generally agree to within 0.02, whereas atmospheric absorptances differ by over 0.1. In terms of the global and annual mean flux absorbed at the surface, the maximum difference is nearly 50 Wm^{-2}. More importantly, surface fluxes computed by models are usually larger than ground-based observations and satellite-based estimates. Satellite- and ground-based values agree well under most circumstances, except for regions affected by strongly absorbing aerosols. Since such an effect is limited to a portion of the continental areas, it does not alter significantly zonal and global mean solar radiation budgets.

The discrepancies between satellite-based estimations and model simulations are on the order of 20 Wm^{-2} to 25 Wm^{-2}, comparable to those found from direct comparisons between model simulations and surface observations. Therefore, two critical issues need to be addressed. First, it is necessary to narrow the large gap in our knowledge of the partitioning of the solar energy between the atmosphere and surface. This requires better and more consistent observations. Second, it is necessary to determine why there is a discrepancy between models and observations, if a discrepancy remains after the observations are better established. As the differences of similar magnitude also exist under clear-sky conditions (Li et al. 1997; Randall et al. 1992), it was argued that the model deficiency stems mainly from clear-sky calculations. The analyses of zonal comparisons further suggest that the use of dated schemes for water vapor absorption (Li 1995) and neglect of absorbing aerosols are the two major factors causing the under (over)-estimation of atmospheric (surface) absorption (Francis et al. 1997). This finding was challenged by claims of a substantial cloud absorption anomaly reported in the mid-1990s that was the subject of a major controversy in the atmospheric radiation community. Note that the CAA was not a new discovery but it was the renewed debate that caught the attention of the climate community at large due to its immense implications for climate modeling. The following section is devoted to the debate.

3. The Cloud Absorption Anomaly (CAA) Debate

3.1. *Historical Debate Prior to the 1990s*

The debate over the existence of a CAA has been ongoing for over half a century. An early discovery of a discrepancy between model calculated and observed cloud absorption was reported by Fritz (1951). Contradictory findings of disagreement (e.g. Robinson 1958; Reynolds et al. 1975; Foot 1988) and agreement (Slingo et al. 1982; Herman and Curry 1984; Hignett 1987; Rawlins 1989) between measured and calculated cloud absorption have since been reported from time to time. Stephens and Tsay (1990) reviewed all reported CAAs within the context of various proposed explanations. It was concluded that broadband absorption measurements were subject to such large uncertainties that some reported anomalies could not be established but an absorption anomaly in the near–infrared (NIR) appeared to be somewhat real. The large droplet theory proposed by Wiscombe et al. (1984) is too small to explain most reported CAA. Absorbing aerosols may contribute to the purported CAA, but would be limited primarily to the visible region where little evidence of a CAA has been reported. The effects of the continuum absorption by water vapor in the NIR and heterogeneous clouds may contribute to a CAA to some degree, but does not seem to be a leading or sole factor. Note that the majority of CAAs reported prior to the 1990s has a magnitude comparable to or smaller than uncertainties incurred in observation and modeling. As both observation techniques and modeling capability advance, some of the reported CAAs may no longer be valid. This is, however, unlikely the case for the renewed CAA debate since the mid-1990s, as reviewed below.

3.2. *Renewed Debate since the Mid-1990s*

The renewed debate was ignited by three studies claiming that the solar radiation absorbed by clouds is substantially underestimated by models (Cess et al. 1995; Ramanathan et al. 1995; Pilewskie and Valero 1995). Except for the study of Pilewskie and Valero (1995), which follows the traditional approach of using aircraft to directly measure cloud absorption, the other two studies deal with atmospheric column absorption using ground and TOA fluxes. Since the absorption bands of water vapor and water droplets are highly overlapping and are of similar magnitudes (Davies et al. 1984), total absorption in the atmospheric column is supposed to be rather insensitive to cloud, except for modification of its vertical distribution. Having a small effect on atmospheric absorption, however, does not imply that clouds absorb little solar radiation, only that whatever absorption occurs, the bulk of it is in place of clear-sky absorption (Stephens 1996). As a result, solar cloud radiative forcing (CRF) (the difference in net solar flux between cloudy and clear skies) at the surface is close to, or slightly larger than that at the TOA, according to our conventional wisdom of radiative transfer.

Using satellite and ground-based measurements made at a handful of tropical and mid-latitude sites, the studies of Ramanathan et al. (1995) and Pilewskie and Valero (1995) found that the ratio of surface to TOA CRF is not only significantly larger (~1.5) than unity, but also invariant in terms of global distribution. The amount of underestimation in solar radiation absorbed by clouds was found to be on the order of 25 Wm^{-2} (daily average), comparable to the average discrepancy between models and observations. If these findings were true, it would cause the biggest change in our understanding of the Earth's energy balance since the late 1960s, when emerging satellite observations helped revise the planetary albedo substantially (Wiscombe 1995). Chou et al. (1995) pointed out that to reach such a high ratio would require an increase in the single-scattering albedo of cloud droplets by a factor of 40 in the NIR, assuming no absorption in the visible region as supported by previous studies (Stephens and Tsay 1990) and also later by Li and Kou (1998). Since absorption in the NIR is proportional to cloud droplet size (King et al. 1990), the finding implies that either the fundamentals of radiative transfer theory originating from the Maxwell equations are seriously flawed, or cloud droplets must be huge; the latter would be at odds with the plethora of in-situ measurements.

Employing quality-controlled radiation data measured around the globe at the surface, together with calibrated ERBE satellite data, Li et al. (1995) reached a nearly opposite conclusion that the CRF ratio is essentially close to unity and varies with season and region (except for some tropical sites where the ratio reaches 1.5). Both the magnitude and variation were explained by extensive model simulation results (Li and Moreau 1996). The high values in certain parts of the tropics were later found to stem from heavy smoke aerosols generated by biomass burning where the strong absorption led to an erroneous estimation of clear-sky fluxes. After correcting the errors, Li (1998) obtained CRF ratios near 1. Under the assumption that clouds were as strongly absorbing as reported, Ackerman and Toon (1996) were unable to simulate the persistent existence of stratocumulus cloud along the west coast of the United States, simply because the excessive absorption "burns out" cloud droplets once they are formed.

Meanwhile, the analysis methods used to support and challenge the CAA were brought under scrutiny (Stephens 1996; Arking et al. 1996; Imre et al. 1996; Barker and Li 1997; Cess et al. 1997; Zhang et al. 1998). Among others, a major concern regarding methodology lies in whether one should use instantaneous or spatially and temporally averaged TOA albedo and surface transmittance data for regression analysis between the two variables. Note that the slope of the regression may serve as a proxy of cloud absorption under the assumption that no exchange of photons occurs horizontally. Instantaneous radiative flux measurements made at individual sites were employed by Cess et al. (1995), whereas monthly mean fluxes averaged over a month and over grid areas of 280 km^2 were used in Li et al. (1995) and Li and Moreau (1996).

To help unravel this issue, we derive here a simple conceptual relationship between the slope(s) and cloud absorption. By energy conservation,

$$R_{toa} + A_{atm} + I_{sfc}(1 - R_{sfc}) + T_{hor} = I_{toa} \tag{1}$$

where I_{toa} denotes the incoming solar radiation at the TOA which is distributed in four ways namely, reflection at the TOA, absorption in the atmosphere and at the surface, and horizontal transportation (positive if losing photons and negative if receiving photons). Radiative transfer modeling has shown that atmospheric absorption is a linear function of TOA albedo (Li et al. 1993):

$$A_{atm} = a + b\, R_{toa}\,. \tag{2}$$

Note that the coefficient b is usually very small, since atmospheric column absorption depends weakly on clouds. While no specific relation between T_{hor} and R_{toa} was established, it is known that T_{hor} is also positively correlated with R_{toa}, in that a region of more cloud deflects (net effect) photons toward an adjacent clear or less cloudy region. To the first order of approximation, we may assume the relation is linear,

$$T_{hor} = c + d\, R_{toa}. \tag{3}$$

Substituting Eqs. (2)-(3) into (1) and rearranging it, we can derive a linear relation between TOA reflected and surface transmitted fluxes:

$$R_{toa} = e - s\, I_{sfc} \tag{4}$$

where the slope is given by:

$$s = (1 - R_{sfc}) \,/\, (1 + b + d). \tag{5}$$

Note that in a plane-parallel model, the coefficient d vanishes to zero, i.e. no horizontal energy exchange is allowed, which is the assumption used in GCMs. Using typical values for R_{sfc} and b, s turn out to be around 0.8, as was found by Cess et al. (1995). In nature, however, there rarely exists a uniform cloud that has no net horizontal exchange of photons. So long as photon exchange takes place, d is a non-zero positive value leading to a slope that is bound to be smaller than that derived from plane-parallel simulations. This was confirmed by Monte Carlo simulations that can modulate the horizontal exchange of photon (Barker and Li 1997). For the same cloud, the slopes resulting from no photon exchange are always significantly larger than those that include the horizontal exchange of photons, despite the fact that the cloud possesses the same optical properties.

These arguments are illustrated more clearly in Fig. 2. The left panel demonstrates how reflection and transmission of solar energy are altered by the horizontal exchange of solar photons, while the right panel shows how the exchange affects the slope. For non-homogeneous clouds, a thicker cloud (or a portion of the cloud that is relatively thick) loses solar energy due to scattering by the cloud edge. As a result, its reflection and transmission are both reduced relative to a plane-parallel cloud of the same cloud thickness and containing the same microphysics. On the contrary, over clear or less cloudy regions, both reflection and transmission are enhanced by the incoming photons scattered by the neighboring thicker clouds. Consequently, the slope of the regression between TOA reflection and surface transmittance is lowered relative to the slope determined by a plane-parallel one-dimensional model, as used in GCMs. This is illustrated in the right panel. It is thus an ill posed comparison between the slopes derived from observations of instantaneous fluxes and GCM-modeled fluxes. On the other hand, the comparison would be meaningful if observations were averaged over a spatial and/or temporal domain that is large enough to effectively remove the effect of photon horizontal exchange. Use of monthly-mean fluxes averaged over the GCM grids may be sufficient in this regard (Li et al. 1995).

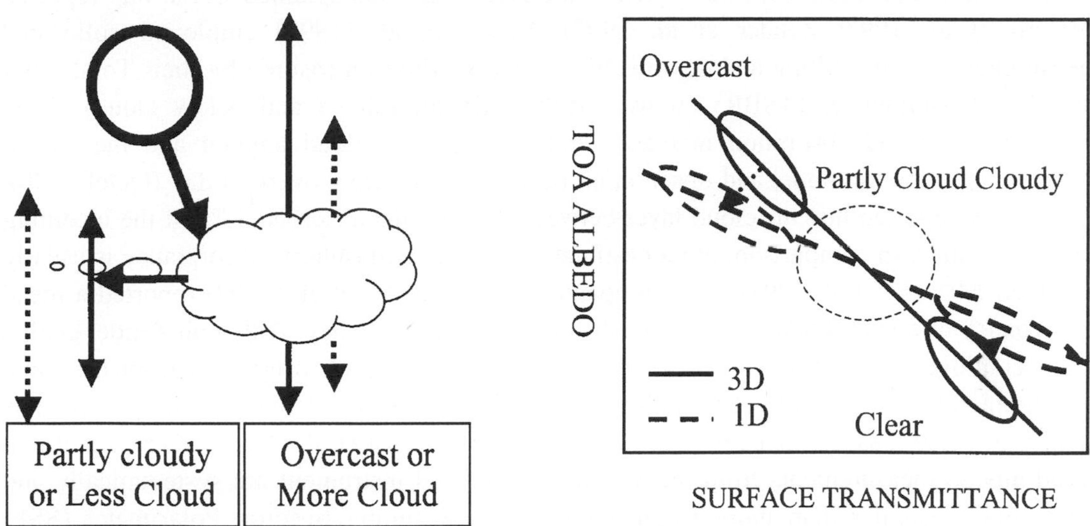

Figure 2. A schematic of the horizontal exchange of solar photons and its impact on the slope of the regression between TOA albedo and surface transmittance. The dashed and solid arrows in the left panel illustrates changes in the amounts of solar radiation reflected and transmitted with and without photon horizontal exchange. The right panel demonstrates resulting changes in albedo and transmittance.

Another closely related debate is concerned with analysis of measurements made by two aircraft flying in stack below and above a cloud (Marshak et al. 1997). Horizontal linkage of photons has also been blamed for apparent CAA in studies using aircraft data. This problem can be remedied by subtracting the apparent cloud absorption in the visible or any non-absorbing band following the original or modified Ackerman and Cox (1981) approach (Marshak et al. 1999). This apparent effect should not be mixed with the real effect of 3-D cloud in altering cloud absorption. Throughout the course of the debate, controversial findings on the magnitude of absorption enhancement have also been reported, ranging from very significant (Byrne et al. 1996), significant (at steep solar zenith angles only) (O'Hirok et al. 1998) to negligible (Markshak et al. 1998, Barker et al. 1998).

Another factor that may lead to the apparent CAA is the influence of aerosols. Li and Trishchenko (2001) demonstrated that absorbing aerosols usually exert much less influence on the TOA radiation budget than the surface radiation budget. The ensuing larger surface aerosol radiative forcing could be attributed to a surface cloud radiative forcing resulting from excessive cloud screening, leading to overestimation of the surface to TOA cloud radiative forcing ratio and thus the apparent CAA.

3.3. *Inconclusive Findings from the ARESE I*

To help resolve the CAA debate, the U.S. Department of Energy sponsored a field experiment in the fall of 1995 dubbed the ARM Enhanced Shortwave Experiment (ARESE), conducted around the ARM Southern Great Plains (SGP) Central Facility (CF) site in north-central Oklahoma. Following the ARESE, an even larger unexplained CAA was reported (Valero et al. 1997; Zender et al. 1997). Valero et al. (1997) employed collocated measurements of upwelling and downwelling radiative fluxes measured by their Total Solar Broadband Radiometer (TSBR) on two stacked aircraft (above and below cloud). They reported that cloud absorption increases dramatically with cloud amount and the largest discrepancy between model and observation occurred on a heavy overcast day (October 30), 1995. They reported that the cloud layer between the aircrafts absorbed 37% of the incoming solar irradiance. In comparison, model estimation of *total* atmospheric absorptance is usually less than 24% (Li et al. 1997). In a companion study, Zender et al. (1997) reported a mean discrepancy of about 100 Wm^{-2} in cloud absorption on the same day. Although Zender et al.'s study employed ground-based surface measurements, their CAA originated from the same measurements, namely, Valero et al.'s airborne measurements of the upwelling flux. The quality of these measurements was called into question by Li et al. (1999). They found that cloud albedo measurements from the Valero's broadband instrument are systematically and substantially smaller than those estimated from the Scanning Spectral Polarimeter (SSP) (Stephens et al. 2000), a well-calibrated airborne radiometer with an accuracy of ±3–5%.

An extensive consistency analysis was carried out by Li et al. (1999) using a variety of data sets acquired from various platforms including ground-based cloud radar data, microwave water vapor data, surface spectral and broadband radiation data, satellite data, airborne spectrometers and broadband radiometers. The spectral albedos were consistent with all other measurements, while Valero's broadband reflected flux data differed considerably from others. Valero et al. (2000) later admitted that their cloud albedo on this heavy overcast day was underestimated but not to the degree found by Li et al. (1999). They maintained that their aircraft data was generally sound and argued that their revised data agreed well with simultaneous GOES-8 satellite measurements. Note that the GOES-based broadband albedo values were estimated from GOES visible reflectances, following vicarious calibration and narrow-to-broadband conversion by Minnis and Smith (1998). Recently, Minnis et al. (2002) found that the calibration gain used in Minnis and Smith (1998) is underestimated by 9% relative to measurements from on-board calibrated sensors of the VIRS and MODIS. This finding reinforces the argument of Li et al. (1999) that the large CAA resulting from the ARESE by Valero et al. (1997) and Zender et al. (1997) is an artifact originating from erroneous observations. The error in the calibration gain also explains why the discrepancy decreases with cloud amount as noted by Li et al. (1999). It is also plausible that the calibration error could have contributed to the CAA finding by Cess et al. (1995) who employed similar GOES-based TOA albedos.

3.4. *Conclusive Findings from ARESE II*

The controversial findings from the first ARESE motivated the second experiment conducted in the spring of 2000. To solve the calibration conundrum resulting from ARESE I, ARESE II devoted much effort into instrument calibration and intercomparison (Michalsky et al. 2001). Several sets of spectral and broadband radiometers were deployed and intercompared and in general, all agreed to within 10 Wm^{-2}. Independent investigations (Li et. al. 2002; Ackerman and Stokes 2003) did not find a cloud absorption anomaly anywhere close to that claimed before, although a small discrepancy on the order of 20 Wm^{-2} (instantaneous value) exists. While one cannot negate a CAA of smaller magnitude, if exists, it must be viewed within the framework of various uncertainties due to measurement uncertainties (~10 Wm^{-2}), mismatch between ground and aircraft observations (<20 Wm^{-2}), model calculations (<20 Wm^{-2}), and errors in input data of unknown magnitude.

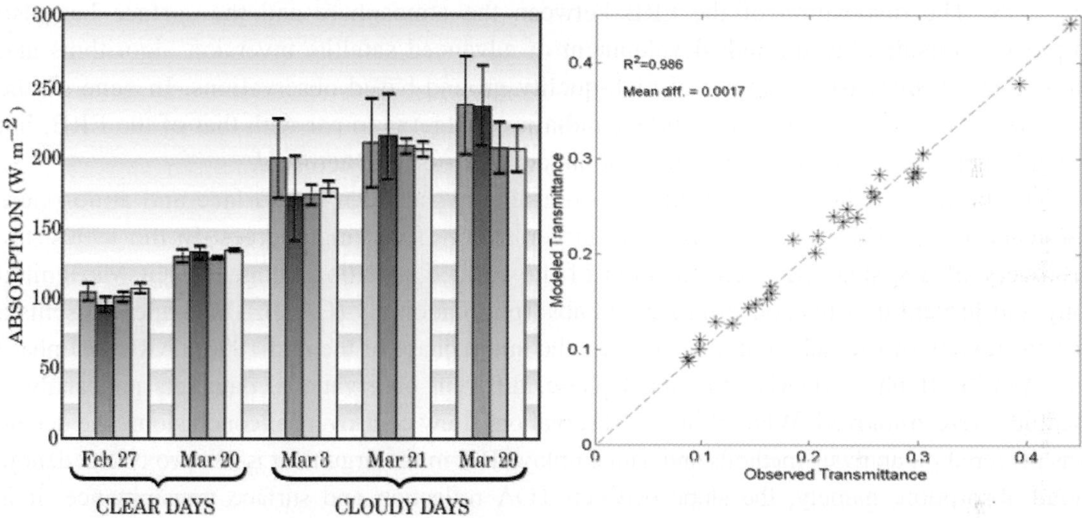

Figure 3 The left panel shows comparisons of atmospheric column absorption from the ground up to 7 km measured by ground and aircraft observations (blue and red) and computed by models (green and yellow) on two clear days and three cloudy days in 2002 (after Ackerman and Stokes, 2003. The right panel shows a comparison between observed and modeled broadband atmospheric transmittance using areal-mean surface albedo (after Li et al. 2002).

Li et al. (2002) found much of the discrepancy in the NIR may be caused by surface heterogeneity surrounding the ARM CF. The ARESE II was conducted over the ARM CF (dry grass and bare soil) where the surface albedo is systematically smaller than that over the surrounding area which was primarily covered by winter wheat. Over the large field of view seen by the downward radiometer, atmospheric transmittance in the NIR as measured on the ground is enhanced substantially by multiple scattering of photons between the "bright" surface (in NIR region) and the cloud base. As a result, use of local surface albedo measured on the ground would systematically underestimate cloud absorption, which is a likely cause for the significant CAA in NIR as reported by O'Hirok et al. (2000) using ARESE I data.

Using an areal-mean albedo, near perfect agreement is reached for both broadband (Fig.3) and spectral comparisons (*Li et al.* 2002). Ackerman and Stokes (2003 thus declared that the CAA is now largely behind us. In parallel to the controversial findings related to the ARESE experiments, a number of other studies using data from different sources reach essentially the same conclusion (e.g. Francis et al. 1997; Chou et al. 1998).

4. Concluding Remarks

This paper reviews the state-of-the-art in solar radiation budget studies. Thanks to satellite observations, we have a rather accurate knowledge of the earth radiation budget at the top of the atmosphere. Prior to the EOS era, the monthly mean ERB at the GCM grid level was known to within 5 Wm^{-2}, but much larger errors existed for instantaneous and daily mean quantities. These errors are lowered by the new generation of ERB sensor (CERES) aboard the EOS. The disposition of the ERB between the atmosphere and the surface has also improved considerably through development of advanced satellite inversion algorithms and backed by an increasing number of high-quality ground-based observations. In general, the accuracy of the satellite-retrieved surface radiation budget is in par with that of the ERB, but much larger errors may occur in regions abundant of absorbing aerosols.

On the other hand, the disposition of solar energy between the surface and atmosphere simulated by GCMs usually suffers from larger errors of varying degrees. In the 1990s, the discovery of a systematic overestimation of the surface radiation budget by GCMs ignited long and fervent debates concerning cloud absorption anomaly (CAA). This paper presents a critical review of the debate in three phases: the initial phase of the mid-1990s, ARESE I phase and ARESE II phase. During the initial phase, different observational data sets and analysis methods were employed. While different observations drew controversial conclusions, the debate was centered on analysis methods and data employed. A major argument is if a proxy can denote cloud absorption, namely, the slope between TOA reflection and surface transmittance. It is shown that the ubiquitous presence of the horizontal exchange of solar photons always leads to a larger slope from models than from observations. In the ARESE I phase, the quality of the measurements used in studying the CAA came under scrutiny. Key measurements used to support the finding of a large CAA were found to be in serious error. A consensus on the CAA was finally reached during ARESE II, namely that there exists no CAA within the context of observation and modeling uncertainties.

The non-existence of the CAA in the magnitude as claimed before does not mean GCMs are free from systematic errors in simulating the solar radiation budget. Many GCMs are subject to numerous errors affecting the ERB, most notably, use of outdated radiative transfer codes and spectroscopic data base from which parameterization schemes are derived, a lack of or inadequate treatments of aerosol absorption, surface albedo, water vapor, etc. Presumbly inspired by the CAA debate, GCMs as a whole now have stronger absorption than before that are more consistent with observations. However, the enhancement stems from various factors but clouds (Barker and Li 1995, Francis et al. 1997, among others).

Acknowledgement. The study is supported by the US Department of Energy's Atmospheric Radiation Measurement grant #DE-FG02-01ER63166, and partly by the National Science Foundation of China (40028503). The author is grateful to the co-authors of his papers quoted here, in particular, Dr. A. Trishchenko, H.W. Barker, L. Moreau, and M. C. Cribb.

References:

Ackerman, S.A. and S.K. Cox, 1981: Aircraft observation of the shortwave fractional absorptance of non-homogeneous clouds, *J. Appl. Meteor*., **20**, 1510-1515,

Ackerman, A.S., and O.B. Toon, 1996: Unrealistic desiccation of marine stratocumulus clouds by enhanced cloud absorption, *Nature*, **380**, 512-517.

Ackerman, T.P., and G.M. Stokes, 2003, The Atmospheric Radiation Measurement program, *Physics Today*, January, 38-44.

Arking, Albert. 1999: Bringing climate models into agreement with observations of atmospheric absorption. *J. Clim.* **12**, 1589-1600.

Arking, A., M.-D. Chou, and W.L. Ridgway, 1996: On estimating the effects of clouds on atmospheric absorption based on flux observations above and below cloud level, *Geophy. Res. Let*, **23,** 829-832.

Arking, A., 1991: The radiative effects of clouds and their impact on climate, *Bull Amer. Meteor. Soc.*, **72,** 795-813.

Arking, A., and S. Vemury, 1984: The Nimbus 7 ERB data set: A critical analysis, *J. Geophys. Res.*, **89,** 5089-5097.

Arking, A., and J. S. Levine, 1967: Earth albedo measurements: July 1963 to June 1964, *J. Atmos. Sci.*, **24,** 721-724.

Abbot, C.G., and F.E. Fowle, 1908: in *Annals of the Astrophysical Observatory of the Smithsonian Institution*, Vol. **2,** Smithsonian Institution, Washington, D.C.

Barker, H., Z. Li, and J.-P. Blanchet, 1994: Radiative characteristics of the Canadian Climate Centre second-generation general circulation model, *J. Climate*. **7:** 1070-1091.

Barker, H.W., and Z. Li, 1995: Improved simulation of clear-sky shortwave radiative transfer in the CCC GCM, *J. Climate*, **8:** 2213-2223.

Barker, H.W., and Z Li. 1997: Interpreting shortwave albedo-transmittance plots: True or apparent anomalous absorption. *Geophy. Res. Let.* **24**, 2023-2026.

Barker, H.W., J.-J. Morcrette, and G.D. Alexander, 1998: Broadband solar fluxes and heating rates for atmospheres with 3 clouds, *Q. J. R. Meteor. Soc.,* **124**, 1245-1271.

Barkstrom, B., E. Harrison, G. Smith, R. Green, J. Kibler, R.D. Cess, and the ERB Science Team, 1989: Earth radiation budget experiment (ERBE) archival and April 1985 results, *Bull. Amer. Meteor. Soc.*, **70,** 1254-1262.

Budyko, M.I., 1956: Heat balance of the Earth's surface, 1958 translated in English, U.S. Weather Bureau, Dept. Commerce, Washington, D.C.

Budyko, M.I., 1982: *The Earth's Climate: Past and Future*, Academic Press, 307 pp.

Byrne, RN, RC J Somerville, and B Subasilar. 1996: Broken-cloud enhancement of solar radiation absorption." *J. Atmos. Sci.* **53**,878 - 886.

Cess.R.D., G.L. Potter, and co-authors, 1989: Interpretation of cloud-climate feedbacks as produced by 14 atmospheric general circulation models, *Science*, **245,** 513-516.

Cess. R.D., M.H. Zhang, and co-authors, 1995: Absorption of solar radiation by clouds: Observations versus models. *Science*, **267**, 496-499.

Cess, RD, MH Zhang, and co-Authors, 1996: Cloud feedback in atmospheric general circulation models: An update. *J. Geophys. Res.* **101**,12791-12794.

Cess, R, F Valero, and co-authors. 1997. Absorption of solar radiation by the cloudy atmosphere: Interpretations of collocated aircraft measurements. *J. Geophys. Res.* **102**, 29,917-29,927.

Chang, F-L, Z Li, and A Trishchenko, 2000: The dependence of TOA anisotropic reflection on cloud properties inferred from ScaRaB satellite data. *J. Appl. Meteor.* **39**, 2480–2493.

Chou, M.-D., A. Arking, J. Otterman, W.L. Ridgway, 1995: The effect of clouds on atmospheric absorption of solar radiation, *Geophy. Res. Letters*, **22,** 1885-1888.

Chou, M.-D., 1989: On the estimation of surface radiation using satellite data. *Theor.Appl. Climat.*, **40**,25-36.

Chou, M.-D., W. Zhao, S.-H. Chou, 1998: Radiation budgets and cloud radiative forcing in the Pacific warm pool during TOGA COARE, *J. Geophy. Res.*, **103**, 16,967-16,977.

Davies, R., W.L. Ridgway, and K.E. Kim, 1984: Spectral absorption of solar radiation in cloudy atmosphere: A 20-cm^{-1} model. *J. Atmos. Sci.*, **4,** 2126-2137.

Dutton, E.G., J.J. Michalsky, and co-authors. 2001: Measurement of broadband diffuse solar irradiance using current commercial instrumentation with a correction for thermal offset errors. *J. Atmos. and Ocean. Tech.* **18**, 297–314.

Foot, J.S., 1988: Some observations of the optical properties of clouds. Part I: Stratocumulus. *Quart. J. Roy. Meteor. Soc.*, **114**, 129-144.

Francis, P.N., J.P. Taylor, P. Hignett, and A. Slingo, 1997: On the question of enhanced absorption of solar radiation by clouds, *Q. J. Meteor. Soc.*, **123**, 419-434.

Fritz, S., 1951: Solar radiant energy, Compendium of Meteorology, T.F. Malone, Ed. Wiley and Sons, 14-29.

Fritz, S., P. Rao and M. Weinstein, 1964: Satellite measurements of reflected solar energy and energy received at the ground. *J. Atmos. Sci.*, **21**,141-151.

Gauthier, C., G. Diak and S. Masse, 1980: A simple physical model to estimate incident solar radiation at the surface from GOES satellite data. *J. Appl. Meteor.*, **19**,1005-1012.

Hartmann, D.L., V. Ramanathan, A. Berroir, and G.E. Hunt, 1986: Earth radiation budget data and climate research, *Rev. Geophy.*, **24,** 439-468.

Hanson, K.J., T.H. Vonder Haar, and V.E. Suomi, 1967: Reflection of sunlight to space and absorption by the earth and atmosphere over the United States during spring 1962. J. *Atmos. Sci.*, **95,** 354-362.

Herman, G.F., and J.A. Curry, 1984: Observational and theoretical studies of solar radiation in Arctic stratus clouds, *J. Climate Appl. Meteor.*, **23**, 5-24.

Hignett, P., 1987: A study of the short-wave radiative properties of marine stratus: Aircraft measurements and model comparisons, *Quart. J. Roy. Meteor. Soc.*, **113**, 1011-1024.

House, F.B., A. Gruber, G.E. Hunt, and A.T. Mecherikunnel, 1986: History of satellite missions and measurements of the earth radiation budget (1957-1984), *Rev. Geophy.*, **24,** 357-377

Houghton, H.G., 1954: On the annual heat balance of the northern hemisphere, *J. Meteor.*, **11,** 1-9.

Hunt, G.E., R. Kandel, and A.T. Mecherikunnel, 1986: A history of pre-satellite investigations of the earth's radiation budget, *Rev. Geophy.*, **24,** 351-356.

Imre, D.G., E.H. Abramson, and P.H. Daum, 1996: Quantifying cloud-induced shortwave absorption: An examination of uncertainties and of recent arguments for large excess absorption, *J. Appl. Meteor.*, **35**, 1991-2010.

Jacobowitz, H., H.V. Soule, H.L. Kyle, F.B. House, and the NIMBUS 7 ERB Experiment Team, 1984: The Earth Radiation Budget (ERB) Experiment: An overview, *J. Geophy. Res.*, **89,** 5021-5038.

Kiehl, JT, JJ Hack, and BP Briegleb. 1994. The simulated earth radiation budget of the National Center for Atmospheric Research Community Climate Model CCM2 and Comparisons with the Earth Radiation Budget Experiment (ERBE). *J. Geophys. Res.* **99**, 20,815-20,827.

Kiehl, J.T., J.J. Hack, M.H. Zhang, and R.D. Cess, 1995: Sensitivity of a GCM climate to enhanced shortwave cloud absorption, *J. Climate*, **8,** 2200-2212.

King, M.D., L.F. Radke, and P.V. Hobbs, 1990: Determination of the spectral absorption of solar radiation by marine stratocumulus clouds from airborne measurements within clouds, *J. Atmos. Sci.*, **47**, 894-907.

Lacis, A.A., and J.E. Hansen, 1974, A parameterization for the absorption of solar radiation in the Earth's atmosphere, *J. Atmos. Sci.,* 31, 118-132, 1974.

Li, Z., and H.G. Leighton, 1992: Narrowband to broadband conversion with spatially autocorrelated reflectance measurements, *J. Appl. Meteor.*, **31,** 421-432.

Li, Z., H.G. Leighton, 1993: Global climatologies of solar radiation budgets at the surface and in the atmosphere from 5 years of ERBE data, *J. Geophy. Res.*, **98,** 4919-4930.

Li, Z., H.G. Leighton, K. Masuda, and T. Takashima, 1993: Estimation of SW flux absorbed at the surface from TOA reflected flux, *J. Climate*, **6,** 317-330.

Li, Z., and L. Garand, 1994: Estimation of surface albedo from space: A parameterization for global application, *J. Geophy. Res.*, **99,** 8335-8350.

Li, Z., 1995: Intercomparison between two satellite-based products of net surface shortwave radiation, *J. Geophy. Res.*, **100**, 3221-3232.

Li, Z, T. Charlock, and C. Whitlock, 1995: Assessment of the global monthly mean surface insolation estimated from satellite measurements using global energy balance archive data, *J. Climate*, **8**, 315-328.

Li, Z., H. Barker, and L. Moreau, 1995: The variable effect of clouds on atmospheric absorption of solar radiation, *Nature*, **376,** 486-490.

Li, Z., and L. Moreau, 1996: Alteration of atmospheric solar absorption by clouds: Simulation and observation, *J. Appl. Meteor.*, **35,** 653-670.

Li, Z., 1996: On the angular correction of satellite-based radiation data: An evaluation of the performance of ERBE ADM in the Arctic, *J. Theor. Appl. Climat.* **54,** 235-248.

Li, Z., L. Moreau, A. Arking, 1997: On solar energy disposition: a perspective from observation and modeling, *Bull. Amer. Meteor. Soc.,* **78,** 53-70

Li, Z., 1998, Influence of absorbing aerosols on the inference of solar surface radiation budget and cloud absorption, *J. Climate,* **11,** 5-17.

Li, Z., and L. Kou, 1998, Atmospheric direct radiative forcing by smoke aerosols determined from satellite and surface measurements, *Tellus (B)*, **50**, 543-554.

Li, Z., A. Trishchenko, H.W. Barker, G.L. Stephens, P.T. Partain, 1999: Analysis of Atmospheric Radiation Measurement (ARM) programs's Enhanced Shortwave Experiment (ARESE) multiple data sets for studying cloud absorption, *J. Geophy. Res.* **104**, 19127-19134.

Li, Z., and A. Trishchenko, 1999, A study towards an improved understanding of the relationship between visible and SW albedo measurements, *J. Atmos. & Ocean. Tech.* **16**, 347-360.

Li, Z., A. Trishchenko, 2001, Quantifying the uncertainties in determining SW cloud radiative forcing and cloud absorption due to variability in atmospheric condition, *J. Atmos. Sci.*, **58**, 376-389.

Li, Z., M. Cribb, A. Trishchenko, 2002, Impact of surface inhomogeneity on solar radiative transfer under overcast conditions, *J. Geophy. Res.*, **107**, 10.1029/2001JD000976.

Loeb, N.G., N.M. Smith, and co-authors, 2002: Angular distribution models for top-of-atmosphere radiative flux estimation from the clouds and the earth's radiant energy system instrument on the tropical rainfall measuring mission satellite, Part I, methodology, *J. Appl. Meteor.*, in press.

London, J., 1957: A study of the atmospheric heat balance, *Final Report*, AFC-TR-57-287, OTSPB129551, 99pp., New York Univ., New York.

Masuda, K., H.G. Leighton, and Z. Li, 1995: A new parameterization for the determination of solar flux absorbed at the surface from satellite measurements, *J. Climate.*, **8,** 1615-1629.

Marshak, A, A Davis, W Wiscombe, and R Cahalan. 1997: Inhomogeneity effects on cloud shortwave absorption: Two-aircraft simulations. *J. Geophys. Res.* **102**, 619-16637.

Marshak, A, A Davis, W Wiscombe, W Ridgway, and R Cahalan. 1998: Biases in shortwave column absorption in the presence of fractal clouds. *J. Climate*, **11**, 431–446.

Marshak, A, W.J. Wiscombe, A.B. Davis, L. Oreopoulos, and R.F. Cahalan. 1999: On the removal of the effect of horizontal fluxes in two-aircraft measurements of cloud absorption. *Quart. J. Roy. Meteor. Soc.* **125**, 2153-2170.

Michalsky, J, E Dutton, and co-authors. 1999: Optimal measurement of surface shortwave irradiance using current instrumentation. *J. Atmos. & Oceanic Tech.* 16: 55–69.

Michalsky, J., et al., 2001: Broadband and spectral shortwave calibration results from ARESE II, paper presented at 11th Annual ARM Science Team Meeting, Dep. of Energy, Atlanta, Georgia, March 19-23, 2001. [Available on-line: www.arm.gov].

Minnis, P., and E.F. Harrison, 1984: Diurnal variability of regional cloud and clear-sky radiative parameters derived from GOES data, III: November 1979 radiative parameters, *J. Climate and Appl. Meteor.*, **23**, 1032-1050.

Minnis, P., W. L. Smith Jr. 1998: Cloud and radiative fields derived from GOES-8 during SUCCESS and the ARM-UAV Spring 1996 Flight Series, *Geophys. Res. Lett.*, **25**, 1113-1116.

Minnis, P. L. Nguyen, D. Doelling, D. Young, W. Miller, D. Kratz, 2002: Rapid calibration of operational and research meteorological satellite imagers. Part I: Evaluation of research satellite visible channels as references. *J. Atmos. & Ocean. Tech.*, **19**, 1233-1249.

Möser, W., and E. Raschke, 1983: Mapping of global radiation and of cloudiness from METEOSAT image data. *Meteor Rundsch.*, **36,** 33-41.

O'Hirok, W, and C Gautier. 1998. A three-dimensional radiative transfer model to investigate the solar radiation within a cloudy atmosphere Part II: Spectral effects. *J. Atmos. Sci.* **55**, 3065–3076.

O'Hirok, W., C. Gautier, and P. Ricchiazzi. 2000: Spectral signature of column solar radiation absorption during the Atmospheric Radiation Measurement Enhanced Shortwave Experiment (ARESE), *J. Geophys. Res.* **105**, 17,471 - 17,480.

Ohmura, A., and H. Gilgen, 1991: *Global Energy Balance Archive (GEBA)*, World Climate Program - Water Project A7, Rep.2: The GEBA Database, Interactive Application, Retrieving Data. Zurich, Verlag der Fachvereine, 60 pp.

Ohmura, A., and H. Gilgen, 1993: Re-evaluation of the global energy balance. *Geophy. Monogra. 75.*, IUGG Vol. 15, 93-110, 1993.

Pilewskie, P., and F. Valero, 1995: Direct observations of excess solar absorption by clouds. *Science*, **257,** 1626-1629.

Pinker, R.T., R. Frouin, and Z. Li, 1995: A review of satellite methods to derive surface shortwave irradiance, *Remote Sens. Environ.*, **51,** 108-124.

Pinker, R.T., and I. Laszlo, 1992: Modelling surface solar irradiance for satellite applications on a global scale. *J. Appl. Meteor.*, **31,** 194-211

Pinker, R.T., and J.A. Ewing, 1985: Modelling surface solar radiation: model formulation and validation. *J. Climate Appl. Meteor.*, **24**,389-401.

Randall, DA, RD Cess, and co-Authors, 1992: Intercomparison and interpretation of surface energy fluxes in atmospheric general circulation models. *J. Geophys. Res.* **97,** 3711-3724.

Ramanathan, V., B. Subasilar, and co-authors, 1995: Warm pool heat budget and shortwave cloud forcing: A missing physics, *Science*, **267,** 499-503.

Ramanathan, V ,1987: The role of earth radiation budget studies in climate and general circulation research, *J. Geophy. Res.*, **92,** 4075-4095.

Randall, D.A, R.D. Cess, and co-authors, 1992: Intercomparison and interpretation of surface energy fluxes in atmospheric general circulation models, *J. Geophy. Res.*, **97,** 3711-3724.

Raschke, E., T.H. Vonder Haar, W.R. Bandeen, and M. Pasternak, 1973: The annual radiation balance of the earth-atmosphere system during 1969-70 from Nimbus 3 measurements, *J. Atmos. Sci.*, **30,** 341-364.

Rawlins, F., 1989: Aircraft measurements of the solar absorption by broken cloud fields: A case study, *Quart. J. Roy. Meteor. Soc.*, **115**, 365-382.

Robinson, 1958, G.D., 1958: Some observations from aircraft of surface albedo and the albedo and absorption of cloud, *Arch. Meteor. Geophy. Bioklimatol.*, B9, 28-41.

Reynolds D.W., T.H. Vonder Harr, and S.K. Cox, 1975: The effect of solar radiation absorption in the tropical atmosphere*, J. Appl. Meteor.*, **14,** 433-444.

Rossow, W.B., and Y.-C. Zhang, 1995: Calculation of surface and top of atmosphere radiative fluxes from physical quantities based on ISCCP data sets, 2. Validation and first results, *J. Geophys. Res.,* **97,** 1167-1197.

Rossow, W.B., and R.A. Schiffer, 1991: ISCCP cloud data products, *Bull. Am. Meteor. Soc.*, **72,** 2-20.

Sasamori, T., J. London, and D.V. Hoyt, 1972: Radiation budget of the southern hemisphere, *Meteor. Monogr.*, **13.**

Schmetz, J., 1989: Towards a surface radiation climatology: Retrieval of downward irradiance from satellite, *Atmos. Res.*, **23,** 287-321.

Simpson, G.C., 1929: The distribution of terrestrial radiation, *Mem. R. Meteor. Soc.*, **3,** 36-41.

Slingo, A., S. Nicholls, and J. Schmetz, 1982: Aircraft observations of marine stratocumulus during JASIN. *Quart. J. Roy. Meteor. Soc.*, **108**, 833-856.

Stephens, G.L., G.G. Campbell, and T.H. Vonder Haar, 1981: Earth radiation budgets, *J. Geophy. Res.*, **86,** 9739-9760.

Stephens, G.L., and S.-C. Tsay, 1990: On the cloud absorption anomaly, *Quart. J. Roy. Meteor. Soc.*, **116,** 671-704.

Stephens, G.L., and T.J. Greenwald, 1991: The earth's radiation budget and its relation to atmospheric hydrology. 2. Observations of cloud effects, *J. Geophy. Res.*, **96,** 15,325-15340.

Stephens, G.L., 1996: How much solar radiation do clouds absorb?, *Science*, **271,** 1131-1133.

Stephens, G.L., R.F. McCoy Jr, R.B. McCoy, P.G., P.T. Partain, and S.D. Miller. 2000. A multipurpose Scanning Spectral Polarimeter (SSP): Instrument description and sample results. *J. Atmos. & Oceanic Tech.*, **17**, 616–627.

Stuhlmann, R., M. Rieland and E. Raschke, 1990: An improvement of the IGMK model to derive total and diffuse solar radiation at the surface from satellite data. *J. Appl. Meteor.*, **29**,586-603.

Stuhlmann, R., and E. Raschke, 1987: Satellite measurements of the earth radiation budget: sampling and retrieval of shortwave exitances - a sampling study, *Beitr. Phys. Atmos.*, **60,** 393-410.

Suttles, J.T., Green, R.N., Minnis, P., and co-authors, 1988: Angular radiation models for Earth-atmosphere system. Vol. 1- shortwave radiation. *NASA Refer. Publ.*, **1184**, 114 pp.

Suttles, J.T., B.A. Wielicki, and S. Vemury, 1992: Top-of-atmosphere radiation fluxes: validation of ERBE scanner inversion algorithm using NIMBUS-7 ERB data, *J. Appl. Meteor.*, **31,** 784-796.

Suttles, J.T., and G. Ohring, 1986: Surface radiation budget for climate applications. *NASA Reference Publication* **1169,** 132 pp.

Tarpley, J.D., 1979: Estimating incident solar radiation at the surface from geostationary satellite data. *J. Appl. Meteor.*, **18,**1172-1181.

Taylor, V.R., and L.L. Stowe, 1984: Reflectance characteristics of uniform earth and cloud surfaces derived from Nimbus-7 ERB, *J. Geophy. Res.*, **89**, 4987-4996.

Trishchenko, Alexander, and Zhanqing Li. 1998: Use of ScaRaB measurements for validating a GOES-based TOA radiation product. *J. Appl. Meteor.* **37**: 591–605.

Valero, FPJ, RD Cess, and co-authors, 1997: Absorption of Solar Radiation by Clouds: Interpretations of Collocated Aircraft Measurement. *J. Geophys. Res*. **104**, 29,917-29,927.

Valero, FP J, P Minnis, and co-authors 2000: The absorption of solar radiation by the atmosphere as determined using consistent satellite, aircraft, and surface data during the ARM Enhanced Short-Wave Experiment (ARESE). *J. Geophys. Res*. **105**, 4743-4758.

Vonder Haar, T.H., and V.E. Suomi, 1971: Measurements of the earth's radiation budget from satellites during a five-year period. Part I: extended time and space means, *J. Atmos. Sci.*, **28,** 305-314.

Ward, D.M., 1995: Comparison of the surface solar radiation budget derived from satellite data with that simulated by the NCAR GCM2, *J. Climate*, **8,** 2824-2842.

Wielicki, B.A., R.D. Cess, M.D. King, D.A. Randall, and E.F. Harrison, 1995: Mission to planet Earth: Role of clouds and radiation in climate, *Bull. Amer. Meteor. Soc.*, **76,** 2125-2153.

Wielicki, B.A., B.R. Barkstrom, and co-authors, 1996: Clouds and the Earth's Radiative Energy System (CERES): An earth observing system experiment. *Bull. Amer. Meteor. Soc.*, **76**, 2125-2153.

Wild, M., A. Ohmura, H. Gilgen, and E. Roeckner, 1995: Validation of general circulation model radiative fluxes using surface observations. *J. Climate*, **8,** 1309-1324.

Wiscombe, W.J., R.M. Welch, and W.D. Hall, 1984: The effects of very large drops on cloud absorption, Part I: Parcel models. *J. Atmos. Sci.*, **41**, 1336-1355.

Wiscombe, W.J., 1995: Atmospheric physics: An absorption mystery. *Nature*, **376**, 466-467.

Zender, C. S., B. Bush, and co-authors, 1997: Atmospheric absorption during the Atmospheric Radiation Measurement (ARM) Enhanced Shortwave Experiment (ARESE), *J. Geophys. Res.* **102**, 29,901-29,916.

Zhang, M.H., W.Y. Lin, and J.T. Kiehl, 1998: Bias of atmospheric shortwave absorption in the NCAR CCM2 and CCM3: Comparison with monthly ERBE/GEBA measurements, *J. Geophys. Res.* **103**, 8919-8925.

Zhang, Y., Z. Li, A. Macke, 2002, Retrieval of surface solar radiation budget under ice cloud sky: uncertainty analysis and parameterization, *J. Atmos. Sci.*, **59**, 2951-2965.

DETECTING SOLAR ULTRAVIOLET IMPACT ON STRATOSPHERIC OZONE

SHUNTAI ZHOU
NOAA/NCEP/Climate Prediction Center
5200 Auth Road, Camp Springs, MD, USA
E-mail: shuntai.zhou@noaa.gov

(Manuscript received 28 October 2002)

Solar cycle effects on stratospheric ozone mixing ratio and total column ozone are studied using 23 years (1979-2001) of Solar Backscatter Ultraviolet (SBUV) data measured by Nimbus-7 and the National Oceanic and Atmospheric Administration (NOAA) satellites. It is shown that the solar cycle has the strongest signal in the upper stratospheric ozone variation, and that the ozone sensitivity to the solar cycle is comparable to that derived from the 27-day solar rotational effect. Stratospheric ozone response to the 27-day solar ultraviolet variation is also studied in three 1000-day periods with strong, moderate and weak solar ultraviolet variations respectively. The temperature effect on ozone variations is removed (or separated) by partial correlation and multiple regression methods. The correlation coefficients, ozone sensitivities and coherency squares are compared for the different periods and for cases with or without temperature effect.

1. Introduction

It is important to understand and quantify solar ultraviolet (UV) impact on stratospheric ozone variability, because solar UV flux-induced ozone variability is one of the fundamental mechanisms of solar-terrestrial connections, which has large influences on stratospheric temperature, radiative transfer, general circulation, and probably the Earth's climate (Haigh 1996; Shindell et al. 1999). In particular, because the maximum percent ozone change induced by solar UV occurs in the same altitude range as that caused by man-made ozone destroying chemicals (Wuebbles 1983; Herman and McQuillan 1985), the long-term solar UV effect must be taken into account in the assessment of anthropogenic effects on the global climate change.

The 11-year solar cycle and the 27-day solar rotational effects on stratospheric ozone have been extensively studied in the past, by using satellite data with various statistical techniques and sophisticated dynamical-radiative-chemical models (Gille et al. 1984; Garcia et al. 1984; Keating et al. 1985; Chandra 1986; Hood 1986; Lean 1987; Hood et al. 1993; Brasseur 1993; Huang and Brasseur 1993; Chandra and McPeters 1994; McCormack and Hood 1996; Hood and Zhou 1998). Because continuous operational satellite ozone measurements began in the end of 1970's, research on the solar cycle impact on ozone is largely limited by the relatively short dataset. The recent studies of solar cycle effect on stratospheric ozone profile and total ozone were based on 15~16 years (1979-1994) of Solar Backscatter Ultraviolet (SBUV) ozone data (Chandra and McPeters 1994; McCormack and Hood 1996; Hood and Soukharev 2000). For instance, Chandra and McPeters (1994) calculated ozone sensitivity to solar cycle variations (percent change in ozone per 1% change in solar UV) for the total column ozone and ozone mixing ratio from 0.5 hPa to 10 hPa. Their results are generally larger than the observed 27-day ozone sensitivity and model predictions by a factor of two. It is not clear whether the ozone response to the long-term solar change differs significantly from the response to the short-term solar change, or the calculated long-term sensitivity was biased due to the short dataset which is a little more than one solar cycle. Since then, the SBUV(/2) data have been extended and reprocessed. It is therefore desirable to take a further look at the solar cycle effect on ozone using the longer reprocessed dataset. In the next section, the solar cycle impact on total ozone and ozone profiles will be revisited based on a cohesive multiple-satellite ozone dataset.

The 27-day solar UV modulation on ozone has attracted much more attention because the magnitude of the solar UV variability on this short time scale is comparable to that of solar cycle variations and the dataset is long enough for studying the solar rotational effect. Although the ozone response to the solar rotational effect may be a good analogy to that of the solar cycle effect, the problem with the short-term solar UV impact is that ozone variations are also affected by atmospheric dynamical changes with similar time scales. Actually, stratospheric ozone is more sensitive to dynamically induced changes in temperature than to changes in solar UV flux. Nevertheless, early correlative and cross-spectral analyses have demonstrated that the solar UV effect on stratospheric ozone can be detected under favorable conditions, such as at low latitudes and during high solar UV variations. Recent studies indicate that the 27-day solar UV effect can also be detected and distinguished from dynamical effects during moderate solar UV variations, by using statistical methods to remove or separate the temperature effect from the solar UV effect (e.g., Zhou et al. 2000). However, to what extent those methods work properly is so far unclear. For instance, whether the solar UV effect is detectable during a solar minimum has not been examined. This issue will be addressed in Section 3, in which two statistical methods are tested for selected periods with strong, moderate and weak solar UV variations, respectively.

2. Solar Cycle Effects

In this study, solar UV flux is represented by the MgII index (280 nm core-to-wing ratio) measured by the National Oceanic and Atmospheric Administration (NOAA) satellites. Both total ozone column and ozone mixing ratio profiles are from the SBUV measurements, which began in November 1978 on Nimbus-7, and continued as SBUV/2 to the present time on a series of NOAA satellites (NOAA-9, 11, 14 and 16). The SBUV(/2) data are almost continuous, with only four months of missing data, covering the last half of solar cycle 21, the complete solar cycle 22 and the first half of solar cycle 23. Monthly mean data are used here for the solar cycle impact on ozone. Recently, Miller et al. (2003) have addressed the inter-satellite instrument differences and developed a cohesive total ozone dataset, using the overlap periods of satellite data to make necessary adjustments. They adjusted the datasets to NOAA-9 as the standard because of its length of record and the amount of overlap with the other instruments. Systematic differences between the satellites also exist in the stratospheric ozone profile data. For instance, Fig. 1a shows the ozone mixing ratio at 5 hPa averaged over 30°S-30°N for all the SBUV (/2) measurements. Those data were internally calibrated for each satellite to be accurate (Herman 1991; Hilsenrath et al. 1995; Bhartia et al. 1996). Due to instrumental differences and satellite drifting problems, the inter-satellite differences are within several percent, as seen in the overlap periods, which is of the same order as solar cycle variations.

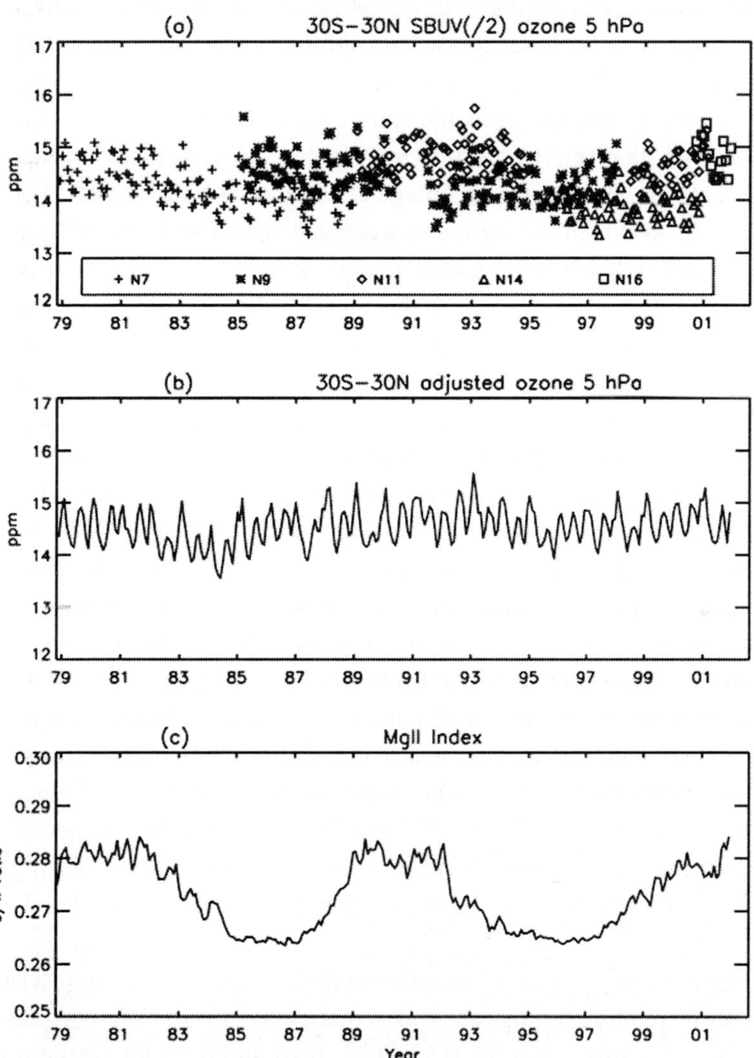

Figure 1. (a) Monthly ozone mixing ratio (ppm) at 5 hPa measured by SBUV and SBUV/2 instruments. (b) The adjusted ozone. (c) The MgII index.

Unlike the total ozone that is calculated by the ratio of radiances at different pairs of frequency channels, the ozone profiles are measured directly by radiances. The ozone profiles are more sensitive than the total ozone to instrument changes because any system bias would otherwise be minimized in the ratio of radiances. In addition, the ozone profiles are more sensitive to local equator-crossing time changes due to the satellite drifting problem, because the lifetime of ozone is short in the upper stratosphere. There is no easy way to remove the systematic difference in the ozone profile data. Here, all the satellite data after 1985 are tentatively adjusted to the NOAA-9 afternoon satellite (1985-1989), by using either direct overlaps (Nimbus-7 and NOAA-11) or indirect extrapolations (NOAA-14 and 16). Prior to 1985, the Nimbus-7 data are used without adjustments. The adjusted ozone mixing ratio at 5 hPa is shown in Fig. 1b. The justification and limitation of this adjustment method will be discussed later.

Because the tropical stratosphere is the source region of ozone, and dynamical effects are smaller in the tropics than in the mid-high latitudes, only tropical area average ozone is considered in this study. The ozone percentage changes relative to the 1979-2001 climatology are shown in Fig. 2, in which the ozone is averaged over the 30°S-30°N tropical area, and linear trends are removed. The total ozone change is within ±5%, which is about the same range of solar UV change represented by the monthly MgII index. The ozone mixing ratio has relatively large variability in the upper stratosphere (within ±10% above 5 hPa). The power spectra of ozone variations indicate that solar cycle has very large amplitudes in the upper stratosphere (1-2 hPa), while the annual cycle dominates 30 hPa and total ozone variations. Between 2 hPa and 10 hPa the semiannual oscillation shows the largest variation. Note that there is almost no quasi-biennial oscillation (QBO) signal at all the levels in the 30°S-30°N average ozone. However, in the equatorial ozone (5°S-5°N) the QBO has the largest power spectrum among other time scales (not shown). It implies that the ozone outside the equatorial region may vary in the opposite phase with the equatorial ozone, so that the QBO signal is averaged out in the 30°S-30°N average. Another point is that the data contamination by El Chichon and Mt. Pinatubo eruptions is apparent at 10 hPa. It reflects the instrument measurement errors due to the increased mid-stratospheric aerosols rather than photochemical ozone loss, because the backscattered radiances increased by the aerosols are interpreted by the instrument as decreased ozone (Fleig et al. 1990; Chandra and McPeters 1994). The total ozone does not appear to have been affected by the aerosols because it is calculated by the ratio of radiances. Neither does the 30 hPa ozone mixing ratio, which is more dependent of the total ozone related "first guess" in the retrieval rather than the measured radiances.

Prior to correlation analyses, a 13-month running average of the time series of ozone and solar UV is taken to remove the annual and semiannual cycles. The correlation coefficients between solar UV and ozone are 0.69, 0.53, 0.37, 0.16 and 0.59 for ozone mixing ratio at 1, 2, 5, 10 and 30 hPa respectively. The coefficient for solar UV and total ozone is 0.57. All of the correlations except for that at 10 hPa are statistically significant at 95% confidence level. The ozone sensitivity to solar cycle variations is estimated by a linear regression of the smoothed ozone and solar UV data. Time lags have not been considered because the ozone

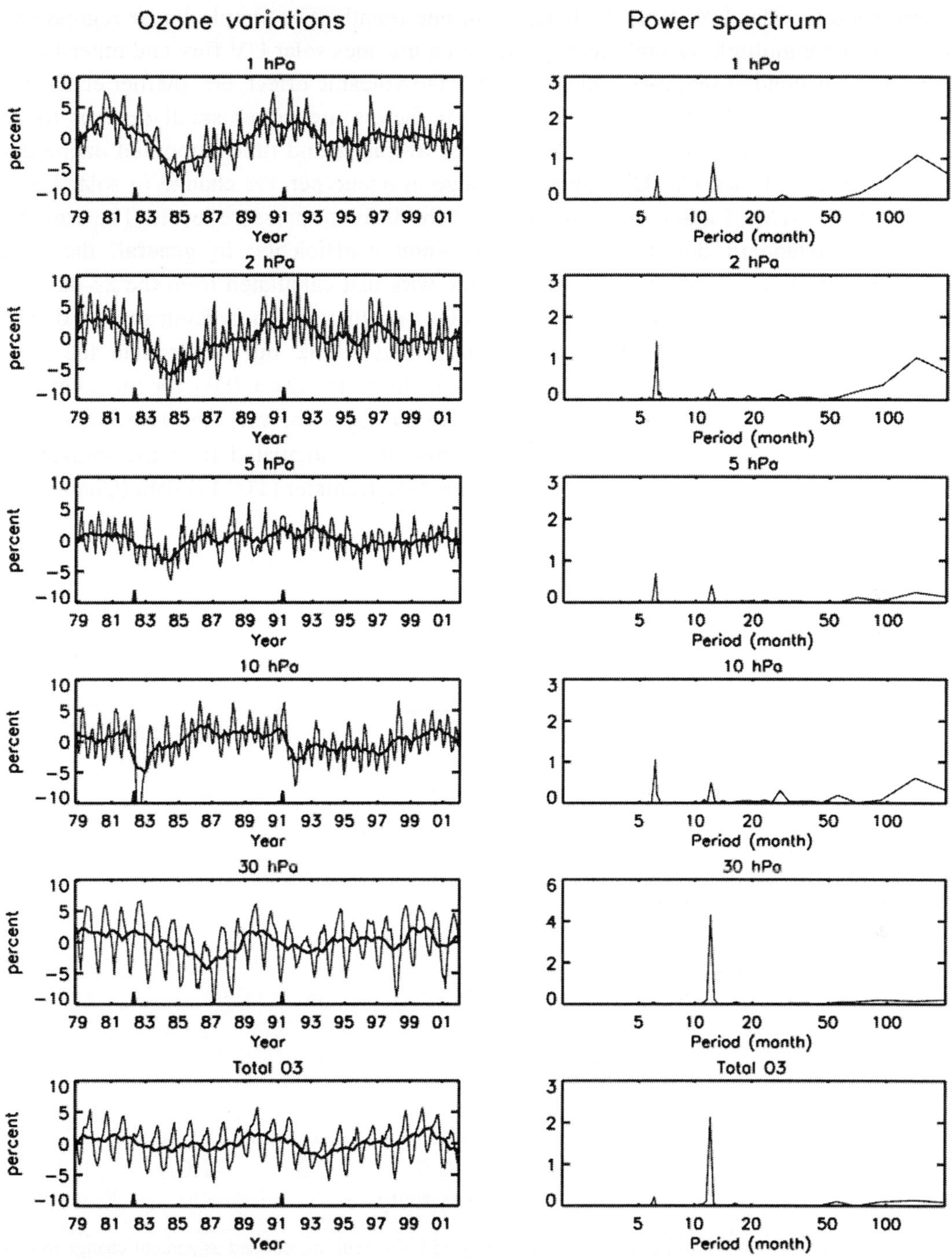

Figure 2. (Left): Percent changes of tropical (30°S-30°N) ozone mixing ratio at 1, 2, 5, 10, 30 hPa and total column ozone. The thick lines are the 13-month running averages. The eruption times of El Chichon and Mt. Pinatubo are marked at the bottom. (Right): Power spectra of ozone variations.

response time to solar UV is much shorter than one month. The simple linear regression is equivalent to the multiple variable regression which includes solar UV flux and other factors, such as a linear trend, a seasonal cycle, the QBO, the volcanic effect, etc. (Miller et al. 1992; McCormack and Hood 1996), because the other factors are either small (except for the volcanic effect at 10 hPa) or excluded in the area averaged and time smoothed ozone data. The ozone sensitivity to solar UV (percent change in ozone per 1% change in solar UV, as represented by the MgII index) is shown in Fig. 3 for different pressure levels. The error bars indicate $\pm 2\sigma$ (standard deviation) of the regression coefficients. In general, the ozone sensitivity in the upper stratosphere is comparable with that calculated from the 27-day solar variation (e.g., Hood and Zhou 1999). It is smaller than that obtained using shorter datasets (Chandra and McPeters 1994; McCormack and Hood 1996), but larger than the model prediction by Brasseur (1993). The total ozone sensitivity is 0.24 ± 0.04 for the solar cycle change, which is also larger than 0.09-0.11 for the 27-day solar variation (Hood and Zhou 1999; Fleming et al. 1995), but slightly smaller than 0.27 calculated from the shorter time periods of SBUV(/2) and the Total Ozone Mapping Spectrometer (TOMS) data (Chandra and McPeters 1994; Fleming et al. 1995).

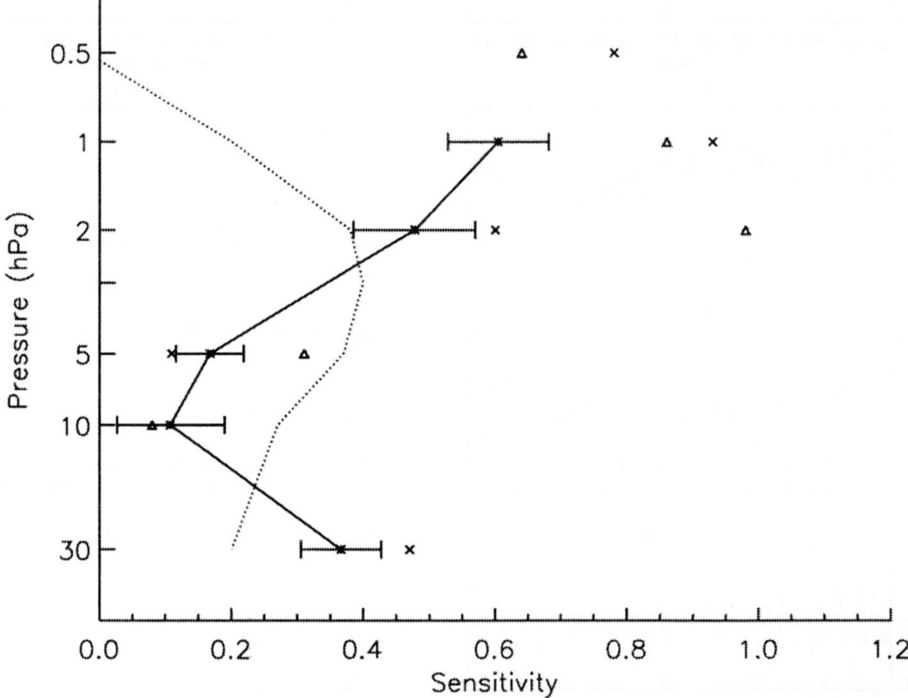

Figure 3. Tropical (30°S-30°N) ozone sensitivity to solar cycle UV variations, defined as percent change in ozone at a given pressure level per 1% change in solar UV as represented by the MgII index. Error bars represent the $\pm 2\sigma$ (standard deviation) of the regression coefficients. The triangle and cross signs indicate ozone sensitivities calculated by Chandra and McPeters (1994) and McCormack and Hood (1996), respectively. The dotted line indicates model results by Brasseur (1993).

3. Solar Rotational Effects

Daily solar MgII index and SBUV/2 ozone were used to study solar UV impact on the stratospheric ozone at the 27-day solar rotational time scale, in which three 1000-day periods in solar cycle 22 were chosen. The first period is from January 23, 1989 to October 19, 1991, the second period from October 20, 1991 to July 15, 1994, and the third period from July 16, 1994 to April 10, 1997. They represent approximately the solar maximum, the declining stage and the solar minimum, respectively. The first two periods are the same as in Zhou et al. (2000), in which they were compared with two similar periods in solar cycle 21, except now the solar MgII index and SBUV/2 ozone data have been reprocessed. NOAA-11 ozone data are used for the first two periods and NOAA-9 ozone data are used for the third period. The temperature effect on ozone is considered in the solar UV-ozone correlation. The temperature data are from the National Centers for Environmental Prediction (NCEP) daily analyses. Missing data are filled by linear interpolations. The ozone and temperature data are averaged over the 30°S-30°N tropical area to minimize dynamical effects. All data are then detrended and deseasonalized by subtraction of a 35-day running average and smoothed by a 7-day running average. This procedure has been commonly used to filter long-term variations, such as annual and seasonal changes, as well as to filter very short-term fluctuations (e.g., Hood 1986; Chandra 1986). Here, it also automatically removes the systematic difference between the two satellite datasets.

Figure 4 shows the percent changes of solar UV, ozone and temperature at 2 hPa. The amplitude of solar UV is relatively large in Period 1 and decreases gradually. However, this is not the case for the ozone and temperature. The short-term ozone variations are mixed with solar modulations and dynamical influences. For instance, the large ozone variation episodes during 1991-1993 were closely related to the large and opposite temperature variations in the same time. If solar UV impact is regarded as a "signal" and temperature effect as a "noise", then the signal-to-noise ratio (snr) can be estimated from the average solar UV and temperature variability for each period. As shown in Table 1, the percent change of solar UV decreases geometrically from Period 1 to Period 3, while the temperature changes are almost constant. As a result, the snr also decreases geometrically from the solar maximum period to the solar minimum period. The temperature effect must be excluded in order to reveal the real relationship of solar UV and ozone. In the following, two statistical methods are introduced to remove (or separate) the temperature effect.

The first method is to calculate partial correlations of solar UV and ozone with the temperature effect removed, directly from cross-correlations of solar UV, ozone and temperature. Although temperature is not independent of solar UV in general, it is very different from ozone response to solar UV in many aspects, so that it can be approximately treated as an independent variable (Zhou et al. 2000). In the case of three variables, the partial correlation coefficient is defined as

$$r_{ij,k} = \frac{r_{ij} - r_{ik} r_{jk}}{\sqrt{(1-r_{ik}^2)(1-r_{jk}^2)}}, \tag{1}$$

where r_{ij}, r_{ik}, and r_{jk} are total correlation coefficients between two variables, and $r_{ij,k}$ is the partial correlation coefficient between variables *i* and *j* with the effect of variable *k* removed (Hald 1952). Here we may consider variables *i, j* as ozone and solar UV, and variable *k* as temperature.

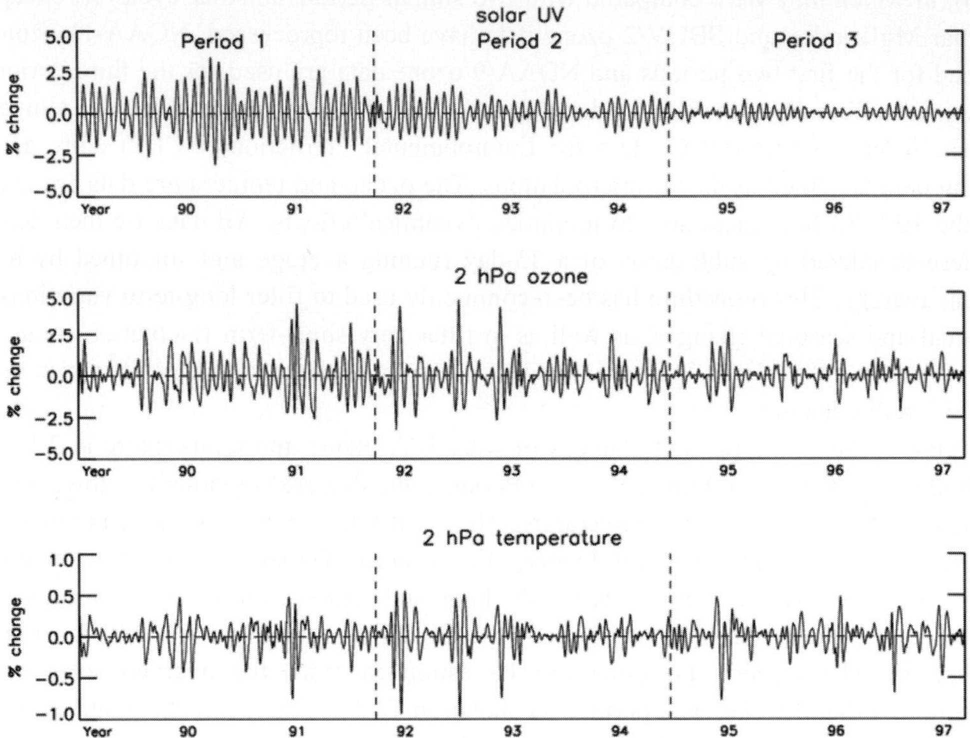

Figure 4. Percent changes of solar UV, the 30°S-30°N average ozone mixing ratio and temperature at 2 hPa in the three periods of solar cycle 22. The changes are calculated by subtracting a 35-day running average and then smoothed by a 7-day running average.

Table 1. Average percent changes in solar UV and temperature and ratio (snr) of solar UV change to temperature change.

P	Period 1 (solar max).			Period 2 (decline)			Period 3 (solar min.)		
(hPa)	UV	T	snr	UV	T	snr	UV	T	snr
1	3.4	0.31	11.2	1.8	0.31	5.7	0.7	0.29	2.4
2	3.4	0.38	9.0	1.8	0.40	4.4	0.7	0.38	1.9
5	3.4	0.43	8.0	1.8	0.48	3.8	0.7	0.45	1.6
10	3.4	0.36	9.5	1.8	0.42	4.3	0.7	0.40	1.8

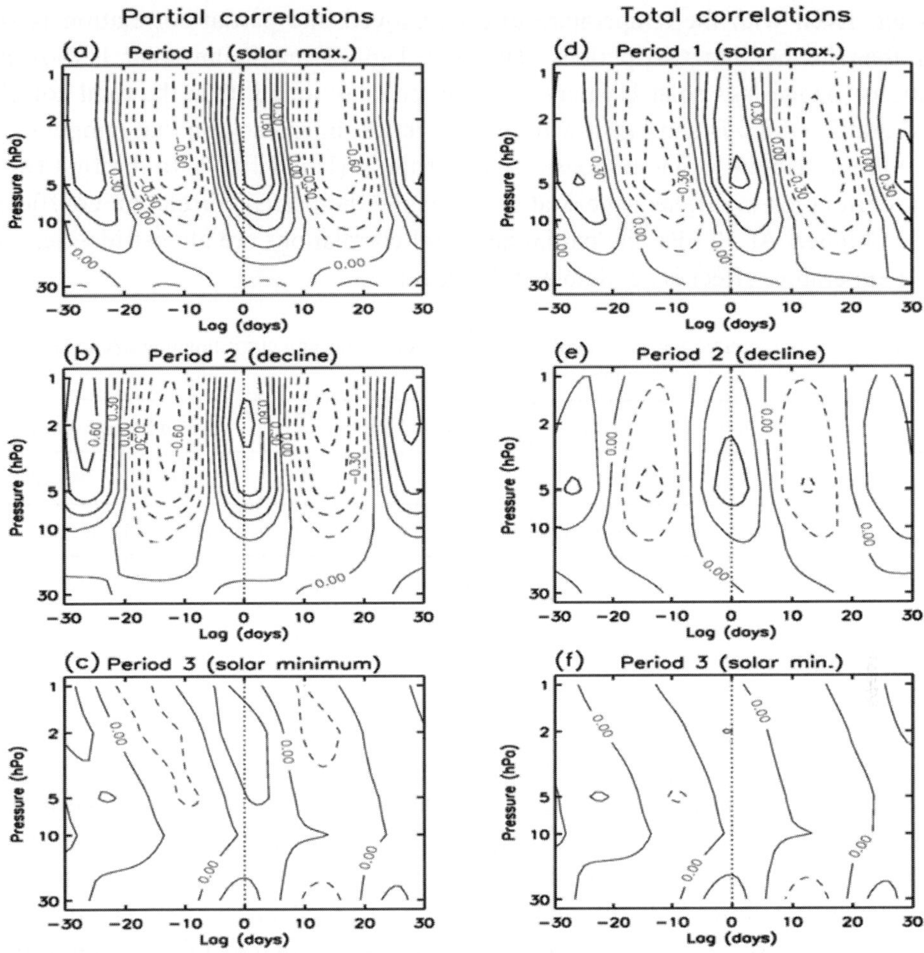

Figure 5. The 27-day partial (left) and total (right) correlation coefficients of tropical (30°S-30°N) ozone and solar UV as a function of pressure and time lag in the three periods. The contour interval is 0.15. The 95 % significance level is ± 0.15.

The partial correlation coefficients of solar UV and the 30°S-30°N average ozone are shown in Fig. 5 and Table 2. In calculating the lag correlations the same time lag relative to solar UV is used for ozone and temperature. The total correlation coefficients of ozone and solar UV (r_{ij}) are also shown for comparison. To determine the statistical significance level, a Monte-Carlo experiment is performed in which three random time series of the same length of actual data (1000 points each), with the same filtering and smoothing procedure, are tested for 800 trials. The 95% significance level is ±0.15 for both the total and partial correlation analyses. In all the three periods the partial correlation coefficients are significantly larger than the total correlation coefficients in the upper stratosphere (1-5 hPa). The difference is remarkably large in Period 2 (declining solar activity) when the snr is intermediate. In Period 3 (solar minimum), because the snr is too small, the total correlation is not statistically significant. Even with the temperature effect removed, the partial correlation is still much smaller than in the first two periods. The time lags of maximum solar UV-ozone partial correlations (ozone lags solar UV) are typically one day longer than the total correlations in Period 1 and 2, which is consistent with the theoretical interpretation that temperature effects can shift the maximum ozone response to earlier times (Hood 1987). In the first two periods, the total correlation coefficients peak at 5 hPa, while the partial correlation coefficients peak somewhat higher. At 10 hPa, the partial and total correlations are almost identical, implying that the temperature effect is negligible at that level.

Table 2. The maximum total and partial correlation coefficients and corresponding ozone lag days to solar UV variations (in parentheses).

P	**Period 1 (solar max.)**		**Period 2 (decline)**		**Period 3 (solar min.)**	
(hPa)	Total	Partial	Total	Partial	Total	Partial
1	0.49 (0)	0.78 (1)	0.16 (0)	0.70 (1)	0.12 (-5)	0.16 (-4)
2	0.56 (0)	0.80 (1)	0.28 (0)	0.80 (0)	0.15 (-1)	0.25 (0)
5	0.64 (2)	0.81 (2)	0.39 (0)	0.66 (1)	0.14 (2)	0.17 (2)
10	0.39 (4)	0.39 (4)	0.20 (1)	0.19 (1)	0.05 (3)	0.05 (3)

The second method to separate the temperature effect from ozone variations is to use multiple linear regressions. It is particularly useful to estimate ozone sensitivities to solar UV and temperature separately (Chandra 1986). Though ozone sensitivity to solar UV was sometimes calculated by using solar UV alone (Hood and Cantrell 1988; Hood and Jirikowic 1991; Fleming et al. 1995), it is appropriate only under certain conditions, such as in the solar maximum, as with the total correlation analysis. To illustrate this, multiple linear regressions are applied to the three periods as in the partial correlation analyses, and ozone sensitivities are compared with those using solar UV alone. The multiple regression equation for ozone is written as

$$O_3 = A_0 + A_1 S + A_2 T \,, \tag{2}$$

where O_3 , S and T denote percent changes in ozone, solar UV and temperature respectively, A_0 is a constant, A_1 is equivalent to ozone sensitivity to solar UV and A_2 is equivalent to ozone sensitivity to temperature.

Figure 6 shows the results of multiple regressions as a function of time lag (ozone lags solar UV or temperature) and pressure level for the three periods, respectively. The characteristics of ozone sensitivity to solar UV are very similar to the partial correlation coefficients of ozone and solar UV (Fig. 5). The largest value is found at 2 hPa, with a zero time lag. The ozone variation is in phase with the solar UV variation, but out of phase with the temperature variation in the upper stratosphere. The largest ozone sensitivity to temperature is negative near the stratopause (1-2 hPa) with a zero time lag. It decreases downward and changes sign below 10 hPa. However, the time lag only increases slightly downward.

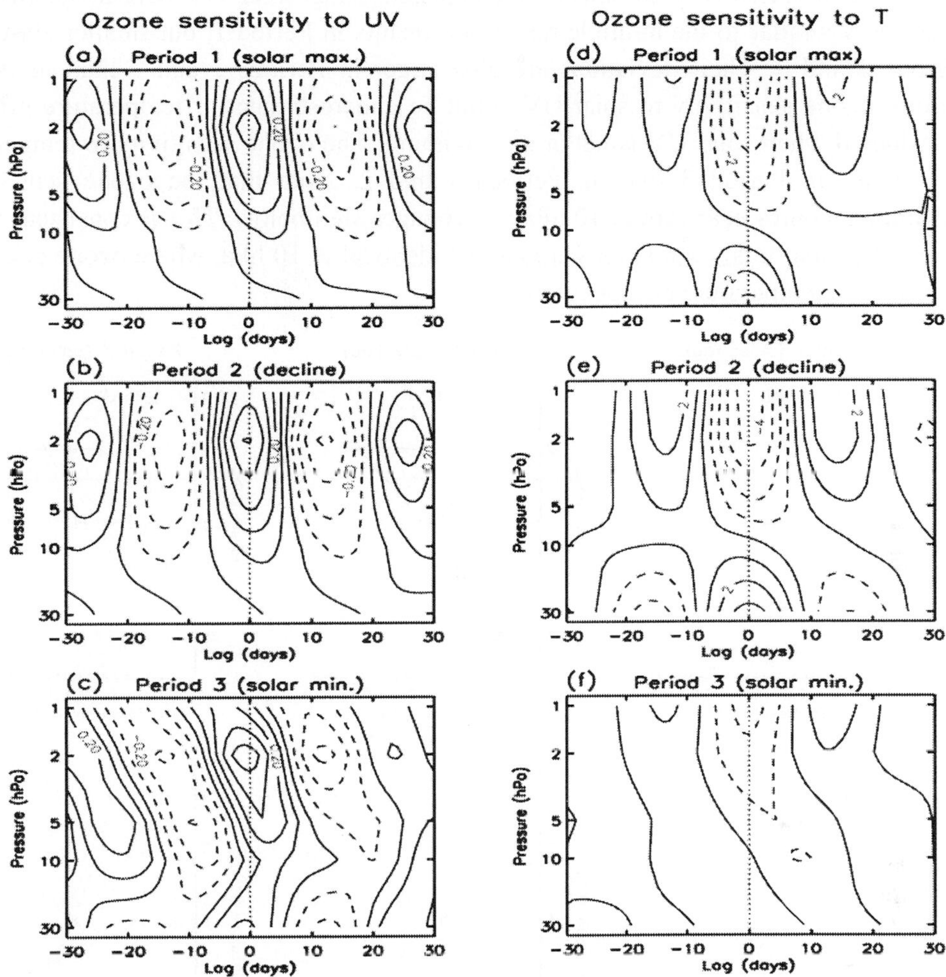

Figure 6. The 27-day ozone sensitivity to solar UV (left panels) and temperature (right panels) as a function of time lag and pressure level in the three periods.

For easy comparisons, the ozone sensitivities to solar UV and temperature at the zero time lag are shown in Fig. 7 (dashed lines). Error bars of $\pm 2\sigma$ are also shown to indicate statistical significance. The ozone sensitivities are not much different between Periods 1 and 2. The maximum ozone sensitivity to solar UV is 0.4~0.6 (±0.03) at 2 hPa, and the maximum ozone sensitivity to temperature is –4.8 (±0.2) at 2 hPa (or –1.8% change in ozone per 1 K change in temperature). Those values are comparable to the results obtained by Chandra (1986), who calculated the ozone sensitivity to solar UV and temperature using the SBUV ozone, the 205 nm solar UV flux and the Stratosphere and Mesosphere Sounder (SAMS) temperature in a 30-month period in solar cycle 21 (December 1978 - May 1981). In Period 3 the ozone sensitivity to solar UV has a similar peak value at 2 hPa, but it is smaller than the first two periods above and below the peak. At 30 hPa, the ozone sensitivity is relatively large in Period 3, which is very different from Periods 1 and 2 where it is near zero. Figure 7 also shows the ozone sensitivity to solar UV calculated using solar UV data alone (triangles). They are very similar to the multiple regression results in Period 1, but smaller above 5 hPa in Periods 2 and 3. Their deviations are also larger in Periods 2 and 3 than in Period 1. Therefore, ozone sensitivity to solar UV could be underestimated if temperature effects are not considered when solar UV is not at its maximum. The ozone sensitivity to temperature is much smaller in Period 3 than in Periods 1 and 2. Note that the ozone sensitivity to temperature becomes near zero at 10 hPa and changes sign below. This is consistent with the fact the partial and total correlations are almost identical at 10 hPa, where ozone changes are not sensitive to temperature changes at all.

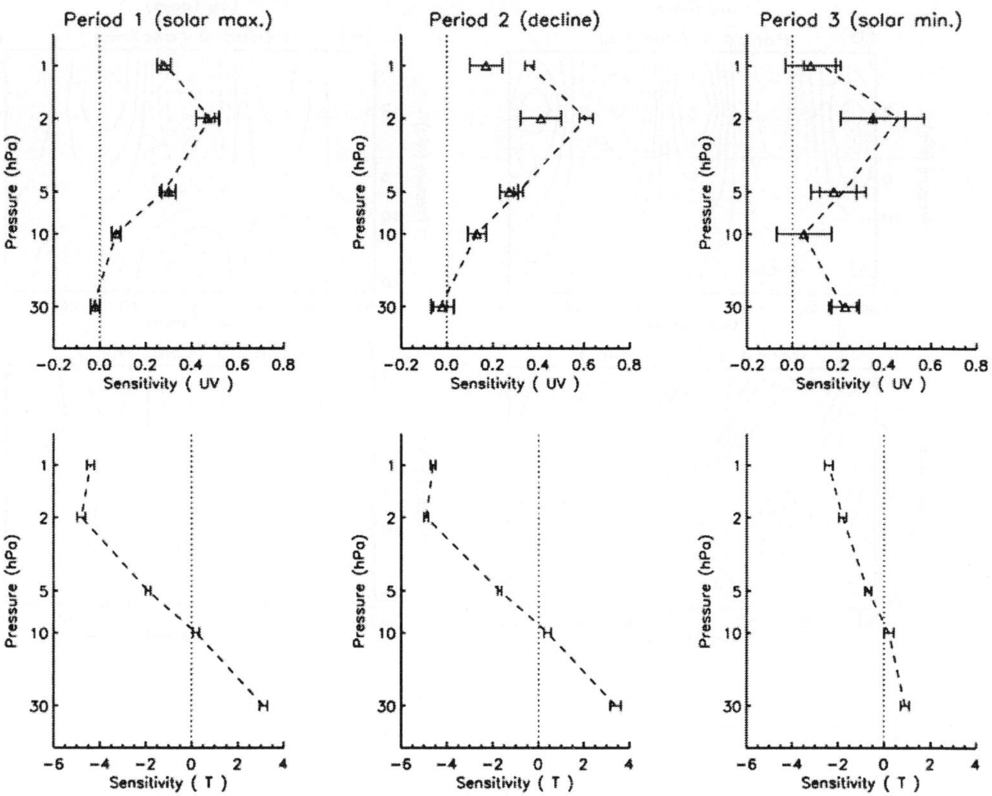

Figure 7. The 27-day ozone sensitivities to solar UV (upper panels) and temperature (lower panels) at the zero time lag in the three periods. The dashed lines are from the multiple regression and the triangles are from solar UV alone. Error bars represent the $\pm 2\sigma$ (standard deviation) of the regression coefficients.

It has been demonstrated that partial correlations and multiple regressions are two useful methods to remove (or separate) temperature effect in studying ozone response to the short-term solar UV variations. However, correlation coefficients and sensitivities only represent a general relationship between two time series. They do not indicate what frequencies of the time series contribute most to the relationship. Therefore, a cross-spectral analysis such as coherency is needed (Hood and Jirikowic 1991). The coherency is analogous to the square of a correlation coefficient, except that the coherency is a function of frequency. Because the ozone variation is very well represented by the linear combination of solar UV and temperature changes, the temperature effect can be removed from the ozone data by subtracting the portion of regressed ozone due to temperature variations. The remainder then includes portions due to solar UV and other secondary factors. Then the coherency square is calculated by using solar UV and the modified ozone. The results of coherency square at 5 hPa are shown in Fig. 8 (solid lines). In Periods 1 and 2, there is a peak at or near the 27-day period, with a confidence level of at least 95%. For a comparison, Fig. 8 also shows the coherency square calculated using unmodified ozone data (dashed lines), which is significantly reduced, though there is a peak at or near the 27-day period in Period 1 and the confidence level for this peak is above 95%. However, in Period 2 the coherency square is too small to meet the 95% confidence level, and the peak is shifted off the 27-day period. In Period 3 the coherency square is very small, not significant, and without a peak near the 27-day period for either modified or unmodified ozone. Therefore, only in the solar maximum period, the 27-day solar UV modulation could be detected by using direct ozone and solar UV data.

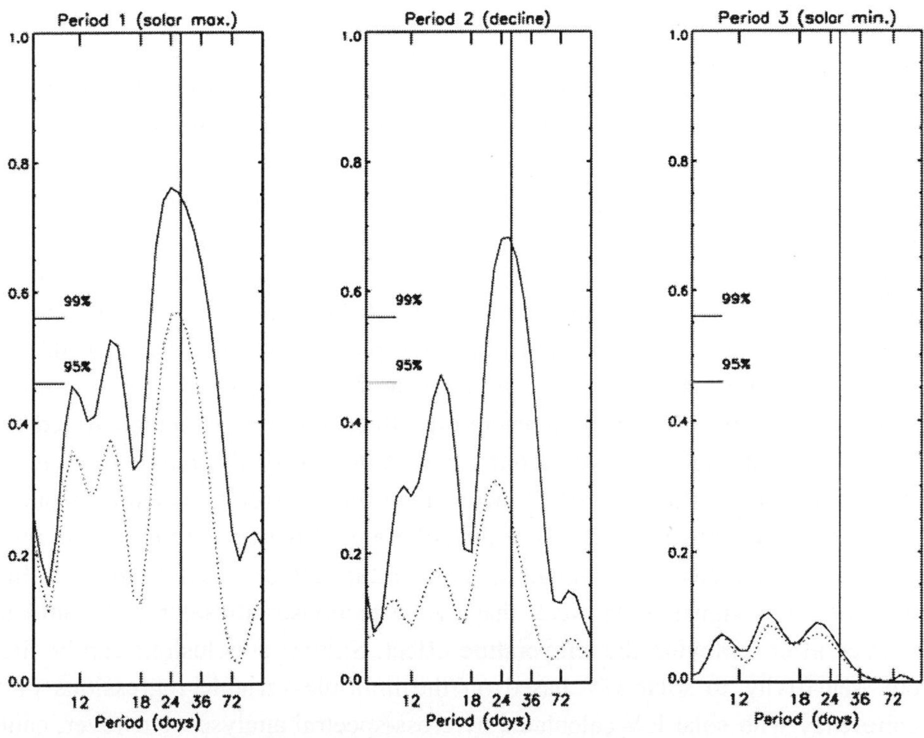

Figure 8. Coherency square calculated from solar UV and modified ozone with temperature effect removed (solid lines) and from solar UV and unmodified ozone (dashed lines). The confidence levels are marked at the left side. The vertical lines indicate the 27-day period.

4. Summary and Discussions

This paper addresses solar UV effects on the tropical total column ozone and stratospheric ozone profiles on both the long-term and short-term time scales. The long-term variation of ozone by the 11-year solar cycle was studied using 23 years of SBUV(/2) total ozone and ozone mixing ratio data. The inter-satellite differences are adjusted to form a cohesive dataset. The adjusted data seem to be reasonable in their power spectra and in the derived ozone sensitivity to solar UV. For instance, the solar cycle signal is mostly clearly seen in the upper stratosphere (1-2 hPa). The ozone sensitivity is improved over the previous analyses based on shorter time periods, and is closer to that derived from the solar rotational effect. For instance, the ozone sensitivity at 2 hPa is 0.48 ± 0.09 for the solar cycle change and 0.44 ± 0.02 for the 27-day solar variation.

However, since ozone profiles are more sensitive to changes of NOAA satellites and the satellite drifting problem, the present inter-satellite adjustment method may need to be improved by more sophisticated methods. From this adjusted dataset, the ozone sensitivity to the solar cycle variation at 1 hPa is about twice as large as that to the 27-day solar variation. Because of the strong dependence of ozone on temperature, ozone measurements are sensitive to the time of a day in the upper stratosphere. For instance, the diurnal cycle effect on temperature is large at and above 1 hPa but negligible below 2 hPa (Miller, personal communication). This may explain the relatively large difference of ozone sensitivity at 1 hPa between the solar cycle and solar rotational effects, because the local time of ozone measurement was different between different satellites. Even within the same satellite the equator-crossing time was gradually delayed when the satellite orbit drifted. In addition, the ozone data only cover a time period of about two solar cycles, the result of ozone response to the solar cycle is less rigorous than that to the 27-day solar variation, in which many solar rotations are considered.

For the short-term variation of stratospheric ozone, here the emphasis is put on how to separate dynamically induced temperature effect from the 27-day solar UV variations. Previous analyses of partial correlations and multiple regressions (Zhou et al. 2000) were extended to three 1000-day periods in solar cycle 22, including the solar minimum. The partial correlation analysis with the temperature effect removed is demonstrated to be most effective for moderate solar UV variations, such as between solar maximum and minimum, although in other periods it is generally better than total correlations using ozone and solar UV only. In the solar maximum, because the solar signal is very strong, ozone responses to solar UV could be detected even though the temperature effect was not removed. In the solar minimum, the solar signal is so weak that ozone responses to solar UV could hardly be detected even after removing the temperature effect. Similar conclusions can be drawn from the ozone sensitivity to solar UV based on the multiple-variable regressions or from the ozone coherency with solar UV calculated by cross-spectral analyses. However, caution must be taken in interpreting the ozone sensitivity to solar UV because it may be sometimes misleading, even after the temperature effect is removed. For instance, at 5 hPa the ozone sensitivity is not much different in all the three periods (Fig. 7), but the coherency is totally different in Period 3 from Periods 1 and 2 (Fig. 8). The ozone variation in the solar minimum period is not coherent with the 27-day solar UV variation at all.

Acknowledgments. The author thanks Ronald Nagatani and Alvin Miller of the National Oceanic and Atmospheric Administration (NOAA) for providing the SBUV(/2) ozone data and helpful discussions. Thanks are also due to Lon Hood of University of Arizona, Steven Lloyd of Johns Hopkins University, and David Rusch of University of Colorado for their constructive comments and suggestions. The MgII data were provided by the NOAA Space Environment Center.

References:

Bhartia, P. K., R. D. McPeters, C. L. Mateer, L. E. Flynn, and C. Wellemeyer, 1996: Algorithm for the estimation of vertical ozone profiles from the backscattered ultraviolet technique. *J. Geophys. Res*., **101**, 18,793-18,806.

Brasseur, G. P., 1993: The response of the middle atmosphere to long-term and short-term solar variability: A two-dimensional model. *J. Geophys. Res.,* **98**, 23,079-23,090.

Chandra, S., 1986: The solar and dynamically induced oscillations in the stratosphere, *J. Geophys. Res.,* **91**, 2719-2734.

_____, and R. D. McPeters, 1994: The solar cycle variation of ozone in the stratosphere inferred from Nimbus 7 and NOAA 11 satellites. *J. Geophys. Res*., **99**, 20,665-20,671.

Fleig, A. J., et al., 1990: Nimbus-7 solar backscatter ultraviolet (SBUV) ozone production user guide. *NASA Ref. Publ*., 1234.

Fleming, E. L., S. Chandra, C. H. Jackman, D. B. Considine, and A. R. Douglass, 1995: The middle atmospheric response to short and long term solar UV variations: analysis of observations and 2D model results. *J. Atmos. Terr. Phys.,* **57**, 333-365.

Garcia, R. R., S. Solomon, R. G. Roble, and D. W. Rusch, 1984: A numerical response of the middle atmosphere to the 11-year solar cycle. *Planet. Space Sci*., **32**, 411-423.

Gille, J. C., C. M. Smythe, and D. F. Heath, 1984: Observed ozone response to variations in solar ultraviolet radiation. *Science*, **225**, 315-317.

Hald, A., 1952: *Statistical theory with engineering applications*. John Wiley & Sons, Inc., 783 pp.

Haigh, J. D., 1996: The impact of solar variability on climate. *Science*, **272**, 981-984.

Herman, J. R., and C. J. McQuillan, 1985: Atmospheric chlorine and stratospheric ozone nonlinearities and trend detection. *J. Geophys. Res*., **90**, 5721-5732.

Herman, J. R., 1991: A new self-calibration method applied to TOMS and SBUV backscattered data to determine long-term global ozone change. *J. Geophys. Res*., **96**, 7531-7545.

Hilsenrath, E., R. P. Cebula, M. T. DeLand, K. Laamann, S. Tayler, C. Wellemeyer, and P. K. Bhartia, 1995: Calibration of the NOAA 11 solar backscatter ultraviolet (SBUV/2) ozone data set from 1989 to 1993 using in-flight calibration data and SSBUV. *J. Geophys. Res*., **100**, 1351-1366.

Huang, T. Y. W., and G. P. Brasseur, 1993: The effect of long-term solar variability in a two dimensional interactive model of the middle atmosphere. *J. Geophys. Res*., **98**, 20,413-20,427.

Hood, L. L., 1986: Coupled stratospheric ozone and temperature responses to short-term changes in solar ultraviolet flux: An analysis of Nimbus 7 SBUV and SAMS data. *J. Geophys. Res.*, **91**, 5264-5276.

_____, 1987: Solar ultraviolet radiation induced variations in the stratosphere and mesosphere. *J. Geophys. Res.*, **92**, 876-888.

_____, and S. Cantrell, 1988: Stratospheric ozone and temperature responses to short-term solar ultraviolet variations: Reproducibility of low-latitude response measurements. *Ann. Geophys.*, **6**, 525-530.

_____, and J. L. Jirikowic, 1991: Stratospheric dynamical effects of solar ultraviolet variations: Evidence from zonal mean ozone and temperature data. *J. Geophys. Res.*, **96**, 7565-7577.

_____, _____, and J. P. McCormack, 1993: Quasi-decadal variability of the stratosphere: Influence of long-term solar ultraviolet variations. *J. Atmos. Sci.*, **50**, 3941-3958.

_____, and S. Zhou, 1998: Stratospheric effects of 27-day solar ultraviolet variations: An analysis of UARS MLS ozone and temperature data. *J. Geophys. Res.*, **103**, 3629-3638.

_____, and _____, 1999: Stratospheric effects of 27-day solar ultraviolet variations: The column ozone response and comparisons of solar cycles 21 and 22. *J. Geophys. Res.*, **104**, 26,473-26,479.

_____, and B. Soukharev, 2000: The solar component of long-term stratospheric variability: Observation, model comparisons, and possible mechanisms (extended abstract), SPARC 2000 Symposium, Mar del Plata, Argentina, November, 2000.

Keating, G. M., G. P. Brasseur, J. Y. Nicholson III, and A. De Rudder, 1985: Detection of the response of ozone in the middle atmosphere to short term solar ultraviolet variations. *Geophys. Res. Lett.*, **12**, 449-452.

Lean, J., 1987: Solar ultraviolet irradiance variations: A review. *J. Geophys. Res.*, **92**, 839-868.

Miller, A. J., R. M. Nagatani, G. C. Tiao, X. F. Niu, G. C. Reinsel, D. Wuebbles, and K. Grant, 1992: Comparisons of observed ozone and temperature trends in the lower stratosphere. *Geophys. Res. Lett.*, **19**, 929-932.

Miller, A. J., R. M. Nagatani, L. E. Flynn, S. Kondragunta, E. Beach, R. Stolarski, R. McPeters, P. K. Bhartia, and M. DeLand, 2003: A cohesive total ozone dataset from the SBUV(/2) satellite system. *J. Geophys. Res.*, in press.

McCormack, J. P. and L. L. Hood, 1996: The apparent solar cycle variation of upper stratospheric ozone and temperature: Latitude and seasonal dependences. *J. Geophys. Res.*, **101**, 20,933-20,944.

Shindell, D., D. Rind, N. Balachandran, J. Lean, and P. Lonergan, 1999: Solar cycle variability, ozone, and climate. *Science*, **284**, 305-308.

Wuebbles, D. J., 1983: Chlorocarbon emission scenarios: Potential impact on stratospheric ozone. *J. Geophys. Res.*, **88**, 1433-1443.

Zhou, S., A. J. Miller, and L. L. Hood, 2000: A partial correlation analysis of the stratospheric ozone response to 27-day solar UV variations with temperature effect removed. *J. Geophys. Res.*, **105**, 4491-4500.

APPLICATIONS OF TRMM SCIENTIFIC DATA

RUNHUA YANG AND HUALAN RUI
Science Systems and Applications Incorporation
10210 Greenbelt Road, Lanham, MD 20706, USA
E-mails: ryang@dao.gsfc.nasa.gov and *rui@daac.gsfc.nasa.gov*

(Manuscript received 15 November 2002)

This article introduces an education-orientated web page about the application of Tropical Rainfall Measuring Mission (TRMM) data in the atmospheric research. We write that on-line page with a two-fold purpose. First, it exhibits examples of the TRMM data usage in the atmospheric study. These examples provide users a means of grasping the concepts of the scientific information in the TRMM data quickly and easily. Secondly, the contents of the sample TRMM data are described and visually depicted. This is intended to familiarize users with TRMM data contents. These examples are picked up from TRMM science reports in both professional journals and distributed web sites. Rather than providing detailed descriptions of these examples, we integrate them into a brief overall page for public access. Six topics are listed at the time of this writing, including: (1) TRMM improves rainfall forecast; (2) TRMM improves rainfall data assimilation; (3) tropical cyclones: new views; (4) cloud and precipitation formation; (5) TRMM for monsoon study; and (6) TRMM for El Niño and La Niña study. Each topic covers two aspects, scientific application and sample data description. In addition, a complete list of TRMM data products, data search and order system, and data processing tools are briefly described. The web page address is at:

http://eosdata.gsfc.nasa.gov/CAMPAIGN_DOCS/hydrology/TDST_SCI/sci_main.html

1. Introduction

Tropical Rainfall Measuring Mission (TRMM) was jointly sponsored by the National Aeronautics and Space Administration (NASA) of the United States and the National Space Development Agency (NASDA) of Japan. Since its launch at November 1997, TRMM has successfully provided five-year long precipitation data at the time of this writing. TRMM rainfall estimates are of good quality in terms of its consistency with the ground validation products and rain gauge net work data (Kummerow et al. 2000). TRMM satellite's low inclination, non-sun-synchronous, and highly precessing orbit allows it to fly over a given location of the tropical region (35°N-35°S) at a different time every day with an approximate 42-day cycle. This kind of sampling allows the documentation of the large diurnal variation of tropical rainfall. Now, with a 5-year long record, TRMM provides a very useful data set for the study of interannual variation of tropical precipitation.

The hydrology group at the Distributed Active Archive Center (DAAC) of Goddard Space Flight Center (GSFC), NASA, provides a full service for TRMM data, including ingesting, archiving, and distributing the data. The group also produces value-added TRMM subset data, data processing software, and scientific information. In this article we introduce readers to an education-orientated web page developed by the hydrology group at DAAC GSFC. We wrote that web page with a two-fold purpose. First, it exhibits examples of the TRMM data usage in the atmospheric research and study. Those examples provide users a means of grasping the concepts of the scientific information in the TRMM data quickly and easily. Secondly, the contents of the sample TRMM data are described and visually depicted. This is intended to familiarize users with TRMM data contents. Those examples are picked up from TRMM science reports in both professional journals and distributed web sites. Rather than providing detailed descriptions of these examples, we integrate them into a brief overall page for public access. Six topics are chosen including: (1) TRMM improves rainfall forecast; (2) TRMM improves rainfall data assimilation; (3) tropical cyclones: new views; (4) cloud and precipitation formation; (5) TRMM for monsoon study; (6) TRMM for El Niño and La Niña study. Each topic covers the two aspects, scientific application and sample data description.

In this short article, we will first cover the six examples of using TRMM scientific data to the atmospheric research and study. Next, the complete list of TRMM data products, data accessing information, and the data analysis tools will be described. Finally, a keynote about the next phase of TRMM mission, Global Precipitation Mission (GPM), will be given.

2. Sample Applications

2.1. *TRMM Improves Rainfall Forecast*

Tropical rainfall forecast is a challenging task. It is common that rainfall forecast for day one is reliable, but for day 2 and day 3, or longer time, forecast skill, which is a quantitative measure of similarity between forecasts and observations, will degrade significantly. To improve tropical rainfall forecast is important, particularly when it comes to hurricane tracks and rainfall accumulations. Moreover, the improvement will have integrated effect on the model global scale behavior of rainfall, and help us understand the model rainfall patterns. One major accomplishment from the use of TRMM satellite data is the short-term rainfall forecast study carried out by researchers at Florida State University (Krishnamurti et al. 2001). In their study, TRMM rainfall data and Special Sensor Microwave Imager (SSM/I) data are included into their multi-analysis process of numerical weather forecast model, and the results show that the global, as well as the regional forecast skill, are higher than that without TRMM data. Also, TRMM data are used as the validation data for deriving the statistical parameters of their multi-model super ensemble system. The resulted forecast correctly give the tracks of major hurricanes for 1999. The rainfall forecasting accuracy is dramatically increased. The scientists attribute this success to a combination of improved analyses available from the super-ensemble approach as well as the availability of accurate rainfall estimates over the tropics from the TRMM satellite.

A case of three-day rainfall forecast made by multi-model super ensemble method (TRMM 2002 News): the initialization process of the model made the use of the precipitation observed by TRMM Microwavc Imager (TMI) and SSM/I at 12Z Oct. 27, 1999. The heavy precipitation along with supercyclone Orissa occurs over South Asia, over Bay of Bengal, and around Indonesia. For day one, the forecast rainfall field looks very similar to the observation in terms of rainfall regions and intensity. For the day two and day three, the predicted precipitation are still fairly close to the observation, though the discrepancies seem to develop around Orissa's center precipitation region.

The researchers at Florida State University currently extend their short-term rainfall forecast to thirty-day forecast experiment using TRMM and SSM/I data in near real time. For the detailed description of this study, please see Krishnamurti et al. (2001).

2.2. *TRMM Improves Rainfall Data Assimilation*

Atmospheric data assimilation is a process that incorporates observational data into numerical atmospheric models with consideration of both observation and model errors. Conventional global assimilated data sets currently contain significant errors in primary hydrological fields, such as precipitation and evaporation, especially in the tropics. Part of these errors is related to relatively coarse precipitation observations. The TRMM-derived rainfall and total precipitable water (TPW) estimates may be used to constrain these fields in assimilation systems, and make improvements on assimilated data sets.

Atmospheric scientists at GSFC NASA have successfully developed analysis techniques to bring TRMM rainfall observation into their global numerical model, called Terra Goddard Earth Observation System (GEOS) data assimilation system (DAS). By assimilating the 6-hour averaged TMI surface rain and other TPW data into the Terra GEOS-DAS, they found that not only the primary hydrological fields, but also key climate parameters, such as clouds and radiation, have been improved significantly (Hou et al. 2001).

A super-typhoon Paka (December 10, 1997) assimilation: a pair of assimilation was carried out by using Terra GEOS-DAS, with and without the addition of TMI and TPW observational data. The one with TMI and TPW data depicted the position of Paka along with the intensive low surface pressure and strong wind convergence, whereas the one without the TMI and TPW data underestimated the intensity significantly.

2.3. *Tropical Cyclones: New Views*

Tropical cyclones bring strong wind and heavy rainfall on their path. The strong tropical cyclones, such as hurricanes or typhoon, cause economic loss and damages on human life when they cross the land. For tropical cyclones initiated over the oceans, it is difficult to monitor them and to forecast their intensity change, since there is no sufficient observation there. But TRMM, for the first time, offers unique opportunities to examine tropical cyclones. With TRMM, scientists are able to make extremely precise radar measurements of tropical storms over the oceans, and identify both accelerators and brakes upon their intensity. The resulting data has provided invaluable insights into the dynamics of tropical storms and rainfall (Simpson et al. 2000; Kummerow et al. 2000).

TRMM marks the first time that tropical cyclones in all ocean basins are able to be viewed from above by high resolution down-looking rain radar. In the first 13 months of operation TRMM samples 84 tropical cyclones with 1189 orbits passing within 750km of a tropical cyclone center (19% of 6227 total orbits). This sample represents over an order of magnitude more data than can be obtained from any other platform. It collects the tropical hurricanes and typhoons observed by TRMM. As pointed out in the reference (Simpson et al. 2000) the ability to forecast intensity changes in tropical cyclones has shown little progress in the past two decades. Now, with TRMM's intensive observations, scientists are able to examine the intensity changes of tropical cyclones. The study on super typhoon Paka is an example. Paka formed in the Northern Hemisphere and remained weak until December 10 when a huge convective burst occurred. In Fig. 1, the upper left panel shows the geosynchronous view from GMS satellite. The large round white area is the top of one of the early "hot towers". The upper right panel shows the TRMM radar superimposed on the geosynchronous image, while the lower left panel is the 85 GHz image from the TMI. Both the radar and the passive microwave show a clear eye, which was hidden on the geosynchronous image. The lower right shows a radar cross-section from A to B on the radar image above. The very high tower leans slightly inward toward the eye. Other radar cross sections show cloud material extruding from the cloud into the eye and almost surely sinking. The convective burst is associated with Paka's first rapid intensity increase from about 27 m/s to above 50 m/s on December 11, 1997. This first rapid deepening has been studied and related to a combination of the convective burst's carrying up high energy air and the storm core moving over the warmer.

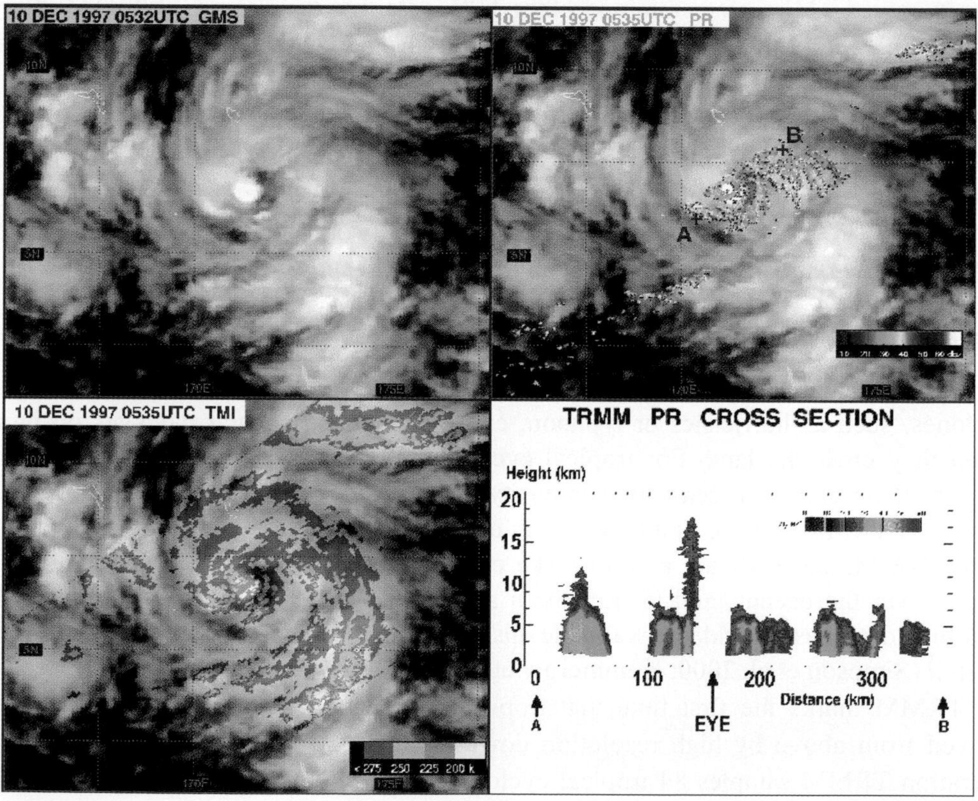

Fig. 1 Paka observed by TRMM (courtesy of Dr. J. Simpson at GSFC/NASA).

2.4. *Cloud and Precipitation Formation*

Cloud and precipitation formation is one of the key issues in the tropical precipitation study. For the first time TRMM provides such rich products that cover many components of cloud and precipitation formation. TRMM data include hydrometeors distribution, latent heat profiles derived from the Goddard Cloud Ensemble Model, separated convective and stratiform rainfall data, cloud electrification, and so on. As the time of this writing, the reported applications to cloud and precipitation include study of tropical rainfall diurnal cycles; study of the electrification of cyclones and hurricanes; and the improvement in cloud and precipitation parameterization used in numerical general circulation models.

An example: scientists (Simpson et al. 2000) systematically investigated the diurnal variation of precipitation for 1998 using TMI, precipitation radar (PR) and TMI/PR combined algorithms. Temporal variations of diurnal cycle of rainfall derived from TRMM 2A12, 2A25, and 2B31 were averaged over the oceans and the land separately over the band of 35°S-35°N.

There is a clear difference in the diurnal cycle between the land and the ocean regions. Over the ocean, a rainfall peak exists consistently in early morning, whereas over the land, the peak exists in early-mid afternoon. For both land and ocean there is seasonal variation on intensity of rainfall diurnal cycle. Over the ocean, relatively large values appeared during the late spring and early fall, whereas over the land, relatively large values exist during the winter and fall.

2.5. *TRMM for Monsoon Study*

Monsoon probably is the most prominent weather phenomenon for the people living in the subtropics because monsoon precipitation, in particular, flood or drought, may have tremendous effect on agriculture and human lives there. As an example, the Yangtze River flood from June through August 1998 destroyed over 30 million acres of farmlands and ruined more than 11 million acres of crops. Over Asia there are two well recognized monsoon systems, namely Indian Monsoon and China Monsoon.

Though Monsoon occurs seasonally resulting from the thermal contrast between the land and the around oceans, its occurring time or onset, affecting area, and the intensity vary yearly. To accurately predict monsoon is of vital importance since possible preventive measures may save life loss and reduce economic loss. Meteorologists have started to forecast monsoon back to a century ago. Mathematical climate models and statistical or empirical models are the main means for the forecast.

Scientists at GSFC NASA study the 1998 devastating natural disaster over Yangtze River region with TRMM data (Lau and Li 1999). They use TRMM TMI data-TRMM 2A12, Precipitation Radar data-TRMM 2A25, and in-situ observations from South China Sea Monsoon Experiment (SCSMEX). The analysis based on these data reveals the dynamic and thermodynamic conditions associated with development of meso-scale convective system that gave rise to the Yangtze River flood in relation to the evolution of South China Sea Monsoon. The top panel of Fig. 2 shows the two defined areas, South China Sea (10°N-24°N, 108°E-

122°E, blue shaded), and Yangtze River area (24°N-38°N, 116°E-140°E, red shaded). The bottom one shows the time series of TRMM TMI rain rates (mm day^{-1}) averaged over the South China Sea (blue line) and the Yangtze River area (red line) for the period of May 1 to June 30 1998.

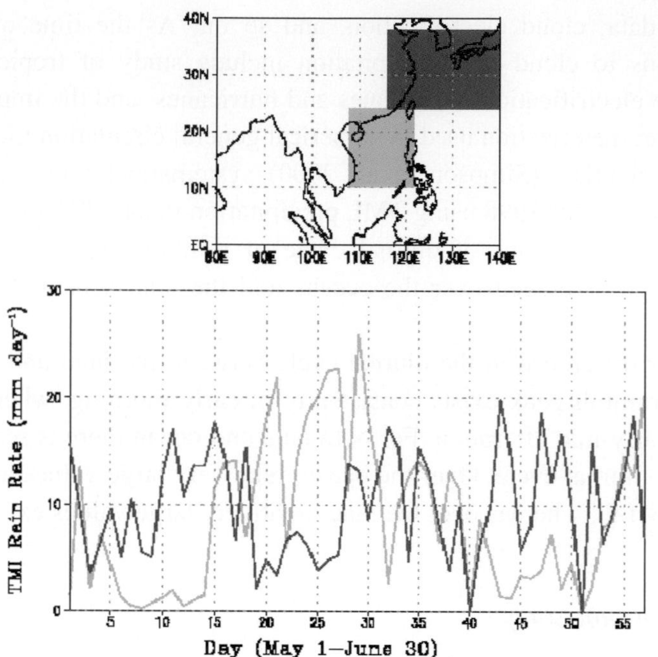

Fig. 2. Top panel: two defined areas, South China Sea (10°N -24°N, 108°E -122°E, blue shaded), and Yangtze River area (24°N-38°N, 116°E-140°E, red shaded). The bottom panel: time series of TRM TMI rain rates (mm/day) averaged over the South China Sea (blue line) and the Yangtz River area (red line) for the period of May 1 to June 30 1998 (Courtesy of Dr. X. Li at GSFC/NASA).

Two main features shown in Fig. 2: first, two lines are nearly out of phase during the whole May denoting the different precipitation variations over the two areas; secondly, high precipitation persisted over the South China Sea for a long period of 25 days (day 15 to day 39, blue line), and when it decreased after day 40 (June 9), the precipitation over the Yangtze River area seemed to increase significantly and persisted for more than 10 days. A question is asked then: "Is there a temporary connection between the evolution of South China Sea Monsoon and Yangtze River area flood?". To answer this question, the scientists investigate the evolution of the South China Sea monsoon and Yangtze River area flood. In their study, Lau and Li (1999) used TRMM TMI rain rate data combined with 850mb wind for the period of May 18-May 23, 1998. The data shows a monsoon depression over the Bay of Bengal developed in May 18, and convection occurred east of China. With the southward shift of the convection, westerly winds associated with the Bay of Bengal depression developed feeding moisture into the South China Sea around 20 May, when the monsoon onset occurred. The similar analysis over the Yangtze River area during the early stage of severe flood, June 11-

16, 1998, shows that strong southerly winds clashed with northerly winds over the Yangtze River area to produce strong horizontal wind shear with strong low level moist convergence, which leads to the first stage of severe flood.

They also use TRMM PR data to reveal the fine features of the cloud and precipitation of the convective systems over the two areas. The vertical structure of precipitation intensity, precipitation types, and melting level are clearly indicated by the PR data.

2.6. *TRMM for El Niño and La Niña Study*

El Niño and La Niña refer to abnormal cold and warm sea surface temperature (SST) events respectively, which initiate along the western coast of Peru and expand west to the central equatorial Pacific Ocean. Scientists have found that El Niño and La Niña have profound effects on weather and climate in the whole tropics and subtropics, as well as in the middle and high latitudes. To understand El Niño/La Niña and further predict them is a challenging task, and needs to treat the ocean and the atmosphere above as an integrated system: SST affects the atmosphere mainly by providing latent heat and sensible heat fluxes, and the atmosphere behavior, such as wind and precipitation, in turn, affects SST. Scientists gain their knowledge of El Niño and La Niña by analyzing observed data and carrying out the experiments with numeric atmospheric, ocean, or coupled ocean-atmosphere general circulation models.

It is well known that the low-frequency variation of tropic precipitation with an irregular 4-5 year period is mainly corresponding to SST cold/warm events. TRMM data by themselves provide a uniquely accurate rainfall data set,but the sampling becomes inadequate for short intervals and small areas. However, by combining TRMM data with other satellite rainfall estimates and rain gauge analysis, the product TRMM 3B43, titled TRMM Other Data Source, extends TRMM-like accuracy to finer space and time resolutions.

Scientists at GSFC explore the use of TRMM precipitation data in their numerical ocean model simulation (Murtugudde et al. 2000). With TRMM precipitation data as an input, their model reproduces the real SST variations over the Indian Ocean, as well as over the eastern equatorial Pacific and the equatorial Atlantic Ocean, for the El Niño period of fall 1997 to spring 1998. TRMM data successfully document the interannual precipitation variation, coordinating the SST changes beneath the atmosphere.

3. TRMM Data Products and Accessing Information

3.1. *TRMM Orbit and Instruments*

The TRMM satellite has low inclination of 35°, non-sun-synchronous, and highly precessing orbit, which allows it to fly over each position on the Earth's surface at different local time. The orbit was maintained at approximately 350 km before August 6, 2001. The average operating altitude for TRMM was changed to 403 km during the period from August 7 to 24, 2001 (referred as boost). The three primary instruments on TRMM are Visible and Infrared Scanner (VIRS), TMI, and PR. The VIRS serves as a very indirect indicator of rainfall and a transfer standard to other measurements that are made routinely using the meteorological

Polar Orbiting Environmental Satellites (POES) and the Geostationary Operational Environmental Satellites (GOES) satellites. The TMI is a passive microwave sensor and able quantify the water vapor, the cloud water, and the rainfall intensity in the atmosphere over a wide swath. The PR is the first spaceborne instrument designed to provide three-dimensional maps of storm structure. The PR measurements yield invaluable information on the intensity and distribution of the rain, on the rain type, on the storm depth and on the height at which the snow melts into rain, and the estimates of the heat released into the atmosphere at different heights. The characteristics of the three rain instruments are shown in the following table.

Table 1. Characteristics of TRMM Instruments

	Visible and Infrared Scanner (VIRS)	**TRMM Microwave Imager (TMI)**	**Precipitation Radar (PR)**
Frequency/ Wavelength	0.63, 1.6, 3.75, 10.8, and 12 μm	Dual polarization: 10.65, 19.35, 37, and 85.5 GHz Vertical polarization: 21 GHz	Vertical polarization: 13.8 GHz
Scanning Mode	Cross track	Conical	Cross track
Ground Resolution*	2.2 km / 2.4 km	4.4 km / 5.1 km at 85.5 GHz	4.3 km / 5.0 km
Swath Width *	720 km / 833 km	760 km / 878 km	215 km / 247 km

* Format: Pre-boost/Post-boost

In addition, a Lightning Imaging Sensor (LIS) and a Clouds and the Earth's Radiant Energy System (CERES) are carried on the TRMM satellite. The LIS is a calibrated optical sensor operating at 0.7774 μm and observes distribution and variation of lightning. The CERES is visible infrared sensor which measures emitted and reflected radiative energy from the surface of the Earth and the atmosphere and its constituents, operating at 0.3 to 0.5 μm in the short wave range and 8.0 to 12.0 μm in the long wave range.

3.2. *TRMM Science Data Products*

The real-time processing and reprocessing of the TRMM data are performed by TRMM Science Data Information System (TSDIS). Raw instrument data are received in near-real-time by TSDIS and then processed by the first tier of TRMM science algorithms to produce calibrated, swath-level, instrument data. With the latter, the second tier of algorithms are used to compute geophysical parameters, such as precipitation rate, also at the swath-level resolution. Finally, the third tier algorithms are used to produce gridded geophysical parameters from the first and second tier instrument data (TSDIS web page, 2003). TRMM instrument algorithms are shown in Fig. 3 (TRMM Data Processing Overview).

TRMM data are provided at five levels (level 0 to level 4), representing the processing done at different stages after GSFC receives the raw data. Level 0 is the time-ordered and quality-checked raw data received from the satellite. Level 1 products are the VIRS calibrated radiance, the TMI brightness temperatures, and the PR return power and reflectivities, at instrument pixel resolution. Level 2 products are derived geophysical parameters (e.g., rain rate, cloud liquid water, latent heat) at the same resolution and location as those of the Level 1 data. Level 3 gridded products are space-time averaged parameters. Level 4 products are analyzed products or those produced from merging measurements from TRMM and other sources. All TRMM satellite standard data products and their short names can be found in Fig. 3.

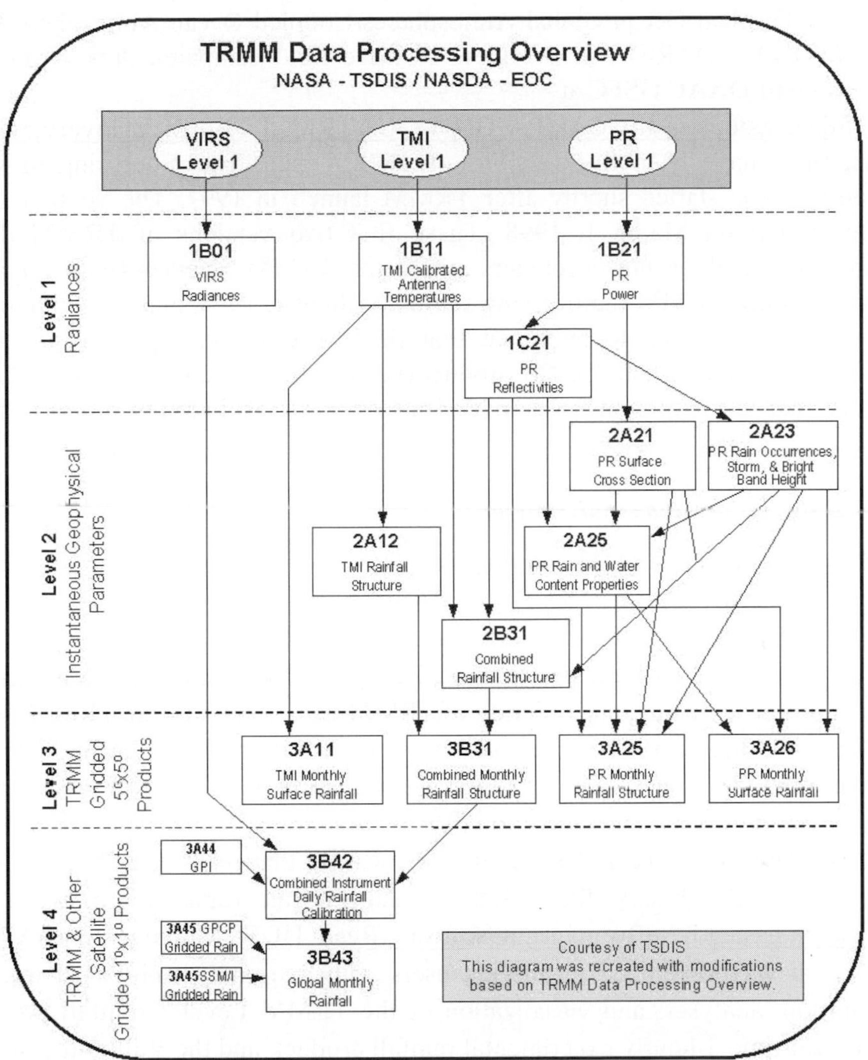

Fig. 3. TRMM data processing overview.

The field campaign program of TRMM is designed to provide ground truth for use in algorithm development of TRMM satellite measurement. To meet this goal, TRMM field campaigns employ ground-based radars and rain gauge networks to provide independent estimates of the TRMM variables, which the TRMM satellite also estimates. Also, the campaigns obtain aircraft measurements with instrumentation similar to the TMI and PR on the TRMM satellite. The NASA DC8 and ER2 aircraft support microwave sensors similar to those aboard the satellite, and the DC8 also supports Airborne Mapping Radar (ARMAR), a prototype of the TRMM satellite radar. The field campaigns consist of TExas-FLorida

Underflight TEFLUN A and TEFLUN B (focus on East Florida), Large-scale Biosphere-Atmosphere Experiment in Amazonia (TRMM-LBA), Kwajalein Experiment (KWAJEX), South China Sea Monsoon Experiment (SCSMEX), Convection And Moisture EXperiment (CAMEX), and Tropical Ocean Global Atmospheres/Coupled Ocean Atmosphere Response Experiment (TOGA COARE). An overview of these field campaign data is provided by Hydrology group of DAAC GSFC at

http://daac.gsfc.nasa.gov/CAMPAIGN_DOCS/TRMM_FE/FE_dataoverview.html.

To date, there have been five versions of TRMM data. The processing of version 1 TRMM products was started shortly after TRMM launch in 1997. The version 2 TRMM reprocessing started on March 1, 1998. These first two versions of TRMM data were provided only to the algorithm developers within the TRMM Science Team for algorithm evaluations. The third TRMM reprocessing started on June 1, 1998, and the version 3 orbital products were first released to public at that time. The fourth reprocessing started on September 1, 1998. The current TRMM product (version 5) started on November 1, 1999. A major product change for version 5 was that of the temporal resolution of product 3B42 from pentad to daily.

3.3. *TRMM Data Access and Analysis Tools*

TRMM data, value-added TRMM subsets, and TRMM ancillary data can be accessed by TRMM search and order system of the GSFC DAAC at

http://lake.nascom.nasa.gov/data/dataset/TRMM/index.html.

The system provides users with a simple and friendly web-based interface for visualization, spatial and temporal searching, and ordering of TRMM data. All first-time TRMM users must register before ordering data. Click the "New User Registration" button at the top of a page and fill the information requested in the form. A notification with new user's account ID and password is emailed to you after completion of registration.

The DAAC Hydrology Team has developed software that reads TRMM HDF data files and writes out to flat binary files. Both C and Fortran versions are available from ftp://lake.nascom.nasa.gov/software/trmm_software/Read_HDF/. Moreover, a TRMM Online Analysis System was developed to provide users with a friendly web-based interface for quick exploration, analyses, and visualization of the TRMM Level-3 rainfall products, the TRMM near-real-time 3-hourly experimental rainfall product, and the Willmott and Matsuura global climate data. Users can plot area averages (area plot) and time series (time plot) for selected areas and time periods. The system can be accessed at

http://daac.gsfc.nasa.gov/CAMPAIGN_DOCS/hydrology/TRMM_analysis.html.

TSDIS developed the TSDIS Science Algorithm Toolkit to assist the TRMM Science Team's algorithm developers. The toolkit provides a library of commonly used routines, constants, and macros for reading and writing data to and from TRMM HDF files. Each of the routines in the toolkit is callable in either C or Fortran. The toolkit also includes routines for reading land/sea data and topographical data. TSDIS Toolkit can be accessed from

http://www-tsdis.gsfc.nasa.gov/tsdis/tsdistk.html.

The TSDIS Orbit Viewer is a menu-driven graphical interface for dynamically generating images from TRMM HDF files. The viewer can display, at the full instrument resolution,

TRMM satellite, Ground Validation, browse, and Coincidence Subsetted Intermediate products, as well as other derived products. Vertical cross sections and 3D images of rain structure can also be created. The software runs on Microsoft Windows and UNIX. The source code and installation instructions for the Orbit Viewer are available from the TSDIS Web Site:

http://www-tsdis.gsfc.nasa.gov/tsdis/TSDISorbitViewer/release.html.

4. Global Precipitation Mission

Global Precipitation Measurement, targeted to launch on 2007, is one of the next generation of systematic measurement missions that will measure global precipitation with improved temporal resolution and spatial coverage. The instruments will be carried on primary spacecraft and the existing international constellation. The basic instruments of the primary spacecraft include dual-frequency precipitation radar and passive microwave radiometer. The instrument of the constellation spacecraft is passive microwave radiometer. The objectives of the GPM are to improve climate prediction; to improve the accuracy of weather and precipitation forecasts; and to provide more frequent and complete sampling of the Earth's precipitation. For more information see: http://gpm.gsfc.nasa.gov

Acknowledgments. This article would not have been written without the support of Dr. Joanne Simpson who is the TRMM project scientist from 1986 to launch in 1997 and a principal member of the TRMM Science Team since then. She gave us permission to excerpt the sample examples from the review of TRMM mission (Simpson et al. 2000) and encouraged us for doing this work. We are also grateful to three reviewers for their insightful comments.

References:

Hou, A. Y., and Coauthors, 2001: Improving global analysis and short-range forecast using rainfall and moisture observations derived from TRMM and SSM/I passive microwave sensors. *Bull. Amer. Meteor. Soc.*, **82**(4), 659-679.

Krishnamurti, T. N., and Coauthors, 2001: Real-time multianalysis-multimodel superensemble forecasts of precipitation using TRMM and SSM/I products. *Mon. Wea. Rew.*, **129**, 2861-2883.

Kummerow, C., and Coauthors, 2000: The status of the Tropical Rainfall Measuring Mission (TRMM) after two years in orbit. *J. Appl. Meteor.*, 39, 1965-1982.

Lau, K. M., and X. Li, 1999: Diagnosis of the 1998 Yangtze River flood using TRMM/SCSMEX data. *TRMM Global Precipitation Mission Meeting,* October 1999, College Park, MD, USA.

Murtugudde, R., J. P. McCreary Jr., and A. J. Busalachi, 2000: Oceanic processes associated with anomalous events in the Indian Ocean with relevance to 1997-1998. *J. Geophy. Res.* **105**(C2), 3295-3306.

Simpson, J., and Coauthors, 2000: The tropical rainfall measuring Mission (TRMM). *Earth Observation and Remote Sensing*, **4C**, 71-90.

TRMM 2000 News: http://trmm.gsfc.nasa.gov/overview_dir/mission_status.html#2000

TSDIS web page, 2003: http://tsdis.gsfc.nasa.gov

PRECIPITATION AND LATENT HEATING ESTIMATION FROM PASSIVE MICROWAVE SATELLITE MESUREMENTS: A REVIEW

SONG YANG

Goddard Earth and Science Technology Center, University of Maryland at Baltimore County
National Aeronautics and Space Administration, Goddard Space Flight Center, Code 912
Greenbelt, MD 20771, USA
E-mail: ysong@agnes.gsfc.nasa.gov

(Manuscript received 19 November 2002)

A brief history of precipitation from satellite passive microwave remote sensing and a theoretical basis for retrieval of precipitation from passive microwave measurements have been described. Two multi-frequency microwave-based physical inversion rain retrieval algorithms are given as examples of satellite-based rain retrieval algorithms. Some applications of rainfall from satellite measurements, especially from the Tropical Rainfall Measuring Mission (TRMM), have been summarized. Discussions on available latent heating retrieval algorithms indicate that the near future TRMM latent heating products would be reasonable. Future efforts on rain/latent heating retrievals are discussed. The quality of both precipitation and latent heating derived from satellite passive microwave measurements is expected to be better in the era of Global Precipitation Measurement mission.

1. Introduction

A thorough understanding of the atmospheric circulation is ultimately connected to an adequate knowledge of the time-space distribution of atmospheric diabatic heating. Precipitation represents a net condensation of water substance in the atmosphere, and the latent heat released in this process is an important component of the atmospheric heat budget. It is well established that the space-time variations of latent heating and other diabatic processes have an important impact on the large-scale circulation of the atmosphere both in the Tropics and Midlatitudes (e.g., Puri, 1987; Hack et al., 1990). In particular, atmospheric circulations are very sensitive to the vertical structure of the latent heating from precipitating clouds. Thus, an accurate determination of the time-space variations of latent heating would be of substantial benefit to atmospheric prediction over a range of time scales.

Despite the importance of the 4-dimensional structure of latent heating in governing atmospheric motion, there has been no direct measurement or dependable method of obtaining the distribution of latent heating. Precipitation naturally becomes the primary scientific target because of its strong relationship to atmospheric heating. Although traditional surface rain gauge networks and ground radar observations can supply rain measurements over land, there are not sufficient measurements, especially in remote regions, to provide a complete picture. In addition, more than two-third of the earth's surface is covered by water, where traditional meteorological measurements are not available. Precipitation estimates from satellite measurements can be used to overcome the sampling problems associated with traditional rain measuring instruments.

Methods for satellite remote sensing of precipitation have been evolving and improving for several decades. The early satellite passive microwave (PMW) radiometer rain retrieval algorithms were based on single frequency measurements from the Nimbus 5 Electrically Scanning Microwave Radiometer (ESMR-5) 19 GHz and ESMR-6 37 GHz (e.g., Wilheit et al., 1977; Weinman and Guetter, 1977). The 1978 launch of the first multi-spectral passive microwave radiometer, i.e., the Nimbus 7 Scanning Multichannel Microwave Radiometer (SMMR), would provide a golden opportunity to develop a new type of multichannel algorithms in which microphysical activity at different levels within a precipitating cloud could be detected by different frequencies. Detailed description about SMMR could be found in Gloersen and Barath (1977) and Gloersen et al. (1984). Several rain algorithms have been developed based on SMMR 37 GHz measurements, such as the "scattering method" of Spencer (1986) and Petty and Katsaros (1992). Prabhakara et al. (1986) estimated rainrates with SMMR 6.6 and 10.7 GHz information while Hinton et al. (1992) applied 6.6, 10.7, 18, 37 GHz measurements in rain retrievals. Those SMMR rain algorithms were based on schemes with regression relationships between surface rainrate and single channel Tb. Olson (1989) developed a rain retrieval algorithm aimed directly at exploiting the SMMR multichannel information. However, it was focused on hurricane rainfall. The introduction of Special Sensor Microwave Imager (SSM/I) measurements from the Defense Meteorological Satellite Program (DMSP) platforms (Hollinger et al., 1990) has stimulated multi-frequency, passive microwave-based rain retrieval algorithms. Algorithms such as developed by Wilheit et al. (1991), Liu and Curry (1992) and Petty (1994) incorporates modeled, statistically, or theoretically derived deterministic functions between measured microwave radiances and rainfall rates. The most advanced rain retrieval algorithm is the physically-based inversion scheme, which retrieves details on the vertical distribution of hydrometeors through quantifiable physical factors affecting the relationship between the satellite-measured brightness temperatures and rainrates (e.g, Olson 1989; Smith et al., 1994a; Yang and Smith, 1999a; Kummerow and Giglio, 1994a-b, and Kummerow et al., 2001). With the successful mission of TRMM launched in November 1997, the TRMM microwave imager (TMI)-based physical rain algorithms have reached a higher level for retrieving rainfall from PMW measurements (Kummerow et al., 2000; Haddard et al., 1997). One article in this book provides additional materials on microwave remote sensing of clouds and precipitation (Liu, 2003).

There are other rain retrieval algorithms based on satellite visible (VIS) and/or infrared (IR) radiometer data. This kind of algorithms is using statistical regression methods to derive relationships between VIS/IR measurements and rainrates, such as from the early simple scheme (e.g., Lethbridge, 1967) to the currently sophisticated scheme (e.g., Adler et al., 2000). The later scheme applied TRMM microwave measured rainrates to calibrate IR-based rainrates for improving the accuracy of IR rain retrievals. Barrett and Beaumont (1994) gave a good review on satellite rainfall monitoring.

The results of more than six precipitation intercomparison projects have clearly shown that instantaneous estimates of rainrates from passive microwave retrieval algorithms are much better than those from the VIS/IR-based algorithms (Wilheit et al., 1994; Ebert and Manton, 1998; Smith et al., 1998). This conclusion was expected because of the ability of microwave radiances to penetrate clouds, and the crude cloud/precipitation profiling capability that the multichannel sensors could provide. The physically-based, precipitation profile inversion approach to microwave rain estimation has been adapted to estimate not only vertical hydrometeor profiles, but also latent heating profiles (Yang and Smith, 1999a-b; Olson et al., 1999).

2. Passive Microwave Remote Sensing

2.1. *Advantages of Microwave measurements for rain retrieval*

There are two strong absorption bands due to water vapor at 22 GHz and 183 GHz and oxygen at 60 GHz and 118 GHz in the microwave spectrum. The overall effect of the combined absorption spectrum is a sequence of relatively transparent microwave windows with local peaks around 0-19, 37, 85, 130, and 220 GHz (Figure 1a). Due to the broad windows, exact frequency selections for rainfall measurements would vary with different satellite mission requirements. The relative large imaginary component of the complex refraction index of water drops in the 10-100 GHZ microwave spectrum makes cloud and rain drops effective emitters/absorbers, while relative small absorption coefficients for ice particles make these particles effective insulators and a negligible source of radiation at microwave frequencies. In addition, both liquid and ice hydrometeors become more effective scatterers as their sizes approach precipitation-size (~1 mm) while their scattering cross-sections increase as a function of frequency. Since air temperatures in the troposphere normally decrease with height, ice particles are generally located above liquid hydrometeors. Therefore, the upwelling microwave radiances emerging from precipitating clouds are essentially controlled by the emission-absorption taking place in the lower-level liquid layers and by the scattering by larger particles in the entire column. The emission-absorption-scattering features of hydrometeors in the microwave spectrum generally enable each frequency to "sense" different layers within the cloud. Thus, a sequence of microwave window channels can provide, approximately, the hydrometeor vertical distribution information within a precipitating cloud. Detailed explanations of the penetration capabilities of microwave radiation in precipitating clouds can be found in Smith and Mugnai (1989), Mugnai et al. (1988) and Smith et al. (1994b).

Figure 1b shows an example of the sensitivity of each microwave window frequency channel to hydrometeor concentrations for a heavy precipitation cloud (See Mugnai et al., 1992). The 10 GHz channel is sensitive to liquid precipitation, while 19 GHz is more sensitive to cloud drops. The 85 GHz is sensitive to precipitation in the top layer of a precipitating cloud, where graupel and snow particles predominate. The 37 GHz provides information regarding precipitation concentrations above the freezing level, but below the top layer of the precipitating cloud. However, we should understand that there is no simple correspondence between hydrometeors and microwave brightness temperatures (Tbs). Shown in Fig. 2 is an example of relationship between surface rainrates and *Tbs* for three microwave frequencies, 18, 37, and 85.6 GHz. Over ocean surfaces, 18 and 37 GHz *Tbs* increase with increasing rainrate at lower rain intensities due to increasing microwave emission/absorption by raindrops. However, as the proportion of large raindrops/ice particles become significant, *Tbs* flatten and then decrease with increasing rain intensity, as scattering by these larger precipitating hydrometeors begins to dominate. *Tbs* at 85 GHz always decrease with rain rate due to efficient scattering by ice-phase precipitation at this frequency. The use of various combinations of these channels in methods for retrieving rainrates over oceans is well documented by many investigators.

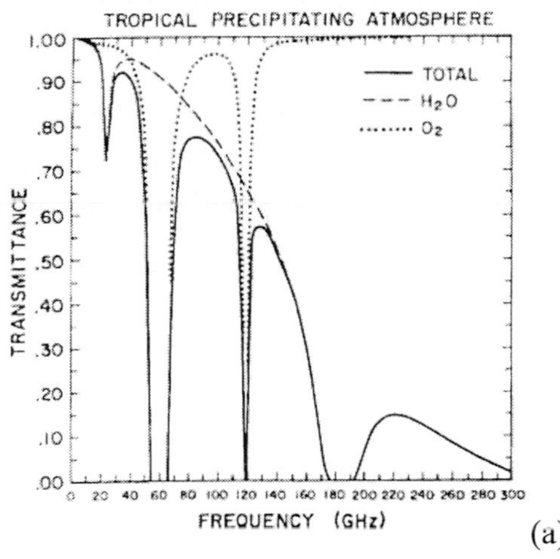

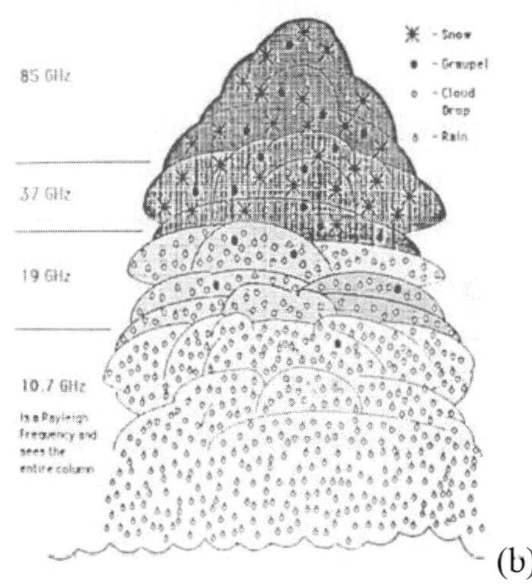

Figure 1. (a) Microwave transmittance due water vapor oxygen and the combined spectrum in a cloud-free tropical precipitation atmosphere (after Smith et al. 1994b); (b) Radiation measured in different passive microwave channel emanates from different levels in the clouds (after Mugnai et al. 1992).

Over land surfaces, microwave emission at the lower microwave frequencies is essentially indistinguishable from the emission from the relatively high-emissivity land surface, making retrieval of rainrates over continental regions based upon these channels unfeasible. Only the scattering signal of precipitation of high frequencies (e.g. 85 GHz) can be utilized to estimate rainrates over land. Since the total information content of multichannel microwave observations is significantly reduced over land, precipitation estimates from passive microwave radiometers over land are much less accurate than those over ocean.

The use of the 10 GHz channel in rain estimation methods could lead to enhanced retrieval accuracy in heavy rain situations (Smith et al., 1994c), although the precipitation information content of the 10 GHz channel is somewhat compromised by resolution in satellite applications. The TMI and the Advanced Microwave Scanning Radiometer (AMSR) both include a 10 GHz channel for better rain estimates.

2.2. *Theoretical basis*

The radiative transfer equation (RTE) (Smith et al., 1994b) for a plane-parallel medium is:

$$\cos\theta \frac{dI_v(z,\theta,\phi)}{dz} = -k_v(z)[I_v(z,\theta,\phi) - J_v(z,\theta,\phi)], \tag{1}$$

where $I_v(z,\theta,\phi)$ is the radiance at frequency v at height z propagating in the direction (θ,ϕ), k_v is the extinction coefficient of the medium. The source function including radiative scattering, $J_v(z,\theta,\phi)$, is given by

$$J_v(z,\theta,\phi) = \{1-\omega_v(z)\} B_v[T(z)] + \frac{\omega_v(z)}{4\pi}\int_0^{2\pi}\int_0^{\pi} P_v(\theta,\phi;\theta',\phi') I_v(z,\theta',\phi') \sin\phi' d\theta' d\phi', \tag{2}$$

where $\omega_v(z)$ is the single-scattering albedo, $P_v(\theta,\phi;\theta',\phi')$ is the phase function for scattering of radiation from direction (θ',ϕ') into direction (θ,ϕ), B_v is the Planck function at temperature $T(z)$ at atmospheric level z. Because the complexity of the phase function in microwave precipitation applications, equation (1) does not lend itself to simple analytical solutions. More detailed background information about the RTE and its microwave applications can be found in the book edited by Janssen (1993).

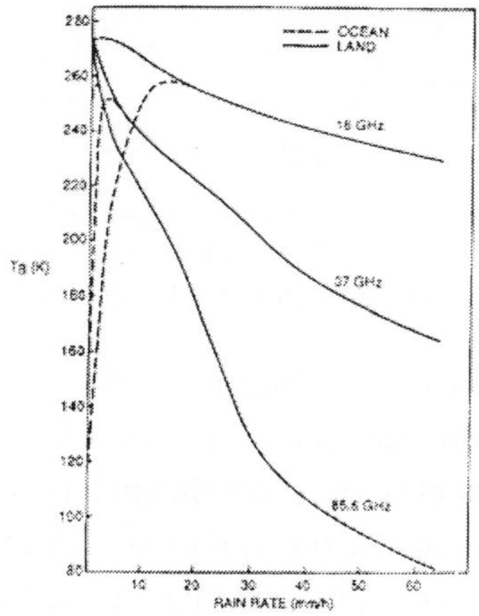

Figure 2. Relationship between microwave *Tbs* and rainrates over land and ocean (after Spencer et al. 1989).

One practical approach for precipitation retrieval applications is to use optimization schemes. Starting from an initial guess of the observed hydrometeor profiles drawn from an ensemble of cloud-resolving model simulations, the hydrometeor profile constituents are perturbed until agreement is obtained between the multi-frequency *Tbs* simulated by applying the RTE (1) to hydrometeor profile constituents and the radiometer-observed Tbs. The optimization scheme is used to iteratively select new perturbations of the hydrometeor constituents that produce better and better agreement between the simulated and observed Tbs. This method has been successfully adopted in the Florida State University (FSU) rain profile algorithm (Smith et al., 1994a; Yang and Smith, 1999a). However, this optimization scheme is difficult to be use in operational applications due to the computationally intensive RTE calculations required.

As an alternative, Bayesian approaches have been developed to circumvent the iterative RTE calculations (Kummerow et al., 1996). In these methods, a large number of cloud-resolving model simulations is used to generate candidate hydrometeor profile solutions and the upwelling microwave brightness temperatures associated with them. Since all RTE calculations are performed beforehand in the creation of the database of candidate hydrometeor profile solutions/associated Tbs, the method is relatively efficient. This technique has been employed in the development of the Goddard profiling algorithm (GPROF; Kummerow et al., 1996; Olson et al., 1996).

There are other schemes such as the IFA-SAP rain algorithm that utilizes a principal component transfer technique (Marzano, 1993). The FSU and GRPOF rain retrieval algorithms will be discussed more in the next section.

3. Inversion-based Profiling Rain Algorithms

3.1. *FSU Rain Algorithm*

The FSU precipitation profile retrieval algorithm is based on the assumption that an explicit 3-dimensional cloud model can be used to provide the microphysical underpinnings for an inversion-based retrieval procedure. The algorithm is a fully physical inversion technique designed to accept any combination of polarized or unpolarized satellite or aircraft passive microwave measurements. The algorithm uses basic multispectral radiative inversion concepts for scattering constituents, designed to retrieve the hydrometeor profiles of cloud, rain, precipitating, and non-precipitating ice. The actual design of the algorithm for satellite applications has been described by Smith et al. (1994a), and its recent improvements by Yang and Smith (1999a, 2000).

The FSU retrieval algorithm configured for SSM/I measurements consists of three key components: (1)cloud-radiation model calculations forming the initial guess database; (2)spatially deconvolved SSM/I measurements at 19, 22, 37, and 85 GHz; and (3)a RTE model configured as a functional in an optimization scheme. The initial cloud microphysics used in the algorithm are based on 3-dimensional, non-hydrostatic mesoscale model output from the University of Wisconsin-Numerical Modeling System (UW-NMS); see Tripoli (1992). This algorithm can be easily adapted to TMI measurements.

To initiate the retrieval processes, raw SSM/I or TMI satellite measurements at different channels are spatially enhanced to the resolution of the 85 GHz. This process is based on an energy-conserving spatial deconvolution scheme developed by Farrar and Smith (1992). Its impact on precipitation retrieval and its ability to partially offset problems due to non-homogenerous beam filling have been described by Farrar et al. (1994), although this procedure could introduce some small noises on the Tbs. Given the measured multispectral *Tbs* at pixel coordinates, a selection process is applied to obtain initial guess profiles from the database needed by the RTE model for carrying out the inversion process. A perturbation process is then guided by a numerical optimization scheme to generate a best solution profile from a set of initial guess profiles.

The rainfall detection module is based on a modified version of the Grody (1991) screening scheme. The algorithm estimates vertical profiles of liquid and ice water contents (i.e., LWC(z) and IWC(z)), for precipitating rain drops and ice particles, as well as LWC's and IWC's for suspended cloud drops and ice particles. The vertical rainrate and mass flux profiles, as well as surface rainrates, are derived from gravity fallout equations consistent with those applied in the cloud model. The equivalent water content profiles of the two precipitating components are finally used to estimate latent heat profiles by evaluating the vertical derivative of the precipitation mass fluxes. The vertical rainrate profile $RR_{rd}(z)$ and ice particle profile $RR_{ip}(z)$ are given by:

$$RR_{rd}(z) = -\rho_0(z) \bullet [\overline{W}_{rd}(z) + \overline{W}(z)] \bullet q_{rd}(z) / \rho_w \tag{3}$$

$$RR_{ip}(z) = -\rho_0(z) \bullet [\overline{W}_{ip}(z) + \overline{W}(z)] \bullet q_{ip}(z) / \rho_i \tag{4}$$

where ρ_0 is basic state air density, ρ_w water density, ρ_i ice density, q_{rd} rain water mixing ratio, $\overline{W}(z)$ cloud scale vertical velocity, and $RR_{rd}(z)$ and $RR_{ip}(z)$ vertically dependent mean terminal velocity.

The latent heat profile (Q) due to condensation and deposition heating is given by:

$$Q(z) = \frac{g}{C_p} \bullet [L_v \frac{\partial R^*_{rd}(z)}{\partial p} + (L_v + L_f) \frac{\partial R^*_{ip}(z)}{\partial p}] \tag{5}$$

where R^*_{rd} and R^*_{ip} are mass flux profiles for rain drops and ice particles, L_v and L_f are latent heats of vaporization and fusion, and C_p is specific heat at constant pressure p. Figure 3 is the flowchart of FSU rain algorithms. Detailed descriptions about FSU rain algorithm are given by Smith et al. (1994a) and Yang and Smith (1999a, 2000).

3.2. *The Goddard Profiling Algorithm (GPROF)*

The GPROF is supported by a large database of vertical hydrometeor and heating profiles derived from the Goddard Cumulus Ensemble (GCE) Model simulations (Kummerow et al., 1994a-6, 1996, 2001; Olson et al., 1996; Tao et al., 1993, 2003; Tao and Simpson, 1993). The vertical profiles are resampled at the nominal resolution of the radiometer (typically ~10 km) to create "footprint-resolution" profiles. RTE calculations are then performed to determine the upwelling microwave *Tbs* at the microwave radiometer frequencies or polarizations corresponding to each simulated footprint. The upwelling *Tbs* are convolved by 2-D Gaussian functions that approximate the radiometer antenna patterns at each frequency, and then the footprint-resolution profiles and corresponding convolved *Tbs* are stored together in a database that is used to support the algorithm.

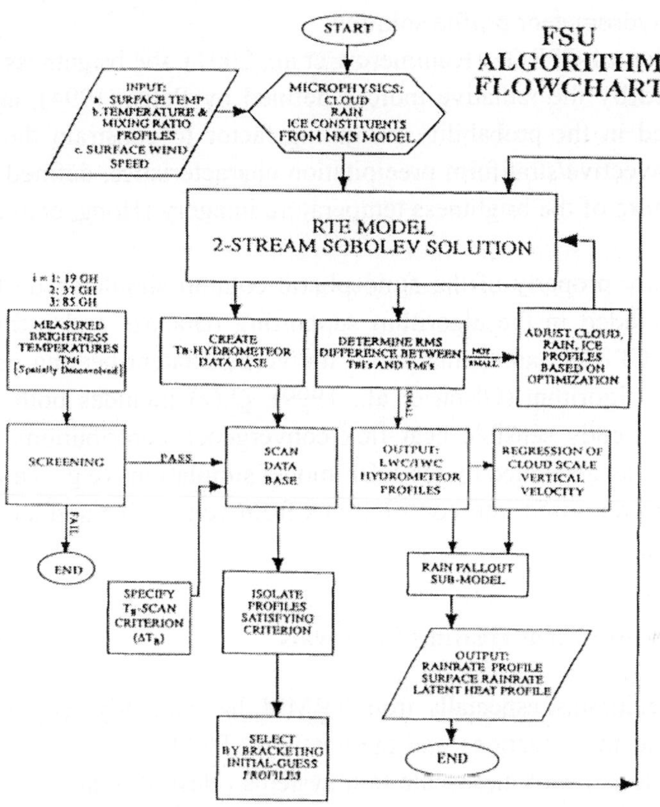

Figure 3. FSU rain algorithm flowchart (after Yang and Smith 1999a).

Given a set of observed microwave brightness temperatures, the large database of simulated profiles is scanned to find those that are radiatively consistent with the observations. An expected value of the observed hydrometeor profile is estimated by taking a weighted average of all radiatively-consistent simulated profiles- the weighting is proportional to the probability that a given simulated profile is the observed. Schematically,

$$\hat{\mathbf{E}}(\mathbf{x}) = \sum_j \mathbf{x}_j \frac{\exp\left\{-0.5\left(\mathbf{TB}_s(\mathbf{x}_j)-\mathbf{TB}_o\right)^T(\mathbf{S}+\mathbf{O})^{-1}\left(\mathbf{TB}_s(\mathbf{x}_j)-\mathbf{TB}_o\right)\right\}}{A} \tag{6}$$

where $\hat{\mathbf{E}}(\mathbf{x})$ is the expected value of the observed hydrometeor profile, $\mathbf{x}_j$ is one of the simulated hydrometeor profiles in the supporting database, $\mathbf{TB}(\mathbf{x}_j)$ is the set of simulated brightness temperatures associated with the profile $\mathbf{x}_j$, $\mathbf{TB}_o$ is the set of observed brightness temperatures, **S** and **O** are covariance matrices for the random errors in the simulated and observed brightness temperatures, respectively, and A is a normalization factor. In (6), the exponential factor is the probability that the hydrometeor profile x_j is observed, given no prior information. The Bayesian approach incorporated in GPROF is motivated by the fact that observed microwave $\mathbf{TB_s}$ do not contain enough specific precipitation information to identify a unique hydrometeor profile solution.

In recent version of GPROF (Kummerow et al., 2001), the brightness temperatures in (6) have been replaced by the radiative indices defined by Petty (1994), and additional terms have been included in the probability weighting factor to constrain the solution profile to have observed convective/stratiform precipitation characteristics, defined by the polarization and horizontal texture of the brightness temperature imagery (Hong, et al., 1999; Olson et al., 2001).

In principle, any property of the atmospheric column simulated by the cloud-resolving models can be included in the algorithm supporting database and estimated using (6). In recent versions of GPROF, an estimation of the vertical latent heating profile, or Q_l(z), has been added to the algorithm (Olson et al., 1999). Q_l(z) includes both hydrometeor phase change and vertical eddy sensible heat flux convergence contributions. The eddy sensible heat flux database was generated by the GCE model simulations (e.g., Tao et al., 2003). Q_l(z) has been further decomposed into contributions from convective and stratiform rain regions (Olson et al., 2002).

4. Precipitation and Latent Heating Retrievals

Satellite retrieved rainrates, especially from TRMM, have already had significant impacts on atmospheric science investigations and applications. TRMM datasets have been applied to identify the characteristics of mesoscale rain systems (Nesbitt et al., 2000), to study the 3-D hurricane structures (Simpson et al., 1998), and to improve hurricane simulations (Pu et al., 2001). The convergence of rain products from different TRMM instruments indicates that TRMM precipitation would show a more accurate global tropical rainfall distributions in terms of rain magnitudes and positions, such as for the well-known climatological regimes: the Intertropical Convergence Zone (ITCZ), Southern Pacific Convergence Zone (SPCZ), storm tracks, and monsoon activities (Adler et al., 2002). It makes the TRMM precipitation so far the best rain datasets to validate and improve general circulation model (GCM) results (Kummerow et al., 2001; Lin et al., 2000). Another very important role of PMW-based precipitation is its positive impacts on GCM data assimilations (e.g., Hou et al., 2000). The better precipitation imported into numerical prediction models have resulted in better short-term predictions, better tropical precipitation and tropical storm forests (Hou et al., 2001; Krishnamurti et al., 2001; Chang and Holt, 1994). The satellite-based rainrates combined with surface observations have being used to provide global monthly rainfall estimates (Adler et al., 2002). These datasets have been applied in climate variability analysis (e.g., Curtis et al., 2002), climate diagnostic studies (Rasmusson and Arkin, 1993). In addition, retrieved rainfall could be input for surface process models (Shinoda and Lukas 1995) and hydrology studics (e.g., van der Linden and Christensen, 2003).

Although research activities focused on passive microwave remote sensing of latent heating distributions have been relatively few, several investigators have published studies in this area (Olson et al, 1999; Shige et al., 2003; Tao et al., 1990, 1993, 2000, 2001; Yang and Smith 1999a-b, 2000). Yang and Smith (1999a-b, 2000) estimated latent heating from SSM/I measurements and validated the latent heating with Q_1/Q_2 budget studies from sounding datasets over the Tropical Ocean-Global Atmosphere (TOGA) Coupled Ocean-Atmosphere Response Experiment (COARE) Intensive Flux Array (IFA). Their results indicate that the estimated vertical structure of latent heating is in general consistent with Q_1 vertical distribution. They also investigated the time-space distributions of latent heating and their variations for different weather regimes. The convection shifts associated with the 1992 El Niño-Southern Oscillation (ENSO) episode were well represented with the monthly mean latent heating field. The vertical structure of monthly mean latent heating over the equatorial regions shows a primary peak around 5km and while a secondary maximum is around 1km. This secondary maximum latent heating is different from published Q_1 budget profiles so that further investigation on the lower level peak is necessary to determine its confidence. Depicted in Fig. 4 is the north-south propagation of the heating field associated with the summer Indian monsoon (top panel), the West Pacific warm pool (middle panel), and the East Pacific (bottom panel). Three interesting features are captured in the heating field over the monsoon area: (1) the seasonal variation of the latent heating in equatorial areas with maximum heating in winter and summer; (2) the northward migration of latent heating in summer associated with the evolution of the monsoon circulation; and (3) the intraseasonal oscillation of latent heating during its northward propagation. All of these features have been well documented (e.g., Yasunari 1979; Lau and Chen 1986). Over the west Pacific warm pool area, the seasonal variation of latent heating and its slow extension northward from winter to summer are obvious. It is also evident that only an annual oscillation of latent heating exists in the east Pacific area. Detailed discussions of the time-space distributions and variations of large-scale latent heating can be found in Yang and Smith (1999b).

There are so far four different latent heating retrieval algorithms: hydrometeor heating (HH), convective/stratiform heating (CSH), Goddard profiling (GPROF), and spectra latent heating (SLH). Table 1 shows previously published studies on latent heating retrieval (Tao et al., 2002). The SLH is similar to CSH except the rain height information from TRMM precipitation radar (PR) is included in SLH retrieval processes. Tao et al. (2001) summarized the strengths and weaknesses of the first three algorithms and compared their horizontal and vertical distributions of heating fields based upon one month TRMM data. The estimated latent heating structures exhibited a general consistency among these algorithms, but some discrepancies were also obvious, especially on the altitude of the maximum heating level. For example, the CSH shows only one maximum heating level, and the level varies from various geographic locations. These features were in agreement with diagnostic budget studies. A broad maximum of heating, often with two embedded peaks, was evident in HH and GPROF algorithms. The feature of the second peak of latent heating profiles needs further investigations. The altitude of primary maximum heating from HH and GPROF was at a little lower level than that from CSH, while the heating structures from CSH changed somewhat depending on applying the TRMM TMI or PR rain products.

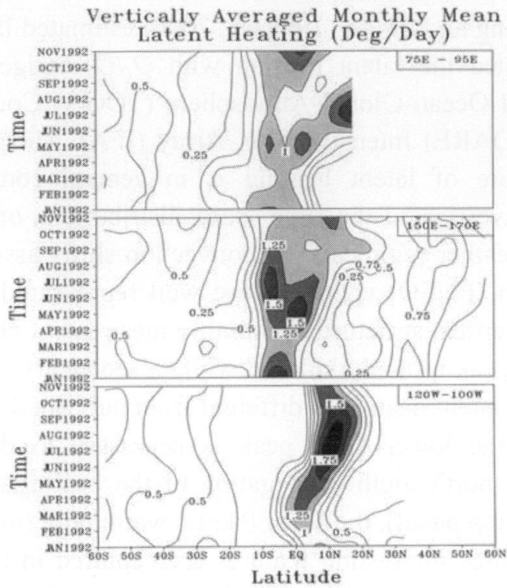

Figure 4. Time-latitude section of vertically averaged latent heating over selected regions (after Yang and Smith 1999b).

Table 1: Summary of Published Studies on Latent heating Retrieval

Author/Algorithm	Input	Cases	Resolution
Tao *et al.* (1990) HH Algorithm	Surface Rainfall, Hydrometeor Profiles (cloud water, rain, cloud ice, snow and graupel) and terminal velocity of rain, snow and graupel	GATE (1974), PRESTORM (1985)	200 - 300 km Daily
Tao *et al.* (1993b) CSH Algorithm	Surface Rainfall and its stratiform percentage	GATE (1974), PRESTORM (1985), Typhoon Thelma (1987)	200 - 300 km Daily
Olson *et al.* (1999) GPROF	Cloud model latent heating profiles and hydrometeor profiles	Hurricane Andrew (1992), TOGA COARE (1992-93)	25 - 50 km Instantaneously
Yang and Smith (1999a, b, 2000) HH Algorithm	Hydrometeor Profiles (cloud water, rain, ice particles) and Cloud Vertical Velocities/ Terminal Velocity of rain and ice particles	TOGA COARE (1992-93) Global Tropical region (1992)	15 - 50 km Instantaneously $2.5^{o} \times 2.5^{o}$ Instantaneously
Tao *et al.* (2000) CSH Algorithm	Surface Rainfall and its stratiform percentage	TOGA COARE (1992-93)	500 x 500 km 3-6 hourly
Tao *et al.* (2001)	CSH, GPROF and HH Algorithms	February 1998	$0.5^{o} \times 0.5^{o}$ Daily
Shige *et al.* (2003) SLH Algorithm	Convective and stratiform Characteristics Precipitation Top Height Rainfall rate at melting layer	TOGA COARE (1992-93)	500 x 500 km Instantaneously

Shown in Fig. 5 are surface rainrates and Q_l estimates from TMI observations of Supertyphoon Paka, derived from applications of the prototype Version 6 GPROF algorithm. The eye wall structure, spiral convective rainbands, and weaker outer bands/intervening stratiform rainfall are well-defined (left panel). The azimuthally-averaged latent heating profile within 1-degree radius of the storm center is shown in the right panel. The mean heating profile indicates a maximum of heating at 6-7 km altitude and weak evaporative cooling close to the ocean surface. It is generally in agreement with Q_l budget studies for severe rain systems (e.g., Houze, 1989). However, there are differences in latent heating profiles at different areas of the Supertyphoon Paka. The strong, deep heating in the inner core (eye wall) region, lesser heating associated with the first major spiral rainbands, and weaker upper-level heating/low-level cooling associated with stratiform precipitation at larger radii are evident (figure not shown here). These kinds of vertical heating distributions are linked to their different cloud structures. The vertical structures of retrieved latent heating seem reasonable, but await more thorough analysis.

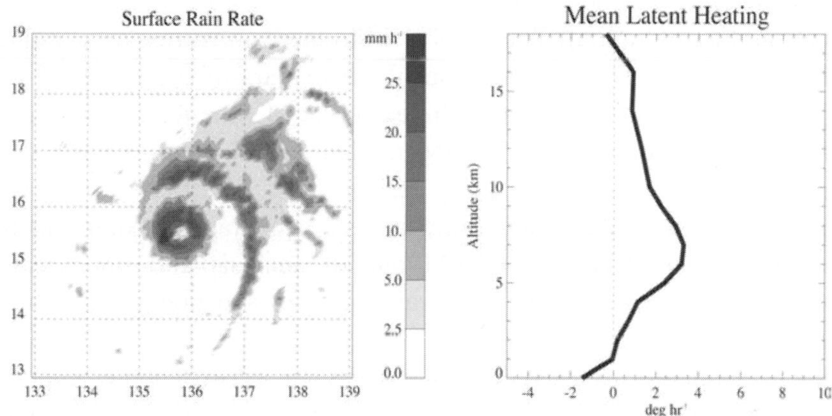

Figure 5. Surface rain distribution (left panel) and azimuthally-averaged latent heating structure within 1-degree radius of Supertyphoon Paka (right panel) on 19 December of 1997.

It is difficult to validate latent heating estimates because latent heating induced temperature changes are quickly balanced by vertical motions and adiabatic cooling or subsidence warming. However, the dynamical response to heating can be detected using other methods, such as dual-Doppler radar analyses and large-scale budget studies based upon rawinsonde data, to quantitatively check retrieved heating profile estimates. Descriptions of recent GPROF modifications and validation tests can be found in studies by Olson et al. (2002) and Yang et al. (2002).

5. Conclusions and Future Work

The satellite passive microwave remote sensing of precipitation has been developed for several decades. It started from the EMSR-5 in early 1970's, evolved into SMMR in late 1970's, SSM/I in late 1980's, and the current TRMM mission. Major progress on the accuracy of rainfall from microwave remote sensing technology emerged with the passive microwave SSM/I measurements in the last decade. Many rainfall comparison/ground validation projects have demonstrated that precipitation from passive microwave radiometer measurements is more accurate than from IR-based measurements. TRMM precipitation products have already stimulated huge interests in the earth science communities for applying TRMM data in many scientific research fields.

The theoretical basis of rain retrievals from passive microwave radiometers has been discussed in this paper. A brief history of developments of PMW-based rain retrievals is also described. The GPROF and FSU rain retrieval algorithms are given as examples of the multi-frequency microwave-based physical inversion rain algorithms. The GPROF rain algorithm has been successfully operated in the TRMM Science Data and Information System (TSDIS) to generate near-real time surface rainfall products. In addition, some precipitation-related applications in earth sciences, such as in weather predictions, GCM model data assimilations, and model validations, have been summarized to show the importance of precipitation derived from satellite PMW remote sensing.

Since uncertainties of rainfall from PMW instrument measurements still exist, more validation studies are necessary to further qualify error bars of the retrieved precipitation. Analysis of TRMM rain datasets revealed the existence of time-dependent regional biases, which indicated that uncertainties of TRMM rainfall would be associated with different weather regimes (Berg et al., 2002). Early results show that the weather regime-based database approach in rain retrieval algorithm is very encouraging (Shin and Kummerow, 2003). Therefore, further efforts in TRMM rainfall analysis and improvement of rain algorithms would lead to a better rain retrieval algorithm and precipitation products from satellite PMW measurements.

Latent heating estimates from satellite PMW measurements is another important TRMM objective. Several latent heating retrieval algorithms developed in recent years have been summarized. Results indicated that the retrieved latent heating from PMW measurements appeared reasonable. However, official TRMM latent heating products are still not available yet. Due to the difficulty of making direct measurements of latent heating, more studies on latent heating retrieval techniques and the accuracy of estimated latent heating are required. Although preliminary latent heating outputs will be soon available in the version 6 TRMM products, more validation studies using sounding network budget analyses and dual-Doppler radar observations are under development. In addition, efforts on latent heating separation for convective and stratiform components and their uncertainties are in progress. Applications of latent heating such as in numerical model prediction and data assimilation are at an early stage, while TRMM latent heating products have already attracted a lot of attentions from modeling community.

The Global Precipitation Measurement (GPM) mission proposed by the National Aeronautics and Space Administration (NASA) and its national/international partners would generate an more accurate global precipitation map from PMW measurements at 3-hour time scale in the near future (Shepherd et al., 2002). Both accuracy and confidence of global precipitation and latent heating products would be better in the GPM era. More impacts of their applications are expected on research and operational communities of earth sciences.

Acknowledgements: The author gratefully thanks William S. Olson from NASA/GSFC for his comments and editing this manuscript. Comments from W-K Tao (NASA/GSFC) and Chris Kummerow (Colorado State University) are appreciated for substantially improving the quality of this paper. This research was supported by NASA Global Water and Energy Cycle (GWEC), TRMM, and GPM projects.

References

Adler, R. F., G. J. Huffman, D. T. Bolvin, S. Curtis, and E. J. Nelkin, 2000: Tropical rainfall distributions determined using TRMM combined with other satellite and raingauge information. *J. Appl. Meteor.*, **39**, 2007-2023.

Adler, R. F., C. Kummerow, D. Bolvin, S. Curtis, and C. Kidd, 2002: Status of TRMM monthly estimates of tropical precipitation. *Meteorological Monographs*, in press.

Barrett, E.C., and M.J. Beaumont: Satellite rainfall mornitoring: An overview. *Remote Sensing Reviews*, **11**, 23-48, 1994.

Berg, W., C. Kummerow, and C.A. Morales, 2002: Differences between east and west Pacific rainfall systems. *J. Climate*, **15**, 3659-3672.

Chang, S.W., and T.R. Holt, 1994: Impact of assimilating SSM/I rainfall rates on numerical prediction of winter cyclones. *Mon. Wea. Rev.*, **122**, 151-164.

Curtis, S., G. J. Huffman, and R. F. Adler, 2002: Precipitation anomalies in the tropical Indian Ocean and their relation to the initiation of El Niño. *Geophys. Res. Letters*, **29**(10), 1441,

Ebert, E.E., and M.J. Manton, 1998: Performance of satellite rainfall estimation algorithms during TOGA-COARE. *J. Atmos. Sci.*, **55**, 1537-1557.

Farrar, M.R., and E.A. Smith, 1992: Spatial resolution enhancement of terrestrial features using deconvolved SSM/I microwave brightness temperatures. *IEEE Trans. Geosci. Rem. Sens.*, **30**, 349-355.

Farrar, M.R., E.A. Smith, and X. Xiang, 1994: The impact of spatial resolution enhancement of SSM/I microwave brightness temperatures on rainfall retrieval algorithms. *J. Appl. Meteor.*, 33, 313-333.

Gloerson, P., and Co-authors, 1984: A summary of results from the first Nimbus 7 SMMR observations. *J. Geophys. Res.*, **89**, 5335-5344.

Gloerson, P., and F.T. Barath, 1977: A scanning multichannel microwave radiometer for Nimbus-G and Seasat-A. *IEEE J. Ocean. Eng.*, **OE-2**, 172-178.

Grody, N.C., 1991: Classification of snow cover and precipitation using the Special Sensor Microwave Imager. *J. Geophys. Res.*, **96**, 7423-7435.

Hack, J.J., and W.H. Schubert, 1990: Some dynamical properties of idealized thermally–forced meridional circulations in the tropics. *Meteor. Atmos. Phys.*, **44**, 101–118.

Haddad, Z.S., and Co-authors, 1997: The TRMM “day-1” radar/radiometer combined rain-profing algorithm. *J. Meteor., Soc. Japan*, **75**, 799-809.

Hinton, B.B., W.S. Olson, D.W.Martin, and B. Auvine, 1992: A passive microwave algorithm for tropical rainfall. *J. Appl. Meteor.*, **31**, 1379-1395.

Hollinger, J.P., J.L. Peirce, and G.A. Goe, 1990: SSM/I instrument evaluation. *IEEE Trans. Geosci. Rem. Sens.* **GE-28**, 781-790.

Hong, Y., C. D. Kummerow, and W. S. Olson, 1999: Separation of convective/stratiform precipitation using microwave brightness temperature. *J. Appl. Meteor.*, **38**, 1195-1213.

Hou, A. Y., and Coauthors, 2001: Improving global analysis and short-range forecast using rainfall and moisture observations derived from TRMM and SSM/I passive microwave instruments. *Bull. Amer. Meteorological Soc.*, **82**, 659-679.

Hou, A.Y., S. Q. Zhang, A. M. da Silva, and W. S. Olson, 2000: Improving assimilated global data sets using TMI rainfall and columnar moisture observations. *J. of Climate*, **13**, 4180-4195.

Houze, R.A., Jr., 1989: Observed structures of mesoscale convective systems and implications for large-scale heating. *Quart. J. Roy. Meteor. Soc.*, **115**, 427-461.

Janssen, M.A., 1993: Atmospheric Remote Sensing by Microwave Radiometry. John Wiley & Sons, Inc. 572pp.

Kedem, B., L. Chiu, and G.R. North, 1990: Estimation of mean rain rate: Application to satellite observations. *J. Geophys. Res.*, **95**, 1965-1972.

Krishnamurti, T. N., and Coauthors, 2001: Real-time multianalysis-multimodel superensemble forecasts of precipitation using TRMM and SSM/I products. *Mon. Wea. Rev.*, **129**, 2861-2883.

Kummerow, C., and L. Giglio, 1994a: A passive microwave technique for estimating rainfall and vertical structure information from space. Part I: Algorithm description. *J. Appl. Meteor.*, **33**, 3-18.

Kummerow, C., and L. Giglio, 1994b: A passive microwave technique for estimating rainfall and vertical structure information from space. Part II: Applications to SSM/I data. *J. Appl. Meteor.*, **33**, 19-34.

Kummerow, C., W.S. Olson, and L. Giglio, 1996: A simplified scheme for obtaining precipitation and vertical hydrometeor profiles from passive microwave sensors. IEEE Trans. *Geosci. Rem. Sens.*, **34**, 1213-1232.

Kummerow, C., and Co-authors, 2000: The status of the Tropical Rainfall Measuring Mission (TRMM) after two years in orbit. *J. Appl. Meteor.*, **39**, 1965-1982.

Kummerow, C., and Co-authors, 2001: The evolution of the Goddard profile algorithm (GPROF) for rainfall estimation from passive microwave sensors. *J. Appl. Meteor.*, **40**, 1801-1820.

Lau, K.M., and P.H. Chen, 1986: Aspects of the 40-50 day oscillation during the northern summer as inferred from out going longwave radiation. *Mon. Wea. Rev.*, **113**, 1889-1909.

Lethbridge, M, 1967: Precipitation probability and satellite radiation data. *Mon. Wea. Rev.*, **95**, 487-490.

Lin, X., D.A. Randall, and L.D. Fowler, 2000: Diurnal variation of the hydrologic cycle and radiative fluxes: comparison between observations and a GCM. *J. Climate*, **13**, 4159-4179.

Liu, G., 2004: Satellite microwave remote sensing of clouds and precipitation. *In this book.*

Liu, G., and J.A. Curry, 1992: Retrieval of precipitation from satellite microwave measurements using both emission and scattering. *J. Geophys. Res.*, **97**, 9959-9974.

Marzano, F.S., 1993: *Telerilevamento e propagazione elettromagnetica in mezzi naturali diffondenti.* Ph.D. Dissertation, Department of Electrical Engineering, University "La Sapienza" of Rome, Italy, 211 pp.

Mugnai, A., and E.A. Smith, 1988: Radiative transfer to space through a precipitating cloud at multiple microwave frequencies. Part I: Model description. *J. Appl. Meteor.*, **27**, 1055-1073.

Mugnai, A., E.A. Smith, and X. Xiang, 1992: Passive microwave precipitation retrieval from space: A hybid statistical-physical algorithm. In URAD'92, *Proceedings of the Special Meeting on Microwave Radiometry and Remote Sensing Applications*, Wave Propagation Laboratory, Boulder, CO, June 1992, NOAA, 237-244.

Nesbitt, S.W., E.J. Zipser, and D.J. Cecil, 2000: A census of precipitation features in the tropics using TRMM: radar, ice scattering, and lightning observations. *J. Climate*, **13**, 4087-4106.

Olson, W. S., Y. Hong, C. D. Kummerow, and J. Turk, 2001: A texture-polarization method for estimating convective/stratiform precipitation area coverage from passive microwave radiometer data. *J. Appl. Meteor.*, **40**, 1577-1591.

Olson, W. S., 1989: Physical retrieval of rainfall rates over the ocean by multispectral microwave radiometry- Application to tropical cyclones. *J. Geophys. Res.*, **94**, 2267 - 2280.

Olson, W. S., C. D. Kummerow, G. M. Heymsfield, and L. Giglio, 1996: A method for combined passive-active microwave retrievals of cloud and precipitation profiles. *J. Appl. Meteor.*, **35**, 1763-1789.

Olson, W. S., C. D. Kummerow, Y. Hong, and W.-K. Tao, 1999: Atmospheric latent heating distributions in the Tropics derived from passive microwave radiometer measurements. *J. Appl. Meteor.*, **38**, 633-664.

Olson, W., and Co-authors, 2002: Estimation of precipitation and latent heating distributiuions by satellite pass/active microwave remote sensing. *Proceedings of the SPIE*, October 23-27, Hangzhou, China.

Petty, G.W., and K.B. Katsaros, 1990: Precipitation observed over the south China Sea by the Nimbus-7 scanning multichannel microwave radiometer during Winter MONEX. *J. Appl. Meteor.*, **29**, 273-278.

Petty, G. W., 1994: Physical retrieval of over-ocean rain rate from multichannel microwave imagery. Part I: Theoretical characteristics of normalized polarization and scattering indices. *Meteorol. Atmos. Phys.*, **54**, 79-100.

Prabhakara, C., D.S. Short, W. Wiscombe, R.S. Fraser, B.E. Vollmer, 1986: Rainfall over oceans inferred from Nimbus 7 AMMR: Application to 1982-83 El Niño. *J. Climate Appl. Meteor.*, **25**, 1464-1474.

Pu, Z., and Co-authors, 2002: The impact of TRMM data on mesoscale numerical simulation of supertyphoon Paka. *Mon. Wea. Rev.*, **130**, 2448-2458.

Puri, K., 1987: Some experiments on the use of tropical diabatic heating information for initial state specification. *Mon. Wea. Rev.*, **115**, 1394-1406.

Rasmusson, E.M., and P.A. Arkin, 1993: A global view of large-scale precipitation variability. *J. Clim.*, **6**, 1495-1522.

Shepherd, J.M., A. Mehta, E.S. Smith and W.J. Admas, 2002: Global Precipitation Measurement – report 1, Summary of the first GPM Partners Planning Workshop. NASA official technical report.

Shige, S., Y. N. Takayabu, W.-K. Tao and D. Johnson, 2003: Spectral retrieved of latent heating profiles from TRMM PR data. Part I: Algorithm development with a cloud resolving model. *J. Applied Meteor.* (submitted).

Shin, D.-B. and C. Kummerow, 2003: Parametric rainfall retrieval algorithms for passive microwave radiometers. *J. Appl. Meteor.,* conditionally accepted.

Shinoda, T., and R. Lukas, 1995: Lagrangian mixed layer modeling of the western equatorial Pacific. *J. Geophys. Res.*, **100**, 2523-2541.

Simpson, J., J. Halverson, H. Pierce, C. Morales, and T. Iguchi, 1998: Eyeing the eye: Exciting early stage science results from TRMM. *Bull. Amer. Meteor. Soc.*, **79**, 1711-1711.

Smith. E.A., and A. Mugnai, 1989: Radiative transfer to space through a precipitating cloud at multiple microwave frequencies. Part II: Influence of large ice particles. *J. Meteor. Soc. Japan*, **67**, 739-755.

Smith, E.A., X. Xiang, A. Mugnai, and G.J. Tripoli, 1994a: Design of an inversion-based precipitation profile retrieval algorithm using an explicit cloud model for initial guess microphysics. *Meteorol. Atmos. Phys.*, **54**, 53-78.

Smith, E.A., C. Kummerow, and A. Mugnai, 1994b: The emergence of inversion-type precipitation profile algorithms for estimation of precipitation from satellite microwave measurements. *Remote Sensing Reviews*, **11**, 211-242.

Smith E.A., X. Xiang, A. Mugnai, R. Hood, and R.W. Spencer, 1994c: Behavior of an inversion-based precipitation retrieval algorithm with high resolution AMPR measurements including a low frequency 10.7 GHz channel. *J. Atmos. Oceanic Tech.*, **11**, 858-873.

Smith, E.A., and Co-authors, 1998: Results of WetNet PIP-2 project. *J. Atmos. Sci.*, **55**, 1483-1536.

Spencer, R.W., 1986: A satellite passive 37 GHz scattering-based method for measuring oceanic rain rates. *J. Climtate Appl. Meteor.*, **25**, 754-766.

Spencer, R. W., H. M. Goodman, R. E. Hood, 1989: Precipitation retrieval over land and ocean with the SSM/I: Identification and characteristics of the scattering signal. *J. Atmos. Oceanic Tech.* **6**, 254–273.

Tao, W.-K., and Co-authors, 2001: Retrieved vertical profiles of latent heating release using TRMM rainfall products for February 1998. *J. appl. Meteor.*, **40**, 957-982..

Tao, W.-K., J. Simpson, S. Lang, M. McCumber, R. Adler and R. Penc, 1990: An algorithm to estimate the heating budget from vertical hydrometeor profiles. *J. Appl. Meteor.*, **29**, 1232-1244.

Tao, W.-K., S. Lang, J. Simpson and R. Adler, 1993: Retrieval Algorithms for estimating the vertical profiles of latent heat release: Their applications for TRMM. *J. Meteor. Soc. Japan* , **71**, 685-700.

Tao, W.-K., S. Lang, J. Simpson, W. S. Olson, D. Johnson, B. Ferrier, C. Kummerow and R. Adler, 2000: Retrieving vertical profiles of latent heat release in TOGA COARE convective systems using a cloud resolving model, SSM/I and radar data, *J. Meteor. Soc. Japan,* **78**, 333-355.

Tao, W.-K., and co-authors, 2002: TRMM Latent Heating Algorithms and Proposed GPM Cloud Heating/Moistening Algorithms. *2nd GPM Workshop* , May 22-24, *Tokyo, Japan.*

Tao, W.-K., and J. Simpson, 1993: The Goddard Cumulus Ensemble Model. Part I: Model description. *Terrestrial, Atmospheric and Oceanic Sciences,* Vol. **4**, 19-54.

Tao, W.-K., and Co-authors, 2003: Microphysics, Radiation and Surface Processes in a Non-hydrostatic Model, *Meteorology and Atmospheric Physics*, **82**, 97-137.

Tripoli, G.J., 1992: An explicit three-dimensional nonhydrostatic numerical simulation of a tropical cyclone. *Meteorol. Atmos. Phys.*, 49, 229-254.

Van der Linden S., and J.H. Christensen, 2003: Improved hydrological modeling for remote regions using a combination of observed and simulated precipitation data. *J. Geophys. Res.*, **108**(D2), 4072, doi: 10.1029/2001JD001420.

Weinman, J.A. and P.J. Guetter, 1977: Determination of rainfall distributions from microwave radiation measured by the Nimbus 6 ESMR. *J. Appl. M eteor.*, **16**, 437-442.

Wilheit, T. T., A. T. C. Chang, M. S. V. Rao, E. B. Rodgers, and J. S. Theon, 1977: A satellite technique for quantitatively mapping rainfall rates over the oceans. *J. Appl. Meteor.*, **16**, 551-560.

Wilheit, T.T., A.T.C. Chang, and L.S. Chiu, 1991: Retrieval of monthly rainfall indices from microwave radiometric measurements using probability distribution functions. *J. Atmos. Oceanic Technol.*, **8**, 1118-1136.

Wilheit, T., and Co-authors, 1994: Algorithms for the retrieval of rainfall from passive microwave measurements. *Remote Sensing Rev.*, **11**, 163-194.

Yang, S., and E.A. Smith, 1999a: Moisture budget analysis of TOGA-COARE using SSM/I retrieved latent heating and large scale Q_2 estimates. *J. Atmos. Oceanic Technol.*, **16**, 633-655.

Yang, S., and E.A. Smith, 1999b: Four dimensional structure of monthly latent heating derived from SSM/I satellite measurements. *J. Clim.*, **12**, 1016-1037.

Yang, S., and E.A. Smith, 2000: Vertical Structure and transient behavior of convective-stratiform heating in TOGA COARE from combined satellite-sounding analysis. *J. Appl. Meteor.*, **39**, 1491-1513.

Yang, S., W.S. Olson, E.A. Smith, C.D. Kummerow, and Shuyi Chen, 2002: Precipitation/Latent Heating Retrieval and Validation. *Proceedings of the SPIE*, October 23-27, Hangzhou, China.

Yasunari, T., 1979: Cloudiness fluctuations associated with the Northern Hemisphere summer monsoon. *J. Meteor. Soc. Japan*, **57**, 227-242.

AEROSOLS AND CLIMATE: A PERSPECTIVE OVER EAST ASIA

ZHANQING LI
Department of Meteorology and the Earth System Science Interdisciplinary Center
University of Maryland, College Park, MD 20742, USA
E-mail: zli@atmos.umd.edu, http://www.atmos.umd.edu/~zli/

(Manuscript received 11 March 2003)

Aerosol is becoming a central theme in the climate research arena, due to many new findings concerning their significant direct and indirect effects on climate (e.g. by altering temperature, cloud, radiation and precipitation) and to the large uncertainties in our estimates of aerosol forcing on climate. Despite the large loading and complex properties of East Asian aerosols, our knowledge of these aerosols and their climatic effects is so meager that they arguably present the last frontier in aerosol and climate research. While their climate effects are notably strong, the magnitude and mechanisms of their influence are far from being clear. Findings concerning how aerosols interact with energy and water cycles in other regions of cleaner environment may not be valid here. More attention needs to be focused on Asian aerosols in order to examine the existing aerosol-climate paradigms and to explore new ones. This paper provides an overview of Chinese aerosols in terms of their physical, chemical and optical properties and their potential impact on regional climate. Both anthropogenic and natural aerosols are addressed. General discussions concerning aerosol observation methods, research tools and approaches are also given. Findings by Chinese scientists are also reviewed to provide the state-of-the-art of Chinese aerosol research.

1. Introduction

While aerosols have been studied for a long time, it was not until the early 1990s that the role of aerosols in climate was widely recognized. Aerosols were identified as a central missing component in most, if not all, general circulation models (GCMs) that simulated climate changes at odds with observations in terms of long-term trends and spatial distributions (Hansen et al. 1997). After introducing sulfate aerosols estimated from industrial emissions (e.g. Langner and Rodhe 1991), model-simulated climate changes in response to the buildup of greenhouse gases became more realistic (Kiehl and Rodhe 1995; Mitchell et al. 1995). Investigations of a wide-range of aerosol effects ensued and many breakthrough findings were reported in the following years, referred to as the "exploratory phase" of aerosol research (Kaufman et al. 2002a). Major findings include the contribution of aerosols to the suppression of precipitation and the slowdown of hydrological cycles by dust storms (Rosenfeld et al. 2001; Ramanathan et al. 2001); to air pollution (Rosenfeld 2000; Rosenfeld and Woodley 2001), and to fire smoke plumes (Kaufman and Fraser 1997; Rosenfeld 2000). Other important findings include a larger reduction of the solar radiation budget at the surface than at the top of the atmosphere due to absorbing aerosols (Li 1998; Satheesh and

Ramanathan 2000; Li and Trishchenko 2001) and strong radiative heating in the atmosphere due to the mixing state of black carbon (Jacobson 2001). It must be borne in mind that these findings were drawn largely from a handful of cases. It remains an open question whether the findings are fortuitous. If so, it is critical to investigate under what circumstances the effects come into play. As we progress from the exploratory phase to quantitative phase in aerosol research, regional and global effects of aerosols must now be quantified. At present, the global averages of aerosol direct and indirect radiative forcing estimates are subject to very large uncertainties, as assessed by the Intergovernmental Panel on Climate Change (Penner et al. 2001).

While carbon dioxide (CO_2)-induced climate warming still dominates climate change at present, the sum of other anthropogenic forcings, especially ozone (O_3), methane (CH_4) and black carbon associated with air pollution, could exceed the effect of CO_2 over the next century (Hansen et al. 2000, 2002). Unlike the uniformly mixed CO_2, aerosol properties and effects exhibit considerable spatial and temporal variability (Fig. 1, left), of which we have a rather poor knowledge and understanding. This is especially the case in the developing world where many absorbing aerosols originate from air pollution, biomass burning, and dust storms (Streets et al. 2001; Lelieveld et al. 2001; Dickerson et al.2002).

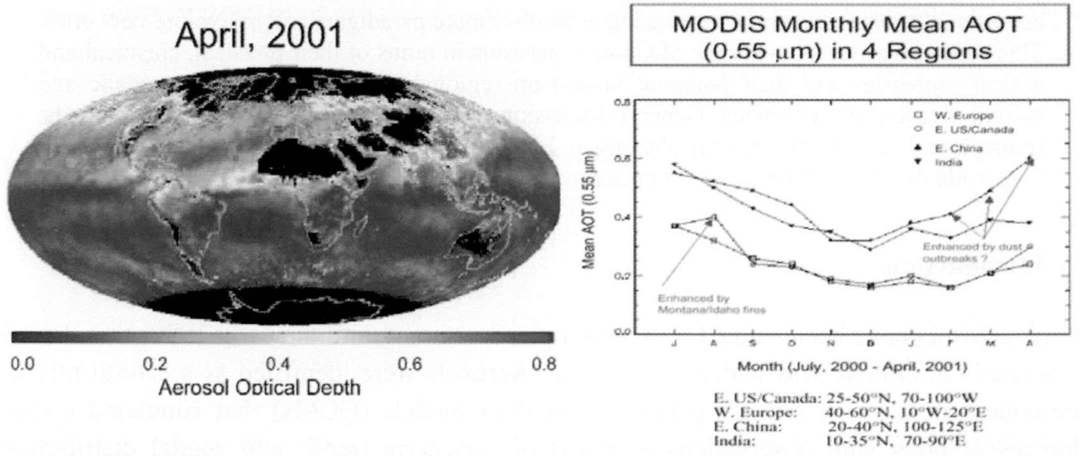

Figure 1. Left: Aerosol optical depth derived from MODIS showing total aerosol loading from all sources. Right: Comparison of mean aerosol optical depths derived over East China, India, West Europe and East United States and Canada from MODIS (Courtesy of Y. Kaufman and A. Chu).

Asia, with 60% of the world's population, is one of the heaviest aerosol-laden regions in the world (Fig. 1 left). At present, aerosol optical depth averaged over India and China is about twice the value in western Europe and eastern United States (Fig. 1 right). As this region is undergoing rapid economic development, increases in the emissions of aerosols and gases may continue for a foreseeable future, especially for black carbon aerosols, a by-product of energy consumption. Demands for energy in Asia are projected to increase rapidly in the future. Coal will remain the dominant source of energy. Dust storms, which are particularly prevalent in northeast Asia in the spring, are another significant source of aerosol. In recent years, dust storms in this region appear to have increased in severity (Husar et al. 2001). The physical properties and chemical composition of the aerosols in this region differ considerably from those found elsewhere. As a result, not only is the magnitude of aerosol forcing large, the mechanisms by which aerosols interact with energy and water cycles may differ from those identified in relatively clean environments elsewhere. Aerosol effects are not limited to climate. Severe air pollution in China may have significantly reduced crop yields (Chameides et al. 1999).

China, as a primary source of both natural and anthropogenic aerosols in eastern Asia, is drawing much scientific attention. A couple of international aerosol experiments (e.g. ACE/Asia, TRACE-P) took place in this region. Chinese scientists have conducted many aerosol-related investigations (Zhang and Mao 2001) but few are known to the world since the majority of publications are written in the Chinese language.

This paper reviews aerosol and climate studies, primarily in the context of East Asian aerosols over China. The state of knowledge concerning both natural (mineral dust) and anthropogenic aerosols and major advances in understanding their climatic effects are presented, based on papers published in both English and Chinese. As a self-complete overview article of certain tutorial value to graduate students and junior scientists, necessary fundamental knowledge on aerosol observation and understanding their climate effects is included.

2. Aerosol Optical Properties and Direct and Indirect Forcing

2.1. *Basic Aerosol Optical Parameters*

Aerosol direct radiative effects can be determined by three basic optical parameters: the aerosol optical depth, the single-scattering albedo and the asymmetry factor. The indirect effect is much more complicated and cannot be fully defined by any set of aerosol parameters. It depends on both aerosol inherent physical and chemical properties and the atmospheric environment. However, aerosol particle size plays an essential role that is also more readily available. The most widely used aerosol size information is the effective radius and the Ångström coefficient.

Aerosol Optical Thickness

A fundamental aerosol optical property is the total column aerosol optical depth, or thickness. It is essentially the aerosol extinction coefficient (σ_e) integrated over a vertical path through the atmosphere from the top of the atmospheric column (TOA) to the ground and is a function of wavelength, λ:

$$\tau(\lambda) = \int_0^{\mathrm{TOA}} \sigma_e(\lambda, h)\, dh. \quad (1)$$

Single- Scattering Albedo

Aerosol absorption properties are usually characterized by aerosol single-scattering albedo, defined as the ratio between the particle scattering coefficient (σ_s) and the total extinction coefficient (the sum of scattering and absorption (σ_a) extinction coefficients):

$$\omega(\lambda)=\frac{\sigma_s(\lambda)}{\sigma_a(\lambda)+\sigma_s(\lambda)}. \tag{2}$$

It is dependent on aerosol chemical composition and particle size distribution.

Phase Function, Asymmetry Parameter and Backscattering Fraction

The aerosol phase function describes the angular distribution of the scattered photons and is given as a function of the angle between the incident radiation and the scattering angle (θ). The asymmetry parameter (g) measures the fraction of forward scattering defined by:

$$g(\lambda)=\frac{1}{2}\int_0^{\pi} p(\lambda,\theta)\cos\theta\sin\theta d\theta\,. \tag{3}$$

It is equal to zero when the scattering is isotropic or symmetric about a scattering angle of 90°. The asymmetry parameter is equal to 1 when the scattering is confined to the forward direction (θ= 0°) and equal to –1 when the scattering is completely in the backward direction (θ= 180°). The backscattering fraction, or coefficient, is given by

$$\beta(\lambda)=\frac{1}{2}[1-g(\lambda)]\,. \tag{4}$$

Effective Radius

The effective radius (r_e) is the area-weighted average radius of an aerosol particle size distribution or, equivalently, the ratio of the third to the second moments of the size distribution:

$$r_e=\frac{\int_0^{\infty} r^3 n(r)dr}{\int_0^{\infty} r^2 n(r)dr}. \tag{5}$$

Ångström Exponent

The Ångström wavelength exponent (α) is an approximate measure of aerosol particle size that can be estimated from aerosol optical depths measured at two wavelengths:

$$\alpha = -\frac{\Delta \ln \tau}{\Delta \ln \lambda}, \tag{6}$$

where Δ refers to the difference between measurements in two wavelength narrowbands. It can be linked to the exponent in an inverse power law, such as the Junge size distribution: $n(r) \sim r^{-4}$.

2.2. *Aerosol Direct Radiative Forcing*

Direct aerosol radiative forcing refers to the difference in the radiation budget with and without the presence of aerosol. While the majority of studies on aerosol direct forcing have been focused on the TOA, increasing attention is now directed toward surface aerosol forcing. Surface forcing is exerted by both scattering and/or absorption due to aerosol particles. Scattering increases the fraction of sunlight reflected to space and consequently reduces the fraction of sunlight reaching the surface. Absorption reduces both TOA reflection and the surface illumination. For conservatively scattering aerosols, the forcing at the TOA is identical to that at the surface. However, for absorbing aerosols, the former is less than the latter, due to the offsetting effects of scattering and absorption (Li and Trishchenko 2001). Because of the offsetting effects, aerosol direct forcing at the TOA may be positive (warming) or negative (cooling), depending on the aerosol single-scattering albedo, backscattering fraction, and surface albedo (R_s), as governed by the following threshold for cooling (Chylek and Coakley 1974):

$$\omega > \frac{2R_s}{\beta(1-R_s^{\ 2})+2R_s}. \tag{7}$$

The dependence on surface albedo may be understood by comparing dark (e.g. ocean) and bright (e.g. snow) surfaces. Over the dark surface, nearly all types of aerosols cool the earth system due to the overwhelming effect of aerosol scattering. Over the bright surface, aerosol absorption is enhanced by multiple reflections between the surface and the aerosol layer, where even a weakly absorbing aerosol may have a net warming effect. The magnitude of aerosol direct radiative forcing is determined by (Haywood and Shine 1995):

$$\Delta F = -\frac{1}{2} S_0 T^2 (1-C)[\omega\beta(1-R_s)^2 - 2(1-\omega)R_s]\tau, \tag{8}$$

where S_0 denotes the solar constant, T is the atmospheric transmittance, and C is the cloud fraction. Note that this equation only accounts for the aerosol effect under clear-sky conditions. Aerosol direct forcing for cloudy skies is usually very small, since the bulk of aerosols are situated below the cloud and cloud scattering outweighs by far aerosol scattering.

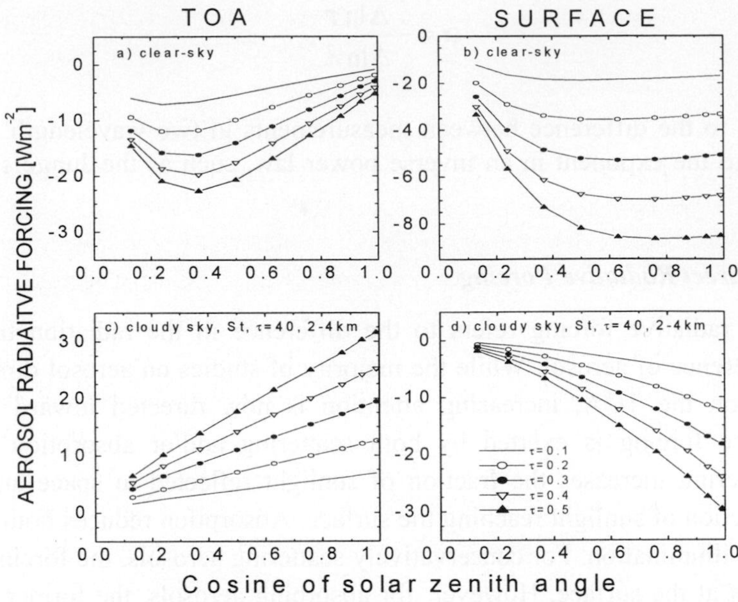

Figure 2. Model simulated aerosol radiative forcing under clear (a and b) and cloudy (c and d) skies at the top of the atmosphere (a and c) and at the surface (b and d) of variable aerosol loading (after Li and Trishchenko 2001).

However, if a significant amount of absorbing aerosol is imbedded inside a cloud layer, multiple scattering by the cloud droplets can substantially strengthen absorption, as demonstrated in Fig. 2. It shows direct aerosol radiative forcing at the top and bottom of the atmosphere for a continental type of model aerosol mixed with a stratus model cloud. For such moderately absorbing aerosols, TOA aerosol forcing under clear and cloudy conditions have opposite signs: cooling and warming for clear and cloudy skies, respectively. The forcing at the surface is always negative, but the magnitude of clear-sky aerosol forcing is much larger than cloudy-sky aerosol forcing. Under clear-sky conditions, surface aerosol forcing is substantially larger than TOA forcing, whereas the magnitude of the forcing is about the same at the surface and at the TOA but have opposite signs. These findings demonstrate the complexity of aerosol direct forcing (Li and Trishchenko 2001).

Observational estimation of aerosol direct radiative forcing is usually only feasible under clear-sky conditions for which the measurements of aerosol optical properties required to compute the forcing are available. Li and Kou (1998) made an attempt to derive aerosol direct radiative forcing inside an atmospheric column under all-sky conditions using coincident surface and satellite observations of solar radiation in the visible wavelengths (400 nm to 700 nm). At these wavelengths, absorption by atmospheric molecules and clouds is negligible so that solar radiation absorbed inside the atmosphere, as determined from satellite and ground observations, is attributed to aerosol forcing inside the atmospheric column. Note that heating by aerosols takes place primarily in the lower troposphere (0-4 km), at the expense of absorption by the surface (Satheesh and Ramanathan 2000). These effects can have important ramifications for cloud dynamics and climate (Hansen et al. 1997; Ackerman et al. 2000).

2.3. *Indirect Aerosol Forcing*

Aerosol indirect forcing refers to the impact of aerosols on cloud and precipitation processes and is further classified into two categories known as the first and second type of aerosol indirect effect. The former is concerned with the enhancement of cloud reflection due to more but smaller cloud droplets formed upon hygroscopic aerosol particles that serve as cloud condensation nuclei (CCN). Given the increasing numbers of cloud droplets and a limited water supply, clouds tend to become less likely to precipitate. As a consequence, clouds last longer, thereby increasing the solar radiation reflected back to space. The effect of prolonged cloud lifetime (primarily for boundary layer clouds) is generally referred to as the indirect effect of the second kind. Many case studies demonstrated the first and second indirect aerosol effects focused on ship tracks over oceans (King et al. 1993, 1995; Coakley et al. 2000), cloud particle size differences over ocean and land (Han et al. 1994), and suppression of precipitation by air pollution and smoke plumes from fires (Rosenfeld 1999, 2000). However, the overall magnitude of the aerosol indirect effect on the global climate system remains uncertain.

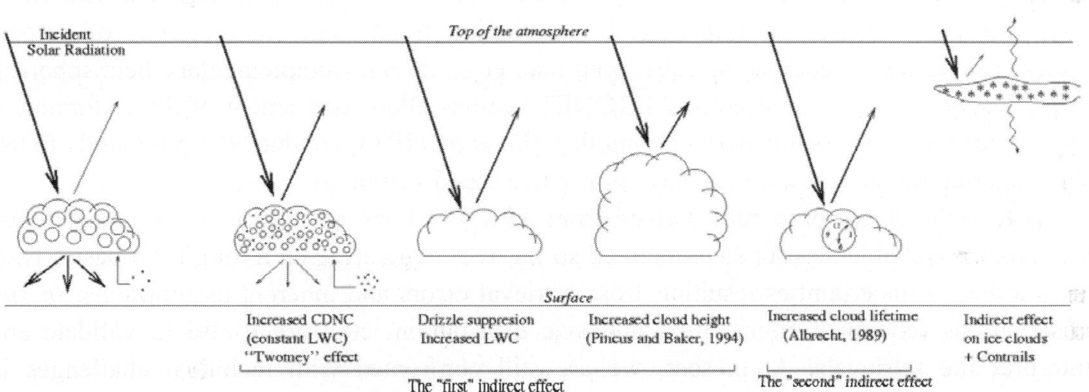

Figure 3. A schematic of aerosol indirect effects (courtesy of Jim Haywood).

The aerosol indirect effect may be determined by 1) taking *in situ* measurements of aerosol size-dependent chemistry, including organics and their dependence on hygroscopicity and CCN concentration; 2) analyzing aerosol data and cloud parameter statistics of the droplet size and cloud albedo as a function of cloud top temperature; and 3) analyzing rainfall data and relating them to cloud microphysics and CCN. Figure 3 is a schematic of four possible aerosol indirect effects: 1) increased CCN, reducing effective radius, 2) drizzle suppression, 3) increased cloud height and 4) increased cloud lifetime. An extensive review of aerosol direct and indirect effects was made by Haywood and Boucher (2000). Satellite retrievals of cloud and aerosol parameters by Nakajima et al. (2001) revealed positive and negative correlations between aerosol and τ and r_e, respectively.

3. Advanced Aerosol Observation Techniques

3.1. *Ground-based Observation*

The most successful ground-based aerosol observation network is the internationally federated AEROsol NETwork (AERONET) employing Cimel scanning sun photometers at multiple wavelengths (Holben et al. 1998, 2000; aeronet.gsfc.nasa.gov). At present, the AERONET includes about 160 instruments distributed with varying spatial density around the world, there are very few AERONET sites in East Asia and none permanent ones in China though. Using AERONET sun photometers, it is possible to characterize the microphysical and optical properties of aerosols. The AERONET system consists of a robotically controlled radiometer that makes direct sun measurements every 15 minutes and diffuse sky scans hourly in eight spectral bands during daylight hours. It has been demonstrated that AERONET data can be employed to characterize aerosol optical, radiative and microphysical properties from spectral direct and diffuse sky radiances (Dubovik and King 2000; Dubovik et al. 2002). The measured and retrieved parameters include aerosol optical thickness, Ångström exponent, particle size distribution (radii from 0.05 to 15 μm), single-scattering albedo, phase function, complex refractive index (n_r, n_i), and spectral flux. These data are valuable for validation of satellite-derived aerosol optical properties, while long-term observations are important in developing trends and climatologies of aerosol optical properties (Holben et al. 2001; Kaufman et al. 2001). Complementary hemispherical solar flux observations at selected AERONET stations allow assessment of direct forcing at the surface by aerosols. It is worth noting that *Qiu et al.* (1983) conducted a pilot study to use sky-scattered radiance to estimate aerosol particle size distribution.

Although aerosol properties derived from AERONET are not direct measurements, they are the most reliable aerosol data obtained so far over large areas on a long term basis. They are subject to uncertainties resulting from retrieval errors and inherent assumptions. *In situ* observations (chemical composition and size distribution, etc.) are useful to validate and interpret the retrievals. At present, we are still confronted with technical challenges in measuring aerosol properties, especially its absorbing properties. One of the most difficult yet most important quantities to measure is black carbon (BC). Inventories of BC are difficult to compile even under the best conditions. Recent analyses of data from the Indian Ocean and the United States (Chen et al. 2001; Dickerson et al. 2002) indicate that measurements of BC and carbon monoxide (much easier to measure) are highly correlated, but their ratio varies with the type of energy consumption. The dominant source of BC in South Asia remains a mystery. Using these data, Dickerson et al. (2002) found that using two different techniques resulted in BC emissions in India that differ by a factor of 2 to 3. Since China has a multitude of small coal-fired sources, a scenario that differs considerably from both the United States and India, a BC inventory must be developed specifically for China. The vertical profile of aerosol extinction can be monitored by a ground-based lidar.

3.2. *Satellite Remote Sensing*

Aerosol optical properties (primarily optical thickness) have been retrieved from several spaceborne sensors such as AVHRR, MODIS, MISR, SeaWIFS, POLDER, and TOMS. King et al. (1999) gave an extensive in-depth review of various aerosol remote sensing methods, while Kaufman et al. (2002a) presented a state-of-the-art review of new aerosol knowledge gained primarily from satellite observations. The essence of aerosol remote sensing is to decompose mixed signals emanating from the surface, atmospheric gases and aerosol particles. Satellite-measured reflectance is composed of contributions from all these atmospheric components and is expressed in the following equation:

$$R(\tau_a,\varpi_0;\mu,\mu_0,\phi)=R_{atm}(\tau_a,\varpi_0;\mu,\mu_0,\phi)+\frac{A_g}{1-A_g r_{atm}(\tau_a,\varpi_0)}\times t_{atm}(\tau_a,\varpi_0;\mu)t_{atm}(\tau_a,\varpi_0;\mu), \quad (9)$$

where R_{atm} is atmospheric reflectance with a "black surface"; $r_{atm}(\tau_a,\varpi_0)$ is the atmospheric spherical albedo, and $t_{atm}(\tau_a,\varpi_0;\mu)$ is the total transmission. Note that each of these variables is a function of aerosol optical thickness and single-scattering albedo and is, implicitly, a function of aerosol size distribution. From this equation, we can demonstrate many issues associated with aerosol remote sensing.

The first step is to remove the presence of clouds because the equation is only valid for clear-sky conditions. Identification of clear scenes from satellite imagery for the purpose of aerosol remote sensing is a non-trivial task. Any cloud contamination can easily confuse the faint signal of aerosol, whereas excessive cloud screening may remove pixels of heavy aerosol loading. Because of this delicate situation, cloud identification schemes used for cloud studies are usually not adequate for aerosol investigations. The majority of cloud screening methods is threshold-based, while more complicated methods, such as the artificial neural networks (Li et al. 2001), may be applied for aerosol detection on a smaller scale or for a specific aerosol type.

Aerosol remote sensing is often an ill-posed problem for we have much more unknown variables than known variables. When retrieving aerosol optical depth, we need to know the single-scattering albedo. Unfortunately, the latter is most difficult to measure by any means. Recently, remote sensing of single-scattering albedo from satellites was proposed using sun-glint measurements that are most sensitive to aerosol absorption (Kaufman et al. 2002b). Given the mixture of reflection from the surface and scattering from aerosols, one may not be able to retrieve aerosol optical depth at all, even if the single-scattering albedo is known perfectly. This occurs when the single-scattering albedo is equal to or close to the threshold given by Eq. (7). In this case, the aerosol absorption and scattering effects are just about the same so that an increase in aerosol optical depth does not increase TOA radiance. Wong and Li (2002) demonstrated the importance of knowing the single-scattering albedo in the retrieval of smoke aerosol optical depth. In addition to aerosol optical properties, the location of aerosol layer is important to study its direct and indirect effects.

Another complication arises from surface albedo. The premise of remote sensing of aerosol properties is that there is a contrast in TOA reflectance between an aerosol-laden and an aerosol-free atmosphere. As the surface reflectance increases, the contrast decreases and the accuracy of the aerosol properties retrieved is reduced. As a result, the majority of aerosol remote sensing methods are only valid over oceans or dark land. Satellite measurements depend on viewing geometries, which introduces another factor that needs accounting for in aerosol retrievals, namely, the bidirectional reflectance distribution function (BRDF). The BRDF is particularly troublesome over land, since it varies with many land parameters such as the land cover type and vegetation condition (e.g. greenness) (Li et al. 1996). Extensive BRDF models are under development using ample data acquired by the Earth Observation System (EOS) sensors. However, the persistent presence of aerosols poses a serious problem in developing the surface BRDF and in retrieving aerosol properties, as the two factors work together.

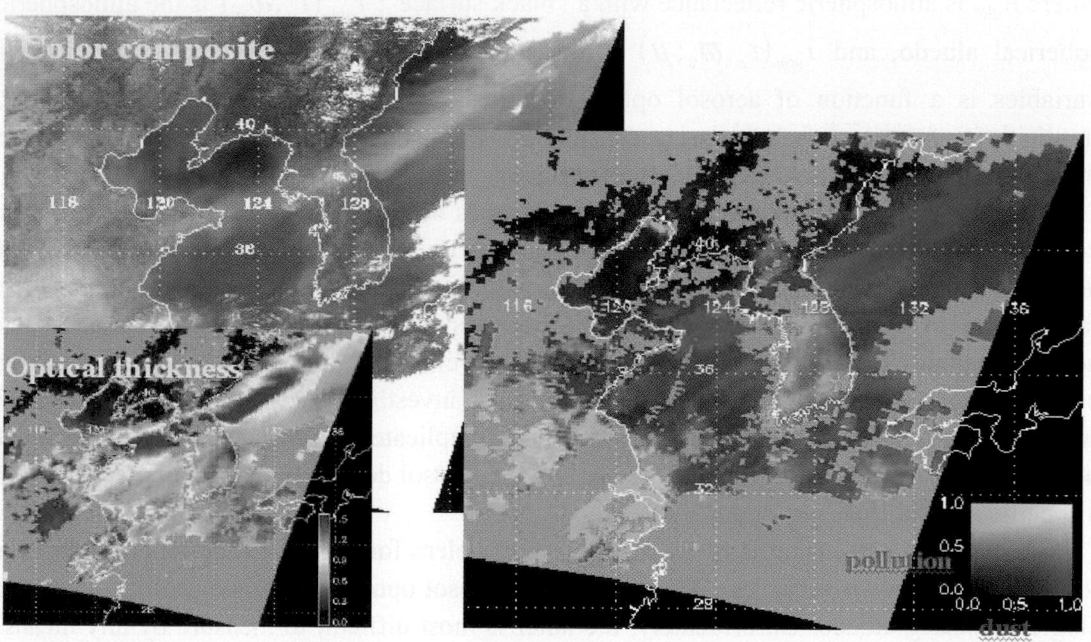

Figure 4. MODIS data from East Asia during a dust episode advected over a pollution layer in March 20, 2001. Top left is a color composite of the MODIS blue, green and red channels. Bottom left is the analysis of the aerosol optical thickness, and the right image is a separation of the fine particle pollution optical thickness (green) and coarse dust optical thickness (red) using the MODIS aerosol spectral signature. A two-dimensional color bar is shown on the bottom-right (after Kaufman et al. 2002a).

The state-of-the-art satellite sensors used for aerosol remote sensing can be classified into three categories: the multi-wavelength sensors (MODIS), the multi-angle sensors (MISR), and the polarization sensors (POLDER). The multi-wavelength sensors take measurements at several spectral channels, which is advantageous because aerosol effects differ from one wavelength to another. By virtue of these spectral differences, one can infer both aerosol size information and surface reflectance (even in the presence of aerosols). Given that aerosol extinction decreases with wavelength, Kaufman et al. (1997) proposed using a longer-wavelength channel (transparent to aerosol) to estimate shorter-wavelength surface albedo from which aerosol optical thickness of fine particles can be estimated. Figure 4 illustrates the rich information that can be extracted from MODIS by virtue of its multispectral coverage. It detects two distinct aerosol types, a dust episode overlying a polluted layer on March 20, 2001 over eastern Asia. Both optical depth and size information are extracted. Aerosol retrievals for this type of sensor require well-characterized low reflectance, or dark, land surfaces.

The dark surface requirement is relaxed or even eliminated by the multi-angle and polarization sensors. The multi-angle sensors observe the same ground target from different viewing angles to separate the ground and atmospheric contributions based on their different directional reflectance behaviors. Since the retrieval relies on varying aerosol reflectance associated with changing atmospheric pathlength, dark targets are not absolutely required, and knowledge of ground surface reflectance is not needed for Lambertian surfaces or surfaces with a known BRDF. Polarization sensors measure the polarization of the incoming signal, which is sensitive to the refractive index of the aerosol particles, and not sensitive to the ground contribution. This allows an approximation of the aerosol type without *a priori* knowledge of surface reflectance. The POLDER algorithm, for example, is applicable to both ocean and land regardless of brightness, due to the negligible polarization contribution from the land surface relative to aerosol. Likewise, polarization of larger particles is so weak that POLDER is most sensitive to fine-mode aerosols such as smoke from biomass burning and anthropogenic aerosols from industrial pollution. Figure 5 shows a global distribution of POLDER-derived aerosol index in the spring of 1997 (Bréon et al. 2002). A high aerosol index is observed over Asia and Africa. Air pollution from fossil and bio-fuel combustion is responsible for the high loading of fine mode aerosols in these regions. Note that it remains a major challenge to derive aerosol optical depth from the POLDER aerosol index. The angular properties used by MISR are also yet to be validated together with the MISR operational aerosol products.

The Total Ozone Mapping Spectrometer (TOMS) is another very useful sensor for aerosol monitoring. The TOMS aerosol index is derived from reflectance measured at 340 nm and 380 nm. The TOMS aerosol index provides a mixed measure of aerosol loading and absorption that is insensitive to surface reflectance (surface albedo at ultraviolet wavelengths is small and invariant). It has proven to be useful for identifying major episodes of absorbing aerosols from such events as biomass burning and dust storms. This sensor is especially useful in monitoring dust storms that are usually swept over desert or semi-arid regions and whose high brightness renders spectral techniques useless. The long-term record of TOMS data (since 1979) allows us to reveal the trend of aerosol change due to dust storms over the East Asia. There is a clear increasing trend in the number of strong dust storms over the past few years. However, because of the dependence of the aerosol index on aerosol altitude (usually not known), weak and/or low-flying dust cannot be seen from the TOMS platform. This, together with contamination by clouds, poses a big challenge for the quantitative determination of aerosol optical depth and single-scattering albedo from TOMS. Note that the index is not necessarily correlated with aerosol loading. Li et al. (2001) found that the high

TOMS aerosol index due to biomass burning corresponded to elevated tenuous smoke far from the heavy smoke surrounding the origin of burning. If the aerosol vertical distribution is known, as will be determined from the future CALIPSO spaceborne lidar, aerosol parameters retrieved from TOMS would be more comparable to ground-based AERONET data (Torres et al. 2002).

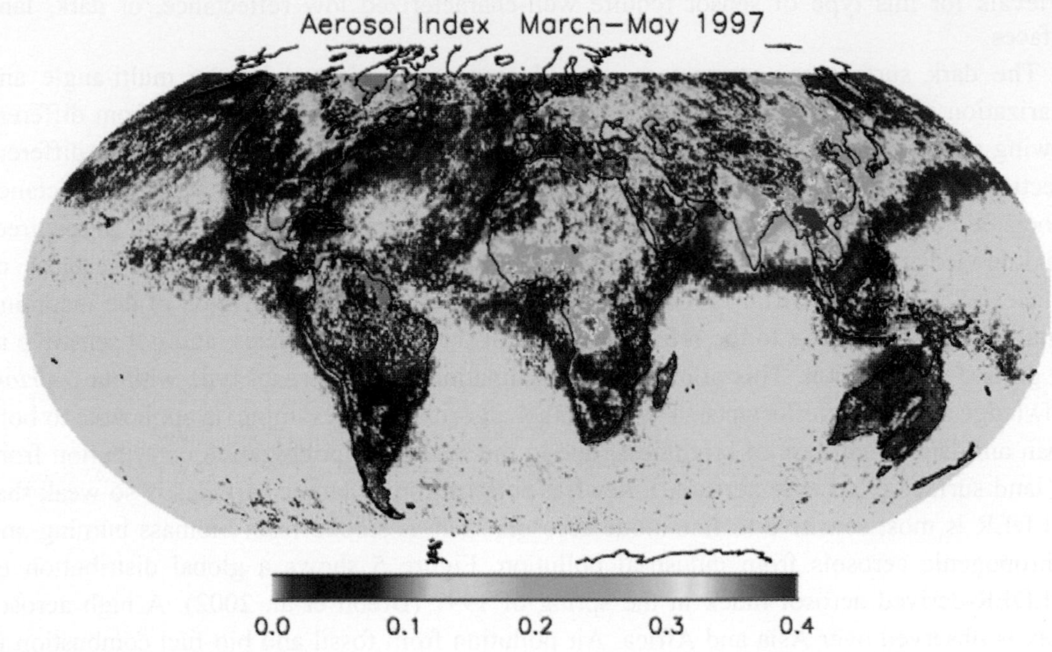

Figure 5. Global distribution of the aerosol index derived from the POLDER (Bréon et al. 2002). The index is most sensitive to fine-mode aerosols.

4. Aerosols and Aerosol Studies in China

Aerosol optical depth over most Chinese cities has increased significantly (*Qiu et al. 1997*). Using routine ground-based solar radiation measurements made across China since the 1960s, Luo et al. (2001) obtained the long-term trend and spatial variations of aerosol optical depth in China (Fig. 6). Similar estimates were also made with the direct solar radiation data by *Qiu et al* (1995). While the retrieved values are likely subject to large uncertainties resulting from numerous assumptions and from cloud screening issues, the patterns of temporal and spatial variations are reasonable. Overall, aerosol optical depth increases steadily from 1960s to 1980s; eruption of the El Chichon volcano might be responsible for the peak observed in the early 1980s. A weak decreasing trend during the 1990s could be an indication of the consequences of emission control enforced by the government. Streets et al. (2001) found a general decrease in sulfur dioxide (SO_2) emissions in China since 1995. On the other hand, emissions of mineral dust in China have an opposite trend. It is thus necessary to separate dust aerosols from anthropogenic aerosols; fortunately, their respective optical properties are distinct.

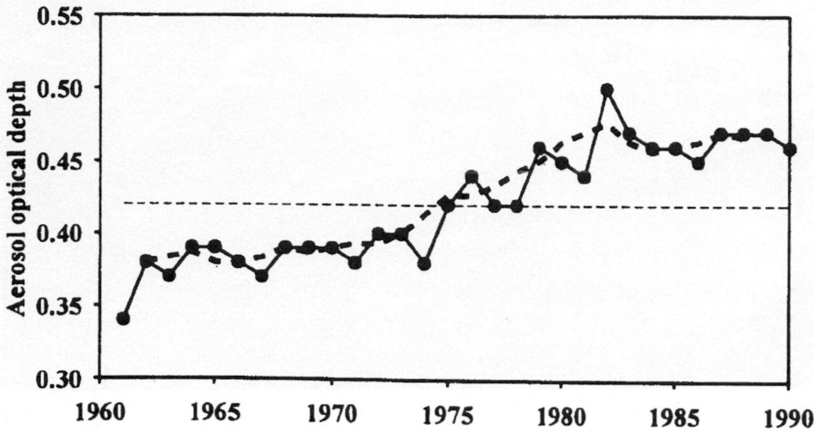

Figure 6. National average of aerosol optical depth over China derived from ground-based radiation measurements (after Luo et al. 2001).

4.1. *Dust Aerosols*

Mineral dust aerosols originate from desert regions, agricultural fields, barren land, and semi-arid regions. Dust storms are of primary concern due to the magnitude of influence and spatial coverage. Major dust storm tracks have been identified, as shown in Fig. 7. (Sun et al. 2001). There are two major sources of dust outbreaks in China: the western Xinjiang territory that includes the Taklamakan Desert, and northwestern Inner Mongolia that includes the Gobi Desert. In Xinjiang, desertification has increased by 400 km^2 every year. Dust emitted from China's nine deserts and semi-arid regions may affect the environment and climate of the mid-latitudes regions in the northern hemisphere. The annual mean dust emission from China is estimated to be around 800 teragrams (Tg). Drought in northern China may have played an important role in dust outbreaks. It remains a critical question, though, whether the aerosol indirect effect has anything to do with the drought trend. The direct cause of dust storms is vegetation cover lost in farming and ranching activities. 27% of the total land area of China is desert or is undergoing desertification with an economical impact of approximately $0.6 billion annually. Dust storms tend to intensify during daytime and weaken at night (*Xu et al.* 1979).

To probe the vertical concentration of dust aerosol, a handful of lidar instruments were deployed in China (*Qiu et al. 1984*). Long-term lidar observations of aerosol profiles have been obtained over Beijing and Hefei (*Zhou et al. 1998b*), supplemented by some short-term campaigns. A unique and seemingly ubiquitous maximum in aerosol concentration between 5 km and 8 km was found both in China and in the United States. This is because dust is usually lifted to the upper atmosphere by cold fronts in the lower atmosphere (below 4 km) and traveled with the westerly wind at higher altitudes.

Figure 7. Major dust storm tracks in China (after Sun et al. 2001) superimposed upon MODIS real-color image (courtesy of S.-C. Tsay).

The radiative forcing of dust aerosol was evaluated using AVHRR and ground observations made during the HEIFE experiment (*Shen and Wei 1999a, 1999b; Wei and Shen 1998*). In Beijing, solar radiative heating of the atmosphere is larger on dust storm days than on dust-free days by 80-318% (*Yi and Han 1989*). *Zhao et al.* (*1983*) monitored aerosols in Beijing with a seven-band sun photometer, from which the aerosol size distribution was retrieved. The radii of dust particles were found to be generally larger than 2.1 μm, while the radii of pollutant aerosols were generally less than 2.1 μm (*Yang 1995*). The average refractive index in Beijing was found to be 1.517–0.034i during heating periods and 1.533–0.016i during non-heating periods. Note that these values are very crude estimates from broadbance solar direct and sky diffuse radiance measurements using shadowband radiometers for the simultaneous retrieval of aerosol optical thickness, size distribution, refractive index, and surface albedo (*Lu et al. 1981; Qiu 1983, 1986*). The imaginary part of the refractive index is dictated primarily by anthropogenic aerosols that are often mixed with mineral dust in Northern China. For pure dust aerosol, its value is much smaller (near zero revealed by the ACE-Asia studies). Dusty weather can reduce visibility to less than 50 m, increase local TOA albedo by 50-100%, and reduce total solar radiation by 10-40% (*Zhou et al. 1994*).

Dust originating in East Asia can travel a long distance, influencing a large area of the Asia-Pacific rim (Arimoto et al. 1996). During the major dust episodes observed in the spring of 1998, dust storms swept from the Takamoukou desert crossed the Pacific and reached the United States within one week. Extensive documentation has shown the impacts of these aerosols (and gases) at the Mauna Loa Observatory in Hawaii (Hubert et al. 2001) and in the Midway Island (Prospero et al. 2002). High loading of transported aerosols was found in the optically important 0.3 to 1.0 μm size range (Perry et al. 1999).

4.2. *Anthropogenic Aerosols*

As the most populated and fastest developing country of vast territory, China is a major source of anthropogenic aerosols. At present, SO_2 emissions from China account for about 20% of the world's total (Lefohn et al. 1999) and is expected to double between 1995-2050 (Streets and Waldhoff 2000). About 50% of the anthropogenic SO_x emitted over East Asia is removed from the continental source regions (Tan et al. 2002). The remaining could be carried to Hawaii or even further (Perry et al. 1999; Propero et al. 2003). The primary energy sources in China are coal (74.8%, or 1.36 billion tons per year in 1997), oil (17.9%), natural gas (1.77%), hydroelectric power (0.09%), and nuclear-generated power (5.44%). The dominance of coal combustion combined with a generally low burning efficiency results in large quantities of black carbon emitted into the atmosphere (Streets et al. 2001). Elemental carbon and organic aerosol concentrations in China are also exceptionally high and may contribute substantially to aerosol direct forcing at both regional and global scales. Other emissions have also increased due to a dramatic rise in automobile use and the rapid expansion of private industry. Major pollutants include SO_2, nitrogen oxides, soot, and suspended particles. The total annual emission of SO_2 amounts to 20.9 Tg (76.2% industrial and 23.8% domestic) and the countrywide average atmospheric concentration of SO_2 is 0.056 mg m^{-3} and 0.037 mg m^{-3} for nitrogen oxides. Note that nitrogen oxide concentrations in Chinese cities may increase rapidly with a drastic increase in private vehicles.

Heavy pollution has a severe adverse impact on the environment, on human life, and on the economy. For example, the large concentration of SO_2 over China has caused widespread acid rain (*Wang 1997*). The total area affected by acid rain is approximately equal to one third of China's total territory, including more than half of Chinese cities, mostly in southern China. In the foreseeable future, the composition of the energy infrastructure in China is unlikely to change dramatically, but the demand for energy will continue to grow, likely leading to more pollution and more anthropogenic aerosols. The problem is exacerbated by the increasing numbers of small, inefficient power plants owned privately or by local governments. Such small enterprises usually do not have coal-cleaning and sulfate-removal facilities essential to produce "clean energy." Currently, the overall consumption of energy per capita in China is very low, about 1/48 the level of the United States; the potential for increase in energy consumption is phenomenal.

Due to air pollution, the imaginary part of the aerosol refractive index is generally high, and decreases from cities to the countryside (*Hu et al. 1991*). The single-scattering albedo is generally low, as is shown in a comparison between Beijing and Washington, DC (Fig. 8). The chemical composition of urban aerosols in China differs considerably from that in North America and Europe, with significantly higher concentrations of sulfate and black carbon (*Wang et al. 1981; Ren et al. 1982*). This is consistent with the findings of *Zhang and Shi* (2000) showing that the fine mode aerosols have increased more rapidly in the past decade, associated with a sharp increase in coal combustion and automobile emissions. Industrial and urban emissions are present year-round and are likely exacerbated by the hot, humid summer monsoon season in the heavily populated southeastern provinces. Note that the differences in single-scattering albedo as shown in Fig.8 are tremendous in terms of the influence of aerosol on the solar radiation budget.

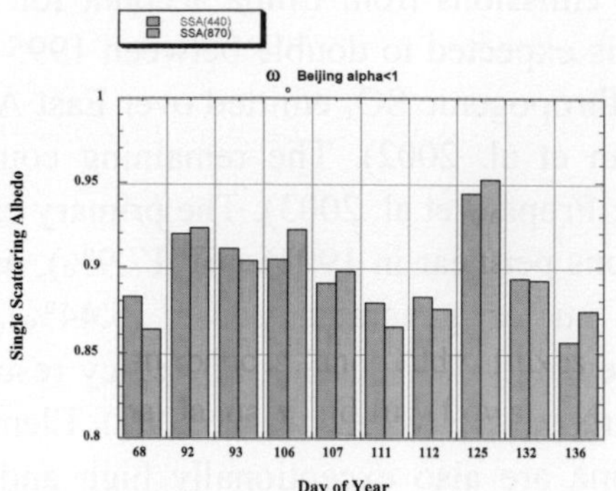

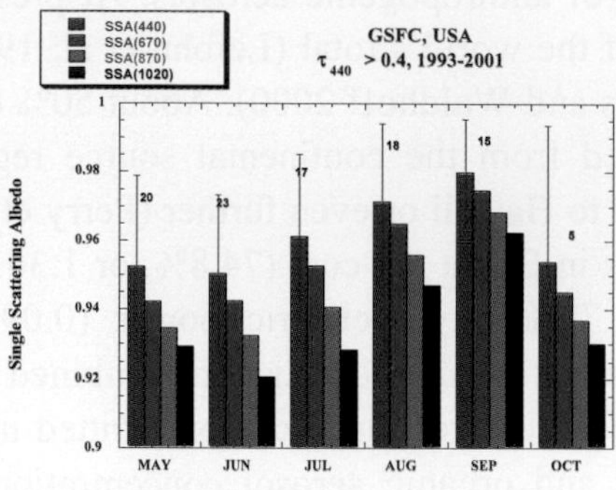

Figure 8. Comparison of single-scattering albedos in Beijing and Washington as retrieved from AERONET measurements (courtesy of B. Holben, 2002).

The highly hygroscopic nature of the pollutants compounds its climate effect due to the influence of humidity (Li et al. 1996, 2000). Observations in Chongqing indicated that aerosol optical depth is doubled as relative humidity increases from 65% to 90%. Because of their high soot content, these aerosols have a warming effect (over 1 °C per day) in the upper boundary layer and a cooling effect in the lower boundary layer, leading to a more stable atmosphere during the daytime. At night, the effect is opposite. Chameides et al. (2002) even argued that the world thickest clouds found in Northwestern Pacific is caused by air-pollution from China.

4.3. *Climate Effects*

Direct aerosol forcing is strong and complex due to high loading and strong absorption of aerosols overlying bright surfaces. One significant effect is a reduction in direct solar radiation at the surface (*Xu, 1990*). According to Luo et al. (2001), this reduction from the 1960s to the 1980s is larger than 20% in all cities, with the largest decrease of 29.2% occurring in Guangzhou, where visibility has diminished by more than 50% during the last two decades! Using the aerosol optical thickness data estimated from surface radiation measurements, Luo (1998) computed aerosol radiative forcing across China and found that the largest forcing (−13 Wm^{-2}) occurred in spring and the smallest (−8 Wm^{-2}), in winter. Note that these estimates are subject to considerable uncertainties due to errors in the retrieval of optical thickness and lack of observations for other aerosol attributes (e.g. single-scattering albedo). The distribution and radiative forcing of anthropogenic aerosols was also modeled, based on an inventory of emission sources and strengths (*Hu and Shi 1998a, 1998b; Zhang and Gao 1997*).

The strong radiative forcing resulting from large aerosol loading impinges substantially upon regional climate (Zhou et al. 1998b; Hu and Shi 1998). One direct effect is surface cooling that can exceed the warming effect of increased greenhouse gases. For example, a generally decreasing trend of surface temperature observed in the heavily industrialized Sichuan Basin (Fig. 9 left) was attributed to the influence of exceptionally high aerosol concentrations. Statistical studies (Xu and Wang 1993; Xu 1997) revealed a strong link between air pollution and China's monsoon regime. Significant correlations were found between an increase in aerosol loading (air pollution), a decrease in solar radiation, and the abnormal location of the subtropical high-pressure system located in the western Pacific. He argued that the "northern drought, southern flood" anomaly is associated with increases in air pollution (Xu 2001).

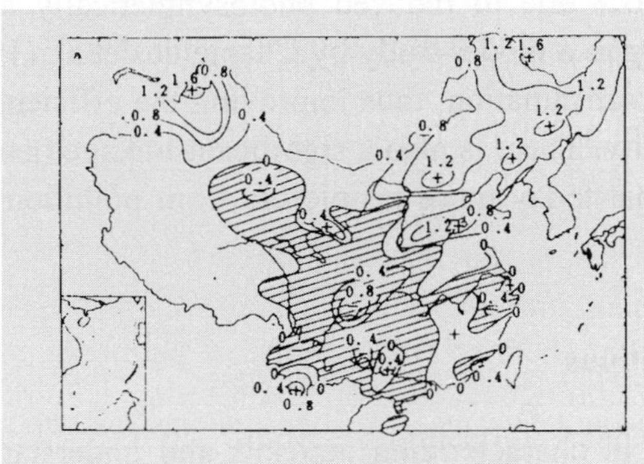

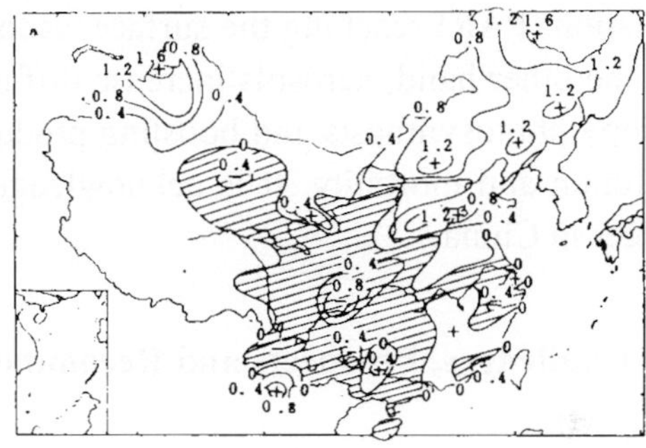

Figure 9. Temperature and precipitation changes in China. Left: Yearly mean temperature difference between 1980s and 1950s (Luo 1998); Right: Rainfall anomaly in 1980-1993 relative to long-term mean (1951-1980) (Xu 1997).

The argument was further substantiated by Menon et al. (2002) using the NASA/GISS general circulation model. They simulated three aerosol scenarios: no aerosol, sulfate aerosol, and sulfate plus black carbon aerosols. Given that the BC concentration in Asia, in particular India and China, is very high, they adopted a very low single-scattering albedo of 0.85, invariant with time and location for the third scenario. Using the aerosol optical depths estimated by Luo et al. (2001), they obtained two distinct climates in the Asian region in terms of changes in temperature and precipitation distribution. It was found that the temperature decrease in East Asia due to sulfate and BC aerosols is much more than that due to sulfate aerosol alone. This is because both types of aerosols lower the amount of solar radiation reaching the ground. Changes in precipitation due to sulfate aerosol are relatively minor. However, when an absorbing aerosol is added, the difference is dramatic. Precipitation decreases in northern China but increases in southern China. The changes in precipitation are induced by changes in atmospheric convection initiated by inhomogeneous aerosol heating in the boundary layer. In the Sichuan basin located in southern China, high aerosol loading and humidity accelerates the upwelling convection leading to excessive precipitation. In northern China, the subsiding air causes a precipitation deficit. While this theory may have oversimplified the workings of the climate in the region (e.g. the monsoon dynamic is not included), it demonstrates a potentially very large role played by aerosols in dictating the regional climate.

There are very few studies on the aerosol indirect effect in China, although the effect could be substantial, especially with regard to suppression of precipitation. If so, aerosols would have a significant impact on the hydrological cycle and on the monsoon regime that governs summer weather and climate across China. These arguments are essentially hypotheses that are yet to be substantiated with more sound physical mechanisms following rigorous scientific investigations.

Of course, the impact of pollution is more far-reaching than just climate. Air pollution in China is likely to reduce crop yield by 5-30% due to reduced photosynthetically active radiation (PAR) reaching the surface, according to a model study by Chameides et al. (1999). On the other hand, aerosols increase diffuse solar radiation, thus improving the efficiency of canopy photosynthesis and boosting productivity. There is also a significant linkage between pollution and mortality. It is acknowledged that today more people die from pollution than cancer in China.

5. Challenges, Prospects and Recommendations

Despite the numerous encouraging advances in characterizing aerosols and understanding their climate effects in China, we are confronted with many daunting tasks such as unraveling the complex relationships between aerosol and climate on both regional and global scales. Previous studies provided preliminary insights into the basic characteristics of aerosols in East Asia, but our knowledge remains very much uncertain for the following reasons.

First, many of the instruments used in the aforementioned studies are outdated with low or un-quantified accuracies. Some advanced instruments were deployed for a short period over few locations. There are no operational aerosol observation stations equipped with rigorously calibrated instruments. China recently purchased 30 Cimel sun photometers mainly for monitoring dust storms. The lack of long-term seasonal and large-scale measurements of aerosol microphysical and radiative properties means there is no baseline against which to gauge changes in aerosol loading and climate impacts into the future. Given the uniqueness of the aerosols in the region, measurements and information acquired over other source regions cannot be substituted to fill the gap. With major advances in ground-based observational techniques and analysis methods, it is highly desirable to establish a long-term routine aerosol observation network equipped with advanced instruments such as those used in the AERONET. At present, there exists a big gap in the coverage of AERONET in China.

Few attempts have been made to utilize satellite data to quantify the spatial and temporal variations of aerosols in China. Only a handful of attempts were made by Chinese scientists to use satellite data to monitor aerosols (Zhao and Yu 1986). Most studies have been limited to qualitatively browsing imagery for major aerosol episodes, such as dust storms (Zhou et al. 1994). A quantitative investigation was made recently to retrieve aerosol optical thickness over major lakes across China using the GMS-5 and MODIS satellite data (Liu 1999; Mao et al. 2000). There are several major challenges in the retrieval of aerosol properties from satellite. A combination of bright surfaces and absorbing aerosols that are often encountered in China makes the retrieval of aerosol optical thickness particularly cumbersome, unless aerosol properties are known reasonably well.

Lack or insufficient knowledge of ground truth information pertaining to aerosol optical properties, surface reflectance and BRDF all hinder remote sensing of aerosol. *In situ* aerosol measurements are required to properly characterize the physical and chemical properties of major aerosol types found in the region. A major piece of missing information is the composition of pollution aerosols. A large uncertainty is probably linked to the source of carbonaceous aerosols. Although its contribution to total optical thickness is relatively small, the black carbon concentration is often a determinant factor for aerosols to warm or cool the atmospheric column.

The largest source of uncertainty lies in estimating aerosol direct and indirect effects and understanding their impact on regional and global climate. Many aspects of aerosol direct forcing are yet to be explored, especially its impact and relation with the Asian monsoon system. In China, aerosols may have an even stronger effect on atmospheric dynamics through which both precipitation intensity and pattern may be severely affected. Aerosols could have a significant impact on the monsoon regime that governs summer weather and climate across China. Physical mechanisms that may link air pollution and the monsoon episode await discovery. Reducing solar illumination at the surface, aerosols may weaken the monsoon atmospheric circulation by lowering the land-ocean thermal contrast, a condition unfavorable for the development of monsoon weather. Since the annual rainfall in northern China comes chiefly from precipitation associated with the summer monsoon, any retreat/weakening of the monsoon system could have tremendous impact on the hydrological cycle. To tackle this problem, one sound approach is to develop a more reliable climatology of aerosol radiative forcing (at the TOA, the surface and inside the atmosphere) across the region over a long period of time, incorporate them into regional climate models capable of generating the monsoon regime (e.g. Tao et al. 2002) and compare the model- simulated and observed monsoon time series.

Given the high concentration and chemical composition of aerosols in China, there may be particularly strong aerosol indirect effects in China. Yet, the aerosol indirect effects are likely more complex than what we have known from studies conducted in relatively clean environments. So far, aerosol indirect effects are mainly confined to alteration of cloud microphysics and lifetimes, although the indirect effect may be coupled with the direct effect and atmospheric dynamics. In China, the indirect effect may be a direct cause of precipitation reduction in northern China, where there is limited water vapor but plentiful CCN, whereas in the southern region, an abundance of both water vapor and CCN may produce excessive precipitation. The latter is corroborated at least in part by acid rain observations that indicate that most precipitation in southern China has rather high values of pH. This hypothesis needs rigorous testing using comprehensive *in situ* cloud and aerosol measurements in cloud microphysical models containing aerosol-cloud interaction schemes like those proposed by Chen and Lamb (1999) and Khain et al. (2000).

In summary, there are plenty of investigations that can be conducted to unravel the complex aerosol-energy-hydrology relationship. China is an excellent testbed for examining aerosol climate effects that may well go beyond the current paradigms. To this end, we need to tackle the following fundamental questions (to name a few):

- What are the spatial and temporal variations of single-scattering albedo?
- What is the aerosol radiative forcing at the surface and at the TOA in the region?
- How do mineral dust aerosols interact with anthropogenic aerosols?
- Do Asian aerosols modify cloud microphysics and precipitation patterns, and if so, how?
- Are the radiative and hydrological effects of anthropogenic aerosols large enough to alter atmospheric circulation patterns and regional climate?

To address these questions, it is recommended that extensive research on the following basic issues be conducted:

(1) Aerosol characterization

- *In situ* observation and ground-based remote sensing (especially AERONET) of aerosol microphysical, chemical, and optical properties;
- Identification of aerosol types (dust, soot, sulfate, organic and carbonaceous aerosols);
- Use measurements of aerosols and trace gases to better understand the origins of the particles and to test and improve emissions inventories;
- Satellite-based mapping of aerosol spatial (including vertical) and temporal variations from such sensors as MODIS, TOMS, SeaWiFS, AVHRR, etc.;
- In situ and remote sensing of broadband and spectral surface albedos, radiative fluxes, precipitable water, etc.

(2) Aerosol direct radiative forcing

- Direct radiative forcing at the top, bottom and within the atmosphere;
- Radiative forcing of different aerosol types (dust, soot, sulfate, mixed);
- Combining observations and modeling studies to estimate aerosol forcing under cloudy conditions.

(3) Aerosol indirect effects

- Indirect radiative forcing; hygroscopic (humidification) effect; hydrological suppression and distribution;
- Acid rain and acidity effects;
- Cloud lifetime.

(4) Impact on regional/global climate

- Absorbing/non-absorbing aerosol in regional/global models for EOS-data assimilation; aerosol-climate-atmospheric chemistry;
- Regional/global climate simulation; feedback effects (absorbing aerosol).

Acknowledgements: The author is grateful to Drs. Y. Kaufman and J. Qiu for their review of the paper, and to the following scientists for inspiring interactions: S.-C. Tsay, M.D. King, B. Holben, A. Chu, H.-B. Chen, Y.-F. Luo, G.-Y. Shi, J.-T. Mao, C.-S. Zhao, D.-R. Lu, etc. The work is partially supported by a Cooperative Research Fund for Overseas Chinese Scholars awarded by the National Science Foundation of China #40028503 and a US DOE/ARM grant DE-FG02-97ER62361.

References (Chinese literatures quotations are in *Italic*):

Ackerman, A. S. *et al.,* 2000, Reduction of tropical cloudiness by soot, *Science* **288**, 1042-1047.

Arimoto, R., R.A. Duce, and co-authors, 1996, Relationships among the aerosol constituents from Asia and the North Pacific during PEM-West A, *J. Geophy. Res.* **101**. 2011-2024.

Bréon, F.-M., D.D. Tanré, and S. Generoso, 2002, Aerosol effect on cloud droplet size monitored from satellite, *Science*, **295**, 834-837.

Chameides, W.L., et al., 1999, Is ozone pollution affecting crop yields in China, *Geophys. Res. Lett.*, **26**, 867-870.

Chameides, W.L., et al., 2002, Correlation between model-calculated anthropogenic aerosols and satellite-derived cloud optical depth: Indication of Indirect effects ?, *J. Geophy. Res.*, **107**, AAC 2, 1-12.

Chen, J-.P., and D. Lamb, 1999: Simulation of cloud microphysical and chemical processes using a multicomponent framework. Part II: Microphysical evolution of a wintertime orographic cloud. *J. Atmos. Sci.*, **56**, 2293-2312.

Chen, L-W. A., B. G. Doddridge, R. R. Dickerson, and P. K. Mueller, Seasonal Variations in Elemental Carbon Aerosol, Carbon Monoxide, and Sulfur Dioxide: Implications for Sources, *Geophys. Res. Lett.,* **28**(9), 1711-1714, 2001.

Chylek, P., and J.A. Coakley, Jr., 1974, Aerosols and climate, *Science*, **183**, 75-77.

Coakley, J.A. Jr., P.A. Durkee, and co-authors, 2000: The appearance and disappearance of ship tracks on large spatial scales, *J. Atmos. Sci.*,. 57, 2765-2778.

Dickerson, R.R., M. O. Andreae, and co-authors, 2002, Analysis of Black Carbon and Carbon Monoxide Observed over the Indian Ocean: Implications for Emissions and Photochemistry, *J. Geophys, Res.,* **107**(19), 16-1 to 16-11.

Dubovik, O.and M. D. King, 2000, A flexible inversion algorithm for retrieval of aerosol optical properties from Sun and sky radiance measurements, *J. Geophys. Res.,* **105**, 20,673-20,696.

Dubovik, O., B.N. Holben, and co-authors, 2002: Variability of absorption and optical properties of key aerosol types observed in worldwide locations, *J. Atmos. Sci.*, **59**, 590-608 .

Han, Q., W.B. Rossow, and A.A. Lacis 1994. Near-global survey of effective droplet radii in liquid water clouds using ISCCP data. *J. Climate*, **7**, 465-497.

Hansen, J., Sato, M. & Ruedy, R. 1997, Radiative forcing and climate response. *J. Geophys. Res.* **102**, 6831-6864.

Hansen, J., M. Sato, R. Ruedy, A. Lacis, V. Oinas, 2000, *Proc. Nat. Acad. Sci., USA*, **97**, 9875,

Hansen, J., 2002, Air pollution as a climate forcing, *Workshop Report*, Honolulu, Hawaii, Apr. 29 – May 3.

Haywood, J.M., and O. Boucher, 2000, Estimate of the direct and indirect aerosol radiative forcing due to tropospheric aerosols: A review, *Rev. Geophy.*, **38**, 513-543.

Haywood J.M. and K.P. Shine, The effect of anthropogenic sulfate and soot on the clear sky planetary radiation budget, *Geophys. Res. Lett.*, **22**, 603-606,

Holben B. N., T. F. Eck, and co-authors, 1998, AERONET - A federated instrument network and data archive for aerosol characterization, *Rem. Sens. Environ.*, **66**, 1-16.

Holben, B.N., D.Tanre, and co-authors, 2001: An emerging ground-based aerosol climatology: Aerosol Optical Depth from AERONET, *J. Geophys. Res.*, **106**, 12 067-12 097.

Hubert, B.J., C.A. Phillips, and co-authors, 2001, Long-term measurements of free-tropospheric sulfate at Mauna Loa: Comparison with global model simulations, *J. Geophy. Res.*, **106**, 5479-5492.

Husar, R.B., et al. Asian dust events of April 1998, 2001, *J. Geophy. Res.*, **106**, 18317-18330.

Jacobson, M.Z., 2001, Strong radiative heating due to the mixing state of black carbon in atmospheric aerosols, *Nature*, **409**, 695-672.

Kaufman, Y. J. and R. S. Fraser, 1997: The effect of smoke particles on clouds and climate forcing, *Science*, 277, 1636-1639.

Kaufman, Y. J., D. Tanré, L. Remer, et al., 1997: Remote sensing of tropospheric aerosol from EOS-MODIS over the land using dark targets and dynamic aerosol models. *J. Geophys. Res.*, **102**, 17051-17067.

Kaufman, Y. J., D. Tanre, and co-authors, 2001, Absorption of sunlight by dust as inferred from satellite and ground-based remote sensing, *Geophys. Res. Lett.*, **28**, 1479-1483.

Kaufman, Y., D. Tanre, and O. Boucher, 2002a, A satellite view of aerosols in the climate system, *Nature*, 419, 215-223.

Kaufman, Y., J.V. Martins, L.A. Remer, M.R. Schoeberl, and M.A. Yamasoe, 2002b, Satellite retrieval of aerosol absorption over the oceans using sunglint, *Geophy. Res. Lett.*, **29**, doi:10.1029/2002GL015403.

Khain, A. P., M. Ovtchinnikov, M. Pinsky, A. Pokrovsky, and H. Krugliak, 2000: Notes on the state-of-the-art numerical modeling of cloud microphysics. *Atmosph. Res.*, **55**, 159-224.

Kiehl, J.T., and H.Rodhe, 1995: Modeling geographical and seasonal forcing due to aerosols, In: Aerosol Forcing of Climate [Charlson, R.J. and J. Heintzenberg, eds.], J. Wiley and Sons Ltd, pp 281-296.

King, M. D., L. F. Radke and P. V. Hobbs, 1993: Optical properties of marine stratocumulus clouds modified by ships. *J. Geophys. Res.*, **98**, 2729-2739.

King, M. D., S. C. Tsay and S. Platnick, 1995: In situ observations of the indirect effects of aerosol on clouds. *Aerosol Forcing of Climate*, R. J. Charlson and J. Heintzenberg, Eds., John Wiley and Sons, 227-248.

King, M. D., Y. J. Kaufman, D. Tanré and T. Nakajima, 1999: Remote sensing of tropospheric aerosols from space: Past, present, and future. *Bull. Amer. Meteor. Soc.*, **80**, 2229-2259.

Lefohn, A.S. J.D. Husar, and R.B. Husar, 1999, Estimating historical anthropogenic global sulfate emission patterns for the period 1985-1990, *Atmos. Environ.*, **33**, 3435-3444.

Lelieveld, J., P. J. Crutzen, and co-authors, 2001, The Indian Ocean Experiment: Widespread air Pollution from South and South-East Asia, *Science,* **291**(5506), 1031-1036.

Li, Z., J. Cihlar, X. Zhang, L. Moreau, L. Hung, 1996: The bidirectional effect in AVHRR measurements over boreal regions, *IEEE Tran. Geosci. & Rem. Sen.*, **34,** 1308-1322.

Li, Z., 1998, Influence of absorbing aerosols on the inference of solar surface radiation budget and cloud absorption, *J. Climate,* **11,** 5-17.

Li, Z., and L. Kou, 1998, Atmospheric direct radiative forcing by smoke aerosols determined from satellite and surface measurements, *Tellus (B)*, **50**, 543-554.

Li, Z., A. Trishchenko, 2001, Quantifying the uncertainties in determining SW cloud radiative forcing and cloud absorption due to variability in atmospheric condition, *J. Atmos. Sci.*, **58**, 376-389.

Li, Z., A. Khananian, R. Fraser, J. Cihlar, 2001, Detecting smoke from boreal forest fires using neural network and threshold approaches applied to AVHRR imagery, *IEEE Tran. Geosci. & Rem. Sen.,***39**, 1859-1870.

Luo, Y., D. Lu, X. Zhou, W. Li, 2001, Characteristics of the spatial distribution and yearly variation of aerosol optical depth over China in last 30 years, *J. Geophys. Res.*, **106**, 14501-14513.

Mao, J., J. Zhang, C. Li, M. Wang, 2000, Remote sensing of atmospheric aerosol over China, submitted.

Menon, S., J. Hansen, L. Nazarenko, Y. Luo, 2002, Climate effects of black carbon in China and India, *Science*, **297**, 2250-2252.

Mitchell, J.F.B., T.C. Johns, J.M. Gregory, and S.F.B. Tett, 1995: Climate response to increasing levels of greenhouse gases and sulphate aerosols, *Nature*, 376, 501-504.

Nakajima, T., A. Higurashi, K. Kawamoto, J. Penner, 2001, A possible correlation between satellite-derived and aerosol microphysics parameters, *Geophy. Res. Lett.*, **28**, 1171-1174.

Penner, J.E., et al., 2001, Aerosols, their Direct and Indirect Effects, Chapter 5 in Climate Change 2001, *The Scientific Basis, Working Group I to the Third Assessment Report of the Intergovernmental Panel on Climate Change (IPCC)*, Cambridge University Press, pp. 289-348.

Perry, Kevin D., T.A. Cahill, R.C. Schnell, and J.M. Harris, 1999, Long-range transport of anthropogenic aerosols to the NOAA Baseline Station at Mauna Loa Observatory, Hawaii. *J. Geophys .Res.* (Atmospheres). **104**, 18,521-18,533.

Propero, J.M., L.S. Savoie, and Arimoto, 2003, Long-term of nss-sulfate and nitrate in aerosols on Midway Island, 1981-2000: Evidence of increase (now Decreasing?) anthropogenic emissions from Asia, *J. Geophys. Res*., submitted.

Rosenfeld D., 1999: TRMM Observed First Direct Evidence of Smoke from Forest Fires Inhibiting Rainfall. *Geophysical Research Letters*. 26, (20), 3105-3108.

Ramanathan, V., P.J. Crutzen, J.T. Kiehl, and D. Rosenfeld, 2001, Aerosols, climate and the hydrological cycle, *Nature*, **294**, 2119-2124.

Rosenfeld, D., 2000, Suppression of rain and snow by urban and industrial air pollution, *Science*, **287**, 1793-1796.

Rosenfeld, D., and W. Woodley, 2001, Pollution and clouds, *Physics World*, 33-37.

Rosenfeld, D., Y. Rudich, and R. Lahav, 2001, Desert dust suppressing precipitation: A possible desertification feedback loop, *Proceedings of the National Academy of Sciences (PNAS)*, *98*, 5975-5980.

Satheesh, S.K., and V. Ramanathan, 2000, Large differences in tropical aerosol forcing at the top of the atmosphere and Earth's surface, *Nature*, 405, 60-63.

Sts, D., and S.T. Waldhoff, 2000, Present and future emissions of air pollutants from China: SO_2, NO_x, and CO, *Atmos. Environ*., **34**, 363-374.

Streets, D. G., S. Gupta, and co-authors, 2001, Black carbon emissions in China, *Atmos. Environ.,* 35, 4281-4296.

Sun, J., M. Zhang, and T. Liu, 2001, Spatial and temporal characteristics of dust storms in China and its surrounding regions, 1960-1999, *J. Geophys. Res.*, 106, 10325-10347.

Tan, Q., Y. Huang, W.L. Chameides, 2002, Budget and export of anthropogenic SO_x from East Asia during continental outflow conditions, *J. Geophy. Res.*, **107**, 10.1029/2001JD000769.

Tao, W.-K., Y. Wang, J. Qian, W. K.-M. Lau, C.-L. Shie and R. Kakar, 2002: Mesoscale Convective Systems during SCSMEX: Simulations with a Regional Climate Model and a Cloud-Resolving Model, *INDO-US Climate Research Program*, (in press).

Torres, O., P.K. Bhartia, J.R. Herman, A. Syniuk, P. Ginoux, and B. Holben, A long term record of aerosol optical depth from TOMS observations and comparison to AERONET measurements, *J. Atm. Sci.,* 59, 398-413, 2002a

Wang W., The problem of environmental acidification in China, *J. Environ. Sci.*, **17**(3), 259, 1997.

Wong, J., and Z. Li, 2002 Retrieval of optical depth for heavy smoke aerosol plumes: uncertainties and sensitivities to the optical properties, *J. Atmos. Sci.,* 59, 250-261.

Xu, Q., 2001, Abrupt change of the mid summer climate in central east China by the influence of atmospheric pollution, *Atmos. Environ*, **35**, 5029.

Publications in Chinese:

Hu et al.胡欢陵，许军，黄正，1991，中国东部若干地区大气气溶胶虚折射率指数特征，大气科学，15(1)，18-23。

Hu and Shi胡荣明，石广玉，1998a，中国地区气溶胶的辐射强迫及其气候响应试验，大气科学，1998a，22（6），919-925。

Hu and Shi胡荣明，石广玉，1998b，平流层气溶胶的辐射强迫及其气候响应的水平二维分析，大气科学，22（1），18-24。

Li et al.李子华，杨军，黄世鸿，2000，考虑湿度影响的城市气溶胶粒子的白天温度效应，大气科学，24（1），87-94。

Li and Xu李子华，涂晓萍，1996，考虑湿度影响的城市气溶胶夜晚温度效应，大气科学，20（3），359-366。

Liu刘莉，1999，GMS5卫星遥感气溶胶光学厚度的试验研究，北京大学硕士学位论文.

Lu et al.吕达仁，周秀骥，邱金桓，消光角散射综合遥感气溶胶分布的原理与数值试验，中国科学，1981，12，1516-1523。

Luo罗云峰，1998，中国地区气溶胶光学厚度特征及其辐射强迫和气候效应的数值模拟，北京大 学博士论文。

Qiu邱金桓，从全波段太阳直接辐射确定大气气溶胶光学厚度I：理论，大气科学，1995，19（4），385-394。

Qiu et al.邱金桓，汪宏七，周秀骥，吕达仁，消光小角散射法遥感气溶胶谱分布的实验研究，大气科学，1983，7（1），33-41。

Qiu et al.邱金桓，赵燕曾，汪宏七，1984，激光探测沙暴过程中的气溶胶消光系数分布，大气 科学，8（2），205-210。

Qiu et al.邱金桓，潘继东，杨理权，杨景梅，董艺珍，1997，中国10个地方大气气溶胶1980-1994年间变化特征研究，大气科学，，21（6），725-733。

Ren et al. 任丽新，J. Winchester，吕位秀，王明星，1982, 北京冬春季大气气溶胶化学成分的研究，大气科学，6（1），11-17。

Shen and Wei沈志宝，魏丽，1999a，黑河地区大气沙尘对地面辐射能量收支的影响，高原气象，18（1），1-8。

Shen and Wei沈志宝，魏丽，1999b，中国西北大气沙尘对地气系统和大气辐射加热的影响，高原气象，18（3），425-435.

Xu徐国昌，陈敏连，吴国雄，1979，甘肃省“4.22”特大沙暴分析，气象学报，37（4），26-35。

Xu 徐群，1990(B)，近29年冬季我国太阳辐射的显著变化，中国科学，1112-1120。

Xu徐群，1997, 近十余年我国盛夏季风雨带南移的重要成因，中国的气候变化与气候影响研究文集，气象出版社，264-271。

Xu and Wang徐群，王冰梅，1993，太阳辐射变化对我国中东部和西非夏季风雨量的影响，应用气象学报，4，38-43。

Wang et al. 王明星，吕位秀，任丽新，J. W. Winchester，1981, 华北地区大气气溶胶的化学成分，大气科学，5（2），136-144。

Wei and Shen魏丽，沈志宝，1998, 大气沙尘辐射特性的卫星观测，高原气象，17（4），347-355。

Yang et al. 杨东贞，王超，温玉璞等，1995，1990年春季两次沙尘暴特征分析，应用气象学报，6（1），18-26。

Yi and Han尹宏，韩志刚，1989, 气溶胶对大气辐射的吸收，气象学报，47（1），118-123

Zhang and Mao, 张军华, 毛节泰, 2001, 我国大气气溶胶研究综述, submitted.

Zhang and Gao张瑛，高庆先，1997, 硫酸盐和碳黑气溶胶辐射效应的研究，应用气象学报，8（增刊），87-91 {1987 in text}

Zhao et al, 赵柏林，王强，毛节泰，秦瑜，1983, 光学遥感大气气溶胶和水汽的研究，中国科学（B），10，951-962。

Zhao and Yu赵柏林，俞小鼎，1986, 海上大气气溶胶的卫星遥感研究，科学通报，21，1645-1649。

Zhou et al. 周军，岳古明，金传佳等，1998a, L300可移动式双波长激光雷达对流层气溶胶探测，中国科学院安徽光学精密机械研究所国家八六三计划大气光学重点实验室，L300型激 光雷达验收报告。

Zhou et al. 周秀骥，李维亮，罗云峰，1998b, 中国地区大气气溶胶辐射强迫及区域气候效应的数值，大气科学， 22(4)，418-427。

APPLICATIONS OF SATELLITE MICROWAVE DATA FOR HURRICANE PREDICTION

TONG ZHU AND FUZHONG WENG

NOAA/NESDIS/Office of Research and Applications

5200 Auth Road, Rm. 601, Camp Springs, MD 20746, USA

E-mail: tong.zhu@noaa.gov

(Manuscript received 6 January 2003)

The measurements obtained from satellite microwave sensors have been increasingly utilized in numerical weather prediction models. In this study, we reviewed a comprehensive method of using a variety of microwave products for hurricane model initialization. In this method, atmospheric temperature profiles are retrieved from the Advanced Microwave Sounding Unit (AMSU) and are then utilized to derive the rotational winds for the hurricane vortex by solving the nonlinear balance equation. In addition, the divergent winds associated with latent heat release are derived from the Omega equation based on the AMSU rain rate as a diabatic heating source. Atmospheric moisture profiles are iteratively retrieved from the AMSU derived total precipitable water. Furthermore, for hurricane simulations over oceans, the sea surface temperature (SST) is an important component in the overall hurricane initialization. This parameter can be uniquely derived from lower frequencies of satellite microwave sensors. The operational microwave SST products have been made available from the Tropical Rainfall Measuring Mission microwave imager.

The retrieved temperatures in four tropical cyclones shown reasonable warm-core structures, and corresponded well with the storm intensities. Trajectory analyses were performed using model output data from a 5-day explicit simulation of Hurricane Bonnie. It was found that the upper-level westerly flow converged into the west eyewall, and then produced middle-level descending warming in the west to the south eyewall. The frontogenesis function was frontolytic in the west eyewall, with the contributions from the deformation term at low levels and from the tilting term at middle and upper levels. It was shown that upper-level strong westerly inflow was a key factor leading to the development of the partial eyewall.

1. Introduction

Tropical cyclones develop over the ocean, where few upper-air observations are available. The monitoring of tropical cyclones began with ship and commercial airline observations in the 1930s. Aircraft reconnaissance and radar observations were available in the 1940s. As compared with satellite observations, the above data have very limited area coverage. During the past three decades, many attempts have been made to incorporate satellite observations into hurricane models. The satellite retrieval products that were successfully incorporated and have potential use for hurricane models are: (a) visible and infrared cloud-track winds, (b) microwave rain rates, (c) ocean surface winds, and (d) microwave temperature profiles. In the following, we give a brief review of these products and the methods to incorporate them into hurricane models.

The first Geostationary Operational Environmental Satellite (GOES) became available in June 1974. Today's geostationary satellites produce visible imagery at 1-km spatial resolution and infrared imagery at 4-km spatial resolution at about 5-min intervals. The observed cloud patterns in such high temporal and spatial resolution provide a visualization of mesoscale meteorological processes. GOES observations are extremely useful for determining tropical cyclone positions and for estimating storm intensities using the method first developed by Dvorak (1972). Furthermore, geostationary sequential satellite imagery can also be used to determine cloud-drift winds, primarily at upper and lower levels (Hubert and Whitney 1971). An obvious limitation of this product results from the fact that vast cloud-free regions are left unsampled. Velden *et al.* (1997) developed an algorithm to determine the upper-tropospheric water vapor wind vector (WVWV) fields from GOES-8/9 water vapor observations. Using a variational bogus data assimilation scheme, Zou and Xiao (2000) incorporated the WVWV into the MM5 model to simulate a mature hurricane.

In recent years, products from the passive microwave observations from polar-orbiting satellites have been widely used in the initialization of hurricane models. Microwave observations have two main advantages over visible and infrared observations: (1) microwave radiation penetrates through clouds, and (2) microwave radiation is sensitive to a wide variety of atmospheric parameters, such as atmospheric temperature and moisture, cloud liquid water, cloud ice water, rain, and surface winds.

Krishnamurti et al. (1991) developed a method to physically initialize the Florida State university (FSU) Global cumulus parameterization Spectral Model (GSM), which highly depends upon the surface rain rates that were derived from the Special Sensor Microwave Imager (SSM/I). Theoretically, the application of more accurate and consistent rain rates should improve the representation of latent heat release and cumulus parameterization in hurricane models. Krishnamurti et al. (1998) indicated that the intensity forecasts of a hurricane are quite sensitive to the initial meso-convective scale precipitation distribution. Recently, a comparison study was conducted by Tibbetts et al. (2000) to evaluate the performance of four different rain-rate algorithms in hurricane track forecasts. It was found that the SSM/I rain rate algorithm was the most consistent and accurate rain rate in this study.

Hou et al. (2000a,b) assimilated the Tropical Rainfall Measuring Mission (TRMM) Microwave Imager (TMI) derived surface rainfall and total precipitable water (TPW) into the Goddard Earth Observing System (GEOS) global analysis. A unique feature of the GEOS data assimilation system is that it uses the incremental analysis update (IAU) developed by Bloom et al. (1996), which improves the spinup problem. The minimization procedure is one-dimensional, but the evaluation of the cost function involves 6-h time integration. The procedure minimizes the least squares differences between the observations and the corresponding values generated by the column model averaged over the 6-h analysis window. Pu et al. (2002) used the GEOS global analysis data with and without assimilation of the TMI surface rainfall to study the impact of TRMM surface rain rate on mesoscale numerical simulation of Supertyphoon Paka. It was found that the GEOS analysis could modify the environment of the storm, and the model initial conditions are more favorable for the development of the storm. Consequently, the simulated typhoon structure and intensity were improved significantly. The experiment with TMI data produced a storm of typhoon intensity after 36 h, while the simulation without TMI data required 60 h to generate a typhoon.

Satellite microwave sounders directly provide information on atmospheric temperature and water vapor profiles, and their products have demonstrated great potential for use in the hurricane model initialization. The first Advanced Microwave Sounding Unit (AMSU) instrument on board the NOAA 15 satellite was launched in 1998. Since then we have seen many studies utilizing these data in monitoring and simulating tropical cyclones. AMSU measurements can provide much improved information on the atmospheric temperature and moisture profiles, as compared to the Microwave Sounding Unit on earlier NOAA satellites. The associated three-dimensional (3D) temperature field has been used to estimate hurricanes' maximum wind (Kidder et al. 2000) and central pressure (Kidder 1978; Velden and Smith 1983; Kidder et al. 2000). In addition, the temperature gradient could be utilized to derive the tangential winds when a hurricane reaches its mature stage with a well-defined circular structure (Grody 1979, Kidder et al. 2000). The above studies demonstrated that AMSU measurements could provide useful information for hurricane modeling. Recently, a new algorithm to construct initial hurricane vortices using the AMSU measurements was developed by Zhu et al. (2002a). In this paper, we test this scheme in more case studies and present some results from a numerical simulation of Hurricane Bonnie (1998).

2. Retrieving hurricane vortices

Zhu et al. (2002a) developed a new algorithm to retrieve three-dimensional temperature, geopotential height, wind, and moisture fields for a hurricane system using the AMSU measurements. There are five procedures to construct a hurricane vortex: (a) retrieving three-dimensional temperature field from AMSU data; (b) integrating hydrostatic equation to obtain geopotential heights; (c) solving the balance equation to obtain the rotational winds; (d) solving the omega equation to obtain the divergent winds; and (e) specifying initial moisture field based on the AMSU-derived total precipitable water. In this section, we will give a brief description of these procedures in the retrieving hurricane vortex for Hurricane Bonnie and test the temperature retrieval algorithm in four tropical cyclones.

2.1. *Retrieving atmospheric temperature*

The AMSU instrument on NOAA-15 contains two modules: A and B. The A module (AMSU-A) has 15 channels and is mainly designed to provide information on atmospheric temperature profiles, while the B module (AMSU-B) allows for profiling the moisture field. AMSU-A has an instantaneous field-of-view of 3.3° and scans $\pm 48.3^{\circ}$ from nadir with 15 different viewing angles on each side. The resolution of AMSU-B measurements is 3 times higher than that of AMSU-A because of its smaller field-of-view. The AMSU-A measures thermal radiation at microwave frequencies ranging from 23.8 to 89.0 GHz. Atmospheric temperature profiles are primarily based on channels 3-14 near 60 GHz, which is an oxygen absorption band. Since the AMSU-A instrument provides a nominal spatial resolution of 48 km at nadir and 170 km at far limb, the temperature perturbations from synoptic to mesoscale can be reasonably depicted. In addition, several AMSU imaging channels at frequencies of 31.4, 89 and 150 GHz are utilized to determine cloud liquid water and ice water contents because they directly respond to the thermal emission of liquid droplets and scattering of ice particles (Weng and Grody 2000).

Since microwave radiance responds linearly to temperatures and since the weighting functions at various AMSU sounding channels are relatively stable, temperatures at any pressure level can be expressed as a linear combination of brightness temperatures measured at various sounding channels (see Janssen 1993), i.e.,

$$T(p) = C_0(p,\theta_s) + \sum_{i=1}^{n} C_i(p,\theta_s) T_b(\nu_i,\theta_s) \ , \tag{1}$$

where p is the pressure; θ_s is the scanning angle; ν_i is the frequency at channel i; and T_b is the AMSU brightness temperature. The empirical coefficients, C_0 and C_i, are determined using a regression equation, in the same form as Eq. (1), by matching the rawinsonde temperature soundings with the AMSU-A brightness temperatures that were co-located at islands over the low- to mid-latitude oceans (Zhu et al. 2002a). Two sets of retrieval coefficients are obtained to retrieve temperatures in clear and heavily raining regions separately, because the large rain droplets contaminate signals in channels 3-5. The heavy precipitation region is identified by using the cloud liquid water path (LWP), which is retrieved from the AMSU-A window channels 1 and 2 (Weng et al. 2000).

Four tropical cyclones are studied to examine how the temperature retrieval algorithm performs. Figure 1a shows the retrieved temperature at 250 hPa for Hurricane Bonnie at 0000 UTC 26 Aug 1998. The maximum temperature in the warm core was about -28°C. The temperature anomaly is often defined as a deviation from the unperturbed environmental temperature. Apparently, the warm core of Bonnie was about 10 - 12°C at 250 hPa (Fig. 1a). At this time the minimum sea-level pressure of Bonnie was 958 hPa. Hurricane Alberto (2000) was a long-lived hurricane that remained at sea through its lifetime. At 2200 UTC 21 Aug 2000, it was a Category 2 (U.S. Saffir-Simpson Hurricane Scale) hurricane with a minimum see-level pressure of 975 hPa. Alberto had a warm core of –38°C, and the warm temperature anomaly was around 8 – 10°C at 250 hPa (Fig. 1b). Hurricane Isaac (2000) was a Category 3 hurricane at 2200 UTC 29 Sep. 2000 with minimum central pressure of 960 hPa. The retrieved warm core was –38°C, and the warm anomaly was about 10°C at 250 hPa (Fig. 1c). We also tested the temperature retrieval algorithm for tropical cyclones in the Pacific Ocean. Typhoon Dan had Category 2 intensity, with a maximum sustained wind speed of 85 kt at 2300 UTC 05 Oct 1999 in the South China Sea. The AMSU measurements showed an 8°C warm core with –35°C at 250 hPa (Fig. 1d). The above results indicate that AMSU retrieved temperatures can depict the strong warm-core signal at 250-hPa height for tropical cyclones. The magnitude of the retrieved temperature anomaly for each storm is closely related to its intensity in terms of the sea-level pressure.

2.2. *Inverting the balanced flow*

After retrieving temperatures from the AMSU data, we are able to obtain the three-dimensional rotational winds using the balance equation. The gradient balance models have been widely used in theoretical studies of tropical cyclones, and they have provided many fundamental insights into the dynamics of hurricane vortices (Eliassen 1951; Emanuel 1986; Shapiro and Willoughby 1982). The balance equation has also been widely used for initializing primitive equation models as well as for bogusing tropical cyclones (Bolin 1956; Baer 1977). The general form of the balance equation, in terms of geopotential height (ϕ) and stream function (ψ), can be written as

$$f\nabla^2\psi + 2(\psi_{xx}\psi_{yy} - \psi_{xy}{}^2) + \psi_x f_x + \psi_y f_y = \nabla^2\phi, \tag{2}$$

where f is the Coriolis parameter, and ∇ is the two-dimensional gradient operator. Inverting the stream function from the given geopotential ϕ using the successive over-relaxation (SOR) method is a Monge-Ampere type of problem, in which an ellipticity condition must be satisfied. The inverting procedure includes (a) specifying an asymmetrically distributed sea level pressure field using a modified scheme of Holland (1980); (b) integrating the hydrostatic equation using the AMSU-derived temperatures to obtain geopotential heights; (c) solving for the stream function from Eq. (2), given lateral boundary conditions; and (d) calculating horizontal winds from the stream function.

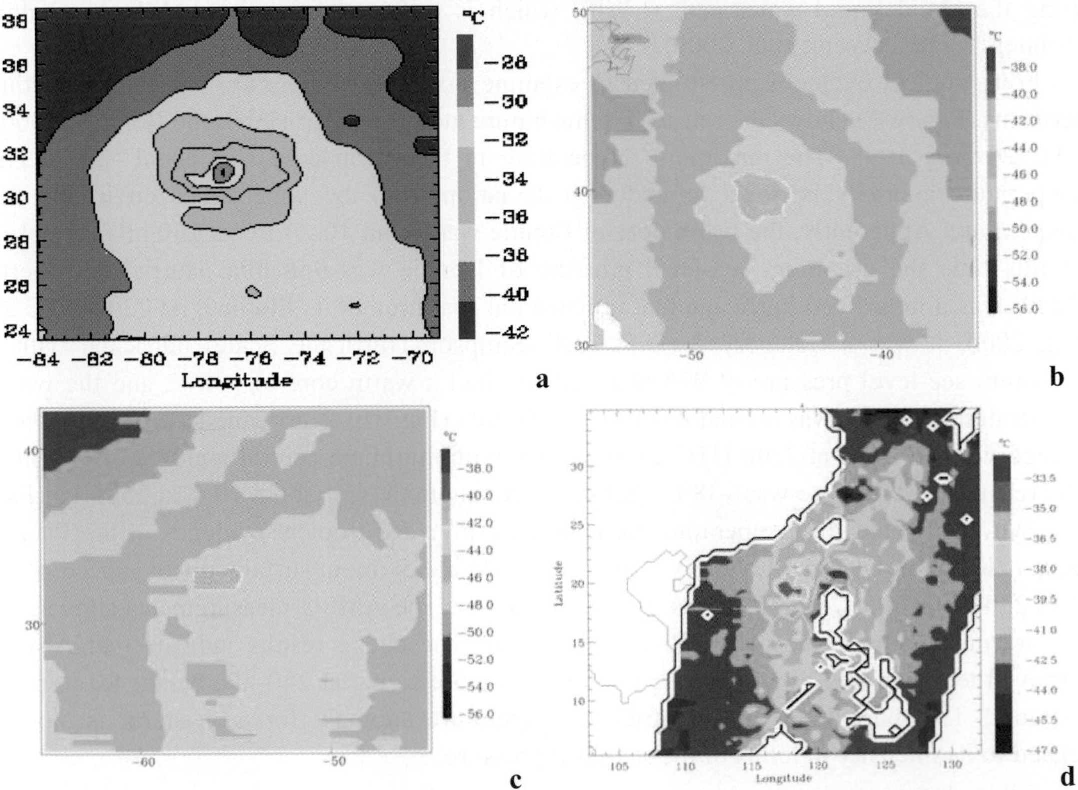

Figure 1. The AMSU-retrieved temperatures at 250 hPa (a) at 0000 UTC 26 Aug 1998 for Hurricane Bonnie, (b) at 2200 UTC 21 Aug 2000 for Hurricane Alberto, (c) at 2200 UTC 29 Sep 2000 for Hurricane Isaac, and (d) at 2300 UTC 05 Oct 1999 for Typhoon Dan.

2.3. *Diagnosing the divergent wind component*

It is evident that horizontal winds derived from Eq. (2) are nondivergent. However, there are pronounced vertical motions associated with latent heat release in the eyewall, which should be included as a divergent component in order to alleviate the initial model spin-up problems. The divergent winds could be obtained through vertical velocity diagnosed from the Omega equation, and it is detailed below.

First, the surface rain rates are derived using NOAA's operational AMSU algorithm (Zhao et al. 2001). Second, the vertical distribution of condensational heating must be determined in accordance with the surface rain rates. A normalized latent heating profile is diagnosed from the explicitly simulated Andrew during its mature stage (Liu et al. 1997). Third, the divergent wind is obtained, following Tarbell et al. (1981), using the balanced winds and the 3D latent heating field. The following Omega and the continuity equations are employed:

$$\nabla^2(\sigma\omega) + f\zeta_a\frac{\partial^2\omega}{\partial p^2} = f\frac{\partial}{\partial p}(V_\varphi + V_\chi)\cdot\nabla\zeta + \frac{R}{p}(\nabla^2 V_\varphi + \nabla^2 V_\chi)\cdot\nabla T - f\frac{\partial}{\partial p}(\zeta\nabla^2\chi) + f\frac{\partial}{\partial p}(\omega\frac{\partial\zeta}{\partial p}) + f\frac{\partial}{\partial p}(\nabla\omega\cdot\nabla\frac{\partial\varphi}{\partial p}) - \frac{R}{C_p P}\nabla^2 Q, \quad (3)$$

$$\nabla^2\chi + \frac{\partial\omega}{\partial p} = 0, \quad (4)$$

where $\sigma \equiv -\frac{RT}{p\theta}\frac{\partial\theta}{\partial p}$ is the static stability and Q is the diabatic heating rate. An iterative procedure is required to solve ω and V_χ from Eq. (3) and (4). Since the latent heating is a major forcing term in the Omega equation, we do see the convergence (divergence) of the lower- (upper-) level flows in the rainfall regions with the divergent winds at about ± 5 m s^{-1} after the above procedure.

After obtaining the divergent winds, the total horizontal winds are just the sum of the rotational and divergent winds. Figure 2 shows the retrieved total flows at 850 hPa at 0000 UTC 22 Aug 1998 when Bonnie reached Category 1 hurricane intensity. At this time, there were Global Positioning System (GPS) dropsondes available to provide verification. In general, the retrieved winds agree with the observed at 850 hPa in both speed and direction. Since the GPS dropsonde data near the storm center were measured around 2130 UTC 21 August, and the satellite passed the spot at 0000 UTC 22 August, the retrieved vortex center is located on the west of the observed by about 0.5°. The retrieved maximum wind is 32 m s^{-1}, in good agreement with the observed. Both the retrieved and observed winds show a region in excess of 30 m s^{-1} on the north of the storm center; whereas on the south winds are relatively weak.

3. Simulation of Hurricane Bonnie

Using the above derived initial hurricane vortex at 0000 UTC 22 Aug 1998, we performed a 5-day simulation of Hurricane Bonnie, and we analyze Bonnie's inner-core structures in this section. The hurricane-scale moisture and SST fields for model initial and boundary conditions are also specified according to satellite microwave measurements.

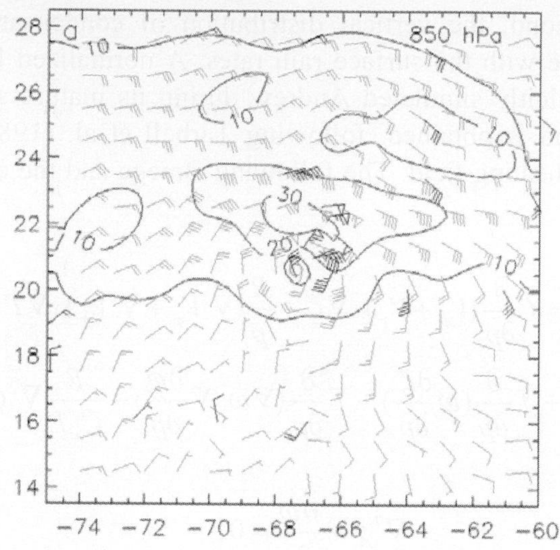

Figure 2. The horizontal winds retrieved from the AMSU (in green) and obtained from the GPS dropsonde measurements (in red) at 0000 UTC 22 August 1998. Solid lines are isotaches at intervals of 5 m s^{-1}. The storm center is denoted by a green hurricane symbol (after Zhu et al. 2002a).

3.1. *A 5-day simulation of Bonnie*

Hurricane Bonnie was the first major hurricane that took place in the Atlantic Ocean after the launch of NOAA-15 with the AMSU instruments on board. Bonnie exhibited two interesting characteristics during its life span. The first one is the evolution of storm intensity: a rapid deepening for a two-day period after reaching Category 1 on 22 August 1998, followed by a near-steady state over the open ocean during the next three days (i.e., 24-27 August). The second distinct characteristic is the evolution of precipitation in the eyewall: from a partial to near-concentric eyewall during 22-27 August. In this section, we present some interesting features associated with the asymmetric eyewall structures in the simulated storm.

Hurricane Bonnie (1998) is explicitly simulated using the PSU/NCAR nonhydrostatic, two-way interactive, movable, triply nested grid model (MM5V3.4, Dudhia 1993 and Grell et al. 1995) with the finest mesh grid length of 4 km. The model physics include the Goddard cloud microphysics scheme (Tao and Simpson, 1993; Simpson and Tao 1993), the Kain-Fritsch (1993) convective parameterization scheme, the Blackadar PBL scheme (Zhang and Anthes 1982) and a cloud-radiation interaction scheme, which are similar to those given in Liu et al. (1997). Although no convective parameterization is applied over the finest mesh, a shallow convective scheme is used. There are 24 σ levels or 23 half-σ layers in the vertical, which is the same as that used by Liu et al. (1997). A 5-day simulation was performed, covering the initial rapid deepening, steady variation and landfalling stages of the storm. See Zhu et al. (2002a,b) for detailed model design.

The model initial conditions and lateral boundary conditions were obtained from the National Centers for Environmental Prediction (NCEP) 2.5°×2.5° global analysis, which was then enhanced by rawinsondes and surface observations. Because the NCEP analysis contained a vortex with a central pressure of 12 hPa weaker than the observed, the above derived hurricane vortex were incorporated into the model initial conditions. The initial moisture field in NCEP analysis was also too dry, particularly in the hurricane eyewall regions. Thus, we iteratively retrieve the 3D moisture field using the Total Precipitable Water (TPW) retrieved from AMSU-A measurements (Grody et al. 2001) as a constraint. The simulated mean RH profiles given in Liu et al. (1999) were used here as a reference to set the maximum and minimum RH values in the vertical. The iteration started from the minimum RH value at a given point, and then computed TPW by vertically integrating its associated specific humidity. The difference between the new TPW and the satellite-retrieved TPW was used to compute a new RH profile. At most points, the calculated TPW values could converge to the satellite-retrieved within 50 iterations.

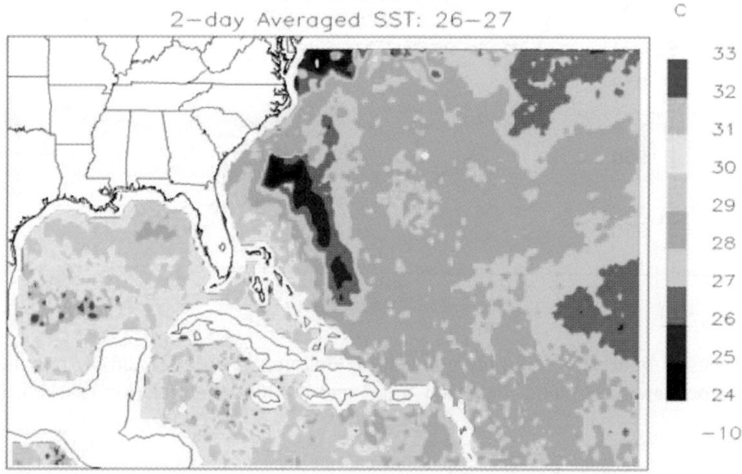

Figure 3. The 2-day averaged SST obtained from Tropical Rainfall Measuring Mission (TRMM) Microwave Imager (TMI) for the period from 0000 UTC 26 to 0000 UTC 27 Aug 1998.

Since this is a 5-day simulation, the oceanic feedback becomes an important factor influencing the intensity of the storm. For this reason, the SST was updated daily using the Tropical Rainfall Measuring Mission (TRMM) Microwave Imager (TMI) Level-1 standard product at 0.25°×0.25° Lat/Lon resolution (Chelton et al. 2000). Although the TRMM/TMI SST can be measured underneath clouds, there were still some missing data near the storm center due to the contamination by heavy rain. They were filled by a three-day running mean of SST at each grid point, using the data up to the current model time. A 2-day averaged SST is given in Fig. 3, which shows more than 4°C SST cooling along and to the north of

Bonnie's track. It can be seen that the cold SST remained for about 2 days after the storm passed. According to our sensitivity test (to be shown in a forthcoming paper) and the coupled ocean-atmosphere model simulations of Hong et al. (2000), this SST cooling could account for about a 20-hPa central pressure change in simulations with or without the SST feedback.

3.2. *Simulation results*

The simulated track compares favorably to the observed (not shown), which shows similar paths in the first 12 h. Then, Bonnie changed its tracking direction from west-northwest to northwest, whereas the simulated one keeps moving west-northwestward. The simulated track deviates from the observed by about 200 km at 24 h. After 48 h, the simulated storm moves closer to the best track, but still with a distance error of 250 km, and it also re-curves somewhat earlier than the observed at 84 h. The simulated landfall occurs to the northeast of that observed by about 150 km and 20 h earlier.

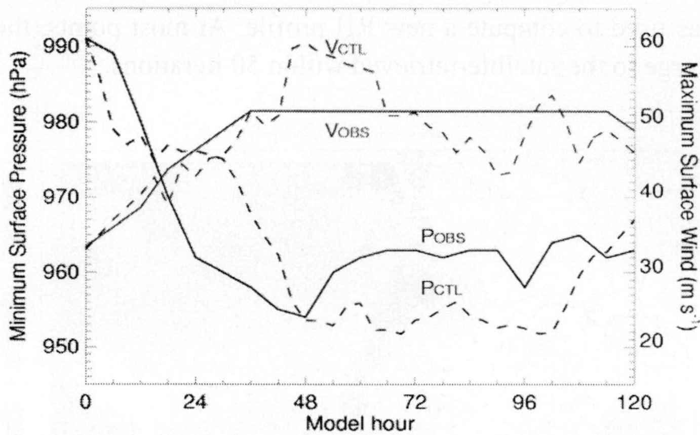

Figure 4. Time series (6 hourly) of the minimum central pressures (P in hPa) and the maximum surface winds (V in m s^{-1}) from the best analysis (solid, P_{OBS} and V_{OBS}) and the model simulation (dashed, P_{CTL} and V_{CTL}) (after Zhu et al. 2003).

Figure 4 compares the simulated hurricane intensity to the observed. They are in general agreement but differ in details during the 5-day period. Initially, the modeled storm (P_{CTL}) deepens more rapidly than the observed (P_{OBS}), indicating the absence of the model spin-up problem. However, the model storm experiences a slow deepening period from 12 to 30 h. This is likely caused by the presence of a too-dry environment that fails to feed the needed energy in the PBL inflow for the continued deepening, since the initial moisture field outside of the hurricane vortex was not modified in the model initial conditions. A deepening rate, similar to the observed, does not occur until the 30-h when the model storm begins to receive comparable energy supply through the air-sea interaction processes. The simulated storm reaches the minimum central pressure of 954 hPa as the observed at 48 h, and maintains its intensity with weak oscillations afterwards. During the maintenance stage, the simulated storm is about 6-8 hPa deeper than the observed. Part of this difference could clearly be attributed to the fact that the simulated storm moves to the southwest of the best track in the first 3 days where local SST is about 1° to 2° warmer than that to its north. Nevertheless, the two distinct development stages of rapid deepening and slow maintenance are reasonably reproduced. The simulated maximum surface winds (V_{CTL}) also compare favorably to the observed (V_{OBS}). The two distinct development stages can also be seen in both the simulated and observed wind fields.

3.3. *Diagnoses of the partial eyewall feature*

The simulated radar reflectivity from the 46-h control simulation was calculated, following Liu et al. (1997); it exhibits highly asymmetric structures of the eyewall and cloud bands (Fig. 5). Specifically, the model produces a partial eyewall in the east and an organized rainband spiraled outward in the northeast semicircle. Of interest is that much weaker or little convective activity occurs in the west, which is in significant contrast with the simulated Hurricane Andrew (1992) shown in Liu et al. (1997). The simulated structures compare favorably to the radar observations by NOAA's WP-3D aircraft during the Convective and Moisture Experiment-3 (CAMEX-3), including the general location, size, and intensity of the eyewall/rainband, although the observed location is about 200 km to the northeast.

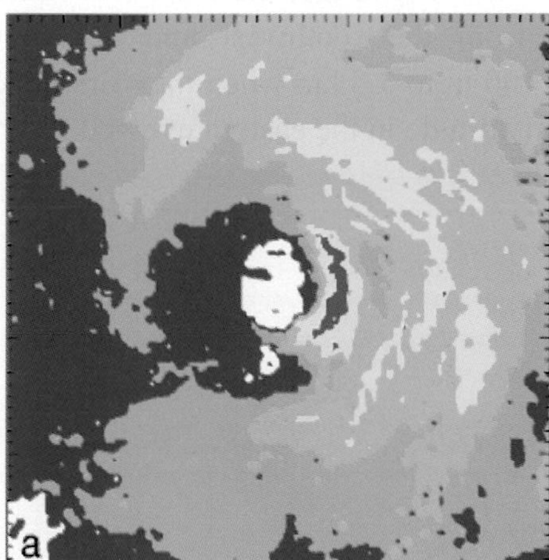

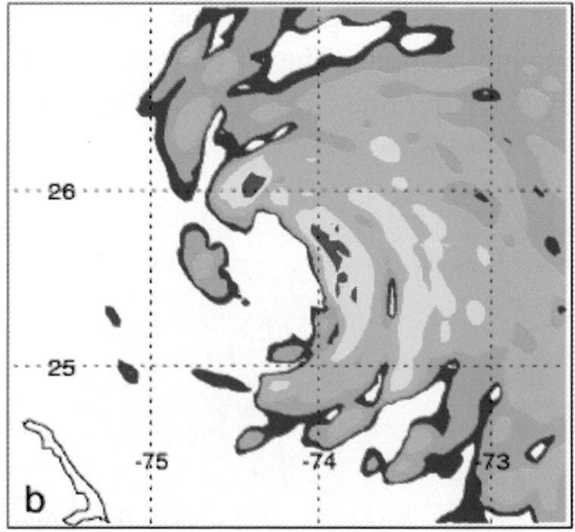

Figure 5. Radar reflectivity (at 15, 21, 28, 35, 41 and 48 dBZ) from (a) NOAA's WP-3D reconnaissance aircraft at 2130 UTC 23 August at the 4544-m altitude; (b) the control simulation at 2200 UTC 23 August near 480 hPa. The two panels have the same color scale and the same domain size of 360 km x 360 km (adapted from Zhu et al 2002a).

Many earlier observations have shown the presence of asymmetric clouds, precipitation and winds in tropical cyclones. The downshear-left asymmetric eyewall pattern has also been clearly demonstrated by idealized model simulations. Frank and Ritchie (1999) simulated tropical-cyclone-like vortices under different vertical wind shear environments using the MM5 model. They found that more intense convection tends to be organized at the left side of the shear vector, looking downshear. There are several hypotheses for the relationship between vertical shear and hurricane asymmetries. These hypotheses all emphasize the processes in lower atmospheric layers, such as frictional drag (Shapiro 1983), and the low-level convergence and divergence (Willoughby et al. 1984, Bender 1997). In the next paragraph, we examine the influence of vertical wind shears on the development of the partial eyewall and other related inner-core structures using the simulation results for Hurricane Bonnie.

Figure 6 presents a few selected trajectories from 42 – 54 h using 15-min model output in order to see the extent of vertical and horizontal displacements associated with the descending and ascending motion in western and eastern eyewall, respectively. The first group of seeds is put in the west eyewall at the altitude of 9 km. The backward trajectory indicates that the air parcels come from the upper-level jet flow region northwest of the storm at about 12-km height. After reaching the west eyewall region, this flow converges and goes downward to 9-km height, then circles cyclonically from the west quadrant to the south quadrant. Once the flow goes out of the eyewall at the south quadrant, it climbed up to 13 km gradually. The second group of seeds is released around 7-km altitude in the west eyewall. This flow comes from the northeast side of environmental region, and it goes into the eyewall at northwest quadrant at 9 km altitude. It takes about 3 h for the flow to complete a half circle in the western eyewall at a speed for about 25 m s^{-1}, and then goes out of the eyewall at the southeast quadrant. Once outside the eyewall, the flow slows down and climbs to 13 km eventually. By examining the low-level trajectories, it is found that there are two sources of moist air that converge into the east eyewall from the southwest and the southeast environmental regions. The air parcels climb from 100 m to 3.5 km when they first reach the east eyewall. After one circle, the flow reaches 6 km altitude in the eastern eyewall.

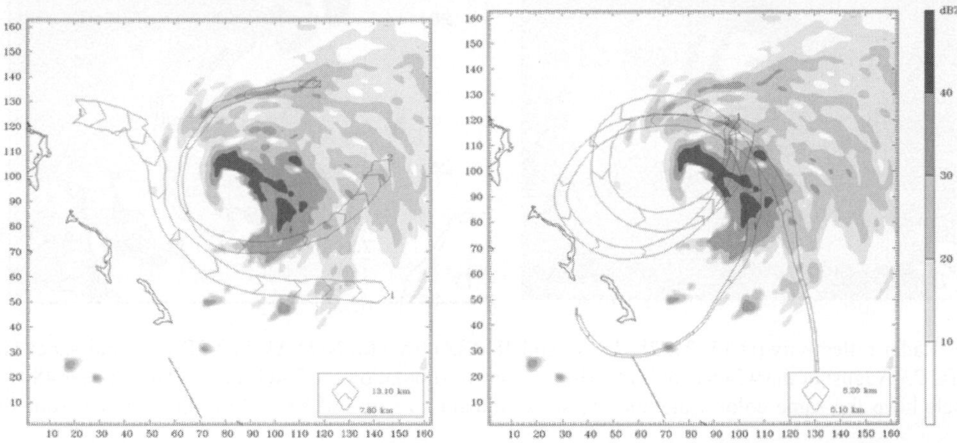

Figure 6. The 6 hourly backward and forward trajectories for parcels released at 48-h simulation (i.e., 0000 UTC 24 August). (a) Number 1 and 2 parcels are released in the western eyewall at z = 9 and 7 km, respectively. (b) Number 3 and 4 parcels are released in the eastern eyewall at z = 3.5 km. Radar reflectivity at $\sigma = 0.44$ level is shaded at intervals of 10 dBZ. A trajectory is shown as a curved ribbon, with its width proportional to the height of the trajectory. The arrowheads are drawn along each trajectory at 1-h intervals.

The above trajectory analysis indicates that an upper-level westerly flow enters the eyewall with deeper and stronger inflow in the northwest. Because the environmental air is cold and dry, downdrafts are initiated in the inflow region, with significant enhancement from evaporative cooling. The dry descent causes adiabatic warming and drying in the middle level, and stabilizes the atmosphere in the western part of the eyewall. Therefore the development of clouds and precipitation in the western eyewall is suppressed, and the partial eyewall feature is generated.

Finally, the frontogenesis function is computed to examine the formation of the partial eyewall feature from kinematic considerations. The frontogenesis function has been widely used to investigate mid-latitude frontogenesis and cyclogenesis (Miller, 1948; Bosart, 1970; Baldwin et al., 1984; Lapenta and Seaman 1990; Doyle and Warner 1993). Emanuel (1997) pointed out that the hurricane eyewall is not necessarily a front in surface temperature, but instead involves the equivalent potential temperature (θ_e) distribution, which is directly related to density in saturated air. At 47 h, the strong warm anomaly or the strong gradient of potential temperature (θ) is located at the west side of eyewall because of subsidence warming, while there is little deep convection. However, the $|\nabla_H \theta_e|$ is weak at the west and strong at the east sides of the eyewall (Fig. 7a). This asymmetric pattern is similar to that of radar reflectivity. In order to focus on the asymmetric convection feature, we analyze frontogenesis in terms of θ_e instead of θ. The temporal rate change of the gradient of θ_e following a parcel is defined as in Lapenta and Seaman (1990) except for the replacement of θ by θ_e and therefore the absence of the condensation term,

$$\begin{aligned}\frac{d}{dt}|\nabla_H\theta_e| = -|\nabla_H\theta_e|^{-1}&\left\{\left[\left(\frac{\partial\theta_e}{\partial x}\right)^2\frac{\partial u}{\partial x}+\left(\frac{\partial\theta_e}{\partial y}\right)^2\frac{\partial v}{\partial y}\right]+\left[\frac{\partial\theta_e}{\partial x}\frac{\partial\theta_e}{\partial y}\left(\frac{\partial v}{\partial x}+\frac{\partial u}{\partial y}\right)\right]\right\}\\ &-|\nabla_H\theta_e|^{-1}\left[\left(\frac{\partial\theta_e}{\partial x}\frac{\partial\theta_e}{\partial z}\right)\frac{\partial w}{\partial x}+\left(\frac{\partial\theta_e}{\partial y}\frac{\partial\theta_e}{\partial z}\right)\frac{\partial w}{\partial y}\right],\end{aligned} \tag{5}$$

where ∇_H is the horizontal gradient operator. The first term on the right-hand side of Eq. (5) represents the contribution of the deformation field to frontogenesis, including the horizontal confluence and shearing effects. The second term is the contribution from the twisting or tilting effect. Note that there is no surface friction term in Eq. (5) because we focus our attention above the maritime boundary layer, especially the middle and upper troposphere where the significant asymmetries occur.

In this analysis, a 1-h averaged dataset is used, which is obtained from a 10-min interval dataset for the period from 46 to 47 h. This is the time when the storm already has asymmetric structures. The typical value of total frontogenesis is approximately 1 K km^{-1} day^{-1} in our analysis, and is about 50 times larger than that in coastal frontogenesis (Lapenta and Seaman 1990; Doyle and Warner 1993), mainly because the magnitudes of the variables in a hurricane system are larger than those in a mid-latitude front system, and our model resolution is higher.

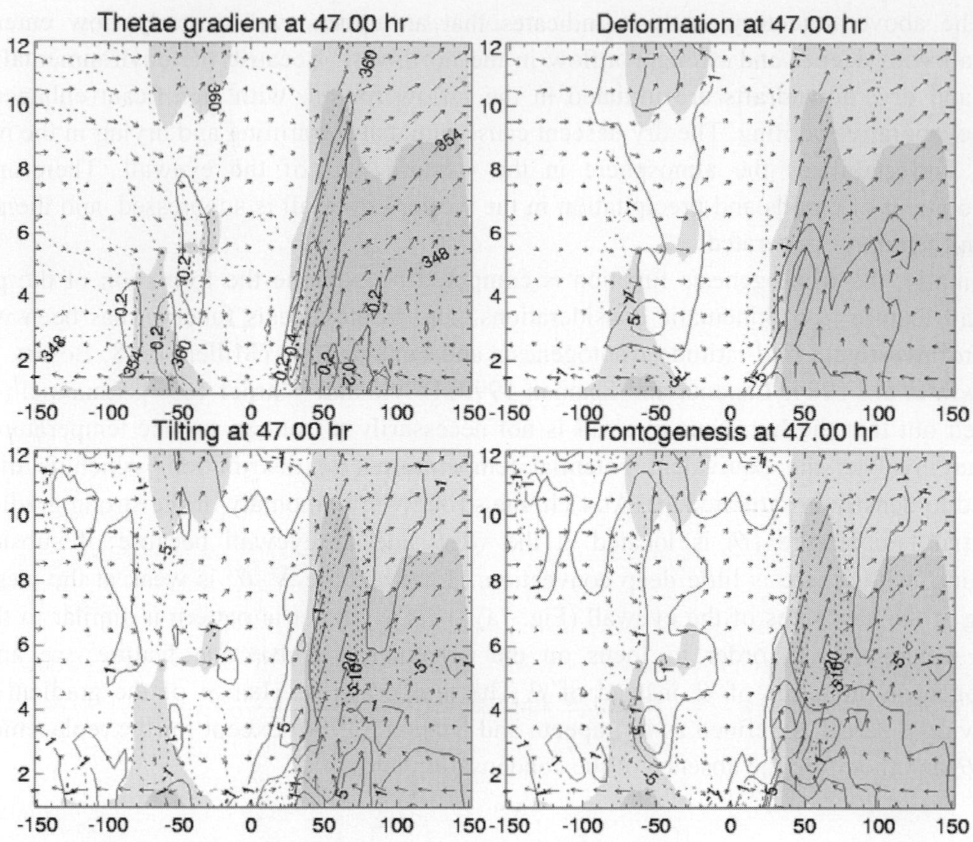

Figure 7. West-east vertical cross sections through the minimum pressure center of (a) potential temperature (dashed) and its gradient (solid); and (b) frontogenesis due to the deformation term; (c) frontogenesis due to the tilting term; and (d) total frontogenesis for the averaged period from 46 to 47 h. Radar reflectivity is shaded at 15 and 30 dBZ. In-plane flow vectors are also superposed.

The west-east vertical cross section of $|\nabla_H \theta_e|$, the deformation term, the tilting term and the total frontogenesis through hurricane eye are shown in Fig. 7. The frontogenesis function due to the deformation term is positive at lower and middle levels of the east side of the storm and middle and upper levels of the west side of the storm (Fig. 7b), as a result of the convergence of environmental flow. The effect of the deformation term is frontolytic at low levels in the west eyewall and at upper levels in the east eyewall, associated with the diffluence zone. The contribution by the tilting term is strong where there is strong vertical motion. At the west side of the eyewall, there are two frontolytic regions at lower and upper levels as a result of the tilting term (Fig. 7c). A deep frontolytic layer exists in the east side of the eyewall, with the maximum at middle and upper levels. A strong frontogenetic zone is produced immediately inside the east eyewall by the tilting term. Overall, the total frontogenesis function (Fig. 7d) is frontolytic in the west eyewall, with contributions from the deformation term at low levels and the tilting term at middle and upper levels. Just inside the [illegible] eyewall, there is a frontogenetic region at middle and upper levels, mainly as a result of [illegible] flow from the upper level. To the east of the center, strong frontogenesis is produced immediately inside the old eyewall with positive contributions from both the deformation and tilting terms. A strong frontolytic zone is created in the eyewall above 5 km by the tilting term. The effect of the west side frontolytic zone, especially in the low level eyewall, is to reduce convection in that region. The total frontogenesis developed inside the east and west eyewalls (around 40-50 km region) suggests that the radius of the inner eyewall is going to keep decreasing. The frontogenesis that develops at the eastern side in about an 80-km area indicates that an outer rainband is going to form there.

4. Summary and conclusions

Tropical cyclones develop over the ocean, where few traditional rawinsonde observations are available. Satellite observations have been proven to be the most useful data source for tropical cyclone monitoring and forecasting. In this paper, we reviewed several satellite products and the methods to incorporate them into numerical models to improve hurricane forecasts. Because microwave radiation penetrates through clouds and is sensitive to a wide variety of atmospheric parameters, satellite microwave observations, such as AMSU data, are extremely useful in the study of tropical cyclones.

Four tropical cyclones were studied to examine the general capability of a newly developed AMSU temperature retrieval algorithm. It was found that the retrieved temperature could depict a strong warm core at 250-hPa height for tropical cyclones. The magnitude of the retrieved temperature anomaly for each storm is closely related to its intensity in term of sea-level pressure. The geopotential heights were calculated by integrating the hydrostatic equation from the bottom upward with the retrieved temperatures. The asymmetric vortex flows were attained by solving the nonlinear balance equation, using the NCEP analysis as the lateral boundary conditions. The divergent wind component was obtained by solving the Omega equation, in which the latent heating profile is specified with a magnitude based on the AMSU-A surface rain rate and a vertical structure based on a previously simulated hurricane. The retrieved temperature and wind fields associated with Hurricane Bonnie compare favorably to the GPS dropsonde observations during CAMEX-3. It should be pointed out that the resolution of the retrieved temperature may not be high enough to resolve the warm-core structure of a small hurricane in lower levels because the AMSU-A field of view increases from 48 to about 170 km at it scans from nadir to extreme limb direction.

Using the above derived initial hurricane vortex, a 5-day simulation of Hurricane Bonnie (1998) was performed with a triply nested mesh and the finest grid size of 4 km. The model initial moisture field of the hurricane vortex was derived from the AMSU-derived TPW, assuming that RH is constant in the eyewall cloud regions. During the model integration, the SST was updated daily according to the TRMM/TMI observation. The simulated track was within 3^0 lat/lon of the best-track during the 5-day integration, with the landfalling point close to the observed. The model also reproduced the hurricane intensity changes during the 5-day period, including the deepening and maintaining stages. The simulated radar reflectivi[illegible] shows pronounced asymmetries in the eyewall and rainbands around 45 h, in agreemen[illegible] the observations from WP-3D radar.

Trajectory analyses indicate that the upper-level westerly flow converged into the west eyewall, and then descended and cyclonically circled around the center from the west to the south sector. The cyclonic flow at middle and lower levels in the eyewall was also forced downward in the west to south quadrant. Because of the strong descending motion, the warm core shifted to the west. The mid-level warming stabilized the atmosphere at the west part of eyewall, and convection was confined to low-levels. When the storm developed the partial eyewall feature, the frontogenesis function was frontolytic in the west eyewall, with contributions from the deformation term at low levels and from the tilting term at the middle and upper levels. It is found that strong upper-level inflow is a key factor that leads to the development of the partial eyewall.

Acknowledgments. This work was supported by NOAA/NESDIS, NSF Grant ATM-9802391, and NASA Grant NAG-57842. The first author wishes to express profound gratitude to his dissertation advisor Dr. Da-Lin Zhang for giving a lot of guidance throughout this study. We want to thank Ms. Huan Meng at NOAA/NESDIS/ORA for helping us to obtain AMSU data.

References:

Baer, F., 1977: The spectral balance equation. *Tellus,* **29,** 107-115.

Baldwin, D., E.-Y., Hsie, and R. A. Anthes, 1984: Diagnostic studies of a two-dimensional simulation of frontogenesis in a moist atmosphere. *J. Atmos. Sci.*, **41**, 2686-2700.

Bender, M., 1997: The effect of relative flow on the asymmetric structure in the interior of hurricanes. *J. Atmos. Sci.*, **54,** 703-724.

Bloom, S. C., L. L. Takacs, A. M. da Silva, and D. V. Ledvina, 1996: Data assimilation using incremental analysis updates. *Mon. Wea. Rev.,* **124**, 1256–1271.

Bolin, B., 1956: An improved barotropic model and some aspects of using the balance equation for three-dimensional flow. *Tellus*, **8**, 61-75.

Bosart, L. F., 1970: Midtropospheric frontogenesis. *Quart J. Roy. Meteor. Soc.*, **96**, 442-471.

Chelton, D.B., F.J. Wentz, C.L. Gentemann, R.A. de Szoeke, and M.G. Schlax, 2000: Satellite microwave SST observations of transequatorial tropical instability waves. *Geophys. Res. Letters*, **27**, 1239-1242.

Doyle, D. J., and T. T. Warner, 1993: A numerical investigation of coastal frontogenesis and mesoscale cyclogenesis during GALE IOP 2. *Mon. Wea. Rev.*, **121**, 1048-1077.

Dudhia, J., 1993: A nonhydrostatic version of the Penn State-NCAR mesoscale model: Validation tests and simulation of an Atlantic cyclone and cold front. *Mon. Wea. Rev.*, **121,** 1493-1513.

Dvorak, V. F., 1972: A technique for the analysis and forecasting of tropical cyclone intensities from satellite pictures. NOAA TMNESS 36, U.S. Dept. of Commerce, Washington, DC, 15 pp.

Eliassen, A., 1951: Slow thermally or frictionally controlled meridional circulation in a circular vortex. *Astrophys. Norv.*, **5,** 19-60.

Emanuel, K. A., 1986: An air-sea interaction theory for tropical cyclone. Part I: Steady-state maintenance. *J. Atmos. Sci.,* **43,** 585-604.

——, 19[illegible]: Some aspects of hurricane inner-core dynamics and energetics. *J. Atmos. Sci.*, **54,** 1014-

Frank, W. M., and E. A. Ritchie, 1999: Effects of environmental flow upon tropical cyclone structure. *Mon. Wea. Rev.*, **127,** 2044-2061.

Grell, G. A., J. Dudhia, and D. R. Stauffer, 1995: A description of the fifth generation Penn State/NCAR mesoscale model (MM5). NCAR Tech Note NCAR/TN-398+STR, 138 pp. [Available from NCAR Publications Office, P. O. Box 3000, Boulder, CO 80307-3000.]

Grody, N., C., 1979: Typhoon "June" winds estimated from scanning microwave spectrometer measurements at 55.45 Ghz. *J. Geophys. Res.*, **84**, 3689-3695.

_____, J. Zhao, R. Ferraro, F. Weng and R. Boers, 2001: Determination of precipitable water and cloud liquid water over oceans from the NOAA-15 Advanced Microwave Sounding Unit (AMSU). *J. of Geophys. Res.,* **106**, 2,943-2,953.

Holland, G. J., 1980:An analytic model of wind and pressure profiles in hurricanes. *Mon. Wea. Rev.*, **108**, 1212-1218.

Hong, X., S. W. Chang, S. Raman, L. K. Shay, and R. Hodur, 2000: The Interaction between Hurricane Opal (1995) and a warm core ring in the Gulf of Mexico. *Mon. Wea. Rev.*, **128**, 1347-1365.

Hou, A. Y., D. V. Ledvina, A. M. Da Silva, S. Q. Zhang, J. Joiner, and R. M. Atlas, 2000a: Assimilation of SSM/I-derived surface rainfall and total precipitable water for improving the GEOS analysis for climate studies. *Mon. Wea. Rev.,* **128**, 509–537.

_____, S. Q. Zhang, A. M. Da Silva, and W. S. Olson, 2000b: Improving assimilated global datasets using TMI rainfall and columnar moisture observations. *J. Climate.,* **13**, 4180–4195.

Hubert, L. F., and L. F. Whitney Jr., 1971: Wind estimation from geostationary satellite pictures. *Mon. Wea. Rev.,* **99,** 665-672.

Janssen, M. A., 1993: *Atmospheric Remote Sensing by Microwave Radiometry.* Wiley Series in Remote Sensing, John Wiley & Sons, Inc., 572pp.

Kain, J. S., and J. M. Fritsch, 1993: Convective parameterization for mesoscale models: The Kain-Fritsch scheme. *The Representation of Cumulus Convection in Numerical Models, Meteor. Monogr.,* No. 46, Amer. Meteor. Soc., 165–170.

Kidder, S. Q., W. M. Gray, and T.H. Vonder Harr, 1978: Estimating tropical cyclone central pressure and outer winds from satellite microwave data. *Mon. Wea. Rev.*,**108**, 144-152.

_____, M. D. Goldberg, R. M. Zehr, M. DeMaria, J. F. W. Purdom, C. S. Velden, N. C. Grody, and S. J. Kusselson, 2000: Satellite analysis of tropical cyclones using the advanced microwave sounding unit (AMSU). *Bull. Amer. Meteor. Soc.* **81,** 1241-1260.

Krishnamurti T. N., Xue J., Bedi H. S., Ingles K., and D. Oosterhof, (1991): Physical initialization for numerical weather prediction over the tropics. *Tellus*, **43**, 53-81.

_____, Han W., and D. Oosterhof, (1998): Sensitivity of hurricane intensity forecasts to physical initialization. *Meteorol. Atmos. Phys*., **65**, 171-181.

Lapenta, W. M., and N. L. Seaman, 1990: A numerical investigation of east coast cyclogenesis during the cold-air damming event of 27-28 February 1982. Part I: Dynamic and thermodynamic structure. *Mon. Wea. Rev.*, **118**, 2668-2695.

Liu, Y., D.-L. Zhang, and M. K. Yau,1997: A multiscale numerical study of Hurricane Andrew(1992). Part I: Explicit simulation and verification. *Mon. Wea. Rev.*, **125**, 3073-3093.

——, ——, and ——,1999: A multiscale numerical study of Hurricane Andrew(1992). Part II: Kinematics and inner-core structures. *Mon. Wea. Rev.*, **127**, 2597-2616.

Miller, J. E., 1948: On the concept of frontogenesis. *J. Meteor.*, **5**, 169-171.

Pu, Z.-X., W.-K. Tao, S. Braun, J. Simpson, Y. Jia, J. Halverson, W. Olson, and A. Hou, 2002: The impact of TRMM data on mesoscale numerical simulation of Supertyphoon Paka. *Mon. Wea. Rev.,* **130**, 2448-2458.

Shapiro, L. J., and H. E. Willoughby, 1982: The response of the balance hurricanes to local sources of heat and momentum. *J. Atmos. Sci.,* **39,** 378-394.

_____, 1983: Asymmetric boundary layer flow under a translating hurricane. *J. Atmos. Sci.*, **40**, 1984-1998.

Simpson, J., and W.-K. Tao, 1993: The Goddard Cumulus Ensemble Model. Part II: Applications for studying cloud precipitating processes and for NASA TRMM. *Terr. Atmos. Oceanic Sci.,* **4**, 73–116.

Tao, W.-K., and J. Simpson, 1993: the Goddard cumulus ensemble model. Part I: Model description. *Terr. Atmos. Oceanic Sci.,* **4,** 35-72.

Tarbell, T. C., T. T. Warner, and R. A. Anthes, 1981: An example of the initialization of the divergent wind component in a mesoscale numerical weather prediction model. *Mon. Wea. Rev*., **109**, 77-95.

Tibbetts, R. T., and T. N. Krishnamurti, 2000: An intercomparison of hurricane forecasts using SSM/I and TRMM rain rate algorithm(s). *Meteor. Atmos. Phys.,* **74,** 37-49.

Velden, C. S., and W. L. Smith, 1983: Monitoring tropical cyclone evolution with NOAA satellite microwave observation. *J. Climate Appl. Meteor*., **22,** 714-724.

_____, C. M. Hayden, S. J. Nieman, W. P. Menzel, and S. T. Wanzong, 1997: Upper-tropospheric winds derived from geostationary satellite water vapor observations. *Bull. Amer. Meteor. Soc.,* **78,** 173-195.

Weng, F., and N. C. Grody, 2000: Retrieval of ice cloud parameters using a microwave imaging radiometer, *J. Atmos. Sci.*, 57, 1069-1081.

_____, R. R. Ferraro, and N. C. Grody, 2000: Effects of AMSU cross-scan asymmetry of brightness temperatures on retrieval of atmospheric and surface parameters. *Microwave Radiometry & Remote Sensing of the Earth's Surface and Atmosphere,* Ed. P. Pampaloni and S. Paloscia, VSP, Netherlands, 255-262.

Willoughby, H. E., F. D. Marks, and R. J. Feinberg, 1984: Stationary and moving convective bands in hurricanes. *J. Atmos. Sci.*, **41**, 3189-3211.

Zhang, D.-L. and R. A. Anthes, 1982: A high-resolution model of the planetary boundary layer – sensitivity tests and comparisons with SESAME-79 data. *J. Appl. Meteor.,* **21,** 1594-1609.

Zhao, L., F. Weng, and R.R. Ferraro, 2001: A physically-based algorithm to derive cloud and precipitation parameters using AMSU measurements, *Preprints, 11th Satellite Conf. on Meteor. & Oceano.,* Amer. Meteor. Soc., Madison, in press.

Zhu, Tong, D.-L. Zhang, and F. Weng, 2002a: Impact of the advanced microwave sounding unit measurements on hurricane prediction. *Mon. Wea. Rev.* **130**, 2416-2432.

_____, _____, _____, 2003: Numerical Simulation of Hurricane Bonnie (1998). Part I: Eyewall Evolution and Intensity Changes. *Mon. Wea. Rev.* In press.

Zou, X., and Q. Xiao, 2000: Studies on the initialization and simulation of a mature hurricane using a variational bogus data assimilation scheme. *J. Atmos. Sci.,* **57,** 836-860.

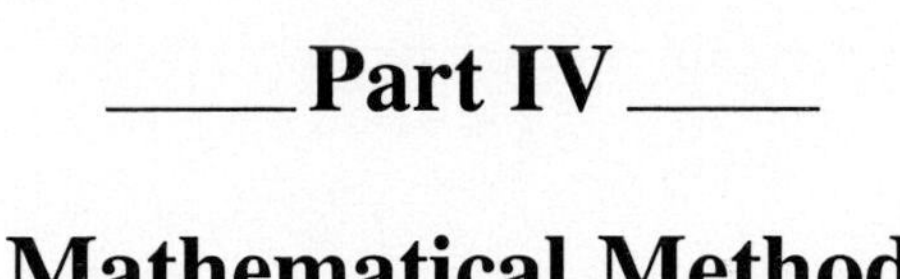

Part IV

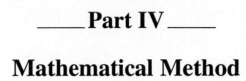

Mathematical Method

THE *RG*-FACTORIZATION IN BLOCK-STRUCTURED MARKOV RENEWAL PROCESSES*

QUAN-LIN LI[1,2] AND YIQIANG Q. ZHAO[1]
[1] *School of Mathematics and Statistics, Carleton University*
Ottawa, Ontario, Canada K1S 5B6
[2] *National Laboratory of Pattern Recognition, Institute of Automation*
Chinese Academy of Sciences, Beijing 100080, P.R. China
E-mail: {qli, zhao}@math.carleton.ca

(Manuscript received 11 November 2002)

In this paper, the censoring technique is used to deal with block-structured Markov renewal processes. Two probabilistic measures, the R- and G-measures, are defined and a censoring invariant property for the R- and G-measures is obtained. The RG-factorization for the transition probability matrix is derived based on the Wiener-Hopf type equations constructed in this paper. For a Markov renewal process of $GI/G/1$ type, the transition probability matrix consists of two sequences of matrices, one is referred to as the boundary sequence and the other the repeating sequence. The RG-facterization of the double transformations for the repeating matrix sequence and four matrix inequalities of the double transformations for the boundary matrix sequence are given.

1. Introduction

We consider a Markov renewal process $\{(X_n, T_n), n \geq 0\}$ on the state space $\Omega \times [0, +\infty)$ with $\Omega = \{(k, j) : k \geq 0, 1 \leq j \leq m_k\}$, where X_n is the state of the process at the nth renewal epoch and T_n is the total renewal time up to the nth renewal, or $T_n = \sum_{i=0}^{n} \tau_i$ with $\tau_0 = 0$ and τ_n being the inter-renewal interval time between the $(n-1)$st and the nth renewal epochs for $n \geq 1$. In this way, the transition probability matrix of the Markov renewal process $\{(X_n, T_n), n \geq 0\}$ is given by

*We present this paper on the factorization of Wiener-Hopf type equations in memory of Professor Jijia Zhang. Over 20 years ago, one of the authors (Y.Q. Zhao) of this paper was studying for his B.S. degree at the Nanjing Institute of Meteorology. A course on partial differential equations (PDEs) was instructed by an excellent teacher from Nanjing University, which attracted a number of teachers and/or researchers to attend his lectures. Professor Jijia Zhang was one of those, who regularly came to the lectures and sat besides this author. During conversations between them following the introduction of Wiener-Hopf type equations to the class, Professor Jijia Zhang pointed out various areas of application in which the Wiener-Hopf type equations could produce significant impact. His comments have influenced this author's research interests and motivated the research in this paper.

$$P(x) = \begin{pmatrix} P_{0,0}(x) & P_{0,1}(x) & P_{0,2}(x) & \cdots \\ P_{1,0}(x) & P_{1,1}(x) & P_{1,2}(x) & \cdots \\ P_{2,0}(x) & P_{2,1}(x) & P_{2,2}(x) & \cdots \\ \vdots & \vdots & \vdots & \ddots \end{pmatrix}, \tag{1}$$

where $P_{i,j}(x)$ is a matrix of size $m_i \times m_j$ whose (r, r')th entry is

$$(P_{ij}(x))_{r,r'} = P\left\{X_{n+1} = (j, r'), T_{n+1} \leq x + T_n \mid X_n = (i, r), T_n\right\}. \tag{2}$$

Because of the block-partitioned structure for $P(x)$, the Markov renewal process $\{(X_n, T_n), n \geq 0\}$ is referred to as a block-structured Markov renewal process. Specifically, $P(+\infty)$, defined as $\lim_{x\to\infty} P(x)$ entry-wise, is referred to as the embedded Markov chain of the Markov renewal process. Throughout this paper, we assume that the Markov renewal process $P(x)$ is irreducible and $P(x)e \lneqq e$ for all $x \geq 0$, where e is a column vector of ones. The state space Ω is partitioned as $\Omega = \bigcup_{i=0}^{\infty} L_i$ with $L_i = \{(i,1), (i,2), \cdots, (i, m_i)\}$. For state (i,k), i is called the level variable and k the phase variable. We also write $L_{\leq k} = \bigcup_{i=0}^{k} L_i$ for the set of all the states in the levels up to k, and $L_{\geq k}$ for the complement of $L_{\leq(k-1)}$.

Many application problems can be naturally modelled as a block-structured Markov renewal process and often it reveals a special structure such as a repeating property, which is referred to as a Markov renewal process of $GI/G/1$ type:

$$P(x) = \begin{pmatrix} D_0(x) & D_1(x) & D_2(x) & D_3(x) & \cdots \\ D_{-1}(x) & A_0(x) & A_1(x) & A_2(x) & \cdots \\ D_{-2}(x) & A_{-1}(x) & A_0(x) & A_1(x) & \cdots \\ D_{-3}(x) & A_{-2}(x) & A_{-1}(x) & A_0(x) & \cdots \\ \vdots & \vdots & \vdots & \vdots & \ddots \end{pmatrix}, \tag{3}$$

where the sizes of the matrices $D_0(x)$, $D_i(x)$, $D_{-i}(x)$ for $i \geq 1$ and $A_j(x)$ for $-\infty < j < \infty$ are $m_0 \times m_0$, $m_0 \times m$, $m \times m_0$ and $m \times m$, respectively.

Motivations of studying a Markov renewal process of $GI/G/1$ type are obvious due to references provided below.

The literature on Markov renewal processes is extensive. References, which are closely related to the study in this paper, include Pyke (1961a, 1961b), Pyke and Schaufele (1964, 1966), Teugels (1968), Hunter (1969) and Çinlar (1968, 1974). For a comprehensive discussion on Markov renewal processes, readers may refer to a survey article by Çinlar (1969). The study of block-structured Markov renewal processes provides a useful modelling tool. Readers may refer to Neuts (1989) for a study of Markov renewal processes of $M/G/1$ type. Other references on block-structured Markov renewal processes include Neuts (1986, 1995), Sengupta (1989), Asmussen and Ramaswami (1990), Ramaswami (1990a, 1990b), Zhao, Li and Alfa (1999) and Hsu, Yuan and Li (2000) among others.

This paper applies the censoring technique to block-structured Markov renewal processes. The same technique has been successfully applied to block-structured Markov chains. Examples include Grassmann and Heyman (1990), Latouche and Ramaswami (1999), Zhao (2000), Zhao, Li and Braun (2001) and Li and Zhao (2002a, 2002b). The main results of this paper are threefold. 1) For a Markov renewal process, a censored process is defined and proved to be again a Markov renewal process. The transition probability matrix of the censored Markov renewal process is expressed. 2) For an irreducible block-structured Markov renewal process, a censoring invariant property for the R- and G-measures is obtained and the RG-factorization for the transition probability matrix is derived based on the Wiener-Hopf type equations constructed in this paper. 3) For a Markov renewal process of $GI/G/1$ type, the RG-facterization of the double transformations for the repeating matrix sequence and four matrix inequalities of the double transformations for the boundary matrix sequence are given. Practical applications, for example the busy period and first passage time distributions, will be supplied in a separate work.

The remainder of the paper is organized as follows. In Section 2, we apply the censoring technique to deal with block-structured Markov renewal processes. We provide a censoring invariant property for the R- and G-measures, and then give the RG-factorization for the transition probability matrix. In Section 3, we discuss Markov renewal processes of $GI/G/1$ type. We derive the RG-facterization of the double transformations for the repeating matrix sequence and four matrix inequalities of the double transformations for the boundary matrix sequence

2. The Censored Markov Renewal Processes

In this section, the censoring technique is applied to deal with irreducible block-structured Markov renewal processes. Based on the censored process, conditions on classification of the states are provided. Two probabilistic measures, the R- and G-measures, are defined. The RG-factorization for the transition probability matrix is derived.

We first define a censored process for a Markov renewal process whose transition probability matrix consists of scalar entries. We then treat a block-structured Markov renewal process as a special case.

Definition 1 *Suppose that* $\{(X_n, T_n), n \geq 0\}$ *is an irreducible Markov renewal process on the state space* $\Omega \times [0, +\infty)$, *where* $X_n \in \Omega = \{0, 1, 2, \cdots\}$ *and* $T_n \in [0, +\infty)$. *Let* E *be a non-empty subset of* Ω. *If the successive visits of* X_n *to the subset* E *take place at the* n_k*th step of state transition, then the inter-visit time* τ_k^E *between the* $(k-1)$*st and the* k*th visits to* E *is given by* $\tau_k^E = \tau_{n_{k-1}+1} + \tau_{n_{k-1}+2} + \cdots + \tau_{n_k}$ *for* $k \geq 1$. *Let* $T_k^E = \sum_{i=1}^{k} \tau_i^E$. *Then the sequence* $\left\{\left(X_{n_k}, T_k^E\right), k \geq 1\right\}$ *is called the censored process with censoring set* E.

Throughout this paper we denote by $(B)_{r,r'}$ the (r, r')th entry of the matrix B and by $B * C(x)$ (or $B(x) * C(x)$) the convolution of two matrix functions $B(x)$ and $C(x)$, i.e., $B * C(x) = \int_0^x B(x-u)\, dC(u)$. We then recursively define $B^{n*}(x) = B * B^{(n-1)*}(x)$ for $n \geq 1$ with $B^{0*}(x) = I$ where I is the identity matrix.

Let $E^c = \Omega - E$. According to the subsets E and E^c, the transition probability matrix $P(x)$ is partitioned as

$$P(x) = \begin{array}{c} \\ E \\ E^c \end{array}\!\!\begin{array}{c} \begin{array}{cc} E & E^c \end{array} \\ \left(\begin{array}{cc} T(x) & U(x) \\ V(x) & Q(x) \end{array}\right) \end{array}. \tag{4}$$

Lemma 1 *If $P(x)$ is irreducible, then each element of $\widehat{Q}(x) = \sum_{n=0}^{\infty} Q^{n*}(x)$ is finite for $x \geq 0$.*

Proof If $P(x)$ is irreducible, then $P(+\infty)$ is irreducible, since $0 \leq P(x) \leq P(+\infty)$. It is obvious that $Q(+\infty)$ is strictly substochastic (substochastic but not stochastic) due to $V(+\infty) \gneq 0$. Hence

$$\widehat{Q}(+\infty) = \sum_{n=0}^{\infty} Q^n(+\infty) = [I - Q(+\infty)]^{-1} < +\infty,$$

where $[I - Q(+\infty)]^{-1}$ is the minimal nonnegative inverse of $I - Q(+\infty)$. Since each element of $\widehat{Q}(+\infty)$ is finite and $0 \leq \widehat{Q}(x) \leq \widehat{Q}(+\infty)$ for $x > 0$, each element of $\widehat{Q}(x)$ is finite. ∎

The matrix $\widehat{Q}(x)$ is referred to as the fundamental matrix of $Q(x)$. In the following, we show that the censored process $\{(X_{n_k}, T_k^E), k \geq 1\}$ is again a Markov renewal process

Theorem 1 *The censored process $\{(X_k^E, T_k^E), k \geq 1\}$ is a Markov renewal process whose transition probability matrix is given by*

$$P^E(x) = T(x) + U * \widehat{Q} * V(x). \tag{5}$$

Proof It follows from the strong Markov property that the censored process $\{(X_k^E, T_k^E), k \geq 1\}$ is a two-dimensional Markov process. To show that it is a Markov renewal process, we need to show that the T_{n+1}^E is independent of X_0^E, X_1^E, $\cdots$, X_{n-1}^E, T_0^E, T_1^E, $\cdots$, T_n^E, given the state of X_n^E. This is clear from the fact that $\{(X_n, T_n), n \geq 0\}$ is a Markov renewal process. The (i,j)th entry of the transition probability matrix of the Markov renewal process $\{(X_k^E, T_k^E), k \geq 1\}$ is

$$\begin{aligned}
\left(P^E(x)\right)_{i,j} &= P\left\{X_{n+1}^E = j, T_{n+1}^E \leq x + T_n^E \mid X_n^E = i, T_n^E\right\} \\
&= P\left\{X_1^E = j, T_1^E \leq x \mid X_0^E = i, T_0^E = 0\right\}.
\end{aligned}$$

To explicitly express $P^E(x)$ in terms of the original transition probability matrix, we consider the following two possible cases:

Case I $n_1 = 1$. In this case, $i, j \in E$, $X_1^E = X_1$, $T_1^E = \tau_1$ and

$$P\left\{X_1^E=j,T_1^E\leq x\mid X_0^E=i,T_0^E=0\right\}=\left(T\left(x\right)\right)_{i,j}. \tag{6}$$

Case II $n_1=k$ for $k\geq 2$. In this case, $i,\ j\in E$, $X_1^E=X_k$, $T_1^E=\sum_{l=1}^{k}\tau_l$ and

$$P\{X_1^E=j,T_1^E\leq x\mid X_0^E=i,T_0^E=0\}=P\{X_k=j,\ X_j\notin E \text{ for } j=1,2, \cdots,k-1,\sum_{l=1}^{k}\tau_l\leq x\mid X_0=i,\tau_0=0\}=\left(U*Q^{(k-2)*}*V\left(x\right)\right)_{i,j}. \tag{7}$$

It follows from (6) and (7) that

$$\begin{aligned}P\left\{X_1^E=j,T_1^E\leq x\mid X_0^E=i,T_0^E=0\right\}&=\left(T\left(x\right)\right)_{i,j}+\sum_{k=2}^{\infty}\left(U*Q^{(k-2)*}*V\left(x\right)\right)_{i,j}\\&=\left(T\left(x\right)\right)_{i,j}+\left(U*\widehat{Q}*V\left(x\right)\right)_{i,j}.\end{aligned}$$

This completes the proof. ∎

Define the double transformation of n_1 and x for the censored Markov renewal process as

$$\widetilde{PE}^*\left(z,s\right)=\left(\widetilde{PE}^*_{i,j}\left(z,s\right)\right)_{i,j\in E},$$

where

$$\widetilde{PE}^*_{i,j}\left(z,s\right)=\sum_{n_1=1}^{\infty}z^{n_1}\int_0^{+\infty}e^{-sx}dP\left\{X_{n_1}=j,T_{n_1}\leq x\mid X_0=i,T_0=0\right\}.$$

The single transformations $\widetilde{T}\left(s\right)$, $\widetilde{U}\left(s\right)$, $\widetilde{V}\left(s\right)$ and $\widetilde{Q}\left(s\right)$ are defined conventionally, for example, $\widetilde{T}\left(s\right)=\int_0^{+\infty}e^{-sx}dT\left(x\right)$.

The following corollary provides a useful result for studying the two-dimensional random vector $\left(X_1^E,T_1^E\right)$, the proof of which is obvious from (6) and (7). $\left(X_1^E,T_1^E\right)$ is important for the study of the Markov renewal process $\left\{\left(X_n,T_n\right),n\geq 1\right\}$. It is worthwhile to notice that an important example is analyzed in Section 2.4 of Neuts (1989).

Corollary 1 *$\widetilde{PE}^*\left(z,s\right)=z\widetilde{T}\left(s\right)+z^2\widetilde{U}\left(s\right)\left[I-z\widetilde{Q}\left(s\right)\right]^{-1}\widetilde{V}\left(s\right)$, where $\left[I-z\widetilde{Q}\left(s\right)\right]^{-1}=\sum_{n=0}^{\infty}z^n\left[\widetilde{Q}\left(s\right)\right]^n$.*

For convenience, we write $P^{[\leq n]}\left(x\right)$ for the censored transition probability matrix $P^E\left(x\right)$ if the censored set $E=L_{\leq n}$, in particular, $P^{[0]}\left(x\right)=P^{[\leq 0]}\left(x\right)$. Based on Definition 1, we have the following two useful properties.

Property 1 For $E_1 \subset E_2$, $P^{E_1}(x) = \left(P^{E_2}\right)^{E_1}(x)$.

Property 2 $P(x)$ is irreducible if and only if $P^E(x)$ is irreducible for all the subsets E of Ω.

Now, we consider the classification of states for an irreducible Markov renewal process. Çinlar (1969) shows that $P(x)$ is recurrent or transient if and only if $P(+\infty)$ is recurrent or transient. Therefore, Lemma 1 in Zhao, Li and Braun (2001) illustrates the following useful relations.

Proposition 1 *1) $P(x)$ is recurrent if and only if $P^E(x)$ is recurrent for some subset $E \subset \Omega$.*

2) $P(x)$ is transient if and only if $P^E(x)$ is transient for some subset $E \subset \Omega$.

The following proposition provides a sufficient condition under which a Markov renewal process $P(x)$ is positive recurrent. The proof is easy according to the fact that $\sum_{j=0}^{\infty} \int_0^{+\infty} x dP_{i,j}(x) e$ is the mean total sojourn time of the Markov renewal process $P(x)$ in state i. It is worthwhile to note that $P(x)$ may not be positive recurrent when $P(+\infty)$ is positive recurrent.

Proposition 2 *The Markov renewal process $P(x)$ is positive recurrent if*

1) $P(+\infty)$ is positive recurrent and

2) $\sum_{j=0}^{\infty} \int_0^{+\infty} x dP_{i,j}(x) e$ is finite for all $i \geq 0$.

Remark 1 *Remark b in Section 3.2 of Neuts (1989) (pp. 140) illustrates that condition 2) in Proposition 2 is stronger than necessary. For example, for a Markov renewal process of $M/G/1$ type, the sufficient condition only requires that $\sum_{j=0}^{\infty} \int_0^{+\infty} x dP_{i,j}(x) e$ for $i = 0$ and $\int_0^{+\infty} x dP_{1,0}(x)$ are finite.*

Based on the above discussion, a probabilistic interpretation for each component in the expression (5) for $P^E(x)$ is available and useful. For the Markov renewal process $P(x)$, let $T_{E^c,E^c}(i,j)$ be total renewal time until the process visits state $j \in E^c$ for the last time before entering E, given that the process starts in state $i \in E^c$. Formally, assume that at the kth transition the process visits state $j \in E^c$ for the last time before entering E, given that the process starts in state $i \in E^c$. Then $T_{E^c,E^c}(i,j) = \sum_{l=1}^{k} \tau_l$. Similarly, let $T_{E,E^c}(i,j)$ be the total renewal time until the process visits state $j \in E^c$ before returning to E, given that the process starts in state $i \in E$, $T_{E^c,E}(i,j)$ the total renewal time until the process enters E and upon entering E the first state visited is $j \in E$, given that the process started at state $i \in E^c$, and $T_{E,E}(i,j)$ the total renewal time until the process enters E and upon returning to E the first state visited is $j \in E$, given that the process started at state $i \in E$.

1. $\left(\widehat{Q}(x)\right)_{i,j}$ is the expected number of visits to state $j \in E^c$ before entering E and $T_{E^c,E^c}(i,j) \le x$, given that the process starts in state $i \in E^c$.

2. $\left(U * \widehat{Q}(x)\right)_{i,j}$ is the expected number of visits to state $j \in E^c$ before returning to E and $T_{E,E^c}(i,j) \le x$, given that the process starts in state $i \in E$.

3. $\left(\widehat{Q} * V(x)\right)_{i,j}$ is the probability that the process enters E and upon entering E the first state visited is $j \in E$ and $T_{E^c,E}(i,j) \le x$, given that the process starts in state $i \in E^c$.

4. $\left(U * \widehat{Q} * V(x)\right)_{i,j}$ is the probability that upon returning to E the first state visited is $j \in E$ and $T_{E,E}(i,j) \le x$, given that the process starts in state $i \in E$.

We now return to the study of a block-structured Markov renewal process. We first define the R- and G-measures for the Markov renewal process $P(x)$. The R- and G-measures will play an important role for studying block-structured Markov renewal processes.

For $0 \le i < j$, $R_{i,j}(k,x)$ is an $m_i \times m_j$ matrix whose (r,r')th entry $(R_{i,j}(x))_{r,r'}$ is the probability that starting in state (i,r) at time 0, the Markov renewal process makes its kth transition in the renewal time interval $[0,x]$ and such a transition is a visit into state (j,r') without visiting any states in $L_{\le(j-1)}$ during intermediate steps; or

$$\begin{aligned}(R_{i,j}(k,x))_{r,r'} =& P\{X_k = (j,r'), X_l \notin L_{\le(j-1)} \text{ for } l = 1,2,\cdots,k-1,\\ & T_k \le x \mid X_0 = (i,r)\}.\end{aligned} \tag{8}$$

Write $R_{i,j}(x) = \sum_{k=1}^{\infty} R_{i,j}(k,x)$. Then the (r,r')th entry of $R_{i,j}(x)$ is the expected number of visits to state (j,r') made in the renewal time interval $[0,x]$ without visiting any states in $L_{\le(j-1)}$ during intermediate steps, given that the process starts in state (i,r) at time 0.

For $0 \le j < i$, $G_{i,j}(k,x)$ is an $m_i \times m_j$ matrix whose (r,r')th entry $(G_{i,j}(x))_{r,r'}$ is the probability that starting in state (i,r) at time 0, the Markov renewal process makes its kth transition in the renewal time interval $[0,x]$ and such a transition is a visit into state (j,r') without visiting any states in $L_{\le(i-1)}$ during intermediate steps; or

$$\begin{aligned}(G_{i,j}(k,x))_{r,r'} =& P\{X_k = (j,r'), X_l \notin L_{\le(i-1)} \text{ for } l = 1,2,\cdots,k-1,\\ & T_k \le x \mid X_0 = (i,r)\}.\end{aligned} \tag{9}$$

Write $G_{i,j}(x) = \sum_{k=1}^{\infty} G_{i,j}(k,x)$. Then the (r,r')th entry of $G_{i,j}(x)$ is the probability that starting in state (i,r) at time 0, the Markov renewal process makes its first visit into $L_{\leq(i-1)}$ in the renewal time interval $[0, x]$ and upon entering $L_{\leq(i-1)}$ it visits state (j, r').

The two matrix sequences $\{R_{i,j}(x)\}$ and $\{G_{i,j}(x)\}$ are called the R- and G-measures of the Markov renewal process $P(x)$, respectively.

The R- and G-measures will be the basic components, in terms of which the analysis of a Markov renewal process will be carried out. One of the reasons that the R- and G-measures are so important in the analysis of a Markov renewal process is that the R- and G-measures can be nicely expressed in terms of the given information, transition probabilities. Therefore, in principle, they can be numerically computed. To see this, partition the transition probability matrix $P(x)$ according to the three subsets $L_{\leq(n-1)}$, L_n and $L_{\geq(n+1)}$ as

$$P(x) = \begin{pmatrix} T(x) & U_0(x) & U_1(x) \\ V_0(x) & T_0(x) & U_2(x) \\ V_1(x) & V_2(x) & Q(x) \end{pmatrix}. \tag{10}$$

Let

$$Q(x) = \begin{pmatrix} T_0(x) & U_2(x) \\ V_2(x) & Q(x) \end{pmatrix} \text{ and } Q^{n*}(x) = \begin{pmatrix} D_{11}(n,x) & D_{12}(n,x) \\ D_{21}(n,x) & D_{22}(n,x) \end{pmatrix}, \quad n \geq 0. \tag{11}$$

Partition $\widehat{Q}(x) = \sum_{n=0}^{\infty} Q^{n*}(x)$ accordingly as

$$\widehat{Q}(x) = \begin{pmatrix} H_{11}(x) & H_{12}(x) \\ H_{21}(x) & H_{22}(x) \end{pmatrix}. \tag{12}$$

We write

$$R_{<n}(x) = \left(R_{0,n}^T(x), R_{1,n}^T(x), R_{2,n}^T(x), \cdots, R_{n-1,n}^T(x)\right)^T, \tag{13}$$

where the superscript T stands for the transpose of a matrix, and

$$G_{<n}(x) = \left(G_{n,0}(x), G_{n,1}(x), G_{n,2}(x), \cdots, G_{n,n-1}(x)\right). \tag{14}$$

For convenience, for a matrix $B = (B_0, B_1, B_2, \cdots)$ or $B = (B_0^T, B_1^T, B_2^T, \cdots)^T$, let $(B)^{\langle i \rangle}$ denote the ith block-entry B_i of the matrix B and $(B)_{r,r'}^{\langle i \rangle}$ the (r, r')th entry in the ith block-entry of B.

Lemma 2 *For $x > 0$ and $n \geq 1$,*

$$R_{<n}(x) = U_0 * H_{11}(x) + U_1 * H_{21}(x) \tag{15}$$

and

$$G_{<n}(x) = H_{11} * V_0(x) + H_{12} * V_1(x). \tag{16}$$

Proof We only prove (15) while (16) can be proved similarly.
For $0 \le i \le n-1$, we consider the following two possible cases for $(R_{i,n}(k,x))_{r,r'}$:
Case I $k = 1$. In this case,

$$\begin{aligned}(R_{i,n}(k,x))_{r,r'} &= P\left\{X_1 = (n,r'), T_1 \le x \mid X_0 = (i,r), T_0 = 0\right\} \\ &= (U_0(x))_{r,r'}^{\langle i \rangle}. \end{aligned} \tag{17}$$

Case II $k \ge 2$. In this case,

$$\begin{aligned}(R_{i,n}(k,x))_{r,r'} =& P\{X_k = (n,r'), X_l \notin L_{\le (j-1)} \text{ for } l = 1,2,\cdots,k-1, \\ & T_k \le x \mid X_0 = (i,r), T_0 = 0\} \\ =& (U_0 * D_{11}(k-1,x) + U_1 * D_{21}(k-1,x))_{r,r'}^{\langle i \rangle}. \end{aligned} \tag{18}$$

Noting that $D_{11}(0,x) = I$ and $D_{21}(0,x) = 0$, it follows from (17) and (18) that

$$\begin{aligned}(R_{i,n}(x))_{r,r'} &= \sum_{k=1}^{\infty} (R_{i,n}(k,x))_{r,r'} \\ &= (U_0(x))_{r,r'}^{\langle i \rangle} + \sum_{k=2}^{\infty} (U_0 * D_{11}(k-1,x) + U_1 * D_{21}(k-1,x))_{r,r'}^{\langle i \rangle} \\ &= \left(\sum_{k=0}^{\infty} (U_0 * D_{11}(k,x) + U_1 * D_{21}(k,x))\right)_{r,r'}^{\langle i \rangle} \\ &= (U_0 * H_{11}(x) + U_1 * H_{21}(x))_{r,r'}^{\langle i \rangle}. \end{aligned}$$

This completes the proof. ∎

For the Markov renewal process $P(x)$, let $Q_n(x)$ be the southeast corner of $P(x)$ from level n, i.e., $Q_n(x) = (P_{i,j}(x))_{i,j \ge n}$. Let $\widehat{Q}_n(x) = \sum_{k=0}^{\infty} Q_n^{k*}(x)$, and $\widehat{Q}_n^{(k,\cdot)}(x)$ and $\widehat{Q}_n^{(\cdot,l)}(x)$ the kth block-row and the lth block-column of $\widehat{Q}_n(x)$, respectively. The following corollary easily follows from Lemma 2.

Corollary 2 *For* $0 \le i < j$,

$$R_{i,j}(x) = (P_{i,j}(x), P_{i,j+1}(x), P_{i,j+2}(x), \cdots) * \widehat{Q}_j^{(\cdot,1)}(x) \tag{19}$$

and for $0 \le j < i$,

$$G_{i,j}(x) = \widehat{Q}_i^{(1,\cdot)}(x) * \left(P_{i,j}(x)^T, P_{i+1,j}(x)^T, P_{i+2,j}(x)^T, \cdots\right)^T. \tag{20}$$

From either Lemma 2 or Corollary 2 it is clear that the determination of the R- and G-measures relies on the entries of the fundamental matrix $\widehat{Q}_k(x)$. Lemma 3 provides a formula for expressing the transformation of the fundamental matrix. The result in this lemma can be viewed as a generalized version of the expression, Lemma 3 in Li and Zhao (2002a), for the minimal nonnegative inverse of a block-structured matrix of infinite size and the proof is similar. This lemma often makes the analysis and computation possible or easier.

It follows from (11) and (12), respectively, that

$$\widetilde{Q}(s) = \begin{pmatrix} \widetilde{T}_0(s) & \widetilde{U}_2(s) \\ \widetilde{V}_2(s) & \widetilde{Q}(s) \end{pmatrix} \quad \text{and} \quad \widetilde{\widehat{Q}}(s) = \begin{pmatrix} \widetilde{H}_{11}(s) & \widetilde{H}_{12}(s) \\ \widetilde{H}_{21}(s) & \widetilde{H}_{22}(s) \end{pmatrix}.$$

Lemma 3 *For $Re(s) \geq 0$,*

$$\widetilde{H}_{11}(s) = \left[I - \widetilde{T}_0(s) - \widetilde{U}_2(s)\widetilde{\widehat{Q}}(s)\widetilde{V}_2(s)\right]^{-1},$$

$$\widetilde{H}_{12}(s) = \left[I - \widetilde{T}_0(s) - \widetilde{U}_2(s)\widetilde{\widehat{Q}}(s)\widetilde{V}_2(s)\right]^{-1}\widetilde{U}_2(s)\widetilde{\widehat{Q}}(s),$$

$$\widetilde{H}_{21}(s) = \widetilde{\widehat{Q}}(s)\widetilde{V}_2(s)\left[I - \widetilde{T}_0(s) - \widetilde{U}_2(s)\widetilde{\widehat{Q}}(s)\widetilde{V}_2(s)\right]^{-1},$$

$$\widetilde{H}_{22}(s) = \widetilde{\widehat{Q}}(s) + \widetilde{\widehat{Q}}(s)\widetilde{V}_2(s)\left[I - \widetilde{T}_0(s) - \widetilde{U}_2(s)\widetilde{\widehat{Q}}(s)\widetilde{V}_2(s)\right]^{-1}\widetilde{U}_2(s)\widetilde{\widehat{Q}}(s).$$

Symmetrically,

$$\widetilde{H}_{11}(s) = \widetilde{\widehat{T}}_0(s) + \widetilde{\widehat{T}}_0(s)\widetilde{U}_2(s)\left[I - \widetilde{Q}(s) - \widetilde{V}_2(s)\widetilde{\widehat{T}}_0(s)\widetilde{U}_2(s)\right]^{-1}\widetilde{V}_2(s)\widetilde{\widehat{T}}_0(s),$$

$$\widetilde{H}_{12}(s) = \widetilde{\widehat{T}}_0(s)\widetilde{U}_2(s)\left[I - \widetilde{Q}(s) - \widetilde{V}_2(s)\widetilde{\widehat{T}}_0(s)\widetilde{U}_2(s)\right]^{-1},$$

$$\widetilde{H}_{21}(s) = \left[I - \widetilde{Q}(s) - \widetilde{V}_2(s)\widetilde{\widehat{T}}_0(s)\widetilde{U}_2(s)\right]^{-1}\widetilde{V}_2(s)\widetilde{\widehat{T}}_0(s),$$

$$\widetilde{H}_{22}(s) = I - \widetilde{Q}(s) - \widetilde{V}_2(s)\widetilde{\widehat{T}}_0(s)\widetilde{U}_2(s).$$

Theorem 2 *For $x \geq 0$ and $n \geq 1$,*

$$R_{<n}(x) = \left[U_0(x) + U_1 * \widehat{Q} * V_2(x)\right] * \sum_{l=0}^{\infty}\sum_{k=0}^{l}\binom{l}{k} T_0^{k*}(x) * \left[U_2 * \widehat{Q} * V_2(x)\right]^{(l-k)*}$$

and

$$G_{<n}(x) = \sum_{l=0}^{\infty}\sum_{k=0}^{l}\binom{l}{k} T_0^{k*}(x) * \left[U_2 * \widehat{Q} * V_2(x)\right]^{(l-k)*} * \left[V_0(x) + U_2 * \widehat{Q} * V_1(x)\right].$$

Proof It follows from Lemmas 2 and 3 that

$$\widetilde{R}_{<n}(s)=\left[\widetilde{U}_0(s)+\widetilde{U}_1(s)\widetilde{\widehat{Q}}(s)\widetilde{V}_2(s)\right]\left[I-\widetilde{T}_0(s)-\widetilde{U}_2(s)\widetilde{\widehat{Q}}(s)\widetilde{V}_2(s)\right]^{-1}$$

and

$$\widetilde{Q}_{<n}(s)=\left[I-\widetilde{T}_0(s)-\widetilde{U}_2(s)\widetilde{\widehat{Q}}(s)\widetilde{V}_2(s)\right]^{-1}\left[\widetilde{V}_0(s)+\widetilde{U}_2(s)\widetilde{\widehat{Q}}(s)\widetilde{V}_1(s)\right].$$

The inverse transform for the above two equations immediately leads to the desired result. ■

The following theorem provides a censoring invariant property for the R- and G-measures of Markov renewal processes. Though this property would be intuitively clear based on the sample path structure of the censored Markov renewal process, a formal proof is also provided. We denote by $R_{i,j}^{[\leq n]}(x)$ and $G_{i,j}^{[\leq n]}(x)$ the R- and G-measures of the censored Markov renewal process $P^{[\leq n]}(x)$.

Theorem 3 *1) For* $0\leq i<j\leq n$, $R_{i,j}^{[\leq n]}(x)=R_{i,j}(x)$.

2) For $0\leq j<i\leq n$, $G_{i,j}^{[\leq n]}(x)=G_{i,j}(x)$.

Proof We only prove 1) while 2) can be proved similarly.

First, we assume that $n=j$ and $P(x)$ is partitioned according to the three subsets $L_{<n}$, L_n and $L_{>n}$ as in (10). It follows from Theorem 1 that

$$\begin{aligned}P^{[\leq n]}(x)&=\begin{pmatrix}T(x)&U_0(x)\\V_0(x)&T_0(x)\end{pmatrix}+\begin{pmatrix}U_1(x)\\U_2(x)\end{pmatrix}*\widehat{Q}(x)*(V_1(x),V_2(x))\\&=\begin{pmatrix}T(x)+U_1*\widehat{Q}*V_1(x)&U_0(x)+U_1*\widehat{Q}*V_2(x)\\V_0(x)+U_2*\widehat{Q}*V_1(x)&T_0(x)+U_2*\widehat{Q}*V_2(x)\end{pmatrix}.\end{aligned}\tag{21}$$

Hence, simple calculations lead to

$$\begin{aligned}R_{<n}^{[\leq n]}(x)&=\left[U_0(x)+U_1*\widehat{Q}*V_2(x)\right]*\sum_{l=0}^{\infty}\left[T_0(x)+U_2*\widehat{Q}*V_2(x)\right]^{l*}\\&=\left[U_0(x)+U_1*\widehat{Q}*V_2(x)\right]*\sum_{l=0}^{\infty}\sum_{k=0}^{l}\binom{l}{k}T_0^{k*}(x)*\left[U_2*\widehat{Q}*V_2(x)\right]^{(l-k)*}.\end{aligned}\tag{22}$$

Therefore, $R_{<n}^{[\leq n]}(x)=R_{<n}(x)$ according to Theorem 2.

If $n>j$, we first censor the matrix $P(x)$ in the set $L_{\leq j}$, $R_{i,j}^{[\leq j]}(x)=R_{i,j}(x)$ based on the fact just proved. Next, we censor the matrix $P(x)$ in the set $L_{\leq n}$. Since according to Property 1 the censored matrix $P^{[\leq j]}$ can be obtained by the censored matrix $P^{[\leq n]}$, $R_{i,j}^{[\leq n]}(x) = R_{i,j}^{[\leq j]}(x)$ based on the fact just proved. Hence, $R_{i,j}^{[\leq n]}(x) = R_{i,j}(x)$ for $j < n$. This completes the proof. ■

Let

$$P^{[\leq n]}(x) = \begin{pmatrix} \phi_{0,0}^{(n)}(x) & \phi_{0,1}^{(n)}(x) & \cdots & \phi_{0,n}^{(n)}(x) \\ \phi_{1,0}^{(n)}(x) & \phi_{1,1}^{(n)}(x) & \cdots & \phi_{1,n}^{(n)}(x) \\ \vdots & \vdots & & \vdots \\ \phi_{n,0}^{(n)}(x) & \phi_{n,1}^{(n)}(x) & \cdots & \phi_{n,n}^{(n)}(x) \end{pmatrix}, \quad n \geq 0,$$

be block-partitioned according to levels. The equations in the following lemma provide a relationship among the entries of censored Markov renewal processes, which are essentially the Wiener-Hopf type equations for the Markov renewal process. The significance of this relationship will become clear very soon.

Lemma 4 *For $n \geq 0$, $0 \leq i, j \leq n$,*

$$\phi_{i,j}^{(n)}(x) = P_{i,j}(x) + \sum_{k=n+1}^{\infty} \phi_{i,k}^{(k)}(x) * \sum_{l=0}^{\infty} \left[\phi_{k,k}^{(k)}(x)\right]^{l*} * \phi_{k,j}^{(k)}(x).$$

Proof Consider the censored matrix $P^{[\leq n]}(x)$ based on $P^{[\leq (n+1)]}(x)$. It follows from Theorem 1 that

$$P^{[\leq n]}(x) = \begin{pmatrix} \phi_{0,0}^{(n+1)}(x) & \phi_{0,1}^{(n+1)}(x) & \cdots & \phi_{0,n}^{(n+1)}(x) \\ \phi_{1,0}^{(n+1)}(x) & \phi_{1,1}^{(n+1)}(x) & \cdots & \phi_{1,n}^{(n+1)}(x) \\ \vdots & \vdots & & \vdots \\ \phi_{n,0}^{(n+1)}(x) & \phi_{n,1}^{(n+1)}(x) & \cdots & \phi_{n,n}^{(n+1)}(x) \end{pmatrix} +$$

$$\begin{pmatrix} \phi_{0,n+1}^{(n+1)}(x) \\ \phi_{1,n+1}^{(n+1)}(x) \\ \vdots \\ \phi_{n,n+1}^{(n+1)}(x) \end{pmatrix} * \sum_{l=0}^{\infty} \left[\phi_{n+1,n+1}^{(n+1)}(x)\right]^{l*} * \left(\phi_{n+1,0}^{(n+1)}(x), \phi_{n+1,1}^{(n+1)}(x), \cdots, \phi_{n+1,n}^{(n+1)}(x)\right).$$

Therefore, from repeatedly using Theorem 1 we obtain

$$
\begin{aligned}
\phi_{i,j}^{(n)}(x) =& \phi_{i,j}^{(n+1)}(x) + \phi_{i,n+1}^{(n+1)}(x) * \sum_{l=0}^{\infty} \left[\phi_{n+1,n+1}^{(n+1)}(x)\right]^{l*} * \phi_{n+1,j}^{(n+1)}(x) \\
=& \phi_{i,j}^{(n+2)}(x) + \phi_{i,n+2}^{(n+2)}(x) * \sum_{l=0}^{\infty} \left[\phi_{n+2,n+2}^{(n+2)}(x)\right]^{l*} * \phi_{n+2,j}^{(n+2)}(x) \\
& + \phi_{i,n+1}^{(n+1)}(x) * \sum_{l=0}^{\infty} \left[\phi_{n+1,n+1}^{(n+1)}(x)\right]^{l*} * \phi_{n+1,j}^{(n+1)}(x) \\
=& \cdots = P_{i,j}(x) + \sum_{k=n+1}^{\infty} \phi_{i,k}^{(k)}(x) * \sum_{l=0}^{\infty} \left[\phi_{k,k}^{(k)}(x)\right]^{l*} * \phi_{k,j}^{(k)}(x),
\end{aligned}
$$

where $P_{i,j}(x) = \phi_{i,j}^{(\infty)}(x)$. This completes the proof. ■

The following lemma provides expressions for the R- and G-measures.

Lemma 5 *1) For* $0 \le i < j$,

$$
R_{i,j}(x) = \phi_{i,j}^{(j)}(x) * \sum_{l=0}^{\infty} \left[\phi_{j,j}^{(j)}(x)\right]^{l*}.
$$

2) For $0 \le j < i$,

$$
G_{i,j}(x) = \sum_{l=0}^{\infty} \left[\phi_{i,i}^{(i)}(x)\right]^{l*} * \phi_{i,j}^{(j)}(x).
$$

Proof Applying Corollary 2 to the censored process $P^{[\le j]}(x)$ gives that

$$
R_{i,j}^{\le j}(x) = \phi_{i,j}^{(j)}(x) * \sum_{l=0}^{\infty} \left[\phi_{j,j}^{(j)}(x)\right]^{l*}, \quad 0 \le i < j,
$$

and

$$
G_{i,j}^{\le i}(x) = \sum_{l=0}^{\infty} \left[\phi_{i,i}^{(i)}(x)\right]^{l*} * \phi_{i,j}^{(j)}(x), \quad 0 \le j < i.
$$

The rest of the proof follows from the invariant property under censoring for the R- and G-measures proved in Theorem 3. ■

Let

$$
\Psi_n(x) = \phi_{n,n}^{(n)}(x), \quad n \ge 0.
$$

The following theorem provides an equivalent form, to the equations in Lemma 4, of the Wiener-Hopf equations stated in terms of the R- and G-measures.

Theorem 4 *1) For* $0 \le i < j$,

$$R_{i,j}(x) * [I - \Psi_j(x)] = P_{i,j}(x) + \sum_{k=j+1}^{\infty} R_{i,k}(x) * [I - \Psi_k(x)] * G_{k,j}(x).$$

2) For $0 \le j < i$,

$$[I - \Psi_i(x)] * G_{i,j}(x) = P_{i,j}(x) + \sum_{k=i+1}^{\infty} R_{i,k}(x) * [I - \Psi_k(x)] * G_{k,j}(x).$$

3) For $n \ge 0$,

$$\Psi_n(x) = P_{n,n}(x) + \sum_{k=n+1}^{\infty} R_{n,k}(x) * [I - \Psi_k(x)] * G_{k,n}(x).$$

Proof We only prove 1) while 2) and 3) can be proved similarly.
It follows from 1) in Lemma 5 that

$$R_{i,j}(x) * [I - \Psi_j(x)] = \phi_{i,j}^{(j)}(x).$$

Using Lemma 4 and Theorem 3 leads to

$$\begin{aligned}
\phi_{i,j}^{(j)}(x) &= P_{i,j}(x) + \sum_{k=j+1}^{\infty} \phi_{i,k}^{(k)}(x) * \sum_{l=0}^{\infty} \left[\phi_{k,k}^{(k)}(x)\right]^{l*} * \phi_{k,j}^{(k)}(x) \\
&= P_{i,j}(x) + \sum_{k=j+1}^{\infty} R_{i,k}^{[\le k]}(x) * [I - \Psi_k(x)] * G_{k,j}^{[\le k]}(x) \\
&= P_{i,j}(x) + \sum_{k=j+1}^{\infty} R_{i,k}(x) * [I - \Psi_k(x)] * G_{k,j}(x).
\end{aligned}$$

This completes the proof. ∎

Based on the Wiener-Hopf equations given in Theorem 4, the RG-factorization for the transition probability matrix , which will play an important role in studying Markov renewal processes, is obtained in the following theorem.

Theorem 5 *For the Markov renewal process* $P(x)$ *given in (1),*

$$I - P(x) = [I - R_U(x)] * [I - \Psi_D(x)] * [I - G_L(x)], \quad x \ge 0, \tag{23}$$

or

$$I - \widetilde{P}(s) = \left[I - \widetilde{R}_U(s)\right]\left[I - \widetilde{\Psi}_D(s)\right]\left[I - \widetilde{G}_L(s)\right], \quad Re(s) \ge 0,$$

where

$$R_U(x) = \begin{pmatrix} 0 & R_{0,1}(x) & R_{0,2}(x) & R_{0,3}(x) & \cdots \\ & 0 & R_{1,2}(x) & R_{1,3}(x) & \cdots \\ & & 0 & R_{2,3}(x) & \cdots \\ & & & 0 & \cdots \\ & & & & \ddots \end{pmatrix},$$

$$\Psi_D(x) = diag(\Psi_0(x), \Psi_1(x), \Psi_2(x), \Psi_3(x), \cdots)$$

and

$$G_L(x) = \begin{pmatrix} 0 & & & & \\ G_{1,0}(x) & 0 & & & \\ G_{2,0}(x) & G_{2,1}(x) & 0 & & \\ G_{3,0}(x) & G_{3,1}(x) & G_{3,2}(x) & 0 & \\ \vdots & \vdots & \vdots & \vdots & \ddots \end{pmatrix}.$$

Proof We only prove (23) for the entries in the first block-row and first block-column. The rest can be proved similarly.

The entry $(0,0)$ on the right-hand side is

$$I - \Psi_0(x) + \sum_{k=1}^{\infty} R_{0,k}(x) * [I - \Psi_k(x)] * G_{k,0}(x),$$

which is equal to $I - P_{0,0}(x)$ according to 3) of Theorem 4.

The entry $(0,l)$ with $l \geq 1$ on the right-hand side is

$$-R_{0,l}(x) * [I - \Psi_l(x)] + \sum_{k=l+1}^{\infty} R_{0,k}(x) * [I - \Psi_k(x)] * G_{k,l}(x),$$

which is equal to $-P_{0,l}(x)$ according to 1) of Theorem 4.

Finally, to see that the entry $(l,0)$ with $l \geq 1$ on the right-hand side is equal to the corresponding entry on the left-hand side, it follows from 2) of Theorem 4 that

$$-[I - \Psi_l(x)] * G_{l,0}(x) + \sum_{k=l+1}^{\infty} R_{l,k}(x) * [I - \Psi_k(x)] * G_{k,0}(x) = -P_{l,0}(x).$$

This completes the proof. ■

In the remainder of this section, we consider three important examples of block-structured Markov renewal processes: 1) Level dependent Markov renewal processes of $M/G/1$ type, 2) level dependent Markov renewal processes of $GI/M/1$ type, and 3) level dependent QBD renewal processes. In the next section, we will consider 4) Markov renewal processes of $GI/G/1$ type. Since the analysis is similar, We only provide details for Markov renewal processes of $M/G/1$ type. For a Markov renewal process $P(x)$ of $M/G/1$ type,

$$Q_k(x) = \begin{pmatrix} P_{k,k}(x) & P_{k,k+1}(x) & P_{k,k+2}(x) & \cdots \\ P_{k+1,k}(x) & P_{k+1,k+1}(x) & P_{k+1,k+2}(x) & \cdots \\ & P_{k+2,k+1}(x) & P_{k+2,k+2}(x) & \cdots \\ & & \ddots & \ddots \end{pmatrix}, \quad k \geq 1.$$

We denote by $(\widehat{Q}_{1,1}^{(k)}(x)^T, \widehat{Q}_{2,1}^{(k)}(x)^T, \ldots)^T$ the first block-column of the matrix $\widehat{Q}_k(x) = \sum_{l=0}^{\infty} [Q_k(x)]^{l*}$. Based on Corollary 2, we have

$$R_{i,j}(x) = \sum_{l=1}^{\infty} P_{i,i+l}(x) * \widehat{Q}_{l,1}^{(j)}(x), \ 0 \leq i < j, \tag{24}$$

$$\underline{G}_k(x) \stackrel{\text{def}}{=} G_{k,k-1}(x) = \widehat{Q}_{1,1}^{(k)}(x) * P_{k,k-1}(x), \ k \geq 1, \tag{25}$$

$G_{i,j}(x) = 0$ for $j \neq i-1$, and

$$\Psi_k(x) = P_{k,k}(x) + R_{k,k+1}(x) * P_{k+1,k}(x), \quad k \geq 0. \tag{26}$$

The following lemma provides the formula for expressing the first block-column in the fundamental matrix $\widehat{Q}_k(x)$ in terms of $\widehat{Q}_{1,1}^{(k)}(x)$. The proof is similar to that of Lemma 4 in Li and Zhao (2002a).

Lemma 6 *For $k \geq 1$ and $j \geq 2$,*

$$\widehat{Q}_{j,1}^{(k)}(x) = \underline{G}_{k+j-1}(x) * \underline{G}_{k+j-2}(x) * \cdots * \underline{G}_{k+1}(x) * \widehat{Q}_{1,1}^{(k)}(x)$$

and

$$[I - \Psi_k(x)] * \widehat{Q}_{1,1}^{(k)}(x) = \widehat{Q}_{1,1}^{(k)}(x) * [I - \Psi_k(x)] = I.$$

The following theorem follows immediately from (24) and Lemma 6.

Theorem 6 *1) For $0 \leq i < j$,*

$$\begin{aligned} R_{i,j}(x) =& [P_{i,i+1}(x) + \sum_{l=2}^{\infty} P_{i,i+l}(x) * \underline{G}_{l+j}(x) * \underline{G}_{l+j-1}(x) \\ & * \cdots * \underline{G}_{j+2}(x)] * \widehat{Q}_{1,1}^{(j)}(x). \end{aligned}$$

2) For $i \geq 0$,

$$\Psi_i(x) = P_{i,i}(x) + \sum_{l=1}^{\infty} P_{i,i+l}(x) * \underline{G}_{i+l}(x) * \underline{G}_{i+l-1}(x) * \cdots * \underline{G}_{i+1}(x).$$

3) The matrix sequence $\{\underline{G}_i(x)\}$ is the minimal nonnegative solution to the system of matrix equations

$$\underline{G}_i(x) = P_{i,i-1}(x) + \sum_{l=0}^{\infty} P_{i,i+l}(x) * \underline{G}_{i+l}(x) * \underline{G}_{i+l-1}(x) * \cdots * \underline{G}_i(x).$$

Remark 2 *For a Markov renewal process of $M/G/1$ type, it is easy to see from Theorem 6 that the two matrix sequences $\{R_{i,j}(x)\}$ and $\{\Psi_i(x)\}$ can be expressed in terms of the matrix sequence $\{\underline{G}_i(x)\}$. Similarly, for a Markov renewal process of $GI/M/1$ type, the two matrix sequences $\{G_{i,j}(x)\}$ and $\{\Psi_i(x)\}$ can be expressed by the matrix sequence $\{\underline{R}_i(x)\}$ defined similarly to (25). It is clear that the RG-factorization (23) can be simplified for Markov renewal processes of $M/G/1$ type and $GI/M/1$ type.*

3. Markov Renewal Processes of $GI/G/1$ Type

In this section, we consider the Markov renewal process of $GI/G/1$ type given in (3). Based on the results in Section 2, we can further simplify expressions for the R- and G-measures and the RG-factorization for the transition probability matrix. Additionally, we obtain the RG-factorization based on the joint transformations for the repeating matrix sequence and the Wiener-Hopf equations for the boundary matrix sequence.

Comparing the two transition probability matrices (1) with (3) leads to $P_{0,j}(x) = D_j$ for $j \geq 0$, $P_{i,0}(x) = D_{-i}$ for $i \geq 1$ and $P_{i,j}(x) = A_{j-i}(x)$ for $i,\ j \geq 1$. It is clear that $Q_n(x) = Q(x)$ for all $n \geq 1$. Therefore, we write $\widehat{Q}^{(\cdot,1)}(x) = \widehat{Q}_j^{(\cdot,1)}(x)$ and $\widehat{Q}^{(1,\cdot)}(x) = \widehat{Q}_i^{(1,\cdot)}(x)$ for all $i,\ j \geq 1$.

It follows from Corollary 2 that

$$R_{0,j}(x) = (D_j(x), D_{j+1}(x), D_{j+2}(x), \cdots) * \widehat{Q}^{(\cdot,1)}(x), \quad j \geq 1,$$

$$R_{i,j}(x) = (A_{j-i}(x), A_{j-i+1}(x), A_{j-i+2}(x), \cdots) * \widehat{Q}^{(\cdot,1)}(x), \ 1 \leq i < j.$$

It is obvious that the matrices $R_{i,j}(x)$ for $1 \leq i < j$ only depend on the difference $j - i$. We write $R_{i,j}(x)$ as $R_{j-i}(x)$ for all $1 < i \leq j$. Therefore, for $k \geq 1$

$$R_k(x) = (A_k(x), A_{k+1}(x), A_{k+2}(x), \cdots) * \widehat{Q}^{(\cdot,1)}(x).$$

Similarly, for $i \geq 1$,

$$G_{i,0}(x) = \widehat{Q}^{(1,\cdot)}(x) * \left(D_{-i}^T(x), D_{-(i+1)}^T(x), D_{-(i+2)}^T(x), \cdots\right)^T$$

and for $1 \leq j < i$,

$$G_{i,j}(x) = \widehat{Q}^{(1,\cdot)}(x) * \left(A_{-(i-j)}^T(x), A_{-(i-j+1)}^T(x), A_{-(i-j+2)}^T(x), \cdots\right)^T.$$

The matrices $G_{i,j}(x)$ for $1 \le j < i$ only depend on the difference $i - j$. We write $G_{i,j}(x)$ as $G_{i-j}(x)$ for all $1 \le j < i$. Therefore, for $k \ge 1$,

$$G_k(x) = \widehat{Q}^{(1,\cdot)}(x) * \left(A_{-k}^T(x), A_{-(k+1)}^T(x), A_{-(k+2)}^T(x), \cdots\right)^T.$$

For the Markov renewal process of $GI/G/1$ type, the following lemma is a consequence of the repeating property of the transition probability matrix and the censoring invariant property provided in Section 2.

Lemma 7 *For $n \ge 1$, $i, j = 1, 2, 3, \cdots, n$,*

$$P_{n-i,n-j}^{[\le n]}(x) = P_{n+1-i,n+1-j}^{[\le (n+1)]}(x) = P_{n+2-i,n+2-j}^{[\le (n+2)]}(x) = \cdots.$$

Proof For $n \ge 1$, $i, j = 1, 2, 3, \cdots, n$, it follows from Theorem 3 that

$$P_{n-i,n-j}^{[\le n]}(x) = A_{i-j}(x) + (A_{i+1}(x), A_{i+2}(x), \cdots) * \widehat{Q}(x) * \left(A_{-(j+1)}^T(x), A_{-(j+2)}^T(x), \cdots\right)^T,$$

which is independent of $n \ge 1$. Thus

$$P_{n-i,n-j}^{[\le n]}(x) = P_{n+1-i,n+1-j}^{[\le (n+1)]}(x) = P_{n+2-i,n+2-j}^{[\le (n+2)]}(x) = \cdots.$$

This completes the proof. ■

Based on Lemma 7, we can define for $1 \le i, j \le n$,

$$\Phi_0(x) = P_{n,n}^{[\le n]}(x), \Phi_i(x) = P_{n-i,n}^{[\le n]}(x), \Phi_{-j}(x) = P_{n,n-j}^{[\le n]}(x). \quad (27)$$

The following theorem is a consequence of Lemma 5, which explicitly expresses the R- and G-measures in terms of the matrices $\Phi_i(x)$ for $-\infty < i < +\infty$.

Theorem 7 *For $i \ge 0$, $R_i(x) = \Phi_i(x) * \sum_{l=0}^{\infty} \Phi_0^{l*}(x)$ and for $j > 1$, $G_j(x) = \sum_{l=0}^{\infty} \Phi_0^{l*}(x) * \Phi_{-j}(x)$.*

Proof We only prove the first equation while the second one can be proved similarly.

It follows from Theorem 3 and (27) that

$$R_i(x) = R_i^{[\le n]}(x) = R_{n-i,n}^{[\le n]}(x) = P_{n-i,n}^{[\le n]}(x) * \sum_{l=0}^{\infty} \left[P_{n,n}^{[\le n]}(x)\right]^{l*} = \Phi_i(x) * \sum_{l=0}^{\infty} \Phi_0^{l*}(x).$$

This completes the proof. ■

The following lemma is a consequence of Theorem 4, which provides Wiener-Hopf equations for the repeating matrix sequence.

Theorem 8 *For* $i \geq 0$,

$$R_i(x) * [I - \Phi_0(x)] = A_i(x) + \sum_{k=1}^{\infty} R_{i+k}(x) * [I - \Phi_0(x)] * G_k(x), \tag{28}$$

for $j \geq 0$,

$$[I - \Phi_0(x)] * G_j(x) = A_{-j}(x) + \sum_{k=1}^{\infty} R_k(x) * [I - \Phi_0(x)] * G_{j+k}(x) \tag{29}$$

and

$$\Phi_0(x) = A_0(x) + \sum_{k=1}^{\infty} R_k(x) * [I - \Phi_0(x)] * G_k(x). \tag{30}$$

Proof We only prove (28) while (29) and (30) can be proved similarly. When n is big enough, it follows from Theorem 7 that

$$R_i(x) * [I - \Phi_0(x)] = \Phi_i(x) * \sum_{l=0}^{\infty} \Phi_0^{l*}(x) * [I - \Phi_0(x)] = \Phi_i(x) = P_{n-i,n}^{[\leq n]}(x),$$

and

$$\begin{aligned} P_{n-i,n}^{[\leq n]}(x) &= P_{n-i,n}^{[\leq (n+1)]}(x) + R_{i+1}^{[\leq (n+1)]}(x) * P_{n+1,n}^{[\leq (n+1)]}(x) \\ &= P_{n-i,n}^{[\leq (n+1)]}(x) + R_{i+1}^{[\leq (n+1)]}(x) * \left[I - P_{n+1,n+1}^{[\leq (n+1)]}(x)\right] * G_1^{[\leq (n+1)]}(x) \\ &= P_{n-i,n}^{[\leq (n+1)]}(x) + R_{i+1}(x) * [I - \Phi_0(x)] * G_1(x) \\ &= \cdots = P_{n-i,n}^{[\leq (n+N)]}(x) + \sum_{k=1}^{N} R_{i+k}(x) * [I - \Phi_0(x)] * G_k(x) \\ &= A_i(x) + \sum_{k=1}^{\infty} R_{i+k}(x) * [I - \Phi_0(x)] * G_k(x). \end{aligned}$$

This completes the proof. ∎

For a Markov renewal process of $GI/G/1$ type, since $R_{i,i+k} = R_k$, $G_{i+k,i} = G_k$ and $\Psi_i = \Phi_0$ for $i \geq 1$, the RG-factorization given in (23) can be simplified.

Let

$$\widetilde{A}_i(s) = \int_0^{+\infty} e^{-sx} dA_i(x), \quad \widetilde{\Phi}_0(s) = \int_0^{+\infty} e^{-sx} d\Phi_0(x),$$

$$\widetilde{R}_i(s) = \int_0^{+\infty} e^{-sx} dR_i(x), \quad \widetilde{G}_i(s) = \int_0^{+\infty} e^{-sx} dG_i(x),$$

$$\widetilde{A}^*(z,s) = \sum_{i=-\infty}^{\infty} z^i \widetilde{A}_i(s),$$

$$\widetilde{R}^*(z,s)=\sum_{i=1}^{\infty}z^i\widetilde{R}_i(s),\quad \widetilde{G}^*(z,s)=\sum_{j=1}^{\infty}z^{-j}\widetilde{G}_j(s).$$

For the Markov renewal process of $GI/G/1$ type, the following theorem provides the RG-factorization based on the double transformations for the repeating matrix sequence.

Theorem 9

$$I-\widetilde{A}^*(z,s)=\left[I-\widetilde{R}^*(z,s)\right]\left[I-\widetilde{\Phi}_0(s)\right]\left[I-\widetilde{G}^*(z,s)\right]. \tag{31}$$

Proof It follows from (28), (29) and (30) that

$$\widetilde{R}_i(s)\left[I-\widetilde{\Phi}_0(s)\right]=\widetilde{A}_i(s)+\sum_{k=1}^{\infty}\widetilde{R}_{i+k}(s)\left[I-\widetilde{\Phi}_0(s)\right]\widetilde{G}_k(s),\quad i\geq 0,$$

$$\left[I-\widetilde{\Phi}_0(s)\right]\widetilde{G}_j(s)=\widetilde{A}_{-j}(s)+\sum_{k=1}^{\infty}\widetilde{R}_k(s)\left[I-\widetilde{\Phi}_0(s)\right]\widetilde{G}_{j+k}(s),\quad j\geq 0,$$

$$\widetilde{\Phi}_0(s)=\widetilde{A}_0(s)+\sum_{k=1}^{\infty}\widetilde{R}_k(s)\left[I-\widetilde{\Phi}_0(s)\right]\widetilde{G}_k(x).$$

Therefore, we obtain

$$\widetilde{R}^*(z,s)\left[I-\widetilde{\Phi}_0(s)\right]+\left[I-\widetilde{\Phi}_0(s)\right]\widetilde{G}^*(z,s)+\widetilde{\Phi}_0(s)$$
$$=I_1+I_2+\sum_{k=1}^{\infty}z^k\widetilde{R}_k(s)\left[I-\widetilde{\Phi}_0(s)\right]z^{-k}\widetilde{G}_k(x),$$

where

$$I_1=\sum_{i=1}^{\infty}\sum_{k=1}^{\infty}z^{i+k}\widetilde{R}_{i+k}(s)\left[I-\widetilde{\Phi}_0(s)\right]z^{-k}\widetilde{G}_k(x)$$
$$=\widetilde{R}^*(z,s)\left[I-\widetilde{\Phi}_0(s)\right]\left[I-\widetilde{G}^*(z,s)\right]-\sum_{k=1}^{\infty}\sum_{i=1}^{k}z^i\widetilde{R}_i(s)\left[I-\widetilde{\Phi}_0(s)\right]z^{-k}\widetilde{G}_k(x)$$

and

$$I_2=\sum_{j=1}^{\infty}\sum_{k=1}^{\infty}z^k\widetilde{R}_k(s)\left[I-\widetilde{\Phi}_0(s)\right]z^{-(j+k)}\widetilde{G}_{j+k}(x)$$
$$=\widetilde{R}^*(z,s)\left[I-\widetilde{\Phi}_0(s)\right]\left[I-\widetilde{G}^*(z,s)\right]-\sum_{k=1}^{\infty}\sum_{j=k}^{\infty}z^j\widetilde{R}_j(s)\left[I-\widetilde{\Phi}_0(s)\right]z^{-k}\widetilde{G}_k(x).$$

Since

$$I_1 + I_2 + \sum_{k=1}^{\infty} z^k \widetilde{R}_k (s) \left[I - \widetilde{\Phi}_0 (s) \right] z^{-k} \widetilde{G}_k (x) = \widetilde{A}^* (z, s) + \widetilde{R}^* (z, s) \left[I - \widetilde{\Phi}_0 (s) \right] \widetilde{G}^* (z, s),$$

we get

$$\begin{aligned} &\widetilde{R}^* (z, s) \left[I - \widetilde{\Phi}_0 (s) \right] + \left[I - \widetilde{\Phi}_0 (s) \right] \widetilde{G}^* (z, s) + \widetilde{\Phi}_0 (s) \\ =&\widetilde{A}^* (z, s) + \widetilde{R}^* (z, s) \left[I - \widetilde{\Phi}_0 (s) \right] \widetilde{G}^* (z, s), \end{aligned}$$

which is equivalent to (31). This completes the proof. ■

The following lemma provides Wiener-Hopf equations for the boundary matrix sequence. The proof is similar to that of Lemma 8.

Lemma 8 *For* $i \geq 1$,

$$R_{0,i} (x) * [I - \Phi_0 (x)] = D_i (x) + \sum_{k=1}^{\infty} R_{0,i+k} (x) * [I - \Phi_0 (x)] * G_k (x), \tag{32}$$

for $j \geq 1$,

$$[I - \Phi_0 (x)] * G_{j,0} (x) = D_{-j} (x) + \sum_{k=1}^{\infty} R_k (x) * [I - \Phi_0 (x)] * G_{j+k,0} (x) \tag{33}$$

and

$$\Psi_0 (x) = P^{[0]} (x) = D_0 (x) + \sum_{k=1}^{\infty} R_{0,k} (x) * [I - \Phi_0 (x)] * G_{k,0} (x). \tag{34}$$

Let

$$\widetilde{D}_i (s) = \int_0^{+\infty} e^{-sx} dD_i (x),$$

$$\widetilde{D}_+^* (z, s) = \sum_{i=1}^{\infty} z^i \widetilde{D}_i (s), \quad \widetilde{D}_-^* (z, s) = \sum_{i=1}^{\infty} z^{-i} \widetilde{D}_{-i} (s),$$

$$\widetilde{R}_{0,i} (s) = \int_0^{+\infty} e^{-sx} dR_{0,i} (x), \quad \widetilde{G}_{i,0} (s) = \int_0^{+\infty} e^{-sx} dG_{i,0} (x),$$

and

$$\widetilde{R}_0^* (z, s) = \sum_{i=1}^{\infty} z^i \widetilde{R}_{0,i} (s), \quad \widetilde{G}_0^* (z, s) = \sum_{j=1}^{\infty} z^{-j} \widetilde{G}_{j,0} (s). \tag{35}$$

Theorem 10 *For $z > 0$ and $s \geq 0$,*

$$\widetilde{R}_0^*(z,s) \geq \widetilde{D}_+^*(z,s)\left[I - \widetilde{\Phi}_0(s)\right]^{-1}, \tag{36}$$

$$\widetilde{R}_0^*(z,s)\left[I - \widetilde{\Phi}_0(s)\right]\left[I - \widetilde{G}^*(z,s)\right] \leq \widetilde{D}_+^*(z,s), \tag{37}$$

$$\widetilde{G}_0^*(z,s) \geq \left[I - \widetilde{\Phi}_0(s)\right]^{-1}\widetilde{D}_-^*(z,s) \tag{38}$$

and

$$\widetilde{G}_0^*(z,s)\left[I - \widetilde{\Phi}_0(s)\right]\left[I - \widetilde{R}^*(z,s)\right] \leq \widetilde{D}_-^*(z,s). \tag{39}$$

Proof We only prove (36) and (37) while (38) and (39) can be proved similarly. It follows from (32) that

$$\widetilde{R}_{0,i}(s)\left[I - \widetilde{\Phi}_0(s)\right] = \widetilde{D}_i(s) + \sum_{k=1}^{\infty}\widetilde{R}_{0,i+k}(s)\left[I - \widetilde{\Phi}_0(s)\right]\widetilde{G}_k(s), \tag{40}$$

and from the second equation of Theorem 7 that

$$\widetilde{G}_k(s) = \left[I - \widetilde{\Phi}_0(s)\right]^{-1}\widetilde{\Phi}_{-k}(s).$$

It is obvious that for $s \geq 0$,

$$\sum_{k=1}^{\infty}\widetilde{R}_{0,i+k}(s)\left[I - \widetilde{\Phi}_0(s)\right]\widetilde{G}_k(s) = \sum_{k=1}^{\infty}\widetilde{R}_{0,i+k}(s)\widetilde{\Phi}_{-k}(s) \geq 0.$$

Hence it follows from (40) that

$$\widetilde{R}_{0,i}(s)\left[I - \widetilde{\Phi}_0(s)\right] \geq \widetilde{D}_i(s)$$

and for $z > 0$,

$$\widetilde{R}_0^*(z,s)\left[I - \widetilde{\Phi}_0(s)\right] \geq \widetilde{D}_+^*(z,s). \tag{41}$$

Since the Markov renewal process is irreducible, the spectral radius

$$sp\left(\widetilde{\Phi}_0(s)\right) \leq sp\left(\widetilde{\Phi}_0(0)\right) = sp\left(\Phi_0(+\infty)\right) < 1$$

for all $s > 0$. Furthermore, the matrix $I - \widetilde{\Phi}_0(s)$ is invertible and $\left[I - \widetilde{\Phi}_0(s)\right]^{-1} \geq 0$. Therefore, it follows from (41) that

$$\widetilde{R}_0^*(z,s) \geq \widetilde{D}_+^*(z,s)\left[I - \widetilde{\Phi}_0(s)\right]^{-1}.$$

It follows from (40) that

$$\widetilde{R}_0^*(z,s)\left[I-\widetilde{\Phi}_0(s)\right] \le \widetilde{D}_+^*(z,s) + \widetilde{R}_0^*(z,s)\left[I-\widetilde{\Phi}_0(s)\right]\widetilde{G}^*(z,s),$$

simple computations lead to

$$\widetilde{R}_0^*(z,s)\left[I-\widetilde{\Phi}_0(s)\right]\left[I-\widetilde{G}^*(z,s)\right] \le \widetilde{D}_+^*(z,s).$$

This completes the proof. ■

Acknowledgements

The authors acknowledge that this work was supported by a research grant from the Natural Sciences and Engineering Research Council of Canada (NSERC). Dr. Li also acknowledges the support from Carleton University.

References

Asmussen, S., 1987: *Applied Probability and Queues.* John Wiley & Sons.

Asmussen, S., 1990: Probabilistic interpretations of some duality results for the matrix paradigms in queueing theory. *Stochastic models*, **6**, 715-733.

Çinlar, E., 1968: Some joint distributions for Markov renewal processes. *Austral. J. Statist*, **10**, 8-20.

Çinlar, E., 1969: Markov renewal processes. *Adv. Appl. Prob.*, **1**, 123-187.

Çinlar, E., 1974: Periodicity in Markov renewal processes. *Adv. Appl. Prob.*, **6**, 61-78.

Çinlar, E., 1975: *Introduction to Stochastic Processes.* Prentice-Hall, Inc..

Grassmann, W. K., and D. P. Heyman, 1990: Equilibrium distribution of block-structured Markov chains with repeating rows. *J. Appl. Prob.*, **27**, 557-576.

Hsu, G. H., X. Yuan, and Q. L. Li, 2000: First passage times for Markov renewal processes and application. *Science in China, series A*, **43**, 1238-1249.

Hunter, J. J., 1969: On the moments of Markov renewal processes. *Adv. Appl. Prob.*, **1**, 188-210.

Latouche, G., and V. Ramaswami, 1999: *Introduction to Matrix Analytic Methods in Stochastic Modeling.* SIAM, Philadelphia, 1999.

Li, Q. L., and Y. Q. Zhao, 2002a: A constructive method for finding $\beta-$invariant measures for transition matrices of $M/G/1$ type. In *Matrix-Analytic Methods Theory and Applications*, Latouche, G., Taylor, P.G., Eds., World Scientific: New Jersey, 237-263.

Li, Q. L., and Y. Q. Zhao, 2002b: Light-tailed asymptotics of stationary probability vectors of Markov chains of $GI/G/1$ type. Technical Report No. 365, Laboratory for Research in Statistics and Probability, Carleton University and University of Ottawa, 2002, 37 pp.

Neuts, M. F., 1981: *Matrix-geometric solutions in stochastic models: An algorithmic approach.* The Johns University Press, Baltimore.

Neuts, M. F., 1986: A new informative embedded Markov renewal process for the $PH/G/1$ queue. *Adv. Appl. Prob.*, **18**, 533-557.

Neuts, M. F., 1989: *Structured Stochastic Matrices of $M/G/1$ Type and Their Applications.* Marcel Decker Inc., New York.

Neuts, M. F., 1995: Matrix-analytic methods in queueing theory. In *Advances of Queueing Theory, Methods, and Open Problems*, Dshalalow, J. H., Ed., CRC: Boca Roton, 265-292.

Pyke, R., 1961a: Markov renewal processes: definitions and preliminary properties. *Ann. Math. Statist.*, **32**, 1231-1242.

Pyke, R., 1961b: Markov renewal processes with finitely many states. *Ann. Math. Statist.*, **32**, 1243-1259.

Pyke, R., and R. Schaufele, 1964: Limit theorems for Markov renewal processes. *Ann. Math. Statist.*, **35**, 1746-1764.

Pyke, R., and R. Schaufele, 1966: The existence and uniqueness of stationary measures for Markov renewal processes. *Ann. Math. Statist.*, **37**, 1439-1462.

Ramaswami, V., 1990a: A duality theorem for the matrix paradigms in queueing theory. *Stochastic Models*, **6**, 151-161.

Ramaswami, V., 1990b: From the matrix-geometric to the matrix-exponential. *Queueing Systems*, **6**, 229-260.

Sengupta, B., 1989: Markov processes whose steady state distribution is matrix-exponential with an application to $GI/PH/1$ queue. *Adv. Appl. Prob.*, **21**, 159-180.

Teugels, J. L., 1968: Exponential ergodicity in Markov renewal processes. *J. Appl. Prob.*, **5**, 387-400.

Zhao, Y. Q., 2000: Censoring technique in studying block-structured Markov chains. *In Advances in Algorithmic Methods for Stochastic Models,* Latouche, G. and Taylor, P. G., Eds, Notable Publications Inc., NJ., 417-433.

Zhao, Y. Q., W. Li, and A. S. Alfa, 1999: Duality results for block-structured transition matrices. *J. Appl. Prob.*, **36**, 1045-1057.

Zhao, Y. Q., W. Li, and W. J. Braun, 2003: Censoring, factorizations and spectral analysis for transition matrices with block-repeating entries. *Methodology and Computing in Applied Probability.* **5**, 35-58.

LARGE SAMPLE PROPERTY OF THE LIKELIHOOD ESTIMATE FOR FRACTIONAL ARIMA MODELS

SHU AN

SAS Institute Inc., SAS Campus Drive, Cary, NC 27513, USA
E-mail: Shu.An@sas.com

PETER BLOOMFIELD AND SASTRY G. PANTULA

Department of Statistics, North Carolina State University, Raleigh, NC 27695-8203, USA
E-mail: bloomfld@eos.ncsu.edu and pantula@stat.ncsu.edu

(Manuscript received November 14, 2002)

In this paper, the authors will investigate the maximum likelihood estimator of fractional autoregressive integrated moving-average time series parameter. Recently, Dahlhaus considered the likelihood function of the processes. Substituting the unknown mean μ by any consistent estimator, Dahlhaus has shown that the maximum likelihood estimator of the time series parameter θ is consistent and asymptotically normal. The authors will estimate μ and θ simultaneously by maximizing the full likelihood function and prove the consistency and asymptotic normality for the estimator of θ. The results also hold for the long range dependence processes.

1. Introduction

Granger and Joyeux (1980) and Hosking (1981) introduced a fractional autoregressive integrated moving-average (fractional ARIMA) process. A fractional ARIMA(p, d, q) $\{y_t\}$ with nonnegative integers p and q and real d is a discrete time stochastic process which may be represented as $\phi(\mathrm{B})\nabla^d(y_t - \mu) = \delta(\mathrm{B})\varepsilon_t$, where B is the backward-shift operator and

$$\nabla^d = (1-\mathrm{B})^d = \sum_{k=0}^{\infty} \binom{d}{k} (-\mathrm{B})^k = 1 - d\mathrm{B} - \frac{1}{2}d(1-d)\mathrm{B}^2 \cdots$$

$\phi(\mathrm{B}) = 1 - \phi_1\mathrm{B} - \cdots - \phi_p\mathrm{B}^p$, $\delta(\mathrm{B}) = 1 - \delta_1\mathrm{B} - \cdots - \delta_q\mathrm{B}^q$, $\mathrm{E}(y_t) = \mu$ and the $\{\varepsilon_t\}$ are independent and identically distributed random variables with mean 0 and variance σ_ε^2.

A fractional ARIMA(p, d, q) process is stationary if $d < \frac{1}{2}$ and $\phi(z) \neq 0$ for $|z| \leq 1$. Also, $\{y_t\}$ is invertible, *i.e.* $\varepsilon_t = \sum_{k=0}^{\infty} b_k y_{t-k}$ in mean square, if $d > -1$ and $\delta(z) \neq 0$ for $|z| \leq 1$. (Granger and Joyeux 1980; Hosking 1981; Bloomfield 1985; Anděl 1986).

The following two properties characterize the fractional ARIMA (p, d, q) process. It is well known that the autocorrelation function of a stationary autoregressive moving-average (ARMA) process converges to zero at a geometric rate and is absolutely summable, and the spectral density function of such stationary process is absolutely continuous. However, when a process is a fractional ARIMA process and is stationary and invertible, its autocorrelation function $\rho_k \sim k^{-(1-2d)}$ as $k \to \infty$, *i.e.* it converges to zero hyperbolically. Therefore, $\sum_{k=0}^{\infty} |\rho_k| = +\infty$ for $d > 0$. It is the slower rate of ρ_k that determines the long memory property of the series. Note that if $d = 0$, it is a stationary ARMA(p, q) model. For $-1 < d < 0$, $\rho(k)$ is absolutely summable. The spectral density function of $\{y_t\}$ is

$$f_\theta(x) = \frac{\sigma_\varepsilon^2}{2\pi} \frac{|\delta(e^{ix})|^2}{|\phi(e^{ix})|^2} |1 - e^{ix}|^{-2d} \sim \frac{\sigma_\varepsilon^2}{2\pi} \left(\frac{1 - \delta_1 - \cdots - \delta_q}{1 - \phi_1 - \cdots - \phi_p} \right)^2 x^{-2d} \qquad \text{as } x \to 0.$$

where $\theta = (\sigma_\varepsilon^2, d, \phi_1, \cdots, \phi_p, \delta_1, \cdots, \delta_q)^\top \in \Theta \subset \mathbb{R}^m$, m is the dimension of θ. Unlike for the stationary ARMA process, the spectral density of a stationary fractional ARIMA process has, for $d > 0$, a singularity at zero frequency.

These differences in the behavior of the autocorrelation and spectral density functions between stationary ARMA and fractional ARIMA processes indicate that the parameter d determines the long term correlation structure, and the small number of δ's and ϕ's explain the short term correlation structure. Both ARMA(p, q) and ARIMA(p, d, q) models can be fitted to the global surface temperature data (Bloomfield, 1992b). ARMA models explain how the current global surface temperature depends on the temperature in last few years (short memory or short term correlation). In addition to the short memory, the nonzero parameter d in a fractional ARIMA(p, d, q) model explores the impact of temperatures for the past hundred years on this year temperature (long memory or long term correlation).

More generally, a stationary process has long-range dependence if its spectral density is of the form $f(x) \sim L(x)|x|^{-2d}$ as $x \to 0$, where $d \in (0, 1/2)$ and $L(\cdot)$ is slowly varying for $|x| \to 0$ or equivalently if its covariance function is $\mathrm{R}_k \sim L_2(k)|k|^{-(1-2d)}$ as $|k| \to \infty$, where $L_2(\cdot)$ is slowly varying for $|k| \to \infty$.

It is easy to see that the fractional ARIMA(p, d, q) process has long range dependence and is a long memory stationary process when $d \in (0, 1/2)$. It is interesting to compare the fractional ARIMA(p, d, q) process with the stationary ARMA(p, q) and the ARMA(p, q) with a unit root process. The spectral density function of a stationary ARMA(p, q) is $f(x) \sim L(x)$, which is slowly varying as $x \to 0$,. The ARMA(p, q) with a unit root process has a spectral density $f(x) \sim L(x)x^{-2}$ as $x \to 0$, which has infinite variance. In fact, if $1/2 \leq d < 1$, a fractional ARIMA(p, d, q) process has infinite variance.

Another widely used class of long range dependence processes is the fractional Gaussian noise $\{y_t\}$ with mean $\mu \in \mathbb{R}$ and covariances

$$R_k = \frac{C}{2} \left(|k+1|^{2H} - 2|k|^{2H} + |k-1|^{2H} \right) \sim C\, H(2H-1)k^{2H-2} \qquad \text{as } k \to \infty,$$

where $\frac{1}{2} < H < 1$, and spectral density

$$f(x, H) = C\, F(H)\, (1 - \cos x) \sum_{k=-\infty}^{\infty} |x + 2\pi k|^{-1-2H} \sim \frac{C\, F(H)}{2} |x|^{1-2H} \qquad \text{as } x \to 0.$$

where $F(H) = \left\{\int_{-\infty}^{\infty}(1-\cos t)|t|^{-1-2H}dt\right\}^{-1}$.

Various examples in which long-range dependence occurs were given by Hampel (1987). Long range dependence models including fractional ARIMA model have been used in various areas; in geophysics and hydrology (Lawrance and Kotegoda 1977), in meteorology (Haslett and Raftery 1989), in economics (Porter-Hudak 1990), and in global warming (Bloomfield 1992a, b; Smith 1992). Also, see the references of Beran (1991) for other applications of the models. Some examples of using autoregressive moving-average models in statistical weather forecasting are explained by Wilks (1995).

Bloomfield (1992b) fitted a fractional ARIMA(p,d,q) model and a few ARMA models to the global surface temperature data. The paper argued that the fractional ARIMA$(0,d,1)$ model outperforms other ARMA models in terms of model simplicity and the precision of the estimated temperature change.

In calculus, fractional derivatives can be defined on a smooth function. Fractional diffusion equations, an application of fractional derivatives on physics, are recently discussed by Sokolov, Klafter and Blumen (2002).

The statistical properties of the maximum likelihood estimator of the parameter θ have received considerable attention recently. The maximum likelihood estimators obtained by maximizing approximate likelihood functions were investigated by Fox and Taqqu (1986), Beran (1986) and Dahlhaus (1989). Dahlhaus (1989) considered the exact maximum likelihood estimate $\tilde{\theta}_N$ obtained by maximizing the full likelihood function evaluated at θ and $\tilde{\mu}_N$, where $\tilde{\mu}_N$ is a consistent estimator of μ. A widely used procedure to obtain the maximum likelihood estimator $\hat{\theta}_N$ of parameter θ is maximizing the profile likelihood function, that is replacing the mean μ in the likelihood function by its maximum likelihood estimator and then maximizing the resulting profile likelihood function with respect to θ. The maximum profile likelihood estimator $\hat{\theta}_N$ equals the overall maximum likelihood estimator of θ (Barndorff-Nielsen 1990). In this article, the authors will prove, under the same assumptions as in Dahlhaus (1989), that the profile likelihood estimator of θ is consistent and asymptotically normal.

In small and moderately-sized samples, like the global surface temperature data, the likelihood based estimates of θ may be biased. The estimate $\hat{\theta}_N$ discussed in this paper is easier than other estimates of θ to be adjusted to reduce the bias. Further research result will be discussed in a separate paper by the authors.

The paper is organized as follows: Section 2 presents some notation and assumptions on a discrete stochastic process $\{y_t\}$, $t \in \mathbb{Z}$. Section 3 establishes the consistency and asymptotic normality of the maximum likelihood estimator $\hat{\theta}_N$. Finally, Appendix gives some lemmas used in order to prove the main results. The proof of the lemmas is available from the authors.

2. Notations and assumptions

This paper deals throughout with parametric problems for which the vector $\mathbf{Y}_N = (y_1, \cdots, y_N)^\top$ contains N observations of a real-valued stationary process with mean $\mu \in \mathbb{R}$ and spectral density $f_\theta(x)$, where $x \in \Pi = [-\pi, \pi]$ and $\theta \in \Theta \subset \mathbb{R}^m$ is a vector of unknown parameters. Let μ_0 and θ_0 be the true value of μ and θ. For a fractional

ARIMA(p,d,q) process, $\theta = (\sigma_\varepsilon^2, d, \phi_1, \cdots, \phi_p, \delta_1, \cdots, \delta_q)^\top$. Let $l_N(\theta,\mu)$ be the $-1/N\times$ log likelihood function, omitting constant terms. In the Gaussian case, $l_N(\theta,\mu)$ takes the form

$$l_N(\theta,\mu) = \frac{1}{2N}\log\det T_\theta + \frac{1}{2N}(\mathbf{Y}_N - \mu\mathbf{1})^\top T_\theta^{-1}(\mathbf{Y}_N - \mu\mathbf{1}),$$

where N by 1 vector $\mathbf{1} = (1,\cdots,1)^\top$ and the covariance matrix is of the form

$$T_N(f_\theta) = \mathrm{var}(\mathbf{Y}_N) = \left\{\int_{-\pi}^{\pi} f_\theta(x) exp(ix(r-s))dx\right\}_{r,s=1,\cdots,N}.$$

Let the vector of matrices

$$T_{\partial,\theta} = \frac{\partial}{\partial\theta}T_N(f_\theta) = \left[T_N\left(\frac{\partial}{\partial\theta_1}f_\theta\right),\cdots,T_N\left(\frac{\partial}{\partial\theta_m}f_\theta\right)\right]^\top$$

and the matrix of matrices

$$T_{\partial^2,\theta} = \frac{\partial^2}{\partial\theta\partial\theta^\top}T_N(f_\theta) = \left\{T_N\left(\frac{\partial^2}{\partial\theta_j\partial\theta_k}f_\theta\right)\right\}_{m\times m}.$$

Let $A_\theta^{(1)} = T_N(f_\theta)^{-1}T_N(\partial/\partial\theta f_\theta)T_N(f_\theta)^{-1}T_N(\partial/\partial\theta f_\theta)^\top T_N(f_\theta)^{-1}$, $A_\theta^{(2)} = T_N(f_\theta)^{-1} T_N(\partial^2/\partial\theta\partial\theta^\top f_\theta)T_N(f_\theta)^{-1}$ and $A_\theta^{(3)} = T_N(f_\theta)^{-1}T_N(\partial/\partial\theta f_\theta)T_N(f_\theta)^{-1}$. It is easy to see that $\partial/\partial\theta(T_\theta^{-1}) = -A_\theta^{(3)}$ and $\partial/\partial\theta(A_\theta^{(3)}) = -2A_\theta^{(1)} + A_\theta^{(2)}$.

Supposing A is an $n\times n$ matrix, the spectral norm of A is defined as $\|A\| = \sup_{\mathbf{x}\in\mathbb{C}^n}(\mathbf{x}^*A^*A\mathbf{x}/\mathbf{x}^*\mathbf{x})^{1/2}$ and the Euclidean norm of A as $|A| = [\,\mathrm{tr}\,(A\,A^*)]^{1/2}$. Some results on these norms are discussed in Davies (1973 Appendix II) and Graybill (1983 Section 5.6).

Dahlhaus (1989) considered the exact maximum likelihood estimator $\tilde\theta_N$ of θ, defined as

$$\begin{aligned} l_N(\tilde\theta_N,\tilde\mu_N) &= \max_{\theta\in\Theta} l_N(\theta,\tilde\mu_N) \\ &= \max_{\theta\in\Theta}\left\{\frac{1}{2N}\log\det T_N(f_\theta) + \frac{1}{2N}(\mathbf{Y}_N - \tilde\mu_N\mathbf{1})^\top T_N(f_\theta)^{-1}(\mathbf{Y}_N - \tilde\mu_N\mathbf{1})\right\}, \end{aligned}$$

where $\tilde\mu_N$ is a consistent estimator of μ_0 (the sample mean $\bar Y_N$, an M-estimator and the maximum likelihood estimator for $\theta = \theta_0$ are examples).

The maximum likelihood estimator $\hat\theta_N$, discussed in this paper, is defined by

$$\begin{aligned} l_N(\hat\theta_N,\hat\mu_N(\hat\theta)) &= \max_{\theta\in\Theta} l_N(\theta,\hat\mu_N(\theta)) \\ &= \max_{\theta\in\Theta}\left\{\frac{1}{2N}\log\det T_N(f_\theta) + \frac{1}{2N}(\mathbf{Y}_N - \hat\mu_N(\theta)\mathbf{1})^\top T_N(f_\theta)^{-1}(\mathbf{Y}_N - \hat\mu_N(\theta)\mathbf{1})\right\}, \end{aligned}$$

where

$$\hat\mu_N(\theta) = \frac{\mathbf{1}^\top T_N(f_\theta)^{-1}\mathbf{Y}_N}{\mathbf{1}^\top T_N(f_\theta)^{-1}\mathbf{1}}$$

is the maximum likelihood estimator of μ_0 at a given θ.

The discussion in this paper is under following assumptions:

(D0) If $\theta \neq \theta'$, the set $\{x | f_\theta(x) \neq f_{\theta'}(x)\}$ is supposed to have positive Lebesgue measure.

There exists a α: $\Theta \to (0,1)$ such that for each $\tau > 0$:

(D1) $f_\theta(x)$ is continuous at all (x, θ), $x \neq 0$, $f_\theta^{-1}(x)$ is continuous at all (x, θ) and $f_\theta(x) = O(|x|^{-\alpha(\theta)-\tau})$.

(D2) $\partial/\partial\theta_j f_\theta(x)$, $\partial^2/\partial\theta_j\partial\theta_k f_\theta(x)$ and $\partial^3/\partial\theta_j\partial\theta_k\partial\theta_h f_\theta(x)$ are all $O(|x|^{-\alpha(\theta)-\tau})$ and continuous at all (x, θ), $x \neq 0$, where $1 \leq j, k, h \leq m$.

(D3) $\partial/(\partial x) f_\theta^{-1}(x)$ and $\partial^2/(\partial x)^2 f_\theta^{-1}(x)$ are continuous at all (x, θ), $x \neq 0$, and $(\partial/\partial x)^k f_\theta^{-1}(x) = O(|x|^{\alpha(\theta)-k-\tau})$ for $k = 1, 2$.

(D4) The above constants for the order statements can be chosen independently of θ (not of τ).

(D5) α is assumed to be continuous. Furthermore, there exists a constant C with $|f_\theta(x) - f_{\theta'}(x)| \leq C\,|\theta - \theta'| f_{\theta'}(x)$ uniformly for all x and all θ, θ' with $\alpha(\theta) \leq \alpha(\theta')$, where $|\cdot|$ denotes the Euclidean norm.

The conditions (D0) to (D5) are satisfied for fractional ARIMA(p, d, q) process with $\alpha(\theta) = 2d$ (Fox and Taqqu 1986 Thereom 3).

In all that follows, K is a constant whose value may vary from equation to equation, unless otherwise stated.

3. Consistency and asymptotic normality

The consistency of $\hat{\theta}_N$ follows the ideas of Walker (1964, section 2).

Theorem 1 *Suppose (D0) – (D5) hold, then* $\hat{\theta}_N \xrightarrow{p} \theta_0$.

Proof: First, the following inequalities and an equality will be proved. For all $\theta_1 \in \Theta - \{\theta_0\}$, there exists a constant $C(\theta_1) > 0$ with

$$\lim_{N\to\infty} \mathrm{E}_{\theta_0}\{l_N(\theta_1, \hat{\mu}_N(\theta_1)) - l_N(\theta_0, \hat{\mu}_N(\theta_0))\} \geq C(\theta_1) \tag{1}$$

$$\lim_{N\to\infty} \mathrm{var}_{\theta_0}\{l_N(\theta_1, \hat{\mu}_N(\theta_1)) - l_N(\theta_0, \hat{\mu}_N(\theta_0))\} = 0 \tag{2}$$

$$\begin{aligned} &|l_N(\theta_2, \hat{\mu}_N(\theta_2)) - l_N(\theta_1, \hat{\mu}_N(\theta_1))| \\ &\quad \leq K\,|\theta_2 - \theta_1| \left\{ 1 + \frac{1}{N}\sum_{i=1}^{N}(y_i - \mu_0)^2 \right\} + K\,|\theta_2 - \theta_1|^{1/2} \frac{1}{N}\sum_{i=1}^{N}(y_i - \mu_0)^2 \end{aligned} \tag{3}$$

where K may depend linearly on $|\theta_2 - \theta_1|$, but not on N.

To prove (1), note that

$$l_N(\theta_1, \hat{\mu}_N(\theta_1)) - l_N(\theta_0, \hat{\mu}_N(\theta_0)) = \frac{1}{2N} \log\det(T_{\theta_1} T_{\theta_0}^{-1})$$

$$+\frac{1}{2N}\{(\mathbf{Y}_N-\mu_0\mathbf{1})^\top(T_{\theta_1}^{-1}-T_{\theta_0}^{-1})(\mathbf{Y}_N-\mu_0\mathbf{1})\} \tag{4}$$

$$-\frac{1}{2N}\left\{\frac{[\mathbf{1}^\top T_{\theta_1}^{-1}(\mathbf{Y}_N-\mu_0\mathbf{1})]^2}{\mathbf{1}^\top T_{\theta_1}^{-1}\mathbf{1}}-\frac{[\mathbf{1}^\top T_{\theta_0}^{-1}(\mathbf{Y}_N-\mu_0\mathbf{1})]^2}{\mathbf{1}^\top T_{\theta_0}^{-1}\mathbf{1}}\right\},$$

and the proof of the Theorem 3.1 in Dahlhaus (1989) covers the first two terms. It remains to consider the last difference. The expectation of the difference is

$$\frac{1}{2N}\left\{\frac{\mathbf{1}^\top T_{\theta_1}^{-1}T_{\theta_0}T_{\theta_1}^{-1}\mathbf{1}}{\mathbf{1}^\top T_{\theta_1}^{-1}\mathbf{1}}-1\right\}.$$

$\mathbf{Y}^\top A\mathbf{Y}\le\mathbf{Y}^\top\mathbf{Y}\,\|A\|$, for $A\ge 0$ and Lemma 5.3 in Dahlhaus (1989) yield that

$$|\mathbf{1}^\top T_{\theta_1}^{-1}T_{\theta_0}T_{\theta_1}^{-1}\mathbf{1}|\le(\mathbf{1}^\top T_{\theta_1}^{-1}\mathbf{1})\|T_{\theta_1}^{-1/2}T_{\theta_0}^{1/2}\|^2=(\mathbf{1}^\top T_{\theta_1}^{-1}\mathbf{1})O(N^{\max\{\alpha(\theta_0)-\alpha(\theta_1),0\}}).$$

Therefore, the above expectation goes to zero as $N\to\infty$, which implies (1).

To establish the assertion (2), first, consider the variance of each term in $l_N(\theta_1,\hat{\mu}_N(\theta_1))-l_N(\theta_0,\hat{\mu}_N(\theta_0))$. By Theorem 5.1 in Dahlhaus (1989), as $N\to\infty$

$$\mathrm{var}_{\theta_0}\left\{\frac{1}{2N}(\mathbf{Y}_N-\mu_0\mathbf{1})^\top(T_{\theta_1}^{-1}-T_{\theta_0}^{-1})(\mathbf{Y}_N-\mu_0\mathbf{1})\right\}=\frac{1}{2N^2}\,\mathrm{tr}\,(T_{\theta_1}^{-1}T_{\theta_0}-I)^2\to 0.$$

The same argument as that used in the proof of (1) shows that the limit of the other two variances are 0 when $N\to\infty$. Now the inequality $\mathrm{var}(x+y+z)\le[(\mathrm{var}x)^{1/2}+(\mathrm{var}y)^{1/2}+(\mathrm{var}z)^{1/2}]^2$ yields (2).

For (3), Dahlhaus (1989 Theorem 3.1) proved that the sum of the first two terms of the right hand side in (4) is bounded by the first term of the right hand side in (3). It remains to show that

$$\frac{1}{2N}\left|\frac{[\mathbf{1}^\top T_{\theta_2}^{-1}(\mathbf{Y}_N-\mu_0\mathbf{1})]^2}{\mathbf{1}^\top T_{\theta_2}^{-1}\mathbf{1}}-\frac{[\mathbf{1}^\top T_{\theta_1}^{-1}(\mathbf{Y}_N-\mu_0\mathbf{1})]^2}{\mathbf{1}^\top T_{\theta_1}^{-1}\mathbf{1}}\right|\le K\,|\theta_2-\theta_1|^{1/2}\frac{1}{N}\sum_{i=1}^N(y_i-\mu_0)^2.$$

Without loss of generality, let $\alpha(\theta_2)\le\alpha(\theta_1)$. Note that

$$\begin{aligned}&\frac{[\mathbf{1}^\top T_{\theta_2}^{-1}(\mathbf{Y}_N-\mu_0\mathbf{1})]^2}{\mathbf{1}^\top T_{\theta_2}^{-1}\mathbf{1}}-\frac{[\mathbf{1}^\top T_{\theta_1}^{-1}(\mathbf{Y}_N-\mu_0\mathbf{1})]^2}{\mathbf{1}^\top T_{\theta_1}^{-1}\mathbf{1}}\\
&\quad=\frac{[\mathbf{1}^\top T_{\theta_2}^{-1}(\mathbf{Y}_N-\mu_0\mathbf{1})]^2-[\mathbf{1}^\top T_{\theta_1}^{-1}(\mathbf{Y}_N-\mu_0\mathbf{1})]^2}{\mathbf{1}^\top T_{\theta_2}^{-1}\mathbf{1}}\\
&\quad\quad+\frac{[\mathbf{1}^\top T_{\theta_1}^{-1}(\mathbf{Y}_N-\mu_0\mathbf{1})]^2}{\mathbf{1}^\top T_{\theta_1}^{-1}\mathbf{1}}\,\frac{\mathbf{1}^\top T_{\theta_1}^{-1}\mathbf{1}-\mathbf{1}^\top T_{\theta_2}^{-1}\mathbf{1}}{\mathbf{1}^\top T_{\theta_2}^{-1}\mathbf{1}}=A+B.\end{aligned}$$

The term B is considered first. By the Cauchy-Schwarz inequality, since $T_{\theta_1}^{-1}\le KI_N$ uniformly in θ_1,

$$\frac{[\mathbf{1}^\top T_{\theta_1}^{-1}(\mathbf{Y}_N-\mu_0\mathbf{1})]^2}{\mathbf{1}^\top T_{\theta_1}^{-1}\mathbf{1}}\le K\sum_{i=1}^N(y_i-\mu_0)^2;$$

and hence, Lemma 1 in Appendix yields $|B| \leq K\,|\theta_2-\theta_1|\,\sum_{i=1}^N (y_i-\mu_0)^2$. For the term A,

$$\begin{aligned}&[\mathbf{1}^\top T_{\theta_2}^{-1}(\mathbf{Y}_N-\mu_0\mathbf{1})]^2-[\mathbf{1}^\top T_{\theta_1}^{-1}(\mathbf{Y}_N-\mu_0\mathbf{1})]^2\\&=[\mathbf{1}^\top T_{\theta_2}^{-1}(\mathbf{Y}_N-\mu_0\mathbf{1})-\mathbf{1}^\top T_{\theta_1}^{-1}(\mathbf{Y}_N-\mu_0\mathbf{1})][\mathbf{1}^\top T_{\theta_2}^{-1}(\mathbf{Y}_N-\mu_0\mathbf{1})+\mathbf{1}^\top T_{\theta_1}^{-1}(\mathbf{Y}_N-\mu_0\mathbf{1})].\end{aligned}$$

Applying the Cauchy-Schwarz inequality yields

$$\begin{aligned}&|\mathbf{1}^\top T_{\theta_2}^{-1}(\mathbf{Y}_N-\mu_0\mathbf{1})-\mathbf{1}^\top T_{\theta_1}^{-1}(\mathbf{Y}_N-\mu_0\mathbf{1})|\\&\leq[\mathbf{1}^\top(T_{\theta_2}^{-1/2}-T_{\theta_1}^{-1}T_{\theta_2}^{1/2})(T_{\theta_2}^{-1/2}-T_{\theta_2}^{1/2}T_{\theta_1}^{-1})\mathbf{1}]^{1/2}[(\mathbf{Y}_N-\mu_0\mathbf{1})^\top T_{\theta_2}^{-1}(\mathbf{Y}_N-\mu_0\mathbf{1})]^{1/2}\\&=[\mathbf{1}^\top(T_{\theta_2}^{-1}-T_{\theta_1}^{-1})\mathbf{1}+\mathbf{1}^\top(T_{\theta_1}^{-1}T_{\theta_2}T_{\theta_1}^{-1}-T_{\theta_1}^{-1})\mathbf{1}]^{1/2}[(\mathbf{Y}_N-\mu_0\mathbf{1})^\top T_{\theta_2}^{-1}(\mathbf{Y}_N-\mu_0\mathbf{1})]^{1/2}.\end{aligned}$$

Since $T_{\theta_2}\leq KT_{\theta_1}$ (assumption D5),

$$\left|\frac{\mathbf{1}^\top(T_{\theta_1}^{-1}T_{\theta_2}T_{\theta_1}^{-1}-T_{\theta_1}^{-1})\mathbf{1}}{\mathbf{1}^\top T_{\theta_2}^{-1}\mathbf{1}}\right|\leq K\sup_{\mathbf{Y}}\left|\frac{\mathbf{Y}^\top(T_{\theta_2}-T_{\theta_1})\mathbf{Y}}{\mathbf{Y}^\top T_{\theta_1}\mathbf{Y}}\right|\leq K|\theta_2-\theta_1|.$$

Therefore, Lemma 1 of Appendix implies

$$\frac{|\mathbf{1}^\top T_{\theta_2}^{-1}(\mathbf{Y}_N-\mu_0\mathbf{1})-\mathbf{1}^\top T_{\theta_1}^{-1}(\mathbf{Y}_N-\mu_0\mathbf{1})|}{(\mathbf{1}^\top T_{\theta_2}^{-1}\mathbf{1})^{1/2}}\leq\left[K\,|\theta_2-\theta_1|\sum_{i=1}^N(y_i-\mu_0)^2\right]^{1/2}.$$

It may be shown by similar arguments that

$$\frac{|\mathbf{1}^\top T_{\theta_2}^{-1}(\mathbf{Y}_N-\mu_0\mathbf{1})+\mathbf{1}^\top T_{\theta_1}^{-1}(\mathbf{Y}_N-\mu_0\mathbf{1})|}{(\mathbf{1}^\top T_{\theta_2}^{-1}\mathbf{1})^{1/2}}\leq K\left[\sum_{i=1}^N(y_i-\mu_0)^2\right]^{1/2}.$$

Consequently, $|A|\leq K|\theta_2-\theta_1|^{1/2}\sum_{i=1}^N(y_i-\mu_0)^2$ and (3) follows.

Next it will be proved, using some probability inequalities, that for any fixed $\theta_1\in\Theta-\{\theta_0\}$, there exists an $\eta>0$ such that for the $C(\theta_1)$ in (2)

$$\lim_{N\to\infty}\mathrm{P}_{\theta_0}\left\{\inf_{\theta\in U_\eta(\theta_1)}l_N(\theta,\hat\mu_N(\theta))-l_N(\theta_0,\hat\mu_N(\theta_0))\geq\frac{C(\theta_1)}{4}\right\}=1,\tag{5}$$

where $U_\eta(\theta_1)=\{\theta\in\Theta;|\theta-\theta_1|<\eta\}$.

In establishing (5), it is sufficient to show

$$\lim_{N\to\infty}\mathrm{P}_{\theta_0}\left\{l_N(\theta_1,\hat\mu_N(\theta_1))-l_N(\theta_0,\hat\mu_N(\theta_0))\geq\frac{C(\theta_1)}{2}\right\}=1,\tag{6}$$

and

$$\lim_{N\to\infty}\mathrm{P}_{\theta_0}\left\{\sup_{\theta\in U_\eta(\theta_1)}|l_N(\theta,\hat\mu_N(\theta))-l_N(\theta_1,\hat\mu_N(\theta_1))|<\frac{C(\theta_1)}{4}\right\}=1.\tag{7}$$

(6) follows immediately from (1) and (2). For (7), define

$$H_N(\mathbf{Y}_N,\eta) = K\,\eta\left\{1+\frac{1}{N}\sum_{i=1}^{N}(y_i-\mu_0)^2\right\}+K\,\eta^{1/2}\frac{1}{N}\sum_{i=1}^{N}(y_i-\mu_0)^2.$$

It is easy to see that $\lim_{\eta\to 0}\ \mathrm{E}_{\theta_0}H_N(\mathbf{Y}_N,\eta) = 0$ uniformly in N, and $\lim_{N\to\infty}\ \mathrm{var}_{\theta_0}H_N(\mathbf{Y}_N,\eta) = 0$. Since $C(\theta_1)>0$, there exists an $\eta>0$ such that $\mathrm{E}_{\theta_0}H_N(\mathbf{Y}_N,\eta)\le C(\theta_1)/8$. Then, (7) follows from (3).

Since the collection of sets $\{U_\eta(\theta_1):\theta_1\in\Theta-\{\theta_0\}\}\bigcup\{U_{\eta_0}(\theta_0):\eta_0>0\}$ constitutes an open covering of Θ and Θ is a compact set, there exist finite $\theta_j\in\Theta, j=0,1,\cdots,k$ such that $\{U_\eta(\theta_j):\theta_j\in\Theta\}$ covers Θ. Then, using (5), as $N\to\infty$

$$\mathrm{P}_{\theta_0}\left\{\inf_{\theta\in\bigcup_{j=1}^{k}U_\eta(\theta_j)} l_N(\theta,\hat{\mu}_N(\theta))-l_N(\theta_0,\hat{\mu}_N(\theta_0))\ge\min_{j=1,\cdots,k}\left[\frac{C(\theta_j)}{4}\right]\right\}\to 1.$$

Therefore

$$\lim_{N\to\infty}\mathrm{P}_{\theta_0}\left\{\inf_{\theta\in\Theta}l_N(\theta,\hat{\mu}_N(\theta))=\inf_{\theta\in U_{\eta_0}(\theta_0)}l_N(\theta,\hat{\mu}_N(\theta))\right\}=1,$$

which inplies $\lim_{N\to\infty}\mathrm{P}_{\theta_0}\{|\hat{\theta}_N-\theta_0|<\eta_0\}=1$, and since η_0 is an arbitrary constant, $\hat{\theta}_N\xrightarrow{p}\theta_0$.

The following theorem establishes the asymptotic normality of $\hat{\theta}_N$.

Theorem 2 *Suppose (D0) – (D5) hold, then* $\sqrt{N}(\hat{\theta}_N-\theta_0)\xrightarrow{D}N(0,\Gamma(\theta_0)^{-1})$, *where*

$$\Gamma(\theta_0)=\left\{\frac{1}{4\pi}\int_{-\pi}^{\pi}\left[\frac{\partial}{\partial\theta}\log f_{\theta_0}(x)\right]\left[\frac{\partial}{\partial\theta}\log f_{\theta_0}(x)\right]^{\top}dx\right\}_{m\times m}$$

Proof: By the mean value theorem,

$$\frac{\partial}{\partial\theta}l_N(\hat{\theta}_N,\hat{\mu}_N(\hat{\theta}_N))-\frac{\partial}{\partial\theta}l_N(\theta_0,\hat{\mu}_N(\theta_0))=\left[\frac{\partial^2}{\partial\theta\partial\theta^{\top}}l_N(\theta_N^*,\hat{\mu}_N(\theta_N^*))\right](\hat{\theta}_N-\theta_0),$$

where $|\theta_N^*-\theta_0|\le|\hat{\theta}_N-\theta_0|$. Since θ_0 is in the interior of Θ, Theorem 1 implies that $\partial/(\partial\theta)l_N(\hat{\theta}_N,\hat{\mu}_N(\hat{\theta}_N))=0$ for N large. Therefore when N is large

$$\frac{\partial}{\partial\theta}l_N(\theta_0,\hat{\mu}_N(\theta_0))=\left[-\frac{\partial^2}{\partial\theta\partial\theta^{\top}}l_N(\theta_N^*,\hat{\mu}_N(\theta_N^*))\right](\hat{\theta}_N-\theta_0).$$

Now the proof of this theorem follows from the verification of the following three statements,

(a) $\sqrt{N}\frac{\partial}{\partial\theta}l_N(\theta_0,\hat{\mu}_N(\theta_0))\xrightarrow{D}N(0,\Gamma(\theta_0))$.

(b) $\left|\frac{\partial^2}{\partial\theta\partial\theta^\top} l_N(\theta_N^*, \hat{\mu}_N(\theta_N^*)) - \frac{\partial^2}{\partial\theta\partial\theta^\top} l_N(\theta_0, \hat{\mu}_N(\theta_0))\right| \xrightarrow{p} 0.$

(c) $\frac{\partial^2}{\partial\theta\partial\theta^\top} l_N(\theta_0, \hat{\mu}_N(\theta_0)) \xrightarrow{p} \Gamma(\theta_0).$

(a) The derivative of $l_N(\theta, \hat{\mu}_N(\theta))$ with respect to θ is

$$\frac{\partial}{\partial\theta} l_N(\theta_0, \hat{\mu}_N(\theta_0)) = \frac{1}{2N}\,\mathrm{tr}\,(T_{\theta_0}^{-1} T_{\partial,\theta_0}) - \frac{1}{2N}(\mathbf{Y}_N - \mu_0\mathbf{1})^\top A_{\theta_0}^{(3)}(\mathbf{Y}_N - \mu_0\mathbf{1})$$
$$+\frac{1}{N}\frac{[\mathbf{1}^\top T_{\theta_0}^{-1}(\mathbf{Y}_N - \mu_0\mathbf{1})][\mathbf{1}^\top A_{\theta_0}^{(3)}(\mathbf{Y}_N - \mu_0\mathbf{1})]}{(\mathbf{1}^\top T_{\theta_0}^{-1}\mathbf{1})} - \frac{1}{2N}\frac{[\mathbf{1}^\top T_{\theta_0}^{-1}(\mathbf{Y}_N - \mu_0\mathbf{1})]^2(\mathbf{1}^\top A_{\theta_0}^{(3)}\mathbf{1})}{(\mathbf{1}^\top T_{\theta_0}^{-1}\mathbf{1})^2}.$$

Multiplying with $\sqrt{N}$, since the first difference converges in distribution to $N(0, \Gamma(\theta_0))$ (Dahlhaus 1989 Theorem 3.2 (ii)), it suffices to prove that the two remaining terms tend to zero in probability.

For the third term, Lemma 2 in Appendix implies that, for $j = 1, \cdots, m$

$$\left|\mathrm{E}_{\theta_0}\left\{\sqrt{N}\frac{1}{N}\frac{[\mathbf{1}^\top T_{\theta_0}^{-1}(\mathbf{Y}_N - \mu_0\mathbf{1})][\mathbf{1}^\top A_{\theta_0}^{(3)}(\mathbf{Y}_N - \mu_0\mathbf{1})]_j}{(\mathbf{1}^\top T_{\theta_0}^{-1}\mathbf{1})}\right\}\right|$$
$$\leq \frac{1}{\sqrt{N}} \parallel T_{\theta_0}^{-1/2}(T_{\partial,\theta_0})_j T_{\theta_0}^{-1/2} \parallel \leq \frac{1}{\sqrt{N}} K N^\tau \to 0 \qquad \text{as } N \to \infty,$$

using Lemma 5.4 in Dahlhaus (1989). Similarly,

$$\mathrm{var}_{\theta_0}\left\{\sqrt{N}\frac{1}{N}\frac{[\mathbf{1}^\top T_{\theta_0}^{-1}(\mathbf{Y}_N - \mu_0\mathbf{1})][\mathbf{1}^\top A_{\theta_0}^{(3)}(\mathbf{Y}_N - \mu_0\mathbf{1})]_j}{(\mathbf{1}^\top T_{\theta_0}^{-1}\mathbf{1})}\right\}$$
$$= \frac{1}{N}\frac{1}{(\mathbf{1}^\top T_{\theta_0}^{-1}\mathbf{1})^2}\left[(\mathbf{1}^\top T_{\theta_0}^{-1}(T_{\partial,\theta_0})_j T_{\theta_0}^{-1}\mathbf{1})^2 + (\mathbf{1}^\top T_{\theta_0}^{-1}\mathbf{1})(\mathbf{1}^\top T_{\theta_0}^{-1}(T_{\partial,\theta_0})_j T_{\theta_0}^{-1}(T_{\partial,\theta_0})_j T_{\theta_0}^{-1}\mathbf{1})\right]$$
$$\leq \frac{K}{N} N^{2\tau} \to 0 \qquad \text{as } N \to \infty.$$

Therefore,

$$\frac{1}{\sqrt{N}}\frac{[\mathbf{1}^\top T_{\theta_0}^{-1}(\mathbf{Y}_N - \mu_0\mathbf{1})][\mathbf{1}^\top A_{\theta_0}^{(3)}(\mathbf{Y}_N - \mu_0\mathbf{1})]}{(\mathbf{1}^\top T_{\theta_0}^{-1}\mathbf{1})} = o_p(1).$$

Similarly, it can be proved that the last term in $\partial/\partial\theta l_N(\theta_0, \hat{\mu}_N(\theta_0))$ is $o_p(1)$.

(b) The second order derivative of $l_N(\theta, \hat{\mu}_N(\theta))$ with respect to θ is

$$\frac{\partial^2}{\partial\theta\partial\theta^\top} l_N(\theta_0, \hat{\mu}_N(\theta_0)) = \Bigg[-\frac{1}{2N}\,\mathrm{tr}\,(A_{\theta_0}^{(3)} T_{\partial,\theta_0}) + \frac{1}{2N}\,\mathrm{tr}\,(T_{\theta_0}^{-1} T_{\partial^2,\theta_0})$$
$$+\frac{1}{N}(\mathbf{Y}_N - \mu_0\mathbf{1})^\top A_{\theta_0}^{(1)}(\mathbf{Y}_N - \mu_0\mathbf{1}) - \frac{1}{2N}(\mathbf{Y}_N - \mu_0\mathbf{1})^\top A_{\theta_0}^{(2)}(\mathbf{Y}_N - \mu_0\mathbf{1})\Bigg]$$

$$-\frac{1}{N}\frac{[\mathbf{1}^\top A^{(3)}_{\theta_0}(\mathbf{Y}_N-\mu_0\mathbf{1})]^2}{(\mathbf{1}^\top T^{-1}_{\theta_0}\mathbf{1})}-\frac{2}{N}\frac{[\mathbf{1}^\top T^{-1}_{\theta_0}(\mathbf{Y}_N-\mu_0\mathbf{1})][\mathbf{1}^\top A^{(1)}_{\theta_0}(\mathbf{Y}_N-\mu_0\mathbf{1})]}{(\mathbf{1}^\top T^{-1}_{\theta_0}\mathbf{1})}$$

$$+\frac{1}{N}\frac{[\mathbf{1}^\top T^{-1}_{\theta_0}(\mathbf{Y}_N-\mu_0\mathbf{1})][\mathbf{1}^\top A^{(2)}_{\theta_0}(\mathbf{Y}_N-\mu_0\mathbf{1})]}{(\mathbf{1}^\top T^{-1}_{\theta_0}\mathbf{1})} \tag{8}$$

$$+\frac{2}{N}\frac{[\mathbf{1}^\top T^{-1}_{\theta_0}(\mathbf{Y}_N-\mu_0\mathbf{1})][\mathbf{1}^\top A^{(3)}_{\theta_0}(\mathbf{Y}_N-\mu_0\mathbf{1})][\mathbf{1}^\top A^{(3)}_{\theta_0}\mathbf{1}]}{(\mathbf{1}^\top T^{-1}_{\theta_0}\mathbf{1})^2}$$

$$+\frac{1}{N}\frac{[\mathbf{1}^\top T^{-1}_{\theta_0}(\mathbf{Y}_N-\mu_0\mathbf{1})]^2(\mathbf{1}^\top A^{(1)}_{\theta_0}\mathbf{1})}{(\mathbf{1}^\top T^{-1}_{\theta_0}\mathbf{1})^2}-\frac{1}{2N}\frac{[\mathbf{1}^\top T^{-1}_{\theta_0}(\mathbf{Y}_N-\mu_0\mathbf{1})]^2(\mathbf{1}^\top A^{(2)}_{\theta_0}\mathbf{1})}{(\mathbf{1}^\top T^{-1}_{\theta_0}\mathbf{1})^2}$$

$$-\frac{1}{N}\frac{[\mathbf{1}^\top T^{-1}_{\theta_0}(\mathbf{Y}_N-\mu_0\mathbf{1})]^2[\mathbf{1}^\top A^{(3)}_{\theta_0}\mathbf{1}]^2}{(\mathbf{1}^\top T^{-1}_{\theta_0}\mathbf{1})^3}.$$

Dahlhaus (1989) Theorem 3.2 (iv) showed that the first four differences in $\partial^2/(\partial\theta\partial\theta^\top)l_N(\theta_N^*,\hat\mu_N(\theta_N^*))$ - $\partial^2/(\partial\theta\partial\theta^\top)l_N(\theta_0,\hat\mu_N(\theta_0))$ converge to zero in probability. The convergence of the remaining 7 differences to zero follows from equicontinuity of the term involved, denoted by $Q_N^{(i)}(\theta)$ $i=1,...,7$, which is proved in Lemma 9 of Appendix .

(c) In order to establish (c), it is sufficient to show that the expectation and variance of $\partial^2/(\partial\theta\partial\theta^\top)l_N(\theta_0,\hat\mu_N(\theta_0))$ converge to $\Gamma(\theta_0)$ and 0, respectively as N goes to infinity. Lemma 2 in Appendix implies that $E_{\theta_0}\partial^2/(\partial\theta\partial\theta^\top)l_N(\theta_0,\hat\mu_N(\theta_0))$ is

$$\frac{1}{2N}\operatorname{tr}\left(A^{(3)}_{\theta_0}T^\top_{\partial\theta_0}\right)-\frac{2}{N}\frac{\mathbf{1}^\top A^{(1)}_{\theta_0}\mathbf{1}}{\mathbf{1}^\top T^{-1}_{\theta_0}\mathbf{1}}+\frac{1}{2N}\frac{\mathbf{1}^\top A^{(2)}_{\theta_0}\mathbf{1}}{\mathbf{1}^\top T^{-1}_{\theta_0}\mathbf{1}}+\frac{1}{N}\frac{(\mathbf{1}^\top A^{(3)}_{\theta_0}\mathbf{1})^2}{\mathbf{1}^\top T^{-1}_{\theta_0}\mathbf{1}}.$$

Theorem 5.1 in Dahlhaus (1989) yields that the first term converges to $\Gamma(\theta_0)$ as N goes to infinity. Lemma 6 in Appendix implies that, as $N\to\infty$, the limit of the last three terms is zero. The variance of the (j,k)th element of the matrix $\partial^2/(\partial\theta\partial\theta^\top)l_N(\theta_0,\hat\mu_N(\theta_0))$, $j,k=1,\cdots,m$, is bounded. And the bound goes to 0 as $N\to\infty$, by Theorem 5.1 in Dahlhaus (1989) and Lemma 6 in Appendix, which completes the proof of Theorem 2.

4. Appendix

The following lemmas are useful in proving consistency and asymptotic normality of $\hat\theta_N$.

Lemma 1 *Suppose $(D1)-(D5)$ hold. If $\alpha(\theta_2)\le\alpha(\theta_1)$, then there exists a constant K independent of θ_1,θ_2 and N with $|(\mathbf{1}^\top T^{-1}_{\theta_1}\mathbf{1}-\mathbf{1}^\top T^{-1}_{\theta_2}\mathbf{1})/\mathbf{1}^\top T^{-1}_{\theta_2}\mathbf{1}|\le K|\theta_1-\theta_2|$.*

Following lemma extends the formulas for the moments of quadratic form with symmetric matrix to the case with an asymmetric matrix.

Lemma 2 *For any $n\times n$ matrix A and $n\times 1$ vector Y, if $Y\sim N(0,V), V>0$, then*

$$\mathrm{E}(Y^\top AY)=\operatorname{tr}(AV)\ \textit{and}\ \operatorname{var}(Y^\top AY)=\operatorname{tr}[(AV)^2+AVA^\top V].$$

Let C_1 and C_2 are $n\times 1$ vectors, then $\mathrm{E}[(C_1^\top Y)(C_2^\top Y)]=C_1^\top VC_2$ *and*

$$\mathrm{var}[(C_1^\top Y)(C_2^\top Y)]=(C_1^\top VC_2)^2+(C_1^\top VC_1)(C_2^\top VC_2).$$

It is well known that for an $n\times n$ matrix $D\geq 0$ and an $n\times 1$ vector Y, $Y^\top DY\leq Y^\top Y\parallel D\parallel$. Next lemma will extend this result to a more general case.

Lemma 3 *Let B be any $n\times n$ matrix, A an $n\times n$ symmetric matrix, and Y an $n\times 1$ vector. Then* $|Y^\top ABAY|\leq(Y^\top A^2Y)\parallel B\parallel$.

Lemma 4 *Suppose a, b are two $n\times 1$ vectors, then* $\|a\,b^\top\|=[(a^\top a)(b^\top b)]^{1/2}$.

The assertion (b) in the proof of Theorem 2 requires the equicontinuity property of some quadratic forms. Lemmas 5-9 establish the equicontinuity property.

For notational simplicity, let $A_\theta^{(13)}=T_\theta^{-1}T_{\partial,\theta}T_\theta^{-1}T_{\partial,\theta}T_\theta^{-1}T_{\partial,\theta}T_\theta^{-1}$, $A_\theta^{(23)}=T_\theta^{-1}T_{\partial^2,\theta}T_\theta^{-1}T_{\partial,\theta}T_\theta^{-1}$, and $A_\theta^{(4)}=T_\theta^{-1}T_{\partial^3,\theta}T_\theta^{-1}$.

Lemma 5 *Suppose $(D1)-(D4)$ hold. Then there exists a constant K independent of θ and N such that for each $\tau>0$ and $j,k,h=1,\cdots,m$*

(a) $\left|\frac{\mathbf{1}^\top B_\theta\ T_{\theta_0}(A_\theta^{(3)})_h\mathbf{1}}{\mathbf{1}^\top T_\theta^{-1}\mathbf{1}}\right|\leq KN^{\frac{1}{2}(1+\alpha(\theta_0))+\tau}$, *where* $B_\theta=(A_\theta^{(i)})_{jk}, i=1,2$ *and* $(A_\theta^{(3)})_j$.

(b) $\left|\frac{\mathbf{1}^\top T_\theta^{-1}T_{\theta_0}B_\theta\mathbf{1}}{\mathbf{1}^\top T_\theta^{-1}\mathbf{1}}\right|\leq KN^{\frac{1}{2}(1+\alpha(\theta_0))+\tau}$, *where* $B_\theta=(A_\theta^{(i)})_{jk}$, $(A_\theta^{(3)})_j$, $(A_\theta^{(i3)})_{jkh}$, $i=1,2$, $(A_\theta^{(4)})_{jkh}$ *and* T_θ^{-1}.

Lemma 6 *Suppose $(D1)-(D4)$ hold. Then there exists a constant K independent of θ and N such that for each $\tau>0$ and $j,k,h=1,\cdots,m$*

(a) $\left|\frac{\mathbf{1}^\top B_\theta\mathbf{1}}{\mathbf{1}^\top T_\theta^{-1}\mathbf{1}}\right|\leq KN^\tau$, (b) $\left|\frac{\mathbf{1}^\top B_\theta B_\theta\mathbf{1}}{\mathbf{1}^\top T_\theta^{-1}\mathbf{1}}\right|\leq KN^\tau$, (c) $\left\|\frac{T_\theta^{-1}\mathbf{1}\mathbf{1}^\top B_\theta}{\mathbf{1}^\top T_\theta^{-1}\mathbf{1}}\right\|\leq KN^\tau$,

where $B_\theta=(A_\theta^{(i)})_{jk}, i=1,2$, $(A_\theta^{(3)})_j$, $(A_\theta^{(i3)})_{jkh}, i=1,2$, *and* $(A_\theta^{(4)})_{jkh}$.

(d) $\left\|\frac{B_\theta\mathbf{1}\mathbf{1}^\top(A_\theta^{(3)})_h}{\mathbf{1}^\top T_\theta^{-1}\mathbf{1}}\right\|\leq KN^\tau$, *where* $B_\theta=(A_\theta^{(i)})_{jk}, i=1,2$, *and* $(A_\theta^{(3)})_j$.

Lemma 7 *Suppose $(D1)-(D4)$ hold. Then there exists a constant K independent of θ and N such that for each $\tau>0$, the following three differences are bounded by* $K|\theta_1-\theta_2|N^{\frac{1}{2}(1+\alpha(\theta_0))+\tau}$.

(a) $\left|\frac{\mathbf{1}^\top A_{\theta_1}^{(3)}T_{\theta_0}A_{\theta_1}^{(3)}\mathbf{1}}{\mathbf{1}^\top T_{\theta_1}^{-1}\mathbf{1}}-\frac{\mathbf{1}^\top A_{\theta_2}^{(3)}T_{\theta_0}A_{\theta_2}^{(3)}\mathbf{1}}{\mathbf{1}^\top T_{\theta_2}^{-1}\mathbf{1}}\right|$, (b) $\left|\frac{\mathbf{1}^\top T_{\theta_1}^{-1}T_{\theta_0}T_{\theta_1}^{-1}\mathbf{1}}{\mathbf{1}^\top T_{\theta_1}^{-1}s\mathbf{1}}-\frac{\mathbf{1}^\top T_{\theta_2}^{-1}T_{\theta_0}T_{\theta_2}^{-1}\mathbf{1}}{\mathbf{1}^\top T_{\theta_2}^{-1}\mathbf{1}}\right|$ and

(c) $\left|\frac{\mathbf{1}^\top T_{\theta_1}^{-1}T_{\theta_0}A_{\theta_1}^{(i)}\mathbf{1}}{\mathbf{1}^\top T_{\theta_1}^{-1}\mathbf{1}}-\frac{\mathbf{1}^\top T_{\theta_2}^{-1}T_{\theta_0}A_{\theta_2}^{(i)}\mathbf{1}}{\mathbf{1}^\top T_{\theta_2}^{-1}\mathbf{1}}\right|$, *where* $i=1,2$ *and* 3.

Lemma 8 *Suppose (D1) – (D4) hold. Then there exists a constant K independent of θ and N such that for each $\tau > 0$, and $j, k = 1, \cdots, m$,*

(a) $\left\| \frac{(A_{\theta_1}^{(3)})_j \mathbf{1}\mathbf{1}^\top (A_{\theta_1}^{(3)})_k}{\mathbf{1}^\top T_{\theta_1}^{-1}\mathbf{1}} - \frac{(A_{\theta_2}^{(3)})_j \mathbf{1}\mathbf{1}^\top (A_{\theta_2}^{(3)})_k}{\mathbf{1}^\top T_{\theta_2}^{-1}\mathbf{1}} \right\| \le K|\theta_1 - \theta_2| N^\tau.$

(b) $\left\| \frac{T_{\theta_1}^{-1} \mathbf{1}\mathbf{1}^\top B_{\theta_1}}{\mathbf{1}^\top T_{\theta_1}^{-1}\mathbf{1}} - \frac{T_{\theta_2}^{-1} \mathbf{1}\mathbf{1}^\top B_{\theta_2}}{\mathbf{1}^\top T_{\theta_2}^{-1}\mathbf{1}} \right\| \le K|\theta_1 - \theta_2| N^\tau$, *where* $B_\theta = (A_{\theta_1}^{(i)})_{jk}$, $i = 1, 2$, $(A_\theta^{(3)})_j$ *and* T_θ^{-1}.

(c) $\left| \frac{\mathbf{1}^\top B_{\theta_1}\mathbf{1}}{\mathbf{1}^\top T_{\theta_1}^{-1}\mathbf{1}} - \frac{\mathbf{1}^\top B_{\theta_2}\mathbf{1}}{\mathbf{1}^\top T_{\theta_2}^{-1}\mathbf{1}} \right| \le K|\theta_1 - \theta_2| N^\tau$, *where* $B_\theta = (A_{\theta_1}^{(i)})_{jk}$, $i = 1, 2$, *and* $(A_\theta^{(3)})_j$.

Next lemma shows that the quadratic forms $Q_N^{(i)}(\theta)$, $i = 1, \cdots, 7$, defined in (8), are equicontinuous. The proofs are closely related to the approach of Pollard (1984 Chapter VII) and Dahlhaus (1989 Section 6).

Lemma 9 *Suppose (D1) – (D4) hold. Then $Q_N^{(i)}(\theta)$, $i = 1, \cdots, 7$, are equicontinuous in probability,* i.e. *for each $\eta > 0$ and $\epsilon > 0$, there exists a $\xi > 0$ such that*

$$\limsup_{N\to\infty} \mathrm{P}_{\theta_0} \left(\sup_{|\theta_1 - \theta_2| \le \xi} \left| Q_N^{(i)}(\theta_1) - Q_N^{(i)}(\theta_2) \right| > \eta \right) < \epsilon.$$

5. Summary

The fractional ARIMA models have been widely applied to problems in many fields. This article concentrates on the asymptotic properties of the maximum likelihood estimator of a fractional ARIMA process. If the mean and time series parameter of the process are estimated simultaneously by maximizing the full likelihood function, then the estimated time series parameters are consistent and asymptotically normally distributed.

6. References

Anděl, J., 1986: Long-memory time series models. *Kybernetika* **22**, 105-123.

Barndorff-Nielsen, O. E., 1990: Likelihood theory. In *Statistical theory and modelling: in honour of Sir David Cox, FRS,* Ed. D. V. Hinkley, N. Reid and E. J. Snell, Ch. 10, pp. 232-264, Chapman and Hall.

Beran, J., 1986: Estimation, testing and prediction for self-similar and related processes. Dissertation, ETH Zürich.

Beran, J., 1991: Statistical methods for data with long-range dependence. Technical Report No. 116, Department of statistics, Texas A & M University.

Bloomfield, P., 1985: On series representations for linear predictors. *Ann. Probab.* **13**, 226-233.

Bloomfield, P., 1992a: Statistical Estimation of Climatic trends. In *The Art of Statistical Science: in Tribute to G. S. Watson,* Ed. K. V. Mardia, Ch. 20, pp. 299-311, John Wiley & Sons.

Bloomfield, P., 1992b: Trends in Global Temperature. *Climatic change* **21**, 1-16.

Dahlhaus, R., 1989: Efficient parameter estimation for self-similar processes. *Ann. Statis.* **17**, 1749-1766.

Fox, R. and M. S. Taqqu, 1986: Large sample properties of parameter estimates for strongly dependent stationary Gaussian time series. *Ann. Statis.* **14**, 517-532.

Granger, C. W. J. and R. Joyeux, 1980: An introduction to long-range time series models and fractional defferencing. *J. Time Series Anal.* **1**, 15-30.

Hampel, F. R., 1987: Data analysis and self-similar processes. *Proceedings of the 46th Session of ISI, Tokyo,* **Book 4**, 235-254.

Haslett, J. and A. E. Raftery, 1989: Space-time modelling with long-memory dependence: Assessing Ireland's wind power resource. Invited paper with discussion. *J. Appl. Stat.* **38**, 1-21.

Hosking, J. R. M., 1981: Fractional differencing. *Biometrika* **68**, 165-176.

Lawrance, A. J. and N. T. Kotegoda, 1977: Stochastic modelling of riverflow time series (with Discussion). *J. Royal Statistical Society* A **140**, 1-47.

McLeod, A. I. and K. W. Hipel, 1978: Preservation of the resealed adjusted range. *Water Resources Res.* **14**, 491-518.

Poeter-Hudak, S., 1990: An application of the seasonal fractionally differenced model to the monetary aggregates. *J. Amer. Statist. Assoc.* **85**, 338-344.

Pollard, D., 1984: *Convergence of Stochastic Processes.* Springer, New York.

Smith, R. L., 1992: Long-range dependence and global warming. Department of Statistics, University of North Carolina, Chapel Hill, N.C.

Sokolov, I. M., J. Klafter, and A. Blumen, 2002: Fractional kinetics. *Physics Today,* **55**(11), 48-54.

Walker, A. M., 1964: Asymptotic properties of least-squares estimates of parameters of the spectrum of a stationary non-deterministic time-series. *J. Austral. Math. Soc.* **4**, 363-384.

Wilks, D. S., 1995: *Statistical Methods in the Atmospheric Sciences: An Introduction.* Academic Press, New York, 464 pp.

ERROR ANALYSIS FOR SCRAMBLED QUASI-MONTE CARLO QUADRATURE

RONG-XIAN YUE
Department of Applied Mathematics, Shanghai Normal University
100 Guilin Road, Shanghai 200234, China
E-mail: yue2@shtu.edu.cn

(Manuscript received 7 July 2002)

Scrambled quasi-Monte Carlo quadrature proposed by Owen (1995) is a hybrid of Monte Carlo and quasi-Monte Carlo methods, which combines the best of these two methods for integration. This article reports some of main results on the scrambled quadrature.

1. Introduction

In many practical problems arising in statistics, finance, physics and engineering, one requires the evaluation of an integral, which, after suitable transformation, may be written as

$$I(f) = \int_{[0,1)^s} f(\boldsymbol{x})d\boldsymbol{x}.$$

When s is small and f is smooth, classical techniques such as those described in Davis and Rabinowitz (1984) can achieve great accuracy with a small number of function evaluations. However, when s is large the integrals are often estimated by Monte Carlo and quasi-Monte Carlo methods by the quadrature rules of the form

$$\hat{I}_n(f;P) = \frac{1}{n}\sum_{i=1}^{n} f(\boldsymbol{x}_i)$$

for some point set $P = \{\boldsymbol{x}_1, \ldots, \boldsymbol{x}_n\}$ which is carefully chosen from $[0,1)^s$.

Simple Monte Carlo method uses n points which are independently drawn from the uniform distribution $[0,1)^s$. It is known that simple Monte Carlo method yields an unbiased estimate $\hat{I}_n$ of the integral I, and the variance of the estimate is σ^2/n where $\sigma^2 = \int[f(\boldsymbol{x}) - I]^2 d\boldsymbol{x}$ provided that $f \in \mathcal{L}^2[0,1)^s$. Hence, the error of simple Monte Carlo integration is of order $n^{-1/2}$ in probability. This is due to gaps and clusters that arise by chance among the points $\boldsymbol{x}_i$.

Quasi-Monte Carlo methods use deterministic sequences of n points which are uniformly scattered in $[0,1)^s$ to avoid, to the extent possible, gaps and clusters. For details on quasi-Monte Carlo methods see Hua and Wang (1981), Niederreiter (1992), Fang and Wang (1994), and Sloan and Joe (1994). Quasi-Monte carlo integration is usually considered to be more accurate than Monte Carlo integration, but for Monte Carlo integration it can be much easier to estimate accuracy. To quantify the uniformity of a list of n points in $[0,1)^s$, one uses a distance between the continuous uniform distribution on $[0,1)^s$ and the discrete uniform distribution taking $\boldsymbol{x}_i$ with probability $1/n$ for $i = 1, \cdots, n$. The most widely studied distance measure is the star discrepancy defined as

$$D_n^*(\boldsymbol{x}_1, \cdots, \boldsymbol{x}_n) = \sup_{0 \le c_r < 1} \left| \prod_{r=1}^{s} c_r - \frac{1}{n} \sum_{i=1}^{n} \prod_{r=1}^{s} 1_{0 \le c_r < 1} \right|.$$

Star discrepancy is related to integration accuracy by the following Koksma-Hlawka inequality

$$|\hat{I}_n - I| \le D_n^*(\boldsymbol{x}_1, \cdots, \boldsymbol{x}_n) V_{\mathrm{HK}}(f),$$

where $V_{\mathrm{HK}}(f)$ is the total variation of f in the sense of Hardy and Krause. It is possible to construct an infinite sequence of points $\boldsymbol{x}_1, \boldsymbol{x}_2, \cdots$ that one can achieve an along which $D_n^* = O(n^{-1}(\log n)^s)$. This proves that one can achieve an asymptotic rate better than that of Monte Carlo estimate, at least for integrants with bounded variation, i.e., $V_{\mathrm{HK}}(f) < \infty$.

A drawback with quasi-Monte Carlo methods is that there is no practical way to estimate the size of $\hat{I}_n - I$ from the function evaluations $f(\boldsymbol{x}_i)$. Estimating the variation $V_{\mathrm{HK}}(f)$ from data appears to be extremely difficult.

To combine the best of Monte Carlo and quasi-Monte Carlo, Owen (1995) proposed a hybrid of these two techniques based on randomly scrambling the digits of the points in a (t, m, s)-net or (t, s)-sequence in base b, which is called a scrambled quadrature rule. The resulting method provides unbiased estimates of $I(f)$.

This article reviews some main results on the scrambled quadrature rules. Section 2 provides definitions of (t, m, s)-nets and (t, s)-sequence in base b and their randomization. Section 3 reports the results on the variance of the estimates for any integrands satisfying certain smooth conditions. Section 4 reports the results on the worst-case error of the quadrature. For proofs the reader is referred to the references cited.

2. Background

2.1. (t, m, s)-nets and (t, s)-sequences

Here we briefly review the quasi-Monte Carlo methods known as (t, m, s)-nets and (t, s)-sequences. See Niederreiter (1992) for details.

Definition 1. Let t and m be nonnegative integers. A finite sequence $\{\boldsymbol{x}_i\}_{i=1}^n$ of points in $[0,1)^s$ with $n = b^m$ is a *(t, m, s)-net in base b* if every b-box of the form $\prod_{j=1}^d [\ell_j / b^{k_j}, (\ell_j + 1)/b^{k_j})$, $k_j \ge 0$, $0 \le \ell_j < b^{k_j}$, of volume b^{t-m} contains exactly b^t points of the sequence.

Clearly, smaller values of t imply stronger equidistribution properties of the net. When $t = 0$, every elementary interval of volume $1/n$ contains one of the n points in the sequence.

Definition 2. An infinite sequence $\{\boldsymbol{x}_i\}_{i\geq 1}$ of points in $[0,1)^s$ is a *(t,s)-sequence in base b* if for all integers $k \geq 0$ and $m \geq t$ the finite sequence $\{\boldsymbol{x}_i\}_{i=kb^m+1}^{(k+1)b^m}$ is a (t,m,s)-net in base b.

An advantage of using nets taken from (t,s)-sequences is that one can increase n through a sequence of values $n = \lambda b^m$, $1 \leq \lambda < b$, so that all of the points used in $\hat{I}_{\lambda b^m}$ are also used in $\hat{I}_{(\lambda+1)b^m}$. Note that the initial λb^m points of a (t,s)-sequence are well equidistributed but are not ordinarily a (t,m,s)-net. Owen (1997a) introduces the following definition to describe such point sequences.

Definition 3. Let s, m, t, b, λ be integers with $s \geq 1$, $m \geq 0$, $0 \leq t \leq m$, $b \geq 2$ and $1 \leq \lambda < b$. A finite sequence $\{\boldsymbol{x}_i\}_{i=1}^n$ of points in $[0,1)^s$ with $n = \lambda b^m$ is called a *(λ, t, m, s)-net in base b* if every b-box of volume b^{t-m} contains λb^t points of the sequence and no b-box of volume b^{t-m-1} contains more than b^t points of the sequence.

Numerical integration by averaging over the points of a (t,m,s)-net has an error $|\hat{I}_n - I| = O(n^{-1}(\log n)^{s-1})$, for integrands of bounded variation in the sense of Hardy and Krause. See Niederreiter (1992) for this result and some sharper versions of it.

2.2. Base b scrambling of the unit cube

The randomization scheme proposed by Owen (1995) can be briefly described as follows:

Suppose that $\{\boldsymbol{a}_i\}_{i=1}^n$ is a (t,m,s)-net in base b. Write the components of $\boldsymbol{a}_i$ as $a_i^r = \sum_{k=1}^{\infty} a_{irk} b^{-k}$. For $i = 1, \ldots, n$, let $\boldsymbol{x}_i = (x_i^1, \ldots, x_i^s)$ with $x_i^r = \sum_{k=1}^{\infty} x_{irk} b^{-k}$, where x_{irk} is a random permutation applied to a_{irk}. The $\boldsymbol{x}_i$'s satisfy the following rules:

(1) Each digit x_{irk} is uniformly distributed on the set $\{0, 1, \ldots, b-1\}$;

(2) For any two points $\boldsymbol{x}_i$ and $\boldsymbol{x}_j$ the s pairs $(x_i^1, x_j^1), \ldots, (x_i^s, x_j^s)$ are mutually independent;

(3) If a_i^r and a_j^r share the same first k digits, but their $k+1^{\text{st}}$ digits are different, then

 (a) $x_{irh} = x_{jrh}$ for $h = 1, \ldots, k$;

 (b) the pair $(x_{ir\,k+1}, x_{jr\,k+1})$ is uniformly distributed on the set $\{(d_i, d_j) : d_i \neq d_j;\ d_i, d_j \in \{0, 1, \ldots, b-1\}\}$ and

 (c) $x_{ir\,k+2}, x_{ir\,k+3}, \cdots, x_{jr\,k+2}, x_{jr\,k+3}, \cdots$ are mutually independent.

We call this a base b scrambling scheme and call the sequence $\{\boldsymbol{x}_i\}_{i=1}^n$ a scrambled version of $\{\boldsymbol{a}_i\}_{i=1}^n$.

The following geometrical description of this scheme may help us visualize the randomization:

Begin by partitioning the unit cube $[0,1)^s$ along the x^1 axis into b parallel b-boxes of the form $[\ell/b, (\ell+1)/b) \times [0,1)^{s-1}$ for $\ell = 0, \ldots, b-1$. Then randomly permute those b-boxes replacing them in one of the $b!$ possible orders, each such order having probability $1/b!$. Next take each such b-box in turn, partition it into b congruent b-boxes of volume b^{-2} along the x^1 axis, and randomly permute those b-boxes. Then repeat this process on b^2 b-boxes of volume b^{-3}, b^3 b-boxes of volume b^4 and so, ad infinitum. In practice this can stop when the b-boxes are narrow compared to machine precision. The full scrambling involves applying the above operations along the other $s-1$ axes $x^2, \cdots, x^s$ as well. All of the many permutations used are to be statistically independent.

Owen(1995, 1997a) proved the following two propositions:

Proposition 1. *If $\{\boldsymbol{a}_i\}$ is a (λ, t, m, s)-net in base b, then the scrambled version $\{\boldsymbol{x}_i\}$ is a (λ, t, m, s)-net in base b with probability* 1.

Proposition 2. *Let $\boldsymbol{a}$ be a point in $[0,1)^s$ and $\boldsymbol{x}$ be the scrambled version of $\boldsymbol{a}$ as described above. Then $\boldsymbol{x}$ has the uniform distribution on $[0,1)^s$.*

3. Variance of Scrambled Quadrature

The variance of the estimate $\hat{I}_n(f;P)$ is defined by

$$\mathrm{var}[\hat{I}_n(f;P)] = E[\hat{I}_n(f;P) - I(f)]^2,$$

where the expectation is taken respect to the scrambling. Owen(1997a, 1997b, 1998) showed that the variance for every integrand in $L^2([0,1)^s)$ under scrambling tends to zero faster than under Monte Carlo sampling. Stronger assumption on the integrand allow one to find the order of the variance.

Theorem 1 (Owen, 1997b, 1998) *Assume that f is of bounded variation in the sense of Hardy and Krause. If $\boldsymbol{x}_1, \cdots, \boldsymbol{x}_n$ are the first n points of a scrambled (t,s)-sequence in base b, then*

$$var[\hat{I}_n(f,P)] = O(n^{-2}(\log n)^{2s}).$$

If $\boldsymbol{x}_1, \cdots, \boldsymbol{x}_n$ are the points of a scrambled (t,m,s)-net in base b, then

$$var[\hat{I}_n(f,P)] = O(n^{-2}(\log n)^{2(s-1)}),$$

and further, if $\partial^s f/\partial \boldsymbol{x}$ is Lipschitz continuous then

$$var[\hat{I}_n(f,P)] = O(n^{-3}(\log n)^{2(s-1)}).$$

Yue (1999) considered the variance of quadrature by scrambling a union of several nets, and found that adding some additional points in sampling may cause a large loss of efficiency. Yue and Mao (1999) further studied the case where integrands are of weaker smoothness than that in Owen (1997b, 1998). It is only required that the integrand satisfies a generalized Lipschitz condition described in Hua and Wang (1981).

Theorem 2 (Yue and Mao, 1999) *Assume that f satisfies a generalized Lipschitz condition on $[0,1)^s$. If $\boldsymbol{x}_1, \cdots, \boldsymbol{x}_n$ are points of a scrambled (t,m,s)-net in base b, then*

$$var[\hat{I}_n(f,P)] = O(n^{-3}(\log n)^{2(s-1)}).$$

If $\boldsymbol{x}_1, \cdots, \boldsymbol{x}_n$ are the first n points a scrambled (t,s)-sequence in base b, then

$$var[\hat{I}_n(f,P)] = O(n^{-2}(\log n)^{(s)}).$$

4. Worst-Case Error Bound

Suppose that the integrand lies in a reproducing kernel Hilbert space, $\mathcal{H}$, say. The reproducing kernel of $\mathcal{H}$ will be denoted by $K : [0,1)^s \times [0,1)^s \to \mathcal{R}$. For basic properties of reproducing kernel Hilbert spaces we refer to Aronszajn (1950), Saitoh (1988), and Wahba (1990). We only recall that $K(\cdot, \boldsymbol{x}) \in \mathcal{H}$ for all $\boldsymbol{x} \in [0,1)^s$, and

$$f(\boldsymbol{x}) = \langle f, K(\cdot, \boldsymbol{x})\rangle \quad \forall f \in \mathcal{H}, \;\; \forall \boldsymbol{x} \in [0,1)^s.$$

Here, $\langle \cdot, \cdot \rangle$ denotes the inner product of $\mathcal{H}$, and $\|f\| = \langle f, f\rangle^{1/2}$.

The worst-case error of the quadrature $\hat{I}_n(f,P)|$ is defined for integration as

$$e(P) = \sup_{f \in \mathcal{H}, \|f\| \le 1} |I(f) - \hat{I}_n(f,P)|.$$

Note that $I(f)$ is a bounded functional on $\mathcal{H}$, and then exists a representor $\xi \in \mathcal{H}$ such that $I(f) = \langle f, \xi \rangle$. Clearly,

$$I(f) - \hat{I}_n(f,P) = \langle f, \xi - \frac{1}{n}\sum_{i=1}^{n} K(\cdot, \boldsymbol{x}_i)\rangle.$$

Therefore,

$$e(P) = \|I(f) - \hat{I}_n(f,P)\| = \|\xi - \frac{1}{n}\sum_{i=1}^{n} K(\cdot, \boldsymbol{x}_i)\rangle.$$

It is customary to call

$$D(P;K) = \|\xi - \frac{1}{n}\sum_{i=1}^{n} K(\cdot, \boldsymbol{x}_i)\rangle.$$

a discrepancy of the point set P. A good set P for multidimensional quadrature is one that has a small discrepancy. Computing the last norm, we obtain an explicit form of the discrepancy

$$[D(P;K)]^2 = \int_{[0,1)^{2s}} K(\boldsymbol{x},\boldsymbol{y}) d\boldsymbol{x} d\boldsymbol{y} - \frac{2}{n}\sum_{i=1}^{n} \int_{[0,1)^s} K(\boldsymbol{x}_i, \boldsymbol{y}) d\boldsymbol{y} + \frac{1}{n^2}\sum_{i=1}^{n}\sum_{j=1}^{n} K(\boldsymbol{x}_i, \boldsymbol{x}_j).$$

Hickernell (1998) derived a family of discrepancy $D(P;K)$. A formula for the mean square L^2-type discrepancy of scrambled $(0,m,s)$-nets in base b is derived in Hickernell (1996). The asymptotic L^2-type discrepancy of scrambled (t,m,s)-nets based on smooth reproducing kernel is investigated in Hickernell and Hong (1999).

Hickernell and Yue (2000) considered the reproducing kernel of the following form

$$K_{\mathrm{sc}}(\boldsymbol{x},\boldsymbol{y}) = \sum_{u}\sum_{\kappa}\sum_{\tau}\sum_{\gamma} \Phi_{u\kappa} \psi_{u\kappa\tau\gamma}(\boldsymbol{x}) \psi_{u\kappa\tau\gamma}(\boldsymbol{y}).$$

This kernel is called a *scrambled-invariant kernel.* Here, the sum is taken over all subsets $u \subseteq \{1, \cdots, s\}$, over all $|u|$-vectors κ of nonnegative integers k_l, over all $|u|$-vectors τ of integers t_l with $0 \le t_l < b^{k_l}$, and over all $|u|$-vectors γ of integers c_l with $0 \le c_l < b$. The functions $\psi_{u\kappa\tau\gamma}$ are the dilated and translated wavelets defined by

$$\psi_{u\kappa\tau\gamma}(\boldsymbol{x}) = b^{(|\kappa|-|u|)/2} \prod_{l \in u} \left(b 1_{\lfloor b^{k_l+1} x_l \rfloor = bt_l + c_l} - 1_{\lfloor b^{k_l} x_l \rfloor = t_l} \right),$$

and the coefficients $\Phi_{u\kappa}$ are of the form

$$\Phi_{u\kappa} = \mu b^{-(\eta+1)|\kappa|}, \quad \mu, \eta > 0.$$

Theorem 3 (Hickernell and Yue, 2000) *If $\boldsymbol{x}_1, \cdots, \boldsymbol{x}_n$ is a (t, m, s)-net in base b, then*

$$[D(P; K_{sc})]^2 = O(n^{-\eta-1} (\log n)^{s-1}).$$

If $\boldsymbol{x}_1, \cdots, \boldsymbol{x}_n$ is the n points of a (t, s)-sequence in base b, then

$$[D(P; K_{sc})]^2 = \begin{cases} O(n^{-\eta-1} (\log n)^{s-1}), & \text{if } 0 < \eta < 1, \\ O(n^{-2} (\log n)^{s}), & \text{if } \eta = 1, \\ O(n^{-2} (\log n)^{s-1}), & \text{if } \eta > 1. \end{cases}$$

Other reproducing kernels which are not scramble-invariant are also considered in Hickernell and Yue (2000).

Digital nets and sequences (Niederreiter (1992), Larcher (1998)), often used for quasi-Monte Carlo quadrature rules. The discrepancy of scrambled digital nets has been considered in Yue and Hickernell (2002). One of the main results is to derive a formula for the discrepancy of a digital (λ, t, m, s)-net in base b that requires only $O(n)$ operations to evaluate.

Theorem 4 (Yue and Hickernell, 2002) *Let $P = \{\boldsymbol{x}_0, \boldsymbol{x}_1, \cdots, \boldsymbol{x}_{n-1}\}$ be a scrambled set from a digital (λ, t, m, s)-net $\{\boldsymbol{z}_0, \boldsymbol{z}_1, \cdots, \boldsymbol{z}_{n-1}\}$ in prime power base b. Then*

$$E[D(p; K_{sc})]^2 = \frac{1}{n} \left[\sum_{i=0}^{b^m-1} K_{sc}(\boldsymbol{z}_i, \boldsymbol{0}) + \sum_{j=1}^{\lambda-1} \frac{2(\lambda - j)}{\lambda} \sum_{i=0}^{b^m-1} K_{sc}(\boldsymbol{z}_{jb^m+i}, \boldsymbol{0}) \right].$$

For further results on the error analysis for the scrambled quasi-Monte Carlo quadrature the reader is referred to Yue and Hickernell (2002), and Heinrich, Hickernell and Yue (2002). And for the application of the scrambled nets in base b to experimental designs see Yue (2001).

Acknowledgments

This work was supported by Shanghai NSF Grant 00JC14057, Shanghai Higher Education STF Grant 01D01-1 and a NSFC grant 10271078.

References

Aronszajn, N. (1950). Theory of reproducing kernels, *Trans. Amer. Math. Soc.* **68**, 337-404.

Davis, P. J. and Rabinowitz, P. (1984), *Methods of Numerical Integration, 2nd Ed.* Academic Press Inc, Orlando FL.

Aronszajn, N. (1950). Theory of reproducing kernels, *Trans. Amer. Math. Soc.* **68**, 337-404.

Davis, P. J. and Rabinowitz, P. (1984), *Methods of Numerical Integration, 2nd Ed.* Academic Press Inc, Orlando FL.

Fang K. T. and Wang Y. (1994). *Number Theoretical Methods in Statistics.* Chapman and Hall, New York.

Heinrich, S., Hickernell, F. J. and Yue R. X. (2002), Optimal quadrature for Haar wavelet spaces, *Math. Comp.* in press.

Hickernell, F. J. (1996). The mean square discrepancy of randomized nets, *ACM Trans. Model. Comput. Simul.* **6**, 274-296.

Hickernell, F. J. (1998). A generalized discrepancy and quadrature error bound, *Math. Comp.* **67**, 299-322.

Hickernell, F. J. and Hong H. S. (1999). The asymptotic efficiency of randomized nets for quadrature, *Math. Comp.* **68**, 767-791.

Hickernell, F. J. and Yue, R. X. (2000). The mean square discrepancy of scrambled (t, s)-sequences, *SIAM, J. Numer. Anal.* **38**, 1089-1112.

Hua L. K. and Wang Y. (1981). *Applications of Number Theory to Numberical Analysis.* Springer-Verlag and Science Press, Berlin, Beijing.

Larcher, G., Digital point sets: Analysis and applications, in *Random and Quasi-Random Point Sets* (P. Hellekaiek and G. Larcher, eds.), Lecture Notes in Statistics 138, Springer-Verlag, New York, 167-222.

Niederreiter, H. (1992). *Random Number Generation and Quasi-Monte Carlo Methods.* SIAM, Philadelphia.

Owen, A. B. (1995). Randomly permuted (t, m, s)-nets and (t, s)-sequences, in *Monte Carlo and Quasi-Monte Carlo Methods in Scientific Computing* (H. Niederreiter and P. J. S. Shiue, eds.), Lecture Notes in Statistics 106, Springer-Verlag, Berlin, 299-317.

Owen, A. B. (1997a). Monte Carlo variance of scrambled equidistribution quadrature. *SIAM, J. Num. Anal.* **34**, 1884-1910.

Owen, A. B. (1997b). Scrambled net variance for integrals of smooth functions. *Ann. Statist.* **25**, 1541-1562.

Owen, A. B. (1998). Scrambled Sobol and Niederreiter-Xing points. *J. Complexity* **14**, 466-489.

Saitoh, S. (1988). *Theory of Reproducing Kernels and Its Applications.* Longman Scientific & Technical, Essex, England.

Sloan, I. H. and Joe, S. (1994). *Lattice Methods for Multiple Integration.* Oxford Science Publications, Oxford.

Wahba, G. (1990). *Spline Models for Observational Data.* SIAM, Philadelphia.

Yue R. X. (1999). Variance of quadrature over scrambled unions of nets. *Statist. Sinica* **9**, 451-473.

Yue R. X. (2001). A comparison of random and quasirandom points for nonparametric response surface design. *Statist. and Probab. Letters* **53**, 129-142.

Yue R. X. and Hickernell, F. J. (2001). Integration and approximation based on scrambled sampling in arbitrary dimensions, *J. Complexity* **17**, 881-897.

Yue R. X. and Hickernell, F. J. (2002). The discrepancy and gain coefficients of scrambled digital nets. *J. Complexity* **18**, 135-151.

Yue R. X. and Mao, S. S. (1999). On the variance of quadrature over scrambled nets and sequences. *Statist. and Probab. Letters* **44**, 267-280.

WEAKLY NONLINEAR EVOLUTION OF SMALL DISTURBANCES IN THE PRESENCE OF LINEAR INSTABILITY

JUN YU
Department of Mathematics and Statistics, University of Vermont
16 Colchester Ave, Burlington, VT 05401, USA
E-mail: Jun.Yu@uvm.edu

(Received 15 November 2002 and in revised form 15 April 2003)

Through examples in channel flow and solid combustion, this paper studies the asymptotic behavior of small disturbances as they evolve in time in the presence of linear instability. Weakly unstable problems where the parameter marginally crosses the neutral value are analyzed and multiple scale expansions are used to derive the dominant evolution equation that governs the weakly nonlinear behavior of solution for long times. For the shallow water waves in an inclined open channel, the evolution equation turns out to be an integro-partial differential equation of first order that can be solved numerically in conjunction with the jump condition that follows from the exact bore conditions. For the channel flow which includes internal dissipation, long wave approximation is made and the evolution equation is a generalized Kuramoto-Sivashinsky equation. For the solid combustion problem, the onset of linear instabilities is studied, and the evolution equations that govern the amplitudes of the normal mode solution are of the Landau-Stuart type. In the first example, the asymptotic theory predicts the solution accurately for both the transient and quasi-steady phases. For the second and third examples, formation of coherent structures is analyzed.

1. Introduction

This paper studies nonlinear evolution of small disturbances in the presence of linear instability. It should be pointed out that linear theory has been widely applied to many problems in physics and science, and it is easier to implement. However, because of the intrinsically nonlinear nature of these problems, linear theory fails for more realistic situations and a nonlinear theory must be developed. Here, we show how some basic solutions become linearly unstable as parameter in the respective system equations passes through certain critical value and how the multiple scale expansion method can be used to incorporate weakly nonlinear terms into the analysis.

In the first example, we consider nonlinear waves on a thin fluid layer flowing down an inclined open channel. A model studied by Dressler (1949), Whitham (1974) and Yu & Kevorkian (1992) will be used. Using dimensionless variables, the equations for mass and momentum conservation for shallow water in a broad, slightly inclined channel are

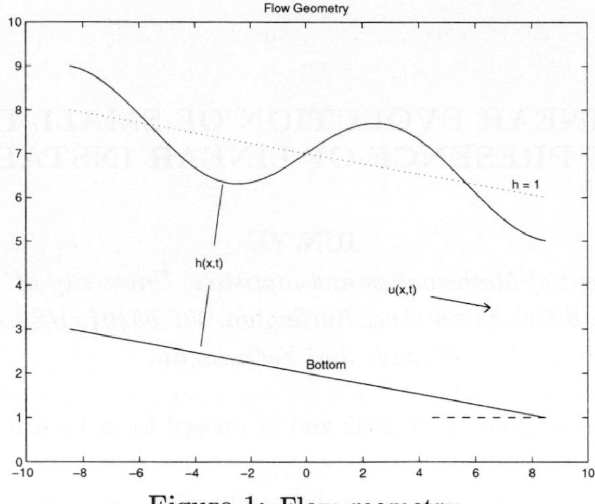

Figure 1: Flow geometry

given by [See Sec. 5.1.1 of Kevorkian (1993)]

$$h_t + uh_x + hu_x = 0, \tag{1.1a}$$

$$h(u_t + h_x + uu_x) = h - \frac{u^2}{F^2}. \tag{1.1b}$$

As shown in Figure 1, u is the flow speed parallel to the channel bottom averaged over h, and h is the height of the free surface normal to the bottom. The Froude number F is the dimensionless speed of undisturbed flow of unit height.

As was pointed out by Whitham (1974), the uniform flow is locally unstable when F is larger than 2. That is, for any small disturbance imposed on the uniform flow, the response predicted by linear theory for (1.1) has an amplitude that grows exponentially in time. However, solutions of the nonlinear equations (1.1) are stable. In fact, various levels of approximation of the flow equations may be used to exhibit bounded quasi-steady solutions; these are time-independent periodic solutions in a coordinate frame moving downstream with a constant progressing speed. In particular, the dimensional form of (1.1) was analyzed by Dressler (1949). He showed that no *continuous* quasi-steady solutions exist, and that it is necessary to have $F > 2$ in order to find *discontinuous* quasi-steady periodic solutions that he called *roll waves.* These results were confirmed by Yu & Kevorkian (1992) for the dimensionless model equations (1.1).

To account for energy dissipation, an additional hu_{xx}/R term was added to (1.1b) so that

$$h_t + uh_x + hu_x = 0, \tag{1.2a}$$

$$h(u_t + h_x + uu_x) = h - \frac{u^2}{F^2} + \frac{hu_{xx}}{R}. \tag{1.2b}$$

where R is the Reynolds number. It can be shown (see for example Needham & Merkin (1984) and Yu *et al.* (2000)) that the linear stability condition $F \leq 2$ now generalizes to $F \leq 2R/(R-k^2)$, where k is the wavenumber of the disturbance. This higher-order model was analyzed by Needham & Merkin (1984) and the stability condition was first derived in their notation. They have also shown that periodic quasi-steady continuous waves exist when the uniform flow is locally unstable.

In these first two examples, we consider the initial conditions

$$h(x,0;\epsilon) = 1 + \epsilon h_0(x), \qquad u(x,0;\epsilon) = F + \epsilon u_0(x), \tag{1.3a,b}$$

where $h_0(x)$ and $u_0(x)$ are prescribed functions and ϵ measures the amplitude. In Section 2.1, we derive a time-dependent asymptotic solution for (1.1) and (1.3). In Section 2.2, we show that our asymptotic solution tends to Dressler's roll wave solution as $t \to \infty$. We also compare our results with numerical integration of the exact problem (1.1) and (1.3). In Section 3, we consider a long wave expansion solution for the initial value problem (1.2) and (1.3).

In the third example, we use perturbation techniques and Laplace transforms to study the onset of linear instability in a simple model of solid combustion. We consider a version of the sharp-interface model of solid combustion introduced by Matkowsky & Sivashinsky (1978). Their free-boundary problem was posed on the whole real line for one-dimensional burning. In Section 4, we consider a one-sided model initially introduced by Frankel (1991). Asymptotic studies on this model (e.g., Gross (1997), Yu & Gross (2001) and Gross (2002)) have distinguished between temporally periodic and non-periodic regimes and have identified dominant spatial patterns, valid for sufficiently large time.

2. Weakly Nonlinear Waves in Channel Flow

We first study the initial-value problem (1.1) and (1.3). We concentrate on weakly unstable problem $0 < F - 2 \ll 1$, as given in Yu & Kevorkian (1992). This is an interesting case, as we are perturbing around a state that is linearly neutral. Therefore, the linearized equation does not have the difficulty of instability. This allows us to explicitly derive all the qualitative features of the more general case $F > 2$, including the steepening of waves, bore formation, and evolution of a quasi-steady pattern.

2.1. *Weakly unstable solution* $(F - 2) \ll 1$

We proceed as in Yu & Kevorkian (1992) to assume $F = 2 + \epsilon\alpha$, where α is a positive $O(1)$ constant and look for a multiple scale expansion solution of (1.1) in the form

$$h(x,t;\epsilon) = 1 + \epsilon h_1(x,t,\tilde{t}) + \epsilon^2 h_2(x,t,\tilde{t}) + O(\epsilon^3), \tag{2.1a}$$

$$u(x,t;\epsilon) = 2 + \epsilon[\alpha + u_1(x,t,\tilde{t})] + \epsilon^2 u_2(x,t,\tilde{t}) + O(\epsilon^3), \tag{2.1b}$$

where $\tilde{t} = \epsilon t$ is the slow time.

There are three key ingredients in this part of analysis. First is the introduction of the characteristic dependent and independent variables of the linearized problem

$$R_1 = h_1 - u_1, \qquad S_1 = h_1 + u_1, \tag{2.2a,b}$$

$$R_2 = h_2 - u_2, \qquad S_2 = h_2 + u_2, \tag{2.2c,d}$$

$$\xi = x - 3t, \qquad \eta = x - t. \tag{2.2e, f}$$

This allows equation of the R-variable to decouple from that of the S-variable at each order so that equations can be solved sequentially. At $O(\epsilon)$ one finds

$$R_1(\xi, \eta, \tilde{t}) = f_1(\eta, \tilde{t}) e^{\xi/2}. \tag{2.3a}$$

$$S_1(\xi, \eta, \tilde{t}) = G(\eta, \tilde{t}) e^{\xi/2} + g_1(\xi, \tilde{t}), \tag{2.3b}$$

where $G(\eta, \tilde{t})$ is defined by

$$\frac{\partial G}{\partial \eta} = \frac{1}{2} f_1(\eta, \tilde{t}). \tag{2.4}$$

Secondly, one needs to apply a consistency condition on solution to $O(\epsilon^2)$ in order to derive equations for f_1 and g_1 defined in (2.3). This is also done sequentially. One finds that

$$f_{1\tilde{t}} + \alpha f_{1\eta} + \frac{\alpha}{4} f_1 + \frac{\alpha}{4} G = 0, \tag{2.5}$$

for f_1 and

$$g_{1\tilde{t}} + (\alpha + \frac{3}{4} g_1) g_{1\xi} - \frac{\alpha}{4} g_1 - \frac{\alpha}{8} e^{\xi/2} (\int^{\xi} g_1 e^{-s/2} ds + C_1) = 0, \tag{2.6}$$

for g_1. Here C_1 is an integration constant.

Finally, the third ingredient is the treatment of the function G in (2.5) as well as the constant C_1 in (2.6). We define

$$f_1^*(\eta, \tilde{t}) = f_1(\eta, \tilde{t}) e^{\eta/2}, \quad G^*(\eta, \tilde{t}) = G(\eta, \tilde{t}) e^{\eta/2}. \tag{2.7a, b}$$

If the initial conditions $u_0(x)$ and $h_0(x)$ in (1.3) are $2l$-periodic functions of x then equations (2.5) and (2.6) become

$$f_{1\tilde{t}}^* + \alpha f_{1\eta}^* = \frac{\alpha}{4} f_1^* - \frac{\alpha}{8} e^{\eta/2} (\int_{-l}^{\eta} f_1^* e^{-s/2} ds + \frac{e^l}{1 - e^l} \int_{-l}^{l} f_1^* e^{-s/2} ds), \tag{2.8a}$$

$$g_{1\tilde{t}} + (\alpha + \frac{3}{4} g_1) g_{1\xi} = \frac{\alpha}{4} g_1 + \frac{\alpha}{8} e^{\xi/2} (\int_{-l}^{\xi} g_1 e^{-s/2} ds + \frac{e^l}{1 - e^l} \int_{-l}^{l} g_1 e^{-s/2} ds). \tag{2.8b}$$

The problem with arbitrary initial conditions may be treated as the limiting case in which the wavelength l approaches to infinity. In this case, (2.5) and (2.6) become

$$f_{1\tilde{t}}^* + \alpha f_{1\eta}^* = \frac{\alpha}{4} f_1^* - \frac{\alpha}{8} e^{\eta/2} \int_{\infty}^{\eta} f_1^* e^{-s/2} ds, \tag{2.9a}$$

$$g_{1\tilde{t}} + (\alpha + \frac{3}{4} g_1) g_{1\xi} = \frac{\alpha}{4} g_1 + \frac{\alpha}{8} e^{\xi/2} \int_{\infty}^{\xi} g_1 e^{-s/2} ds. \tag{2.9b}$$

Once the expressions for f_1^* and g_1 are derived, the solution for h and u to order ϵ is then available through (2.7), (2.3), (2.2) and (2.1) as

$$h(x, t; \epsilon) = 1 + \epsilon \frac{g_1(\xi, \tilde{t}) + [G^*(\eta, \tilde{t}) + f_1^*(\eta, \tilde{t})] e^{(\xi - \eta)/2}}{2} + O(\epsilon^2), \tag{2.10a}$$

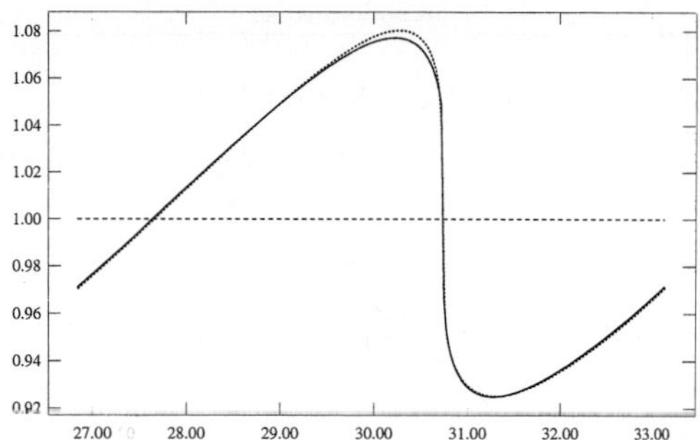

Figure 2: Numerical and asymptotic solutions for the initial conditions $h(x,0;\epsilon) = 1 - \epsilon \sin x$; $u(x,0;\epsilon) = F$ with $F = 2.1$, ($\epsilon = 0.1$, $\alpha = 1$) at $t = 10 = 1/\epsilon$ (after Yu & Kevorkian (1992))

$$u(x,t;\epsilon) = 2 + \epsilon\{\alpha + \frac{g_1(\xi,\tilde{t}) + [G^*(\eta,\tilde{t}) - f_1^*(\eta,\tilde{t})]e^{(\xi-\eta)/2}}{2}\} + O(\epsilon^2). \qquad (2.10b)$$

To illustrate the above ideas, we assume the initial conditions (1.3a,b) in the simple form ($l = \pi$)

$$h(x,0;\epsilon) = 1 - \epsilon \sin x, \qquad u(x,0;\epsilon) = F, \qquad (2.11a,b)$$

so that $h_0(x) = -\sin x$ and $u_0(x) = 0$. In Figure 2 we show a typical comparison between the theoretical (dotted curve) and numerical (solid curve) values of h in a 2π-interval in x for $t = 10$, $\epsilon = 0.1$ and $F = 2.1$. More comparisons with smaller values of ϵ are shown in Yu & Kevorkian (1992) to demonstrate that the solution (2.10) is accurate to $O(\epsilon)$ for times of $O(\epsilon^{-1})$. A more systematic approach to verify the asymptotic validity of our solution is to use the technique discussed in Bosley (1996). That is, if we assume that the maximum error between the asymptotic and exact results is given to leading order by

$$\text{Error} = K\epsilon^{\gamma}, \qquad (2.12)$$

then we obtain a good estimate of the exponent γ by plotting log(Error) versus log(ϵ) for several different values of ϵ. In Figure 3 we summarize errors from the three comparisons given in Yu & Kevorkian (1992). The numerical estimate of γ is 1.88, which is almost 2.

2.2. *Quasi-steady state solution*

We summarize here results of the steady state limit of the asymptotic solution correct to $O(\epsilon)$ given by (2.10). As was pointed out in Yu & Kevorkian (1992), the amplitude of the disturbance along the η =const. ray decays exponentially. Therefore, as $t \to \infty$ these terms vanish and (2.10a,b) become

$$h = 1 + \epsilon\frac{g^*}{2}, \qquad u = 2 + \epsilon(\alpha + \frac{g^*}{2}), \qquad (2.13a,b)$$

where g^* is the steady state value of g_1. There are two key elements in this section of study. First, a change of variables $\tilde{t}^* = \tilde{t}$ and $\zeta = \xi - \beta\tilde{t}$ must be made with β defined in the expansion of the progressing speed $c = 3 + \epsilon\beta + O(\epsilon^2)$. By letting $\tilde{t}^* \to \infty$ while

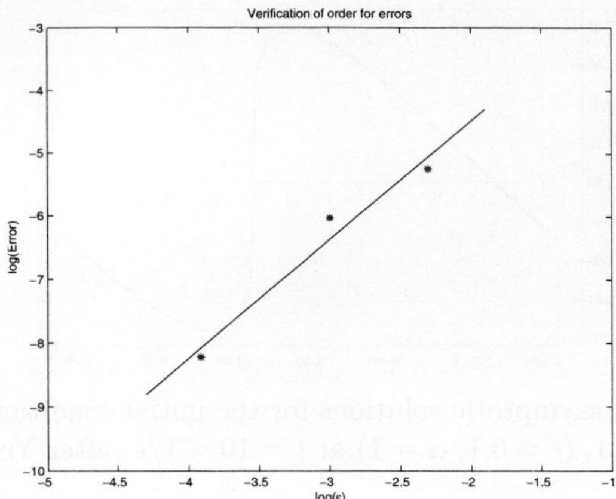

Figure 3: Verification of the order for errors

holding ζ fixed, we are observing wave behavior in a coordinate frame that travels with the progressing speed to $O(\epsilon)$. In this limit, (2.8b) becomes

$$(\alpha - \beta + \frac{3}{4}g^*)\frac{dg^*}{d\zeta} = \frac{\alpha}{4}g^* + \frac{\alpha}{8}e^{\zeta/2}(\int_{-l}^{\zeta} g^* e^{-s/2} ds + \frac{e^l}{1-e^l}\int_{-l}^{l} g^* e^{-s/2} ds). \tag{2.14}$$

Through a consistency argument, equation (2.14) reduces to

$$\frac{dg^*}{d\zeta} = \frac{1}{4}[g^* - \frac{4}{3}(\beta - 2\alpha)]. \tag{2.15}$$

The solution of (2.15) subject to $g^*(-l) = 0$ is given by

$$\zeta = -l + 4\log|1 + \frac{3g^*}{4(2\alpha - \beta)}|, \tag{2.16}$$

and the right segment of the solution is a $2l$-shift of (2.16), i.e.,

$$\zeta = l + 4\log|1 + \frac{3g^*}{4(2\alpha - \beta)}|. \tag{2.17}$$

Secondly, a correct jump condition across a discontinuity of the solution must be obtained from (1.1). This is given by

$$\alpha - \beta + \frac{3}{8}(g^{*+} + g^{*-}) = 0. \tag{2.18}$$

in term of g^*, where the plus and minus superscripts indicate values on either side of the discontinuity. Substituting $g^* = g^{*-}$ and $g^* = g^{*+}$ into (2.16) and (2.17), respectively and equating the two results, one finds a second equation for g^{*-} and g^{*+}:

$$-l + 4\log|1 + \frac{3g^{*-}}{4(2\alpha - \beta)}| = l + 4\log|1 + \frac{3g^{*+}}{4(2\alpha - \beta)}|. \tag{2.19}$$

Solving equations (2.18) and (2.19) gives

$$g^{*+} = \frac{4}{3}(\beta - \frac{2\alpha}{1+e^{-l/2}}), \qquad g^{*-} = \frac{4}{3}(\beta - \frac{2\alpha}{1+e^{l/2}}). \tag{2.20a, b}$$

The location of the discontinuity can now be obtained as

$$\zeta_s = -4\log[\frac{2\alpha-\beta}{\alpha}\cosh(\frac{l}{4})]. \tag{2.20c}$$

Finally, we use the law of mass conservation to determine the unknown constant β in the solution that we introduced in the expansion of c. This is given by

$$\beta = \frac{3}{2}\overline{h}_0 + 2\alpha(1 - \frac{2}{l}\tanh\frac{l}{4}). \tag{2.21}$$

The steady state of the asymptotic solution correct to $O(\epsilon)$ over a $2l$-interval is now available from (2.20)-(2.21), (2.16)-(2.17) and (2.13). Comparisons between Dressler's (numerical) and our asymptotic steady state solutions were given in Yu & Kevorkian (1992) and they indicated that both numerical values of wave height and the location of the discontinuity were very well predicted by our asymptotic solution.

In summary, we have demonstrated the use of multiple scale expansions to obtain asymptotic solution for weakly nonlinear waves in channel Flow. We have shown that even with the presence of linear instability, our theory predicts the solution accurately for long time (over the time interval $[0, T(\epsilon)]$, where $T = O(\epsilon^{-1})$). In the limit $t \to \infty$, our theory predicts the quasi-steady pattern of waves accurately and provides explicit formulae for the quasi-steady solution.

3. Channel Flow with Internal Dissipation

We now study the initial-value problem (1.2) and (1.3). Unlike the situation in Section 2 where all wavenumbers are equally unstable, we must focus on long waves if $0 < F - 2 \ll 1$ (see Yu *et al.* (2000)). We restrict the Froude number to the one-parameter family $F = 2 + \alpha\epsilon$ for $k = O(\epsilon^{\frac{1}{2}})$. Here, α is a positive constant that is independent of ϵ. The strongly unstable problem ($F - 2 = O(1)$) is not discussed because a multiple scale expansion is only valid over a time interval that is marginally longer than $O(1)$ in this case.

The fact that $k = O(\epsilon^{\frac{1}{2}})$ implies that the appropriate spatial variable is $\tilde{x} = \epsilon^{\frac{1}{2}}x$, and in addition to the time t, we allow for the four slow times $\tilde{t} = \sqrt{\epsilon}t$, $\bar{t} = \epsilon t$, $t^* = \epsilon^{\frac{3}{2}}t$ and $\bar{\bar{t}} = \epsilon^2 t$. The multiple scale expansions for h and u are then assumed in the form

$$h(x,t;\epsilon) = 1 + \epsilon h_1(\tilde{x},t,\tilde{t},\bar{t},t^*,\bar{\bar{t}}) + \epsilon^{3/2}h_2(\tilde{x},t,\tilde{t},\bar{t},t^*,\bar{\bar{t}}) + O(\epsilon^2), \tag{3.1a}$$

$$u(x,t;\epsilon) = 2 + \epsilon[\alpha + u_1(\tilde{x},t,\tilde{t},\bar{t},t^*,\bar{\bar{t}})] + \epsilon^{3/2}u_2(\tilde{x},t,\tilde{t},\bar{t},t^*,\bar{\bar{t}}) + O(\epsilon^2). \tag{3.1b}$$

By a similar procedure as given in Section 2.1, the solution for h and u to order ϵ is in the form

$$h(x,t;\epsilon) = 1 + \epsilon f(\xi,t^*) + O(\epsilon^{3/2}), \tag{3.2a}$$

$$u(x,t;\epsilon) = 2 + \epsilon[\alpha + f(\xi, t^*) + g_1(\eta, \tilde{t})e^{-t}] + O(\epsilon^{3/2}), \tag{3.2b}$$

where $\xi = \tilde{x} - 3\tilde{t}$, f satisfies

$$f_{t^*} + \frac{3}{2}(\alpha + f)f_\xi + \frac{1}{R}f_{\xi\xi\xi} + \sqrt{\epsilon}(\alpha f_{\xi\xi} + \frac{2}{R}f_{\xi\xi\xi\xi}) = 0, \tag{3.3}$$

subject to the initial condition $f(x,0) = h_0(x)$ and g_1 satisfies

$$g_{1\tilde{t}} = \frac{\alpha}{2}g_1 + \frac{1}{R}g_{1\eta\eta}, \tag{3.4}$$

subject to $g_1(x,0) = u_0(x) - h_0(x)$.

Equation (3.3) is a generalized Kuramoto-Sivashinsky (GKS) equation for f in terms of ξ and t^*. Note that for $F > 2$ the coefficient of $f_{\xi\xi}$ is positive and represents a negative diffusion. This destabilizing effect is balanced by the positive diffusion of the $\frac{2}{R}f_{\xi\xi\xi\xi}$ term.

Formation of a coherent structure

As was pointed out in Yu *et al.* (2000), deriving amplitude equations for long waves on thin liquid films has a long history beginning with the work of Benney (1966). Equation (3.3) involves a balance between dispersion, dissipation, and nonlinearity for long waves. The Kuramoto-Sivashinsky equation corresponds to vanishing dispersion in (3.3) in which there is a long wave instability. The Korteweg-de Vries (KdV) equation, corresponding to the absence of dissipation, has a one-parameter family of solitary waves. Our interest is in the strongly dispersive case. It has become well-known due to the work of Toh and Kawahara (1985, 1985, 1988) that solitary waves of the KdV equation are destabilized over a long time scale by negative diffusive perturbation present in (3.3) and stabilized by the dissipative fourth-derivative term. In this way, there is a mechanism for the appearance of a coherent state since a solitary pulse slowly evolves into one with a unique equilibrated amplitude.

Notice that in the evolution equation (3.3) the backward diffusion operator and stable fourth order dissipation operator are small compared to the dispersive term of the KdV. We follow Ablowitz & Segur (1981) and introduce a slow time variable $T = \epsilon^{\frac{1}{2}}t^*$ and a spatial variable $\theta = \xi - \frac{G(\epsilon^{\frac{1}{2}}t^*)}{\epsilon^{\frac{1}{2}}}$. A perturbation expansion, $f = f_0 + \epsilon^{\frac{1}{2}}f_1 + ...$, shows that the leading order solution may be a slowly varying solitary wave in the form

$$f_0 = \frac{8}{R}A^2 \text{sech}^2(A\theta), \tag{3.5}$$

where the maximum height $\frac{8}{R}A^2$ is given by

$$\frac{8}{R}A^2 = \frac{7\alpha}{5} = \frac{7(F-2)}{5\epsilon}. \tag{3.6}$$

This is the unique solitary wave for which the energy input by the backwards diffusion operator balances the energy dissipated by the fourth order dissipative operator.

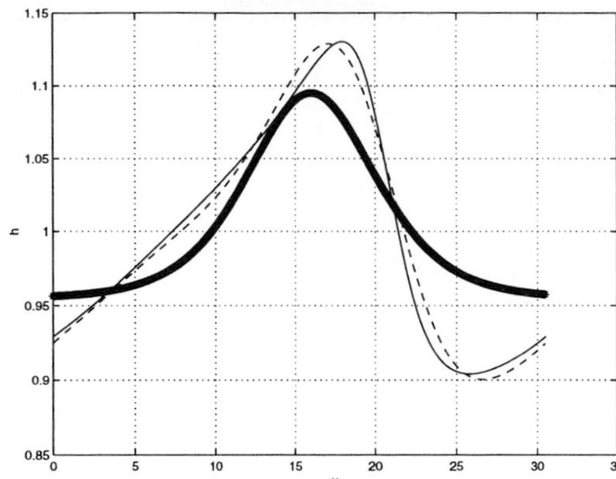

Figure 4: Numerical and asymptotic long-wave solutions with $F = 2.1$, $R = 2$ and $t = 32$ and superimposing the unique equilibrated solitary wave (bold curve) (after Yu *et al.* (2000))

In order to verify the accuracy of the asymptotic results, we perform numerical computation for initial conditions similar to (2.11) so that

$$f(x,0) = -\sin(kx), \qquad g_1(x,0) = \sin(kx). \tag{3.7a, b}$$

The wavenumber k in the initial conditions is of order $\sqrt{\epsilon}$, i.e., $k = k^*\sqrt{\epsilon}$. We choose $\epsilon = 0.1$, $R = 2$ and $k^* = 0.651$ so that $F = 2.1$ is marginally larger than the critical value $2R/(R-k^2) = 2.043$. We show three curves (dashed, solid and bold) in Figure 4. The dashed curve is for the theoretical values of h, obtained by solving (3.3), over one wavelength in x and for $t = 32$. The solid curve is for the numerically integrated values from the exact problem (1.2) and (1.3). The bold curve in Figure 4 is for the unique equilibrated solitary wave given by (3.5), with amplitude $\frac{7}{5}(F-2) = 0.14$ using (3.6). (Notice that equation (3.3) is unaltered under the substitution: $\bar{f} = f - \beta$, $\tau = t^*$ and $\bar{\xi} = \xi - \frac{3}{2}\beta t^*$, where β is a constant.) Comparisons between the theoretical and numerical values of h over one wavelength in x and for other values of times are shown in Yu *et al.* (2000). We observe that, unlike the case for $R = \infty$ discussed in Section 2, our asymptotic results here do not remain valid for long enough times to accurately predict the traveling wave phase. However, as shown in Figure 4, the asymptotic and numerical values of h approach the unique equilibrated solitary wave given by the bold curve.

4. Weak Nonlinear Disturbances in Solid Combustion

The geometry under consideration is a strip of material extending from $x = 0$ to $x = l$ and from $y = 0$ to $y = \infty$. Suppose the material is undergoing an exothermic reaction so that $y = f(t,x)$ defines the reaction front. The burned region is $y < f(t,x)$, and $y > f(t,x)$ is the unburned (see Figure 5). Using dimensionless variables, the temperature distribution $T(x,y,t)$ must satisfy the heat equation in the unburned region (see Gross (1997) and Yu & Gross (2001)):

$$T_t = T_{xx} + T_{yy}, \qquad 0 < x < l, \quad y > f(t,x), \quad t > 0. \tag{4.1}$$

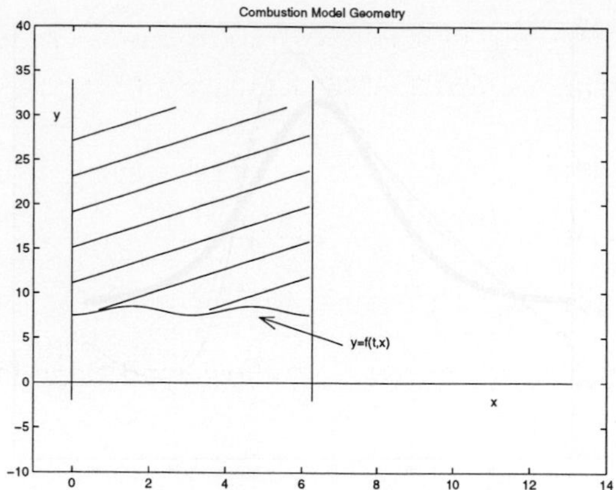

Figure 5: Combustion model geometry

At the interface position $\Gamma(t) = \{(x,y)|y = f(t,x)\}$, we impose two conditions

$$T|_\Gamma = 1 + \nu K(V), \qquad \frac{\partial T}{\partial \mathbf{n}}\Big|_\Gamma = -V. \tag{4.2a, b}$$

Here $V = f_t/\sqrt{1+f_x^2}$ is the front velocity in the direction $\mathbf{n}$ normal to the front, $K(V)$ is the boundary kinetics and ν is inversely proportional to the activation energy. Impose zero-slope contact conditions for the interface

$$f_x(t,0) = 0, \qquad f_x(t,l) = 0, \tag{4.3a, b}$$

and adiabatic (insulated) conditions at the walls

$$T_x(t,0) = 0, \qquad T_x(t,l) = 0. \tag{4.4a, b}$$

In accordance with the normalization of the temperature, we also require that

$$\lim_{y\to\infty} T(t,x,y) = 0. \tag{4.5}$$

Finally, the initial temperature distribution is prescribed as

$$T(0,x,y) = g(x,y). \tag{4.6}$$

Equations (4.1)-(4.6) form a free-boundary, initial-value problem that defines the temperature T and the front position f uniquely. One may use (4.1), (4.2a), (4.4)-(4.6) to solve for the temperature T for any arbitrary front position f and then determine the front position by equations (4.2b) and (4.3).

4.1. *The onset of linear instabilities*

We first study the behavior of the linear problem governing perturbations to the basic solution of (4.1)-(4.6) consisting of a flat constant-speed front ($f(t,x) = t$, $T =$

e^{t-y}). We change the coordinates into the front-attached ones so that $\tau = t$, $\xi = x$ and $\eta = y - f(t,x)$. Perturbing about the basic solution by letting

$$T = e^{-\eta} + \epsilon w, \qquad f = \tau + \epsilon\phi, \tag{4.7a, b}$$

linearizing and considering the kth Fourier mode:

$$w(\tau,\eta,\xi) \sim W(\tau,\eta)\cos(k\xi), \qquad \phi(\tau,\xi) \sim \Phi(\tau)\cos(k\xi), \tag{4.8a, b}$$

one obtains the governing equation and boundary conditions for W and Φ (see Gross (1997) and Yu & Gross (2001)):

$$W_\tau = -k^2 W + W_{\eta\eta} + W_\eta - e^{-\eta}(\Phi' + k^2\Phi), \tag{4.9a}$$

$$W|_{\eta=0} = \nu\Phi', \quad W_\eta|_{\eta=0} = -\Phi', \quad \lim_{\eta\to\infty} W = 0. \tag{4.9b, c, d}$$

The initial conditions are

$$W(0,\eta) = g^*(\eta), \qquad \Phi(0) = \Phi_0, \tag{4.9e, f}$$

where Φ_0 is a constant and $g^*(\eta)$ is related to $g(x,y)$ in (4.6) through (4.8) and (4.7). Also, $\lim_{\eta\to\infty} g^* = 0$ as required by (4.9d). Equations (4.9a,b,d,e) define an initial and boundary value problem for $W(\tau,\eta)$ in a semi-infinite domain $\eta > 0$, $\tau > 0$ of the (η,τ) plane and conditions (4.9c,f) are for $\Phi(\tau)$ in the interval $\tau > 0$.

Proceeding as in Yu & Gross (2001), we make a change of variable

$$u = (W - \nu\Phi')e^{(k^2+\frac{1}{4})\tau+\eta/2}, \tag{4.10}$$

in (4.9a,b,d,e) and use the method of images to obtain

$$u(\tau,\eta) = \int_0^\tau dt \int_0^\infty \frac{H(t,x)}{\sqrt{4\pi(\tau-t)}}[e^{-(\eta-x)^2/4(\tau-t)} - e^{-(\eta+x)^2/4(\tau-t)}]dx$$
$$+ \int_0^\infty \frac{G(x)}{\sqrt{4\pi\tau}}[e^{-(\eta-x)^2/4\tau} - e^{-(\eta+x)^2/4\tau}]dx, \tag{4.11}$$

where $H(\tau,\eta) = -[\nu\Phi'' + \nu k^2\Phi' + e^{-\eta}(\Phi' + k^2\Phi)]e^{(k^2+\frac{1}{4})\tau+\eta/2}$ and $G(\eta) = [g^*(\eta) - \nu\Phi'(0)]e^{\eta/2}$.

Equation (4.9c) implies that Φ must satisfy

$$\int_0^\tau M(t)\int_0^\infty L(\tau-t,x)e^{x/2}dxdt + \int_0^\tau N(t)\int_0^\infty L(\tau-t,x)e^{-x/2}dxdt$$
$$+ \int_0^\infty [g^*(x) - \nu\Phi'(0)]e^{x/2}L(\tau,x)dx = -e^{(k^2+\frac{1}{4})\tau}\Phi'(\tau), \tag{4.12}$$

where $M(\tau) = -(\nu\Phi'' + \nu k^2\Phi')e^{(k^2+\frac{1}{4})\tau}$, $N(\tau) = -(\Phi' + k^2\Phi)e^{(k^2+\frac{1}{4})\tau}$ and $L(\tau,x) = \frac{x}{2\sqrt{\pi\tau^3}}e^{-x^2/4\tau}$. Considering typical initial conditions in the form

$$g^*(\eta) = g_0 e^{-\alpha\eta}, \tag{4.13}$$

where g_0 is a constant and $\alpha = \alpha_r + i\alpha_i$ is a complex number with $\alpha_r > 0$, one finds that the Laplace transform of (4.12) yields

$$\tilde{\Phi}(s - \frac{1}{4} - k^2) = \frac{1}{p(s)}\Big(\frac{-\nu s + (1-\nu)\sqrt{s} - \frac{1}{2}(1-\frac{\nu}{2})}{\sqrt{s}+\frac{1}{2}}\Phi(0) - \frac{g_0}{\sqrt{s}+\alpha-\frac{1}{2}}\Big), \tag{4.14}$$

where $p(s) = -\nu(\sqrt{s})^3 + (1-\frac{\nu}{2})(\sqrt{s})^2 + [\nu(\frac{1}{4}+k^2) - 1]\sqrt{s} + \frac{1}{4} + \frac{\nu}{8} + (\frac{\nu}{2}-1)k^2$.

Notice that $p(s)$ in (4.14) is a third-degree polynomial in $\sqrt{s}$. Denoting the three roots as r_1, r_2 and r_3, one finds that (4.14) becomes

$$\tilde{\Phi}(s - \frac{1}{4} - k^2) = -\frac{1}{\nu}\Big(\frac{-\nu s + (1-\nu)\sqrt{s} - \frac{1}{2}(1-\frac{\nu}{2})}{(\sqrt{s}-r_1)(\sqrt{s}-r_2)(\sqrt{s}-r_3)(\sqrt{s}+\frac{1}{2})}\Phi(0)$$
$$- \frac{g_0}{(\sqrt{s}-r_1)(\sqrt{s}-r_2)(\sqrt{s}-r_3)(\sqrt{s}+\alpha-\frac{1}{2})}\Big). \tag{4.15}$$

As shown in Yu & Gross (2001), a partial fraction expansion of (4.15) can be carried out and the front perturbation Φ can be obtained by an inverse Laplace transform. Neutral stability occurs when $p(s)$ has complex roots $r_{2,3} = \beta_r \pm i\beta_i$ and a typical term in Φ associated with $r_{2,3}$ has the following explicit expression:

$$e^{(\beta_r^2 - \beta_i^2 - \frac{1}{4} - k^2)\tau}\mathrm{erfc}(-\beta_r\sqrt{\tau})\begin{Bmatrix}\cos(2\beta_r\beta_i\tau)\\ \sin(2\beta_r\beta_i\tau)\end{Bmatrix}, \tag{4.16}$$

where the growth rate may be defined by

$$\text{Growth Rate} = \beta_r^2 - \beta_i^2 - \frac{1}{4} - k^2. \tag{4.17}$$

Letting the growth rate equal to zero, one obtains the neutral stability curve as shown in Figure 6. In Figure 7 we plot the growth rate vs k for $\nu = 0.35$, 0.335 and 0.32. As expected, the growth rate for $\nu = 0.35$ is always negative as it lies inside of the stable region. For $\nu = 0.335$, the growth rates are positive between $0.155 < k < 0.962$ and attain a maximum value of 0.1027 at $k = 0.595$. For $\nu = 0.32$, the growth rates are positive for $k < 1.255$ and attain a maximum value of 0.3203 at $k = 0.5876$.

4.2. *Weakly nonlinear analysis*

Proceeding as in Gross (2002), we focus on normal-mode solutions for perturbations of the temperature and interface position:

$$w = e^{\lambda\tau}\cos(k_j\xi)g(\eta;\lambda,\nu), \quad \phi = e^{\lambda\tau}\cos(k_j\xi), \tag{4.18}$$

where $k_j = j\pi/L$, $j = 0, 1, 2, \ldots$. As shown in Figure 6, when $L = \pi$ the two modes with wave numbers $k_0 = 0$ and $k_1 = 1$ lose stability simultaneously as the material parameter ν drops below the critical value $\nu_c = 1/3$. We consider a small deviation from the neutral stability curve in the unstable direction, namely $\nu = \nu_c - \epsilon^2$, and study the evolution of the linearized eigenmodes modulated by slowly-varying, complex-valued

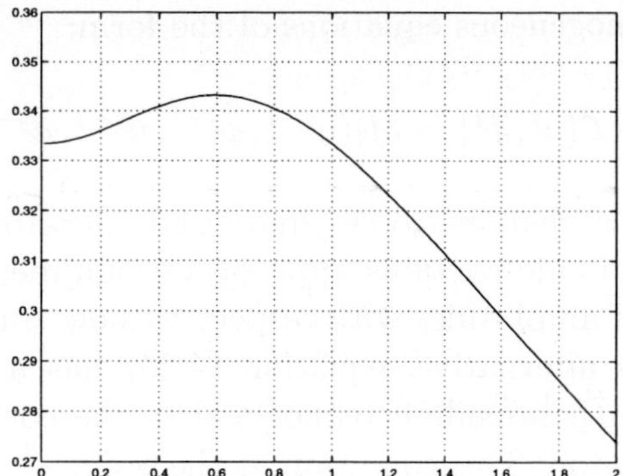

Figure 6: Neutral stability curve (ν vs k) (adopted from Yu & Gross (2001))

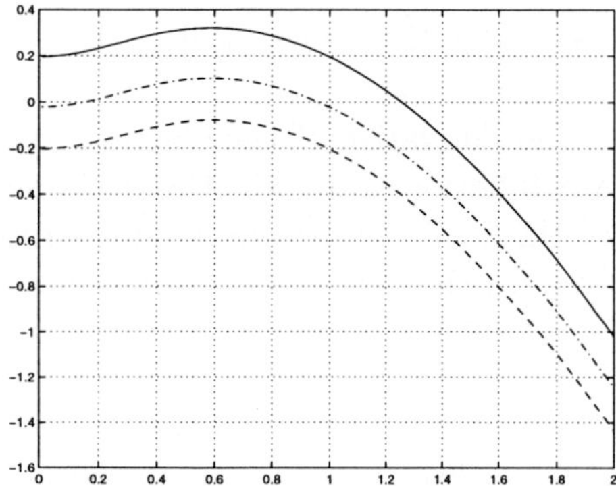

Figure 7: Growth rates vs k for $\nu = 0.35$ (dotted), 0.335 (dashdotted) and 0.32 (solid) (after Yu & Gross (2001))

amplitude functions $A_j(\epsilon\tau, \epsilon^2\tau)$, $j = 0, 1$. The choice of the small parameter ϵ allows for the possibility of a Hopf bifurcation where the magnitude of the solution is on the order of the square root of the bifurcation parameter. Therefore, we assume the following expansions for the temperature and interface position perturbed about the basic solution (see also (4.7)):

$$T(\xi, \eta, \tau, \epsilon\tau, \epsilon^2\tau) = e^{-\eta} + \epsilon \sum_{j=0}^{1} A_j(\epsilon\tau, \epsilon^2\tau) e^{iw_j\tau} \cos(k_j\xi) g(\eta; iw_j, \frac{1}{3})$$

$$+\epsilon^2 w_2(\xi, \eta, \tau, \epsilon\tau, \epsilon^2\tau) + \ldots + CC, \tag{4.19a}$$

$$f(\xi, \tau, \epsilon\tau, \epsilon^2\tau) = \tau + \epsilon \Big\{ \sum_{j=0}^{1} A_j(\epsilon\tau, \epsilon^2\tau) e^{iw_j\tau} \cos(k_j\xi) + \frac{1}{2} B(\epsilon\tau, \epsilon^2\tau) \Big\}$$

$$+\epsilon^2 \phi_2(\xi, \tau, \epsilon\tau, \epsilon^2\tau) + \ldots + CC, \tag{4.19b}$$

where $\lambda = iw_j$ is a purely imaginary discrete eigenvalue, $g(\eta; iw_j, \frac{1}{3})$ is the corresponding eigenfunction and CC stands for complex-conjugate terms.

In order to derive constraints for the slowly-varying amplitudes A_j and B, we insert (4.19) into the nonlinear equation and find the consecutive terms of the perturbation series satisfy linear inhomogeneous equations of the form:

$$\frac{\partial u^j}{\partial \tau} + \mathcal{L}(u^j, \phi^j) = M_j(u^{j-1}, \phi^{j-1}, u^{j-2}, \phi^{j-2}, ...), \tag{4.20}$$

where the left-hand side is identical to the linearized operator in the eigenmode equation, while the right-hand side contains appropriate nonlinear terms plus terms with appropriate derivatives of amplitudes with respect to slow times, $t_1 = \epsilon\tau$ and $t_2 = \epsilon^2\tau$. According to Fredholm's alternative, equation (4.20) has a nonsecular (bounded in time) solution if the right-hand side is orthogonal to the corresponding eigenfunction of the adjoint linear operator. This condition at the order ϵ^3 of the perturbation series gives the system of Landau-Stuart equations

$$\frac{dA_j}{dt_2} = \chi_j A_j + \beta_{j,1} A_j^2 \bar{A}_j + \beta_{j,2} A_j A_{1-j} \bar{A}_{1-j}, \quad j = 0, 1 \tag{4.21}$$

where χ_j is determined by the derivative of the eigenvalue with respect to ν and the Landau coefficients $\beta_{j,n}$ are functions of a kinetics parameter. Equations in (4.21) determine the dynamics of the weakly unstable modes subject to both nonlinear interaction and self-interaction. In the case when a single mode loses stability, equation (4.21) reduces to (see Gross (1997))

$$\frac{dA}{dt_2} = \kappa A + \beta A^2 \bar{A}. \tag{4.22}$$

It can be shown that for the kinetics function given by

$$K(V) = \frac{V^p - V^{-0.7}}{p + 0.7}, \tag{4.23}$$

equation (4.22) has circular limit cycles for $0.47 < p < 1.65$. The existence of limit cycles indicates that the temperature and interface position form coherent periodic pattern due to nonlinear self-interaction.

5. Concluding Remarks

We have done both asymptotic and numerical studies on three problems from channel flow and solid combustion. First, we have shown derivation of asymptotic solution of the model equations (1.1) for initial conditions (1.3) with ϵ small. It turns out that for $t < T$ the solution is time-dependent; a time on the order of $10T$ is required before the solution tends to a quasi-steady state. It is worth to point out that quasi-steady state solution is uniquely determined by its wavelength and the progressing speed. Secondly, we have derived an asymptotic solution of equations (1.2) for initial conditions (1.3). This solution to $O(\epsilon^{3/2})$ has the form (3.2) in which the non-decaying part, f obeys the GKS equation (3.3). We have verified that our results remain accurate, as required by a multiple scale analysis, over the time interval $[0, T(\epsilon)]$, where $T = O(\epsilon^{-3/2})$. Thirdly, We have carried out a linear stability study for a simple solid combustion model. Neutral stability curve and growth rates of a typical linearly unstable term are shown in

Figures 6 - 7. We have also demonstrated a weakly nonlinear analysis to obtain Landau-Stuart type of evolution equations (4.21) and (4.22) for slowly-varying amplitude of the linearized eigenmode.

Through these examples, we see that multiple scale expansion method is a useful perturbation technique for analyzing weakly nonlinear problems in the presence of linear instability. In fact, whenever the linear theory predicts growth one must take into account the cumulative effect of small nonlinearities in order to obtain a correct description of the solution over long times. In all three cases given in this paper, our theory predicts accurately the transient behavior of the small disturbances. The evolution equations also allow us to better understand the dynamics of the problems, in particular, the formation of coherent structures. In the case given in Section 2, our theory even predicts the solution accurately for quasi-steady phase. With the success in these examples, we anticipate that the multiple scale expansion method will become a very powerful tool for understanding the complexities of a wide variety of problems in physics and science, including the transition to turbulence in fluid flow.

References

Ablowitz, M. J., and H. Segur, 1981: Solitons and Inverse Scattering Transform. *SIAM.*

Benney, D. J., 1966: Long Waves on Liquid Films. *J. Math. Phys.*, **45**, 150-155.

Bosley, D. L., 1996: A Technique for the Numerical Verification of Asymptotic Expansions. *SIAM Rev.*, **38**(1), 128-135.

Dressler, R. F., 1949: Mathematical Solution of the Problem of Roll-Waves in Inclined Open Channels. *Comm. on Pure and Appl. Math.*, **2**, 149-194.

Frankel, M. L., 1991: Free Boundary Problems and Dynamical Geometry Associated with Flames. *IMA Volumes in Mathematics and Its Applications*, **35**, 107-127.

Gross, L. K., 1997: *Weakly Nonlinear Analysis of Interface Propagation.* Ph. D. thesis, Rensselaer Polytechnic Institute.

——, 2002: Weakly Nonlinear Dynamics of Interface Propagation. *Studies in Appl. Math.*, **108**, 323-350.

Kawahara, T., and S. Toh, 1985: Nonlinear Dispersive Periodic Waves in the Presence of Instability and Damping. *Phys. Fluids*, **28**, 1636-1638.

——, and ——, 1988: Pulse Interactions in an Unstable Dissipative-Dispersive Nonlinear System. *Phys. Fluids*, **31**, 2103-2111.

Kevorkian, J., 1993: Partial Differential Equations: Analytical Solution Techniques. *Chapmen and Hall.*

Matkowsky, B. J., and G. I. Sivashinsky, 1978: Propagation of a Pulsating Reaction Front in Solid Fuel Combustion. *SIAM J. Appl. Math.*, **35**, 465-478.

Needham, D. J., and J. H. Merkin, 1984: On Roll Waves down an Open Inclined Channel. *Proc. R. Soc. Lond.*, **A394**, 259-278.

Toh, S., and T. Kawahara, 1985: On the Stability of Soliton-Like Pulses in a Nonlinear Dispersive System with Instability and Dissipation. *J. Phys. Soc. Jpn.*, **54**, 1257-1269.

Whitham, G. B., 1974: Linear and Nonlinear Waves. *Wiley-Interscience.*

Yu, J., and L. K. Gross, 2001: The Onset of Linear Instabilities in a Solid Combustion Model. *Studies in Appl. Math.*, **107**, 81-101.

——, and J. Kevorkian, 1992: Nonlinear Evolution of Small Disturbances into Roll Waves in an Inclined Open Channel. *J. Fluid Mech.*, **243**, 575-594.

——, ——, and R. Haberman, 2000: Weak Nonlinear Long Waves in Channel Flow with Internal Dissipation. *Studies in Appl. Math.*, **105**, 143-163.

REDUCTION OF EDGE EFFECTS IN THE CONTINUOUS WAVELET TRANSFORM BY THE EMPIRICAL MODE DECOMPOSITION METHOD

XUN ZHU

The Johns Hopkins University Applied Physics Laboratory
11100 Johns Hopkins Road, Laurel, MD 20723-6099, USA
E-mail: xun.zhu@jhuapl.edu

(Manuscript received 12 November 2002)

The recently developed empirical mode decomposition (EMD) method is applied to developing an algorithm for the continuous wavelet transform based on the fast Fourier transform that greatly reduces edge effects for a finite data series. By introducing an anti-symmetric padding on both sides of the intrinsic mode functions derived from the EMD method, the spectral characteristics of the original data series near the edges are appropriately preserved in the local power spectrum. Numerical experiments show that the proposed algorithm based on the anti-symmetric padding is better than the traditional zero-padding technique in deriving a more accurate power spectrum of various scales near the edges.

1. Introduction

The continuous wavelet transform (CWT) has become a common tool for analyzing time-scale (t-a) or space-scale (x-a) localization of power in a signal that is dependent on time or space (e.g., Farge 1992; Kumar and Foufoula-Georgiou 1997; Torrence and Compo 1998). The local power spectrum derived from CWT can be used to identify transient spectral features of the signal. Therefore, CWT can be considered an extension of the familiar Fourier transform that provides either the power of a stationary signal or an averaged power spectrum of the whole signal.

Numerical implementation of CWT is usually done by convolving a complex bandpass filter, the so-called mother wavelet, with a real data series. This can always be efficiently accomplished by the fast Fourier transform (FFT) method (e.g., Press et al. 1992, also Eq. (3) below). Since FFT requires the length of the data series to be 2^N, with N being an integer, one is often required to either extend or cut the original data series. In addition, Fourier transform to a finite data series while computing the convolution implies periodic boundary conditions to the data. Both will introduce spurious features to the derived power spectrum near the edges of the original data series. Since the width of the mother wavelet is determined by the scale parameter (a) of CWT, the extent of the spurious features near edges also depends on the scale.

In this paper, an anti-symmetric padding technique is introduced to reduce edge effects in the power spectrum of CWT when the traditional zero-padding is used. The padding extends an original data series from N to $N^* = 2^n$ by a factor $\gamma\,(1.5 \le \gamma \le 3)$ before performing FFT to implement CWT. The scale dependence of edge effects is treated by applying the anti-symmetric padding to an individual wave packet of the so-called intrinsic mode function (IMF) that only contains a single scale. The IMF is derived from the recently developed empirical mode decomposition (EMD) method (Huang et al. 1998) that decomposes a given signal into a set of slowly varying wave packets corresponding to Wentzel-Kramers-Brillouin (WKB) solutions of a dispersive-dissipative wave equation (Zhu et al. 1997).

2. An Algorithm for the Continuous Wavelet Transform

The CWT represents a time-scale (t-a) or a space-scale (x-a) analysis of a signal with respect to a particular function called the mother wavelet $\psi(x)$. The mother wavelet $\psi(x)$ can be considered as a bandpass filter in the frequency or wavenumber domain. Furthermore, we choose the standard Morlet wavelet as the mother wavelet for which its Fourier transform, $\hat{\psi}(\omega)$, can be derived analytically as

$$\psi(x) = \frac{1}{\sqrt{2\pi}\sigma} e^{-x^2/(2\sigma^2)} (e^{i\pi x} - e^{-\pi^2\sigma^2/2}), \tag{1a}$$

$$\hat{\psi}(\omega) \equiv FT\{\psi(x)\} = e^{-\sigma^2(\omega-\pi)^2/4} - e^{-\sigma^2(\omega^2+\pi^2)/2}, \tag{1b}$$

where σ is the shaping parameter for the mother wavelet. The CWT of a signal $s(x)$ is defined as (e.g., Daubechies 1992; Farge 1992)

$$W(x,a) = \frac{1}{\sqrt{a}} \int_{-\infty}^{\infty} s(\xi)\psi^*\left(\frac{\xi - x}{a}\right) d\xi, \tag{2}$$

where the translation parameter x and dilation parameter a (> 0) represent the position and scale, respectively. The asterisk denotes a complex conjugation. From Parseval's theorem (e.g., Bracewell 1986) CWT can also be computed from a frequency domain

$$W(x,a) = \frac{1}{2\pi\sqrt{a}} \int_{-\infty}^{\infty} FT\{s(\xi)\} FT^*\left\{\psi\left(\frac{\xi - x}{a}\right)\right\} d\omega = \sqrt{a} FT^{-1}\left\{\hat{s}(\omega)\hat{\psi}^*(a\omega)\right\}. \tag{3}$$

In the above, we have applied the similarity and shift theorems to the second identity. The second terms in (1) assure the admissibility condition of $\hat{\psi}(\omega) = 0$ for the mother wavelet. The admissibility condition is a necessary and sufficient condition for reconstructing the signal from the CWT. When $\sigma > 3$, its numerical effect on the CWT in practical applications of digital signals is negligible. In this paper, we set $\sigma = 4$. Equation (1a) suggests that the mother wavelet is a sinusoidal wave of frequency π modulated by a Gaussian envelope of width 2σ. In the frequency domain, (1b) and (3) indicate that the mother wavelet acts as a bandpass filter with its width inversely proportional to the shaping parameter σ. On the basis of (2), we can introduce a local power spectrum denoted by $P(x, a)$ that describes a space-scale localization of the original signal $s(x)$

$$P(x,a)=|W(x,a)|^2. \tag{4}$$

In many applications of spectral analyses (e.g., Cohen 1995), the local power spectrum P is often expressed in terms of the time-frequency (t-ω) or the space-wavenumber (x-k). In this paper, we present our analysis in the wavenumber domain because a spatially slowly varying wave packet can be more naturally related to a physically based equation (Zhu et al. 1997). To relate the scale parameter a to the wavenumber k in power spectral analysis often used in meteorological applications we consider the following sinusoidal signal with a specified wavenumber K

$$s(x)=e^{iKx} \quad \text{and} \quad \hat{s}(\omega)=2\pi\delta(\omega-K), \tag{5}$$

where $\delta(\omega)$ is the Dirac-delta function. Substituting (1b) and (5) into (3) and (4) we obtain

$$P(x,a)\approx ae^{-\sigma^2(aK-\pi)^2}, \tag{6}$$

which peaks at the scale parameter

$$a=\frac{\pi}{2K}\left(1+\sqrt{1+\frac{2}{\pi^2\sigma^2}}\right). \tag{7}$$

When $\sigma=4$ the local power spectrum peaks at $a\approx\pi/K$ for the sinusoidal wave (5). In general, we will express the local power spectrum as a function of space and wavenumber $P(x, k)$ with $k=\pi/a$. Note that the Morlet wavelet is neither space-limited nor band-limited. However, it decays rapidly in both space and wavenumber domains from their peak values at $x = 0$ and $k = \pi$. Hence, we expect $P(x,k)$ based on the Morlet wavelet (1) to provide a reasonably good localization in both space and wavenumber domains.

Given a discrete data series, the convolution (2) can be efficiently computed by FFT through (3) (e.g., Press et al. 1992). Assuming a signal sequence $s_i=s(x_i)$ ($i = 1,\ldots, N$) sampled at an evenly spaced interval Δ, the resulting FFT sequence $\hat{s}_j=\hat{s}(\omega_j)$ ($j = 1,\ldots, N$) will have a corresponding frequency interval ($\Omega=2\pi/(N\Delta)$) inversely proportional to the total length of the sampling series ($N\Delta$). Since $\hat{\psi}(\omega)$ is analytically given by (1b), the filtering sequence $\hat{s}_j\hat{\psi}^*(a\omega_j)$ in (3) can be easily computed at the required frequency grids ω_j for any values of scale parameter a. The inverse FFT of $\hat{s}_j\hat{\psi}^*(a\omega_j)$ at the given frequency grids ω_j results in the CWT $W(x_i,a)$ at the original spatial grids x_i. To implement this algorithm by FFT, we require the total length of the discrete data series to be a power of 2: N

$= 2^n$. A common practice when performing FFT for a given data series of length other than 2^n is to extend the data series by padding it with zeros (e.g., Press et al. 1992). However, such a zero-padding extension to the original data series will produce spurious features in the local power spectrum $P(x, k)$ near its edges (x_1 and x_N). To reduce such an edge effect, we introduce a new technique that extends the data series by an antisymmetric mapping of the data series with respect to the edges.

To illustrate the padding by an anti-symmetric mapping we consider the following simple example of a digital signal:

$$s_i = 250(1 + x_i) + 45e^{x_i/1.8} \sin\left(24\pi \frac{x_i}{x_N}\right)$$

$$+45e^{(x_N - x_i)/1.8} \sin\left(15\pi \frac{x_i}{x_N}\right) + 45e^{x_i/1.8} \sin\left(11.75\pi \frac{x_i}{x_N}\right), \quad i = 1, \ldots, N, \tag{8}$$

where x_i ($i = 1,\ldots, N$) is an evenly spaced digital series with $\Delta = 5$ km and $N = 1002$. The digital signal (8) consists of a background linear trend and three sinusoidal wave components with amplitude modulations. Since the background linear trend should not contribute to the local power spectrum, we first subtract the background linear trend $Ax_i + B$ from the original data series. The coefficients A and B are determined in the algorithm by fitting the series to a straight line using a least-squared method (Press et al. 1992).

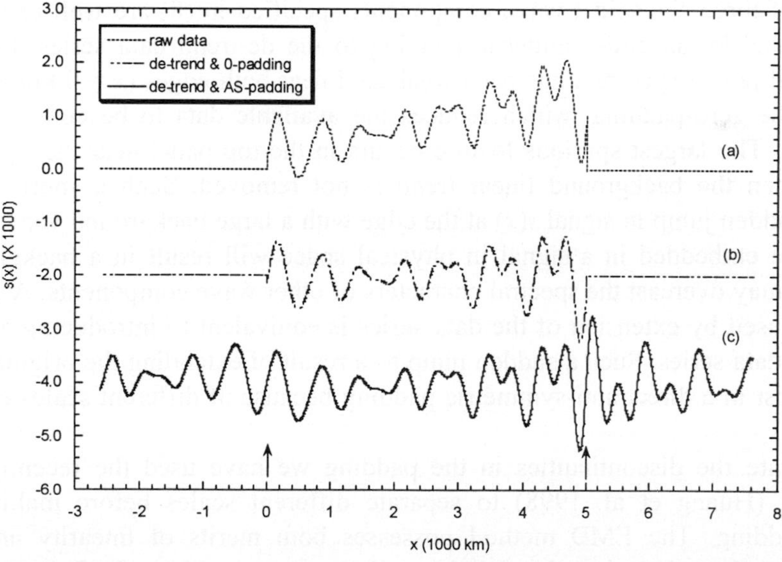

Figure 1. Data series of a superposition of a linear trend and three sinusoidal waves with modulated amplitudes: (a) the original digital signal with a zero-padding extension, (b) de-trend digital signal with a zero-padding extension, and (c) de-trend digital signal with an anti-symmetric padding extension. The arrows indicate where the edges are in the original data series. For a clear display, curves (b) and (c) have been shifted downward by 2 and 4 units, respectively.

We show in Fig. 1 the original and de-trend data series that have been extended to $N^* = 2048$ by padding zeros beyond both edges. By examining (1a) and (2) we know that the CWT of the data series with zero-padding will exhibit spurious features near the edges as a result of big jumps in the signal at the edges. For the de-trend data series we generally expect a relatively weaker $P(x, k)$ in x near the edges because part of the signal to be convolved with the wavelet (1a) is replaced by zeros. As a result, the spatial extent of the spurious feature in $P(x, k)$ depends on the scale parameter a. A larger a that corresponds to a smaller k will produce spurious features of $P(x, k)$ in x farther away from the edges.

The above analysis also indicates that the spurious feature in CWT near an edge is unavoidable because one requires unavailable data series beyond the edge to compute $W(x, k)$ by (2). Note that when the duration of the digital signal (N) is finite, the Fourier transform presumes the signal to be periodic with the largest wavelength of $N\Delta$. Therefore, the use of (4) to calculate CWT by Fourier transform without extension of the original signal has implicitly assumed a periodic extension of the data series beyond the edges. The spectral character near one edge will be unrealistically mapped into the region near the other edge, which usually results in worse edge effects than the zero-padding approach (e.g., Shimomai et al. 1996).

One way to reduce the spurious edge effect is to extend the same spectral structure beyond the edges. We suggest that the de-trend data series be extended beyond an edge by an appropriate anti-symmetric padding of the data series with respect to the edge. One example of such an anti-symmetric padding to the de-trend data series is also shown in Fig. 1 for the test data series. In Fig. 2, we show the local power spectra $P(x, k)$ for three different data series presented in Fig. 1. We see that both the peak wavenumber and the amplitude modulations of three sinusoidal wave components specified in (8) are well characterized in the bottom panel by an anti-symmetric padding to the de-trend data series. In the middle panel, the local power spectrum has been weakened near both edges ($x = 0$ km and $x = 5000$ km), due to the zero-padding, which reduces the available data to be convolved with the wavelet in (2). The largest spurious feature occurs in the top panel near the right edge ($x = 5000$ km), when the background linear trend is not removed. Such a spurious feature is caused by a sudden jump in signal $s(x)$ at the edge with a large background value. Note that a localized pulse embedded in a signal in physical space will result in a background power spectrum that may overcast the spectral characters of other wave components. A sudden jump at the edge caused by extension of the data series is equivalent to introducing an unrealistic pulse into the data series. Such a sudden jump as a result of extending the original data series would also exist in a direct anti-symmetric padding because of different scales existed in the signal.

To eliminate the discontinuities in the padding we have used the recently developed EMD method (Huang et al. 1998) to separate different scales before making the anti-symmetric padding. The EMD method possesses both merits of linearity and internally determined basis by expressing any signal as a linear superposition of a finite number (L) of analytic signals belonging to their IMFs (Huang et al. 1998)

$$s(x) = \mathrm{Re}\left\{\sum_{\ell=1}^{L} S_\ell(x)\exp[i\int^x k_\ell(\tau)d\tau]\right\}, \tag{9}$$

where Re{ } denotes the real part of the complex signal.

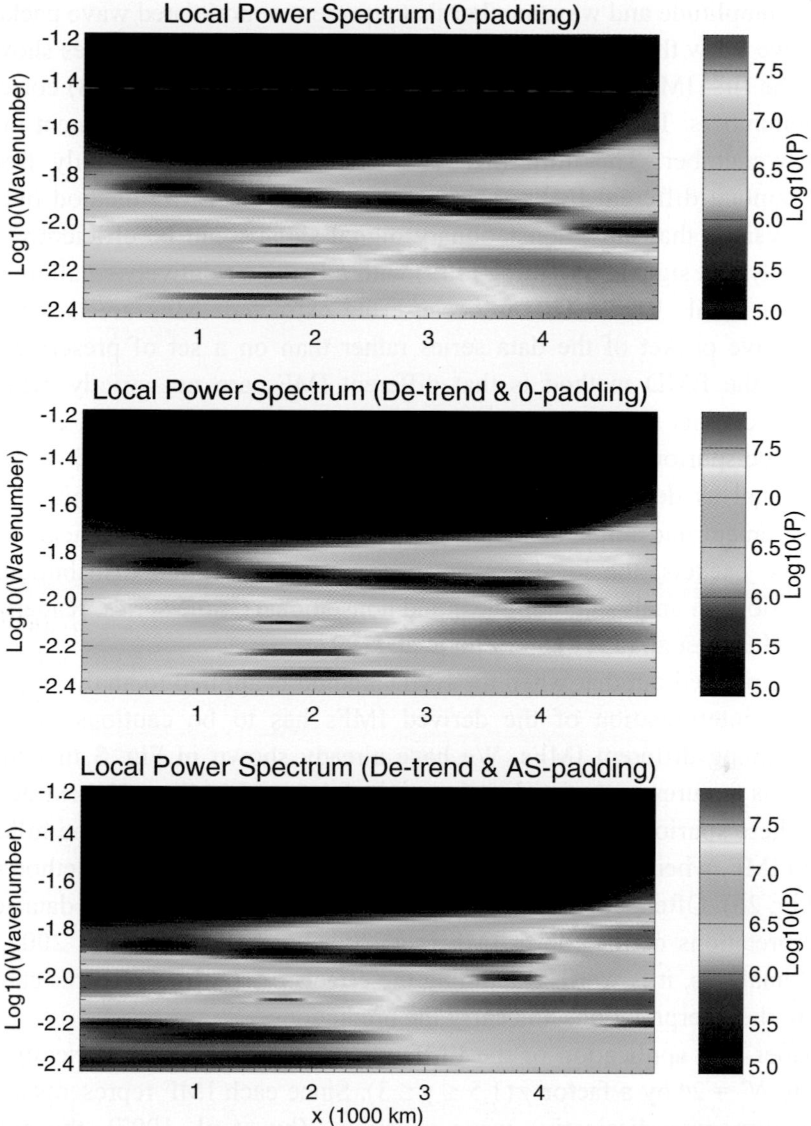

Figure 2. Logarithm of local power spectrum, $\log_{10}[P(x,k)]$, of the test data series of Fig. 1. The color scheme is made to "saturated" below the value of 5 to accentuate significant wave energy in the neighborhood of the local wavenumber. The three panels correspond to (top) the original data series with zero-padding, (middle) the de-trend data series with zero-padding, and (bottom) the de-trend data series with anti-symmetric padding, respectively.

The process of extracting IMFs from the original signal in the EMD method is similar to the multi-resolution approach (e.g., Daubechies 1992, pp. 10-13), where successive coarser components of IMFs are derived from the original signal. Zhu et al. (1997) showed that each IMF derived from the EMD method is closely related to the WKB solutions of an equation describing dispersive-dissipative wave phenomena. Hence, $S_\ell(x)$ and $k_\ell(x)$ in (9) are the slowly varying amplitude and wavenumber that represent a modulated wave packet.

In Fig. 3, we show the three IMFs derived from the de-trend data series shown in Fig. 1. We note that the first IMF represents the first two wave components in (8) corresponding to the high wavenumbers. The second IMF represents the third wave component in (8) that has the lowest wavenumber. The third IMF is the residual resulting mainly from the non-orthogonality among different IMFs. The major merit of the EMD method over the multi-resolution analysis is that most nonstationary digital signals can be efficiently decomposed into a set of analytical signals according to (9) with L being a relatively small number (Huang et al. 1998; Zhu et al. 1997). This is because the extracted IMFs are based on the local characteristic wave packet of the data series rather than on a set of prescribed bases. The shortcoming of the EMD method is that different IMFs are not strictly orthogonal even though there are distinct scale separations among them. As a consequence, coarser resolution IMFs will include spurious residuals of the finer resolution IMFs extracted sequentially from the original signal as demonstrated in Fig. 3. For a more precise definition of the IMF, detailed development and implementation of the EMD method, the physical interpretations related to wave packets, the local power spectrum, the marginal distributions, and their various applications in analyzing nonlinear and nonstationary time series, readers are referred to the work of Huang et al. (1998) and Zhu et al. (1997).

It should be pointed out that when the EMD method is applied to analyzing an individual data series, the interpretation of the derived IMFs has to be cautious due to the non-orthogonality among different IMFs. We have already shown in Fig. 3 that the third IMF contains spurious features not existed in the original data series. Since the extraction of IMFs is sequential, these spurious features could be carried to and amplified in the following IMFs, say, the fourth IMF, when more stringent criterions are used in the EMD method (e.g., Huang et al. 1998, Fig. 26). Often, applications of the EMD method to individual data series rely on physical interpretations of the fourth IMF (e.g., Salisbury and Wimbush 2002, Xie 2002). Based on our analyses, it is worth examining the effects of spurious features carried in those IMFs to assure the interpretations and carry the predictions.

In this paper, our application of the EMD method is to anti-symmetrically extend each IMF from N to $N^* = 2^n$ by a factor $\gamma\,(1.5 \le \gamma \le 3)$. Since each IMF represents a wave packet solution of a dispersive-dissipative wave equation (Zhu et al. 1997), the anti-symmetric padding to each IMF effectively removes the sudden jumps at the edges. Most importantly, the padded IMFs (Fig. 3) are recombined into a "pre-processed" data series (the solid line in Fig. 1) before CWT is performed. Therefore, the non-orthogonality of different IMFs in this application does not affect the spectral properties of the original data series.

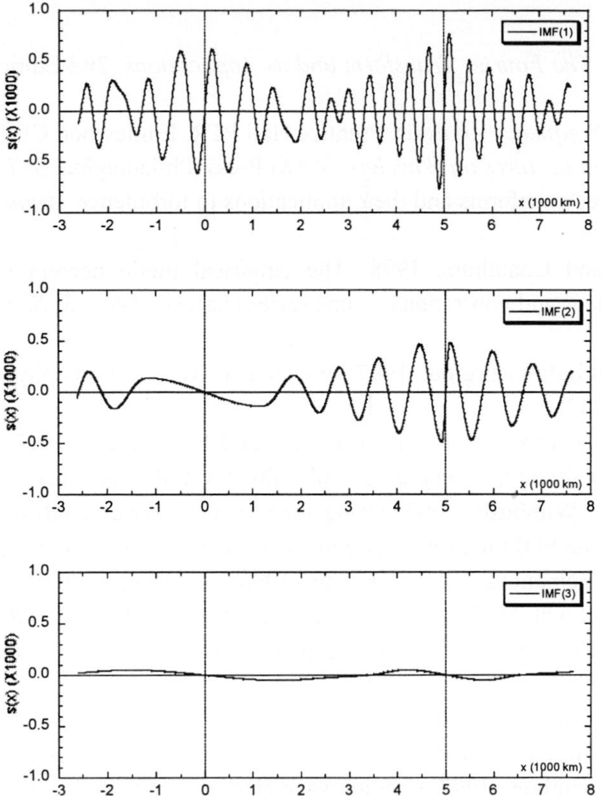

Figure 3. Extended IMFs derived from the de-trend data series shown in Fig. 1.

3. Concluding Remarks

In this paper, we introduce an anti-symmetric padding technique that reduces edge effects in the power spectrum of CWT. The technique was originally developed by the author (Zhu et al. 1997) to deal with a more general problem of extending any given data series into 2^n to accommodate FFT. When FFT is performed to compute CWT, the issue of scale dependence of edge effects in the derived power spectrum as a result of the traditional zero-padding becomes more apparent. The current technique reduces the edge effects and makes physical interpretations of the derived power spectrum near edges more reliable. The FORTRAN program of CWT algorithm described in the last section, which introduces the anti-symmetric padding technique, is available from the author upon request.

Acknowledgments. This research was supported by NSF grant ATM-0091514.

References:

Bracewell, R. N., 1986: *The Fourier Transform and its Applications.* 2nd Edition. McGraw-Hill, New York, 474 pp.

Cohen, L., 1995: *Time-Frequency Analysis*. Prentice Hall PTR, Englewood Cliffs, NJ, 299 pp.

Daubechies, I., 1992: *Ten Lectures on Wavelets*. SIAM Press, Philadelphia, 357 pp.

Farge, M., 1992: Wavelet transforms and their applications to turbulence. *Annu. Rev. Fluid Mech.*, **24**, 395-457.

Huang, N., Z. Shen, and Coauthors, 1998: The empirical mode decomposition and the Hilbert spectrum for nonlinear and nonstationary time series analysis. *Proc. R. Soc. London, A.,* **454**, 903-995.

Kumar, P., and E. Foufoula-Georgiou, 1997: Wavelet analysis for geophysical applications. *Rev. Geophys.*, **35**, 385-412.

Press, W. H., S. A. Teukolsky, W. T. Vetterling, B., and P. Flannery, 1992: *Numerical Recipes in Fortran. The Arts of Scientific Computing.* 2nd Edition. Cambridge University Press, 963 pp.

Salisbury, J. I., and M. Wimbush, 2002: Using modern time series analysis techniques to predict ENSO events from the SOI time series. *Nonlinear Processes in Geophysics*, **9**, 341-345.

Shimomai, T., M. D. Yamanaka, and S. Fukao, 1996: Application of wavelet analysis to wind disturbance observed with MST radar techniques. *J. Atmos. Terr. Phys.,* **58**, 683-696.

Torrence, C., and G. P. Compo, 1998: A practical guide to wavelet analysis. *Bull. Amer. Meteor. Soc.*, **79**, 61-78.

Xie, L., L. J. Pietrafesa, and K. Wu, 2002: Interannual and decadal variability of landfalling tropical cyclones in the southeast coastal states of the United States. *Adv. Atmos. Sci.*, **19**, 677-686.

Zhu, X., Z. Shen, and Coauthors, 1997: Gravity wave characteristics in the middle atmosphere derived from the empirical mode decomposition method. *J. Geophys. Res.,* **102**(D14), 16,545-16,561.

AUTHOR INDEX